W9-BMV-391

Glencoe

BIOLOGY

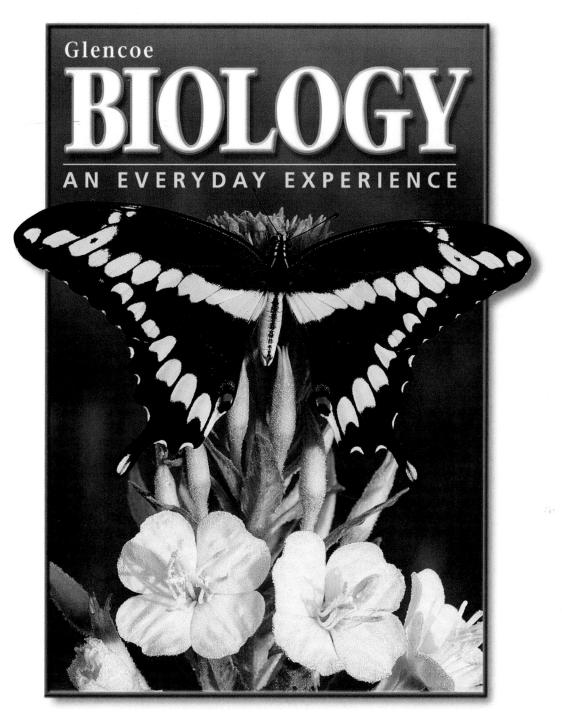

AN EVERYDAY EXPERIENCE

Albert Kaskel Paul J. Hummer, Jr. Lucy Daniel

Glencoe McGraw-Hill

New York, New York Columbus, Ohio Chicago, Illinois Peoria, Illinois Woodland Hills, California

A GLENCOE BIOLOGY PROGRAM

Biology: An Everyday Experience, *Student Edition*
Biology: An Everyday Experience, *Teacher Wraparound Edition*
Biology: An Everyday Experience, *Teacher Classroom Resources*
Biology: An Everyday Experience, *Study Guide*
Biology: An Everyday Experience, *Transparency Package*
Biology: An Everyday Experience, *Laboratory Manual, Student Edition*
Biology: An Everyday Experience, *Laboratory Manual, Teacher Edition*
Biology: An Everyday Experience, *ExamView®Pro*
Biology: An Everyday Experience, *Tech Prep Applications*

CONTENT CONSULTANTS

David M. Armstrong, Ph.D.
Director, University of Colorado Museum
University of Colorado
Boulder, CO

Mary D. Coyne, Ph. D.
Professor of Biological Sciences
Department of Biological Sciences
Wellesley College
Wellesley, MA

Joe W. Crim, Ph.D.
Associate Professor of Zoology
Department of Zoology
University of Georgia
Athens, GA

Marvin Druger
Professor of Biology and Science Education
Department of Biology
Syracuse University
Syracuse, NY

David G. Futch, Ph.D.
Associate Professor of Biology
Department of Biology
San Diego State University
San Diego, CA

Carl Gans, Ph.D.
Professor of Biology
Department of Biology
University of Michigan
Ann Arbor, MI

John Just, Ph.D.
Associate Professor of Biology
School of Biological Sciences
University of Kentucky
Lexington, KY

Richard Storey, Ph. D.
Associate Professor of Biology
Department of Biology
Colorado College
Colorado Springs, CO

James F. Waters, Ph.D.
Professor of Zoology
Department of Biology
Humboldt State University
Arcata, Ca

READING CONSULANT

Barbara S. Pettegrew, Ph.D.
Director of Reading/Study Center
Assistant Professor of Education
Otterbein College
Westerville, OH

SPECIAL CONSULTANT

Alton Biggs
Allen High School
Allen, Texas

REVIEWERS

John A. Beach
Fairless High School
Navarre, OH

Tony Beasley
Science Supervisor
Davidson County School Board
Nashville, TN

Brenda Carrillo
McCollum High School
San Antonio, TX

Renee M. Carroll
Taylor County High School
Perry, FL

Dixie Duncan
Williams Township School
Whiteville, NC

Margorae Freimuth
Argenta-Oreanna High School
Argenta, IL

Raymond P. Gipson
Blue Ridge High School
Morgan Hill, CA

Karen S. Hewitt
Coldspring High School
Coldspring, TX

Marilyn B. Jacobs
Huffman Eastgate High School
Huffman, TX

Rex J. Kartchner
St. David High School
St. David, AZ

Barbara B. Kruse
Alamosa High School
Alamosa, CO

Lynn M. Smith
Waterville Hill School
Waterville, ME

Ouida E. Thomas
B.F. Terry High School
Rosenberg, TX

Glencoe/McGraw-Hill

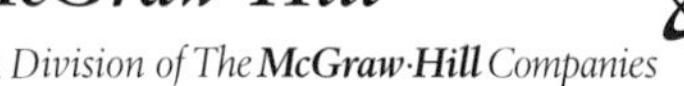

A Division of The **McGraw-Hill** *Companies*

Send all inquiries to:
Glencoe/McGraw-Hill
8787 Orion Place
Columbus, OH 43240-4027

ISBN 0-07-829749-4
Printed in the United States of America.

2 3 4 5 6 7 8 071 09 08 07 06 05 04 03 02

AUTHORS

Albert Kaskel has thirty-one years experience teaching science. He has extensive teaching experience in the city of Chicago and Evanston Township High School, Evanston, Illinois. His teaching experience includes all ability levels of biological science, physical science, and chemistry. He holds an M.Ed. degree from DePaul University. He received the Outstanding Biology Teacher Award for the State of Illinois in 1984.

Paul J. Hummer, Jr. taught science for twenty-eight years in the Fredrick County, Maryland schools. He is currently an Assistant Professor in the education department at Hood College, Frederick, Maryland. He received his B.S.Ed. from Lock Haven State University, Lock Haven, PA, and his M.A.S.T from Union College in Schenectady, NY. He has experience teaching various ability levels of biology as well as general science and physics. He received the Presidential Award for Excellence in Science and Mathematics Teaching in 1984.

Lucy Daniel taught biology at Rutherfordton-Spindale High School, Rutherfordton, North Carolina. She has thirty-five years of teaching experience in biology, life science, and general science. She holds a B.S. degree from the University of North Carolina at Greensboro, a M.A.S.E. from Western Carolina University at Cullowhee, and a Ph.D. from East Carolina University at Greenville, North Carolina. She received the Presidential Award for Excellence in Science and Mathematics Teaching in 1984.

Biology and You

Biology *is* an Everyday Experience

This is your life. Biology is not just another science class. It's a subject you already know well, because it's about life.

It's 10:00, do you know where your dinner is? Every day, you eat and drink to stay alive. Where do your food and water come from, and how does your body use them?

Appreciate the environment—it's the only one you have. Trees don't just stand there, they help supply you with oxygen, prevent soil erosion, and make some of your food.

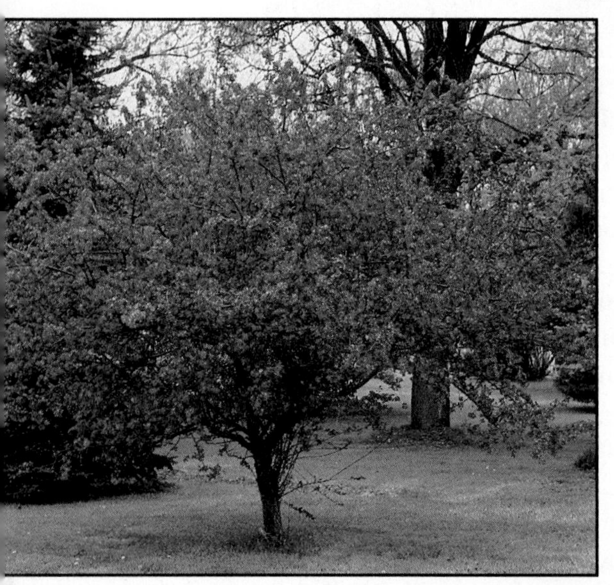

The wonderful world of technology. The field of medicine is closely tied to biology. Advances in medical technology may present you with difficult decisions.

Biology happens! Biology is about every living thing in your world and the relationships among them. The more you learn about biology, the more you will realize that biology *is* an everyday experience.

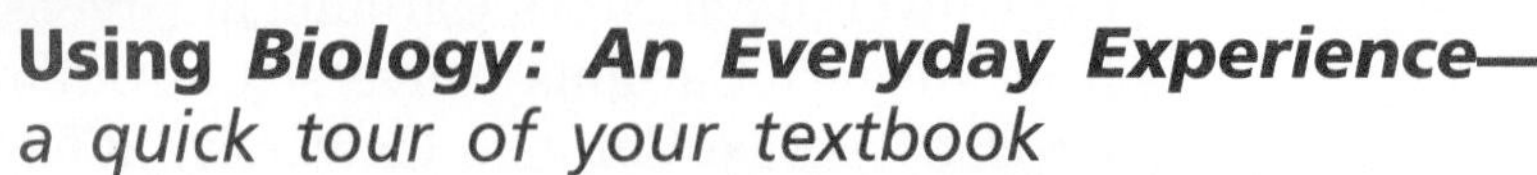

Using *Biology: An Everyday Experience*—*a quick tour of your textbook*

Biology: An Everyday Experience not only presents information, it asks thought-provoking questions. Labs bring the text to life as you use scientific methods to solve problems. You will see how biology affects you as a consumer and learn about careers in biology. Take time now to see what your textbook offers.

from beginning to end, you'll see how biology connects to the world around you

1 What would happen if. . . there were no mosquitoes? Have you ever thought about it? Each unit opener begins with a thought-provoking question like this. The unit introduction then shows you how even small differences in the relationships between living things can change your world in dramatic ways. So, what would happen if there were no mosquitoes? Read the opener to Unit 8 to find out.

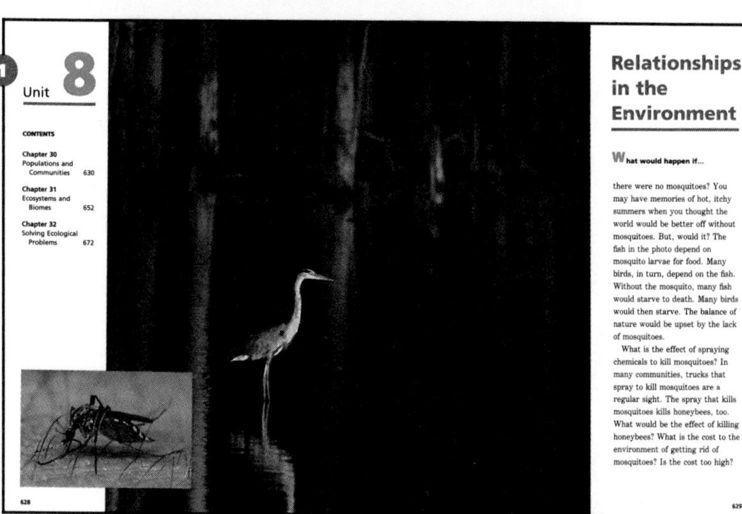
1

Unit 8

CONTENTS

Chapter 30 Populations and Communities 630

Chapter 31 Ecosystems and Biomes 652

Chapter 32 Solving Ecological Problems 672

628

Relationships in the Environment

What would happen if...

there were no mosquitoes? You may have memories of hot, itchy summers when you thought the world would be better off without mosquitoes. But, would it? The fish in the photo depend on mosquito larvae for food. Many birds, in turn, depend on the fish. Without the mosquito, many fish would starve to death. Many birds would then starve. The balance of nature would be upset by the lack of mosquitoes.

What is the effect of spraying chemicals to kill mosquitoes? In many communities, trucks that spray to kill mosquitoes are a regular sight. The spray that kills mosquitoes kills honeybees, too. What would be the effect of killing honeybees? What is the cost to the environment of getting rid of mosquitoes? Is the cost too high?

629

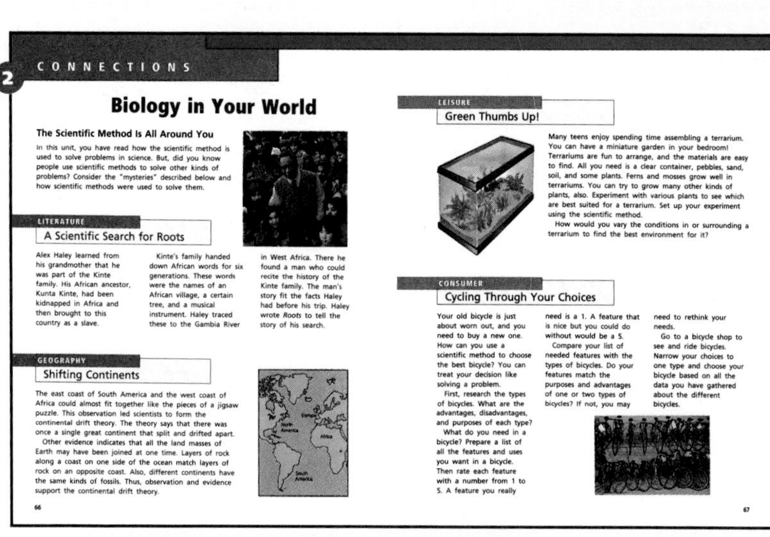
2

CONNECTIONS

Biology in Your World

The Scientific Method Is All Around You

In this unit, you have read how the scientific method is used to solve problems in science. But, did you know people use scientific methods to solve other kinds of problems? Consider the "mysteries" described below and how scientific methods were used to solve them.

LITERATURE

A Scientific Search for Roots

Alex Haley learned from his grandmother that he was part of the Kinte family. His African ancestor, Kunta Kinte, had been kidnapped in Africa and then brought to this country as a slave.

Kinte's family handed down African words for six generations. These words were the names of an African village, a certain tree, and a musical instrument. Haley traced these to the Gambia River in West Africa. There he found a man who could recite the history of the Kinte family. The man's story fit the facts Haley had before his trip. Haley wrote *Roots* to tell the story of his search.

GEOGRAPHY

Shifting Continents

The east coast of South America and the west coast of Africa could almost fit together like the pieces of a jigsaw puzzle. This observation led scientists to form the continental drift theory. The theory says that there was once a single great continent that split and drifted apart.

Other evidence indicates that all the land masses of Earth may have been joined at one time. Layers of rock along a coast on one side of the ocean match layers of rock on an opposite coast. Also, different continents have the same kinds of fossils. Thus, observation and evidence support the continental drift theory.

66

LEISURE

Green Thumbs Up!

Many teens enjoy spending time assembling a terrarium. You can have a miniature garden in your bedroom! Terrariums are fun to arrange, and the materials are easy to find. All you need is a clear container, pebbles, sand, soil, and some plants. Ferns and mosses grow well in terrariums. You can try to grow many other kinds of plants, also. Experiment with various plants to see which are best suited for a terrarium. Set up your experiment using the scientific method.

How would you vary the conditions in or surrounding a terrarium to find the best environment for it?

CONSUMER

Cycling Through Your Choices

Your old bicycle is just about worn out, and you need to buy a new one. How can you use a scientific method to choose the best bicycle? You can treat your decision like solving a problem.

First, research the types of bicycles. What are the advantages, disadvantages, and purposes of each type?

What do you need in a bicycle? Prepare a list of all the features and uses you want in a bicycle. Then rate each feature with a number from 1 to 5. A feature you really need is a 1. A feature that is nice but you could do without would be a 5.

Compare your list of needed features with the types of bicycles. Do your features match the purposes and advantages of one or two types of bicycles? If not, you may need to rethink your needs.

Go to a bicycle shop to see and ride bicycles. Narrow your choices to one type and choose your bicycle based on all the data you have gathered about the different bicycles.

67

2 What do biology and Alex Haley have in common? A lot, as you'll discover when you read the close to Unit 1. Each unit is closed with mini essays that make connections between biology and consumer issues, leisure activities, art, literature, and history.

clearly organized to get you started and keep you going

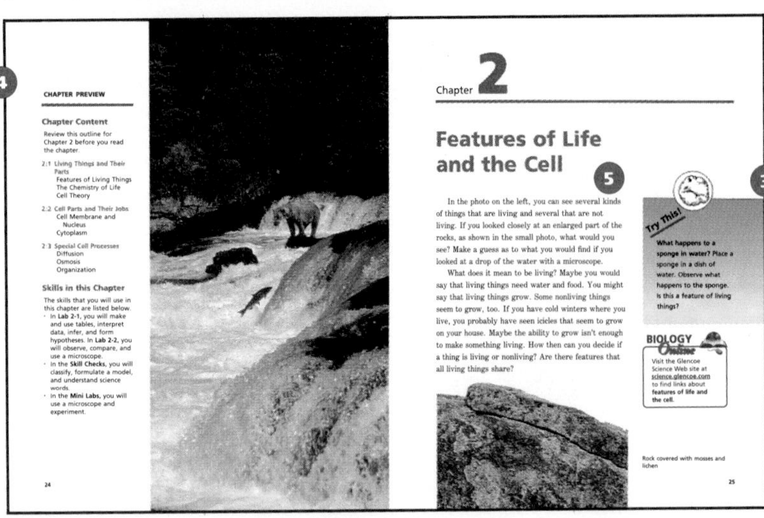
4

CHAPTER PREVIEW

Chapter Content

Review this outline for Chapter 2 before you read the chapter.

2:1 Living Things and Their Parts
Features of Living Things
The Chemistry of Life
Cell Theory

2:2 Cell Parts and Their Jobs
Cell Membrane and Nucleus
Cytoplasm

2:3 Special Cell Processes
Diffusion
Osmosis
Organization

Skills in this Chapter

The skills that you will use in this chapter are listed below.

- In **Lab 2-1**, you will make and use tables, interpret data, infer, and form hypotheses. In **Lab 2-2**, you will observe, compare, and use a microscope.
- In the **Skill Checks**, you will classify, formulate a model, and understand science words.
- In the **Mini Labs**, you will use a microscope and experiment.

24

Chapter 2

Features of Life and the Cell

5

In the photo on the left, you can see several kinds of things that are living and several that are not living. If you looked closely at an enlarged part of the rocks, as shown in the small photo, what would you see? Make a guess as to what you would find if you looked at a drop of the water with a microscope.

What does it mean to be living? Maybe you would say that living things need water and food. You might say that living things grow. Some nonliving things seem to grow, too. If you have cold winters where you live, you probably have seen icicles that seem to grow on your house. Maybe the ability to grow isn't enough to make something living. How then can you decide if a thing is living or nonliving? Are there features that all living things share?

3

Try This!

What happens to a sponge in water? Place a sponge in a dish of water. Observe what happens to the sponge. Is this a feature of living things?

BIOLOGY Online

Visit the Glencoe Science Web site at science.glencoe.com to find links about **features of life and the cell.**

Rock covered with mosses and lichen

25

3 Try This!
Each chapter begins with an easy activity to do right at your desk, or at home. It gets you ready for learning.

4 Listed for you in the **Chapter Preview** are the chapter contents. They tell what topics are covered and how they are organized. Study this before you dive into the chapter material. Also listed are skills that you will practice. A skill is something you get better at with practice. *Biology: An Everyday Experience* gives you all the practice you need to master skills that are important for success in biology, your other classes, and your everyday life.

5 When was the last time you thought about what it means to be alive? Do small living things have the same life processes as large ones? These are the kinds of ideas you will ponder as you read the chapter openers. Each chapter opener has two photographs that are talked about in the introduction. As you read, think about what the photos mean and how they relate to the chapter.

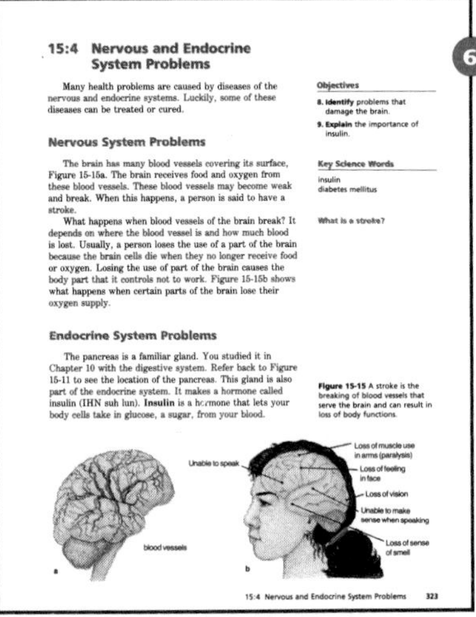
15:4 Nervous and Endocrine System Problems

6

Many health problems are caused by diseases of the nervous and endocrine systems. Luckily, some of these diseases can be treated or cured.

Objectives

8. **Identify** problems that damage the brain.
9. **Explain** the importance of insulin.

Key Science Words

insulin
diabetes mellitus

Nervous System Problems

The brain has many blood vessels covering its surface, Figure 15-15a. The brain receives food and oxygen from these blood vessels. These blood vessels may become weak and break. When this happens, a person is said to have a stroke.

What is a stroke?

What happens when blood vessels of the brain break? It depends on where the blood vessel is and how much blood is lost. Usually, a person loses the use of a part of the brain because the brain cells die when they no longer receive food or oxygen. Losing the use of part of the brain causes the body part that it controls not to work. Figure 15-15b shows what happens when certain parts of the brain lose their oxygen supply.

Endocrine System Problems

The pancreas is a familiar gland. You studied it in Chapter 10 with the digestive system. Refer back to Figure 15-11 to see the location of the pancreas. This gland is also part of the endocrine system. It makes a hormone called insulin (IHN suh lun). **Insulin** is a hormone that lets your body cells take in glucose, a sugar, from your blood.

Figure 15-15 A stroke is the breaking of blood vessels that serve the brain and can result in loss of body functions.

15:4 Nervous and Endocrine System Problems 323

6 Chapters are organized into two to four numbered sections. Each numbered section has several subsections that have red headings. The Objectives at the beginning of the numbered section tell you what major topics you'll be covering and what you should expect to learn about them. The Key Science Words are also listed in the order in which they appear in the section.

lots of ways to help you master important ideas and skills

7 Here's a chance to sharpen your skills. The **Skill Checks** and **Mini Labs** are good ways to practice skills. Each skill exercise requires only pencil and paper or simple materials you can find at home or get from your teacher. If you have trouble with the skill exercises in the Skill Checks or Mini Labs, there is a reference to the **Skill Handbook** at the back of the book. Here, you can find complete information about a particular skill.

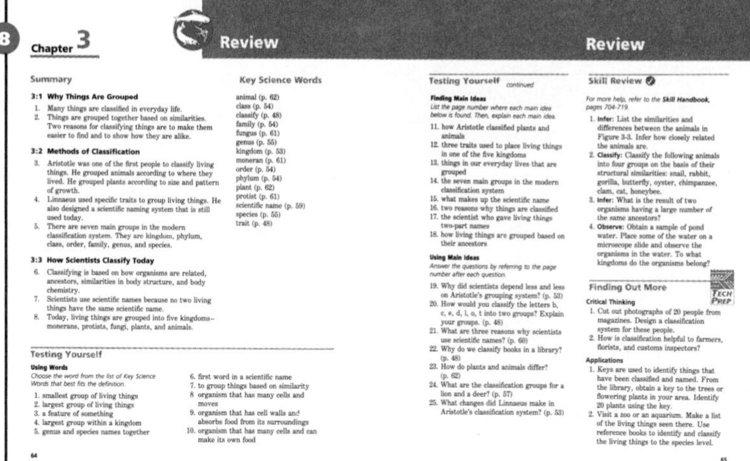

8

Chapter 3 Review

Summary

3:1 Why Things Are Grouped

1. Many things are classified in everyday life.
2. Things are grouped together based on similarities. Two reasons for classifying things are to make them easier to find and to show how they are alike.

3:2 Methods of Classification

3. Aristotle was one of the first people to classify living things. He grouped animals according to where they lived. He grouped plants according to size and pattern of growth.
4. Linnaeus used specific traits to group living things. He also designed a scientific naming system that is still used today.
5. There are seven main groups in the modern classification system. They are kingdom, phylum, class, order, family, genus, and species.

3:3 How Scientists Classify Today

6. Classifying is based on how organisms are related, ancestors, similarities in body structure, and body chemistry.
7. Scientists use scientific names because no two living things have the same scientific name.
8. Today, living things are grouped into five kingdoms–monerans, protists, fungi, plants, and animals.

Key Science Words

animal (p. 62)
class (p. 54)
classify (p. 48)
family (p. 54)
fungus (p. 61)
genus (p. 55)
kingdom (p. 53)
moneran (p. 61)
order (p. 54)
phylum (p. 54)
plant (p. 62)
protist (p. 61)
scientific name (p. 59)
species (p. 55)
trait (p. 48)

Testing Yourself

Using Words

Choose the word from the list of Key Science Words that best fits the definition.

1. smallest group of living things
2. largest group of living things
3. a feature of something
4. largest group within a kingdom
5. genus and species names together
6. first word in a scientific name
7. to group things based on similarity
8. organism that has many cells and moves
9. organism that has cell walls and absorbs food from its surroundings
10. organism that has many cells and can make its own food

64

Review

Testing Yourself *continued*

Finding Main Ideas

List the page number where each main idea below is found. Then, explain each main idea.

11. how Aristotle classified plants and animals
12. three traits used to place living things in one of the five kingdoms
13. things in our everyday lives that are grouped
14. the seven main groups in the modern classification system
15. what makes up the scientific name
16. two reasons why things are classified
17. the scientist who gave living things two-part names
18. how living things are grouped based on their ancestors

Using Main Ideas

Answer the questions by referring to the page number after each question.

19. Why did scientists depend less and less on Aristotle's grouping system? (p. 53)
20. How would you classify the letters b, c, e, d, l, o, t into two groups? Explain your groups. (p. 48)
21. What are three reasons why scientists use scientific names? (p. 60)
22. Why do we classify books in a library? (p. 48)
23. How do plants and animals differ? (p. 62)
24. What are the classification groups for a lion and a deer? (p. 57)
25. What changes did Linnaeus make in Aristotle's classification system? (p. 53)

Skill Review

*For more help, refer to the **Skill Handbook**, pages 704-719.*

1. **Infer:** List the similarities and differences between the animals in Figure 3-3. Infer how closely related the animals are.
2. **Classify:** Classify the following animals into four groups on the basis of their structural similarities: snail, rabbit, gorilla, butterfly, oyster, chimpanzee, clam, cat, honeybee.
3. **Infer:** What is the result of two organisms having a large number of the same ancestors?
4. **Observe:** Obtain a sample of pond water. Place some of the water on a microscope slide and observe the organisms in the water. To what kingdoms do the organisms belong?

Finding Out More

TECH PREP

Critical Thinking

1. Cut out photographs of 20 people from magazines. Design a classification system for these people.
2. How is classification helpful to farmers, florists, and customs inspectors?

Applications

1. Keys are used to identify things that have been classified and named. From the library, obtain a key to the trees or flowering plants in your area. Identify 20 plants using the key.
2. Visit a zoo or an aquarium. Make a list of the living things seen there. Use reference books to identify and classify the living things to the species level.

65

8 **Chapter Reviews** give you an opportunity to reinforce and apply your knowledge.

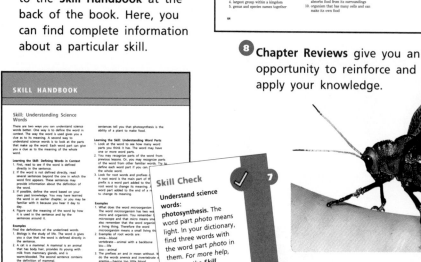

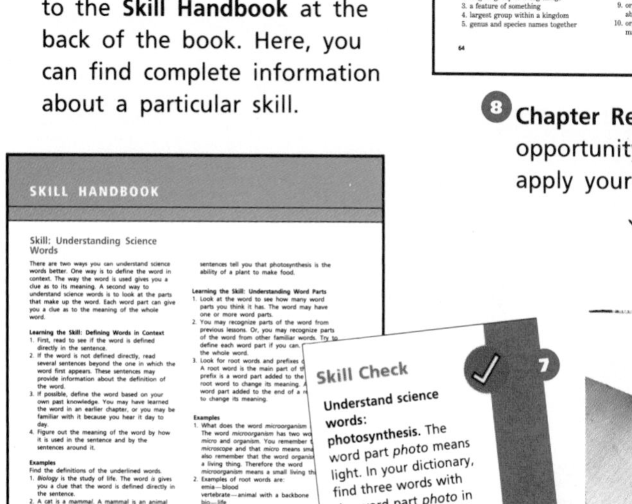

SKILL HANDBOOK

Skill: Understanding Science Words

There are two ways you can understand science words better. One way is to define the word in context. The way the word is used gives you a clue as to its meaning. A second way to understand science words is to look at the parts that make up the word. Each word part can give you a clue as to the meaning of the whole word.

Learning the Skill: Defining Words in Context

1. First, read to see if the word is defined directly in the sentence.
2. If the word is not defined directly, read several sentences beyond the one in which the word first appears. These sentences may provide information about the definition of the word.
3. If possible, define the word based on your own past knowledge. You may have learned the word in an earlier chapter, or you may be familiar with it because you hear it day to day.
4. Figure out the meaning of the word by how it is used in the sentence and by the sentences around it.

Examples

Find the definitions of the underlined words.

1. *Biology* is the study of life. The word *is* gives you a clue that the word is defined directly in the sentence.
2. A cat is a *mammal*. A mammal is an animal that has body hair, provides its young with milk from mammary glands, and is warm-blooded. The second sentence contains the definition of mammal.
3. All living things can *reproduce*. *Reproduce* is a familiar word. It is defined in Chapter 2. It is also one you probably hear from day to day.
4. Green plants carry out *photosynthesis*. Thus, they are able to make their own food. These sentences tell you that photosynthesis is the ability of a plant to make food.

Learning the Skill: Understanding Word Parts

1. Look at the word to see how many word parts you think it has. The word may have one or more word parts.
2. You may recognize parts of the word from previous lessons. Or, you may recognize parts of the word from other familiar words. Try to define each word part if you can. [illegible] the whole word.
3. Look for root words and prefixes [illegible] A root word is the main part of th[illegible] prefix is a word part added to the [illegible] root word to change its meaning. [illegible] word part added to the end of a r[illegible] to change its meaning.

Examples

1. What does the word *microorganism* [illegible] The word *microorganism* has two wo[illegible] *micro* and *organism*. You remember [illegible] *microscope* and that *micro* means sm[illegible] also remember that the word organis[illegible] a living thing. Therefore the word *microorganism* means a small living th[illegible]
2. Examples of root words are:
   emia—blood
   vertebrate—animal with a backbone
   bio—life
   zoo—animal
3. The prefixes *an* and *in* mean without. W[illegible] do the words *anemia* and *invertebrate* [illegible]
   anemia—having too little blood
   invertebrate—animal without a backbone
4. The suffix *logy* means the study of. What [illegible] biology? Zoology?
   biology—the study of life
   zoology—the study of animals

708 Skill Handbook

Skill Check 7

Understand science words: photosynthesis. The word part *photo* means light. In your dictionary, find three words with the word part *photo* in them. *For more help, refer to the **Skill Handbook**, pages 706-711.*

9 If you have trouble remembering and understanding a lesson, help is on the way. Each major concept has an Idea Map that you can use as a study guide. Ideas are summarized in an easy-to-read format. The Idea Maps will help make your study of biology a success.

10 At the end of each major section are five **Check Your Understanding** questions. The first three questions reinforce what you have learned in the section. The fourth question challenges you to think critically about what you have read. The fifth question connects biology with reading, writing, or math.

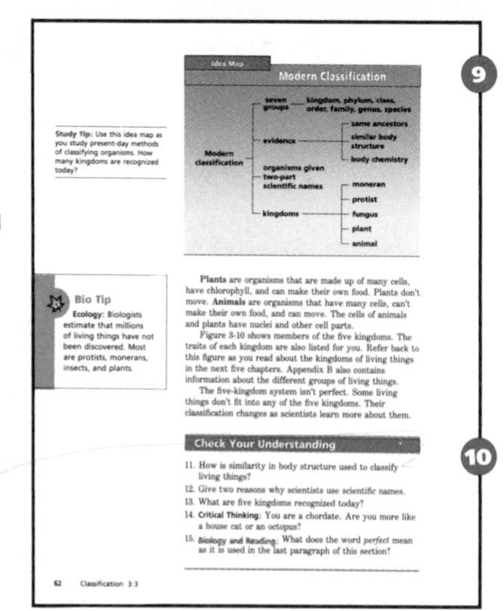

9

Idea Map

Modern Classification

Study Tip: Use this idea map as you study present-day methods of classifying organisms. How many kingdoms are recognized today?

Bio Tip

Ecology: Biologists estimate that millions of living things have not been discovered. Most are protists, monerans, insects, and plants.

Plants are organisms that are made up of many cells, have chlorophyll, and can make their own food. Plants don't move. **Animals** are organisms that have many cells, can't make their own food, and can move. The cells of animals and plants have nuclei and other cell parts.

Figure 3-10 shows members of the five kingdoms. The traits of each kingdom are also listed for you. Refer back to this figure as you read about the kingdoms of living things in the next five chapters. Appendix B also contains information about the different groups of living things.

The five-kingdom system isn't perfect. Some living things don't fit into any of the five kingdoms. Their classification changes as scientists learn more about them.

Check Your Understanding

10

11. How is similarity in body structure used to classify living things?
12. Give two reasons why scientists use scientific names.
13. What are five kingdoms recognized today?
14. **Critical Thinking:** You are a chordate. Are you more like a house cat or an octopus?
15. **Biology and Reading:** What does the word *perfect* mean as it is used in the last paragraph of this section?

62 Classification 3:3

really experience biology by observing, experimenting, asking questions

11 Every chapter has two step-by-step labs. Procedures are clear and easy to follow. Sample data tables are given to help you organize the information you collect. At the end of each lab are questions that help to reinforce what you learned in the lab. Doing a lab has never been so easy!

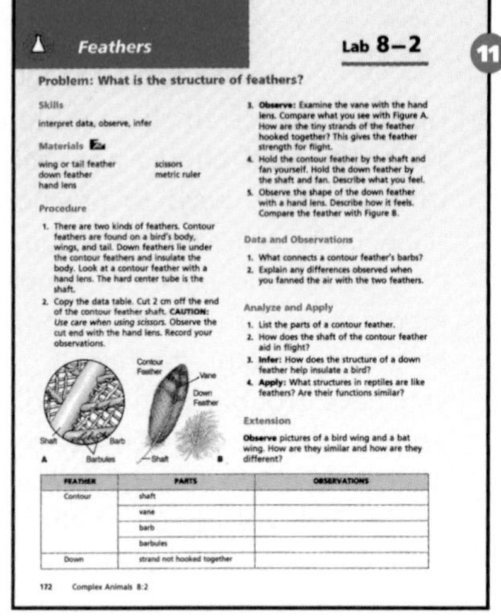

11

Feathers — Lab 8–2

Problem: What is the structure of feathers?

Skills

interpret data, observe, infer

Materials

wing or tail feather
down feather
hand lens
scissors
metric ruler

Procedure

1. There are two kinds of feathers. Contour feathers are found on a bird's body, wings, and tail. Down feathers lie under the contour feathers and insulate the body. Look at a contour feather with a hand lens. The hard center tube is the shaft.
2. Copy the data table. Cut 2 cm off the end of the contour feather shaft. **CAUTION:** *Use care when using scissors.* Observe the cut end with the hand lens. Record your observations.
3. **Observe:** Examine the vane with the hand lens. Compare what you see with Figure A. How are the tiny strands of the feather hooked together? This gives the feather strength for flight.
4. Hold the contour feather by the shaft and fan yourself. Hold the down feather by the shaft and fan. Describe what you feel.
5. Observe the shape of the down feather with a hand lens. Describe how it feels. Compare the feather with Figure B.

FEATHER	PARTS	OBSERVATIONS
Contour	shaft	
	vane	
	barb	
	barbules	
Down	strand not hooked together	

Data and Observations

1. What connects a contour feather's barbs?
2. Explain any differences observed when you fanned the air with the two feathers.

Analyze and Apply

1. List the parts of a contour feather.
2. How does the shaft of the contour feather aid in flight?
3. **Infer:** How does the structure of a down feather help insulate a bird?
4. **Apply:** What structures in reptiles are like feathers? Are their functions similar?

Extension

Observe pictures of a bird wing and a bat wing. How are they similar and how are they different?

172 Complex Animals 8:2

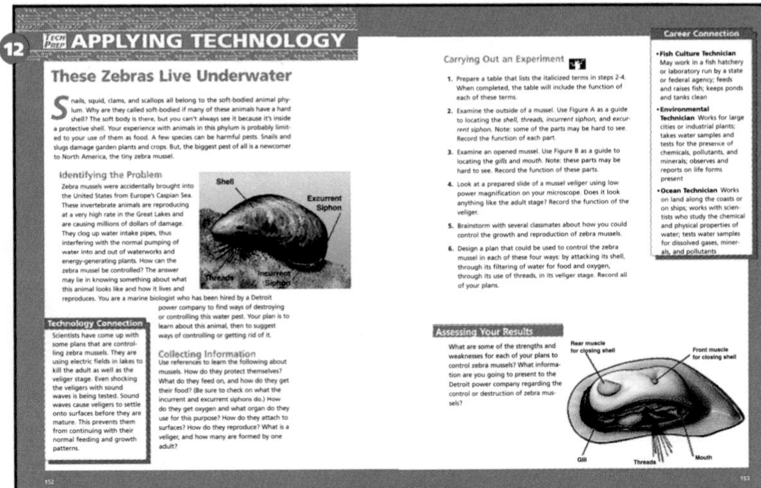

12

TECH PREP APPLYING TECHNOLOGY

These Zebras Live Underwater

Snails, squid, clams, and scallops all belong to the soft-bodied animal phylum. Why are they called soft-bodied if many of these animals have a hard shell? The soft body is there, but you can't always see it because it's inside a protective shell. Your experience with animals in this phylum is probably limited to your use of them as food. A few species can be harmful pests. Snails and slugs damage garden plants and crops. But, the biggest pest of all is a newcomer to North America, the tiny zebra mussel.

Identifying the Problem

Zebra mussels were accidentally brought into the United States from Europe's Caspian Sea. These invertebrate animals are reproducing at a very high rate in the Great Lakes and are causing millions of dollars of damage. They clog up water intake pipes, thus interfering with the normal pumping of water into and out of waterworks and energy-generating plants. How can the zebra mussel be controlled? The answer may lie in knowing something about what this animal looks like and how it lives and reproduces. You are a marine biologist who has been hired by a Detroit power company to find ways of destroying or controlling this water pest. Your plan is to learn about this animal, then to suggest ways of controlling or getting rid of it.

Technology Connection

Scientists have come up with some plans that are controlling zebra mussels. They are using electric fields in lakes to kill the adult as well as the veliger stage. Even shocking the veligers with sound waves is being tested. Sound waves cause veligers to settle onto surfaces before they are mature. This prevents them from continuing with their normal feeding and growth patterns.

Collecting Information

Use references to learn the following about mussels. How do they protect themselves? What do they feed on, and how do they get their food? (Be sure to check on what the incurrent and excurrent siphons do.) How do they get oxygen and what organ do they use for this purpose? How do they attach to surfaces? How do they reproduce? What is a veliger, and how many are formed by one adult?

152

Carrying Out an Experiment

1. Prepare a table that lists the italicized terms in steps 2-4. When completed, the table will include the function of each of these terms.
2. Examine the outside of a mussel. Use Figure A as a guide to locating the *shell, threads, incurrent siphon,* and *excurrent siphon.* Note: some of the parts may be hard to see. Record the function of each part.
3. Examine an opened mussel. Use Figure B as a guide to locating the *gills* and *mouth.* Note: these parts may be hard to see. Record the function of these parts.
4. Look at a prepared slide of a mussel *veliger* using low power magnification on your microscope. Does it look anything like the adult stage? Record the function of the veliger.
5. Brainstorm with several classmates about how you could control the growth and reproduction of zebra mussels.
6. Design a plan that could be used to control the zebra mussel in each of these four ways: by attacking its shell, through its filtering of water for food and oxygen, through its use of threads, in its veliger stage. Record all of your plans.

Career Connection

- **Fish Culture Technician** May work in a fish hatchery or laboratory run by a state or federal agency; feeds and raises fish; keeps ponds and tanks clean
- **Environmental Technician** Works for large cities or industrial plants; takes water samples and tests for the presence of chemicals, pollutants, and minerals; observes and reports on life forms present
- **Ocean Technician** Works on land along the coasts or on ships; works with scientists who study the chemical and physical properties of water; tests water samples for dissolved gases, minerals, and pollutants

Assessing Your Results

What are some of the strengths and weaknesses for each of your plans to control zebra mussels? What information are you going to present to the Detroit power company regarding the control or destruction of zebra mussels?

153

12 Each unit has a two-page Applying Technology feature. Read the short background paragraph, do the activity, and read how it relates to technology and careers.

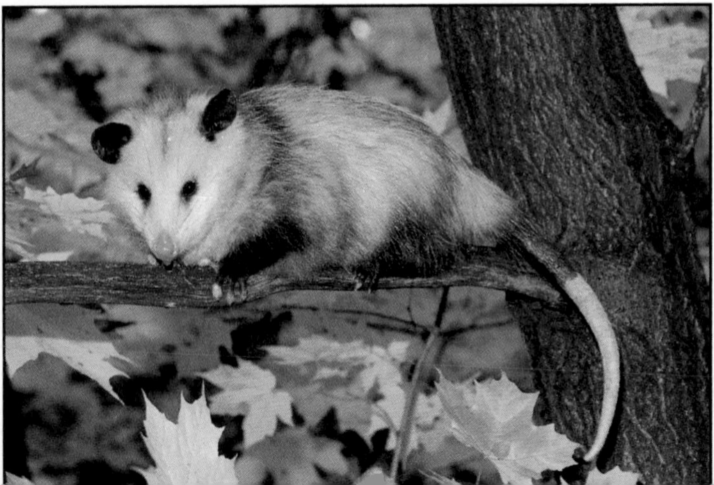

explore how biology impacts technology and news-making issues and offers career choices

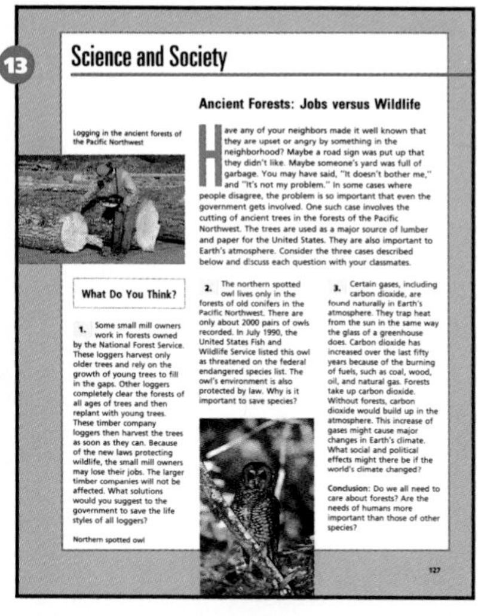

13

Science and Society

Ancient Forests: Jobs versus Wildlife

Logging in the ancient forests of the Pacific Northwest

Have any of your neighbors made it well known that they are upset or angry by something in the neighborhood? Maybe a road sign was put up that they didn't like. Maybe someone's yard was full of garbage. You may have said, "It doesn't bother me," and "It's not my problem." In some cases where people disagree, the problem is so important that even the government gets involved. One such case involves the cutting of ancient trees in the forests of the Pacific Northwest. The trees are used as a major source of lumber and paper for the United States. They are also important to Earth's atmosphere. Consider the three cases described below and discuss each question with your classmates.

What Do You Think?

1. Some small mill owners work in forests owned by the National Forest Service. These loggers harvest only older trees and rely on the growth of young trees to fill in the gaps. Other loggers completely clear the forests of all ages of trees and then replant with young trees. These timber company loggers then harvest the trees as soon as they can. Because of the new laws protecting wildlife, the small mill owners may lose their jobs. The larger timber companies will not be affected. What solutions would you suggest to the government to save the life styles of all loggers?

2. The northern spotted owl lives only in the forests of old conifers in the Pacific Northwest. There are only about 2000 pairs of owls recorded. In July 1990, the United States Fish and Wildlife Service listed this owl as threatened on the federal endangered species list. The owl's environment is also protected by law. Why is it important to save species?

3. Certain gases, including carbon dioxide, are found naturally in Earth's atmosphere. They trap heat from the sun in the same way the glass of a greenhouse does. Carbon dioxide has increased over the last fifty years because of the burning of fuels, such as coal, wood, oil, and natural gas. Forests take up carbon dioxide. Without forests, carbon dioxide would build up in the atmosphere. This increase of gases might cause major changes in Earth's climate. What social and political effects might there be if the world's climate changed?

Conclusion: Do we all need to care about forests? Are the needs of humans more important than those of other species?

Northern spotted owl

127

13 Are animal experiments necessary? Who decides which person in need of an organ transplant gets one? As you read the Science and Society features, you'll find that the answers to these and other questions are not so easy. The Science and Society features bring you closer to current issues and let you see the impact of technology on society. They prepare you for the day when you may need to participate in making decisions that affect your community and your environment.

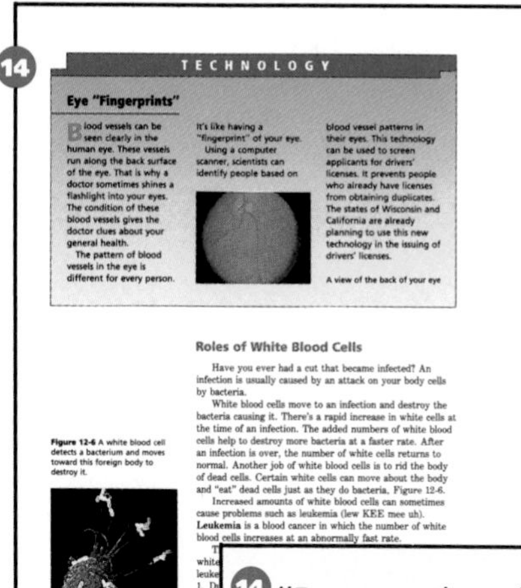

14

TECHNOLOGY

Eye "Fingerprints"

Blood vessels can be seen clearly in the human eye. These vessels run along the back surface of the eye. That is why a doctor sometimes shines a flashlight into your eyes. The condition of these blood vessels gives the doctor clues about your general health.

The pattern of blood vessels in the eye is different for every person. It's like having a "fingerprint" of your eye.

Using a computer scanner, scientists can identify people based on blood vessel patterns in their eyes. This technology can be used to screen applicants for drivers' licenses. It prevents people who already have licenses from obtaining duplicates. The states of Wisconsin and California are already planning to use this new technology in the issuing of drivers' licenses.

A view of the back of your eye

Roles of White Blood Cells

Have you ever had a cut that became infected? An infection is usually caused by an attack on your body cells by bacteria.

White blood cells move to an infection and destroy the bacteria causing it. There's a rapid increase in white cells at the time of an infection. The added numbers of white blood cells help to destroy more bacteria at a faster rate. After an infection is over, the number of white cells returns to normal. Another job of white blood cells is to rid the body of dead cells. Certain white cells can move about the body and "eat" dead cells just as they do bacteria, Figure 12-6.

Increased amounts of white blood cells can sometimes cause problems such as leukemia (lew KEE mee uh). **Leukemia** is a blood cancer in which the number of white blood cells increases at an abnormally fast rate.

Figure 12-6 A white blood cell detects a bacterium and moves toward this foreign body to destroy it.

250 Blood 12:2

14 "Eyes are the windows to the soul." Did you know that your eyes are also like fingerprints? The patterns made by the vessels in the back of your eye are unique and can be used to identify you. No other person has eyes like you. This and other recent discoveries appear in the Technology features. The Technology features tie together biology and applications of the most recent research in biology. They make biology meaningful to *you.*

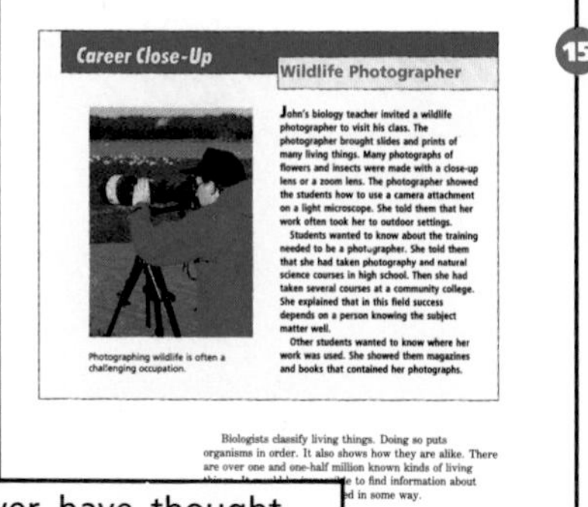

15

Career Close-Up

Wildlife Photographer

John's biology teacher invited a wildlife photographer to visit his class. The photographer brought slides and prints of many living things. Many photographs of flowers and insects were made with a close-up lens or a zoom lens. The photographer showed the students how to use a camera attachment on a light microscope. She told them that her work often took her to outdoor settings.

Students wanted to know about the training needed to be a photographer. She told them that she had taken photography and natural science courses in high school. Then she had taken several courses at a community college. She explained that in this field success depends on a person knowing the subject matter well.

Other students wanted to know where her work was used. She showed them magazines and books that contained her photographs.

Photographing wildlife is often a challenging occupation.

Biologists classify living things. Doing so puts organisms in order. It also shows how they are alike. There are over one and one-half million known kinds of living

15 You may never have thought about a career in biology, but think again. What kinds of interesting careers are there for you in biology? What about wildlife photography? How about being an athletic trainer? Each career feature focuses on one career, telling you the daily ins and outs of the job. Even if you want to work as soon as you graduate from high school, you'll find there are many careers that will let you do just that. Information about career training is given so that if you're interested, you can start planning now.

CONTENTS

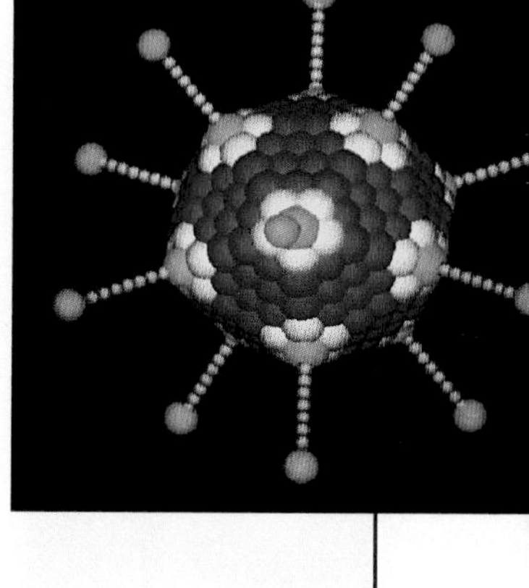

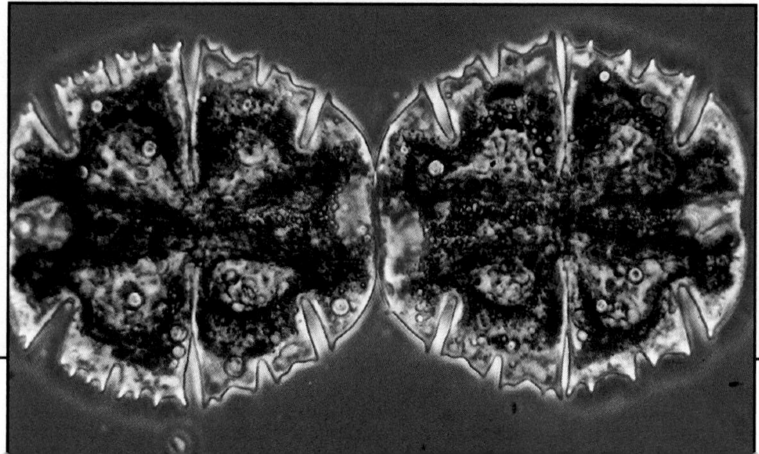

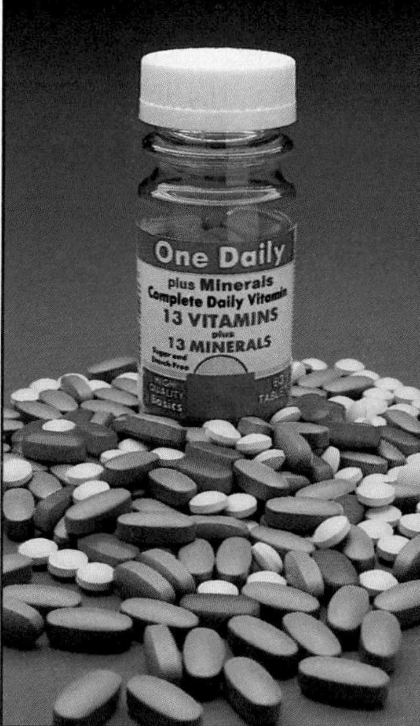

One Daily
plus Minerals
Complete Daily Vitamin
13 VITAMINS
13 MINERALS

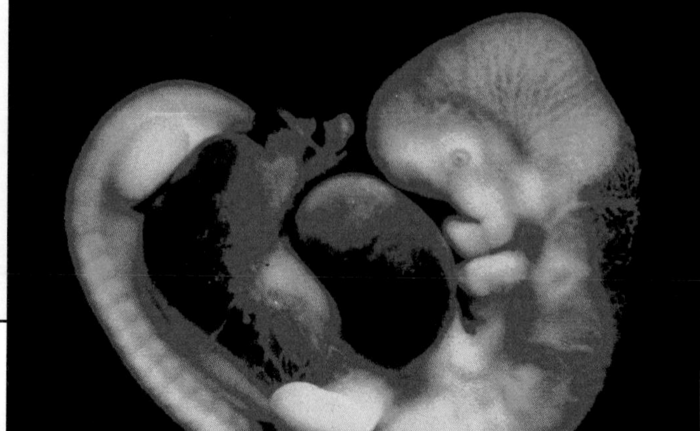

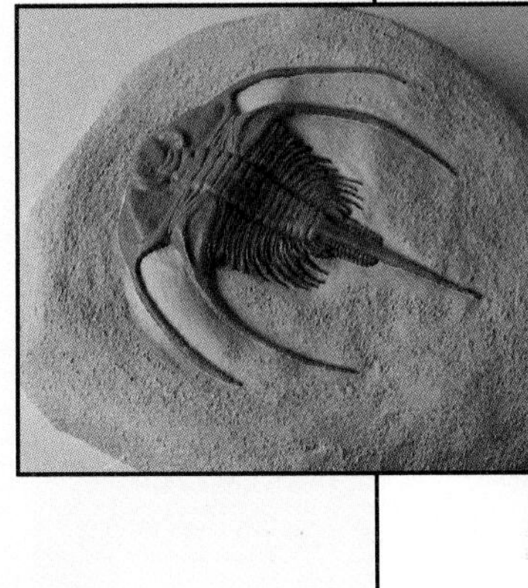

LABS

Have you ever seen an animal capture and eat its prey using tentacles that surround its mouth? You may do this in one of the 64 **Labs** found in your textbook. In another **Lab,** you may watch as protein is digested. In other **Labs,** you test painkillers for the presence of aspirin, build models of viruses, and test the effects of pollution on yeast cells. **Labs** let you use what you have learned to *do* biology.

APPLYING TECHNOLOGY

Can crops be watered with ocean water? Does light butter have less fat than regular butter? These are examples of **Applying Technology** features found in your textbook. You may investigate the problem of zebra mussels in the Great Lakes, or you may try to date rock layers from the fossils they contain. In the **Applying Technology** features, you can apply technology to solving biological problems and perhaps find a future career in the process.

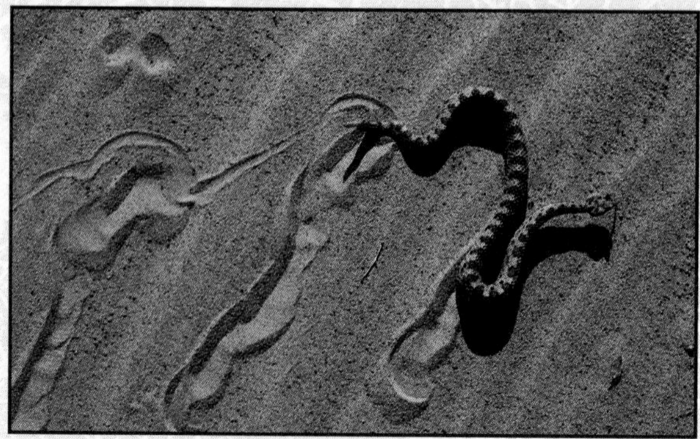

SCIENCE AND SOCIETY

Today it is possible for a couple to choose the sex of their child. Should this be done? How will a country's population be affected if all couples chose the sexes of their children? This is an example of a problem in which science and society interact as people line up on both sides of the debate. **Science and Society** features discuss real cases and present questions to help you form your *own* opinion about a current topic of debate in biology. Should animals be used to test the safety of new drugs? Should athletes use steroids to help them reach their goals? These and many other issues involving the environment, health, and conservation are presented in **Science and Society.**

TECHNOLOGY

What do biologists do with new information about living things? The chemicals in the cells of each human are unique. No two people have the exact same chemical make-up. Learning to put that fact to practical use in solving crimes is one example of technology. Your textbook's **Technology** features show examples of biological knowledge at work. In **Technology,** you learn how bacteria can clean up oil spills and how eye-blinking behavior might help pilots.

CAREER CLOSE-UP

What will you do with everything you learn in your biology course? Maybe you'll begin thinking about a career in a field involving biology. **Career Close-ups** are designed to give you some examples of career paths you might follow. You might become a wildlife photographer and travel around the world photographing living things. Or, you could be an animal breeder working to develop more nutritious beef. Perhaps you would like to become a mushroom farmer. The number of career choices in biology is almost endless.

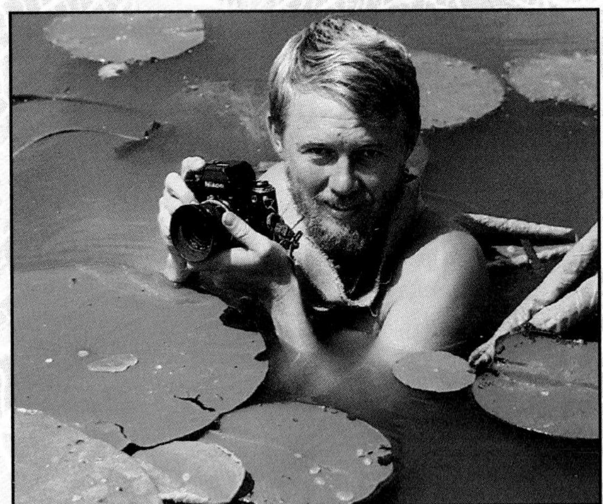

MINI LABS

Would you like to learn how to make a bird feeder or try to design a child-proof drug package? The **Mini Lab** features in your textbook will tell you how.

SKILL CHECKS

What does the word *photosynthesis* mean? What other words are made of the same word parts? The **Skill Check** features will help you understand science words, as well as give you practice using other skills.

Unit 1

CONTENTS

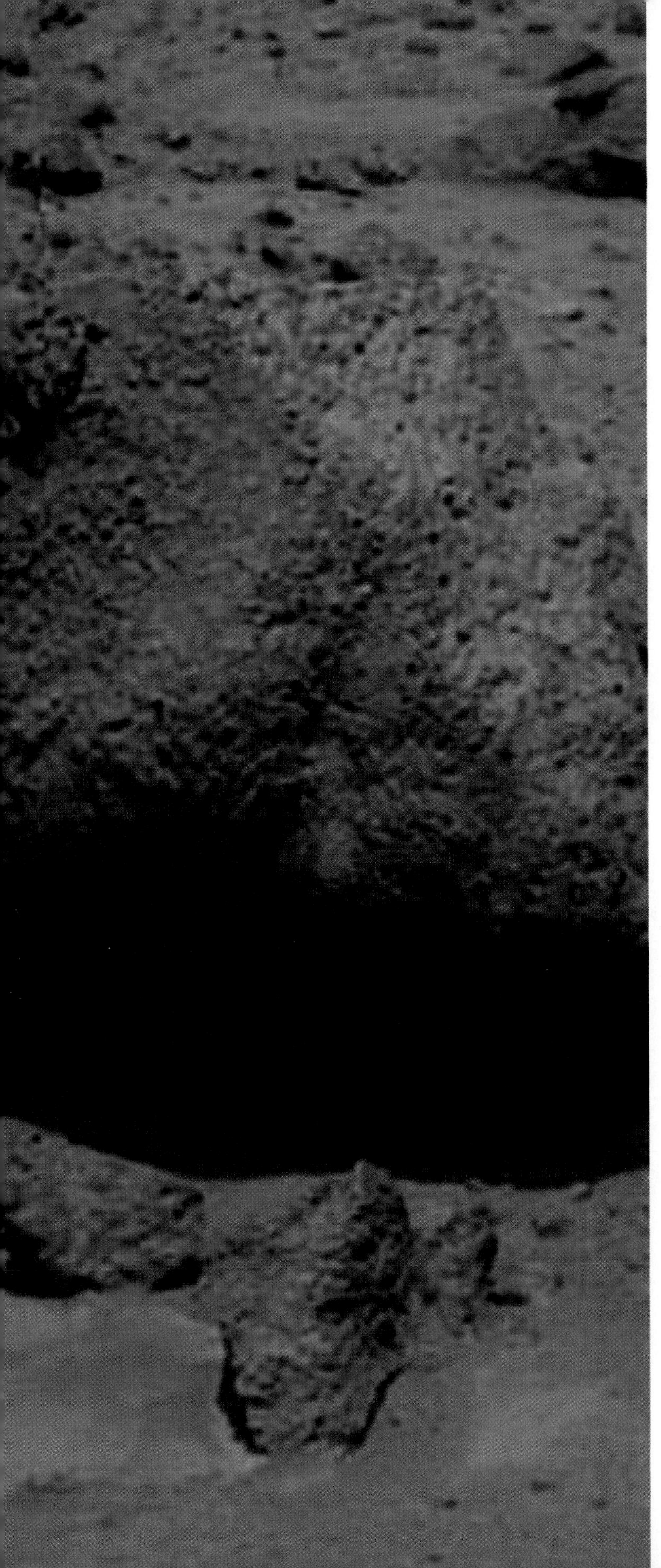

Kinds of Life

What would happen if...

there were life on Mars? Living things as we know them need water, warmth, and protection from radiation from the sun. Does Mars provide these things? Life-forms on Mars would have to survive without any water because all of the water on Mars is frozen. Mars is farther away from the sun than Earth. As a result, the temperature on Mars is always below freezing. The atmosphere of Mars is too thin to block out harmful radiation from the sun. Thus any life on Mars would have to be able to withstand radiation.

The Pathfinder probe, carrying a surface rover named Sojourner, landed on the surface of Mars on July 4th, 1997. Sojourner, shown here, is studying Martian soil and rocks, looking for different chemicals and for living things.

If life exists on Mars, it probably is not life as we know it. But are there other types or definitions of life?

CHAPTER PREVIEW

Chapter Content

Review this outline for Chapter 1 before you read the chapter.

Skills in this Chapter

The skills that you will use in this chapter are listed below.

- In **Lab 1-1,** you will observe and recognize and use spatial relationships. In **Lab 1-2,** you will form hypotheses, interpret data, and separate and control variables.
- In the **Skill Checks,** you will make and use tables, calculate, and understand science words.
- In the **Mini Labs,** you will design an experiment.

Chapter 1

The Study of Life

Look at the photo of the classroom on the left. How many things can you name that are living or were once living? Think about the living or once-living things in your own classroom. The desks, pencils, and books in your classroom are plant products. Chalk comes from tiny living things found in the ocean. Plastics contain materials that come from the remains of plants and animals. Chairs, pens, rulers, and calculators are some of the products that can be made from plastics.

Consider the things that came from living or once-living things in the other parts of the school, such as the gym, cafeteria, or library. You can see many things that you need and use every day are living or were once living. **Biology** is the study of living and once-living things. *Bio-* means life and *-ology* means the study of. A person who studies living and once-living things is called a biologist.

Try This!

How are an animal and a candle flame alike? Observe a living animal and a candle flame. Record how they are similar and how they are different. How is the candle flame like the living animal? How is it different?

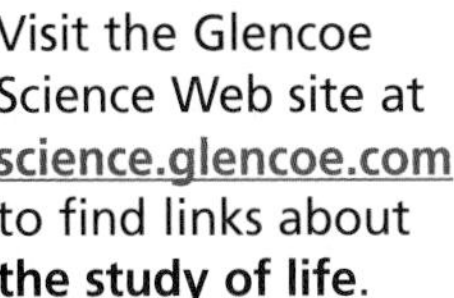

BIOLOGY Online

Visit the Glencoe Science Web site at science.glencoe.com to find links about **the study of life**.

1:1 Biology in Use

Objectives

1. **Describe** three jobs that depend on the use of biology.
2. **List** five tools of biologists.
3. **Compare** the features of the light microscope, stereomicroscope, and electron microscope.

Key Science Words

biology
light microscope
stereomicroscope

The use of biology is not limited to biologists. Many people use biology daily in their hobbies and work. We will look at how you and other people use biology.

Who Uses Biology?

Look at the photos. Are the people shown in the photos using biology in any way?

The beekeeper in Figure 1-1a is removing honey from the hive. A beekeeper must know how bees work together to produce honey. He must know where to place the hives to get the most honey.

How do the police use biology?

Crime lab technicians (tek NISH uhnz) work with different kinds of evidence. Evidence may include cloth fibers, bits of paper, hair, skin, and blood. These things must be studied under a microscope.

If you wear eyeglasses or contact lenses, an optometrist (ahp TAHM uh trust) probably fitted them. The optometrist has studied the eyes and how they work.

Mushrooms are grown in buildings with carefully controlled temperature, moisture, and air flow. Mushroom growers must choose the best soil conditions for their crops. They must know how to identify and treat diseases of mushrooms.

Many people use biology in different ways. You use biology if you have ever asked yourself any of these questions: How can I tell when my plant needs water? What can I do about the fleas on my dog? How can I tell if this plant is poison ivy? The answers to these questions have to do with living things. You use biology all the time and may not even realize it.

Figure 1-1 Beekeepers (a) and optometrists (b) use biology in their work.

a

b

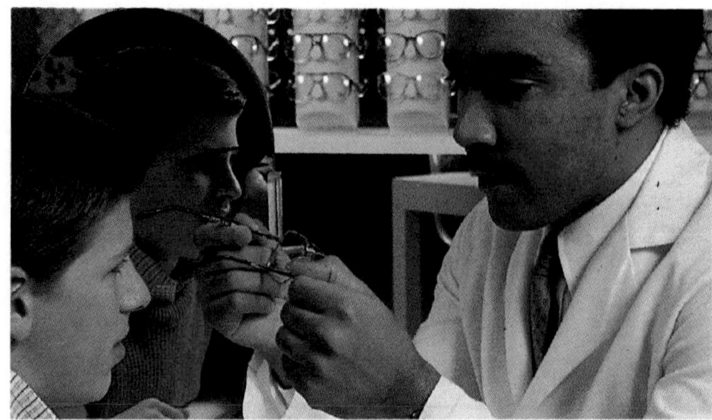

Tools of Biology

Almost every day you use tools to measure and observe. You may use some of the same tools that a biologist uses. Have you ever used a ruler or a magnifying glass? Maybe you have used a pair of binoculars or a microscope.

A hand lens is a type of magnifying glass. It makes an object appear three to five times larger than it is.

A light microscope is one of the most important tools of biology, Figure 1-2. It magnifies small objects similar to the way binoculars magnify far-away objects. Both light microscopes and binoculars have lenses for magnifying objects.

In a **light microscope**, light passes through the object being looked at and then through two or more lenses. What you look at appears larger.

Skill Check

Understand science words: microscope. The word part *micro* means small. In your dictionary, find three words with the word part *micro* in them. *For more help, refer to the* ***Skill Handbook,*** *pages 706-711.*

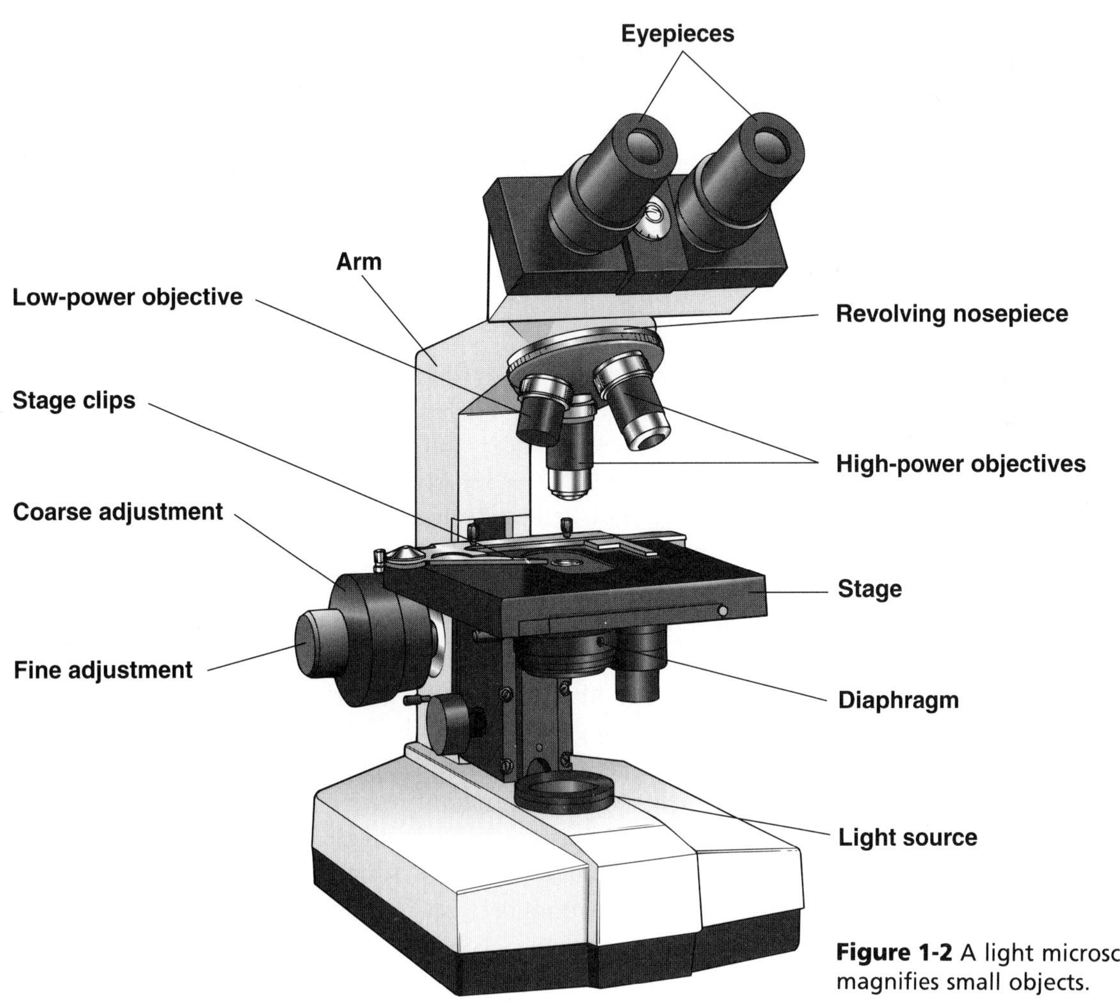

Figure 1-2 A light microscope magnifies small objects.

a

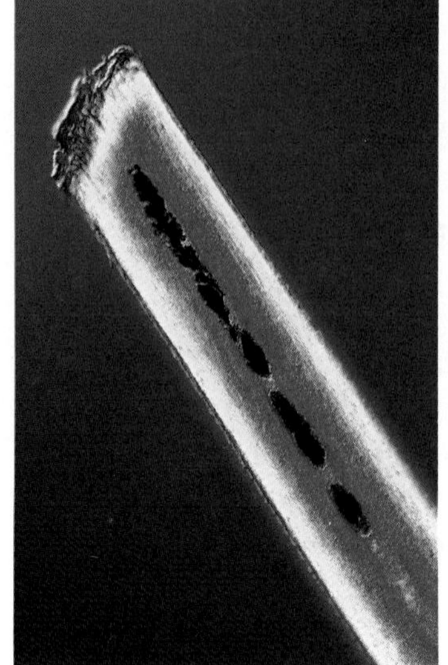

b

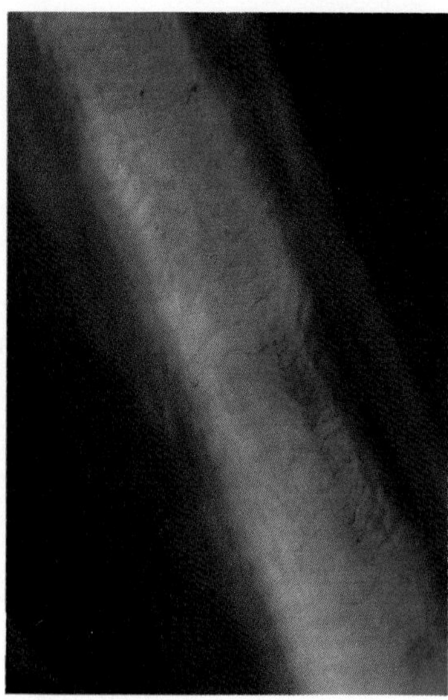

c

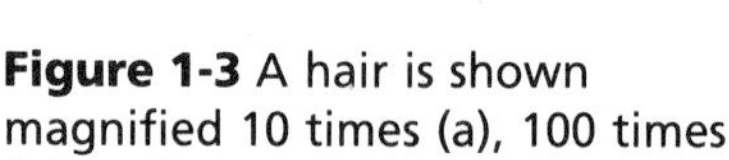

Figure 1-3 A hair is shown magnified 10 times (a), 100 times (b), and 430 times (c).

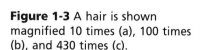

The microscope has a lens in the eyepiece. The eyepiece lens magnifies 10 times and is marked 10 ×. If you looked through just the eyepiece, a hair would look like the one in Figure 1-3a.

In addition to the eyepiece lens, a light microscope has one or more objective lenses that magnify the object further. One objective is a low-power lens. It magnifies 10 times and is marked 10 ×. To find the *total magnification* when looking through the eyepiece lens and the objective lens, you multiply the magnification of the two lenses together.

eyepiece lens × objective lens = total magnification

10 times × 10 times = 100 times larger

If you looked at the same hair with the low-power objective and the eyepiece together, you would see something that looks like Figure 1-3b. This hair is enlarged 100 times.

The second objective lens is a high-power lens. High-power objectives can magnify 40, 43, or 100 times. How would an objective that magnifies 43 times be marked? If you looked at the hair with a high-power objective and the eyepiece together, it might look like the hair in Figure 1-3c. This hair is magnified 430 times. The highest-powered light microscopes can magnify objects 2000 times.

Stereomicroscopes (STER ee oh MI kruh skohps) are often used in biology classes. A **stereomicroscope** is used for viewing large objects and things through which light cannot pass, such as insects. The stereomicroscope has an eyepiece lens for each eye, as shown in Figure 1-4.

Skill Check

Make and use tables: Make a table that compares the magnifying powers of a hand lens, a stereomicroscope, low power and high power of a light microscope, and an electron microscope. *For more help, refer to the **Skill Handbook,** pages 715-717.*

What kinds of objects would you view with a stereomicroscope?

Lab 1–1 Microscopes

Problem: How do you use the microscope?

Skills

observe, recognize spatial relationships

Materials

microscope
microscope slide
coverslip
water
dropper
forceps
small piece of magazine print

Procedure

1. Place a drop of water in the center of a glass slide.
2. With forceps, place a small piece of magazine print in the drop of water. Make sure the piece of magazine has a letter e in it.
3. Hold a clean coverslip by the edges at one side of the drop of water. Carefully lower the coverslip onto the drop of water. You have just made a wet mount slide.
4. Place the slide on the stage of the microscope. Put the stage clips into place. Move the slide so that the magazine print is directly over the hole in the stage.
5. Turn the low-power objective into place.
6. Looking from the side, turn the coarse adjustment so that the low-power objective is near the coverslip.
7. Open the diaphragm so that the most light enters the microscope. If your microscope has a mirror, adjust it until you see a circle of light. This circle of light is called the field of view.
8. Look through the eyepiece. Try to keep both eyes open. Raise the body tube until you can see the letters. **CAUTION:** *Never lower the body tube while looking through the eyepiece.*
9. Use the fine adjustment to bring the letters into sharp focus. Find a letter e.
10. Adjust the slide so that the letter e is in the center of your field of view.
11. On a separate paper, make a drawing of the letter e. Label your drawing with your name and the magnification.
12. **Recognize spatial relationships:** Move the slide to the left, to the right, toward you, and away from you. Note the direction in which the e appears to move.
13. Turn the nosepiece to bring the high-power objective into position. Focus with the fine adjustment. **CAUTION:** *Use only the fine adjustment with the high-power objective.*
14. Draw the letter e as it appears under high power. Label your drawing with your name and the magnification.
15. Remove the wet mount. Rinse and dry the slide and coverslip. Set your microscope on low power and put it away.

Data and Observations

1. What happens when you move the slide to the left? To the right? Toward you? Away from you?
2. Describe the appearance of the letter e under high power.

Analyze and Apply

1. What is the purpose of the coverslip?
2. Why should the coverslip be held by the edges?
3. **Apply:** Why must a specimen be very thin to be viewed under the microscope?

Extension

Observe: View other objects, such as a hair or thread under the microscope.

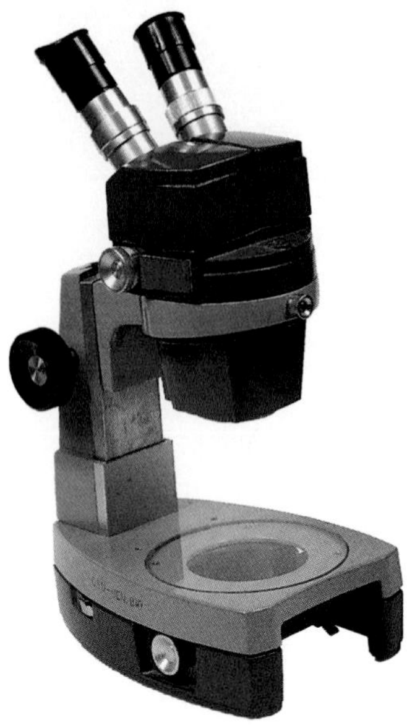

Figure 1-4 A stereomicroscope is used for viewing objects through which light will not pass.

How does having two eyepiece lenses help? The advantage of having a lens for each eye is that it makes the object appear three-dimensional. A disadvantage of the stereomicroscope is that its magnifying power is not as great as that of a light microscope. The stereomicroscope magnifies only 10 to 60 times.

An electron microscope is a very high-powered microscope that uses a beam of electrons instead of light to magnify objects. The magnified object is seen on a monitor similar to a TV screen. An advantage of an electron microscope is that it can magnify more than 500 000 times. Objects can be viewed in great detail. A disadvantage is that objects can't be viewed in color. Also, objects must be stained with expensive dyes before they can be viewed under the electron microscope. The staining process requires a great deal of skill and training.

New information about biology is published every day. How could a biologist remember all of it? Many biologists use computers to store, find, and process information. The computer helps biologists search for information and organize it. It saves many hours of library research.

How do biologists use cameras, nets, and gardens?

Biologists use many other tools that you may not associate with biology. For example, biologists use cameras. A wildlife biologist may take photographs of deer to understand the deer's daily habits. Biologists who study fish need boats to take them to where the fish are. They also use nets to catch the fish so they can observe them. Did you ever imagine that a garden could be a biologist's tool? Biologists who study how plants grow use gardens all the time for their work. As you see, biologists use more than laboratory tools to study living things.

Check Your Understanding

1. How does a beekeeper use biology?
2. What are three tools used by biologists?
3. A light microscope has an eyepiece that magnifies 10 × and an objective that is marked 43 ×. What is the total magnification for these two lenses?
4. **Critical Thinking:** How would a farmer use biology to grow corn?
5. **Biology and Math:** If a microscope eyepiece magnifies 4× and the total magnification is 400×, how much does the objective lens magnify?

1:2 Measurements Used in Biology

It is 40 kilometers to the nearest city. The soft drink is in a 2-liter bottle. The mass of the bag of potato chips is 184 grams. A football field is 300 feet long. A photographer uses a 35 millimeter camera. The medicine comes in 25-milligram tablets.

In which of these sentences is the unit of measurement different? The football field measurement is different. It is in English units. The other measurements are in metric units. Scientists today use a modern form of the metric system. They use the International System of Units, or SI system. The **International System of Units** is a measuring system based on units of 10. Scientists all over the world use it. Measurements in the SI system are similar to measurements in our money system. Let's see how the two systems compare.

Objectives

4. **Explain** how SI units are grouped.
5. **List** the SI units of measure for length, volume, mass, time, and temperature.
6. **Measure** objects using SI units.

Key Science Words

International System of Units
meter
volume
kilogram
Celsius

Length

Length is the distance from one point to another. What does our money system have in common with a system for measuring length? The basic unit in our money system is the dollar. As you know, the dollar can be divided into 10 dimes. Each dime can be divided into 10 cents as shown in Figure 1-5.

The SI unit of length is the **meter** (m). Figure 1-5 shows that like a dollar, a meter can be divided into 10 smaller units. The meter is divided into 10 units called decimeters (DES uh meet urz) (dm). Each decimeter can be divided into 10 centimeters (SENT uh meet urz) (cm).

Figure 1-5 The SI system is based on units of 10, just as our money system is.

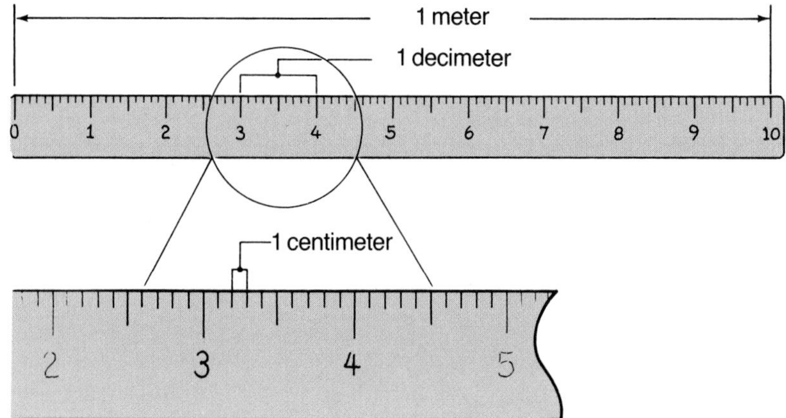

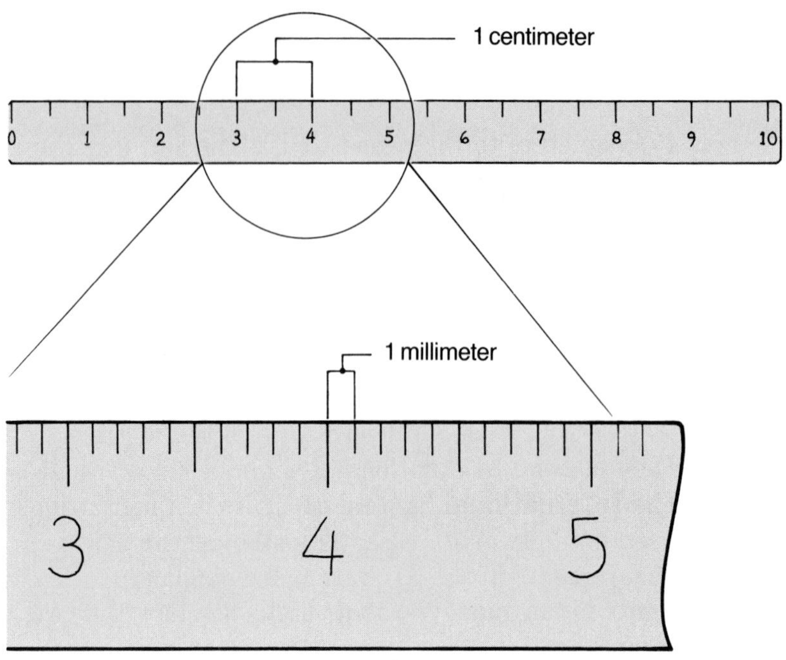

Figure 1-6 There are 10 millimeters in one centimeter.

In our money system, the smallest amount of money is one cent. However, the smallest length you can measure is not one centimeter. Figure 1-6 shows that each centimeter can be further divided into 10 units called millimeters (MIHL uh meet urz) (mm). Very small measurements are made in millimeters.

Do we have any units in our money system that are *larger* than the dollar? Yes, we group dollars together into tens, hundreds, and thousands. We also put meters into larger groups. Table 1-1 shows how metric units are grouped. The prefixes shown have special meanings. Notice that the first three prefixes change meters to larger units of measurement. For example, 1000 meters make one *kilo*meter (KIHL uh meet ur) (km). Ten *deci*meters make one meter. One hundred *centi*meters make one meter. The second three prefixes change meters to smaller units of measurement. How many *milli*meters make one meter?

Skill Check

Calculate: Use a calculator and a meter stick to find the area of your classroom and your desk. *For more help, refer to the **Skill Handbook,** pages 718-719.*

What is the width of your hand in millimeters?

TABLE 1–1. HOW METRIC UNITS ARE GROUPED

Prefix (word part)	Symbol	Meaning
kilo-	k	1000
hecto-	h	100
deka-	da	10
deci-	d	0.1 (1/10)
centi-	c	0.01 (1/100)
milli-	m	0.001 (1/1000)

Volume

Volume is the amount of space a substance occupies. Units of volume are based on units of length. If we multiply length by width by height, we get volume. The SI unit of volume is the cubic meter (m^3). This volume is rather large for everyday use, so scientists often use the liter (L) instead. A liter is 1/1000 as large as a cubic meter. You use the liter when you do labs in science class, or when you buy soft drinks at the grocery store. One liter equals 1000 milliliters (mL).

Into how many smaller units can a liter be divided?

Like meters, liters can be put into large groups. One kiloliter means 1000 liters. What does five kiloliters mean?

Weight and Mass

The photos in Figure 1-7 show an astronaut on Earth and an astronaut on the moon. The astronaut does not weigh the same in both places. Weight is a measure of the force of gravity on an object. The force of gravity on the moon is less than that on Earth. Therefore, an astronaut weighs less on the moon than on Earth. Even though the astronaut weighs less on the moon than on Earth, his or her mass does not change. Mass is how much matter is in something. Let's look at mass in another way. Consider the masses of two different objects, a box of bricks and a box of cotton. Which has more mass? Mass is measured by comparing an object of unknown mass to an object of known mass. The instrument used to measure mass is a balance.

a

Figure 1-7 An astronaut's mass is the same on Earth (a) as it is on the moon (b).

b

The SI unit of mass is the **kilogram** (kg). There are 1000 grams (g) in a kilogram. Grams are used to measure small objects. An egg has a mass of about 60 grams. A gram may be divided into 1000 units called milligrams. Milligrams are often used to measure medicines or ingredients in foods such as breakfast cereals.

Time and Temperature

Time is the period between two events. The SI unit for measuring time is the second (s). A stopwatch or a clock with a second hand is used to measure time.

For what kind of measurement is a kilogram used?

Temperature is a measure of the amount of heat in something. The SI scale for measuring temperature is the Kelvin scale (K). Scientists measure temperature with the **Celsius** (SEL see us) scale because it is easier to use. Temperature scales are divided into units called degrees. On the Celsius scale, the freezing point of water is 0 degrees and the boiling point is 100 degrees. Figure 1-8 shows the temperature in Celsius degrees of other familiar things.

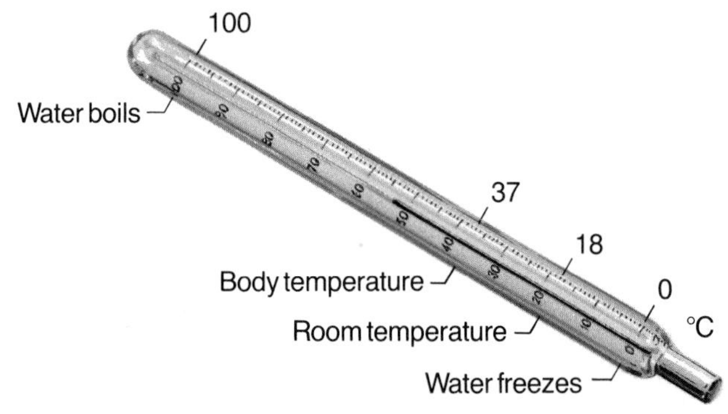

Figure 1-8 The Celsius scale

Check Your Understanding

6. What does the prefix *centi* mean?
7. What are some units used for measuring volume?
8. What is your body temperature in degrees Celsius?
9. **Critical Thinking:** How many centimeters make one kilometer?
10. **Biology and Math:** Suppose a cheetah can run 30 meters per second. How far can it run in 8 seconds?

1:3 Scientific Method

Much of the work of biology is to solve problems. Problems are not solved by flipping a coin or taking a guess as to the outcome. Scientists use a series of steps called a **scientific method** to solve problems. The following steps are often used: recognizing the problem, researching the problem, forming a hypothesis (hi PAHTH uh sus), testing the hypothesis, and drawing conclusions.

Objectives

7. Explain the steps of the scientific method.

8. Compare the difference between a hypothesis and a theory.

9. Explain how technology is used to solve everyday problems.

Key Science Words

scientific method
hypothesis
experiment
variable
control
data
theory
technology

Recognizing and Researching the Problem

Have you ever tried to turn on a light and found that it didn't work? Maybe you have turned the key in a car's ignition and the car didn't start. If you have had problems like those, you probably have used a scientific method to solve them.

A biology class raised guppies for a class project. Paula thought the guppies would have more young if the light in the aquarium was turned off part of the time. Other students thought the light should be left on all the time. But, Paula had observed that her guppies at home had more young than the ones at school. She also knew that she kept her aquarium light turned off part of the time. The class had a problem. What should they do about the aquarium light?

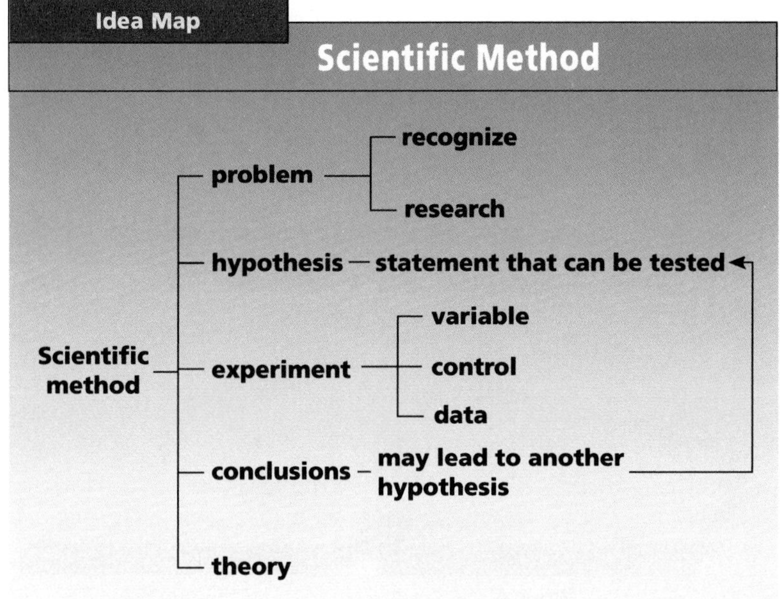

Study Tip: Use this idea map as a study guide to scientific method. Notice that a conclusion may lead to a second hypothesis. The second hypothesis then leads to more experiments and further conclusions.

Mini Lab

Does Aspirin Help Plants Grow?

Design an experiment: Your aunt tells you that plant cuttings will grow faster if aspirin is added to the water. Design and conduct an experiment to find out if this is true. *For more help, refer to the* ***Skill Handbook,*** *pages 704-705.*

Paula was able to recognize the problem. Before Paula could work on the solution to the problem, she needed to do some research at the library. She found out what temperature to keep the water and the amount of water needed. However, there was little information about how much light a guppy needs.

Forming a Hypothesis

A **hypothesis** is a statement that can be tested. Paula's hypothesis was: "If the light is turned off part of the time, then the guppies will have more young." Paula knew that her guppies received less light than the ones at school. She also observed two important things. Her tank at home had the same amount of water in it as the one at school. Also, the water temperature was the same in each tank. Paula was pretty sure that the conditions for the guppies were the same at home and at school, except for the amount of light. However, she needed to do a test to be absolutely sure.

Testing a Hypothesis

What is a hypothesis?

Paula's next step was to test her hypothesis. Testing a hypothesis using a series of steps with controlled conditions is called an **experiment**. In her experiment, Paula divided 16 guppies into two equal groups. She put the groups into separate tanks and labeled them group A and group B. She kept the temperature in the tanks the same. She put the same amount of water in each tank. The light for group B was left on 24 hours a day. The light for group A was turned on for 12 hours and off for 12 hours each day.

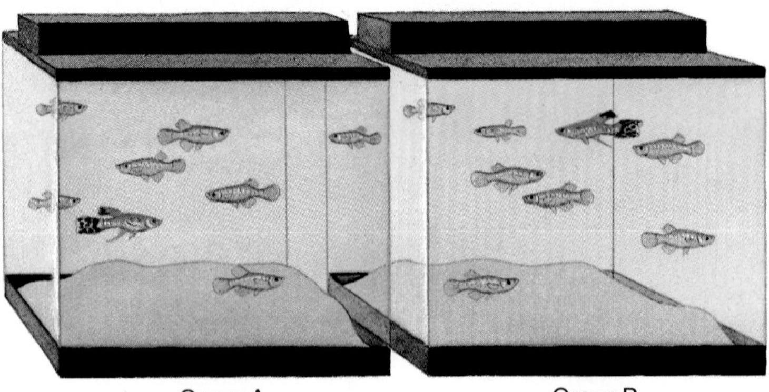

Figure 1-9 In Paula's experiment, the conditions for group A and group B were the same except for the amount of light received.

Lab 1–2

Scientific Method

Problem: How is a scientific method used to solve problems?

Skills

form hypotheses, interpret data, separate and control variables

Materials

2 100-mL beakers	2 paper towels
40 pinto beans	50-mL graduated cylinder
balance	2 self-sealing plastic bags
salt	masking tape
water	

Procedure

1. Solve this problem using a scientific method: Do seeds grow better when soaked in plain water or salt water?
2. Research the problem. Find out what seeds need in order to grow.
3. Develop a **hypothesis** based on the problem.
4. Conduct the experiment.
   a. Copy the data table.
   b. Label two beakers A and B. Fill each beaker with 50 mL of water.
   c. Add 20 beans to beaker A. Add 20 beans and 2 g salt to beaker B. Label each beaker with your name and set them aside overnight.
   d. On the following day, pour out the water from both beakers.
   e. Wrap the seeds from beaker A with a damp paper towel. Place the paper towel into a plastic bag and close the bag. Label the bag A.
   f. Repeat step e with the seeds in beaker B.

	SEEDS USED	SEEDS THAT GREW
Water		
Salt water		

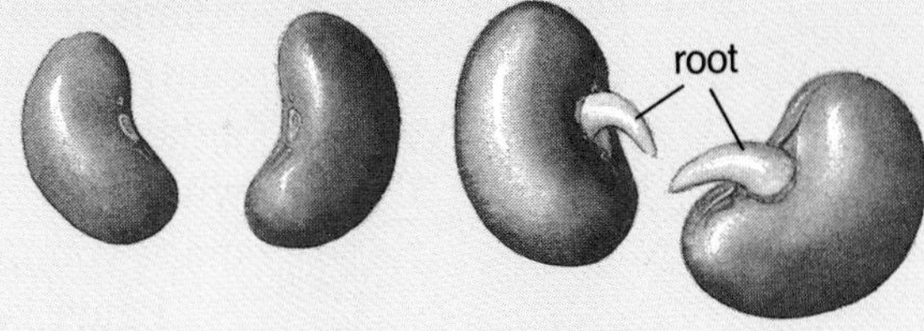

5. Collect the data.
   a. On the next day check the seeds for growth. Look for small white roots.
   b. Count the number of seeds that are growing in each bag. Record these numbers in your data table.
6. Draw conclusions. Based on your data, decide if you need a new hypothesis.
7. If soil and containers are available, you may want to plant the seeds that are growing.

Data and Observations

1. Under what condition, plain water or salt water, did more seeds grow?
2. Under what condition, plain water or salt water, did fewer seeds grow?

Analyze and Apply

1. **Check your hypothesis:** Is your original hypothesis supported by your data? Why or why not?
2. **Infer:** How can you be sure that the conclusion you reached is correct?
3. What variable was being tested in this experiment? What was the control?
4. **Apply:** How do salt water and plain water affect seed growth?

Extension

Design an experiment using distilled water in place of the salt water. Compare your results.

What are five steps of the scientific method described in this section?

A **variable** is something that causes the changes observed in an experiment. Some variables in Paula's experiment were water temperature, amount of food given to each group, and number of hours the light was left on. For meaningful results, the effect of only one variable can be tested at a time. Paula tested the effect of having the light on for different lengths of time.

An experiment must also have a control. A **control** is a standard for comparing results. In Paula's experiment, the control was group B, in which the light was left on 24 hours a day. Paula planned to compare her results for group A with her results for group B to see if group A produced more young.

Paula kept a record of the information she gathered. Recording information is called collecting data. **Data** are the recorded facts or measurements from an experiment.

Paula recorded the number of adult guppies she put into each tank. Every day, she recorded the temperature of the water in each of the tanks. Once a week for four weeks, she recorded the number of young produced by each group, Figure 1-11. She removed the new young each week so she could count more easily.

Drawing Conclusions

Paula's last step was to conclude whether or not the problem was solved. At the end of her experiment, the guppies in group A had more young than the guppies in group B, Figure 1-11. Could Paula conclude that 12 hours

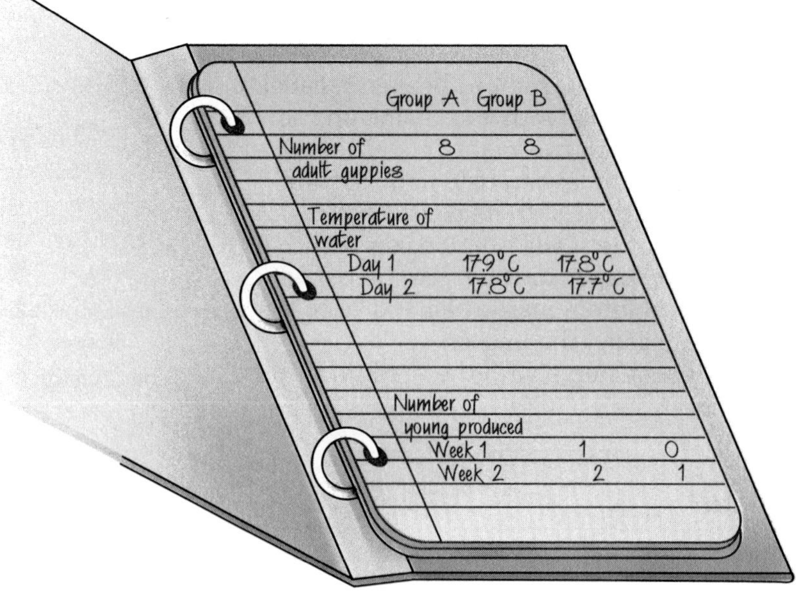

Group A Group B
Number of adult guppies 8 8
Temperature of water
Day 1 17.9°C 17.8°C
Day 2 17.8°C 17.7°C
Number of young produced
Week 1 1 0
Week 2 2 1

Figure 1-10 Paula recorded the data that she collected.

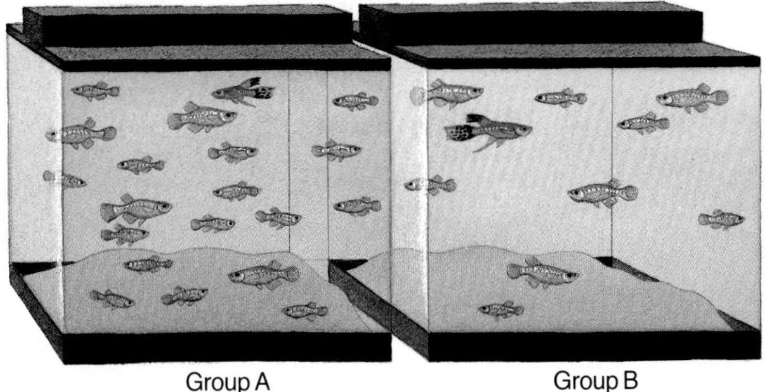

Figure 1-11 At the end of the experiment, group A had more young.

of light and 12 hours of dark caused the difference? She might. One way to be sure of a conclusion is to repeat the experiment. Repeating experiments is how scientists confirm that their hypotheses are correct. To confirm her hypothesis, Paula repeated the experiment three times with different groups of guppies. Each time the same thing happened. The guppies had more young when they were exposed to 12 hours of light and 12 hours of dark each day.

Paula used a scientific method to solve a problem. Let's review the steps that she used.

1. Recognize the problem.
2. Research the problem.
3. Form a hypothesis.
4. Test the hypothesis with an experiment. The experiment should have one variable being tested and a control group.
5. Draw conclusions. If your first hypothesis is wrong, you may need to form another hypothesis and test that.

Theory

If a hypothesis is tested with further experiments and the results are similar, it may become a theory. A **theory** is a hypothesis that has been tested again and again by many scientists, with similar results each time. You may have said at some time, "I think I know why the bus is late. My theory is that it got into a traffic jam on Fifth Street." But, this statement is not a theory. It hasn't been tested. A scientific theory is not a guess. It is the best explanation science has to offer about a problem. Theories can be used to predict future events. In the next chapter you will learn about the cell theory.

Bio Tip

Science: Nonscientists usually use *theory* to mean an idea or thought as opposed to a fact. Scientists use *theory* to describe an idea that has so much supporting evidence that it is almost certainly true.

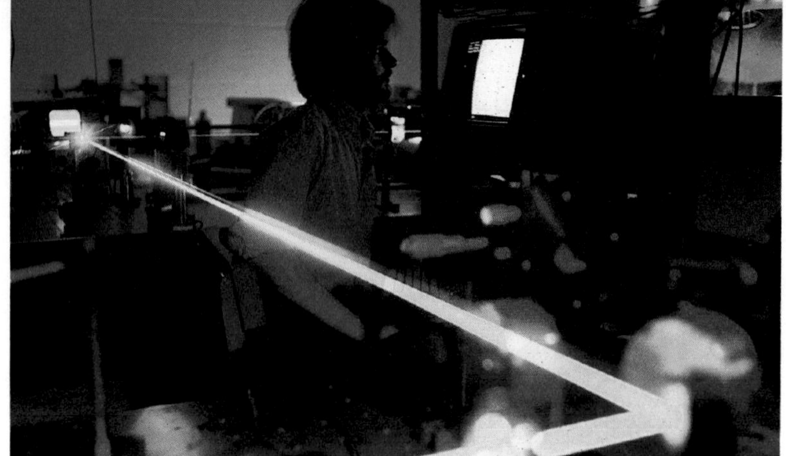

Figure 1-12 Lasers are products of technology.

Technology

Scientists have played a part in solving many problems in farming, industry, and medicine. The use of scientific discoveries to solve everyday problems is called **technology**. Doctors transplant human organs based on what they know about the human body. Over two million artificial devices are used each year to take the place of limbs or body organs lost to illnesses and accidents. These devices are products of technology.

Improved running shoes, laser surgery, and drugs that clear fats from the walls of blood vessels are also products of technology. These kinds of technological advances promote well-being. Other kinds of advances may be as helpful, but have harmful effects as well. Fertilizers help farmers grow better crops, but they may contribute to water pollution. Chemicals added to food keep foods fresh, but may have harmful effects on people if consumed for long periods of time.

Bio Tip

Consumer: A laser is a tool that produces a narrow, intense beam of light that can be used as a kind of scalpel. It can be used to remove unhealthy cells and in certain types of eye surgery.

How do doctors use technology to solve medical problems?

Check Your Understanding

11. What is the variable in an experiment? What is the control?
12. What are five steps in the scientific method described in this section?
13. Why is a theory more accepted than a hypothesis?
14. **Critical Thinking:** Give an example of how you have used technology to solve an everyday problem.
15. **Biology and Reading:** What three steps of the scientific method described in this section should be done before performing an experiment?

Science and Society

Technology: Helpful or Harmful?

Technology has provided many products that we believe make our lives better. Imagine walking three or more miles to and from school each day or trying to keep clean without soap and toothpaste. Look inside your cabinets and closets at home. Almost everything in them is a product of recent technology. Even the clothes you wear may contain fibers invented in a laboratory. Besides providing useful products, our use of technology has raised questions that do not have easy answers. Consider the three situations described below and try to answer the questions.

What Do You Think?

1. Nuclear energy is an important source of power. Many communities use nuclear energy to generate electricity for lighting homes and businesses. Nuclear power plants do not pollute the air as coal-burning power plants do. Use of nuclear energy, however, does present problems. There is no guaranteed safe way to dispose of used nuclear fuel. Also, many nuclear power plants are being shut down because they are not considered safe. Who is responsible for disposing of nuclear wastes safely? Should we continue to use and build nuclear power plants?

2. Farmers use fertilizers to grow more and better crops. More crops mean a better living for the farmer, a better food supply, and lower prices at the grocery store. Without fertilizers, we might not be able to buy many kinds of fruits and vegetables. Fertilizers, however, can drain from farmland into lakes and streams. Fish can't live in water with fertilizers in it. In some areas, people can't drink the water because of the fertilizers in it. Should laws be passed on when and how much fertilizers can be used?

3. A person is rushed to the hospital after an automobile accident. The doctors do not know if there is brain damage, or what the chances of survival are. A quick decision must be made about whether or not to use a life-support system called a respirator. A respirator is a device that helps a person breathe when that person can't breathe on his or her own. The respirator is used. Later, it is discovered that the person shows little brain activity. The person remains in a coma for months. There is little hope for recovery, and the family wants the respirator disconnected. Who should decide when life-support systems are needed?

Conclusion: How does technology help people? Is it possible to misuse technology?

Nuclear power plant

Chapter 1

Review

Summary

1:1 Biology in Use

1. Many jobs depend on the use of biology.
2. Tools that biologists use include microscopes, binoculars, cameras, nets, and computers.
3. The light microscope, stereomicroscope, and electron microscope are important tools in biology.

1:2 Measurements Used in Biology

4. The International System of Units (SI) is a measuring system based on units of 10 that all scientists use.
5. Meter, cubic meter, kilogram, second, and Celsius degrees are SI measurements.
6. SI units are used to measure length, volume, mass, time, and temperature.

1:3 Scientific Method

7. The steps of the scientific method are: recognizing the problem, researching the problem, forming a hypothesis, testing the hypothesis, and drawing conclusions.
8. A theory is a hypothesis that has been tested again and again by many scientists. Results are the same each time.
9. Technology is used to solve many everyday problems.

Key Science Words

biology (p. 5)
Celsius (p. 14)
control (p. 18)
data (p. 18)
experiment (p. 16)
hypothesis (p. 16)
International System of Units (p. 11)
kilogram (p. 14)
light microscope (p. 7)
meter (p. 11)
scientific method (p. 15)
stereomicroscope (p. 8)
technology (p. 20)
theory (p. 19)
variable (p. 18)
volume (p. 13)

Testing Yourself

Using Words

Choose the word from the list of Key Science Words that best fits the definition.

1. one thousand grams
2. measured in liters
3. series of steps in solving a problem
4. can be divided into 100 units called centimeters
5. study of living things
6. magnifies objects 430 ×
7. recorded measurements in an experiment
8. a statement that can be tested
9. using discoveries in science to solve everyday problems
10. steps used to test a hypothesis
11. something that causes changes in an experiment
12. magnifies objects through which light does not pass

Testing Yourself *continued*

Finding Main Ideas

List the page number where each main idea below is found. Then, explain each main idea.

13. how time is measured
14. what a scientific method is
15. what an experiment is
16. how people use biology
17. instrument used to measure mass
18. how to find volume
19. tools a biologist uses
20. what a hypothesis is
21. how prefixes are used
22. part of an experiment that is not changed

Using Main Ideas

Answer the questions by referring to the page number after each question.

23. What must an optometrist and a mushroom grower know about biology? (p. 6)
24. How is a stereomicroscope different from a light microscope? (p. 8)
25. How are mass and weight different? (p. 13)
26. How are biology and technology related? (p. 20)
27. Which SI units could you use to measure
    (a) the distance to your friend's house? (p. 11)
    (b) the volume of one serving of orange juice? (p. 13)
    (c) the mass of a can of soup? (p. 14)
28. How are a variable and a control different? (p. 18)
29. Why do biologists use computers? (p. 10)

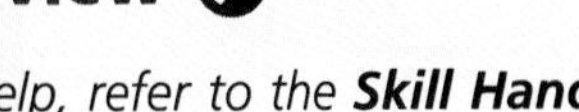

Skill Review

For more help, refer to the ***Skill Handbook,*** *pages 704-719.*

1. **Calculate:** An ostrich has a mass of 155 kg. A hummingbird has a mass of 10 g. How many times greater than the mass of the hummingbird is the mass of the ostrich?
2. **Design an experiment:** Design an experiment to show the effect of sunlight on the growth of bean seedlings.
3. **Observe:** Look around your classroom and name the tools that you observe.
4. **Make and use tables:** Make a table of the different kinds of measurements and the SI units for each.

Finding Out More

Critical Thinking

1. Design a controlled experiment to test the following hypothesis: Radish seeds will germinate faster in light than in darkness.
2. Why is it important to use both a number and a unit when measuring an object?

Applications

1. Report to your class on new technologies in medicine, farming, or industry by reading the newspaper or watching television.
2. Find out which metric units are used by a hospital or pharmacy in preparing medicines. List several examples.

CHAPTER PREVIEW

Chapter Content

Review this outline for Chapter 2 before you read the chapter.

2:1 **Living Things and Their Parts**
- Features of Living Things
- The Chemistry of Life
- Cell Theory

2:2 **Cell Parts and Their Jobs**
- Cell Membrane and Nucleus
- Cytoplasm

2:3 **Special Cell Processes**
- Diffusion
- Osmosis
- Organization

Skills in this Chapter

The skills that you will use in this chapter are listed below.

- In **Lab 2-1,** you will make and use tables, interpret data, infer, and form hypotheses. In **Lab 2-2,** you will observe, compare, and use a microscope.
- In the **Skill Checks,** you will classify, formulate a model, and understand science words.
- In the **Mini Labs,** you will use a microscope and experiment.

Chapter 2

Features of Life and the Cell

In the photo on the left, you can see several kinds of things that are living and several that are not living. If you looked closely at an enlarged part of the rocks, as shown in the small photo, what would you see? Make a guess as to what you would find if you looked at a drop of the water with a microscope.

What does it mean to be living? Maybe you would say that living things need water and food. You might say that living things grow. Some nonliving things seem to grow, too. If you have cold winters where you live, you probably have seen icicles that seem to grow on your house. Maybe the ability to grow isn't enough to make something living. How then can you decide if a thing is living or nonliving? Are there features that all living things share?

Try This!

What happens to a sponge in water? Place a sponge in a dish of water. Observe what happens to the sponge. Is this a feature of living things?

BIOLOGY Online

Visit the Glencoe Science Web site at science.glencoe.com to find links about **features of life and the cell.**

Rock covered with mosses and lichen

2:1 Living Things and Their Parts

Objectives

1. **Describe** eight features common to all living things.
2. **List** the elements that make up living things.
3. **State** the major ideas of the cell theory.

Key Science Words

reproduce
development
consumer
producer
cellular respiration
cell
adaptation

The word *living* is not easy to define. However, biologists recognize that all living things share certain features. Let's look at some of those features.

Features of Living Things

Feature 1 Living things reproduce.

Reproduce means to form offspring similar to the parents. If living things could not reproduce, there might be nothing alive on Earth today.

Living things reproduce in different ways. For example, a one-celled living thing may split to produce two living things, like the amoeba in Figure 2-1.

Sunflowers grow from seeds made by the parent plants. A sunflower produces many seeds, each of which can become a new sunflower. Bear cubs have a male and a female parent and are born alive.

Feature 2 Living things grow.

A living thing grows by using materials and energy from its environment to increase its size. Young oak saplings grow into large oak trees over a period of many years. A kitten grows into a cat.

Feature 3 Living things develop.

Development is all the changes that occur as a living thing grows. When a tadpole first begins to develop, it does not look like an adult frog, Figure 2-2a. It has no front or hind legs. As the animal grows and develops, legs begin to

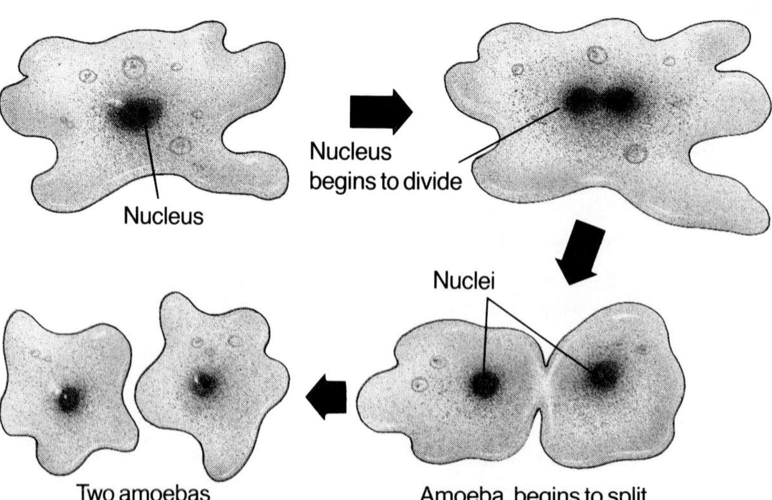

Figure 2-1 An amoeba divides to reproduce.

a

b

Figure 2-2 A tadpole (a) does not look like an adult frog. A developing bean seed (b) does not look like a young bean plant.

form. A young plant in a seed looks different from a fully developed plant, as shown by Figure 2-2b. The young plant does not have roots, stems, or flowers. Many of these parts develop as the plant grows.

Feature 4 Living things need food.

Many animals get food by hunting and eating other animals. Some animals eat only plants. Other animals eat both plants and animals. Animals are consumers. **Consumers** are living things that eat, or consume, other living things. Green plants make their own food. Green plants are producers. **Producers** are living things that make, or produce, their own food.

Skill Check

Classify: Look around you and make a list of things that are living or were once living. How are living things different from things that were once living? *For more help, refer to the **Skill Handbook**, pages 715-717.*

Feature 5 Living things use energy.

Energy is the ability to do work. Moving your arms or legs, blinking your eyes, breathing, and even sleeping require energy. Animals get energy from the food they eat. Plants get energy from the food they make during photosynthesis.

How do living things get energy from food? All living things carry out cellular respiration. **Cellular respiration** is the process by which food is broken down and energy is released. Many living things use oxygen in the process of cellular respiration. Water and a gas called carbon dioxide are given off. The energy given off during cellular respiration is used for other life processes. For example, energy is used to keep human body temperature at 37°C.

What is given off during cellular respiration?

Feature 6 Living things are made of cells.

The **cell** is the basic unit of all living things. Some living things have many cells. Others have only one cell.

Respiration

Lab 2–1

Problem: Do all things give off carbon dioxide?

Materials

5 small baby food jars
red liquid
wax pencil
guppy
yeast suspension
dropper
paper clip
straw

Skills

make and use tables, interpret data, infer, form hypotheses

Procedure

1. Copy the data table.
2. Number the five baby food jars 1 to 5. Fill each of the jars 2/3 full with red liquid.
3. Record in your notebook the color in each jar at the start of the activity.
4. Add the things listed here to the jars. Jar 1—Add nothing. Jar 2—Use the straw to blow gently into the red liquid for 2 minutes. Jar 3—Add 1 guppy. **CAUTION:** *Always use care when handling live animals.* Jar 4—Add 5 drops of yeast solution. Jar 5–Add 1 paper clip. Record what is added to each jar on your table.
5. In your notebook, write a **hypothesis** about which things you think will give off carbon dioxide. You will know when carbon dioxide is given off because the red liquid will turn orange or yellow.
6. Wait 15 minutes. Record in your table the color of the liquid in each jar using red, orange, or yellow as the choices.
7. Return the guppy to your teacher.

Data and Observations

1. In which jars did the color change?
2. In which jars was carbon dioxide present?

Analyze and Apply

1. **Check your hypothesis:** Is your hypothesis supported by your data? Why or why not?
2. What was the purpose of Jar 1?
3. **Interpret data:** Which things gave off carbon dioxide?
4. Which things didn't give off carbon dioxide?
5. What process in living things causes carbon dioxide to be given off?
6. **Apply:** Which of these things give off carbon dioxide: a tree? a bicycle? a fly? a shoe?

Extension

Infer: Look around your classroom. Make a table listing the things that now give off carbon dioxide, things that gave off carbon dioxide when they were living, and things that never gave off carbon dioxide.

JAR	1	2	3	4	5
What was added					
Color at start					
Color of liquid after 15 minutes					
Color of liquid in Jar 2 after 2 minutes					

Figure 2-3 Plants respond to light by growing toward it.

Feature 7 Living things respond.

The plants in Figure 2-3 are responding to the light by growing toward it. When you call your dog, it responds to the sound of your voice. It comes to you. These are examples of living things responding to changes in the environment. The environment is made up of all the living and nonliving things that surround another living thing.

You respond to changes in the environment, too. When you get cold, you shiver. When you run, your heart beats faster. When your body gets warm, you sweat. These are examples of how your body responds to changes.

Feature 8 Living things are adapted to their environments.

A trait that makes a living thing better able to survive is called an **adaptation.** Polar bears are adapted to living in the Arctic. They have thick fur and layers of fat to keep warm. Squirrels are adapted to living in trees. Their bushy tails help them keep their balance.

How can you tell if something is living? It must have all of the features just described. If it does not, it is not living. For example, a burning candle has some features that living things have, Figure 2-4. It gives off carbon dioxide gas and water. It uses oxygen and wax and changes shape. A candle responds. The candle flame flickers when a draft passes over it.

Does having these features mean that a burning candle is alive? The answer is no. A candle does not reproduce. It does not have cells. It does not grow. It is not adapted to its environment. The candle is not living, even though it has a few of the features that a living thing has.

Figure 2-4 A burning candle has some features of living things. What are they?

Skill Check

Formulate a model: Construct a model of a simple molecule, such as water or carbon dioxide. Use polystyrene balls and toothpicks. Label each atom in the molecule. *For more help, refer to the* ***Skill Handbook****, pages 706-711.*

The Chemistry of Life

Living things are made of matter. Matter is anything that has mass and takes up space. Matter is made of tiny particles called atoms. Atoms make up elements. An element is a substance that is made up of only one kind of atom. Oxygen and hydrogen are examples of elements.

Matter is often made of two or more different elements joined together to form a compound. Joining different elements can change the way they act. For example, water is a compound made of the elements oxygen and hydrogen. Oxygen and hydrogen are gases. Yet, when they join they make water. As you know, water is a liquid. The smallest part of a compound is a molecule. A molecule of water is the smallest amount of water you can have and still have water.

What is the chemical formula for water?

Scientists use a certain symbol for each different element. When written together, the symbols make a chemical formula. A chemical formula is a way to write the name of a compound using symbols. The symbol for oxygen is O. The symbol for hydrogen is H. The chemical formula for a molecule of water is H_2O. The number 2 shows that there are two hydrogen atoms for every oxygen atom in a molecule of water.

Seven elements make up over 99 percent of the matter in living things. Those seven elements are carbon, hydrogen, oxygen, nitrogen, phosphorus (FAHS fuh rus), calcium, and sulfur, Figure 2-5. They are often called the building blocks of living matter.

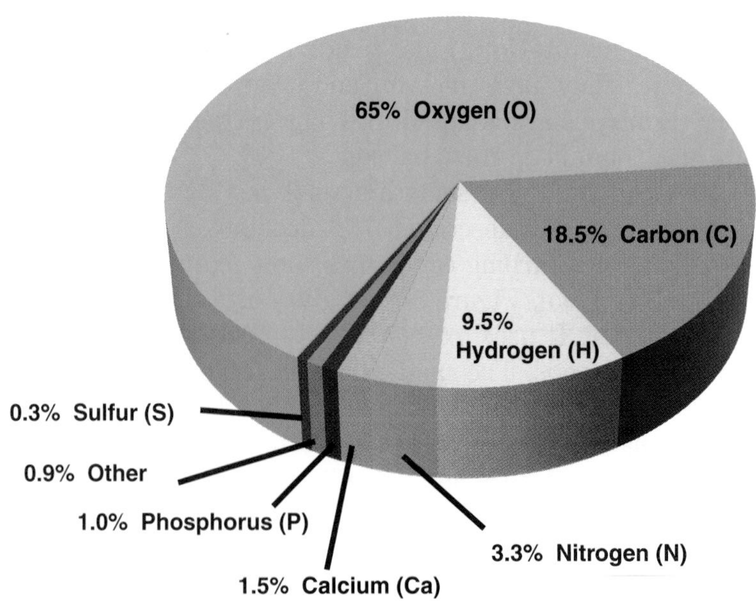

Figure 2-5 Living things are made up of the elements shown here. Which element makes up most of a living thing?

Cell Theory

In 1665, an English scientist named Robert Hooke looked at thin slices of cork under a microscope. Hooke saw that cork had a lot of empty spaces. Hooke used the word *cells* to describe the empty spaces in cork.

Today, biologists know that Hooke did not see living cells. He saw the walls of cells that were alive at one time.

By the nineteenth century, microscopes had been improved. Scientists could see that the cell had parts. First, Robert Brown discovered the central part of the cell, the nucleus (NEW klee us). Then, two German biologists, Matthias Schleiden and Theodor Schwann, did experiments to see what kinds of living things had cells. They formed hypotheses that all plants and animals were made of cells.

The experiments of Schleiden, Schwann, and other scientists led to the development of the cell theory. The major ideas of the cell theory are listed below.

1. All living things are made of one or more cells.
2. Cells are the basic units of structure and function in living things.
3. All cells come from other cells.

Mini Lab

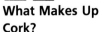

What Makes Up Cork?

Use a microscope: To see what Robert Hooke saw, slice a thin section from a piece of cork and look at it under the microscope. What do you see in the center of each cell? *For more help refer to the **Skill Handbook**, pages 712-714.*

What events led to the cell theory?

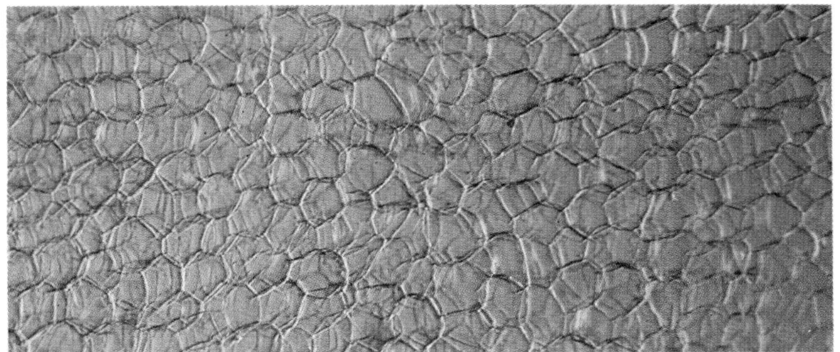

Figure 2-6 These cork cells are similar to what Robert Hooke saw.

Check Your Understanding

1. Name eight features of living things.
2. What six elements make up most of living matter?
3. What is the cell theory?
4. **Critical Thinking:** Why is cellular respiration necessary to living things?
5. **Biology and Writing:** Write a paragraph about a nonliving thing not mentioned in this section. Describe the features of living things that the nonliving thing has.

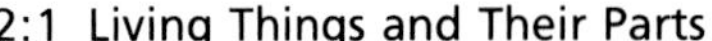

2:2 Cell Parts and Their Jobs

Objectives

4. List the parts of the cell.

5. Describe the functions of the cell parts.

Key Science Words

cell membrane
nucleus
nuclear membrane
nucleolus
chromosome
cytoplasm
ribosome
mitochondria
vacuole
centriole
chloroplast
cell wall

Cells are microscopic units that make up all living things. Cells are alive. They do everything needed to stay alive. They carry on cellular respiration. They grow and reproduce. A cell has many different parts to do all of these jobs. As you study the parts of the cell, refer to Figure 2-7. Figure 2-7 shows an animal cell and a plant cell and the cell parts of each.

Cell Membrane and Nucleus

All cells are surrounded by a cell membrane. The **cell membrane** gives the cell shape and holds the cytoplasm. It also helps control what moves into and out of the cell.

The **nucleus** is the cell part that controls most of the cell's activities. It determines how and when proteins will be made. Proteins are complex substances with several different jobs. Some form cell parts. Others regulate activities of the cell. The nucleus also passes traits from parents to offspring.

The **nuclear membrane** is a structure that surrounds the nucleus and separates it from the rest of the cell. The nuclear membrane has openings that allow certain materials to move into and out of the nucleus.

Inside the nucleus is a smaller body called the nucleolus (new KLEE uh lus). The **nucleolus** is the cell part that helps make ribosomes (RI buh sohmz). You will read about ribosomes in the next section. Some cells have more than one nucleolus.

Also inside the nucleus are threadlike structures called chromosomes (KROH muh sohmz). **Chromosomes** are cell parts with information that determines what traits a living thing will have. Examples of traits are hair color, eye color, and sizes and shapes of leaves.

Cytoplasm

What is found between the cell membrane and the nucleus?

The clear, jellylike material between the cell membrane and the nucleus that makes up most of the cell is called **cytoplasm** (SITE uh plaz um). Most of the cell's chemical reactions take place in the cytoplasm. Cytoplasm is mostly water, but it also has other chemicals. In addition, other cell parts that carry on special functions are found in the cytoplasm.

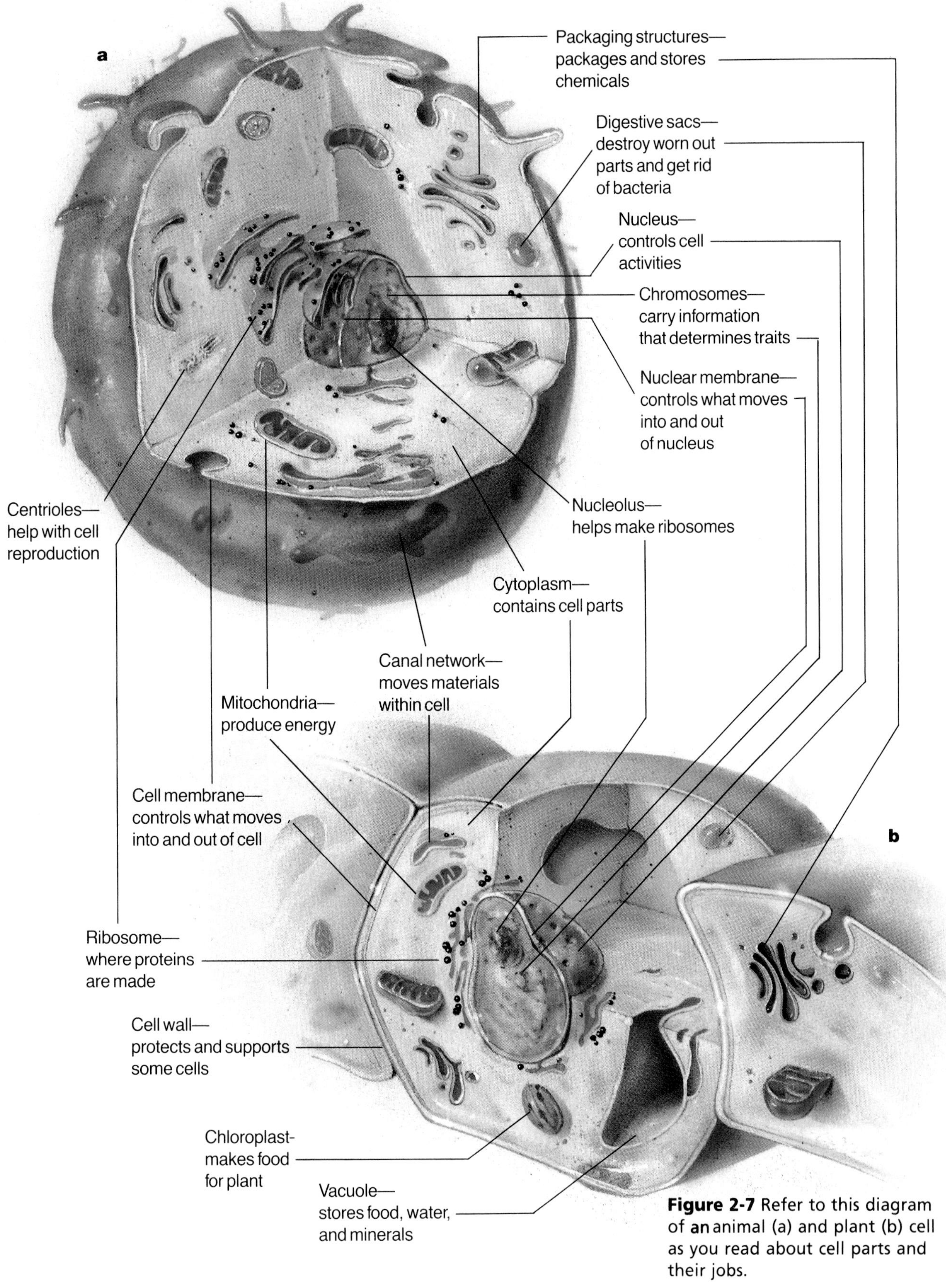

Figure 2-7 Refer to this diagram of an animal (a) and plant (b) cell as you read about cell parts and their jobs.

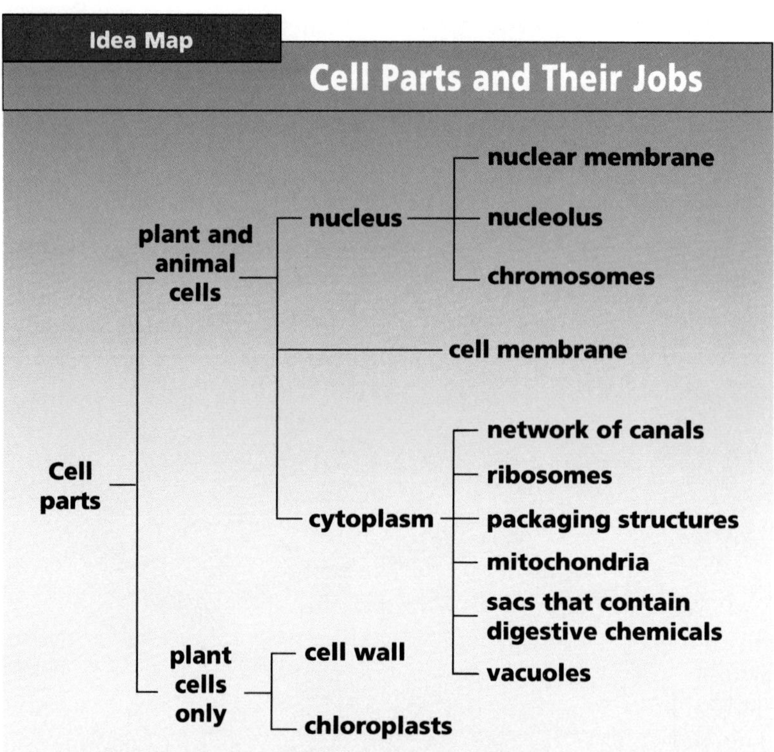

Study Tip: Use this idea map as you study cell parts and their jobs. Remember that only plant cells have cell walls and chloroplasts, and only animal cells have centrioles. All cells have the other cell parts shown in the idea map.

First, the cytoplasm contains a network of canals that help move material around inside the cell. The canals connect the nuclear membrane and the cell membrane.

What is the job of the ribosomes?

A second cell part found in the cytoplasm is the ribosome. **Ribosomes** are cell parts where proteins are made. Large numbers of ribosomes can be found along the canal network, where they are made. Ribosomes can also be found throughout the cytoplasm.

A third cell part found in the cytoplasm is a structure that packages and stores chemicals to be released from the cell. Large numbers of these packaging structures are found in cells that make saliva. Why do you suppose this is so? Large amounts of saliva are needed to break down the foods you eat.

Fourth, the cytoplasm contains rod-shaped bodies called mitochondria (mite uh KAHN dree uh). The **mitochondria** are cell parts that produce energy from food that has been digested. Mitochondria are often called "powerhouses" of the cell because they produce so much energy.

Small sacs that contain digestive chemicals are a fifth structure found in the cytoplasm. The chemicals made in these sacs break down large molecules. They get rid of

Bio Tip

Health: Hair is not made of cells. It is made of protein that is secreted by hair follicle cells. Hair follicle cells surround the root of the hair.

disease-causing bacteria that enter the cell. They also destroy worn-out cell parts and form products that can be used again.

Sixth, most cells have vacuoles (VAK yuh wolz) within the cytoplasm. A **vacuole** is a liquid-filled space that stores food, water, and minerals. Vacuoles also store wastes until the cell is ready to get rid of them. In most plant cells, the vacuole takes up a large amount of space within the cell. The fluid inside the vacuole helps to support the plant.

Centrioles (SEN tree olhz), a seventh structure within the cytoplasm, are located near the nucleus in animal cells but not in plant cells. **Centrioles** are cell parts that help with cell reproduction. They exist in pairs in the cell.

The cytoplasm of plant cells contains an eighth cell part, chloroplasts (KLOR uh plasts). **Chloroplasts** are cell parts that contain the green pigment, chlorophyll. Chlorophyll traps energy from the sun. Plants use this energy to make food. Chloroplasts give plants their green color.

The cells of plants, algae, fungi, and some bacteria have cell walls. Animal cells do not have cell walls. The **cell wall** is a thick outer covering outside the cell membrane. It protects and supports the cell.

The cell wall often remains after the rest of the cell has died. Wood is made of the walls of dead cells. What did Robert Hooke see when he looked at cork cells?

Skill Check

Understand science words: chloroplast. The word part *chloro* means green. In your dictionary, find three words with the word part *chloro* in them. *For more help, refer to the **Skill Handbook**, pages 706-711.*

Figure 2-8 A plant cell magnified about 5000 times (a) and an animal cell magnified about 1900 times (b) are shown here. Note the labeled structures.

a

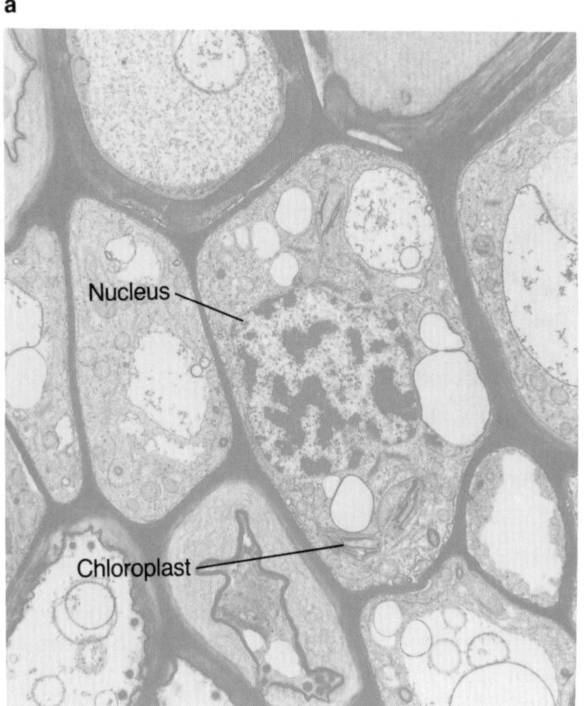

b

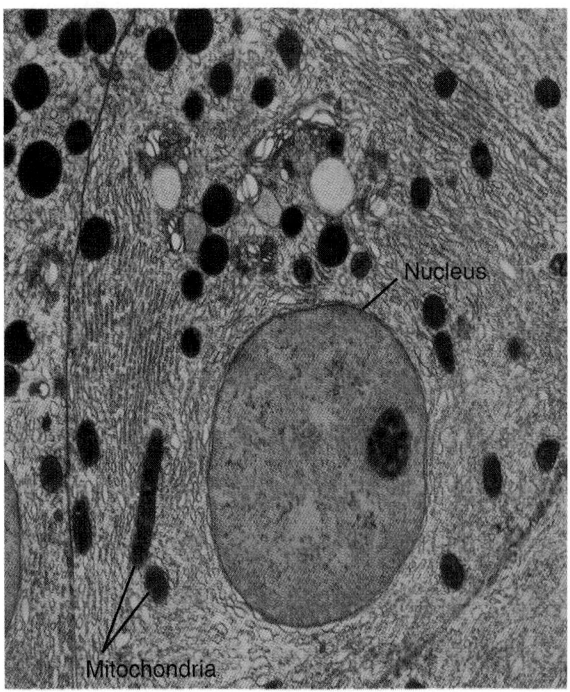

Cells

Lab 2–2

Problem: How do animal and plant cells differ?

Materials

light microscope
2 glass slides
2 coverslips
dropper
methylene blue stain
toothpick, flat type
Elodea leaf
water

CELL PART	CHEEK CELL PARTS PRESENT	*ELODEA* CELL PARTS PRESENT
Cytoplasm		
Nucleus		
Chloroplast		
Cell wall		
Cell membrane		

Skills

observe, compare, use a microscope

Procedure

1. Copy the data table.
2. Put a drop of stain on a slide. Gently scrape the inside of your cheek with a toothpick. **CAUTION:** *Do not scrape hard enough to injure your cheek.*
3. Rub the toothpick in the stain. Break the toothpick in half and discard it as your teacher directs.
4. Cover the slide with a coverslip.
5. **Use a microscope:** Look at the cheek cells under low power, then high power.
6. Locate the nucleus, cytoplasm, and cell membrane. Fill in the table by putting a check mark in the box if the cell part can be seen.
7. Draw and label the nucleus, cytoplasm, and cell membrane of a cheek cell.
8. Prepare a slide of an *Elodea* leaf. Put an *Elodea* leaf in a drop of water on a slide. Add a coverslip.
9. Look at the *Elodea* cells under low power, then high power.
10. Locate the cell wall, chloroplasts, nucleus, and cytoplasm. Fill in the table.
11. Draw and label the cell wall, chloroplasts, nucleus, and cytoplasm of an *Elodea* cell.

Data and Observations

1. Describe the shape of a cheek cell.
2. Describe the shape of an *Elodea* cell.
3. **Compare:** What parts did you see in both cells?
4. What parts are found in plant cells that are absent in animal cells?

Analyze and Apply

1. What do the cell parts found only in plant cells do?
2. Is the nucleus always found in the center of the cell?
3. Which part of an animal cell gives shape to the cell?
4. Which parts of a plant cell give shape to the cell?
5. Why are stains such as methylene blue used when observing cells under the microscope?
6. **Apply:** Why don't animal cells have chloroplasts? (HINT: How do animals get energy?)

Extension

Observe other plant and animal cells under the microscope. How are they different from cheek cells and *Elodea* cells? How are they alike?

Your cells and the cells of every living thing around you are complex structures with many parts. Each part has a function that is important to the life of the cell. All of the living things in Figure 2-9 are made of cells. What other living things can you name that are made of cells?

Figure 2-9 The maple tree (a), ladybug (b), horse (c), and snake (d) are all made of cells.

Check Your Understanding

6. Which cell part is being described?
   (a) helps keep cytoplasm inside
   (b) controls most of the cell's activities
   (c) a liquid-filled space for storage
   (d) green parts of plants that trap energy from the sun
   (e) clear, jellylike material in which most of the cell's chemical reactions take place
7. Name two cell parts found in plant cells that are not found in animal cells.
8. Why are mitochondria called "powerhouses" of the cell?
9. **Critical Thinking:** How do mitochondria and chloroplasts differ?
10. **Biology and Writing:** Write three sentences describing the cell parts that make up a wooden table.

2:3 Special Cell Processes

Objectives

6. **Explain** the process of diffusion in a cell.
7. **Describe** the process of osmosis in a cell.
8. **Communicate** how cells, tissues, organs, and organ systems are organized.

Key Science Words

diffusion
osmosis
tissue
organ
organ system
organism

Cells need certain substances to stay alive. These substances include food and oxygen. How does a substance, such as oxygen, get through the cell membrane and into the cell?

Diffusion

Imagine you have a box of marbles. The box is divided in half by a strip of cardboard with an opening. Figure 2-10 shows the box with marbles in it. Notice that all of the marbles are on one side of the cardboard strip.

The marbles are packed very tightly into the box. Some of the marbles roll to the other side of the box through the opening in the cardboard strip. The marbles keep rolling until each side of the box has about equal numbers of marbles. How did this happen?

If you walk near the chemistry lab in your school, you might smell a gas made during a chemistry experiment. How did the gas move from the lab into the hall where you are? The gas moved from where there was a large amount of it, in the lab, to where there was a small amount of it, in the hall. Gas molecules, like marbles, can move about. The movement of a substance from where there is a large amount of it to where there is a small amount of it is called **diffusion** (dif YEW zhun).

What is diffusion?

How does oxygen diffuse into cells? Doesn't the cell membrane get in the way? Think about the box divided by the cardboard strip. Marbles could roll to the other side of the box through the opening in the cardboard strip.

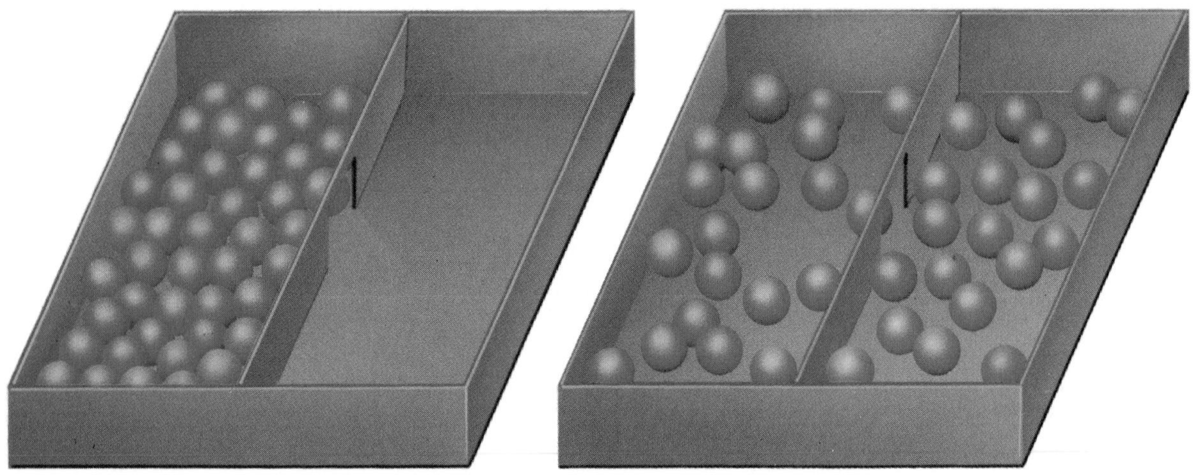

Figure 2-10 Before the marbles roll through the opening in the cardboard strip, they are all on one side of the box (left). The box on the right shows what happens after the marbles roll from one side to the other.

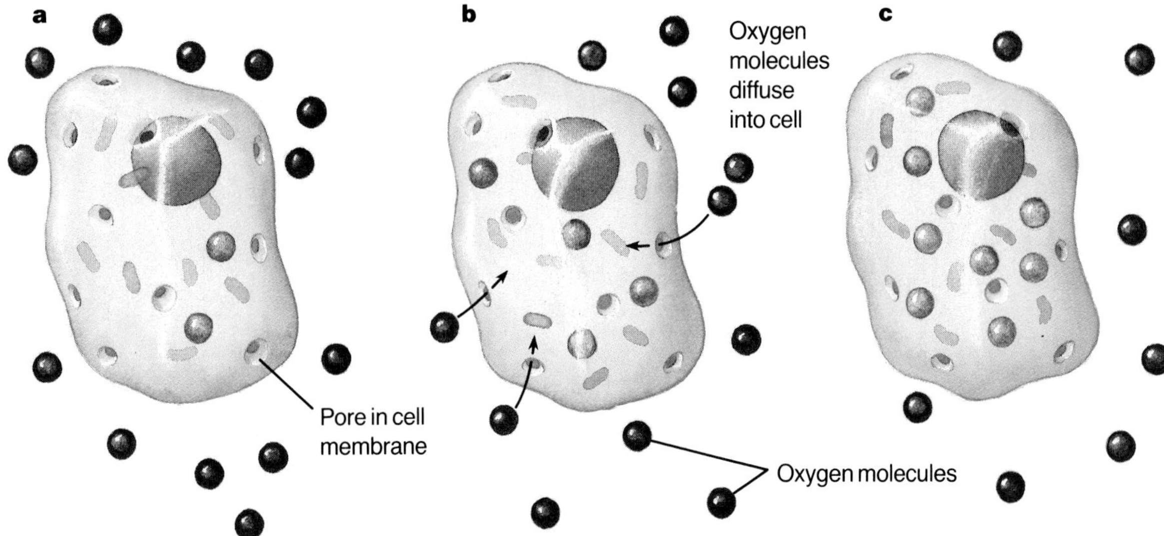

Figure 2-11 Oxygen diffuses into the cell through pores in the membrane.

Membranes also have openings, called pores. Some molecules, such as oxygen, can pass through these pores, Figure 2-11a. The movement of molecules through the pores of a membrane is like the movement of marbles through the opening in the cardboard strip.

In Figure 2-11:

1. There is a large number of oxygen molecules outside the cell.
2. There is a small number of oxygen molecules inside the cell.
3. The oxygen moves from the outside to the inside by diffusion, as shown in Figure 2-11b. The oxygen moves from where there is a large amount of it to where there is a small amount of it. If we look at the cell later, we see about equal numbers of oxygen molecules inside and outside the cell, Figure 2-11c. Remember that at a later time, there were about equal numbers of marbles on both sides of the box.

Osmosis

Water molecules also move across membranes. Like oxygen, they move from where there is a large number of molecules to where there is a small number of molecules. The movement goes on until the number of water molecules is about the same on both sides of the membrane. The movement of water across the cell membrane is called **osmosis** (ahs MOH sus).

Mini Lab

What Is Diffusion?

Experiment: Put 10 drops of vanilla extract inside a balloon. Blow up the balloon, tie it, and place it inside a box. Close the box. After two hours, open the box. What do you smell? *For more help, refer to the* ***Skill Handbook,*** *pages 704-705.*

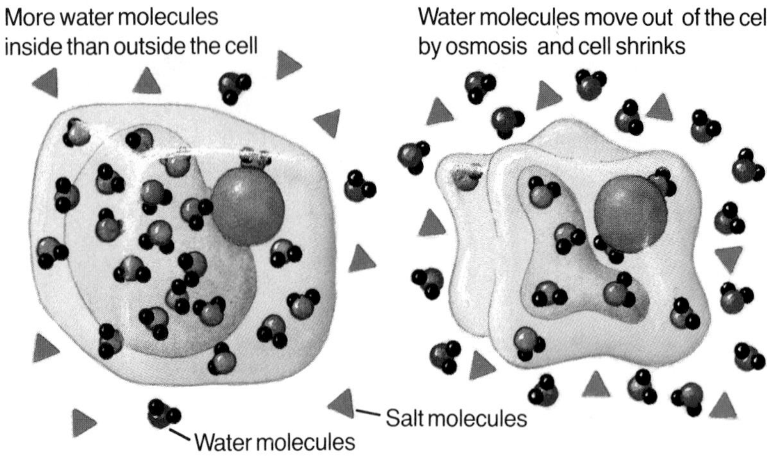
More water molecules inside than outside the cell
Water molecules move out of the cell by osmosis and cell shrinks
Salt molecules
Water molecules

Figure 2-12 Water molecules move into and out of cells by osmosis. If there are more water molecules inside than out, the water molecules will move out of the cell.

Why is osmosis important to cells?

Osmosis is important to cells because they are surrounded by water molecules. The number of water molecules inside and outside the cell must be about the same. When there are more water molecules outside the cell, they move into the cell. When there are more water molecules inside the cell, they move out.

Suppose you placed some cucumber slices into salt water. What do you think would happen? (HINT: Cucumber cells have nearly pure water inside. Salt water is not pure water because it has salt in it. So, there are more water molecules in a cucumber than in salt water.) The water molecules inside the cucumber would move out, as shown in Figure 2-12. When cells of a plant lose too much water, the plant wilts. What do you think would happen if your cells lost too much water?

Bio Tip

Consumer: Osmosis affects the taste of foods. When steak is salted before it is cooked, water moves out of its cells, making the steak dry and tasteless.

Organization

Living things are organized in special ways. Some living things are made of only one cell. That one cell carries out all the activities necessary for life. It responds to the environment, reproduces, and uses energy.

In living things made of many cells, the cells are organized into groups, Figure 2-13. The cells of each group do special jobs. For example, cells that line the small intestine make chemicals for digestion. Groups of cells, such as those that line the intestine, are called tissues. A **tissue** is a group of similar cells that work together to carry out a special job. Bone, muscle, blood, and nerve are kinds of tissues found in animals. Bark and outer surfaces of leaves are kinds of tissues in plants.

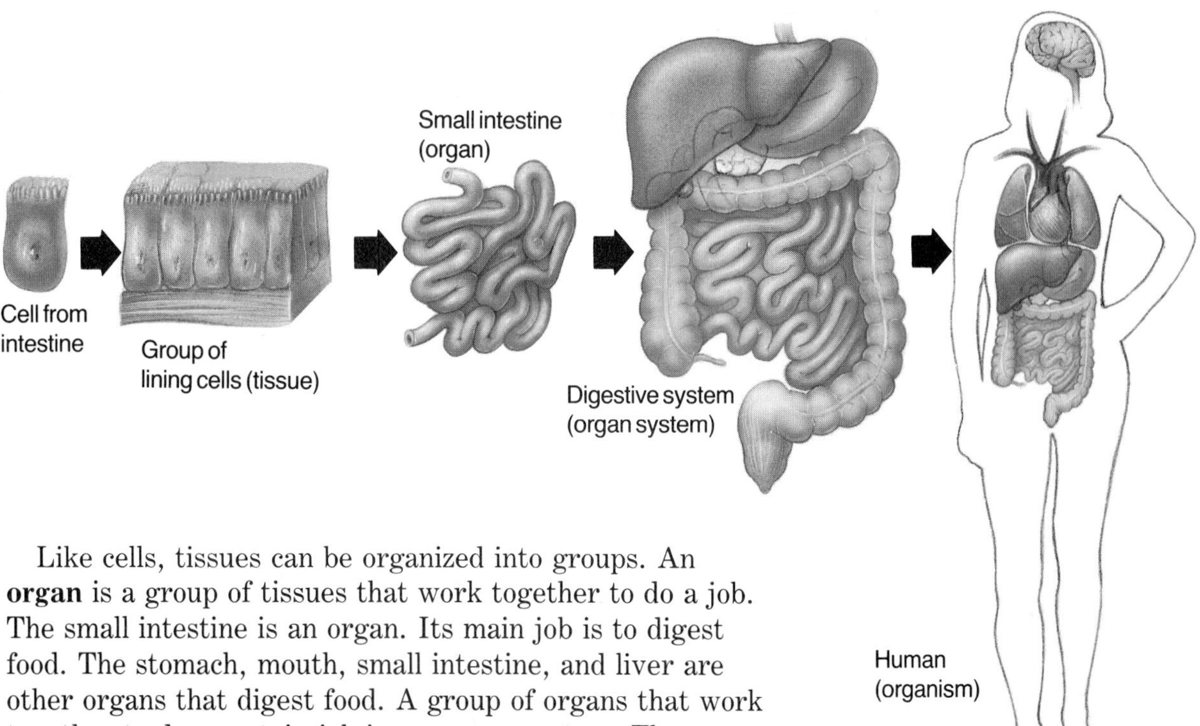

Figure 2-13 Living things (organisms) are organized into cells, tissues, organs, and organ systems.

Like cells, tissues can be organized into groups. An **organ** is a group of tissues that work together to do a job. The small intestine is an organ. Its main job is to digest food. The stomach, mouth, small intestine, and liver are other organs that digest food. A group of organs that work together to do a certain job is an **organ system.** The organs that work together to digest food are called the digestive system.

All the organ systems working together make up an organism (OR guh niz um). An **organism** is a living thing. Many organisms are made of only one cell. Yet, they have all the features of living things. Other organisms have organs and organ systems. Humans are organisms with several organ systems. Among other systems, humans have circulatory, reproductive, and nervous systems. These systems will be explained in Chapters 11, 15, and 24.

What is the relationship among cells, tissues, organs, and organ systems?

Check Your Understanding

11. How does diffusion make it possible for you to smell different odors?
12. In osmosis, which substance diffuses?
13. How are tissues and organs different?
14. **Critical Thinking:** What would happen if you placed a slice of cucumber in distilled water? (HINT: Distilled water is purer than the water inside the cucumber cells.)
15. **Biology and Reading:** What are the living parts that make up an organ system?

Salty Plants—How Will They Grow?

All plants need water in order to live and grow. Farmers usually depend on rainfall to provide water to their growing crops. But, what if there is too little rain available? Then, farmers must supply their crops with water through irrigation. The problem with irrigation, however, is that sometimes there isn't a large enough water supply available. Solution: there is plenty of water in Earth's oceans. Why not use it to irrigate plants?

Identifying the Problem

You are a botanist who has been hired by a large group of farmers along the west coast of the United States. Your job is to determine whether they can use water from the Pacific Ocean to water their crops. Your plan is to conduct an experiment to see whether bean plants can survive when watered with salt water. You will use only one variable, which will be the adding of salt water (seawater) to one group of plants. Also, remember that you will need a control. It will be the adding of freshwater to a second group of plants. Form a hypothesis as to how you believe salt water will affect the growth of plants.

Technology Connection

The need for freshwater has resulted in technology that removes salt from ocean water. Desalination plants built near oceans are helping to supply freshwater to areas that are in need. Cities along the coast of California, as well as in Israel and Saudi Arabia, have already built these plants. They remove the salt from ocean waters by either distillation or reverse osmosis techniques.

Collecting Information

Do some library research to determine what the difference is between "fresh" water and "sea or salt" water. Once you have this information, you are almost ready to begin your experiment. One last item to think about during your research is whether osmosis might be occurring between the cells of plants and the water within the soil in which they are growing.

Carrying Out an Experiment

1. Obtain two trays containing young plants.
2. Label one tray *experimental* and the other tray *control.*
3. Design a data table in which you record your daily observations. Your table should include the number of days that you conduct the experiment. Mark today as *Day 1.*
4. Add only salt water to the experimental tray as needed and freshwater (tap water) to the control tray as needed. Keep the soil in both trays moist.
5. Place the trays in a lighted area.
6. Observe the plants each day for a total of at least ten days. Record all observations in your table. Diagrams may be helpful to show any differences between the two groups.

Assessing Your Results

Compare and contrast the size and general appearance of plants receiving freshwater with those receiving salt water. Based on your observations, what conclusions might you make about watering or irrigating bean plants with salt water? Based on your observations, what advice will you give to the farmers who hired you to solve their problem? Based on your understanding of osmosis, what conclusions might you make about the effect of salt water on plants? What was your control in this experiment and what was the variable? Why did you need a control? Was your hypothesis supported by your experimental findings? Explain.

Career Connection

- **Agricultural Supply Salesperson** Sells farming equipment, agricultural chemicals, and other supplies to farmers; works in stores or in large dealerships
- **Crop Scientist** Works to increase the yield of crops and to improve farming methods. Their concerns may be about weed control, watering problems, and crop development.
- **Water Resources Engineer** Ensures that water is available for drinking, farming, and industrial use; works on problems associated with conserving water and preventing flooding

Chapter 2

Review

Summary

2:1 Living Things and Their Parts

1. Living things have eight features in common. They reproduce, grow, develop, need food, use energy, have cells, respond, and are adapted to their environments.
2. Living things are made of matter. Six elements make up over 97 percent of the matter in living things.
3. The cell theory states: all living things are made of one or more cells; cells are the basic units of structure and function in living things; all cells come from other cells.

2:2 Cell Parts and Their Jobs

4. The main parts of the cell are the cell membrane, nucleus, chromosomes, cytoplasm, ribosomes, mitochondria, and vacuoles.
5. The jobs of the cell parts include protection, the making of energy, the moving of materials into and out of the cell, and reproduction.

2:3 Special Cell Processes

6. Materials move into or out of a cell by diffusion.
7. The movement of water molecules across the cell membrane is osmosis.
8. Living things are organized into cells, tissues, organs, and organ systems.

Key Science Words

adaptation (p. 29)
cell (p. 27)
cell membrane (p. 32)
cellular respiration (p. 27)
cell wall (p. 35)
centriole (p. 35)
chloroplast (p. 35)
chromosome (p. 32)
consumer (p. 27)
cytoplasm (p. 32)
development (p. 26)
diffusion (p. 38)
mitochondria (p. 34)
nuclear membrane (p. 32)
nucleolus (p. 32)
nucleus (p. 32)
organ (p. 41)
organism (p. 41)
organ system (p. 41)
osmosis (p. 39)
producer (p. 27)
reproduce (p. 26)
ribosome (p. 34)
tissue (p. 40)
vacuole (p. 35)

Testing Yourself

Using Words

Choose the word from the list of Key Science Words that best fits the definition.

1. cell part with information that determines a living thing's trait
2. cell parts that produce energy when food is broken down
3. tissues that work together to do the same job
4. green part inside a plant cell
5. movement of substances, such as oxygen, into a cell
6. basic unit of living things
7. process in which food is broken down and energy is released
8. group of organs working together
9. diffusion of water into or out of a cell

Testing Yourself *continued*

Finding Main Ideas

List the page number where each main idea below is found. Then, explain each main idea.

10. what living things respond to
11. what an organ system is
12. the function of centrioles
13. how molecules pass through a membrane
14. how living things get energy from food
15. what a molecule is
16. the six elements that make up over 97 percent of living matter
17. the scientist who first observed cells
18. the job of chromosomes
19. why cucumbers wilt in salt water

Using Main Ideas

Answer these questions by referring to the page number after the question.

20. How are growing and developing different? (p. 26)
21. Why are humans called organisms? (p. 41)
22. Why will a stalk of celery put in salt water begin to wilt? (p. 40)
23. How do your school and a cell compare? Match the job of each cell part with a part of your school that has a similar job. (pp. 32-35)

(a) nucleus	1. main office
(b) vacuole	2. cafeteria
(c) chloroplasts	3. furnace room
(d) mitochondria	4. lockers

24. How is osmosis a special kind of diffusion? (p. 39)
25. What is the function of pores in a cell membrane? (p. 39)
26. What characteristics of living things does a cell have? (pp. 26, 27, 29)

Skill Review 

For more help, refer to the ***Skill Handbook,*** *pages 704-719.*

1. **Make and use tables:** Make a table to show the features of living things shown by a cat, guppy, bird, and maple tree. In your table give examples for each feature of living things shown by these organisms.
2. **Classify:** Classify the following as elements or compounds: hydrogen, oxygen, water, carbon, carbon dioxide, nitrogen, and phosphorus.
3. **Formulate a model:** Construct and label a three-dimensional model of a plant or animal cell.
4. **Infer:** Compare the growth of a single-celled organism with that of a many-celled organism. What is the difference between the two types of growth?

Finding Out More

Critical Thinking

1. Explain why some cells have more mitochondria than others.
2. Make a time line showing the events that led to the cell theory.

Applications

1. What kinds of tissues and organs are you eating when you eat a green salad?
2. Report on ways that desert organisms such as cacti and camels conserve and use water as adaptations for survival.

CHAPTER PREVIEW

Chapter Content

Review this outline for Chapter 3 before you read the chapter.

3:1 Why Things Are Grouped
- Classifying in Everyday Life
- How Grouping Helps Us

3:2 Methods of Classification
- Early Classification
- The Beginning of Modern Classification

3:3 How Scientists Classify Today
- Classifying Based on How Organisms Are Related
- Other Evidence Used in Classifying
- Scientific Names Come from Classification
- Why Scientific Names Are Used
- Classification of Kingdoms

Skills in this Chapter

The skills that you will use in this chapter are listed below.

- In **Lab 3-1,** you will classify and observe. In **Lab 3-2,** you will classify, communicate, and make a table.
- In the **Skill Checks,** you will classify and understand science words.
- In the **Mini Labs,** you will infer and classify.

Tubeworms

Chapter 3

Classification

What features do biologists use to group living things? You know that most plants are green and do not move around. You also know that most animals are not green and do move around. The tubeworm on the left lives in the water attached to one spot. It is not green. Is a tubeworm a plant or an animal? The euglena in the smaller photo is green and moves around in the water. Is it a plant or an animal, or does it belong to some other group?

Biologists have a system for grouping living things. Each living thing has a specific name that biologists all over the world understand. In this chapter, you will learn why we group living things and about the system used to group them.

Try This!

Why do we have two names? Choose a first name that you hear often. Make a list of all the people you know who have the same first name. Can you see why a two-name system is needed?

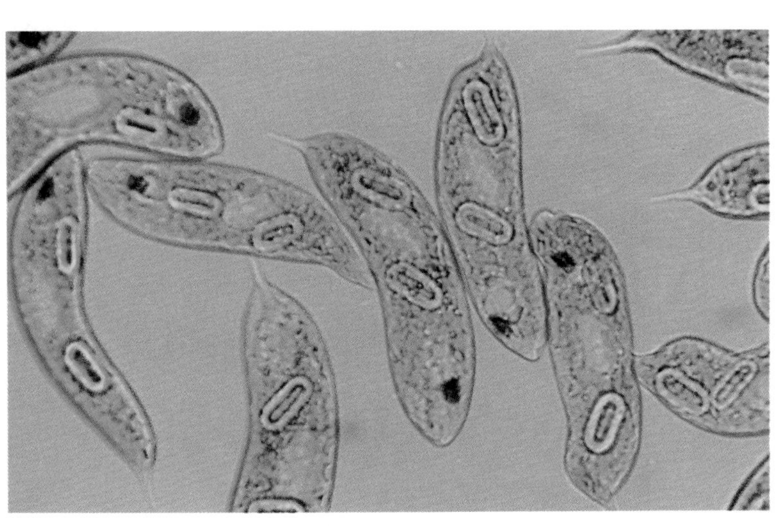

BIOLOGY Online

Visit the Glencoe Science Web site at science.glencoe.com to find links about **classification.**

Euglena, 450x

3:1 Why Things Are Grouped

Objectives

1. **Give examples** of items in daily life that are grouped.
2. **Explain** how and why we classify things.

Key Science Words

classify
trait

How often do you classify things? You probably classify things more often than you think. To **classify** means to group things together based on similarities.

Classifying in Everyday Life

If you were opening a sports store, how would you group the equipment in Figure 3-1? You might put the shoes in one place and the balls in another. You might want items for a certain sport grouped together.

Where would you find lettuce and bread in a grocery store?

Many things in our daily lives have been grouped for us. Think about how food is grouped in a grocery store. Frozen foods, meats, produce, bakery items, and canned foods are found in separate areas.

What subjects are you taking this year? Don't you group your courses? Spanish, French, and German are language courses. Typing and bookkeeping are business courses.

How Grouping Helps Us

There are several reasons to classify things. One reason is to put things in order so that they become easier to find.

How are books arranged in a library?

Your school library must have thousands of books. They are arranged by a system of numbering that makes it easier to find a certain book. Think of trying to find a classroom with the room numbers out of order. Could you find a phone number in a phone book if the names were not in alphabetical order?

A second reason we classify things is to show that they share certain traits. A **trait** is a feature that a thing has. In a library, you see biographies grouped together. What trait do they share?

Figure 3-1 Many everyday objects, such as this sports equipment, are classified.

Lab 3–1 ...assifying

Problem: How can some common objects be classified?

Skills

classify, observe

Materials

envelope that contains 12 items

Procedure

1. Copy the diagram.
2. Place the objects from your envelope on your desk.
3. Make a list of the objects. Your teacher will help you with their names.
4. These objects will be called Kingdom Objects. List the objects under Kingdom Objects in your diagram.
5. Sort the 12 objects into two groups on your desk. The objects in each group must have a common trait.
6. Call each of these two groups a phylum. Use the common trait as the name of the phylum. Place the phylum name for each group on your diagram.
7. On the diagram, write the name of each object that you placed in Phylum 1 and in Phylum 2.
8. Sort the objects of Phylum 1 into two groups. Each of the two groups must have a common trait.
9. Call each of these two groups a class. Use the common trait as the name of the class. Place the class names for Phylum 1 on your diagram under Classes A and B.
10. Sort the objects in Phylum 2 into two groups. Each of these two groups must have a common trait.
11. Each of these groups represents a class. Place the class names for Phylum 2 on your diagram under Classes C and D.
12. Put the items back into the envelope and return the envelope to your teacher.

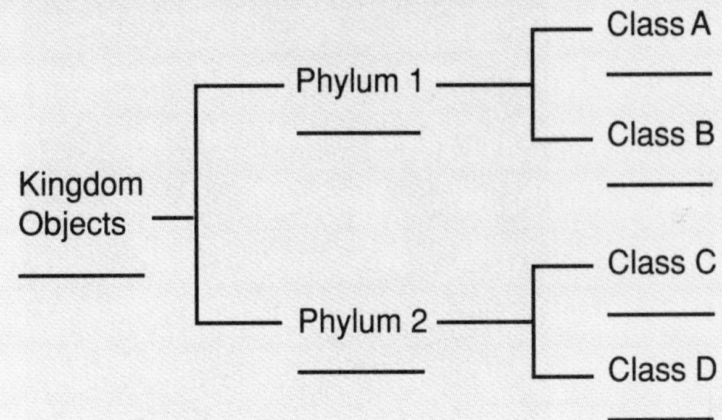

Data and Observations

1. What traits did you use to classify the kingdom objects into two phyla?
2. What is the name of Phylum 1? Phylum 2?
3. What are the names of Class A and Class B?
4. What are the names of Class C and Class D?
5. What trait do all four classes have in common?
6. What traits do Classes A and B have in common?
7. What traits do Classes C and D have in common?

Analyze and Apply

1. At what level of classification, phylum or class, did all objects share more traits?
2. At what level of classification, phylum or class, did all objects share fewer traits?
3. **Apply:** How could you use the classification method in this lab to group a collection of compact discs?

Extension

Classify the items in your bedroom closet using the same method that you used in this lab.

Career Close-Up

Wildlife Photographer

Photographing wildlife is often a challenging occupation.

John's biology teacher invited a wildlife photographer to visit his class. The photographer brought slides and prints of many living things. Many photographs of flowers and insects were made with a close-up lens or a zoom lens. The photographer showed the students how to use a camera attachment on a light microscope. He told them that his work often took him to outdoor settings.

Students wanted to know about the training needed to be a photographer. He told them that he had taken photography and natural science courses in high school. Then he had taken several courses at a community college. He explained that in this field success depends on a person knowing the subject matter well.

Other students wanted to know where his work was used. He showed them magazines and books that contained his photographs.

Biologists classify living things. Doing so puts organisms in order. It also shows how they are alike. There are over one and one-half million known kinds of living things. It would be impossible to find information about them if they were not grouped in some way.

Check Your Understanding

1. What is meant by a trait? Give two examples of traits.
2. Give an example of something at school that is classified.
3. Give two reasons why things are classified.
4. **Critical Thinking:** Group the parts of an animal cell based on their shapes.
5. **Biology and Math:** What are two other ways to write the number one and one-half million?

3:2 Methods of Classification

For hundreds of years, people have been grouping living things. The job has not been easy. What makes the job of classifying hard is that scientists do not always agree on how to group living things.

Objectives

3. **Explain** Aristotle's classification of living things.
4. **Compare** Linnaeus' system of classification with Aristotle's.
5. **Name** the levels of classification that are used today.

Early Classification

Over 2000 years ago, a Greek scientist named Aristotle (AIR uh staht ul) was one of the first people to classify living things. He noticed that living things fit into two main groups—plant and animal. Most plants were green and didn't move. Most animals weren't green and did move.

Key Science Words

kingdom
phylum
class
order
family
genus
species

Aristotle next divided all animals into three groups. He based his groups on where the animals lived as shown in Figure 3-2. Animals that lived in water went into one group. Animals that lived on land went into a second group. Animals that lived in the air and could fly made up a third group.

How did Aristotle divide his animal group?

Aristotle then worked out a system for grouping plants. He based it on size of the plant and pattern of growth. You can see in Figure 3-2 that plants were placed into three smaller groups. Tall plants with one trunk were put into a tree group. Medium plants with many trunks were put into a shrub group. A privet hedge is an example of a shrub. Small plants with soft stems went into an herb group. Herbs included grasses and wildflowers.

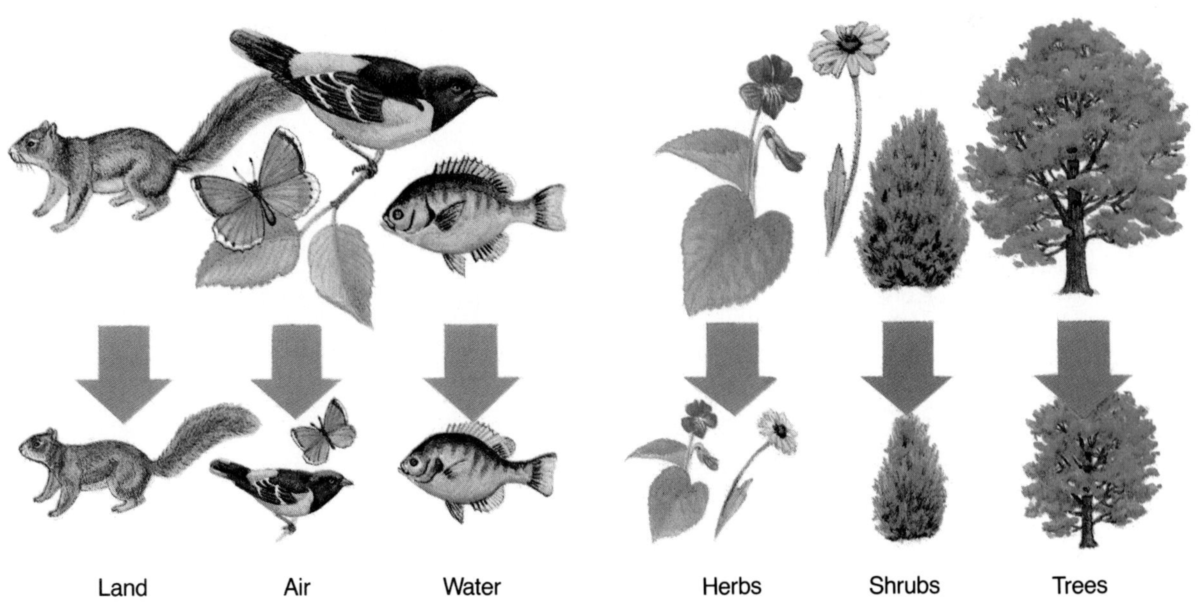

Figure 3-2 Aristotle grouped animals into three groups and plants into three groups.

Common Names

Lab 3–2

Problem: How many places are named for animals or plants?

Skills

classify, communicate, make a table

Materials

state map

Procedure

1. **Make a table** in your notebook in which to record your observations. Use the headings Name and Type of Place in your table.
2. Use the map on this page or your state map to find each city or town that has the name of a plant or animal as part of its name. Record the names in your table.
3. Place a (C) after the name so you will know it is a city or town.
4. Use the map to find state or national parks, rivers, lakes, and historical places that have a plant or animal name as part of their names.
5. Record the names of the parks, rivers, lakes, and historical places in your table. Put a (P) after parks, (R) after rivers, (H) after historical places, and (L) after lakes.

Data and Observations

1. Were animal names or plant names used more often on your map?
2. Which animal name appeared most often on your map?
3. Which plant name appeared most?

Analyze and Apply

1. Why do you suppose humans use certain animal and plant names for cities, rivers, and parks?
2. Do cities, parks, or bodies of water seem to have more plant and animal names?
3. Why do humans use the common names of plants and animals rather than the scientific names to name places?
4. **Apply:** Would you expect to find a town called Giraffesville on your map? Explain.

Extension

Communicate: Make a list of other common themes that are used in naming places, for example Indian names. Share your list with the class.

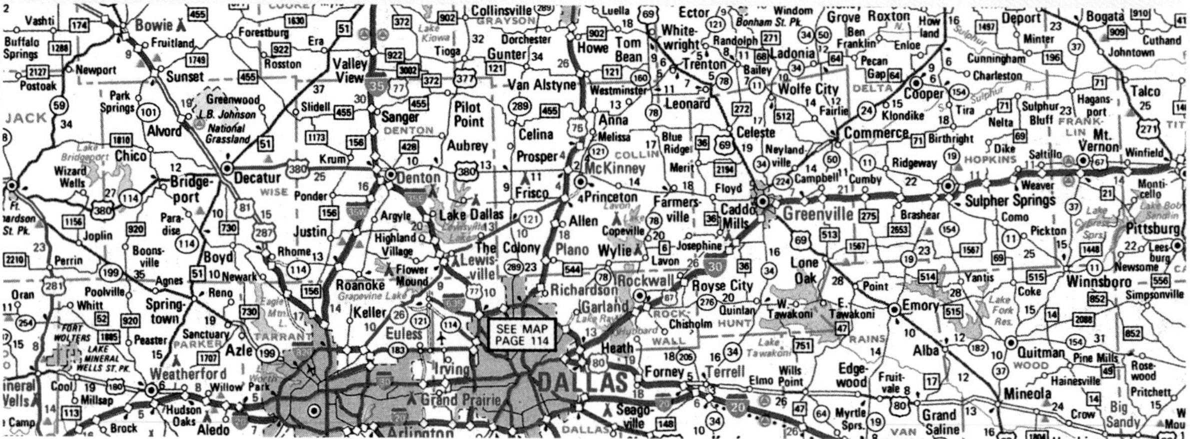

The Beginning of Modern Classification

How do the classification systems of Aristotle and Linnaeus compare?

Scientists used Aristotle's system of classification for hundreds of years. However, as scientists found more and more living things, Aristotle's system became less useful. Many of the newly discovered living things did not fit into his system. Many scientists began to question the system because different traits were used to group plants and animals. A useful classification system should use the same traits for classifying all groups.

In 1735, Carolus Linnaeus (lin NAY us) developed a new classification system. He first placed living things into two main groups. He called these groups kingdoms. A **kingdom** is the largest group of living things. Plants made up one kingdom. Animals made up the other kingdom. Linnaeus placed living things with similar traits into the same group and called this group a species. He used very specific traits for his groups. For example, he used flower parts to group plant species. He placed similar species into a larger group called a genus (JEE nus).

Through his work, Linnaeus made a number of important changes in Aristotle's system.

1. He classified plants and animals into more groups.
2. He based his system on specific traits.
3. He gave organisms names that described their traits. These names had two parts. All living things still have two-part names.

Mini Lab

How Did Aristotle Classify Animals?

Infer: Make a list of animals that you have seen or that you know about from reading magazines and watching television. Classify the animals the way Aristotle would have. *For more help, refer to the **Skill Handbook**, pages 706-711.*

Idea Map

Early Classification

- Early classification
  - Aristotle
    - plant
      - trees
      - shrubs
      - herbs
    - animal
      - water
      - land
      - air
  - Linnaeus
    - plant kingdom — genus — species
    - animal kingdom — genus — species

Study Tip: Use this idea map as you study the development of classification systems. Aristotle and Linnaeus developed early classification systems.

TABLE 3–1. COMPARING CLASSIFICATION GROUPS AND ADDRESS INFORMATION	
Country	Kingdom
State	Phylum
County	Class
Town	Order
Neighborhood	Family
Street	Genus
House number	Species

Today, there are seven groups for classifying organisms—kingdom, phylum, class, order, family, genus, and species. Why so many? Having more groups makes it easier to place an organism in the proper group. Think of how difficult it would be to find a long-lost cousin's house if all you knew was the country your cousin lived in. Table 3-1 shows the different kinds of address information that could be helpful in finding your cousin's house. How do the classification groups compare to the different kinds of address information?

How can classification groups be compared to addresses?

The kingdom is still the largest group of living things, just as it was in Linnaeus' time. A kingdom can be compared to the country your cousin lives in. Kingdoms are divided into groups called phyla (FI luh) (singular phylum). A **phylum** is the largest group within a kingdom. A phylum can be compared to a state. Phyla are divided into even smaller groups called classes. A **class** is the largest group within a phylum. Classes are divided into groups called orders. An **order** is the largest group within a class. Orders are divided into families. A **family** is the largest group within an order. Notice in Table 3-1 that classes can be compared to counties, orders to towns, and families to neighborhoods.

Figure 3-3 The praying mantis (a), walking stick (b), and rhinoceros beetle (c) have names that describe their appearance.

a

b

c

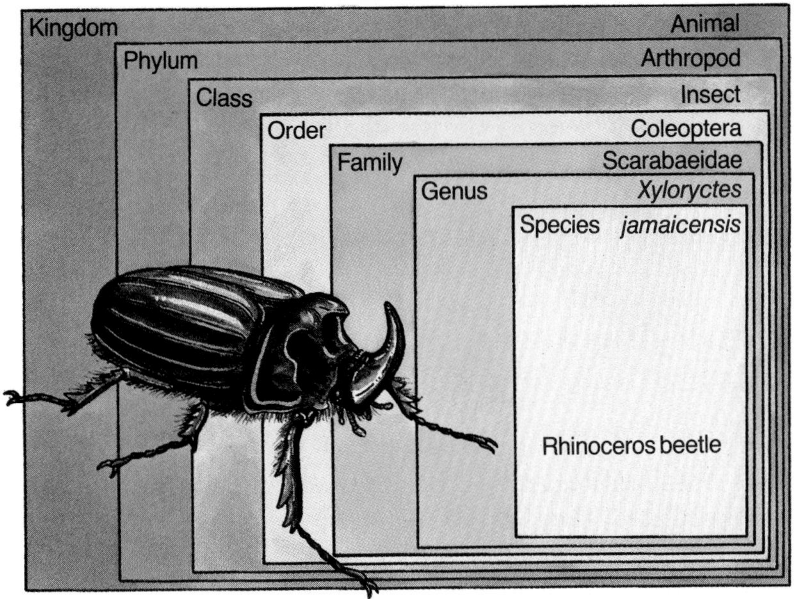

Figure 3-4 Each classification group is a subset of the next-largest group. Of what group is the kingdom a subset?

The names given to the different groups often describe the living things in them. Let's look at an example at the family level. Members of three different insect families are shown in Figure 3-3. Each insect belongs to a different family.

Another way to think of a kingdom is as a set. Think of the other groups, from largest to smallest, as subsets. All seven groups are shown in Figure 3-4. You can see that like classes and orders, families are divided into groups. The largest group within a family is a **genus.** The smallest group of living things is a **species.**

Skill Check

Understand science words: phylum. The word part *phyl* means tribe or race. In your dictionary, find three words with the word part *phyl* in them. *For more help, refer to the **Skill Handbook,** pages 706-711.*

Check Your Understanding

6. What traits were used by Aristotle to divide plants into groups?
7. Linnaeus made important changes in the way living things are classified. Name two.
8. There are seven main classification groups. Name these groups in order from largest to smallest.
9. **Critical Thinking:** A euglena is a one-celled organism that moves on its own and has chloroplasts. Why would Aristotle have trouble classifying this organism?
10. **Biology and Reading:** Who developed a two-kingdom system of classification for plants and animals?

3:3 How Scientists Classify Today

Objectives

6. **List** traits used in classifying living things.
7. **State** the function of scientific names.
8. **List** the five kingdoms scientists recognize today.

Key Science Words

scientific name
moneran
protist
fungi
plant
animal

How do scientists know to which groups an organism belongs? They look at many traits. They compare the traits of one organism with those of another. Scientists also compare organisms living today with those that lived long ago. Let's look at how scientists classify living things.

Classifying Based on How Organisms Are Related

Living things that are closely related are in many of the same classification groups. For example, if two plants are closely related, they will be in almost all of the same groups. If they are not closely related, they will not be in very many of the same groups.

Table 3-2 shows a list of groups to which the house cat shown in Figure 3-5 belongs. It also shows the traits for each group. You may not recognize some of the group names. Many are from the Greek or Latin language.

Compare the lion shown in Figure 3-6 to the house cat. In how many groups are they found together?

TABLE 3–2. CLASSIFYING THE HOUSE CAT

Group	Group name	Group trait
Kingdom	Animal	Has many cells; eats food
Phylum	Chordate	Rodlike structure along the back for support
Class	Mammal	Nurses young; has hair
Order	Carnivore	Eats flesh; has large teeth
Family	Felidae	Sharp claws; large eyes
Genus	*Felis*	Small cats
Species	*catus*	Tame

Figure 3-5 To what classification groups does a house cat belong?

TABLE 3–3. COMPARING THE HOUSE CAT AND LION			
	House Cat	**Lion**	**Comparison**
Kingdom	Animal	Animal	Same kingdom
Phylum	Chordate	Chordate	Same phylum
Class	Mammal	Mammal	Same class
Order	Carnivore	Carnivore	Same order
Family	Felidae	Felidae	Same family
Genus	*Felis*	*Panthera*	Different genus
Species	*catus*	*leo*	Different species

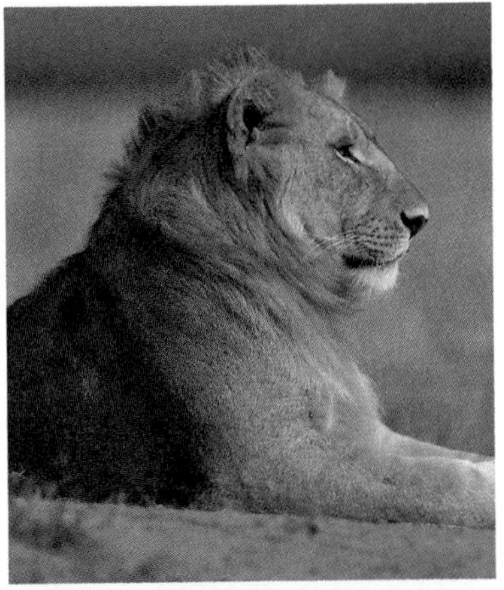

Table 3-3 shows that five of the seven groups for these two animals are the same. Only the genus and species groups are different. The lion and the house cat are very similar. They have many of the same traits.

Look at the classifications in Table 3-4 for the house cat and deer. These two animals do not share many traits. Notice that only their first three classification groups are the same.

TABLE 3–4. COMPARING THE HOUSE CAT AND DEER			
	House Cat	**Deer**	**Comparison**
Kingdom	Animal	Animal	Same kingdom
Phylum	Chordate	Chordate	Same phylum
Class	Mammal	Mammal	Same class
Order	Carnivore	Artiodactyla	Different order
Family	Felidae	Cervidae	Different family
Genus	*Felis*	*Odocoileus*	Different genus
Species	*catus*	*virginianus*	Different species

Let's compare a house cat with an octopus. They have fewer groups in common than house cats and deer. House cats and deer have backbones and hair. What about an octopus? Table 3-5 shows that for the house cat and the octopus, only the kingdom is the same. The house cat and the octopus are both animals. After the kingdom level, these animals belong to different groups.

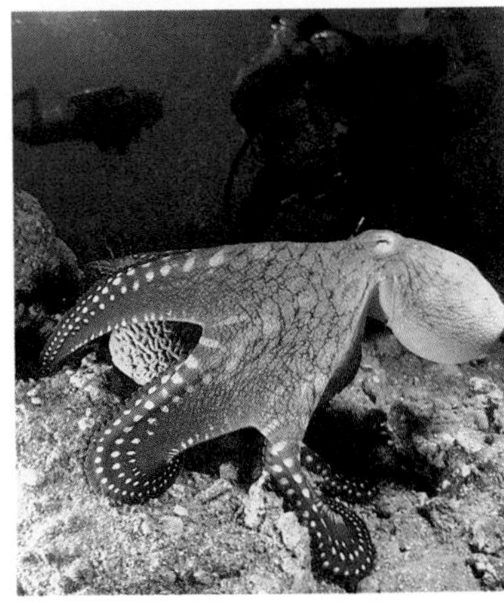

Figure 3-6 How are a lion (top), deer (middle), and octopus (bottom) like a house cat?

TABLE 3–5. COMPARING THE HOUSE CAT AND OCTOPUS			
	House Cat	**Octopus**	**Comparison**
Kingdom	Animal	Animal	Same kingdom
Phylum	Chordate	Mollusk	Different phylum
Class	Mammal	Cephalopod	Different class
Order	Carnivore	Octopoda	Different order
Family	Felidae	Octopodidae	Different family
Genus	*Felis*	*Octopus*	Different genus
Species	*catus*	*vulgaris*	Different species

Mini Lab

What Kingdom?

Classify: Write a description of an imaginary organism so that a person reading the description could classify the organism as a moneran, protist, fungus, plant, or animal. *For more help, refer to the **Skill Handbook,** pages 715-717.*

Other Evidence Used in Classifying

Classification can be based on a living thing's ancestors. An ancestor is a related organism that lived some time in the past. For example, horses and donkeys have many of the same ancestors. They have more of the same ancestors than horses and goats do. Horses and goats have more of the same ancestors than horses and fish do. Of these pairs, horses and donkeys have the most ancestors in common and are the most related.

Similar body structures often show that living things have common ancestors. This is important in classification. You can see in Figure 3-7 that the front limbs of a human (a), a cat (b), a horse (c), a bird (d), and a bat (e) are similar in their bone structures. These similarities show a common ancestor. Compare the front limbs of animals (a) through (e) with the front limbs of an animal that lived long ago (f). All the limbs are similar even though they do different things. They have similar bones arranged in similar patterns.

Why is the horseshoe crab classified with spiders?

Another way to group living things is by body chemistry. A good example is the horseshoe crab. At first, the horseshoe crab was thought to be like other crabs. The blood of the horseshoe crab, however, is more like that of the spider. As a result, the horseshoe crab is now grouped with spiders.

For many years scientists have debated about how to classify the giant panda. Some classified it with raccoons, while others classified it with bears. Now its body chemistry shows it to be more closely related to bears.

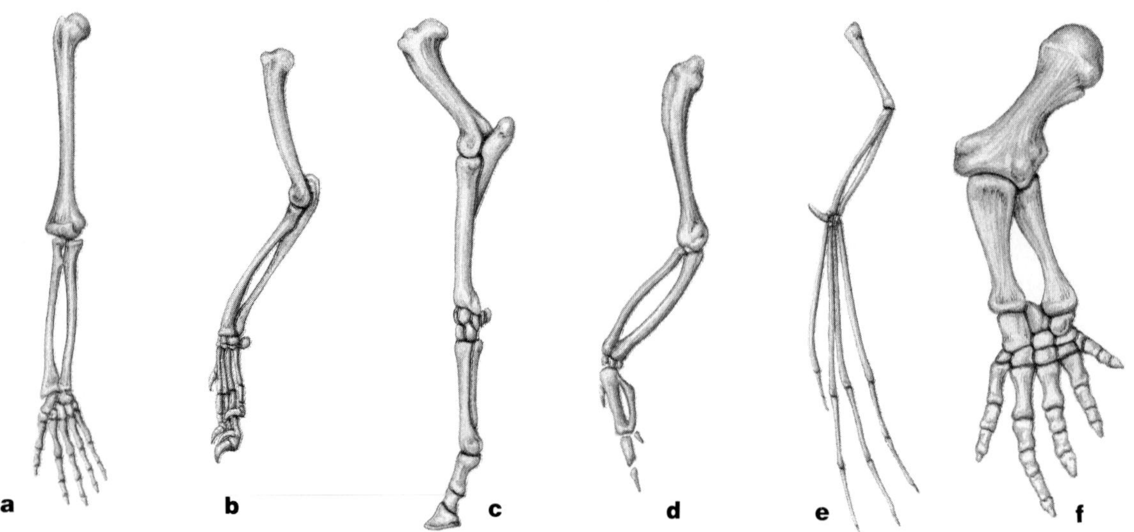

Figure 3-7 Similarity in body structures shows that living things may have had a common ancestor.

Scientific Names Come from Classification

Tables 3-2 to 3-5 showed that living things are classified down to genus and species. The genus and species names together make up the **scientific name.** In Section 3:2, you learned that Linnaeus was the first to give organisms a scientific name with two parts.

The scientific name for a cat is *Felis catus* (FEE lus • CAT us). The name comes from the genus *Felis* and species *catus.* Notice that the genus is always capitalized and the species is not. Both the genus and the species are in italics.

What names make up a scientific name?

A wolf's scientific name is *Canis lupus* (KAY nus • LEW pus). The name includes the genus *Canis* and species *lupus.* The scientific name for humans is *Homo sapiens.* What are the genus and the species to which humans belong?

Have you ever called a plant or an animal by its scientific name? You have, but you may not have realized it. Figure 3-8 shows some living things that you know about. Table 3-6 shows a list of their scientific names. You probably have used parts of these scientific names many times.

TABLE 3–6. SCIENTIFIC NAMES OF SOME PLANTS AND ANIMALS

Scientific name	Name you probably use
Pinus sylvestris	Scotch pine
Rosa carolina	rose
Elephas maximus	Asian elephant
Gorilla gorilla	gorilla
Giraffa camelopardalis	giraffe

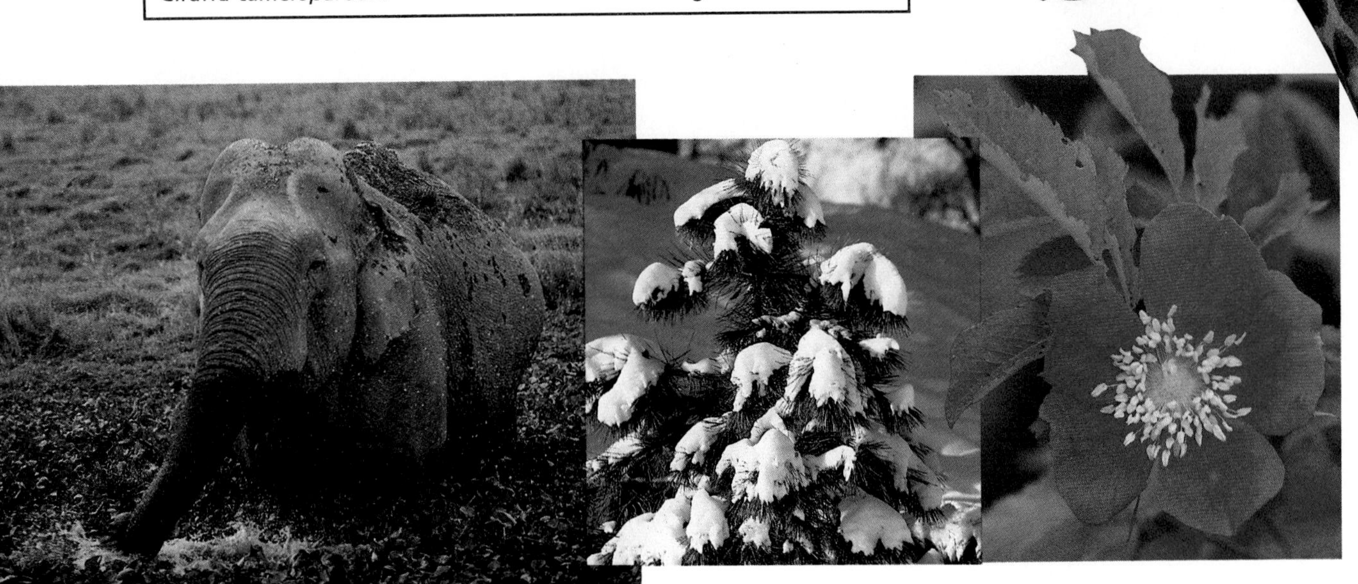

Figure 3-8 Many living things have common names that sound like their scientific names.

Skill Check

Classify: Collect 10 leaves from trees in your neighborhood. Bring the leaves to class and use a key or field guide to find the scientific names of the trees from which the leaves came. *For more help, refer to the* ***Skill Handbook,*** *pages 715-717.*

Why Scientific Names Are Used

You call a robin, seal, apple tree, and house cat by their common names. Common names are names used in everyday language. Scientists prefer to use scientific names. There are several reasons for using scientific names instead of common names.

1. No mistake can be made about which living thing is being described. That's because two different living things don't have the same scientific name. Two different living things, however, may have the same common name. For example, hawk is the common name for several kinds of birds. Only one kind of hawk, however, is named *Buteo jamaicensis* (BOO tee oh • juh may uh KEN sis). This hawk has broad wings and a reddish tail. You may know it as the red-tailed hawk.
2. Scientific names seldom change.
3. Scientific names are written in the same language around the world. Using scientific names allows scientists to communicate no matter what their everyday language is. The language in which scientific names are written is Latin. Latin is used because it does not change.

Figure 3-9 Only one of these hawks is *Buteo jamaicensis* (a).

a

b

c

Classification of Kingdoms

Remember that early scientists grouped living things into two kingdoms—plant and animal. As scientists learned about more living things, they found that some living things did not fit into either kingdom. A new system of classification was needed to group all the living things being discovered.

TECHNOLOGY

Six Kingdoms Instead of Five

Scientists use classification as a tool and kingdom as a category for organizing life forms. Do scientists always agree as to the number of kingdoms to be used? No! They cannot agree as to whether there should be five or six kingdoms, thus creating a controversy.

The problem lies with the moneran kingdom, which includes different kinds of bacteria. Some scientists wish to put all bacteria into Kingdom Monera. Others want to divide bacteria into two kingdoms called Archaea and Eubacteria.

What is the basis for this two-kingdom grouping? Archaea live in very unusual places on Earth, such as in deep sea vents, salt lakes, and the hot sulfur springs shown here. These life forms have unique cell walls and membranes, different kinds of proteins and fats, and they don't need oxygen to live. Many give off smelly marsh gas. The life forms that would be placed into the eubacteria kingdom do not show most of these traits and would die if exposed to some of the same harsh conditions under which the Archaea now thrive.

Hot sulfur spring

Today most scientists use this system to classify living things into five kingdoms. The five kingdoms include: monerans (muh NIHR uns), protists (PROH tihsts), fungi, plants, and animals.

Monerans are one-celled organisms that don't have a nucleus. They lack most of the cell parts that other cells have. The kingdom has only two phyla, bacteria and blue-green bacteria.

Protists are mostly single-celled organisms that have a nucleus and other cell parts. Some have chlorophyll and can make their own food. Others must take in food from the surroundings. Some can move and some can't. The organisms in this kingdom are difficult for scientists to classify because they are so different from one another and from most other living things.

Mushrooms, molds, and yeasts are examples of fungi, (singular fungus). **Fungi** are organisms that have cell walls and absorb food from their surroundings. Fungi don't have chlorophyll, so they can't make their own food.

What are two differences between monerans and protists?

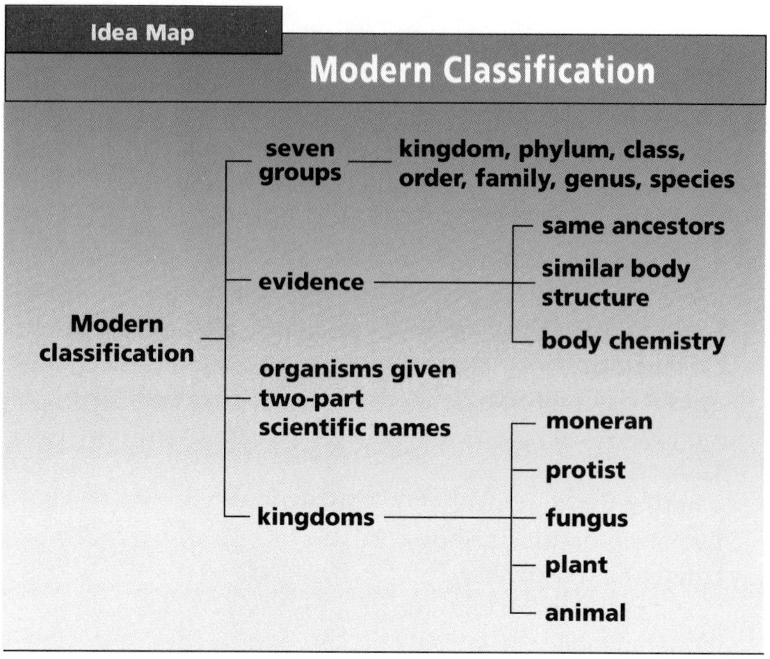

Idea Map

Modern Classification

Modern classification
- seven groups — kingdom, phylum, class, order, family, genus, species
- evidence
  - same ancestors
  - similar body structure
  - body chemistry
- organisms given two-part scientific names
- kingdoms
  - moneran
  - protist
  - fungus
  - plant
  - animal

Study Tip: Use this idea map as you study present-day methods of classifying organisms. How many kingdoms are recognized today?

Bio Tip

Ecology: Biologists estimate that millions of living things have not been discovered. Most are protists, monerans, insects, and plants.

Plants are organisms that are made up of many cells, have chlorophyll, and can make their own food. Plants don't move. **Animals** are organisms that have many cells, can't make their own food, and can move. The cells of animals and plants have nuclei and other cell parts.

Figure 3-10 shows members of the five kingdoms. The traits of each kingdom are also listed for you. Refer back to this figure as you read about the kingdoms of living things in the next five chapters. Appendix B also contains information about the different groups of living things.

The five-kingdom system isn't perfect. Some living things don't fit into any of the five kingdoms. Their classification changes as scientists learn more about them.

Check Your Understanding

11. How is similarity in body structure used to classify living things?
12. Give two reasons why scientists use scientific names.
13. What are five kingdoms recognized today?
14. **Critical Thinking:** You are a chordate. Are you more like a house cat or an octopus?
15. **Biology and Reading:** What does the word *perfect* mean as it is used in the last paragraph of this section?

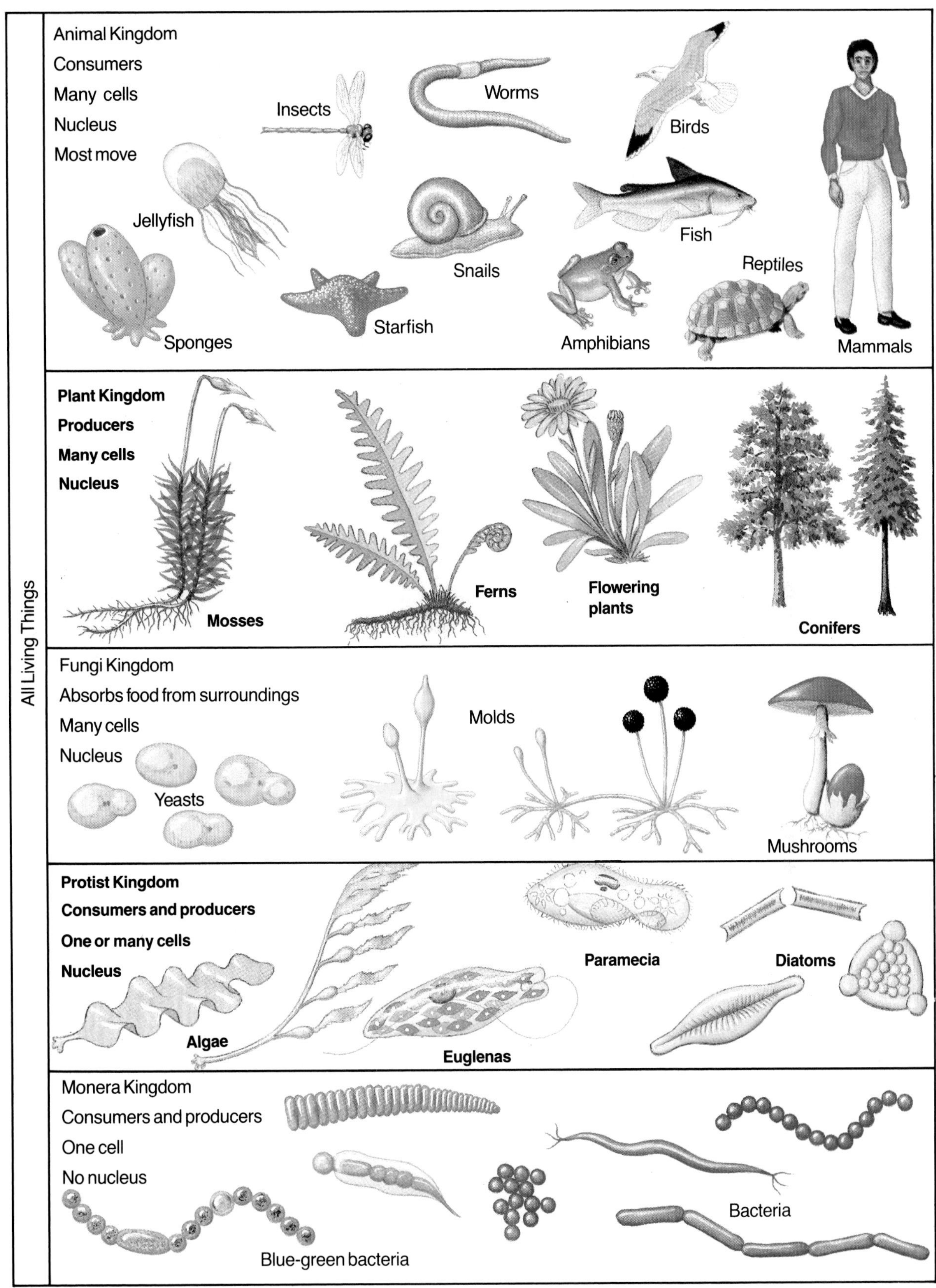

Figure 3-10 Living things are classified into five kingdoms.

Chapter 3

Review

Summary

3:1 Why Things Are Grouped

1. Many things are classified in everyday life.
2. Things are grouped together based on similarities. Two reasons for classifying things are to make them easier to find and to show how they are alike.

3:2 Methods of Classification

3. Aristotle was one of the first people to classify living things. He grouped animals according to where they lived. He grouped plants according to size and pattern of growth.
4. Linnaeus used specific traits to group living things. He also designed a scientific naming system that is still used today.
5. There are seven main groups in the modern classification system. They are kingdom, phylum, class, order, family, genus, and species.

3:3 How Scientists Classify Today

6. Classifying is based on how organisms are related, ancestors, similarities in body structure, and body chemistry.
7. Scientists use scientific names because no two living things have the same scientific name.
8. Today, living things are grouped into five kingdoms–monerans, protists, fungi, plants, and animals.

Key Science Words

animal (p. 62)
class (p. 54)
classify (p. 48)
family (p. 54)
fungus (p. 61)
genus (p. 55)
kingdom (p. 53)
moneran (p. 61)
order (p. 54)
phylum (p. 54)
plant (p. 62)
protist (p. 61)
scientific name (p. 59)
species (p. 55)
trait (p. 48)

Testing Yourself

Using Words

Choose the word from the list of Key Science Words that best fits the definition.

1. smallest group of living things
2. largest group of living things
3. a feature of something
4. largest group within a kingdom
5. genus and species names together
6. first word in a scientific name
7. to group things based on similarity
8. organism that has many cells and moves
9. organism that has cell walls and absorbs food from its surroundings
10. organism that has many cells and can make its own food

Testing Yourself *continued*

Finding Main Ideas

List the page number where each main idea below is found. Then, explain each main idea.

11. how Aristotle classified plants and animals
12. three traits used to place living things in one of the five kingdoms
13. things in our everyday lives that are grouped
14. the seven main groups in the modern classification system
15. what makes up the scientific name
16. two reasons why things are classified
17. the scientist who gave living things two-part names
18. how living things are grouped based on their ancestors

Using Main Ideas

Answer the questions by referring to the page number after each question.

19. Why did scientists depend less and less on Aristotle's grouping system? (p. 53)
20. How would you classify the letters b, c, e, d, l, o, t into two groups? Explain your groups. (p. 48)
21. What are three reasons why scientists use scientific names? (p. 60)
22. Why do we classify books in a library? (p. 48)
23. How do plants and animals differ? (p. 62)
24. What are the classification groups for a lion and a deer? (p. 57)
25. What changes did Linnaeus make in Aristotle's classification system? (p. 53)

Skill Review ✓

For more help, refer to the ***Skill Handbook,*** *pages 704-719.*

1. **Infer:** List the similarities and differences between the animals in Figure 3-3. Infer how closely related the animals are.
2. **Classify:** Classify the following animals into four groups on the basis of their structural similarities: snail, rabbit, gorilla, butterfly, oyster, chimpanzee, clam, cat, honeybee.
3. **Infer:** What is the result of two organisms having a large number of the same ancestors?
4. **Observe:** Obtain a sample of pond water. Place some of the water on a microscope slide and observe the organisms in the water. To what kingdoms do the organisms belong?

Finding Out More

TECH PREP

Critical Thinking

1. Cut out photographs of 20 people from magazines. Design a classification system for these people.
2. How is classification helpful to farmers, florists, and customs inspectors?

Applications

1. Keys are used to identify things that have been classified and named. From the library, obtain a key to the trees or flowering plants in your area. Identify 20 plants using the key.
2. Visit a zoo or an aquarium. Make a list of the living things seen there. Use reference books to identify and classify the living things to the species level.

Biology in Your World

The Scientific Method Is All Around You

In this unit, you have read how the scientific method is used to solve problems in science. But, did you know people use scientific methods to solve other kinds of problems? Consider the "mysteries" described below and how scientific methods were used to solve them.

LITERATURE

A Scientific Search for Roots

Alex Haley learned from his grandmother that he was part of the Kinte family. His African ancestor, Kunta Kinte, had been kidnapped in Africa and then brought to this country as a slave.

Kinte's family handed down African words for six generations. These words were the names of an African village, a certain tree, and a musical instrument. Haley traced these to the Gambia River in West Africa. There he found a man who could recite the history of the Kinte family. The man's story fit the facts Haley had before his trip. Haley wrote *Roots* to tell the story of his search.

GEOGRAPHY

Shifting Continents

The east coast of South America and the west coast of Africa could almost fit together like the pieces of a jigsaw puzzle. This observation led scientists to form the continental drift theory. The theory says that there was once a single great continent that split and drifted apart.

Other evidence indicates that all the land masses of Earth may have been joined at one time. Layers of rock along a coast on one side of the ocean match layers of rock on an opposite coast. Also, different continents have the same kinds of fossils. Thus, observation and evidence support the continental drift theory.

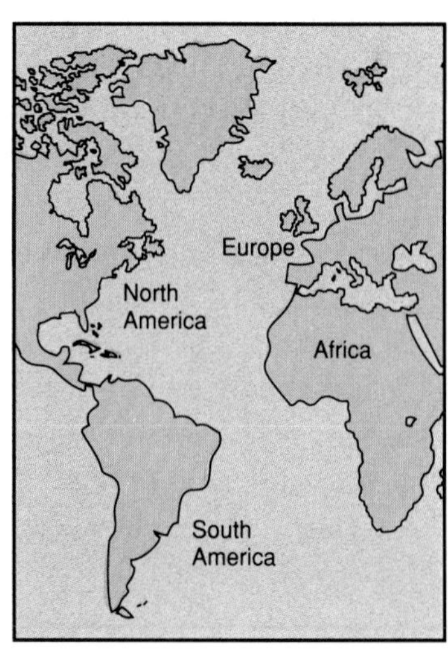

LEISURE

Green Thumbs Up!

Many teens enjoy spending time assembling a terrarium. You can have a miniature garden in your bedroom! Terrariums are fun to arrange, and the materials are easy to find. All you need is a clear container, pebbles, sand, soil, and some plants. Ferns and mosses grow well in terrariums. You can try to grow many other kinds of plants, also. Experiment with various plants to see which are best suited for a terrarium. Set up your experiment using the scientific method.

How would you vary the conditions in or surrounding a terrarium to find the best environment for it?

CONSUMER

Cycling Through Your Choices

Your old bicycle is just about worn out, and you need to buy a new one. How can you use a scientific method to choose the best bicycle? You can treat your decision like solving a problem.

First, research the types of bicycles. What are the advantages, disadvantages, and purposes of each type?

What do you need in a bicycle? Prepare a list of all the features and uses you want in a bicycle. Then rate each feature with a number from 1 to 5. A feature you really need is a 1. A feature that is nice but you could do without would be a 5.

Compare your list of needed features with the types of bicycles. Do your features match the purposes and advantages of one or two types of bicycles? If not, you may need to rethink your needs.

Go to a bicycle shop to see and ride bicycles. Narrow your choices to one type and choose your bicycle based on all the data you have gathered about the different bicycles.

Unit 2

CONTENTS

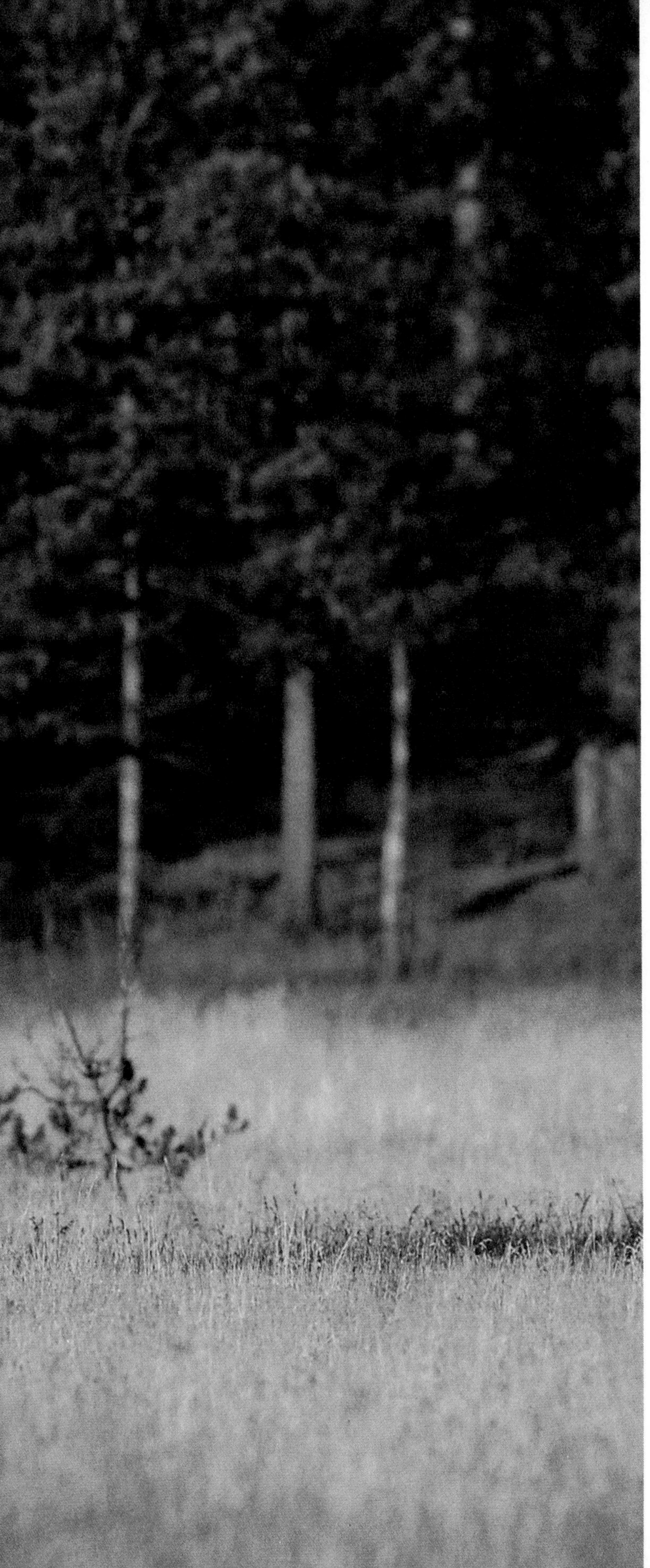

Kingdoms

What would happen if...

there were no fungi? Fungi use and break down substances in rotting logs and other dead organisms. This action releases nutrients into the soil for plants to use. Without fungi, the nutrients in the soil would soon be used up. Then, no plants would grow. Without plants, animals such as the elk in the photo would have no food to eat.

Fungi are also needed to produce many things we've come to depend on. Yeasts, molds, and mushrooms are all fungi. Carbon dioxide gas produced by yeast causes bread to rise. Without yeast, our bread would be flat. Without a certain type of mold, there would be no penicillin. This important antibiotic keeps many deadly diseases under control. Other molds are used to flavor some types of cheeses.

Fungi may be simple and small, but they are a necessary part of our world.

CHAPTER PREVIEW

Chapter Content

Review this outline for Chapter 4 before you read the chapter.

Skills in this Chapter

The skills that you will use in this chapter are listed below.

- In **Lab 4-1**, you will observe, recognize and use spatial relationships, and formulate a model. In **Lab 4-2**, you will make and use tables, observe, measure in SI, and use a microscope.
- In the **Skill Checks**, you will interpret diagrams and understand science words.
- In the **Mini Labs**, you will infer and use a microscope.

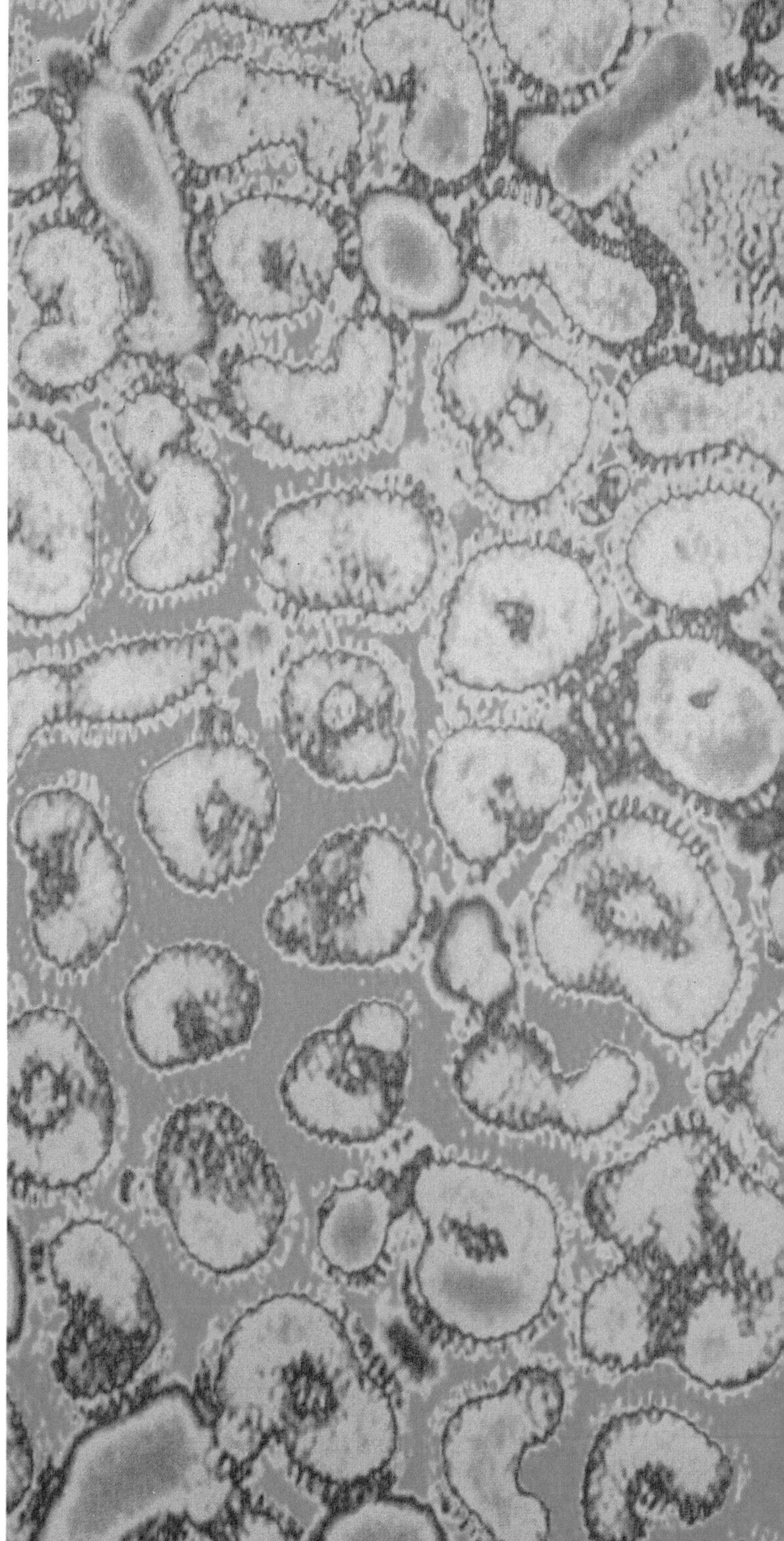

Flu virus

Chapter 4

Viruses and Monerans

Have you ever suffered from the flu? If so, you have had a disease caused by a virus. There are several kinds of viruses that cause the flu. You can get the flu more than once a year because you can be infected by different flu viruses. The photo on the left shows one of the viruses that can cause the flu. It is enlarged 120 000 times. What does this tell you about the size of viruses? In the smaller photo you can see bacteria that cause pneumonia. They are enlarged 75 000 times. Are bacteria or viruses larger?

In this chapter you will read about viruses and monerans. Viruses and monerans affect organisms that are billions of times larger than they are.

Try This!

What methods do people use to fight viruses? Interview several people to learn about folk remedies used to treat the common cold. Make a list of the folk remedies. After you have studied viruses, decide which folk remedies would be the most helpful in fighting a cold.

BIOLOGY Online

Visit the Glencoe Science Web site at science.glencoe.com to find links about **viruses and monerans.**

Pneumonia bacteria

4:1 Viruses

Objectives

1. **Identify** the traits of viruses.
2. **Tell** how viruses reproduce.
3. **List** examples of how viruses affect living things.

Key Science Words

virus
host
parasite
interferon
vaccine

Viruses are neither living nor nonliving things, and yet they have some traits of both. Viruses are difficult for scientists to classify. What are these unusual things and what are their unusual traits?

Traits of Viruses

Viruses are so small that they can be seen only with an electron microscope. Figure 4-1 shows three kinds of viruses highly magnified. Viruses have many different shapes. Some are round and some are rod-shaped. Other viruses are many-sided.

A **virus** is made of a chromosome-like part surrounded by a protein coat. The chromosome-like part carries the hereditary material. The protein coat is responsible for the different shapes of viruses. The structures of some viruses are shown in Figure 4-1.

Viruses are not made of cells and have no cell parts. They do not grow or respond to changes in the environment as living things do. The trait that viruses share with living things is the ability to reproduce. However, viruses reproduce only inside living cells.

Figure 4-1 All viruses are made of a chromosome-like part and a protein coat.

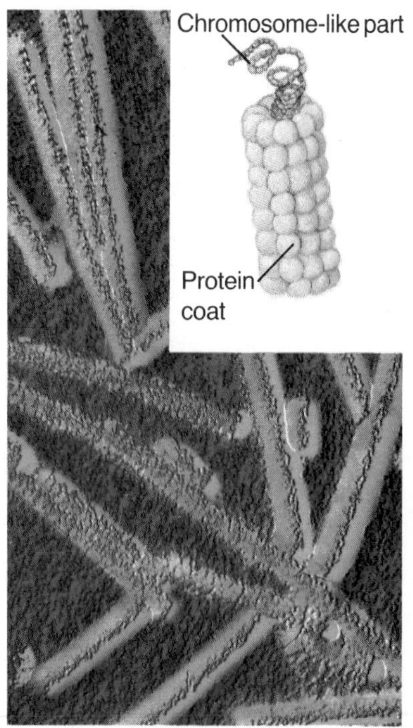

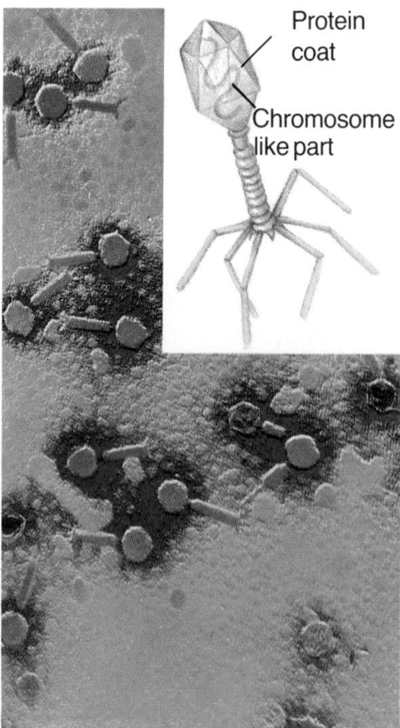

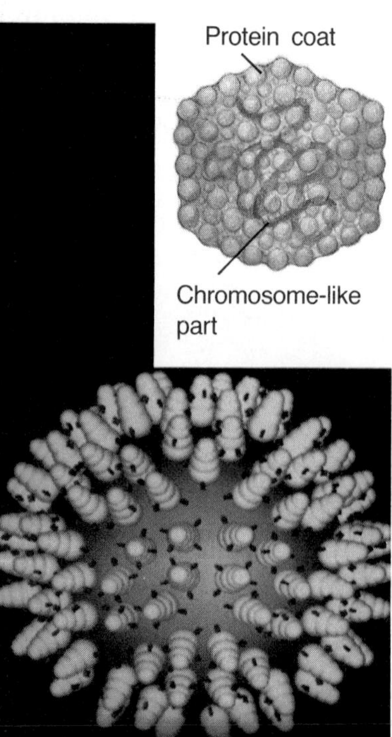

Parts of viruses that look like crystals were stored in labs for as long as 50 years. Scientists then placed the viruses inside living cells. The viruses grew and reproduced. You can see why some scientists consider viruses to be living and others consider them nonliving. For this reason, viruses are not grouped into any kingdom.

One trait used to group viruses is the kind of cell they infect. Plants, animals, fungi, monerans, and protists all serve as hosts to different kinds of viruses. A **host** is an organism that provides food for a parasite (PAR uh site). A **parasite** is an organism that lives in or on another living thing and gets food from it. In this example, viruses act like parasites.

Each kind of virus infects a certain host. For example, the tobacco mosaic virus in Figure 4-2 will infect tobacco plants, but not corn or wheat plants. Some viruses will infect only certain parts of their hosts. The rabies virus will infect only the nervous system of mammals. Common cold viruses will infect only the cells along the air pathway to the lungs. The shape of the protein coat and size of the virus are two other traits that scientists use to classify viruses.

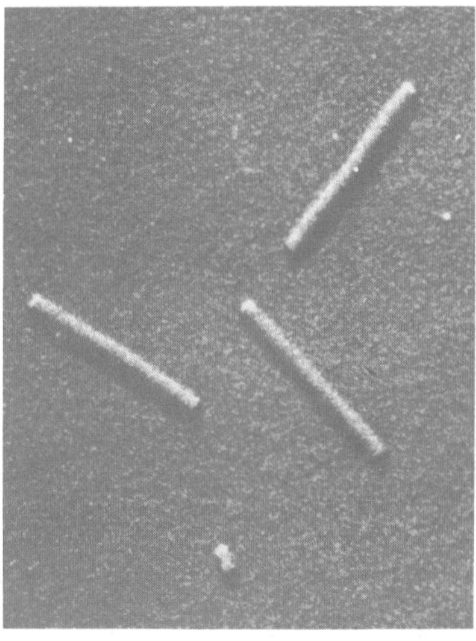

Figure 4-2 Tobacco mosaic virus, magnified 100 000 times

What traits are used to classify viruses?

Some scientists think that viruses came into being before cells existed. Others say that viruses are parts of early cells. Still another idea is that they may have come from disease-causing bacteria.

Viruses and Disease

In spite of their small size, viruses cause many serious diseases in humans and other living things. Some of these diseases are listed in Table 4-1. Viruses may spread from one infected organism to another. In humans, viruses are spread by insects, air, water, food, and other people.

TABLE 4–1. DISEASES CAUSED BY VIRUSES

Animal Diseases	Human Diseases	Plant Diseases
Rabies in dogs	Common cold, flu	Mosaic disease in tobacco
Foot and mouth disease in cattle	German measles, mumps	Bushy stunt in tomato
Newcastle disease in chickens	Measles, chickenpox	Maize dwarf mosaic
Distemper in dogs	Mononucleosis, cold sores	Mosaic disease in alfalfa
Cowpox in cows	Hepatitis, warts	Curly top in sugar beet
Feline leukemia in cats	Polio, smallpox	Dwarfism in rice
	Herpes, AIDS	

Bio Tip

Health: An influenza virus can survive five minutes on human skin and 1.5 days on a kitchen counter top.

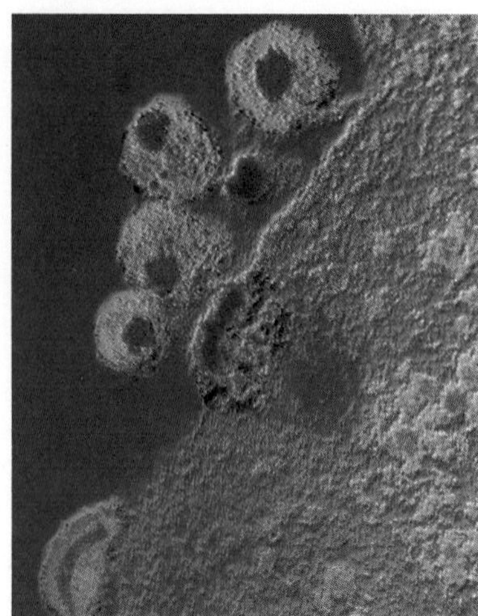

Figure 4-3 AIDS virus being released from an infected cell. The viruses are color enhanced in red and green.

Acquired immune deficiency syndrome (AIDS) is a viral disease that destroys the body's immune system. When the immune system does not work properly, the body can't fight infections. People with AIDS die from diseases that most others can resist. Even a common cold can be life-threatening. The AIDS virus, Figure 4-3, is spread in four known ways. It is spread by sexual intercourse, blood products, the sharing of contaminated needles, and from a pregnant woman to her developing baby. Scientists studying AIDS have not yet discovered a cure.

Plant viruses may be spread by the wind or by insects. Chewing or sucking insects, such as aphids (AY fudz), whiteflies, and beetles, spread most plant viruses. Many plant viruses are unable to enter a plant without the help of an insect. While feeding, the insect breaks through the cell wall of the plant. This break in the cell wall lets the virus get into the host cell. Viruses may infect one or two leaves or an entire plant. Insects or wind then carry the viruses to other plants.

How do viruses get into plant cells?

How do viruses cause disease? Some reproduce rapidly and cause the cell to break open and release viruses, as shown in Figure 4-5. When this type of virus comes in

Figure 4-4 Chewing and sucking insects break cell walls and allow viruses to enter plants.

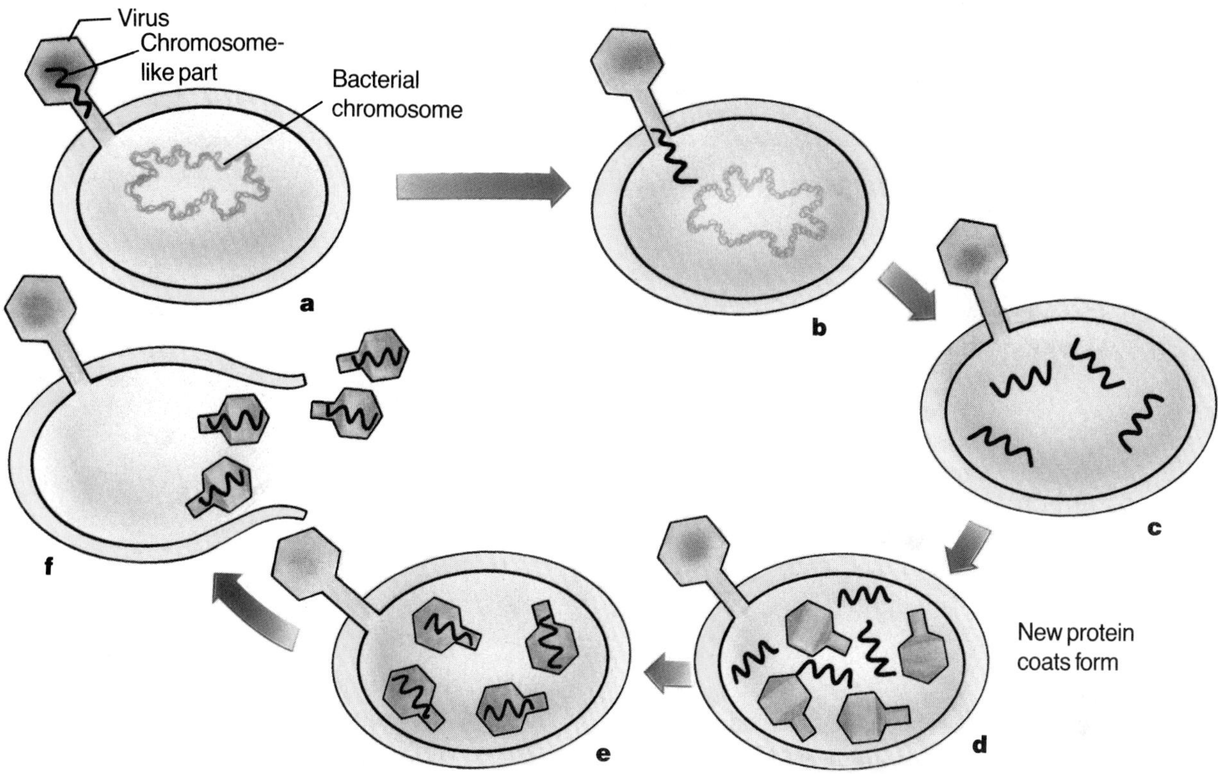

Figure 4-5 Life cycle of a bacterial virus

contact with a host cell, it attaches itself to the cell (a). The chromosome-like part enters the cell (b). Then the chromosome-like part takes over the cell (c). The virus changes the hereditary material in the host cell so that the host cell produces more viruses instead of performing its usual work (d-e). The cell breaks open and releases the new viruses, which then invade other cells (f). Tissue damage and disease result. Viruses that cause polio, mumps, rabies, and flu in humans act this way.

Other viruses remain hidden in the cell a long time without reproducing. Cold sores are blisters around the lips. They are caused by a virus. No symptoms appear until something such as a fever or sunburn causes the virus to become active. The cold sores may disappear for long periods of time, but the virus remains in the body. As long as the virus does not reproduce, no cold sores appear. When the virus begins reproducing again, the cold sores reappear.

Other viruses cause the host cells to reproduce both themselves and the viruses. Groups of infected cells become lumps called tumors. Some tumors are harmless. A wart is an example of a harmless tumor. Examples of harmful tumors are some kinds of cancer.

Skill Check

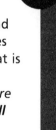

Interpret diagrams: Look at Figure 4-5 and tell how a virus causes disease. Describe what is taking place in each lettered step. *For more help, refer to the* ***Skill Handbook,*** *pages 706-711.*

Viruses

Lab 4–1

Problem: What are some shapes of viruses?

Skills

observe, formulate a model, recognize and use spatial relationships

Materials

bolt, 3.7 cm × 0.7 cm
2 nuts to fit bolt
2 pieces of wire, 14 cm long (#22 gauge)
polystyrene ball, 4.5 cm in diameter
pipecleaners, cut in 2-cm lengths

Procedure

1. **Observe:** Look at Figure A of the bacterial virus enlarged 260 000 times. Note the three parts labeled for you.
2. **Formulate a Model:** To build a model of the virus, attach two nuts onto the bolt as shown in Figure C.
3. Twist the wires around the bolt as in Figure D. Make them as tight as possible.
4. Fold down the wire ends and bend them as in Figure E.
5. Figure B shows a flu virus enlarged 300 000 times. Build a model of this virus using the polystyrene ball and pipecleaners.

Data and Observations

1. Which virus part is represented by the top of the bolt and the nuts in your model?
2. Which virus part is represented by the threaded part of the bolt and the wires?

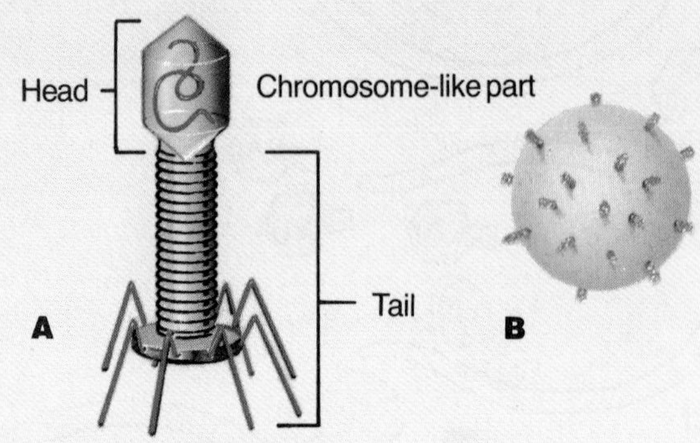

Analyze and Apply

1. What chemical in a real virus would make up the threaded part of the bolt and the wires?
2. Which part of a virus would attach to the host's cell membrane?
3. Which parts would not enter a cell?
4. Which part does enter a cell?
5. If your flu virus model were a real virus, what would you expect to find inside the ball?
6. What chemical would you expect to make up the covering on the ball?
7. What are the basic parts of a virus?
8. **Apply:** How does the structure of a virus differ from that of a bacterial cell?

Extension

Formulate a model: Using references, look up the shapes of some other viruses. Build models with common household items.

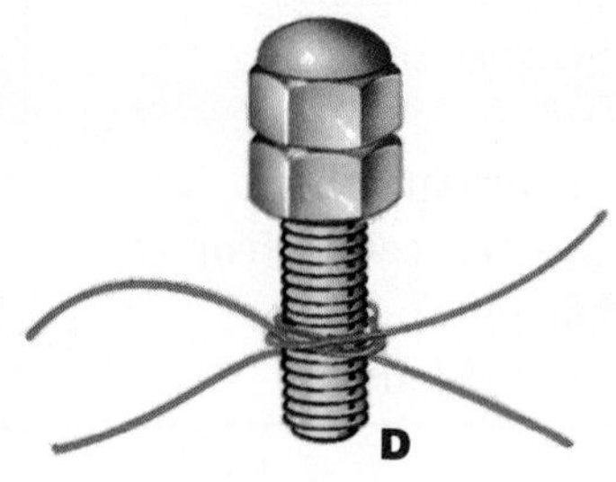

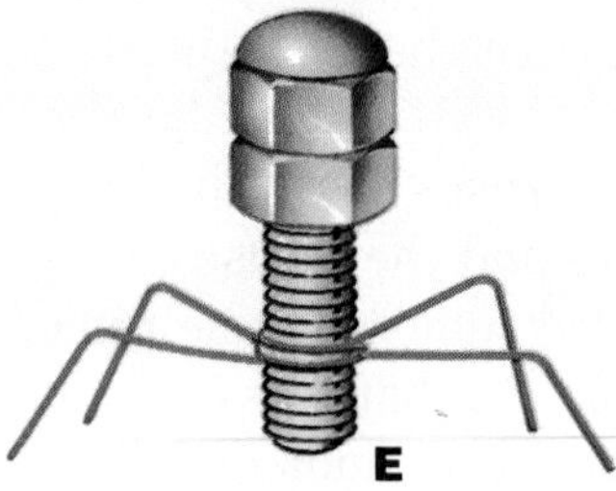

Controlling Viruses

Diseases caused by viruses are hard to treat or cure. There are no known drugs that destroy viruses. However, humans and other animals are protected against some viruses in several ways. Certain white blood cells can surround and destroy a virus. If the virus is not captured and destroyed by these white blood cells, other white blood cells make chemicals called antibodies (ANT ih bohd eez). Antibodies help destroy viruses or harmful bacteria by attaching to them. The virus or bacterium may be destroyed directly by the antibody or it may be held captive by the antibody until white blood cells can surround and destroy it. Each antibody acts only on one specific kind of virus or bacteria.

When human cells are first attacked by a virus, the cells produce interferon (ihnt ur FIHR ahn). **Interferon** is a chemical substance that interferes with the way viruses reproduce. When infected cells burst and release more viruses, they also release interferon. It is believed that the interferon warns nearby cells that a viral infection is taking place. The cells then produce their own chemicals to fight the viruses. Unlike antibodies, interferon will affect any type of virus that invades the body. The interferon produced in one species will work only in that species. For example, interferon produced by a horse will not work in humans.

What does a human cell do when it is attacked by a virus?

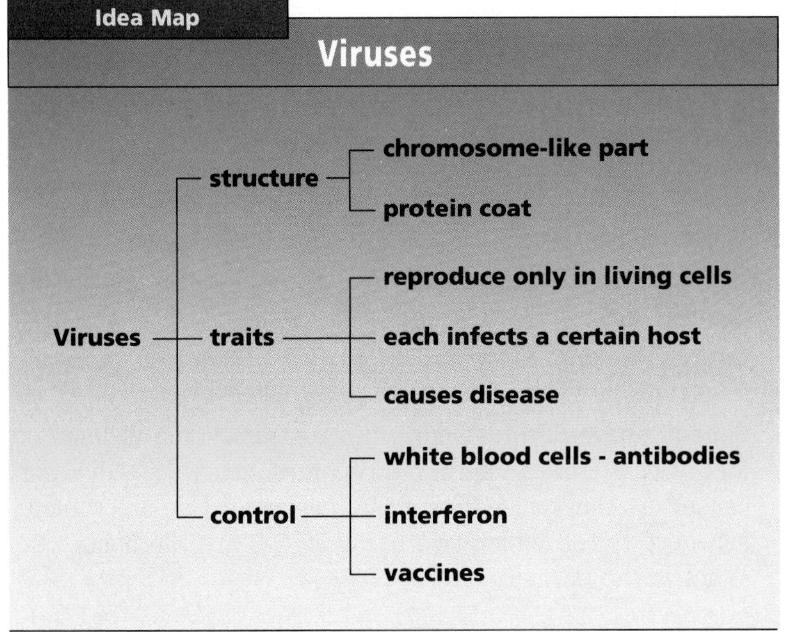

Study Tip: Use this idea map as a study guide to viruses. The features of viruses and how viruses are controlled are shown.

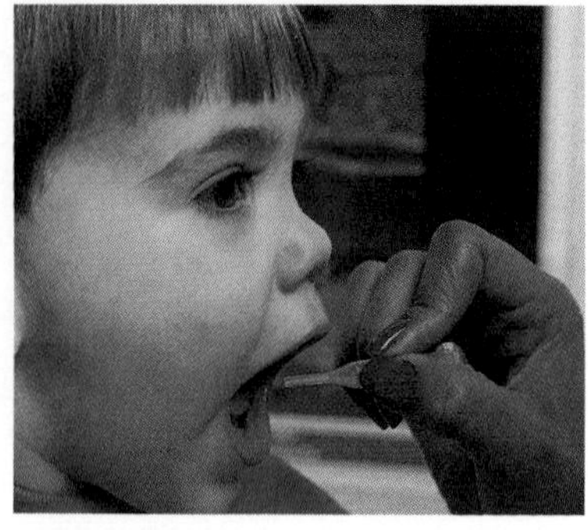

a

b

Figure 4-6 A person takes polio vaccine for protection against the disease (a). Burning infected plants prevents the spread of plant diseases caused by viruses (b).

Vaccines are substances made from weakened or dead viruses. They can be used to treat some viral or bacterial diseases. There are vaccines for polio, rubella, measles, and influenza. Figure 4-6a shows a person getting the polio vaccine. The body reacts to a vaccine by producing antibodies to protect against the disease. In addition to people, pets and farm animals receive vaccines for protection against disease.

It is usually impossible to treat viral infections in plants. Therefore, farmers try to prevent the spread of viral infections by burning infected plants.

Check Your Understanding

1. List two ways viruses differ from living things.
2. Where must viruses be found if they are to reproduce?
3. What are three ways the human body protects itself against viruses?
4. **Critical Thinking:** How is the body's reaction to a vaccine similar to its reaction when it is attacked by a virus?
5. **Biology and Writing:** A mosquito contains some viruses and bites a horse. The horse becomes sick with a disease caused by the viruses from the mosquito. Use a complete sentence to tell which two living things are the hosts. What is the parasite?

4:2 Monera Kingdom

Monerans are one-celled organisms that lack a nucleus, but do have nuclear material within the cell wall. They also lack many of the cell parts found in plant and animal cells. Two groups of monerans, bacteria and blue-green bacteria, are found everywhere around you. They live as single cells or in groups of cells.

Objectives

4. Identify the traits of bacteria.

5. Explain how bacteria affect other living things.

6. Compare the traits of blue-green bacteria and other bacteria.

Key Science Words

bacteria
colony
capsule
flagellum
fission
asexual reproduction
endospore
saprophyte
decomposer
Koch's postulates
communicable disease
antibiotic
biotechnology
pasteurization
blue-green bacteria

Traits of Bacteria

What are bacteria? **Bacteria** are very small, one-celled monerans. They are larger than viruses, but are too small to be seen without a microscope. They are so small that 300 could fit side by side across the period at the end of this sentence.

Bacteria can be found almost everywhere. They live in water, air, soil, food, and on almost every object. They are on your skin and even inside your body. Bacteria have been found in arctic ice as well as in hot springs. Bacteria are so widespread because almost any material may be food for some kind of bacteria.

Bacteria are classified into three groups according to shape. Some bacteria are round. Some are shaped like rods. Others are spiral, Figure 4-7. Bacteria can be found as single cells, in pairs, or in clusters. Some singles and pairs may join together in a cluster or chain. The cells in a rapidly growing cluster or chain can make a colony. A **colony** is a group of similar cells growing next to each other that do not depend on each other.

a

b

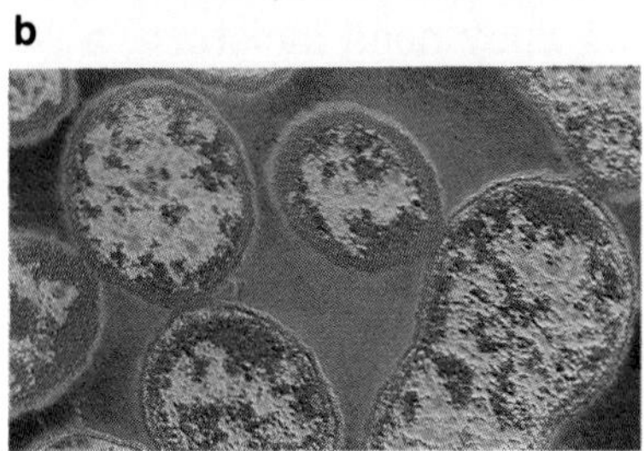

c

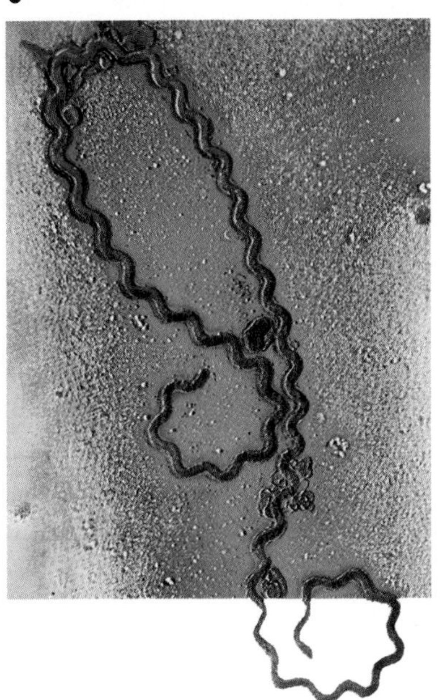

Figure 4-7 Rod-shaped bacteria (a), round bacteria (b), and spiral bacteria (c) highly magnified

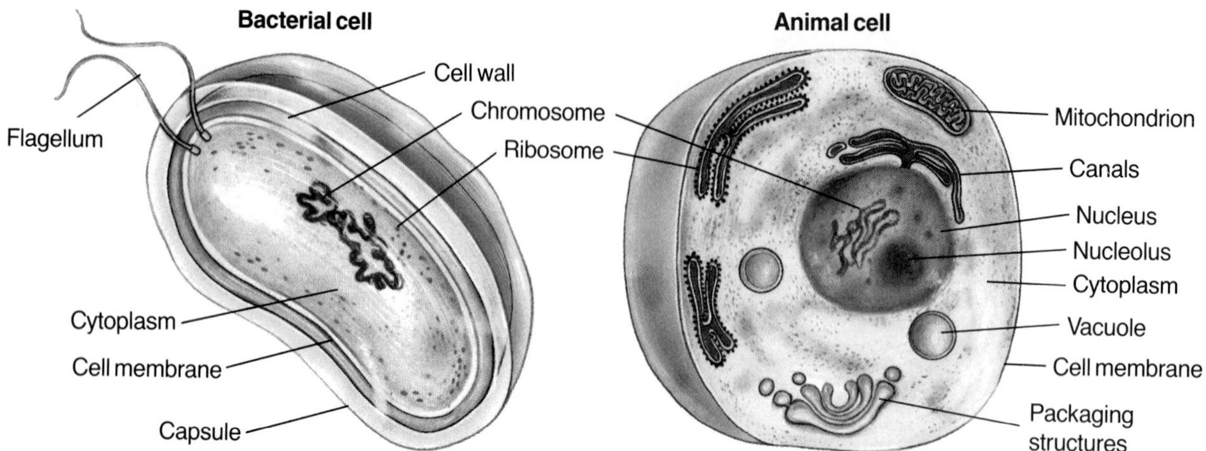

Figure 4-8 Structures of bacterial and animal cells. What structures are found in a bacterial cell but not in an animal cell?

Under a microscope, bacteria look very different from other cells you have studied. Notice that the bacterial cell in Figure 4-8 does not have a nucleus. It has one main chromosome. The animal cell does have a nucleus. The bacterial cell also does not have most cell parts seen in the animal cell.

Even without certain cell parts, bacteria carry on the same jobs as other cells. Bacteria still reproduce, grow, and carry out cellular respiration.

Bacterial cells have a cell wall as well as a cell membrane. Some bacteria have a sticky outer layer called a **capsule,** Figure 4-8. The capsule keeps the cell from drying out and helps the cell stick to food and other cells. Some bacteria move with a long, whiplike thread called a **flagellum** (fluh JEL um).

Bacteria reproduce by fission (FIHSH un). **Fission** is the process of one organism dividing into two organisms. Fission is a kind of asexual (ay SEK shul) reproduction. **Asexual reproduction** is the reproducing of a living thing from only one parent. Bacteria reproduce by asexual reproduction. The circular chromosome of the bacterial cell makes a copy of itself, and the cell divides, Figure 4-9. The time that it takes a bacterial cell to grow and divide into two cells varies. If growing conditions are right, it can take only about 20 minutes.

What do bacteria need to live? They need moisture, a certain temperature, and food. A few bacteria can live at 0°C and others can live at temperatures of 75°C. Those that cause disease in humans live at 37°C, normal body temperature. Most bacteria grow best in darkness. Most need oxygen to live. Others cannot live in the presence of oxygen.

Bio Tip

Consumer: In Southeast Asia, rice is grown without using fertilizers because certain bacteria take nitrogen gas out of the air and use it to make nutrients for the rice plants.

Figure 4-9 A bacterial cell divides to form two new cells. What is this form of reproduction called?

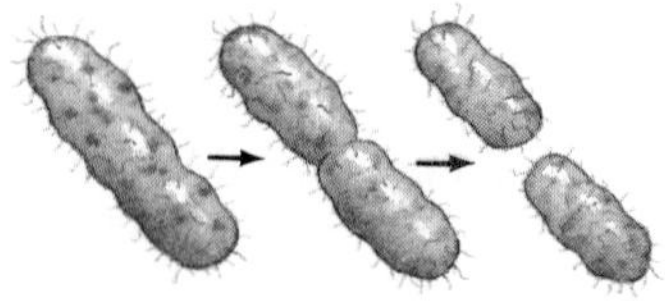

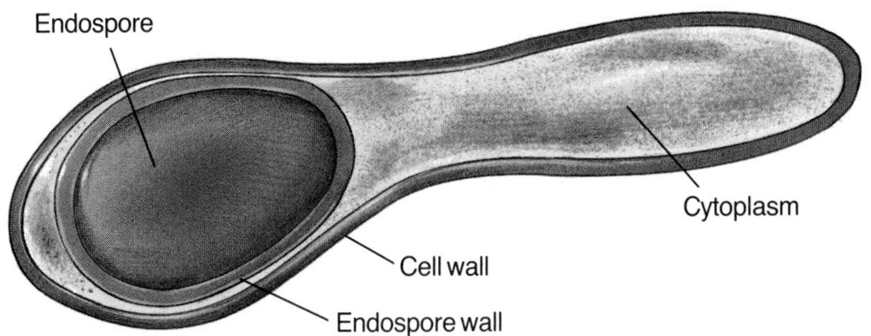

Figure 4-10 An endospore has a thick, protective wall that allows the bacterium to withstand harsh conditions.

If living conditions are not right for bacteria to grow, they can survive by forming endospores (EN duh sporz), Figure 4-10. An **endospore** is a thick-walled structure that forms inside the cell, enclosing all the nuclear material and some cytoplasm. Some endospores can survive for many years. They can withstand boiling, freezing, and extremely dry conditions without damage. When conditions return to normal, endospores develop into bacteria. Endospores are not used for reproduction. They let bacteria survive when living conditions are not ideal.

What do bacteria use for food?

Most bacteria feed on other living things or on dead things. Bacteria that feed on living things are parasites. Bacteria that live on dead things are called saprophytes (SAP ruh fites). **Saprophytes** are organisms that use dead materials for food. They get energy by breaking down, or decomposing, dead materials. **Decomposers** are living things that get their food from breaking down dead matter into simpler chemicals. Decomposers are important because they return minerals and other materials to the soil, where other organisms can use them.

Some bacteria can make their own food. Some use the energy of the sun to make food. Others use the energy in substances containing iron and sulfur to make food.

Bacteria and Disease

Some bacteria are parasites in humans. They can cause a variety of diseases. Human diseases caused by bacteria include strep throat, tuberculosis, certain kinds of pneumonia (noo MOH nyuh), gonorrhea (gahn uh REE uh), and meningitis.

Plants and animals other than humans also may have bacterial diseases. Fire blight and crown gall, Figure 4-11, are plant diseases caused by bacteria. The bacteria that cause fire blight are spread by rain.

Figure 4-11 Crown gall is a plant disease caused by bacteria.

Some bacterial diseases that occur in animals can be passed on to humans. Anthrax, a disease found in livestock, is one example. It is usually passed on to people who work with animals, such as butchers and handlers of wool and leather.

How do we know that bacteria cause disease? In 1876, a German doctor named Robert Koch used a scientific method to show that anthrax was caused by a bacterium. He made a hypothesis that a bacterium caused the disease. He experimented to test the hypothesis. The experiments supported his hypothesis, so he formed a theory about the cause of disease.

Koch's postulates (KAHKS • PAHS chuh lutz) are steps for proving that a disease is caused by a certain microscopic organism. Here are Koch's postulates.

How can scientists prove that bacteria cause disease?

1. The organism must be present in a living thing when the disease occurs.
2. The organism must be taken from the host and grown in the laboratory, Figure 4-12a, b.
3. When the organisms from the laboratory are injected into healthy hosts, they must cause the same disease in the healthy hosts, Figure 4-12c.
4. The organism must be removed from the new hosts, grown in the laboratory, and shown to be the same as the organism from the first host, 4-12d.

Figure 4-12 Using Koch's postulates to prove the cause of a disease

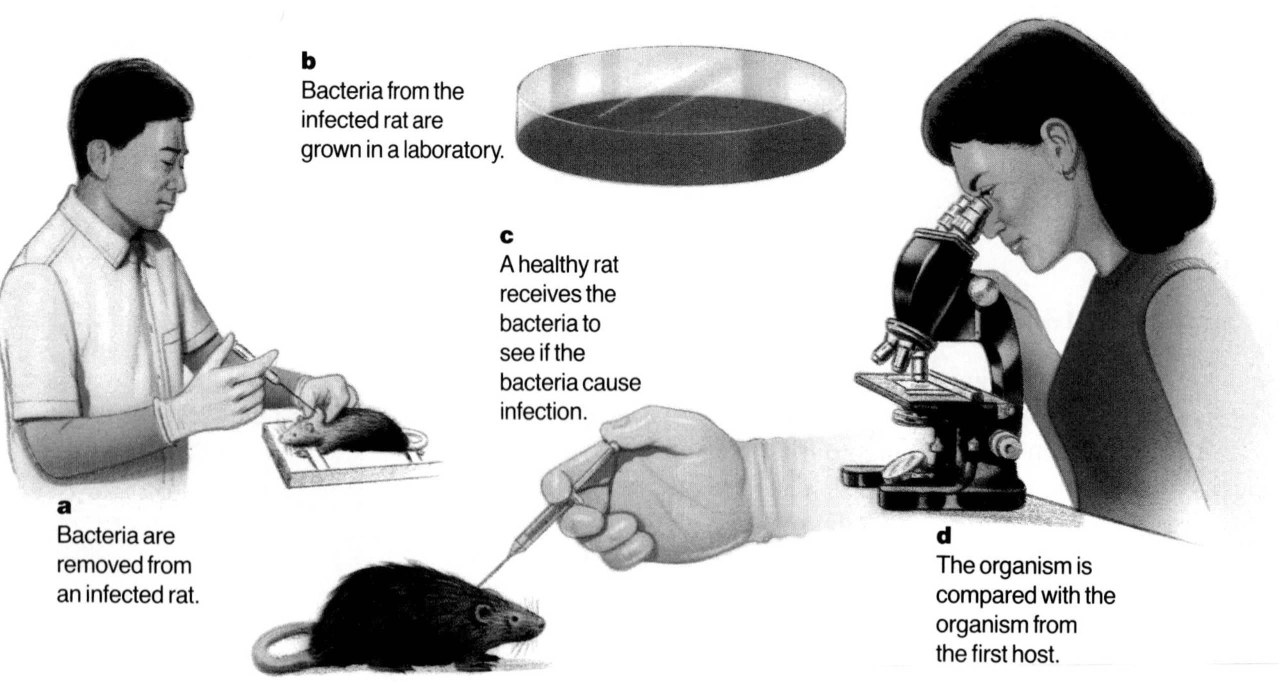
a
Bacteria are removed from an infected rat.
b
Bacteria from the infected rat are grown in a laboratory.
c
A healthy rat receives the bacteria to see if the bacteria cause infection.
d
The organism is compared with the organism from the first host.

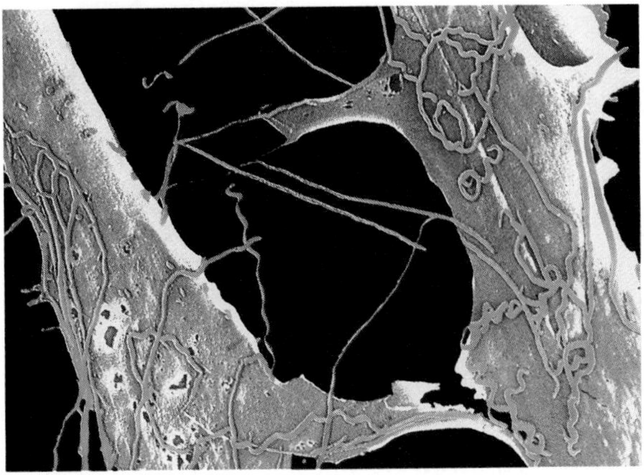

a

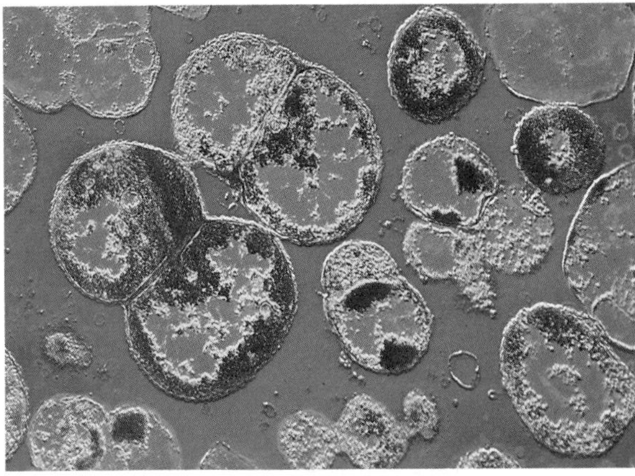

b

Figure 4-13 The spiral bacteria that cause syphilis (a) and the round bacteria that cause gonorrhea (b)

Scientists still use Koch's postulates today. The procedure helps identify bacteria and other disease-causing organisms.

Some diseases caused by bacteria and viruses are called communicable (kuh MYEW nih kuh bul) diseases. **Communicable diseases** are ones that can be passed from one organism to another. They may be spread in several ways. Some diseases are spread by air when a person sneezes or coughs. Pneumonia, strep throat, and tuberculosis are spread through the air. Communicable diseases can also be spread by touching anything an infected organism has touched. Common items that might have bacteria are clothes, food, silverware, or toothbrushes. Another way of spreading diseases is by drinking water that contains bacteria. Some of the more serious diseases are those spread by sexual contact. Syphilis (SIHF uh lus) and gonorrhea are sexually transmitted diseases caused by bacteria, Figures 4-13a, b.

Insects also spread diseases. Flies, fleas, cockroaches, and mosquitoes carry disease organisms.

Helpful Bacteria

You may be surprised to learn that there are more helpful bacteria than harmful ones. Most kinds of bacteria are helpful to humans. Bacteria are needed to decompose dead matter. Bacteria get energy from dead matter as they break it down into materials that other living things can use. In this way, the chemicals that are found in dead matter are recycled so that other organisms can use them to grow.

Mini Lab

What Prevents the Growth of Bacteria?

Infer: Pour 100 mL beef broth into each of 4 beakers: A, B, C, D. Add 1 tsp. salt to B, 1 tsp. sugar to C, and 1 tsp. vinegar to D. Observe after 2 days. What prevents bacterial growth? *For more help, refer to the* ***Skill Handbook,*** *pages 706-711.*

Mini Lab

What Makes Yogurt?

Use a microscope: Thin plain yogurt with water. Place a drop on a slide. Add a drop of methylene blue and a coverslip. Observe under high power. What do you see in the yogurt? *For more help, refer to the **Skill Handbook,** pages 712-714.*

Some bacteria grow in other living things and help them. Bacteria in the stomach of a cow break down grass and hay. In this way, the grass and hay can be used as food by both the bacteria and the cow. Without the action of bacteria, plants would have little food value for cows, sheep, and deer. Bacteria in your intestine make vitamins that you need.

Bacteria are used to make many products that are useful to us. Some bacteria break down the fibers of plants used to make linen and rope. Some are used in making leather from skins. Some of today's most useful antibiotics (an ti bi AHT iks) are produced by bacteria, Figure 4-14a. **Antibiotics** are chemical substances that kill or slow the growth of bacteria. Some antibiotics have a specific effect, working only on certain types of bacteria. Others destroy many types of bacteria.

Some foods owe their taste or texture to bacteria. Many dairy products are made by adding certain species of bacteria to milk. For example, cottage cheese, Swiss cheese, yogurt, sour cream, and buttermilk are all made by bacteria growing in milk. The flavors of coffee and cocoa are due to bacteria, Figure 4-14b. Sauerkraut is made by the action of bacteria on chopped cabbage. Bacteria also help change alcohol to vinegar.

Figure 4-14 Bacteria produce some of today's most useful antibiotics (a). Bacteria are used in making some food products (b).

Bacteria are very important in the field of biotechnology (bi oh tek NAHL uh jee). **Biotechnology** is the use of living things to solve practical problems. Bacteria are being used to produce natural gas and detergents. Some are used to produce human insulin. Insulin is a chemical that the body normally makes to control the level of sugar in the blood. Some people cannot make insulin, so they must use the insulin produced by bacteria.

Skill Check

Understand science words: biotechnology. The word part *bio* means life. In your dictionary, find three words with the word part *bio* in them. *For more help, refer to the **Skill Handbook,** pages 706-711.*

Controlling Bacteria

If food is put into containers and heated to a high enough temperature, the bacteria inside can be destroyed. The containers can then be sealed while they are hot. Food can remain in them for long periods of time without spoiling, as long as the seals are not broken and the cans remain airtight. The process of sealing food in airtight cans or jars after killing the bacteria is called canning. Endospores can be killed by the canning process. They are a major cause of food poisoning.

The process of heating milk to kill harmful bacteria is called **pasteurization** (pas chuh ruh ZAY shun). The milk you buy in stores is pasteurized. Even pasteurized milk that has not been opened will finally spoil. Can you explain why?

Cooling food to a low temperature slows the growth of bacteria. You keep food in the refrigerator to slow bacterial growth. Freezing food is also used to slow bacterial growth. Frozen food can be stored safely for several months. Items kept in a freezer last longer than those kept in the refrigerator. Why?

Figure 4-15 Canning is a way to prevent bacteria from spoiling food.

TECHNOLOGY

Bacteria are helpful in removing pollutants from soil and groundwater. After an oil spill in Prince William Sound in Alaska, the natural populations of oil-eating bacteria were not large enough to take care of the spill. Researchers added fertilizers to the oil-covered beaches to increase the number of bacteria that lived on the beaches.

At Los Alamos National Laboratory, bacteria that break down explosive wastes are being grown. A researcher has collected soil samples that contain TNT and nitroglycerin. The samples also contain bacteria that have adapted to the explosives and break them down. The bacteria are being cultured in large quantities and transferred back to the contaminated area.

Bacteria are used to mine gold, copper, and other valuable minerals from mines where the concentrations are so low they cannot be mined by traditional methods. Bacteria that release bound-up minerals are added to the ore. The minerals can then be mined.

Oil spill

Figure 4-16 When water is removed from foods, bacteria cannot grow in them.

If water is taken from food, bacteria can't live in the food. Bacteria need the water for growth. For this reason, uncooked noodles and dry cereals can be left open without spoiling. These foods have very little water. Removing the water from food is called dehydration (dee hi DRAY shun). What other foods do you know of that are dehydrated?

Why do you put iodine, hydrogen peroxide, or alcohol on a cut, scrape, or other skin injury? These chemicals are called antiseptics (an ti SEP tiks). The word *antiseptic* means against infection. Antiseptics are chemicals that kill bacteria on living things. Disinfectants (dihs un FEK tunts) are stronger chemicals that are used to destroy bacteria on objects that are not living.

What do we do to get rid of bacteria on or in our bodies? We wash with soap, we brush our teeth, and we use mouthwash. Each time this is done, we wash, brush, kill, or rinse away millions of bacteria. The bacteria can cause body and breath odor and tooth decay.

Some medicines help stop bacterial infections after you become sick. You may have been given an antibiotic or other medicine to fight a bacterial infection when you were sick. Vaccines prevent bacterial infections from happening. They protect the body from invading organisms.

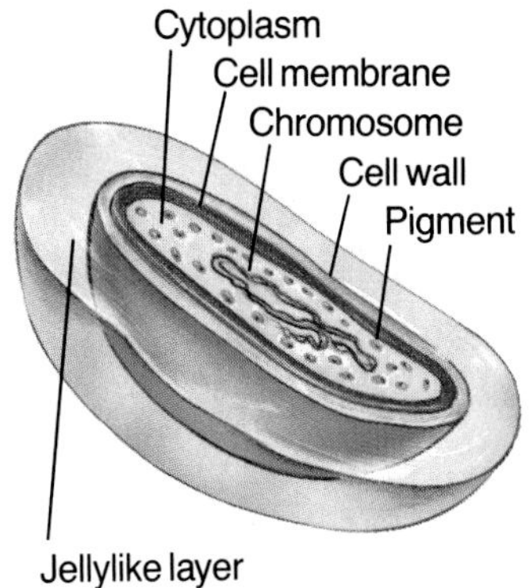

Figure 4-17 Structure of a typical blue-green bacterium. Does it look more like a bacterial cell or an animal cell?

Blue-green Bacteria

Blue-green bacteria are small, one-celled monerans that contain chlorophyll and can make their own food. For many years they were called blue-green algae and were classified as plants. Their cell structure is more like that of bacteria than of plants. They do not have nuclei or other typical plant cell parts, Figure 4-17. Each has a cell wall outside a cell membrane. The cell wall of a blue-green bacterium is made of materials different from the cell wall of a plant cell.

Blue-green bacteria contain colored pigments. They are usually a blue-green color, but some blue-green bacteria may appear red, black, brown, or purple. The blue-green bacteria in the Red Sea grow rapidly at certain times of the year, giving the sea its red color and its name.

Where do blue-green bacteria live? They are found in ponds, lakes, moist soils, swimming pools, and sometimes around leaky faucets. Some can even live in salt water and on snow. Others grow in the acid water of hot springs.

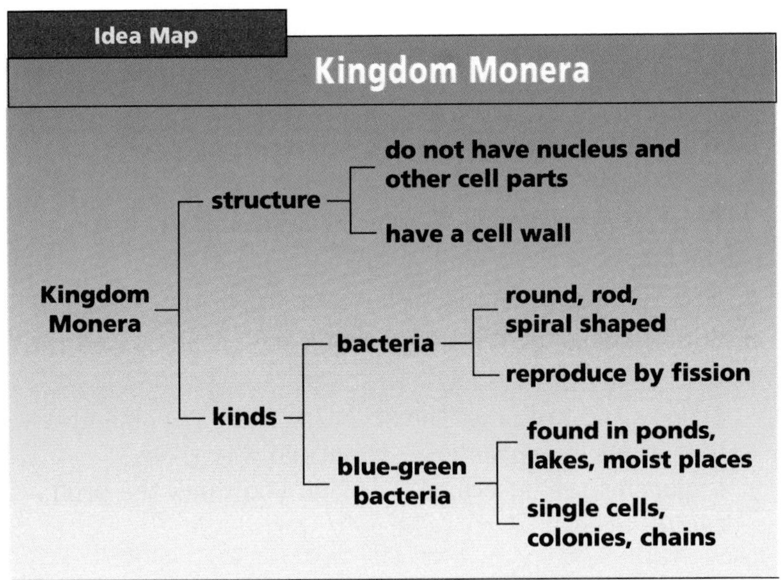

Study Tip: Use this idea map as a study guide to the monerans. Can you give an example of each kind of moneran?

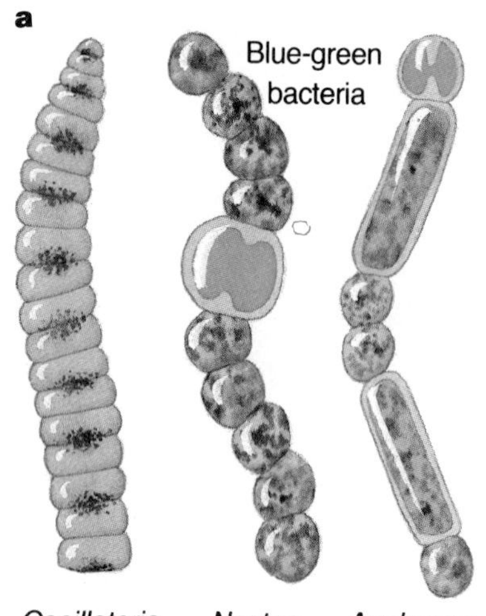

a

b

Figure 4-18 Blue-green bacteria (a) can grow very rapidly and cover the surface of a pond (b).

Blue-green bacteria occur as single cells, colonies, and long, threadlike chains. Many have an outer jellylike layer that holds the cell to other cells. Some have gas bubbles that let the cells float at the surface, where they get sun.

Blue-green bacteria serve as food for animals that live in water. They produce oxygen as they make their own food. They are also important in recycling nitrogen that can then be used by plants.

Blue-green bacteria may grow too rapidly and cover an entire pond. When this happens, they can use up the oxygen in the pond and kill other living things. Many blue-green bacteria produce a bad odor. The odor is one clue that the water is unfit for drinking. Substances that pollute water are often used by blue-green bacteria for food. The amount of pollution in water is measured by counting the number of blue-green bacteria.

Check Your Understanding

6. What are two ways that bacterial cells differ from animal cells?
7. What are three ways in which communicable diseases can be spread?
8. How do blue-green bacteria differ from other bacteria?
9. **Critical Thinking:** Why don't foods such as uncooked rice and raisins spoil?
10. **Biology and Reading:** The scientific name for blue-green bacteria is *cyanobacteria.* From the way these organisms are described, what do you think the prefix *cyano-* means?

Lab 4–2

Monerans

Problem: How do blue-green bacteria and bacteria compare?

Skills

make and use tables, observe, measure in SI, use a microscope

Materials

microscope
prepared slide of *Anabaena*
prepared slide of bacteria
petri dish
metric ruler

Procedure

1. Copy the data table.
2. Use the low-power objective to locate a strand of *Anabaena*, Figure A.
3. **Observe:** Switch to the high-power objective of your microscope and observe the *Anabaena*. Look for cell parts inside the cells.
4. Complete the first three rows of the data table for *Anabaena*.
5. Using the petri dish as a guide, draw a circle on a sheet of paper.
6. Look through the eyepiece of the microscope. Draw one cell in the circle. (Pretend that the circle is the view through the eyepiece. Draw the cell the same size it appears through the eyepiece. This is called drawing to scale.)
7. **Measure in SI:** Measure the length of your *Anabaena* diagram in millimeters.
8. Record this number in your data table.
9. To find the actual length of the cell, multiply the length of your diagram by 0.0035. Use a calculator to help you. Record your answer.
10. Repeat steps 2 through 9 with the bacteria slide. See Figure B.

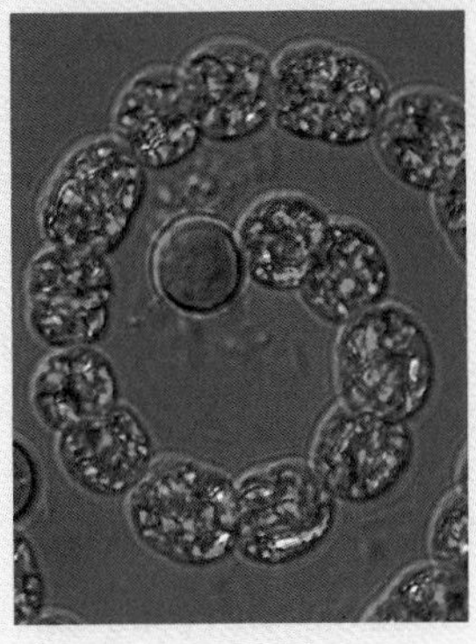

A

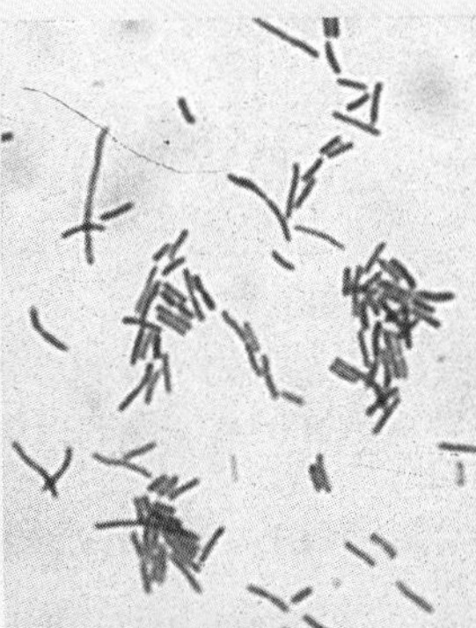

B

Data and Observations

1. Which organism is smaller?
2. Which organism lives as a single cell?

	ANABAENA	*BACTERIUM*
Shape of cells		
Single cell or filament		
Color		
Length of diagram		
Actual length of cell		

Analyze and Apply

1. Which organism can make its own food?
2. To which kingdom do *Anabaena* and bacteria belong? How do you know?
3. **Apply:** How do blue-green bacteria and other bacteria compare?

Extension

Observe other bacteria and blue-green bacteria under the microscope. Compare these with *Anabaena* and the bacteria in this lab.

Chapter 4

Review

Summary

4:1 Viruses

1. Viruses are not cells and have no cell parts. They are made of a chromosome-like part that carries the hereditary material. They are surrounded by a protein coat.
2. Viruses reproduce only inside living cells.
3. Viruses cause diseases in many living things.

4:2 Monera Kingdom

4. All bacteria are one-celled, have no nucleus, and are either round, rod-shaped, or spiral. Other bacteria cannot make their own food or make oxygen.
5. Some bacteria cause serious diseases. They may be spread by air, food, water, sexual contact, or insects.
6. Blue-green bacteria can make their own food and make oxygen. They occur as single cells, colonies, or long chains.

Key Science Words

antibiotic (p. 84)
asexual reproduction (p. 80)
bacteria (p. 79)
biotechnology (p. 85)
blue-green bacteria (p. 87)
capsule (p. 80)
colony (p. 79)
communicable disease (p. 83)
decomposer (p. 81)
endospore (p. 81)
fission (p. 80)
flagellum (p. 80)
host (p. 73)
interferon (p. 77)
Koch's postulates (p. 82)
parasite (p. 73)
pasteurization (p. 85)
saprophyte (p. 81)
vaccine (p. 78)
virus (p. 72)

Testing Yourself

Using Words

Choose the word from the list of Key Science Words that best fits the definition.

1. organism that lives on or in another living thing and causes harm to it
2. substance that interferes with how a virus reproduces
3. process of heating milk to kill harmful bacteria
4. group of similar cells growing next to each other
5. the use of living things to solve practical problems
6. chemical that kills or slows the growth of bacteria
7. monerans that contain chlorophyll and can make their own food
8. dividing in half to form two new cells
9. whiplike thread used for movement
10. chromosome-like part surrounded by a protein coat

Finding Main Ideas

List the page number where each main idea below is found. Then, explain each main idea.

11. how viruses are classified
12. how human viruses are spread
13. how humans and other animals protect themselves from viruses

Testing Yourself *continued*

14. where bacteria are found
15. how bacteria are classified
16. how communicable diseases are spread
17. two plant diseases caused by bacteria
18. methods by which bacteria can be controlled
19. how decay-causing bacteria are important
20. how blue-green bacteria are harmful

Using Main Ideas

Answer the questions by referring to the page number after each question.

21. What are some traits of viruses? (p. 72)
22. How do insects spread plant viruses? (p. 74)
23. In what two ways do viruses cause disease? (p. 74)
24. How does interferon act against viruses? (p. 77)
25. What are the differences between a bacterial cell and an animal cell? (p. 80)
26. How can bacteria survive when growing conditions are not favorable? (p. 81)
27. How do antiseptics and disinfectants differ from one another? (p. 86)
28. How are Koch's postulates used today? (p. 83)
29. In what ways are bacteria useful to humans? (p. 83)
30. How do blue-green bacteria differ from other bacteria? (p. 87)
31. Where can blue-green bacteria be found? (p. 87)

Skill Review ✓

For more help, refer to the ***Skill Handbook,*** *pages 704-719.*

1. **Formulate a model:** Place several green marbles in a bag filled with water. What organism does this represent?
2. **Interpret diagrams:** Look at Figure 4-8 and compare the parts of a bacterium with those of an animal cell. What cell parts does an animal cell have that a bacterium does not have?
3. **Infer:** Blue-green bacteria were once classified as algae. They are now classified as bacteria. The organisms have not changed. Why do you think their classification has changed?
4. **Make and use tables:** Make a table showing the cause of: measles, AIDS, pneumonia, cold, crown gall, flu, syphilis, warts, gonorrhea, tooth decay, fire blight, herpes.

Finding Out More

TECH PREP

Critical Thinking

1. Bacteria that live on teeth produce an acid that causes decay. Why do people who do not brush regularly have more cavities than those who do?
2. Few people get the diseases measles and polio. Why are the vaccines still required in many states?

Applications

1. Report on the economic effects of viral diseases on crops.
2. Make a collage of ads for products such as deodorants, soaps, and toothpastes that control bacteria.

CHAPTER PREVIEW

Chapter Content

Review this outline for Chapter 5 before you read the chapter.

Skills in this Chapter

The skills that you will use in this chapter are listed below.

- In **Lab 5-1,** you will make and use tables, interpret data, observe, classify, and use a microscope. In **Lab 5-2,** you will observe, infer, use a microscope, and design an experiment.
- In the **Skill Checks,** you will understand science words and interpret diagrams.
- In the **Mini Labs,** you will use a microscope and observe.

Chapter 5

Protists and Fungi

Many different protists live in streams, ponds, and oceans. Some have chlorophyll and can make their own food, while others must find food. Some have cell walls and others do not. Most protists are one-celled, but some seaweeds are several meters long. In the small photograph on this page, you can see some of the many kinds of protists that are found in a drop of water from a pond.

The mushroom growing near the pond on the left is a fungus. Mushrooms and other fungi were once grouped with plants. Then it was discovered that fungi have traits that are different from plants. Fungi do have cell walls, but they lack chlorophyll. They absorb food from their surroundings. This is why fungi are grouped in a separate kingdom.

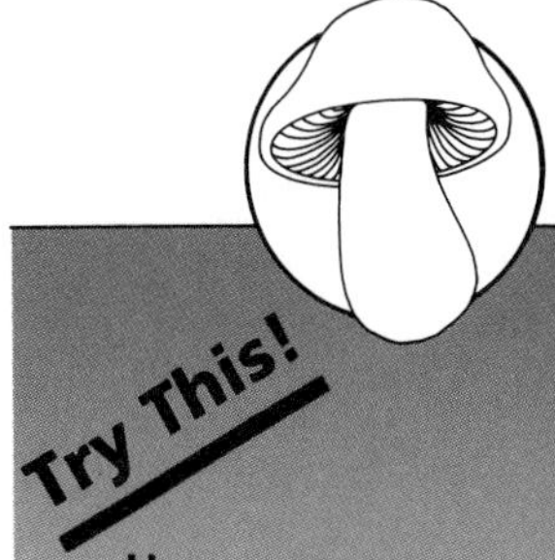

Try This!

Have you ever made mushroom prints? Remove the stalks from several different mushrooms. Place each cap, open side down, on a separate sheet of paper. Cover each one with an inverted bowl or jar and leave overnight. Uncover and remove the caps. What do you see?

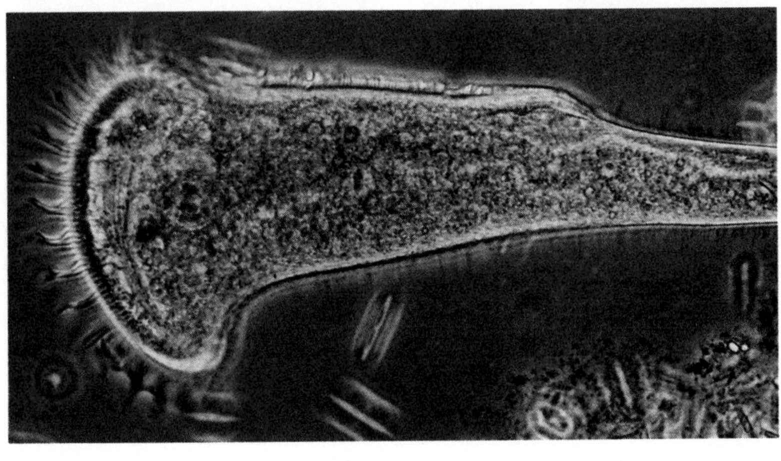

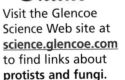

BIOLOGY Online

Visit the Glencoe Science Web site at science.glencoe.com to find links about **protists and fungi.**

Stentor is a protist that lives in pond water.

5:1 Protist Kingdom

Objectives

1. **Identify** the general characteristics of protists.
2. **Compare** the traits of animal-like protists, plantlike protists, and funguslike protists.
3. **Give an example** of each kind of protist.

Key Science Words

protozoan
cilia
sporozoan
spore
algae
multicellular
slime mold

Protists live almost everywhere. In Chapter 3, protists were described as one-celled organisms that have a nucleus and other cell parts with membranes. Protists are larger than bacteria, but most are so small they can be seen only with a microscope. Protists may be producers or consumers. Most can move about in search of food, light, or a place to live.

Although there are a number of different protist phyla, they can be put into three groups. You can see in Figure 5-1 an example of an animal-like protist, plantlike protist, and a funguslike protist.

Animal-like Protists

How many cells do protozoans have?

Protozoans (proht uh ZOH uhnz) make up the largest group of protists. **Protozoans** are one-celled animal-like organisms with a nucleus. The name *protozoan* comes from the Greek words meaning "first animal." At one time, these organisms were classified in the animal kingdom because scientists thought they looked like tiny animals.

Protozoans, like animals, are consumers. The single cell takes in and digests food. It also reproduces. Most protozoans can move about. They are classified in different phyla depending on how they move. Table 5-1 lists the traits of four protozoan phyla.

Figure 5-1 Protists can be grouped into animal-like protists (a), plantlike protists (b), and funguslike protists (c).

a

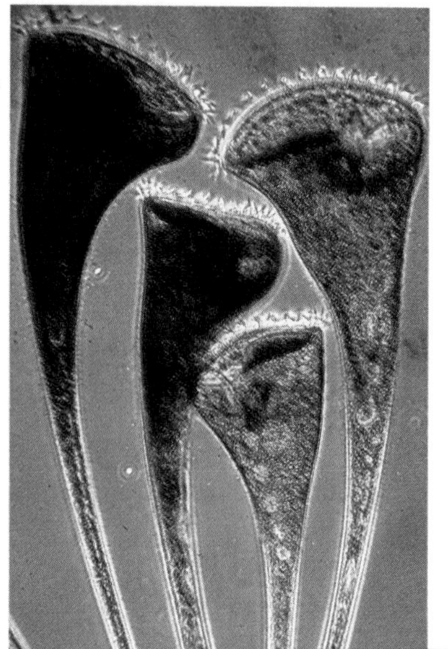

b

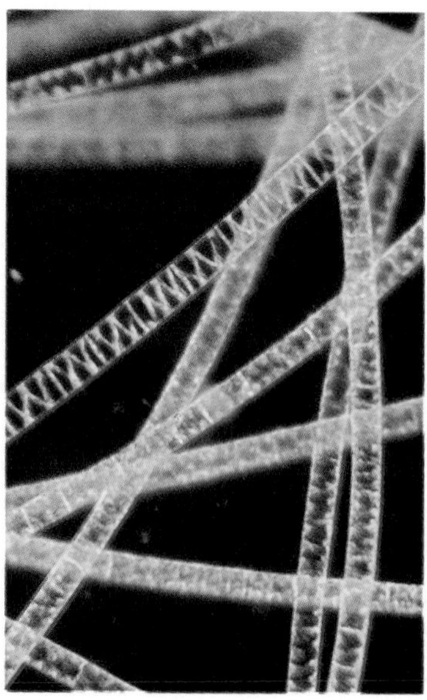

c

TABLE 5–1. ANIMAL-LIKE PROTISTS	
Traits of Phylum	**Example**
move with false feet	amoeba
move with cilia	*Paramecium*
do not move; all parasites	*Plasmodium* *(plaz MOH dee uhm)*
move with one or more flagella	trypanosome (trip AN uh some)

Let's look at some of the protozoans. The protozoan in Figure 5-2 is an amoeba (uh MEE buh). The amoeba is a protist that moves by changing shape. It has parts made of cytoplasm that can reach out from the main body. These parts are called false feet. They help the amoeba move about and trap food. Figure 5-3 shows how. First, the amoeba's false feet surround a small protist or a bacterium. Soon the food is trapped inside the amoeba, where it will be digested.

How does an amoeba move?

Like bacteria, amoebas reproduce by fission. A parent cell divides into two new cells. Each of the cells has the same hereditary material as the parent cell.

Most kinds of amoebas are not harmful to humans, but some cause disease. One kind of amoeba causes a severe type of diarrhea called amoebic dysentery (uh MEE bihk • DIHS un ter ee). People get this disease by drinking water or eating food that contains the amoebas. This usually occurs in areas where conditions are not sanitary.

Figure 5-2 Structures of an amoeba. Do amoebas have cell walls?

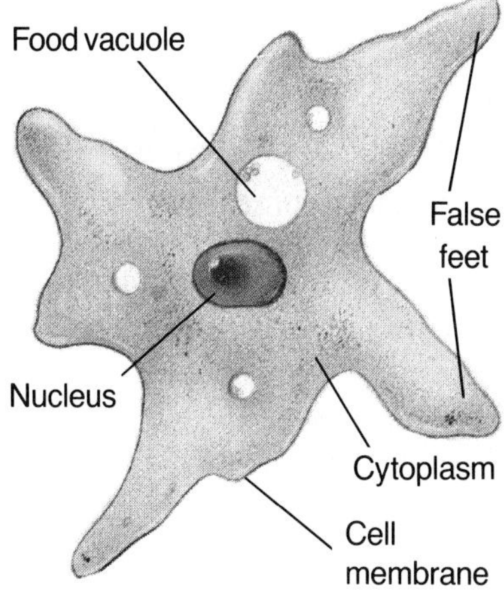

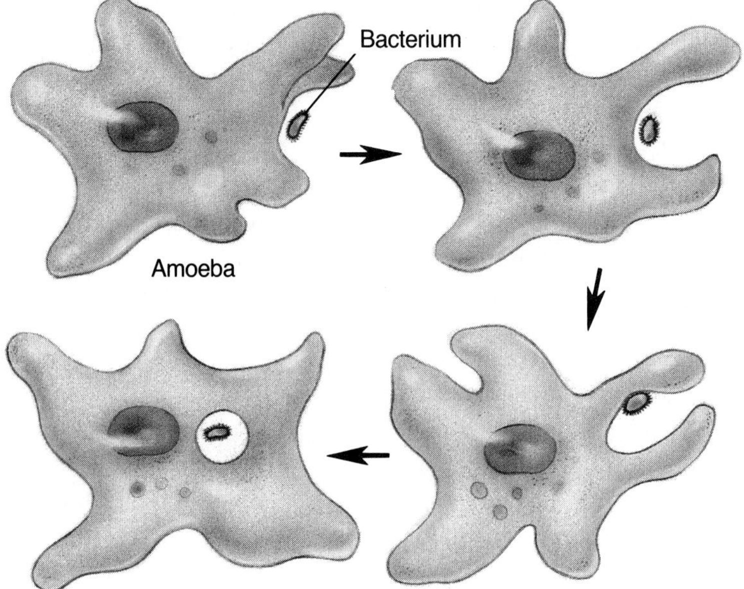

Figure 5-3 Amoebas capture food with their false feet.

Figure 5-4 A paramecium (a) and some of its structures (b). What structures does a paramecium use for movement?

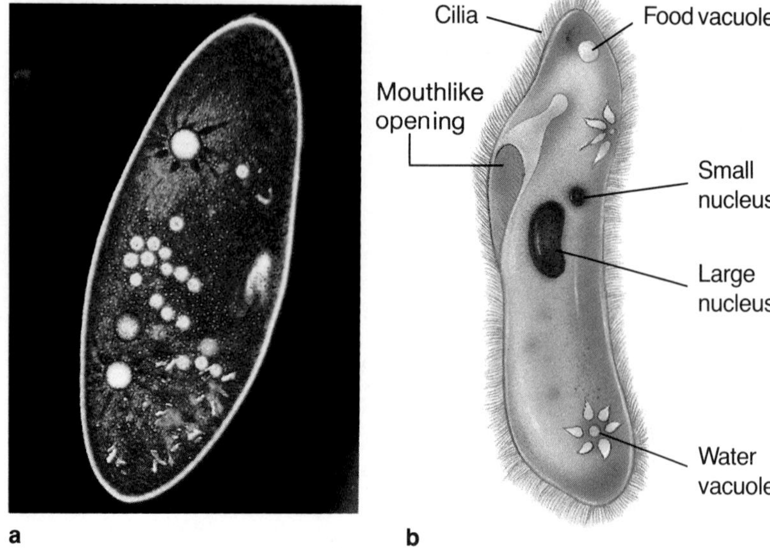

Mini Lab

Is There Life in a Mud Puddle?

Use a microscope: Use a dropper to collect a sample of water from a mud puddle. Place a drop on a slide and add a coverslip. Focus on low power and then on high. Draw what you see. *For more help, refer to the* ***Skill Handbook,*** *pages 712-714.*

An example of another type of protozoan is the paramecium. A paramecium is a protozoan that has a definite shape and moves by means of cilia. **Cilia** are short, hairlike parts on the surface of the cell, as seen in Figure 5-4. The cilia move back and forth very fast, causing the organism to move through the water. Protists with cilia are found in most ponds and streams. Most members of this group do not cause diseases in humans.

When a paramecium finds food, such as a bacterium, it sweeps the food into its body with its cilia. The food enters through a mouthlike opening. A vacuole then forms around the food. The food is digested inside the food vacuole and used by the cell.

Some protozoans move by means of long, whiplike flagella. These protists are found in all kinds of water and in the soil. Some even live inside animals, such as the protists that live in the digestive systems of termites and digest wood for the termite.

Some protozoans are parasites and cause disease. Trypanosomes cause sleeping sickness in humans. They are carried from one person to another by a certain kind of fly that is found in Africa. Once inside the bloodstream, these protozoans produce poison that causes weakness or death in humans.

Which protist reproduces by forming spores?

Sporozoans (spor uh ZOH uhnz) are protozoans that reproduce by forming spores. **Spores** are special cells that develop into new organisms. Sporozoans do not have cilia, flagella, or false feet, so they can't move to get food. They live as parasites in humans and other animals. Remember that parasites cause harm to other living things.

Perhaps you have heard of malaria. The sporozoan named *Plasmodium* causes this disease. It is carried by female *Anopheles* (uh NAHF uh leez) mosquitoes. Humans get the disease when they are bitten by infected mosquitoes. The disease is common in tropical areas where these mosquitoes live.

Plantlike Protists

Algae (AL jee) (singular, alga) are plantlike protists. They are like plants because they have chlorophyll and make their own food. Algae also have other pigments. The other pigments help capture light and pass its energy on to chlorophyll. Algae are producers. They produce large amounts of oxygen, which is used by other living things.

Algae live in fresh and salt water. Some also live in moist soils and on tree bark. Like animals, many algae can move about. They do not have a cell wall. Is this trait like that of a plant or an animal?

Algae may be one-celled or multicellular. **Multicellular** means that an organism has many different cells that do certain jobs for the organism. All algae cells have nuclei. Algae are grouped into different phyla depending on their color and their structures. Note the features of each phylum in Table 5-2.

Skill Check

Understand science words: multicellular. The word part *multi* means many. In your dictionary, find three words with the word part *multi* in them. *For more help, refer to the* ***Skill Handbook****, pages 706-711.*

TABLE 5–2. PLANTLIKE PROTISTS

Traits of Phylum	Example
move with flagellum, one-celled, no cell wall, found mostly in fresh water, green color	euglena
found mostly in salt water, one-celled, gold-brown color	diatom
found mostly in salt water, red or brown color, one-celled	dinoflagellate
green color, found in fresh and salt water and in soil, one-celled, multicellular, or in colonies	green algae
brown color, found in salt water, multicellular	kelp

The euglena shown in Figure 5-5 lives in fresh water. A euglena is a one-celled alga that moves with a flagellum. The euglena swims by moving its flagellum back and forth in the water. This causes the cell to move in an unusual wiggling motion. Euglenas have a red eyespot near their front end. The eyespot is sensitive to light and helps the euglena find the light it needs to make food. Are euglenas producers or consumers?

Figure 5-5 A euglena moves with a flagellum and has chloroplasts for making food. It has a red eyespot for detecting light.

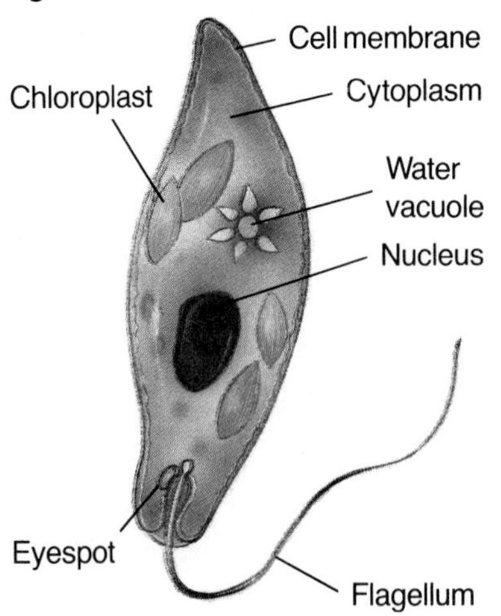

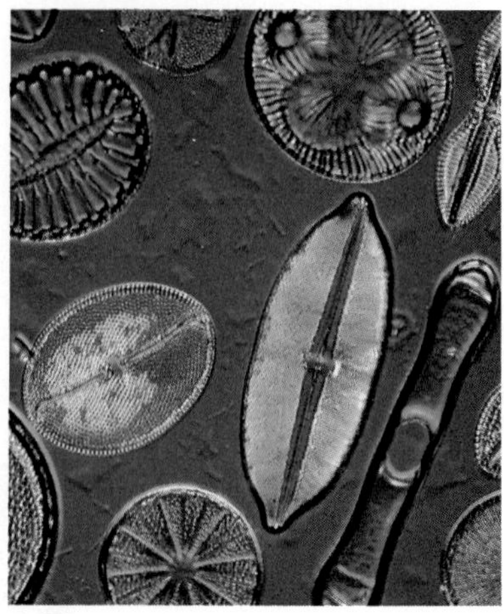

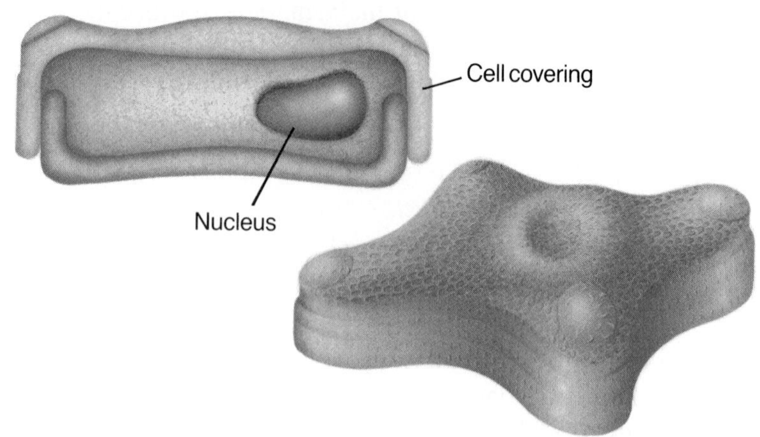

Figure 5-6 Diatoms have many shapes. Each cell has a two-part cell covering.

Diatoms (DI uh tahmz) are beautiful, one-celled protists that have many shapes, Figure 5-6. They are shaped like boats, rods, disks, or triangles. Diatoms are producers. Would you expect them to have chlorophyll? Diatoms can move slowly through the water. They are one of the most important food sources for animals that live in water.

Diatoms have an interesting structure. They have a cell covering made of two parts. The covering overlaps like a box with a lid, as shown in Figure 5-6. It is made of the same material as glass. When diatoms die, they fall to the bottom of the lake or ocean. The coverings do not decay, but form thick layers. The coverings in these layers are used in toothpastes, scouring powders, and filters because they are gritty and are good cleansers.

Dinoflagellates (di noh FLAJ uh luhts) are algae that are usually found in oceans. Only a few types live in fresh water. They are usually red or brown. They have chlorophyll, but their red and brown pigments hide it. They move by beating their two flagella.

Red dinoflagellates are responsible for red tides. They reproduce in such large numbers that the ocean water turns red. Chemicals produced by these protists during red tides can kill thousands of fish. Humans can become ill if they eat shellfish that have absorbed these chemicals. Red tides are common off the coast of Florida and in other warm waters.

Green algae come in many different forms. They may be one-celled, multicellular, or they may form colonies—groups of cells that live together. *Volvox*, the organism shown in Figure 5-7, is a green alga that lives as a colony containing hundreds of one-celled organisms with flagella. The cells are arranged in a single layer with their flagella facing outward. When the flagella beat, the colony spins

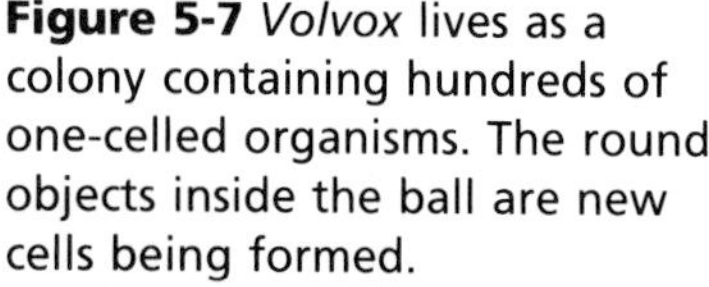

Figure 5-7 *Volvox* lives as a colony containing hundreds of one-celled organisms. The round objects inside the ball are new cells being formed.

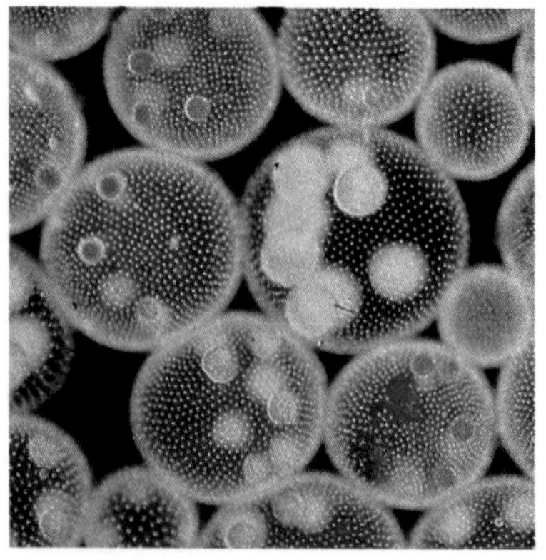

through the water. A *Volvox* colony reproduces when some of the cells in the colony come together to form small groups inside the colony. The colony bursts open and releases the small groups of cells.

Multicellular algae have many different shapes. Some form long strands that have many cells. Others have a leaflike shape that looks like lettuce.

Algae are important to us in many ways. They release oxygen into the water and air. The oxygen is used by other living things. Algae are important as food for animals such as fish, snails, and crayfish. Have you ever heard of a seaweed called kelp? Kelp is one kind of brown algae used by many people for food, Figure 5-8. Some scientists think that green algae will be an even more important source of food for humans in the near future.

Are algae products found in any foods that you eat? If you have eaten ice cream or jelly, you probably have eaten certain substances that come from algae. These substances are used to thicken foods.

Figure 5-8 Kelp is a brown alga that many people eat.

Bio Tip

Ecology: Slime molds have been known to cover an entire golf course with their slimy masses. The masses creep over the ground, engulfing microorganisms and digesting them.

Funguslike Protists

Slime molds are funguslike protists that are consumers. They live in cool, damp places, such as the forest floor. They feed on bacteria growing on rotting logs and decaying leaves. A few slime molds are parasites.

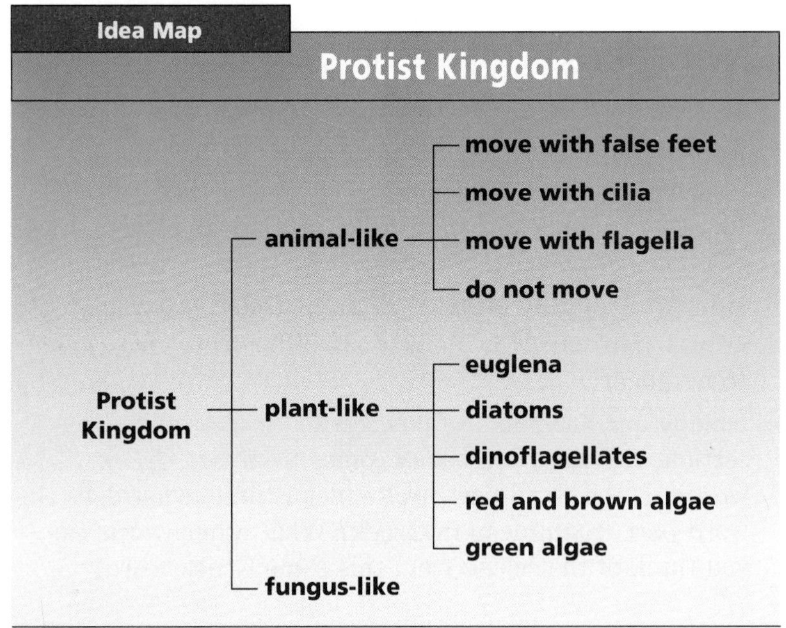

Study Tip: Use this idea map as a study guide to the protist kingdom. The features of the three protist groups are shown.

Figure 5-9 The life cycle of a slime mold has three stages: the feeding and growth stage (a), the spore-forming stage (b), and the amoebalike stage (c).

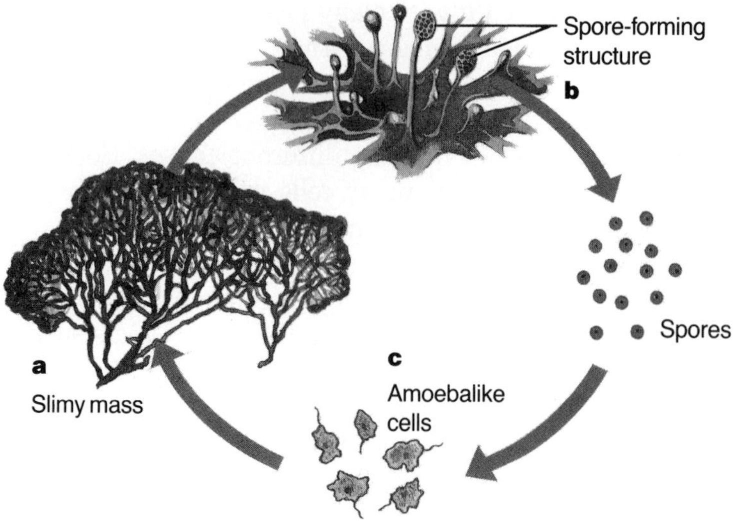

Skill Check

Interpret diagrams: Look at the slime mold in Figure 5-9. Identify the two stages that are like other organisms. What organisms in this chapter do the stages resemble? *For more help, refer to the* ***Skill Handbook****, pages 706-711.*

There are three stages in the life cycle of a slime mold, Figure 5-9. The first stage looks like a slimy mass. It moves much like an amoeba, by flowing across a surface. The mass may have beautiful colors such as red, yellow, or violet, or it may have no color.

In the second stage, the slime mold stops growing and moving. It produces spores inside a structure on a stalk. The slime mold in this second stage is like a fungus. This stage usually starts when the slime mold has no food.

During the third stage, the spores develop into little amoebalike cells with flagella. Then each of these cells loses its flagella and grows into a slimy mass again.

Check Your Understanding

1. Explain how the following move: a *Volvox* colony, a paramecium, a euglena.
2. List two examples of animal-like protists and two examples of plantlike protists.
3. How are plantlike protists different from protozoans?
4. **Critical Thinking:** How do protists differ from viruses and from monerans?
5. **Biology and Reading:** As you learned from reading this section, the word *protozoan* comes from two Greek words. If the word part *proto-* means first, what does the word part *-zoan* mean in Greek? What other words can you think of that come from this same Greek word?

Lab 5–1 Protists

Problem: What are the traits of protists?

Skills

make and use tables, interpret data, observe, classify, use a microscope

Materials

microscope	cultures of:
3 coverslips	amoeba
3 droppers	paramecium
3 glass slides	euglena

Procedure

1. Copy the data table.
2. Put a drop of amoeba culture on a glass slide. Gently add a coverslip.
3. **Observe:** Examine the amoeba on low, then high power. Record in your table the parts you see.
4. Repeat steps 2 and 3 with the paramecium and euglena cultures.
5. Wash and dry the glass slides, coverslips, and droppers. Return them to their proper places.

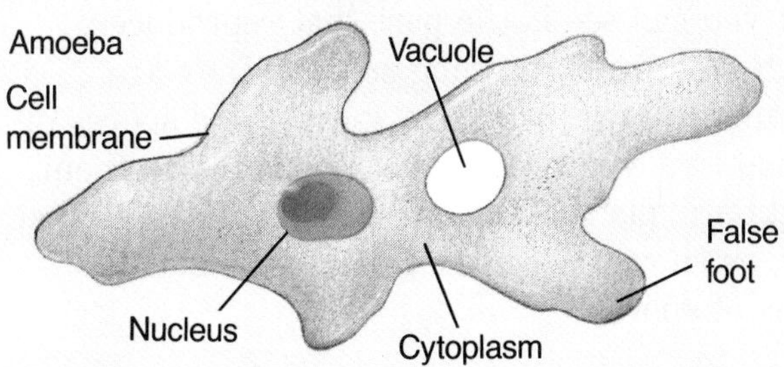

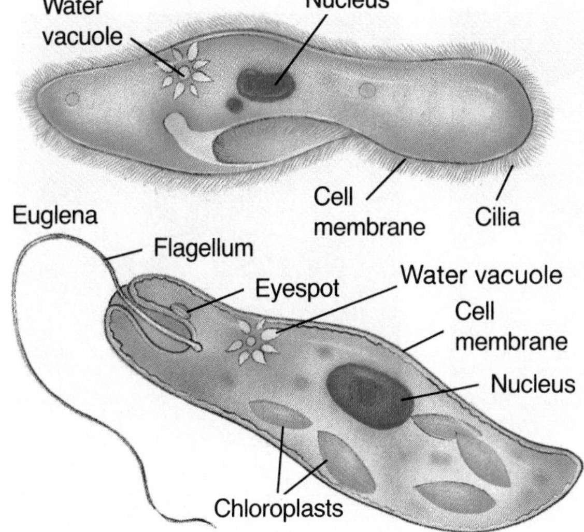

Data and Observations

1. How does the amoeba move?
2. Which protist is green?

	Amoeba	Paramecium	Euglena
Cell membrane			
Parts used for movement			
Cell wall			
Vacuoles			
Nucleus			
Cytoplasm			
Green color (chlorophyll)			

Analyze and Apply

1. **Interpret data:** What traits do protists have in common?
2. Which protist moves the slowest? Which moves the fastest?
3. **Apply:** Which protists are like animals? Why? Which protists are like plants? Why?

Extension

Observe a drop of pond water under the microscope. Identify the protists you see.

5:2 Fungus Kingdom

Objectives

4. List the traits of fungi and compare the traits of the major kinds of fungi.

5. Tell how fungi can be harmful and helpful.

Key Science Words

hyphae
sporangium fungi
sporangia
club fungi
sac fungi
budding
mutualism
lichen

Mushrooms, molds, mildews, yeasts, rusts, and smuts are all fungi. Fungi are consumers and decomposers. They cannot make their own food. Some are parasites, but most of them are saprophytes. Remember from Chapter 4 that saprophytes live on dead matter. They break down waste and dead materials for food and return them to the soil.

Traits of Fungi

You read in Chapter 3 that a fungus is an organism that has a cell wall and does not make its own food. Cells of fungi often have more than one nucleus. Fungi range in size from one-celled yeasts to large, multicellular mushrooms.

The bodies of most fungi are made of a network of threadlike structures called **hyphae** (HI fee). Hyphae contain cytoplasm and are usually divided by cross walls. The hyphae grow and branch until they cover and digest the food source on which the fungus is growing. Look at the bracket fungi growing on the tree in Figure 5-10b. The hyphae have spread throughout the tree bark. The part of the fungus that you can see is the part that reproduces.

Most fungi reproduce by forming spores. They are classified according to how they form spores. The spores of fungi are so small that you need a microscope to see them. Mushrooms and puffballs release large clouds of spores that are easy to see, as shown in Figure 5-10c. Each cloud contains millions of spores.

Figure 5-10 Hyphae (a) of a fungus usually have cross walls. Bracket fungi (b) grow on trees. Mushrooms (c) reproduce by releasing spores.

a

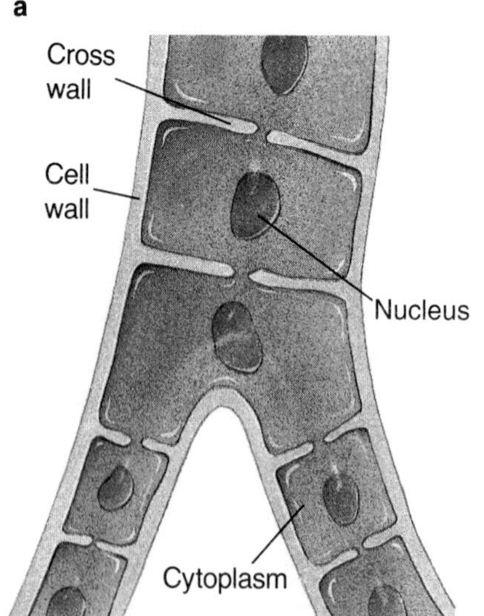

b

c

Chemicals released by hyphae digest dead materials.
Hyphae absorb the digested food.

Figure 5-11 Fungi digest food outside their bodies.

Fungi also reproduce from pieces of hyphae. Sometimes, wind or water carry pieces of broken hyphae to new places. If enough moisture and food are present, the pieces grow into new fungi.

What do fungi use for food?

Most fungi can use once-living things for food. The bracket fungi in Figure 5-10b are feeding on a dead tree. Mildew grows on different items, using them as food. Have you ever seen mildew growing on old clothing? As fungi grow on a once-living thing, they decompose it.

A large amount of waste and dead material is deposited on Earth every day. If it were not for decomposers, the waste would build up and become a problem for all living things. Many of the decomposers that help break down dead materials and return them to the soil are fungi.

Fungi digest food outside their bodies. First, the hyphae release chemicals into the material surrounding them. The chemicals break down the food into small molecules. Then the hyphae absorb the digested food.

Not all fungi feed on dead material. Some feed on living things. They are parasites. Most fungi that are parasites grow on plants. Some attack crops such as corn, wheat, potatoes, and soybeans.

Some fungi are parasites of animals. They cause problems such as athlete's foot by living on human skin cells.

Many fungi grow in and on the roots of living plants, helping the plants get water and minerals. In return, the fungi receive food from the plant. Many trees and other plants can't live without these fungi.

Bio Tip

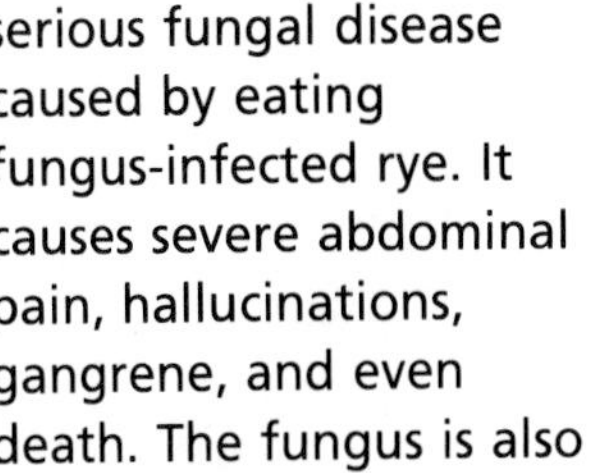

Health: Ergotism is a serious fungal disease caused by eating fungus-infected rye. It causes severe abdominal pain, hallucinations, gangrene, and even death. The fungus is also a source of lysergic acid diethylamide, also known as LSD.

Mini Lab

What Do Fungi Need?

Observe: Sprinkle a slice of bread with water. Place it in a plastic bag. Place a dry slice in a bag. Seal the bags and set in a warm place for a week. What do fungi need to grow? *For more help, refer to the* ***Skill Handbook,*** *pages 704-705.*

Kinds of Fungi

Have you ever seen bread mold—a black, fuzzy substance growing on bread? If so, you have seen a sporangium (spuh RAN jee uhm) fungus. **Sporangium fungi** are fungi that produce spores in sporangia. **Sporangia** are structures, found on the tips of hyphae, that make spores. You can see that the phylum gets its name from these structures.

Let's return to our bread mold. Bread mold produces spores within sporangia that stick up above the bread, as shown in Figure 5-12. Each spore can form a new bread mold. Hyphae grow into the bread. The mold takes in food and water by means of its hyphae. Molds grow best in warm, moist, dark places, but they can grow also in cold places, such as the refrigerator.

Growth of molds on food can be prevented by cleanliness, drying, and the use of chemicals. Chemicals are added to breads to delay the growth of molds. Fruits such as raisins and apricots are preserved by drying.

Bread mold

Spores

Sporangium

Hyphae

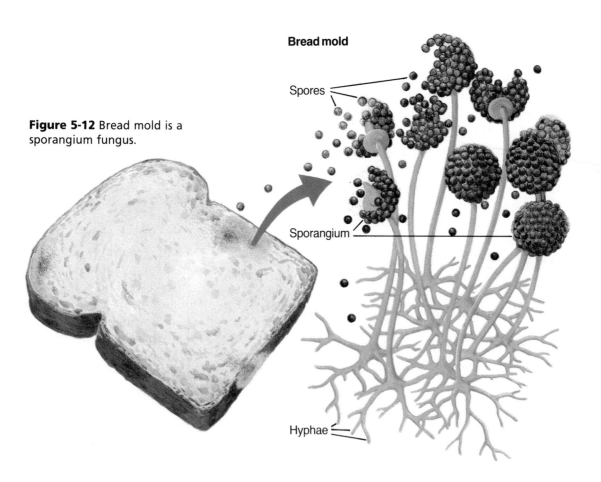

Figure 5-12 Bread mold is a sporangium fungus.

Career Close-Up

Mushroom Farmer

Did you ever wonder where the mushrooms on your pizza came from? Most mushrooms are grown indoors, but some may be grown outdoors or in caves. Equipment is used to control temperature, humidity, and ventilation.

To grow mushrooms, a compost mixture is placed in shallow beds. Spores from mushroom caps are sown on the compost and covered with a thin layer of soil. A mass of hyphae will soon grow through the compost. In seven or eight weeks, mushrooms will appear.

After three months, the mushrooms can be harvested and put into boxes. They will be shipped to markets the same day.

No special schooling is needed for this job, but a knowledge of fungi is very helpful.

Mushroom farmers grow mushrooms in shallow beds of compost.

Fungi with club-shaped parts that produce spores are called **club fungi,** Figure 5-13. They form networks of branched hyphae underground. Mushrooms are club fungi. Mushrooms get food in much the same way as bread mold.

Mushrooms are an important food crop. Some, however, cause illness and a few types cause death. You should never eat wild mushrooms unless an expert says they are safe.

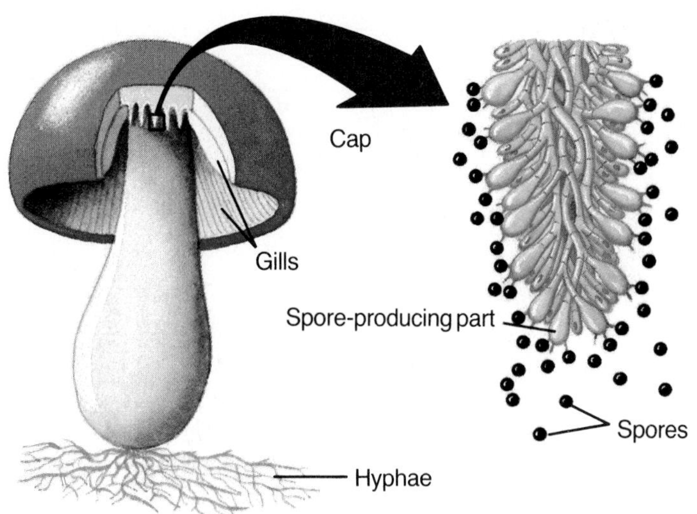

Figure 5-13 Club fungi produce spores in club-shaped structures within gills. Where are the gills located?

Figure 5-14 Sac fungi produce spores in saclike structures.

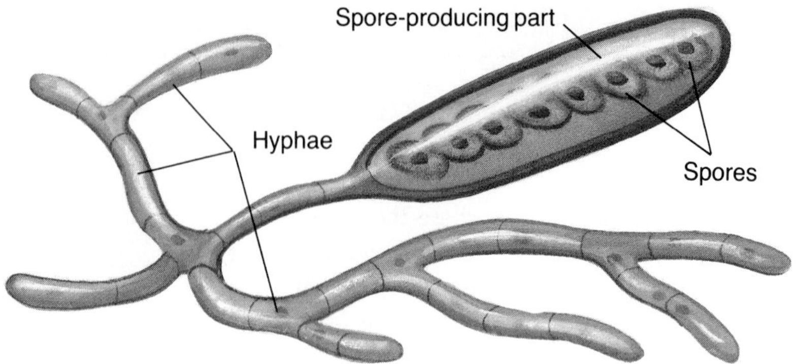

Shelf fungi, rusts, smuts, and puffballs are club fungi, too. Many of these fungi are harmful to plants. Rusts and smuts can destroy entire crops of corn, wheat, and oats.

Yeasts, cup fungi, and powdery mildews are called sac fungi. **Sac fungi** produce spores in saclike structures, Figure 5-14. Yeasts are unusual sac fungi because they have only one cell. Yeasts reproduce by budding. **Budding** is reproduction in which a small part of the parent grows into a new organism. The bud grows out of the parent as shown in Figure 5-15b. Budding produces offspring that are identical to the parent. If conditions are not right for growth, the yeast cell can form a spore. When growing conditions improve, the spore forms a new yeast cell.

How are yeasts helpful to humans?

Sac fungi are useful to humans. Yeasts are used for making bread and alcohol. *Penicillium* (pen uh SIL ee uhm) is a fungus used to make the antibiotic penicillin. It is shown in Figure 5-15a. One species of *Penicillium* is used

Figure 5-15 *Penicillium* (a) is a sac fungus used to make penicillin. Yeasts reproduce by budding (b). Is budding sexual or asexual reproduction?

a

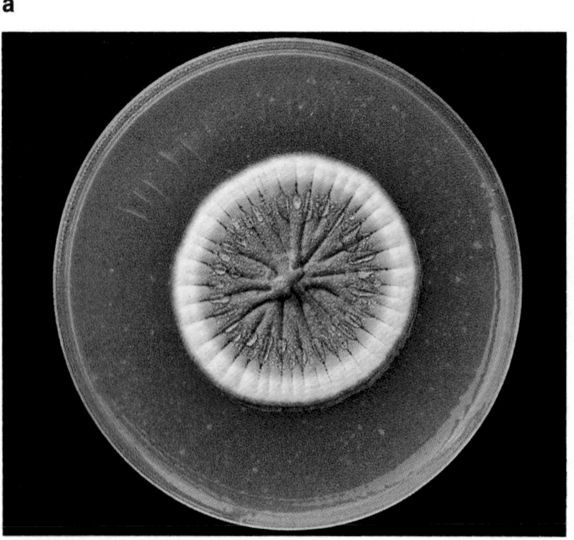

b

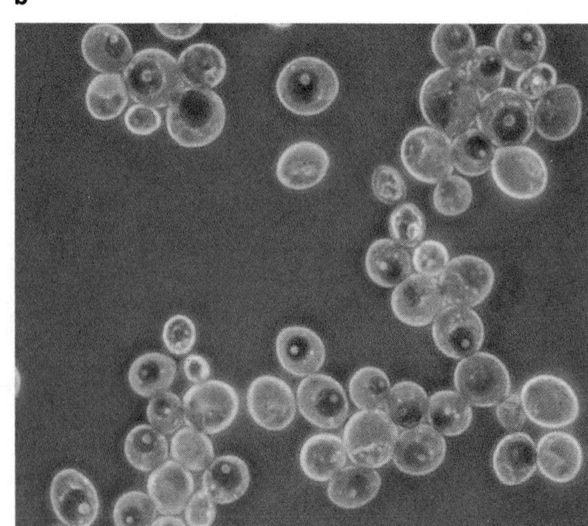

TABLE 5–3. HELPFUL AND HARMFUL FUNGI
Helpful Fungi
1. Used as food—mushrooms
2. Used to make blue and Roquefort cheese
3. Used to make wine, beer, and whiskey—yeast
4. Used to make bread rise—yeast
5. Used to make soy sauce
6. Used to make penicillin and other antibiotics
7. Help to break down materials and get rid of wastes
8. Enrich the soil
Harmful Fungi
1. Cause foods, such as fruit and bread, to spoil
2. Cause plant diseases, such as rusts, smuts, Dutch elm disease, and mildew
3. Cause human diseases, such as athlete's foot, ringworm, thrush, and lung infections
4. Destroy leather, fabrics, plastics

Bio Tip

Consumer: Most lemon flavoring (citric acid) today is made from a black mold called *Aspergillus niger.*

to make blue cheeses. While many sac fungi are helpful, one causes Dutch elm disease. This disease has destroyed thousands of elm trees in the United States. Table 5-3 shows some examples of how fungi are helpful and harmful.

Lichens

Some fungi live neither as parasites nor as saprophytes. They get food from other organisms without causing harm. In turn, they may give support or protection. A living arrangement in which both organisms benefit is called **mutualism.**

A **lichen** (LI kun) is a fungus and an organism with chlorophyll that live together. The organism with chlorophyll can be either a green alga or a blue-green bacterium. A lichen looks like a single organism but it is not. The two organisms in the lichen are so closely tangled together that they cannot be easily separated. You may have seen a British-soldier lichen, as shown in Figure 5-16. The green alga provides the food for the fungus. The fungus provides support, and holds water and minerals for the alga. The alga uses the water and minerals to make food for both itself and the fungus. Both organisms benefit.

Lichens are very sensitive to changes in the environment. If the air becomes too polluted, the green organism can't live. If the green organism dies, the fungus also dies.

Figure 5-16 British soldier lichen

Study Tip: Use this idea map as a study guide to the fungus kingdom. The features and kinds of fungi are shown.

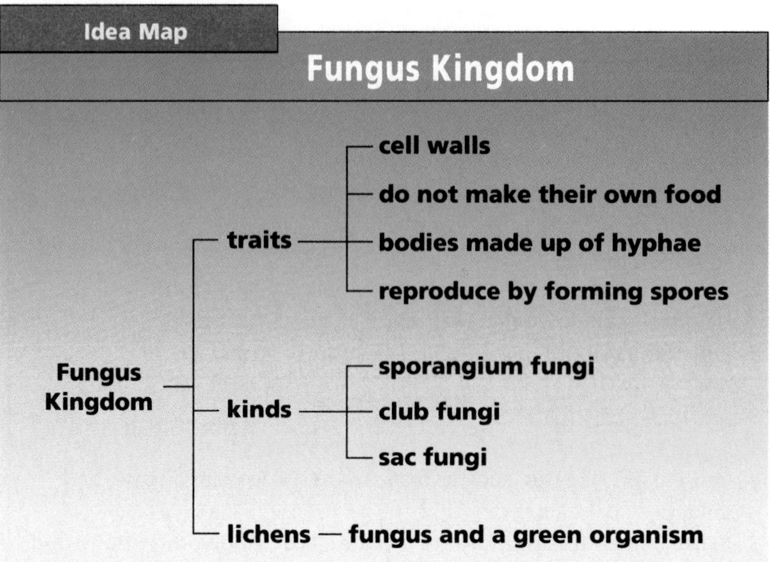

Lichens grow in places where neither fungi nor organisms with chlorophyll can be found alone. They grow on bare rocks, trees, and even on Arctic ice. They are often the first organisms to appear on rocks, such as lava, after a volcano erupts. Lichens release acids that break down the rock. Soil begins to form when the broken down rock mixes with lichens that have died. This process provides a place where other living things can grow.

Some types of lichens provide food for animals. In the Arctic, there is very little plant life. Reindeer moss is eaten by reindeer and caribou (KAIR uh boo), much as cattle graze on grass elsewhere. Caribou are deerlike animals that live in the Arctic.

Check Your Understanding

6. List two ways in which fungi reproduce.
7. How do sporangium fungi differ from club fungi?
8. Name three uses of sac fungi.
9. **Critical Thinking:** How does budding in yeast differ from fission in bacteria?
10. **Biology and Writing:** Combine the following three simple sentences into one compound sentence. Some fungi are saprophytes. Some fungi are parasites. Some fungi are mutualistic.

Lab 5–2 Fungi

Problem: Why is a mushroom a fungus and not a plant?

Skills

observe, infer, use a microscope, design an experiment

Materials

scalpel
microscope
hand lens
mushroom
prepared slide of mushroom gill

Procedure

1. Obtain a mushroom and study its structure. Use Figure A to find the parts of the mushroom.
2. Draw and label the parts in your notebook.
3. Cut the mushroom lengthwise through the stipe and cap. **CAUTION:** *Always be careful when using a scalpel.*
4. **Observe:** Examine the cut areas with a hand lens. Locate the hyphae that form the reproductive part. Figure B will help.
5. Use the hand lens to observe the gills on the underside of the cap.
6. **Use a microscope:** Obtain a prepared slide of a mushroom gill. Examine it under low power on the microscope. Locate the structures that produce spores.
7. Locate the spores.
8. Draw and label the spores and the structures that produce them.

Data and Observations

1. What mushroom part supports the cap?
2. Where are the gills located?
3. Where are the reproductive parts of the mushroom located?
4. Describe the color of the mushroom.

Analyze and Apply

1. **Infer:** How would the location of the mushroom's reproductive parts help spread the spores?
2. What is a mushroom's food source?
3. Describe how a mushroom gets its food. Is it a parasite or a saprophyte?
4. **Apply:** How do mushrooms appear in your yard year after year when there is no sign of them between seasons? (HINT: Think about how mushrooms grow.)

Extension

Design an experiment to show what foods a fungus can use for growth.

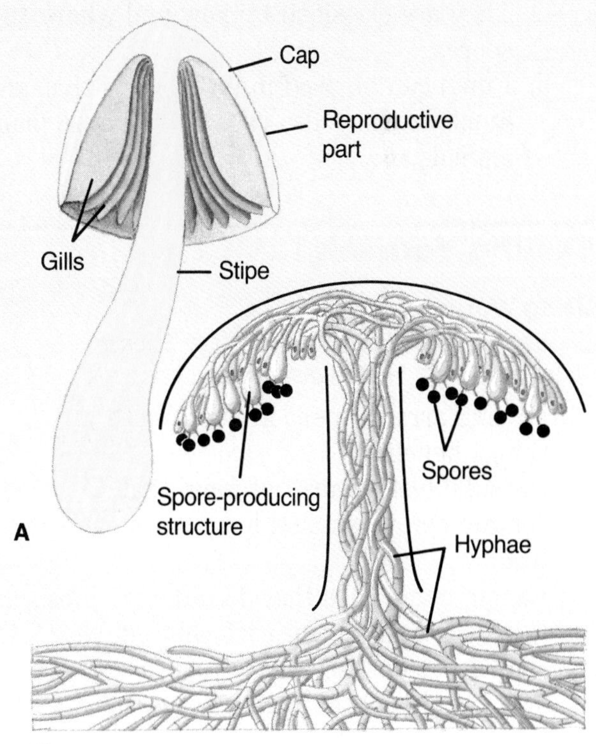

Chapter 5

Review

Summary

5:1 Protist Kingdom

1. Protists are usually one-celled, have a nucleus and other cell parts with membranes, and may be either producers or consumers.
2. There are three kinds of protists—animal-like protists, plantlike protists, and funguslike protists. They have different methods of movement, feeding, and reproduction.
3. Amoebas, paramecia, and trypanosomes are animal-like protists. Euglena, diatoms, and *Volvox* are plantlike protists. Slime molds are funguslike protists.

5:2 Fungus Kingdom

4. Fungi have many cells with cell walls and nuclei. They are consumers and feed as saprophytes or parasites. They form mutualistic relationships with green organisms. There are three phyla of fungi. They are classified by how and where they produce spores.
5. Fungi can be used in many ways that are helpful to humans. They also cause diseases in plants and animals.

Key Science Words

algae (p. 97)
budding (p. 106)
cilia (p. 96)
club fungi (p. 105)
hyphae (p. 102)
lichen (p. 107)
multicellular (p. 97)
mutualism (p. 107)
protozoan (p. 94)
sac fungi (p. 106)
slime mold (p. 99)
sporangia (p. 104)
sporangium fungi (p. 104)
spore (p. 96)
sporozoan (p. 96)

Testing Yourself

Using Words

Choose the word from the list of Key Science Words that best fits the definition.

1. a living arrangement in which both things benefit
2. threadlike structures of fungi that contain cytoplasm and have cross walls
3. having many cells that do different jobs
4. fungus and organism with chlorophyll that live together
5. special cell that forms a new organism
6. a way of producing offspring that are identical to the parent
7. tiny, hairlike parts on a cell surface
8. plantlike protist group
9. animal-like protist group
10. fungi with club-shaped parts
11. funguslike protist
12. fungi that have saclike structures
13. structures, on tips of hyphae, that make spores
14. a protozoan that forms spores

Review

Testing Yourself *continued*

Finding Main Ideas

List the page number where each main idea below is found. Then, explain each main idea.

15. how yeasts are different from other fungi
16. the name of protists that do not move
17. the structure of a diatom
18. what fungi are made of
19. the three kinds of fungi
20. why sac fungi are important
21. the cause of red tides
22. the life cycle of a slime mold
23. what most slime molds use for food
24. two human diseases caused by protists

Using Main Ideas

Answer the questions by referring to the page number after each question.

25. What is the role of each organism in a lichen? (p. 107)
26. How does an amoeba get food? (p. 95)
27. How does a person get malaria? (p. 97)
28. How do fungi digest food? (p. 103)
29. How are plantlike protists different from funguslike protists? (pp. 97, 99)
30. How do bracket fungi get food from trees? (p. 102)
31. Why should you not eat wild mushrooms? (p. 105)
32. How do protozoans get food? (p. 94)
33. Where are lichens found? (p. 108)
34 How can growth of molds on food be prevented? (p. 104)
35. How are lichens helpful? (p. 108)
36. How do certain dinoflagellates cause red tides? (p. 98)

Skill Review 

For more help, refer to the ***Skill Handbook,*** *pages 704-719.*

1. **Make and use tables:** Make a table to show the three kinds of fungi, where they produce spores, and examples of each.
2. **Classify:** Make a list of the traits of animal-like protists, plantlike protists, and funguslike protists. How are these traits used to classify protists?
3. **Infer:** Mushrooms often appear on a lawn soon after a rain. Using your knowledge of fungi, infer how this happens.
4. **Observe:** Observe a small amount of blue cheese under a stereomicroscope. What makes the cheese blue?

Finding Out More

Critical Thinking

1. What advantages does a lichen have over its fungus and green-organism parts?
2. How would life on Earth be affected if all the algae died?

Applications

1. Make a list of things you can do to keep fungi from growing in your home. Where are fungi most likely to be found in the home?
2. Use the library to learn about fungicides. Report on how they are used to keep fruits and vegetables safe from fungal infection.

CHAPTER PREVIEW

Chapter Content

Review this outline for Chapter 6 before you read the chapter.

Skills in this Chapter

The skills that you will use in this chapter are listed below.

- In **Lab 6-1,** you will make and use tables, interpret data, use a microscope, measure in SI, and infer. In **Lab 6-2,** you will make and use tables, observe, and use a microscope.
- In the **Skill Checks,** you will understand science words and interpret diagrams.
- In the **Mini Labs,** you will experiment and observe.

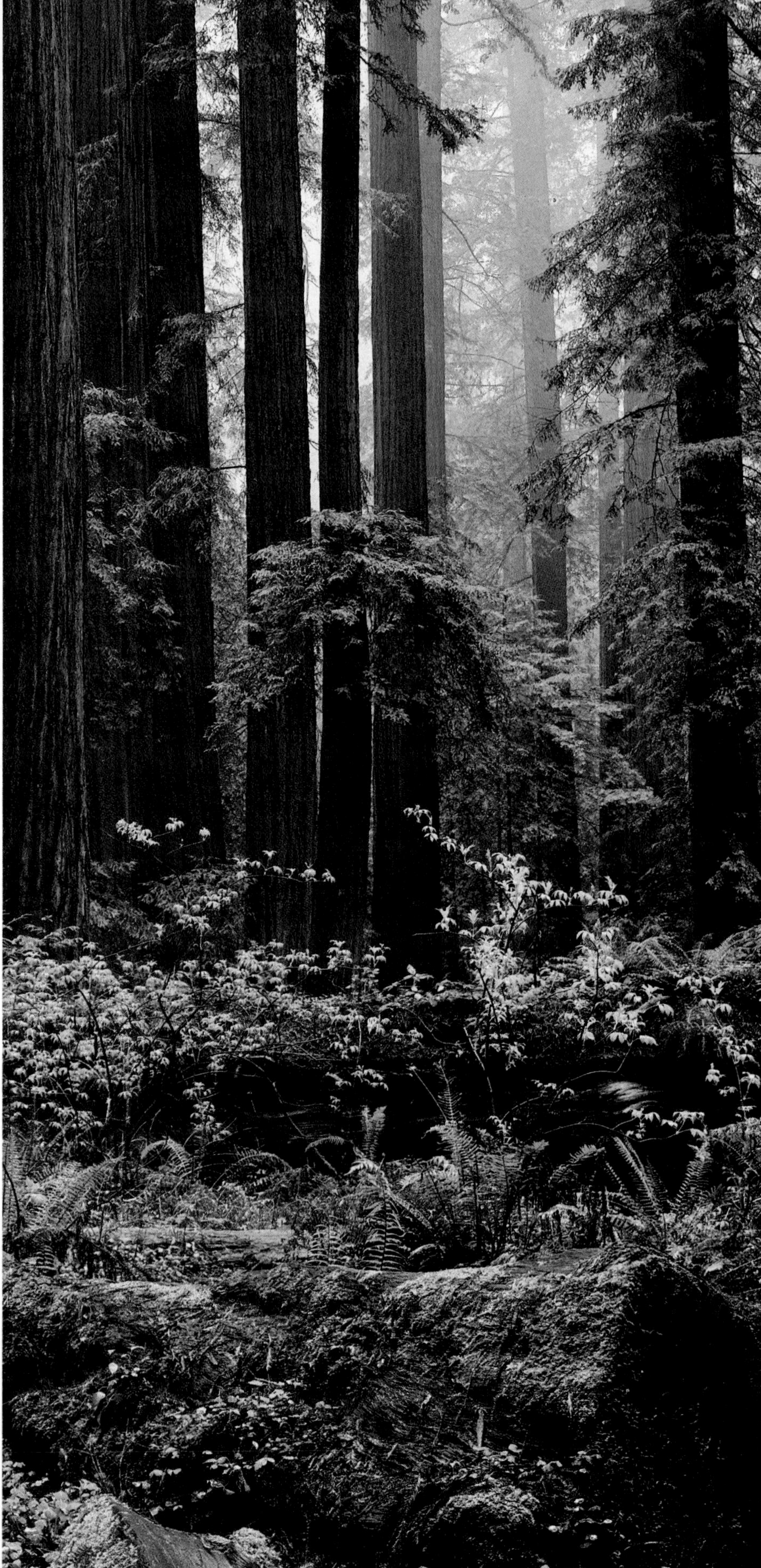

Chapter 6

Plants

If you were to walk through the forest pictured on the left, you would first see the giant redwood trees. Because of their small size, you might not notice the plants growing around the roots of the trees. Now, look at the tiny water plants in the photo below. These plants live on the surfaces of lakes or ponds.

There are many kinds of plants. They vary in size, shape, color, and life span. Some, like the small water plants, are only a few millimeters long. Others, such as the redwood tree, can grow to 100 meters tall. Some plants change colors and lose their leaves. Others stay green for many years. Some plants live for just a few months. Plants grow almost everywhere on Earth. They are adapted to many kinds of environments. In this chapter, you will read about plants and where they grow.

Try This!

Where does the moisture come from? Cover a small houseplant with a wide-mouth jar. Observe the plant for a day or two. What do you see in the jar? What is its source?

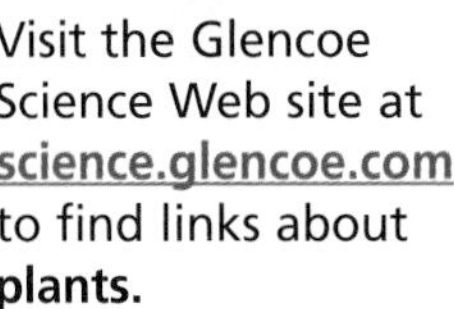

Visit the Glencoe Science Web site at science.glencoe.com to find links about **plants.**

Duckweed

6:1 Plant Classification

Objectives

1. **Describe** two features common to most plants.
2. **Compare** vascular and nonvascular plants.

Key Science Words

chloropyhll
photosynthesis
vascular plant
nonvascular plant

There are over 350 000 different kinds of plants. How could you ever know them all? Scientists have grouped plants to make it easier to study and learn them. Just as with all other living things, plants are classified based on their traits.

Plant Features

Although it may seem all plants are different, most have two traits in common. First, almost all plants have green cell parts called chloroplasts (KLOHR uh plasts). A chloroplast is shown in Figure 6-1. Chloroplasts are green because they contain chlorophyll (KLOHR uh fihl). **Chlorophyll** is a chemical that gives plants their green color and traps light energy. Plants trap light energy for photosynthesis (foht oh SIHN thuh sus). **Photosynthesis** is the process in which plants use water, carbon dioxide, and energy from the sun to make food. In the process, they release oxygen. This food-making process of plants occurs only in the chloroplasts.

What feature of cells is common to all plants?

Photosynthesis is the main process that separates plants from animals. Animals can't make their own food. They don't have chlorophyll. Unlike animals, plants can't move about to search for food. They have to take in the materials and energy needed for photosynthesis.

The second trait of plants is a stiff cell wall. The cell wall gives structure to each cell. The cell wall helps to support a plant on land. Fungi, monerans, and many protists also have cell walls, but animal cells don't.

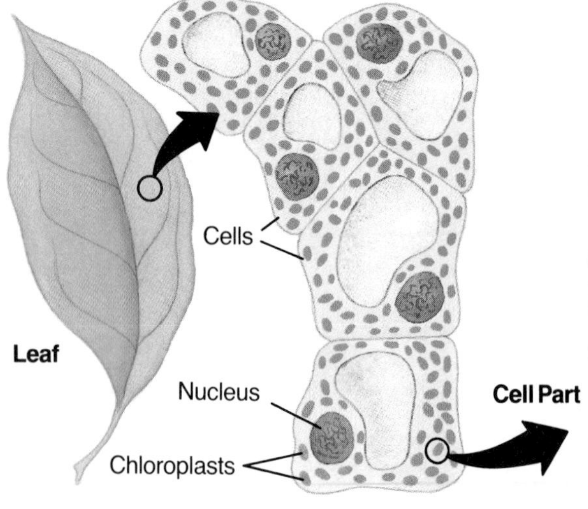

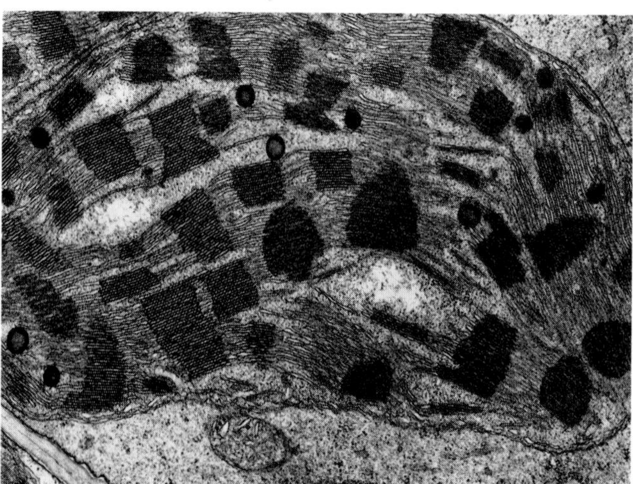

Figure 6-1 A common cell part of almost all plants is the chloroplast, shown in the photo magnified 15 000 times.

How Are Plants Grouped?

Plants are grouped in a way similar to how you might group office buildings. If you were to group office buildings by whether they have elevators, you would be able to put them into two groups. Some office buildings have elevators, some don't.

Notice that the taller building in Figure 6-2 has an elevator. The building with only three floors does not have an elevator. Why do we build tall buildings with elevators rather than just stairways? People can get to offices in a tall building more easily if there are elevators in the building. Supplies can also be moved more easily from one floor to another.

Just as there are two groups of buildings, there are also two groups of plants. The two groups of plants are based on whether or not they have cells that form tubes through the length of the plant. Some plants have tubelike cells and some don't. The tubelike cells join end to end and look like water pipes or soda straws joined together. The function of the tubelike cells in plants is much like that of the elevators in a tall building. The elevators carry people up and down within the building. The people can easily reach all the offices. One set of tubelike cells allows water and minerals to move up and down within a plant. A separate set of tubelike cells allows food made by the leaves to move to other parts of the plant. The tubelike cells run through the roots, stems, and leaves of this kind of plant, Figure 6-2. They carry supplies up and down the plant. This is the way that water and food reach all the plant's cells.

Skill Check

Understand science words: photosynthesis. The word part *photo* means light. In your dictionary, find three words with the word part *photo* in them. *For more help, refer to the **Skill Handbook,** pages 706-711.*

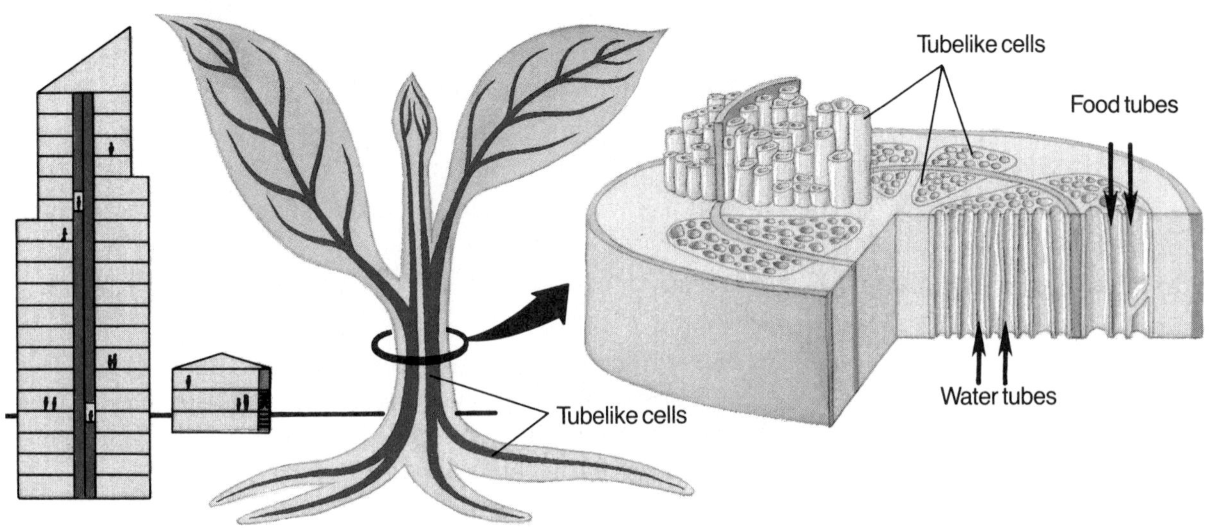

Figure 6-2 The elevator in a tall building acts like the tubelike cells in vascular plants.

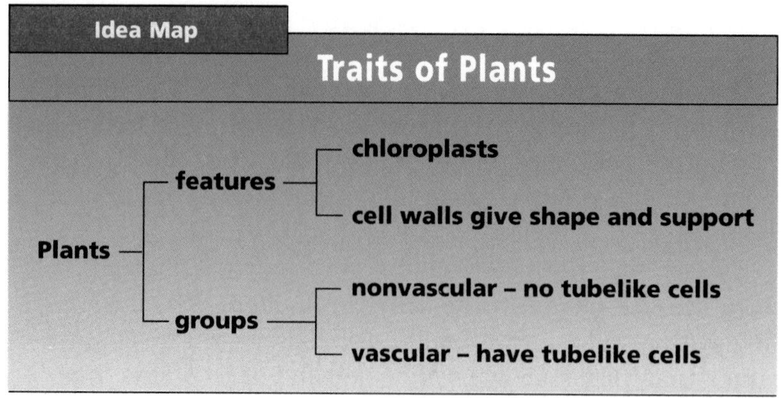

Study Tip: Use this idea map as a study guide to the traits of plants. Notice how one trait is used to classify plants into two main groups.

Mini Lab

How Does Water Move Up in a Plant?

Experiment: Fill a beaker with water. Add red food coloring. Place a white carnation in the beaker. Explain what happens. *For more help, refer to the* ***Skill Handbook****, pages 704-705.*

What plant part common to vascular plants is absent in nonvascular plants?

Plants that have tubelike cells in their roots, stems, and leaves to carry food and water are **vascular** (VAS kyuh lur) **plants.** Roots, stems, and leaves are three kinds of organs. In vascular plants, roots anchor plants in the ground and take in water and minerals from the soil. Stems carry water to all parts of the plant and hold the leaves up to the sunlight. Leaves are the main organ for food making. Most of the chloroplasts are in the leaves of a plant.

Plants that don't have tubelike cells are shorter, just as office buildings without elevators are shorter than those with elevators. Plants that don't have tubelike cells in their stems and leaves are **nonvascular** (nahn VAS kyuh lur) **plants.** Nonvascular plants grow close to the ground in moist areas. They don't have roots. They take up water by osmosis through hairlike cells. You have read in Chapter 2 that osmosis is the movement of water across a cell membrane. The water also carries minerals from the soil to the plant.

Check Your Understanding

1. Why is chlorophyll important to plants?
2. What is the main difference between vascular and nonvascular plants?
3. Why are vascular plants usually taller than nonvascular plants?
4. **Critical Thinking:** How does photosynthesis differ from cellular respiration?
5. **Biology and Reading:** Go back and read again the paragraph that introduces the word *photosynthesis*. Then read the paragraph that follows. What is the main idea of these two paragraphs?

6:2 Nonvascular Plants

Objectives

3. **Compare** mosses and liverworts with other plants.
4. **Sequence** the steps in the life cycle of a moss.
5. **List** three ways mosses are important to other living things.

Key Science Words

moss
sexual reproduction
egg
sperm
fertilization

You may have hiked on the mossy banks of a stream. Mosses are one kind of nonvascular plant. Most nonvascular plants live near water. They grow close to the ground where they can take up water even though they don't have tubelike cells.

Mosses and Liverworts

You may have seen mosses as a lush, green, leafy mat on the floor of a forest or on rocks by a stream. A **moss** is a small, nonvascular plant that has both stems and leaves but no roots. If you looked closer to the ground you may have seen smaller patches of green plants that looked wet and slippery. These are liverworts, Figure 6-3.

Mosses and liverworts (LIV uhr wuhrts) are two very similar kinds of nonvascular plants. Unlike vascular plants, mosses and liverworts don't have roots. They are fixed to the surface of the ground or tree trunks by hairlike cells that take up water. Perhaps you have seen mosses and liverworts. If you have, you may have noticed that they are only a few centimeters tall and are common in wet or damp areas.

How do leaves and stems of mosses and liverworts differ from those in vascular plants?

Mosses have fine, soft stems that often grow upright in mats. These mosses look like short-cropped hair. Some mosses form mats but have stems that creep along the ground. These mosses look like longer hair that is tangled and wavy. The leaves of mosses are only one or two cells thick and will soon dry out if taken from their moist environment. Moss leaves grow from all sides of the stems. Leaf shapes of mosses vary and moss species can be classified by their leaves.

Figure 6-3 Liverworts (a) and mosses (b) are nonvascular land plants.

a

b

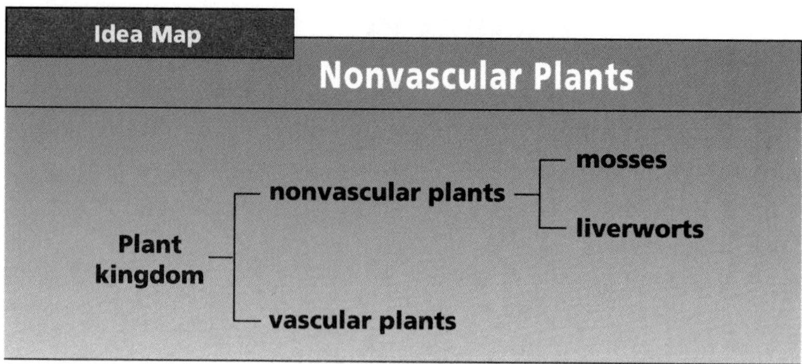

Study Tip: Use this idea map as a study guide to nonvascular plants. What do mosses and liverworts have in common?

Many liverworts don't have roots, stems, or leaves. The body of these liverworts is often a flat, slippery layer of green cells that lies close to the ground. In other species, liverworts look more like mosses. They have creeping stems and small leaves. The main difference between mosses and these leafy liverworts is the arrangement of the leaves. In liverworts, the leaves grow in two or three flattened rows along the stem. Remember that moss leaves grow all around the stems.

How are mosses important to small animals?

Mosses and liverworts are useful to other living things, including you. They are food for some animals such as worms and snails. They also help hold the soil in place to keep it from washing away. Some mosses that live on rocks cause them to break down to form soil.

You may have used sphagnum (SFAHG num) moss in hanging baskets or flowers, Figure 6-4. Have you ever seen a gardener using sphagnum moss? Sometimes called peat moss, sphagnum moss is added to soil to increase the amount of water the soil can hold. Sphagnum moss is an upright moss that forms large, deep mats in bogs. In some areas, peat is cut, dried, and burned as fuel.

Figure 6-4 Moss helps keep moisture around potted flowers.

The Life Cycle of a Moss

Mosses and liverworts need a constant supply of water to survive. They also need water for sexual reproduction. **Sexual reproduction** is the forming of a new organism by the union of two reproductive cells. A female reproductive cell called an **egg** joins with a male reproductive cell called a **sperm.** The joining of the egg and sperm is called **fertilization.** Sexual reproduction occurs in most plants and animals. In many living things, the sperm swims to the egg. The sperm of mosses and liverworts must swim to the eggs to fertilize them. Sperm and eggs of mosses form at the tips of leafy stems.

Lab 6–1

Nonvascular Plants

Problem: What are some traits of a nonvascular plant?

Skills

make and use tables, interpret data, use a microscope, measure in SI, infer

Materials

moss
light microscope
3 microscope slides
coverslip
hand lens or stereomicroscope
forceps
paper towel
metric ruler
dropper

Procedure

1. Copy the data table.
2. Place the moss plant on the paper towel and examine it with the hand lens.
3. **Measure in SI:** Find the hairlike structures that act as roots for the moss. Measure their lengths in millimeters. Complete the table.
4. Observe the stem part with the hand lens. Complete the table for this part.
5. Observe the leaf parts with the hand lens. Complete the table for these parts.
6. Use the forceps to remove one of the leaves. Make a wet mount slide and observe the cells under low power.
7. Find the chloroplasts in the cells. Determine how thick the leaf is by moving the fine focus up and down.

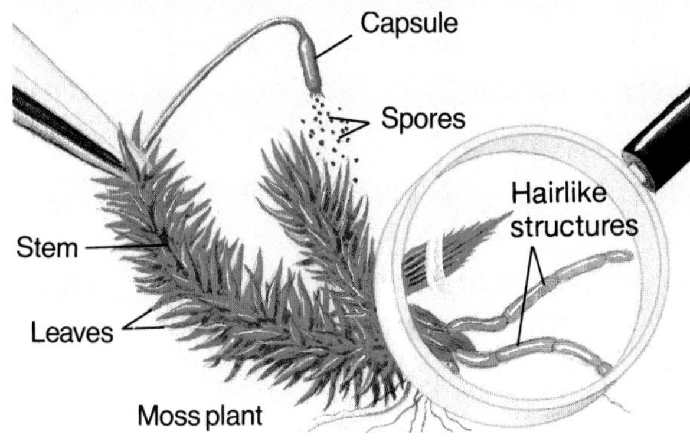

8. Find a spore capsule at the end of a brownish stalk. Observe the spore capsule with the hand lens. Complete the table.
9. Place the spore capsule on a slide. Add two drops of water. Place a second slide on top of the capsule. Press firmly down with your thumb on the top slide so that the capsule is crushed. Carefully remove the top slide. Place a coverslip over the crushed capsule.
10. Examine the released spores under low power. Draw and label the spores.

Data and Observations

1. Which parts of the moss are green?
2. What shapes are moss leaves?

PART	COLOR	SHAPE	LENGTH (mm)
Rootlike hairs			
Stem			
Leaf			
Spore case			

Analyze and Apply

1. **Infer:** How are the green parts used by a moss?
2. Describe the overall size of the plant.
3. How many cell layers are in a moss leaf?
4. **Apply:** Why do moss plants live in moist areas?

Extension

Design an experiment to find out if mosses grow better in wet or dry places.

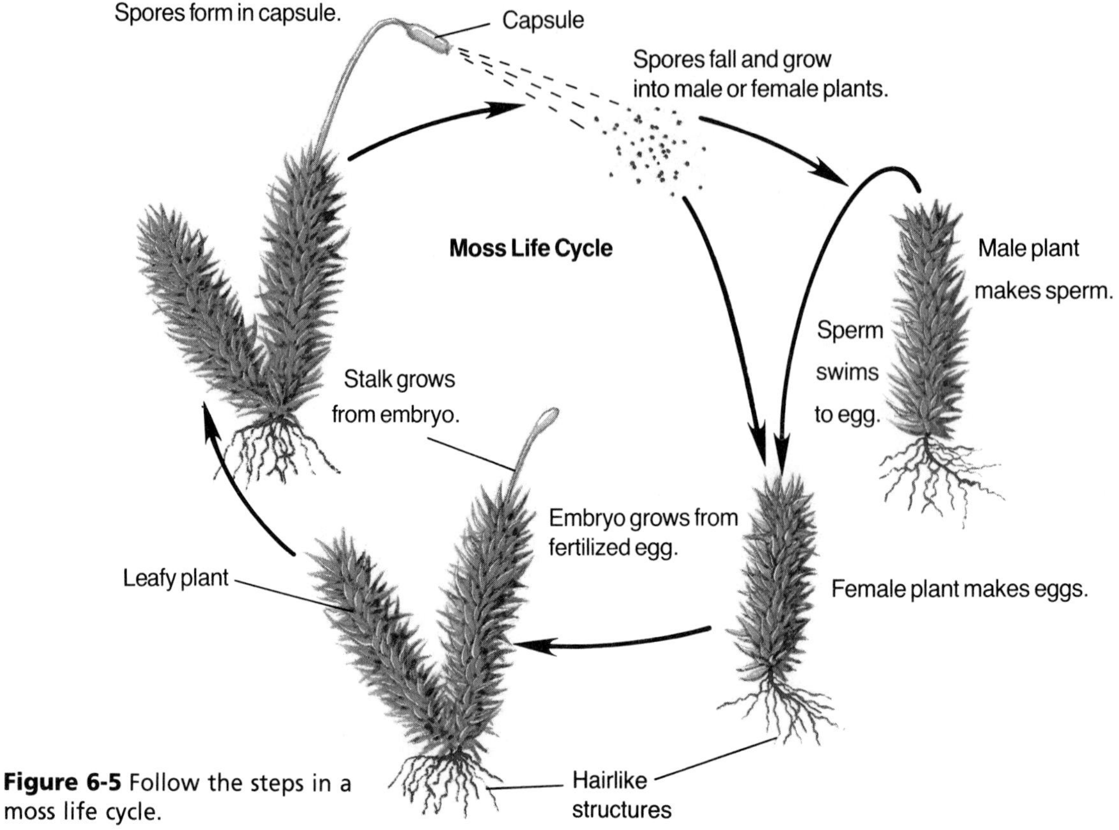
Spores form in capsule.
Capsule
Spores fall and grow into male or female plants.
Moss Life Cycle
Male plant makes sperm.
Sperm swims to egg.
Stalk grows from embryo.
Embryo grows from fertilized egg.
Female plant makes eggs.
Leafy plant
Hairlike structures

Figure 6-5 Follow the steps in a moss life cycle.

What forms inside a moss capsule?

Follow the arrows in Figure 6-5, which shows the steps in the life cycle of a moss. Notice in the figure that after fertilization, a stalk grows from the fertilized egg. Brown capsules form at the tip of each stalk. The capsules contain spores. Many plants make spores. Moss spores are as small and light as powdered sugar. Spores are blown away from the parent plant by wind, and where they land they grow into new leafy plants.

Check Your Understanding

6. How do mosses and liverworts take up water?
7. How are mosses important?
8. How do mosses reproduce?
9. **Critical Thinking:** How might a mat of mosses protect a forest floor?
10. **Biology and Reading:** From what you have learned about mosses and liverworts, what would be the effect on these plants if the climate became drier?

6:3 Vascular Plants

Most of the plants that you see every day are vascular plants. You have read that vascular plants are plants with tubelike cells in roots, stems, and leaves. The tubelike cells carry food and water throughout the plant. There are two types of tubelike cells in vascular plants. **Xylem** (ZI lum) cells carry water and dissolved minerals from the roots to the leaves. **Phloem** (FLOH em) cells carry food that is made in the leaves to all parts of the plant.

Objectives

6. **Compare** the traits of ferns, conifers, and flowering plants.
7. **Explain** how ferns, conifers, and flowering plants reproduce.
8. **Describe** how conifers and flowering plants are important to other living things.

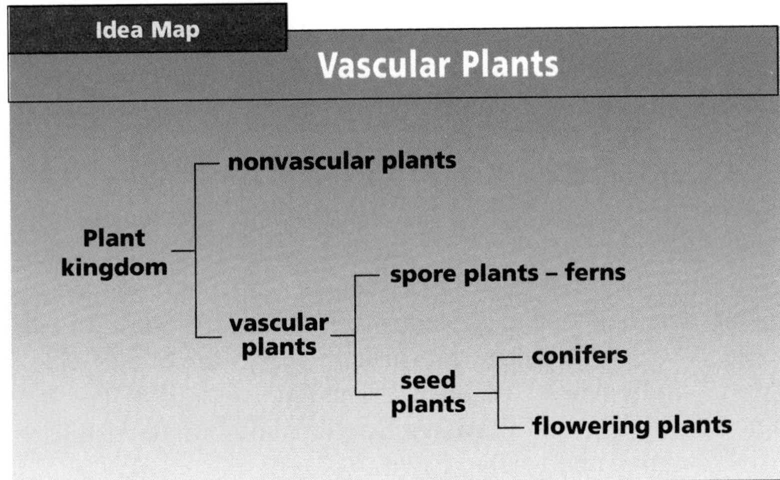

Key Science Words

xylem
phloem
fern
seed
embryo
conifer
pollen
flowering plant
flower

Study Tip: Use this idea map as a study guide to vascular plants. How do ferns differ from other vascular plants?

Ferns

A **fern** is a vascular plant that reproduces with spores. Because of their vascular system, ferns can grow taller than mosses and liverworts. However, one stage of the life cycle of a fern does not have a vascular system at all. For this reason ferns often grow in moist, shaded areas. Some ferns have adapted to much drier climates. Some live in water. Some species called tree ferns live in tropical forests and can grow to a height of 25 meters. There are over 12 000 known species of ferns. About 8000 of these species live in tropical regions.

How are ferns different from conifers and flowering plants?

You might think that a tree fern is unusual. However, over 300 million years ago, most of the plants that grew on Earth were tree ferns or other vascular spore plants. We know that these plants lived at that time because scientists have found ancient, preserved pieces of their stems. These ancient tree ferns lived in swamps. When they died and fell into the swamps, they formed the coal that we use for fuel today.

Figure 6-6 A fern produces thousands of spores each year.

The leaves of a fern grow from a stem that is horizontal and lies underground. The stem stores food and water. Small roots grow down from the underground stem. The roots anchor the plant and take up water and minerals from the soil into the xylem cells.

A fern leaf is often divided into many tiny leaflets that give it a feathery appearance. Many ferns lose their leaves at the end of a growing season. The underground stems remain and produce new leaves each spring. Ferns can be named by looking at the shapes of their leaves.

Ferns are similar to nonvascular plants in that they reproduce with spores. The spores of ferns are found on the underside of the leaves, Figure 6-6. If you have a fern in your school or home, look for brown or orange spots on the leaves. These roundish structures are called spore cases. The spore cases hold the spores.

What part of a fern plant produces eggs and sperm?

Figure 6-7 shows the life cycle of a fern. Look for the spore cases on the fern leaf. When a spore case opens, the tiny spores are carried by wind and water. If a spore lands in a moist place, it grows into a small, flat, heart-shaped plant. The heart-shaped plant produces sperm cells and egg cells. The sperm cells swim through the water to egg cells on the underside of the plant. Each egg cell is fertilized by one sperm. The fertilized egg develops into a new fern with roots, stems, and leaves, Figure 6-7.

Skill Check

Interpret diagrams: Look at the fern life cycle in Figure 6-7. What does the spore grow into? Where do the sperm and eggs develop? How do the sperm get to the eggs? *For more help, refer to the* ***Skill Handbook,*** *pages 706-711.*

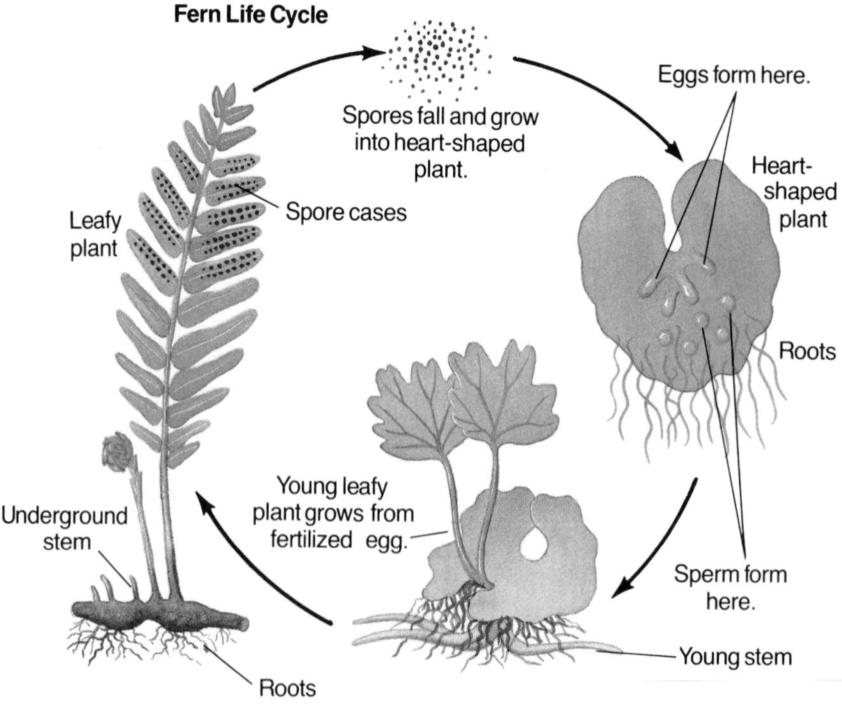

Figure 6-7 Follow the steps in the fern life cycle.

Lab 6–2

Ferns

Problem: What are some traits of ferns?

Skills

make and use tables, observe, use a microscope

Materials

stereomicroscope
2 microscope slides
ethyl alcohol
whole fern plant (for the class)
fern leaf with spore cases
scalpel
dropper
paper and pencil

Procedure

1. **Observe:** Look at a whole fern plant. Locate the horizontal stem at the base.
2. Note how the leaves grow from the stem.
3. Look for veins that indicate xylem and phloem cells in the leaf.
4. Look at the underside of each leaf for brownish-yellow spots. Each spot is made up of a group of spore cases.
5. Draw a table like the one shown. Draw and label your fern plant. Label all the parts as shown in Figure A.
6. Remove a fern leaf. Notice how the cells of the veins connect the stem to the leaf.
7. Obtain a part of a leaf with spore cases.
8. Place a drop of water on a clean slide.
9. Use the scalpel to scrape one of the brownish-yellow spots into the drop of water on the slide. **CAUTION:** *Always cut away from yourself when using a scalpel.*
10. **Use a microscope:** Observe the slide under low power of the stereomicroscope. Look for club-shaped spore cases like the one shown in Figure B.
11. Draw and label a single spore case.
12. Now, make a slide of spores by removing one or two spore cases from your first slide to a clean, dry slide. Add a drop of ethyl alcohol.
13. Observe the slide under low power. Watch what happens to the spore cases.
14. Draw a few of the spores like those shown in Figure C.

Data and Observations

1. What kinds of cells made it difficult to tear fern leaves from the plant?
2. What do fern roots grow from?

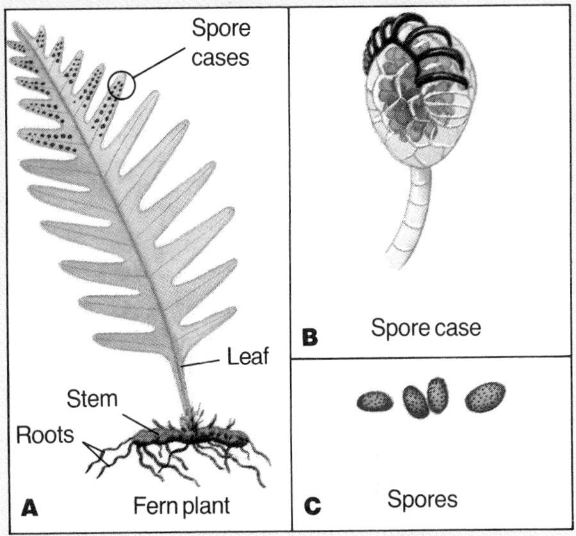

Analyze and Apply

1. What parts of the fern carry on photosynthesis?
2. When are spores produced in a fern life cycle?
3. Why are ferns usually larger than mosses and liverworts?
4. **Apply:** How are the leaves of ferns similar to the leaves of flowering plants?

Extension

Compare the roots of ferns with the rootlike hairs in mosses and liverworts.

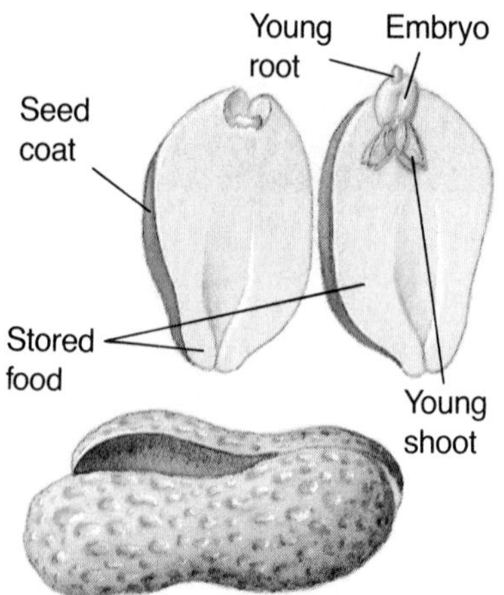

Figure 6-8 A seed has a seed coat to protect the embryo with its food supply. What kind of seed is shown here?

Conifers

The most common land plants are seed plants. Seed plants reproduce by forming seeds. A **seed** is the part of a plant that contains a new, young plant and stored food. The young plant in a seed is the embryo (EM bree oh). An **embryo** is an organism in its earliest stages of growth. A seed has a hard outer covering called the seed coat that protects the embryo, Figure 6-8. The food supply and the seed coat help the embryo survive for long periods of time when conditions are not right for growth.

There are two kinds of seed plants. One kind, called a **conifer** (KAHN uh fur), is a plant that produces seeds in cones. Conifers generally keep their leaves throughout the year. This is why they are sometimes called evergreens. A long time ago there were many different kinds of conifers. Conditions on Earth changed and many of the plants died out. Today, conifers are found mainly in northern areas of the world.

You have probably seen many examples of conifers. Most conifers are evergreen trees with small, needle-shaped leaves. An evergreen tree sheds its leaves in the same way that fur falls from the coat of a dog or cat—not all at once! If you are familiar with pine, spruce, and fir trees and think they are evergreen conifers, you are correct. Check your knowledge of these conifers with Figure 6-9. The larch, dawn redwood, and bald cypress are conifers that lose all their leaves each fall.

Figure 6-9 The pine (a), the spruce (b), and the fir (c) are all conifers with needlelike leaves.

a

b

c

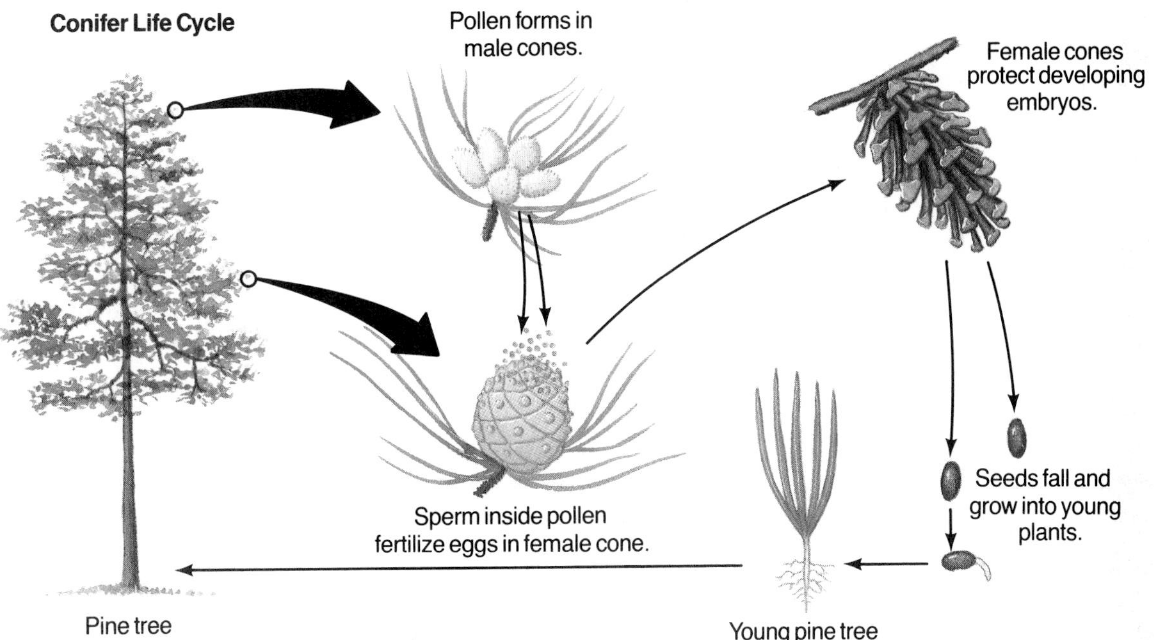
Conifer Life Cycle
Pollen forms in male cones.
Female cones protect developing embryos.
Seeds fall and grow into young plants.
Sperm inside pollen fertilize eggs in female cone.
Pine tree
Young pine tree

Figure 6-10 Follow the steps in the life cycle of a conifer.

Let's look at the life cycle of a pine tree. Have you ever noticed that two different kinds of cones grow on pine trees? Pine trees produce male and female cones. The small cones shown in Figure 6-10 are male cones, which produce pollen. **Pollen** are the tiny grains of seed plants in which sperm develop. You see male cones early in the spring. They open and shed pollen into the air. Wind carries the pollen to the larger female cones. The female cones, as shown in Figure 6-10, contain egg cells. The sperm cells fertilize the egg cells and the seeds containing the embryo form between the woody scales of the cone.

When the seeds are ripe, the cones become dry and the woody scales open. The seeds fall to the ground. When conditions are right, each seed grows into a young plant, Figure 6-10. Some conifers don't have woody cones. The conifer known as a yew produces its seeds inside red, fleshy cuplike parts. Birds eat these juicy cuplike parts and later drop the seeds away from the tree.

The roots, stems, and leaves of conifers are not like the soft stems of ferns. The roots and stems of conifers are hard and woody. The xylem cells have thicker cell walls. Also, the leaves of conifers are tough and needlelike. Examine the different kinds of conifer leaves shown in Figure 6-11. Can you see why some of them are called needles? Other leaves of conifers are scalelike in their shape, but feel quite soft.

Figure 6-11 Conifer leaves are either needlelike as in pines, flattened needles as in hemlock, or scalelike as in arborvitae, a common garden shrub.

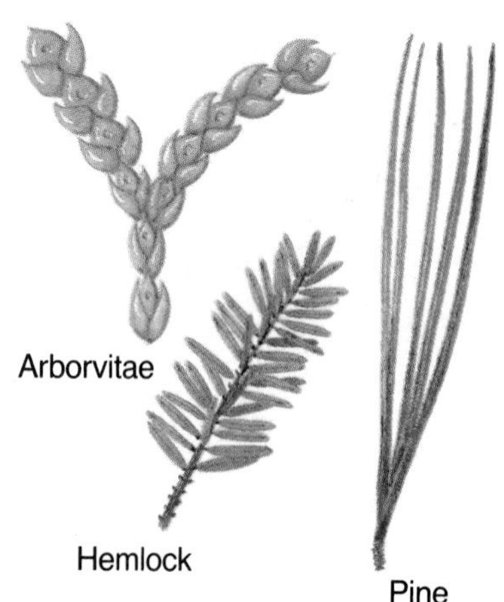
Arborvitae
Hemlock
Pine

Figure 6-12 A bald cypress is a conifer that lives in swamps.

Conifers can live in many more places than most other plants. Some can live in very wet areas. For example, the bald cypress shown in Figure 6-12 lives in swamps. Other conifers live in very dry places. Junipers are common conifers of the desert. Conifers grow also on high mountain slopes or close to the sea.

Take a look around at everything that is made of wood. Wood is a plant material. Wood is made up mainly of xylem cells. Your pencil, desk, and many parts of your school are probably made of wood. What about your home? What parts of your home are made of wood? Conifers supply three-fourths of the lumber that is used in the world.

You are reading words that are printed on plant material. You pay for things with plant material. You even write on plant material. Why are all these statements true? Paper is made from wood. Cardboard is also made of paper. Think of all the things you buy that are in cardboard boxes. Almost all of the world's paper comes from conifers, Figure 6-13. Conifers also are a source of turpentines, disinfectants, and fuel.

Conifers are very important to other animals for food and shelter. The bark, buds, and seeds of conifers are eaten by insects, birds, squirrels, rabbits, and many other animals.

Mini Lab

What Conditions Cause Pine Cones to Open?

Observe: Place open and closed pine cones into two plastic bags. Add a little water to one. Leave the other in a warm, dry place. Which conditions caused the cones to open? *For more help, refer to the* ***Skill Handbook,*** *pages 704-705.*

Figure 6-13 Conifers supply most of the world's paper products. Which of these products have you used?

Science and Society

Logging in the ancient forests of the Pacific Northwest

Ancient Forests: Jobs versus Wildlife

Have any of your neighbors made it well known that they are upset or angry by something in the neighborhood? Maybe a road sign was put up that they didn't like. Maybe someone's yard was full of garbage. You may have said, "It doesn't bother me," and "It's not my problem." In some cases where people disagree, the problem is so important that even the government gets involved. One such case involves the cutting of ancient trees in the forests of the Pacific Northwest. The trees are used as a major source of lumber and paper for the United States. They are also important to Earth's atmosphere. Consider the three cases described below and discuss each question with your classmates.

What Do You Think?

1. Some small mill owners work in forests owned by the National Forest Service. These loggers harvest only older trees and rely on the growth of young trees to fill in the gaps. Other loggers completely clear the forests of all ages of trees and then replant with young trees. These timber company loggers then harvest the trees as soon as they can. Because of the new laws protecting wildlife, the small mill owners may lose their jobs. The larger timber companies will not be affected. What solutions would you suggest to the government to save the life styles of all loggers?

2. The northern spotted owl lives only in the forests of old conifers in the Pacific Northwest. There are only about 3000 pairs of owls recorded. In July 1990, the United States Fish and Wildlife Service listed this owl as threatened on the federal endangered species list. The owl's environment is also protected by law. Why is it important to save species?

Northern spotted owl

3. Certain gases, including carbon dioxide, are found naturally in Earth's atmosphere. They trap heat from the sun in the same way the glass of a greenhouse does. Carbon dioxide has increased over the last fifty years because of the burning of fuels, such as coal, wood, oil, and natural gas. Forests take up carbon dioxide. Without forests, carbon dioxide would build up in the atmosphere. This increase of gases might cause major changes in Earth's climate. What social and political effects might there be if the world's climate changed?

Conclusion: Do we all need to care about forests? Are the needs of humans more important than those of other species?

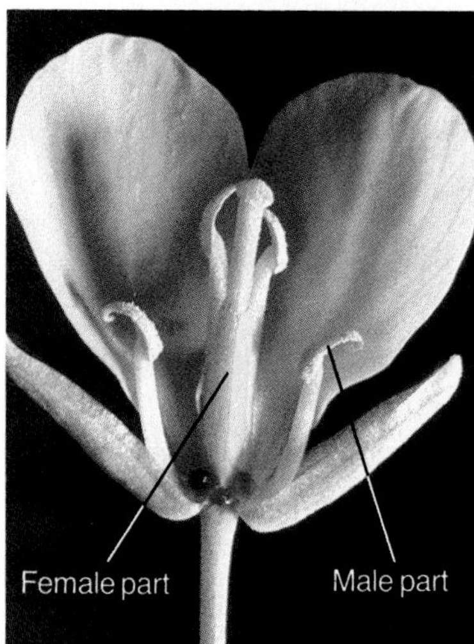

Figure 6-14 A flower is the reproductive part of a flowering plant.

Flowering Plants

The second kind of seed plant produces flowers and forms fruits. This seed plant is a flowering plant. A **flowering plant** is a vascular plant that produces seeds inside a flower. The **flower** is the reproductive part of the plant, Figure 6-14. Flowers have male parts that produce pollen. They also have female parts that produce eggs. A sperm cell produced in a pollen grain must join with an egg for a seed to form. Pollen is carried to the egg cells by wind, insects, or other animals. The female flower parts develop into a fruit that protects the seeds.

There are many more kinds of flowering plants in the world than nonflowering plants. Why might that be? The way the seeds are protected in flowering plants helps to make sure that the new plants survive. Flowering plants are adapted to live in many different environments.

Just because there are many more kinds of flowering plants than nonflowering plants, don't think that you will see mostly flowers when you take a walk outside. Not all flowering plants have big, sweet-smelling flowers. Many flowering plants, such as the maple in Figure 6-15a, have small, non-showy flowers. You may not have known they were flowering plants. Many broadleaved trees, vegetables, grasses, roadside weeds, and thorn bushes have flowers. So you see that not just roses and lilies are examples of flowering plants. Think about grass in lawns, fields, and parks. All grasses have flowers. The flowers on a grass plant are hard to see.

Another reason for not seeing flowers everywhere when you take a walk is that most flowering plants produce their

Bio Tip

Consumer: Did you know that cauliflower and broccoli are heads of tightly packed flower buds? Next time you buy these vegetables, look more closely to find the parts of the flowers.

a

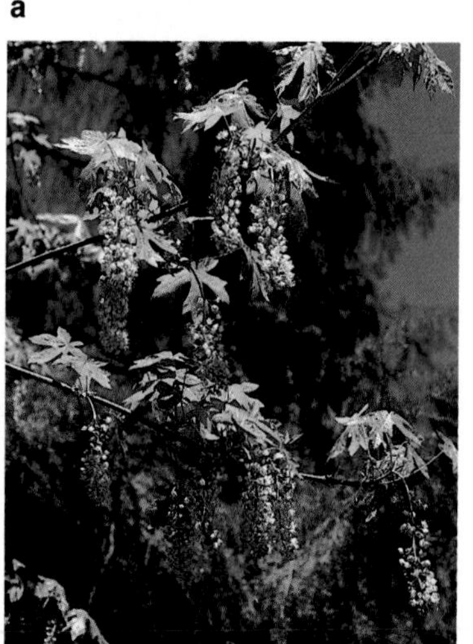

b

Figure 6-15 The flowers of maples (a) are not as showy as those of magnolia trees (b).

flowers only at certain times of the year. For example, many forest trees bloom only in the spring. The garden lily and tomato plants flower in the summer. Some plants even flower at times other than spring and summer, Figure 6-16.

Figure 6-16 Chrysanthemums bloom in late summer. Name some flowers that bloom in the spring.

How are flowering plants important to us? You know that showy flowers are often used for decoration, but plants have many more uses. Oranges and potatoes are plant parts you can eat. Bread and cereals are foods that are made from plants. Without plants, there would be no fruits, vegetables, breads, or cereals to eat. What parts of the plants in Figure 6-17 have you eaten?

If there were no plants, could you eat meat instead? No, if there were no plants, there would be no animals. Animals depend on plants or other producers to live.

In what ways do animals need plants?

In what other ways do all animals need plants? Remember that plants make their own food by photosynthesis. Oxygen given off by the plant is needed by both plants and animals for cellular respiration. Much of the oxygen gas that we have on Earth is produced by plants. In turn, the carbon dioxide produced during cellular respiration in plants and animals is needed by plants to make food. You can now see why every living thing depends on every other living thing.

Figure 6-17 Which part of a plant do you eat—stems, leaves, seeds, or fruits?

Check Your Understanding

11. What are three examples of conifers?
12. Where are sperm and egg cells formed in seed plants?
13. How are flowering plants important to us?
14. **Critical Thinking:** Which gas is more important to life on Earth, carbon dioxide or oxygen?
15. **Biology and Reading:** From reading the previous section, what are the three major groups of vascular plants?

Chapter 6

Review

Summary

6:1 Plant Classification

1. Most plants have chlorophyll in chloroplasts that traps light energy for photosynthesis. All plants have stiff cell walls.
2. Nonvascular plants do not have tubelike cells. Vascular plants have tubelike cells that carry food and water.

6:2 Nonvascular Plants

3. Mosses and liverworts are nonvascular plants. They have stems and leaves but no roots.
4. Mosses reproduce by sperm and eggs formed at the tips of leafy stems. The fertilized eggs develop into a stalk with a capsule. Spores are produced in a capsule. New moss plants grow from the spores.
5. Mosses are food for animals. They help form soil and keep it moist. Mosses are used in gardens.

6:3 Vascular Plants

6. Ferns, conifers, and flowering plants all have xylem and phloem cells. Conifers have seeds inside cones. Flowering plants have flowers and seeds.
7. Ferns reproduce by eggs and sperm and by spores on the leaves. Conifers produce sperm in pollen. Pollen and eggs are formed in cones. Flowering plants produce sperm in pollen and eggs in flowers.
8. Conifers are used for lumber and paper. All living things depend on plants for food and oxygen.

Key Science Words

chlorophyll (p. 114)
conifer (p. 124)
egg (p. 118)
embryo (p. 124)
fern (p. 121)
fertilization (p. 118)
flower (p. 128)
flowering plant (p. 128)
moss (p. 117)
nonvascular plant (p. 116)
phloem (p. 121)
photosynthesis (p. 114)
pollen (p. 125)
seed (p. 124)
sexual reproduction (p. 118)
sperm (p. 118)
vascular plant (p. 116)
xylem (p. 121)

Testing Yourself

Using Words

Choose the word from the list of Key Science Words that best fits the definition.

1. chemical that gives plants their green color
2. when a sperm and an egg cell join
3. kind of plant with tubelike cells to carry water and food
4. type of reproduction when a sperm fertilizes an egg
5. female sex cell

Review

Testing Yourself *continued*

6. a structure that contains a small, new plant and a supply of stored food
7. plant that grows from the seeds of woody cones
8. vascular plant that produces seeds inside a flower
9. powderlike structures in seed plants in which sperm develop

Finding Main Ideas

List the page number where each main idea below is found. Then, explain each main idea.

10. what's carried in tubelike cells
11. what plants use to make food
12. how ferns reproduce
13. how plants are grouped
14. what mosses look like
15. how flowering plants are important to other organisms

Using Main Ideas

Answer the questions by referring to the page number after each question.

16. What are several ways that conifers are important? (p. 126)
17. How do mosses reproduce? (p. 118)
18. Why do mosses and liverworts grow in wet or damp places? (p. 117)
19. What are two reasons you may not find flowers on flowering plants? (p. 128)
20. How do nonvascular plants survive without xylem and phloem? (p. 116)
21. What is one way of grouping vascular plants? (pp. 121, 124)
22. What are the uses of mosses? (p. 118)
23. How are plants different from animals? (p. 114)
24. How are spores different from pollen? (pp. 120, 125)

Skill Review 

For more help, refer to the ***Skill Handbook,*** *pages 704-719.*

1. **Infer:** Why are the spores of a fern more likely to be carried farther by wind than the spores of a moss?
2. **Observe:** Place a flower on a sheet of white paper. Carefully take the flower apart and compare it with the one in Figure 6-14.
3. **Interpret diagrams:** Look at the diagram in Figure 6-10. Why must pollen be formed before fertilization?
4. **Make and use tables:** Make a table to show the importance of mosses and liverworts, ferns, conifers, and flowering plants.

Finding Out More

Critical Thinking

1. How do you think certain medicines from plants were discovered?
2. Compare the methods of reproduction between spore plants and flowering plants.

Applications

1. In an old aquarium or glass jar, set up a terrarium with mosses, liverworts, and small ferns.
2. Report on common poisonous plants. Explain what part of the plant is poisonous, the effects of the poison, and whether there are any known antidotes.

CHAPTER PREVIEW

Chapter Content

Review this outline for Chapter 7 before you read the chapter.

Skills in this Chapter

The skills that you will use in this chapter are listed below.

- In **Lab 7-1,** you will observe, make and use tables, and interpret data. In **Lab 7-2,** you will observe, infer, predict, and make and use tables.
- In the **Skill Checks,** you will understand science words, interpret diagrams, and infer.
- In the **Mini Labs,** you will infer and observe.

Chapter 7

Simple Animals

In the photo on the opposite page, you can see some of the beautiful and unusual animals that are found in the sea. For many years people thought that sponges were plants. The sponges didn't move around by themselves and they didn't seem to capture food, both animal-like traits. Can you tell from the photo how a sponge captures its food? Early scientists couldn't either.

The small photo shows the mouth of a coral. It is surrounded by structures that contain the stinging cells the animal uses to capture food. With careful observation, scientists were able to watch such animal-like traits as food-getting. Today there is no doubt that these organisms are animals.

Try This!

How does a squid move? Inflate a balloon and release it. Observe the movement and direction of the balloon. Compare it with the movement of a squid.

BIOLOGY Online

Visit the Glencoe Science Web site at science.glencoe.com to find links about **simple animals.**

The mouth of a coral is surrounded by tentacles.

7:1 Animal Classification

Objectives

1. **List** four traits of animals.
2. **Identify** nine major phyla of animals and give an example of each.

Key Science Words

vertebrate
invertebrate
symmetry

Scientists have discovered and classified about 1 1/2 million different kinds of living things. Of this number, about one million are animals. The different kinds of animals are grouped according to traits they have in common.

Traits of Animals

What are some of the traits that make animals different from other living things? First, animals are consumers and can't make their own food. They must take in food from their surroundings. Some animals eat plants, some eat other animals, and others eat both plants and animals. Most animals digest and store food in their bodies.

Second, most animals can move from place to place. Moving around helps them find food. Animals that don't move have developed other ways to get food.

Third, animals are multicellular organisms. The cells of most animals are organized into tissues and organs that form systems. Many animals have nervous systems, digestive systems, and reproductive systems. Nearly all animals have muscles and systems for getting rid of wastes.

All animals belong to one of two groups—the vertebrates and the invertebrates (in VERT uh brayts). **Vertebrates** are animals with backbones. People and the animals most closely related to us, such as fish, amphibians, reptiles, birds, and other mammals, are vertebrates. **Invertebrates** are animals without backbones. Worms and insects are examples of invertebrates.

Finally, most animals have some kind of symmetry (SIH muh tree). **Symmetry** is the balanced arrangement of body parts around a center point or along a center line. Only a few very simple animals do not show symmetry. These animals grow in a variety of shapes. The sponge in Figure 7-1a does not show symmetry.

Some invertebrates show radial (RAYD ee ul) symmetry. In radial symmetry, the body parts are arranged in a circle around a center point. The sea anemone (uh NEM uh nee) in Figure 7-1b is an example of an animal that has radial symmetry.

All vertebrates and some invertebrates have bilateral (bi LAT uh rul) symmetry. In an animal with bilateral symmetry, the body can be divided lengthwise into two equal sides, a right side and a left side. The bee in Figure

Skill Check

Understand science words: invertebrate. The word part *in* means not. In your dictionary, find three words with the word part *in* in them. *For more help, refer to the **Skill Handbook**, pages 706-711.*

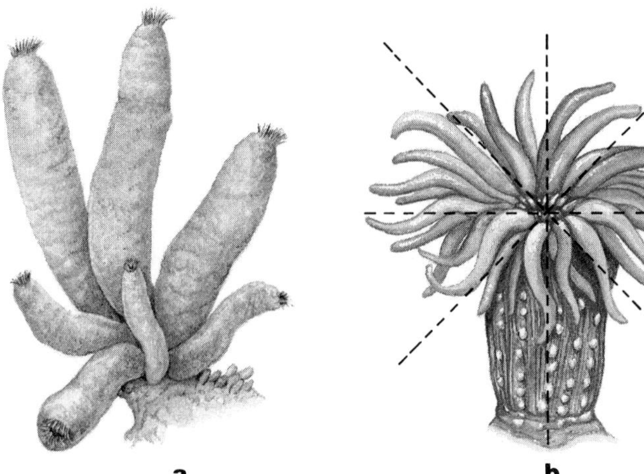

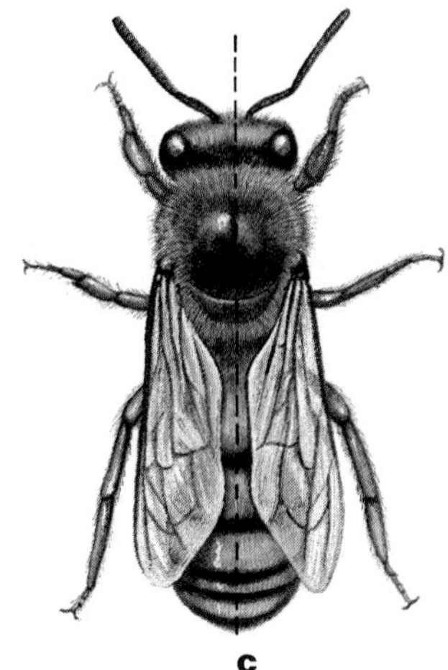

Figure 7-1 A sponge does not have symmetry (a). A sea anemone has radial symmetry (b). A bee has bilateral symmetry (c).

7-1c has bilateral symmetry. Notice that it has a head end and a back end. The head end studies where the animal is going and finds food. The bee also has an upper and a lower side. The lower side moves the animal along a surface.

Let's review the traits of animals.

1. Animals can't make food, but must catch and eat it.
2. Most animals can move from place to place.
3. Animals have many cells. The cells make up tissues and organs that form systems.
4. Most animals have symmetry.

How are an animal's cells organized?

How Animals Are Classified

Scientists group living things into one of five kingdoms. If an organism shows most of the traits just listed, then it is placed in the animal kingdom. If not, then it is placed in one of the other four kingdoms.

How do scientists know to which phylum an animal belongs? Consider the sponges. They have certain traits. If an animal shows these traits, it belongs in the phylum with sponges. If it does not, then it belongs in a different phylum.

Here is an example with which you are familiar. A new student arrives at your school. How does the office staff know whether the student should be placed in the ninth, tenth, eleventh, or twelfth grade? The student can be placed in a certain grade only if he or she has passed a certain number of courses. If the student has passed the number of courses required in the ninth grade, the student is put in the tenth grade. The number of courses passed is the "trait" used to place students into the proper grades.

Skill Check

Interpret diagrams: Make a list of the invertebrates pictured in this chapter. Indicate the type of symmetry shown by each one. *For more help, refer to the **Skill Handbook**, pages 706-711.*

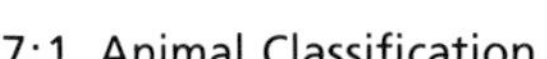

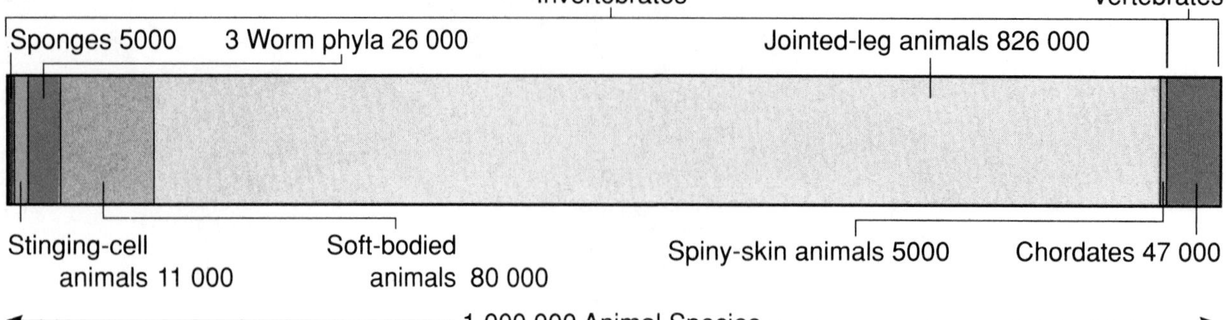

Figure 7-2 Animals are placed in nine major phyla. Which phylum has the most complex animals?

The nine major groups into which animals are placed are shown in Figure 7-2. Each major group is a phylum. Also shown is the number of animal species in each phylum.

The figure shows that some phyla include many different kinds of animals. Other phyla are small. Animals with jointed legs make up the largest phylum. Insects, spiders, and crayfish belong to this phylum. Sponges make up the smallest of the phyla shown.

Which phylum contains the simplest animals?

Figure 7-2 also shows the simplest phyla at the left and the most complex at the right. Sponges are the simplest of all animals. Chordates (KOR dayts) are the most complex. Most chordates are vertebrates. Eight of the nine phyla shown are grouped together as invertebrates. You can see from Figure 7-2 that there are many more invertebrates than vertebrates on Earth.

Check Your Understanding

1. List four traits of animals.
2. Which animal phylum is the largest? Give two examples of members of this phylum.
3. How can an animal body be divided if it has radial symmetry?
4. **Critical Thinking:** An astronaut discovers an organism on a distant planet. It can move around, is multicellular, has bilateral symmetry, and makes its own food. Can this new organism be classified as an animal? Explain.
5. **Biology and Reading:** An amoeba has no symmetry. A starfish has radial symmetry. A human has bilateral symmetry. Draw an amoeba, a starfish, and a person. Draw the line or lines of symmetry for each.

7:2 Sponges and Stinging-cell Animals

Objectives

3. List the traits of sponges.

4. Describe the traits of stinging-cell animals.

Key Science Words

sponge
pore
stinging-cell animal
tentacle

Sponges and stinging-cell animals do not have backbones. They are invertebrates. Sponges are the simplest invertebrates.

Sponges

Sponges are simple invertebrates that have pores. A **pore** is a small opening. You may know a sponge as an object that you use when you take a bath or wash cars. Look at the sponges in Figure 7-3. Some look like small trees. Others look like vases. Most sponges have no definite shape. You will also notice that they come in many colors.

What are the traits of sponges? Unlike most other animals, sponges do not move about freely on their own. Most live attached to rocks in shallow oceans. Some sponges live in fresh water. The body of a sponge has many pores. Water enters through these pores and leaves through an opening in the top center of the sponge's body. Many canals carry the water throughout the sponge's body.

How is water carried through the walls of a sponge's body?

Since sponges can't move, how do they get food? All sponges must live in water. Water has small organisms in it that sponges can use for food. The pores in the sponge allow water and any food in the water to flow into the animal. Once inside, the food is trapped by food-getting cells.

Figure 7-3 Sponges come in a variety of sizes, shapes, and colors.

Figure 7-4 Arrows show the direction in which water flows through a sponge. Sponges have three kinds of cells.

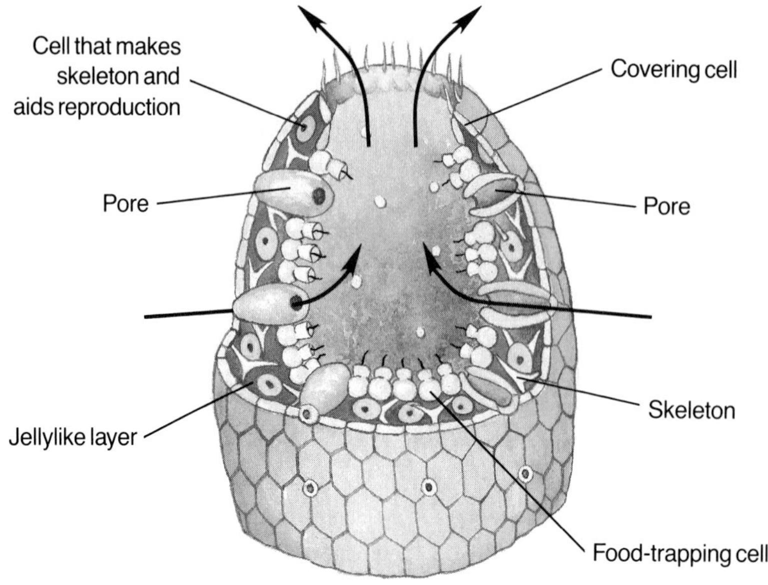

Sponges are only two cell layers thick, have no muscle or nerve cells, and have no tissues, organs, or organ systems. Sponges have three main kinds of cells. Each cell type has a certain job. One cell type traps food and moves water through the sponge by beating its flagellum. Other cells cover and protect the sponge. The third cell type produces chemicals that make the skeleton. This third cell type also aids in reproduction and carries food to other cells. The three kinds of cells are shown in Figure 7-4.

> **Bio Tip**
>
> **Consumer:** Most of the sponges sold today are artificial. They are made from cellulose or synthetic materials. Natural sponges hold much more water than artificial sponges, but they are not commonly used because they are expensive.

Sponges can reproduce both sexually and asexually. In sexual reproduction, the same sponge produces both sperm and eggs, but at different times. It takes two sponges for sexual reproduction to take place. Having both sperm and eggs ensures that sponges near each other can reproduce. Sperm cells produced by one sponge are carried to another sponge by the water. There, the sperm fertilize the eggs. New sponges usually develop on rocks on the ocean floor.

Sponges also reproduce asexually. Small pieces of a sponge's body may break off and form separate, new sponges. Sponges may also form buds. Each bud can develop into a new sponge.

You may have used a natural sponge for cleaning and washing because they hold a lot of water. The skeletons of sponges have many fibers. When the animal dies and decays, the fibers remain. The spaces between these fibers can be filled with water. Because the fibers are elastic, the water can be squeezed out.

Stinging-cell Animals

Stinging-cell animals are animals with stinging cells and hollow, saclike bodies that lack organs. Most of these animals live in the ocean. A few live in freshwater lakes and streams. Many stinging-cell animals are beautiful and delicate. They look very different from the animals you may see every day.

It is hard at first to compare hydras, corals, sea fans, jellyfish, and sea anemones with one another. Look at the animals in Figure 7-6. Animals in the stinging-cell animal phylum do not look much alike. Yet, all of them share certain traits.

Many stinging-cell animals have armlike parts called **tentacles** (TENT ih kulz) that surround the mouth. The jellyfish in Figure 7-6a has tentacles. If you were to look closely at corals and sea fans, you would see that these animals also have tentacles. A coral is really a group of many small animals. The same is true of sea fans. A hard structure on the outside of the animals supports and protects them and is made of chemicals produced by the animals. The chemicals harden into different shapes.

Figure 7-5 shows that stinging-cell animals have radial symmetry and saclike bodies made of two cell layers. A jellylike layer lies between the two cell layers. Inside the body of each animal is a body cavity. The cavity has one opening called the mouth. The only way into and out of the body is through the mouth.

Many stinging-cell animals fasten themselves to the ocean bottom or to rocks with a structure called a disc. They do not move from that place. How then do they get food? They catch it with their tentacles.

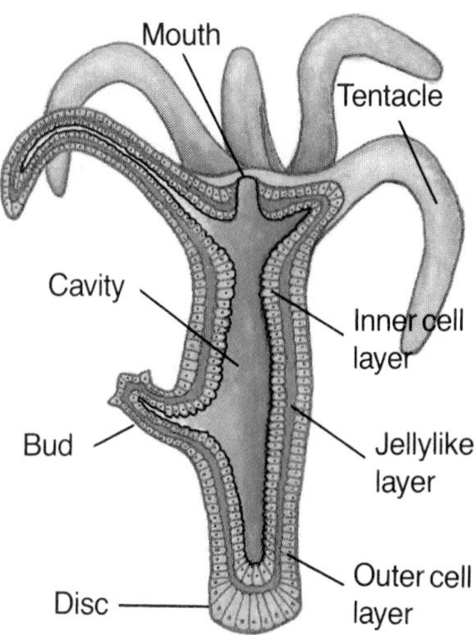

Figure 7-5 Stinging-cell animals have saclike bodies with two cell layers.

How many cell layers does a stinging-cell animal's body contain?

Figure 7-6 Some of the many stinging-cell animals are a jellyfish (a), coral (b), and sea fan (c).

a

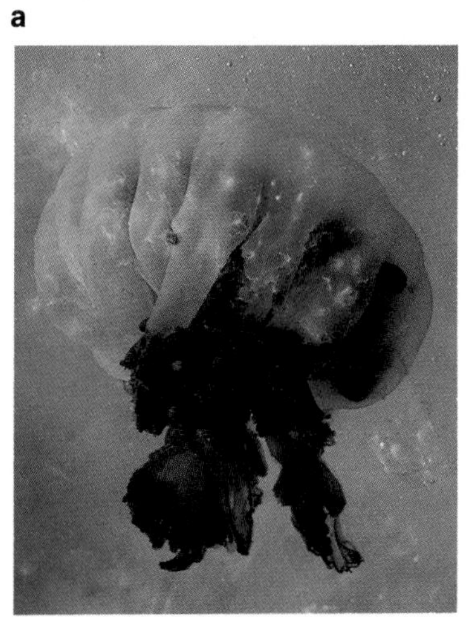

b

c

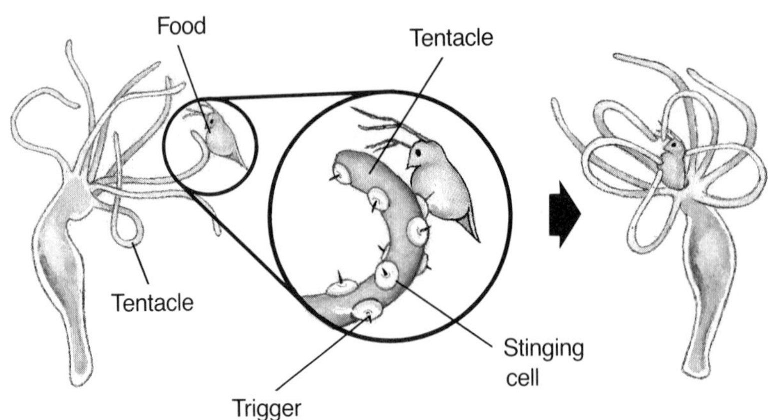

Figure 7-7 Hydras catch food by stinging a small animal with a poison dart. The stunned animal is then stuffed into the hydra's mouth.

Mini Lab

How Does a Sea Anemone Attach Itself?

Infer: Wet a suction cup and press it against a small rock. Pick the rock up by holding the suction cup. How are you able to lift the rock with the suction cup? *For more help, refer to the **Skill Handbook**, pages 706-711.*

The tentacles have thousands of stinging cells with hairlike triggers. When an animal brushes against the trigger, a stinging dart containing poison shoots into the water. The dart hits the animal and stuns it with the poison. The tentacles then push the newly caught meal through the mouth and into the body cavity, where it is digested. Undigested food leaves the animal through the mouth. Figure 7-7 shows a hydra with food it has caught.

Stinging cells of a jellyfish can cause a very painful sting in people. Anyone who lives near an ocean probably knows not to bother jellyfish that have washed onto shore.

Stinging-cell animals have muscle cells and nerve cells. Animals in this phylum use muscles and nerves to move their tentacles. Jellyfish use muscles to swim.

Stinging-cell animals reproduce sexually by forming eggs and sperm. These are released into the water, where the sperm fertilize the eggs. Some stinging-cell animals, like the hydra, can also reproduce asexually by forming buds. These buds break off and become separate animals.

Check Your Understanding

6. List three traits of sponges.
7. How do stinging-cell animals get food?
8. List three traits of stinging-cell animals.
9. **Critical Thinking:** Why is radial symmetry useful to a stinging-cell animal?
10. **Biology and Reading:** Most of the sponges sold today are made from cellulose, a plant material. How do you think this kind of sponge got its name?

Lab 7–1 Stinging-cell Animals

Problem: What are the traits of a stinging-cell animal?

Skills

observe, make and use tables, interpret data

Materials

hydra culture
dropper
stereomicroscope
daphnia or brine shrimp
small dish
hand lens
flashlight

Procedure

1. Copy the data tables.
2. Use a dropper to place a hydra and some water from the culture into a small dish.
3. **Observe:** Examine the hydra with a hand lens. Count the number of tentacles. Find the disc by which the animal attaches itself. Find the mouth. Record your observations in Table 1.
4. Tap the dish gently with your finger. Observe how the hydra moves and changes shape. Record your observations.
5. Place the dish on the stage of a stereomicroscope and focus on the hydra.
6. Draw the hydra and label the structures.
7. Drop a daphnia or a small amount of brine shrimp into the dish. Observe and record how the hydra takes in food.
8. Shine a flashlight on the dish. Observe and record the hydra's reaction.
9. Return the hydra to the culture.

Data and Observations

1. How does the hydra react to light?
2. Where is the disc located?

Analyze and Apply

1. How does the hydra capture food?
2. **Interpret data:** Why does the hydra react the way it does when you tap the dish?
3. What is the advantage to the hydra of having a mouth surrounded by tentacles?
4. **Apply:** How does a hydra differ from a sponge in food-getting?

Extension

Observe a hydra when it is near other small animals and protists to determine what a hydra will eat.

TABLE 1.

PART	OBSERVATION
Location of mouth	
Number of tentacles	
Location of disc	

TABLE 2.

STIMULUS	REACTION
Touch (tapping on dish)	
Food	
Light	

7:3 Worms

Objectives

5. **List** the three main phyla of worms and give an example of each.
6. **Identify** the main features of flatworms, roundworms, and segmented worms.

Key Science Words

flatworm
tapeworm
cyst
planarian
roundworm
hookworm
anus
segmented worm

When you hear the word *worm*, you probably think of an earthworm, but there are many different kinds of worms. Worms are invertebrates. They are classified into three main phyla—flatworms, roundworms, and segmented worms, based on their structures.

Flatworms

Worms are more complex than sponges and stinging-cell animals. **Flatworms** are the simplest worms. They have a flattened body and three layers of cells. These layers include an outer layer, an inner layer, and a thick, middle layer. Organs and systems develop from cells of the middle layer.

Most flatworms are parasites. Remember that parasites live in or on other living things and get food from them. The organism that provides the food is the host. Some flatworms can cause serious problems.

A **tapeworm** is a kind of flatworm that has a flattened, ribbonlike body divided into sections. Tapeworms can live in the intestine of almost any kind of vertebrate. A tapeworm has suckers and hooks on one end that are used to hold onto the intestine of its host. Tapeworms have no mouth or organs to digest food. They absorb food that the host has already digested.

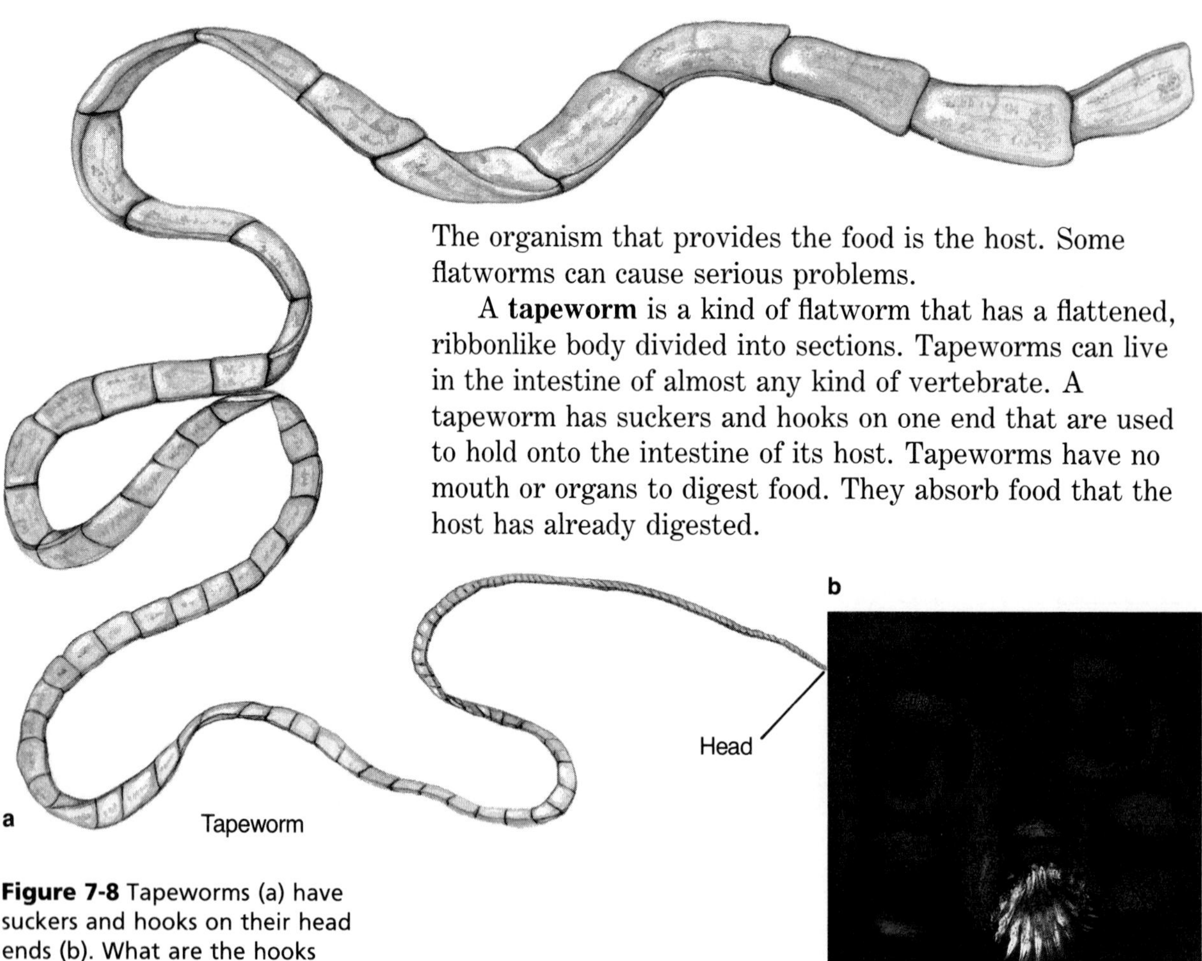

Figure 7-8 Tapeworms (a) have suckers and hooks on their head ends (b). What are the hooks used for?

The tapeworm in Figure 7-8 is shown life size. Although some are much smaller, this one happens to be very large. Is it any wonder that a worm of this size uses a lot of the host's food?

Tapeworms have life cycles with many stages. Follow the steps of the pork tapeworm life cycle in Figure 7-9.

1. A pig eats tapeworm eggs that are on the ground. The eggs hatch in the pig's intestine.
2. The young worms enter the pig's bloodstream. Then they travel to the muscles and burrow into them.
3. The young worms form cysts (SIHSTZ) in the muscles. A **cyst** is a young worm with a protective covering.
4. If a person eats raw or undercooked meat that contains the cysts, the tapeworms get inside the person's intestine. The worms come out of their cysts, attach to the inside of the intestine, and begin to grow.
5. Tapeworms produce both eggs and sperm in each of their body sections. The sperm fertilize the eggs.
6. The body sections containing fertilized eggs break off and leave the host's body in solid waste through the intestine.

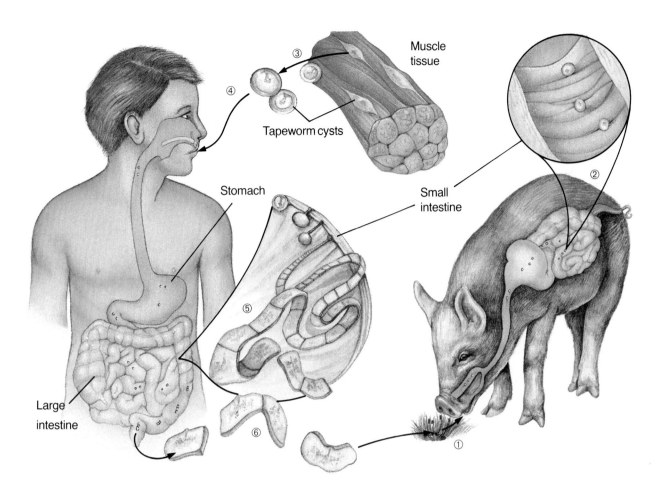

Figure 7-9 Life cycle of a pork tapeworm

Skill Check

Infer: Someone with tapeworms eats a lot of food but still feels hungry, tired, and loses weight. What is the cause? *For more help, refer to the* ***Skill Handbook,*** *pages 706-711.*

Figure 7-10 A planarian (a) can reproduce asexually by regrowing new parts (b). How many parents are involved in this type of reproduction?

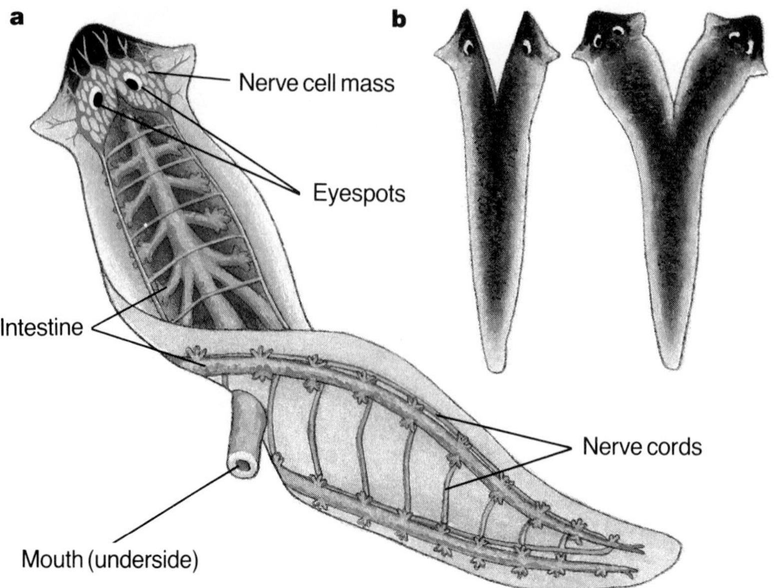

Tapeworms in humans are not as common in the United States as in some other countries. In the United States, human wastes are treated with chemicals at sewage treatment plants. Meat is inspected for cysts by government inspectors. However, completely cooking meat is the best way to keep from getting tapeworms.

Although most flatworms are parasites, some aren't. Those that are not parasites are said to be free-living. Some of these flatworms live in water, but a few are found on land. A **planarian** (pluh NAIR ee un) is a common freshwater flatworm that is not a parasite. It is less than one centimeter long.

A planarian has the features shown in Figure 7-10a.

1. It has a flat body with muscles.
2. It has a triangular head with two spots that look like eyes. The eyespots have nerve cells that detect light.
3. It has two nerve cords. It also has two masses of nerve cells near the head. These structures allow the planarian to respond to the environment.

What flatworm has a mouth?

4. The mouth is near the middle of the body on the underside.
5. Planarians have an intestine that breaks down food. Undigested food leaves the intestine through the mouth.
6. Each animal has both male and female reproductive organs and produces both sperm and eggs. It reproduces sexually by exchanging sperm with another planarian. Planarians also reproduce asexually by pinching into two parts. Each part then forms a new animal, Figure 7-10b.

Roundworms

Roundworms are worms that have long bodies with pointed ends. They have three layers of cells. Many of these worms are small and can't be seen without a microscope. Many are parasites. Their hosts are people, dogs, cats, or even plants.

Some roundworms live in great numbers in the soil. A square meter of soil may contain four million worms! These worms are not animal parasites, but some of them may be plant parasites. Most roundworms, however, are free-living.

Hookworms are found in soil in the southeastern part of the United States. A **hookworm** is a roundworm that is a parasite of humans. Hookworms enter the body through the skin of the feet. Once inside, they move to the intestine. There they attach themselves and feed on the host's blood.

Let's take a closer look at a roundworm. Look at Figure 7-11 as you read about the features of this group.

1. These worms have long, rounded bodies. Each end is pointed. A set of muscles runs the length of the body. When roundworms move, they whip about.
2. One end of a roundworm has a mouth. The other end has an anus (AY nus). An **anus** is an opening through which undigested food leaves the body.
3. A roundworm has a tube within its body that connects the mouth and the anus. The tube is an intestine that digests food. Food and wastes do not mix as they do in the stinging-cell animals and flatworms, which have only one opening.
4. Males and females are separate animals.

Mini Lab

Where Do You Find Roundworms?

Observe: Cover a funnel with wire mesh. Stand it in a jar of alcohol. Place moist soil on the mesh. What do you see in the alcohol when the soil begins to dry out? Why? *For more help, refer to the* ***Skill Handbook,*** *pages 704-705.*

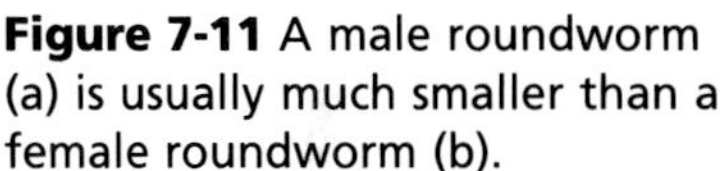

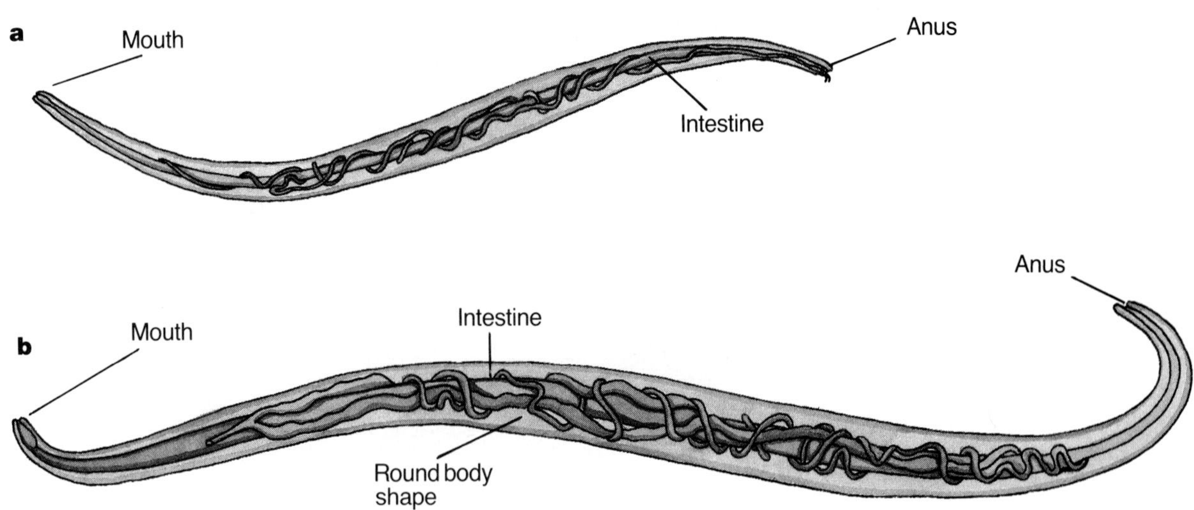

Figure 7-11 A male roundworm (a) is usually much smaller than a female roundworm (b).

a

b

Figure 7-12 Earthworms (a) and leeches (b) are segmented worms.

Segmented Worms

The **segmented worms** are worms with bodies divided into sections called segments. These worms have three cell layers. They are the most complex of all the worms.

You may be familiar with some of the segmented worms. They live almost everywhere—in fresh water, salt water, and on land. Examples are shown in Figure 7-12. The leech is a segmented worm that is a parasite. Most leeches live in streams and lakes. They attach to the skin of host animals and suck their blood.

The common earthworm is a segmented worm. Figure 7-13 shows some of the earthworm's features. Look at this figure as you read about the earthworm's traits.

1. The body wall has layers of muscles. Each segment has bristles that help the earthworm move by gripping the soil.
2. Earthworms have a mouth and an anus.
3. The intestine is a tube. It has parts for holding food, grinding food, and digesting food.
4. Most segments have organs to get rid of wastes.
5. Two blood vessels run along the body. They meet to form five pairs of simple hearts. The hearts pump blood through the body, carrying oxygen and food to all cells.

Bio Tip

Health: Leeches are used in medicine to drain excess blood from reattached body parts following surgery. Leeches produce a chemical that prevents blood from clotting. This chemical is used for treating cardiovascular disorders.

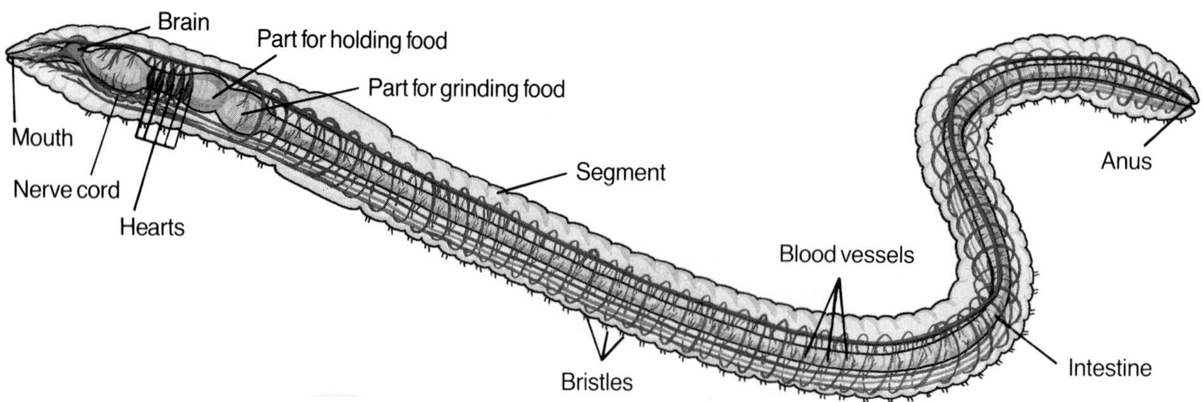

Figure 7-13 Segmented worms, such as earthworms, have a complex structure.

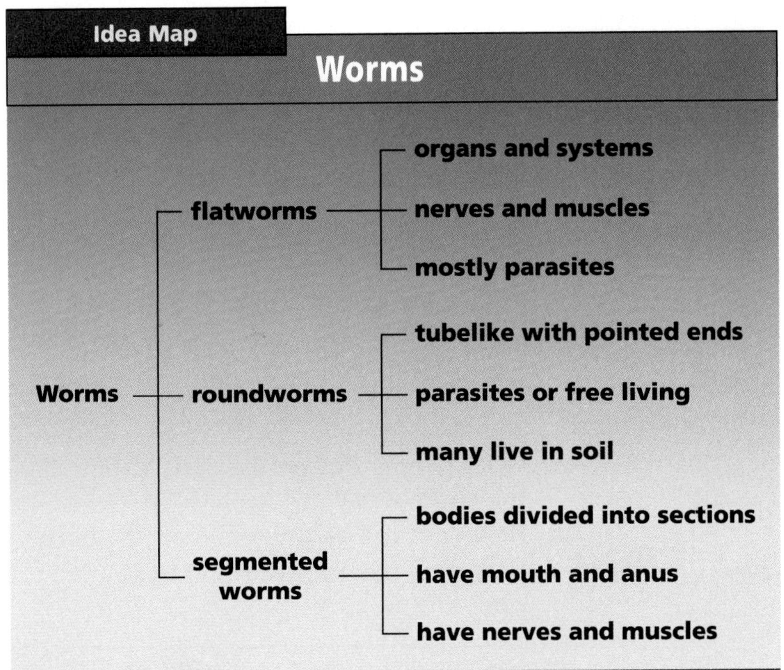

Study Tip: Use this idea map as a study guide to worms. The traits of each worm group are shown.

6. Nerves run the length of the body. There is a simple brain at the front end.
7. Each earthworm has both male and female sex organs. Earthworms reproduce when two worms exchange sperm. Other segmented worms may have separate sexes.

Why do earthworms seem to eat soil?

Earthworms take in soil through their mouths. Soil contains decayed matter, such as dead leaves, insects, and seeds. These things are food for the earthworm. The soil itself is not food and passes through the animal's intestine. Soil leaves the intestine through the anus. Earthworms move large amounts of soil from place to place by passing it through their bodies. They enrich the soil and loosen it, helping plants grow.

Check Your Understanding

11. Give two examples of flatworms.
12. In what way are tapeworms harmful?
13. List three traits of roundworms.
14. **Critical Thinking:** How can you prevent hookworms from entering your body?
15. **Biology and Reading:** List two things that earthworms do that are helpful.

Earthworms

Lab 7–2

Problem: What traits does an earthworm have that help it live in soil?

Skills

observe, infer, predict, make and use tables

Materials

live earthworms in a covered container
hand lens
shallow pan
toothpick
paper towels
beaker of water
vinegar
cotton swab
flashlight

Procedure

1. Copy the data table.
2. Open a container of earthworms and shine the flashlight on the worms. What do the worms do? Record what you see.
3. Wet your hands. Carefully place an earthworm on a moist paper towel in a shallow pan. **CAUTION:** *Do not let the worm dry out.* Keep your hands wet and moisten the worm by sprinkling it with water. Hold the worm gently between your thumb and forefinger. Observe its movements and record them in your table.
4. Rub your fingers gently along the body. Examine the bristles along the earthworm with a hand lens. See the figure for help.
5. **Observe:** With a toothpick, gently touch the earthworm on the front and back ends. Record your observations.
6. **Make and use tables:** Dip a cotton swab in vinegar. Place it in front of the earthworm on the paper towel. Record your observations.

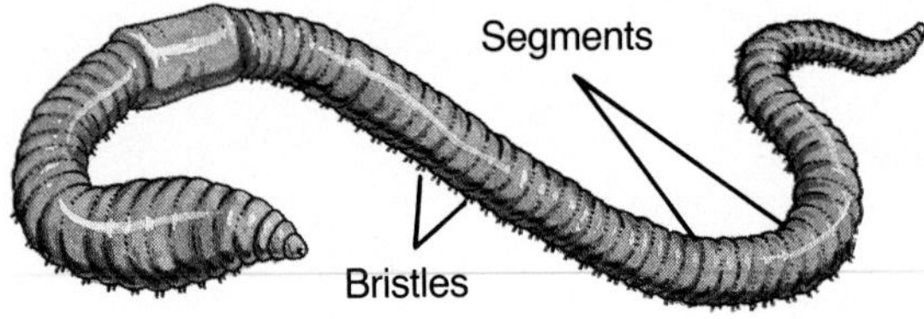

7. Return the earthworm to the covered container.

Data and Observations

1. What happens when the light is shined on the earthworms?
2. How does the earthworm react to touch?
3. How does the earthworm react to vinegar?

CONDITION	REACTION
Light	
Holding	
Touch	
Vinegar	

Analyze and Apply

1. **Infer:** How does the earthworm's reaction to light help it to live in the soil?
2. How does the earthworm's reaction to being touched help it to live in the soil?
3. **Predict:** How would you expect an earthworm to react to chemicals like vinegar in the soil?
4. How is the front end of the earthworm helpful for living in soil?
5. How is the long, soft body helpful for living in the soil?
6. How are the bristles useful for living in the soil?
7. **Apply:** How is an earthworm similar to a hookworm?

Extension

Observe a planarian's response to light and touch.

7:4 Soft-bodied Animals

The soft-bodied animals are more complex than the other invertebrates that you have studied. Soft-bodied animals come in all sizes and live in many different environments. They are placed together in one phylum because of the traits they share.

Objectives

7. Identify the major features of soft-bodied animals.

8. Explain how soft-bodied animals are classified.

Traits of Soft-bodied Animals

Key Science Words

soft-bodied animal
mantle

Soft-bodied animals are animals with a soft body that is usually protected by a hard shell. Animals in the soft-bodied animal phylum may be familiar to you, no matter where you live. Some live along the ocean on rocks, buried in the sand, or in the water. You may have gathered their shells while walking along a beach, Figure 7-14a. Some of these animals are found around streams and ponds. Others live on land.

What are the traits of soft-bodied animals?

Soft-bodied animals have certain traits. The body is covered by a thin, fleshy tissue called a **mantle.** The mantle makes the shell. The shell is the outermost covering. It protects the soft body. Soft-bodied animals have a muscular foot for moving from place to place, Figure 7-14b. Most have a head with a mouth. Many soft-bodied animals have a structure like a tongue covered with teeth inside the mouth. This structure scrapes food from surfaces, such as rocks. A snail has this structure.

The soft-bodied animals are grouped into classes based on three traits. The three traits are the kind of foot, whether or not a shell is present, and the number of shells.

Figure 7-14 The shells of soft-bodied animals can be found on many beaches (a). Clams use a muscular foot for movement (b).

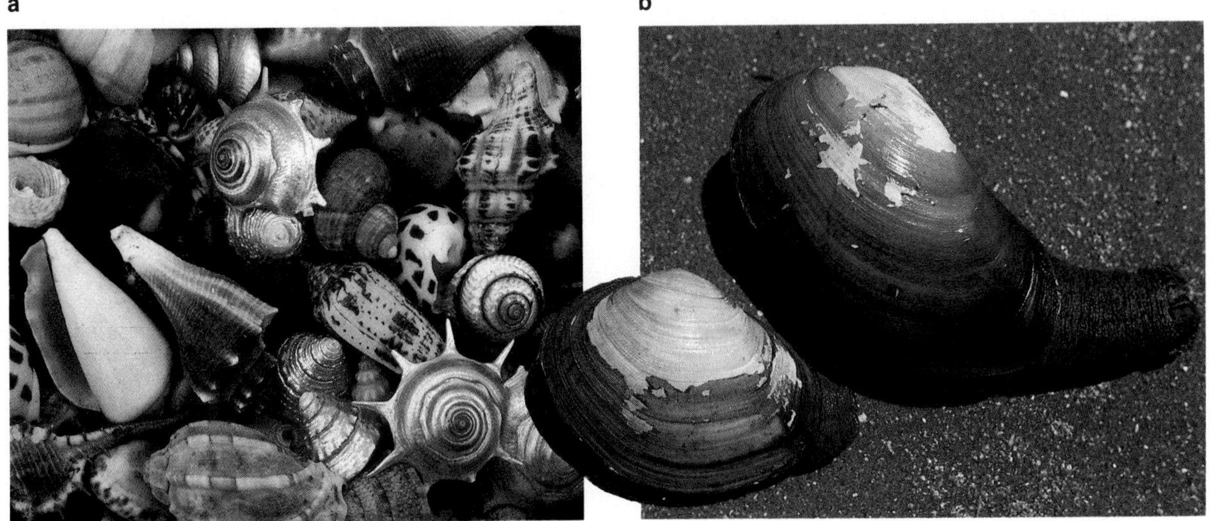

a

b

Figure 7-15 A snail (a) has a muscular foot. The foot of an octopus (b) is divided into eight arms that surround the head.

What are the three classes of soft-bodied animals?

Classes of Soft-bodied Animals

The first class of soft-bodied animals includes snails and slugs, shown in Figure 7-15a. They live on land and in the water. They glide slowly along by means of a wide, muscular foot. The foot puts down a trail of slime that helps the animal glide. Snails have a single shell and slugs have no shell. Sea and land slugs look like snails without shells. Notice the two pairs of tentacles on the snail's head in Figure 7-15a. They are sense organs. Each of the larger tentacles has an eye that detects light.

The second class of soft-bodied animals includes clams, oysters, and scallops. See Figure 7-14b. Unlike the snails, animals in the second class have two shells that fit together. Many clams spend their lives under water buried in sand or mud. Their foot is shaped like a shovel and is used to burrow in the sand.

How do buried clams get food? Figure 7-16 shows you. These animals usually have two tubes that stick out of the sand into the water. The animal takes in water through one tube. Any food in the water is filtered out by the animal. The water is then forced out through its other tube.

The octopus in Figure 7-15b and squid and cuttlefish belong to the third class of soft-bodied animals. Both squid and cuttlefish have shells inside their bodies. The octopus has no shell. In these animals, the muscular foot is divided into arms, or tentacles, surrounding the head. These animals have very well-developed eyes. They are rapid swimmers and move by shooting a jet of water.

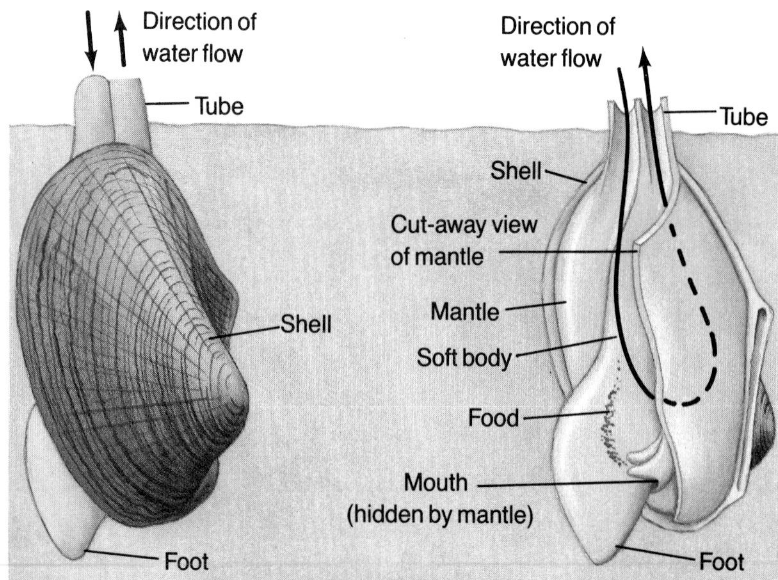

Figure 7-16 Clams filter food out of water. The water is taken into the clam through one tube and forced out through another tube.

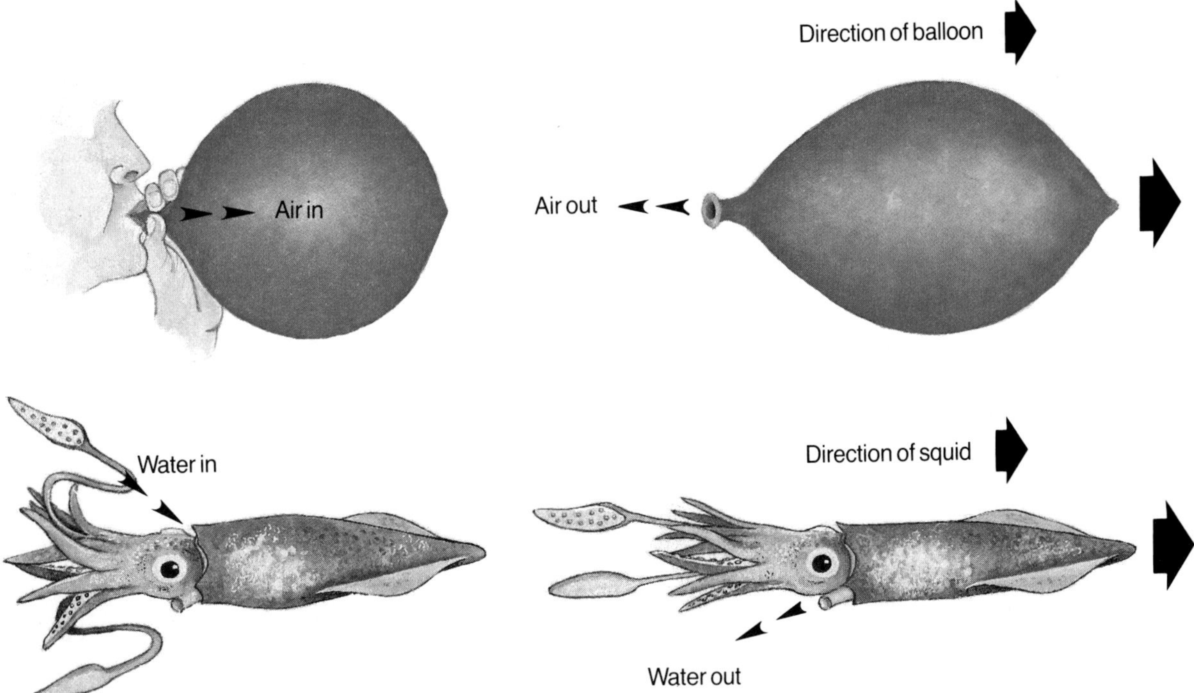

Figure 7-17 Squid move rapidly by shooting a jet of water out of their bodies. Can you think of something else that moves this way?

The way a squid moves is shown in Figure 7-17. The movement is similar to that of a balloon. If you let go of a balloon filled with air, it flies in a direction opposite to that of the air coming out of it. The squid also swims in a direction opposite to that of the water coming out of it.

Let's review the features of soft-bodied animals.

1. All are invertebrates. No backbone is present.
2. They have a soft body covered by a mantle.
3. Most have one or two external shells or an internal shell. A few have no shell.
4. Most have a foot by which they move about.

Check Your Understanding

16. What are four traits of soft-bodied animals?
17. How do clams get food?
18. What traits are used to classify soft-bodied animals?
19. **Critical Thinking:** How does the shell of a snail differ from that of a clam? How does the shell of an octopus differ from that of a clam?
20. **Biology and Math:** Giant squid may grow as long as 49 feet. Estimate how many meters long these squid grow.

These Zebras Live Underwater

Snails, squid, clams, and scallops all belong to the soft-bodied animal phylum. Why are they called soft-bodied if many of these animals have a hard shell? The soft body is there, but you can't always see it because it's inside a protective shell. Your experience with animals in this phylum is probably limited to your use of them as food. A few species can be harmful pests. Snails and slugs damage garden plants and crops. But, the biggest pest of all is a newcomer to North America, the tiny zebra mussel.

Identifying the Problem

Zebra mussels were accidentally brought into the United States from Europe's Caspian Sea. These invertebrate animals are reproducing at a very high rate in the Great Lakes and are causing millions of dollars of damage. They clog up water intake pipes, thus interfering with the normal pumping of water into and out of waterworks and energy-generating plants. How can the zebra mussel be controlled? The answer may lie in knowing something about what this animal looks like and how it lives and reproduces. You are a marine biologist who has been hired by a Detroit power company to find ways of destroying or controlling this water pest. Your plan is to learn about this animal, then to suggest ways of controlling or getting rid of it.

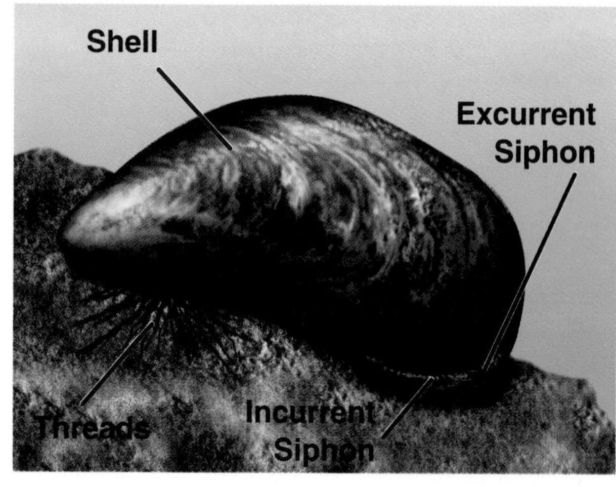

Technology Connection

Scientists have come up with some plans that are controlling zebra mussels. They are using electric fields in lakes to kill the adult as well as the veliger stage. Even shocking the veligers with sound waves is being tested. Sound waves cause veligers to settle onto surfaces before they are mature. This prevents them from continuing with their normal feeding and growth patterns.

Collecting Information

Use references to learn the following about mussels. How do they protect themselves? What do they feed on, and how do they get their food? (Be sure to check on what the incurrent and excurrent siphons do.) How do they get oxygen and what organ do they use for this purpose? How do they attach to surfaces? How do they reproduce? What is a veliger, and how many are formed by one adult?

Carrying Out an Experiment   

1. Prepare a table that lists the italicized terms in steps 2-4. When completed, the table will include the function of each of these terms.
2. Examine the outside of a mussel. Use Figure A as a guide to locating the *shell, threads, incurrent siphon,* and *excurrent siphon.* Note: some of the parts may be hard to see. Record the function of each part.
3. Examine an opened mussel. Use Figure B as a guide to locating the *gills* and *mouth.* Note: these parts may be hard to see. Record the function of these parts.
4. Look at a prepared slide of a mussel *veliger* using low power magnification on your microscope. Does it look anything like the adult stage? Record the function of the veliger.
5. Brainstorm with several classmates about how you could control the growth and reproduction of zebra mussels.
6. Design a plan that could be used to control the zebra mussel in each of these four ways: by attacking its shell, through its filtering of water for food and oxygen, through its use of threads, in its veliger stage. Record all of your plans.

Career Connection

- **Fish Culture Technician** May work in a fish hatchery or laboratory run by a state or federal agency; feeds and raises fish; keeps ponds and tanks clean
- **Environmental Technician** Works for large cities or industrial plants; takes water samples and tests for the presence of chemicals, pollutants, and minerals; observes and reports on life forms present
- **Ocean Technician** Works on land along the coasts or on ships; works with scientists who study the chemical and physical properties of water; tests water samples for dissolved gases, minerals, and pollutants

Assessing Your Results

What are some of the strengths and weaknesses for each of your plans to control zebra mussels? What information are you going to present to the Detroit power company regarding the control or destruction of zebra mussels?

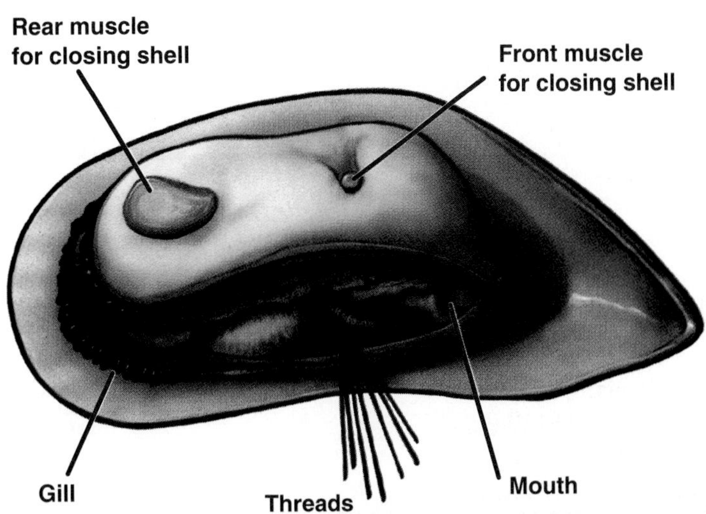

Chapter 7

Review

Summary

7:1 Animal Classification

1. Animals can't make their own food. They can move about, are multicellular, and usually have cells organized into tissues, organs, and organ systems.
2. Animals are divided into nine major phyla based on traits they have in common.

7:2 Sponges and Stinging-cell Animals

3. Sponges are invertebrates that live attached to a surface in water.
4. Stinging-cell animals have stinging cells and hollow bodies with one opening.

7:3 Worms

5. Worms are classified into three major phyla—flatworms, roundworms, and segmented worms.
6. Flatworms have flattened bodies. Roundworms have rounded bodies. Segmented worms have segmented bodies.

7:4 Soft-bodied Animals

7. Soft-bodied animals are invertebrates that have a soft body covered by a mantle. They have one, two, or no shells and usually have a foot for movement.
8. Soft-bodied animals are grouped into classes by the kind of foot, whether or not a shell is present, and the number of shells.

Key Science Words

anus (p. 145)
cyst (p. 143)
flatworm (p. 142)
hookworm (p. 145)
invertebrate (p. 134)
mantle (p. 149)
planarian (p. 144)
pore (p. 137)
roundworm (p. 145)
segmented worm (p. 146)
soft-bodied animal (p. 149)
sponge (p. 137)
stinging-cell animal (p. 139)
symmetry (p. 134)
tapeworm (p. 142)
tentacle (p. 139)
vertebrate (p. 134)

Testing Yourself

Using Words

Choose the word from the list of Key Science Words that best fits the definition.

1. young tapeworm with protective covering
2. the simplest worm
3. simplest animal
4. the arrangement of body parts
5. armlike part that surrounds a stinging-cell animal's mouth
6. thin, fleshy tissue that makes a shell
7. earthworm phylum
8. flatworm that lives on the digested food of another animal

Testing Yourself *continued*

9. animal without a backbone
10. flatworm that is not a parasite

Finding Main Ideas

List the page number where each main idea below is found. Then, explain each main idea.

11. the phylum that has the largest number of animal types
12. how to avoid getting a tapeworm
13. what a planarian uses to detect light
14. how animals such as the hydra and sea anemone capture their food
15. why sponges hold water easily
16. what an anus is
17. how a squid moves
18. how a hookworm enters the body
19. how clams buried in the sand get their food
20. what an earthworm eats

Using Main Ideas

Answer the questions by referring to the page number after each question.

21. How do you determine if a living thing belongs to the sponge phylum? (p. 137)
22. What features make animals different from other living things? (p. 134)
23. How are some stinging-cell animals harmful to humans? (p. 140)
24. How does radial symmetry differ from bilateral symmetry? (p. 134)
25. Why are earthworms important to farmers? (p. 147)
26. In which phylum are leeches found? (p. 146)
27. Why are leeches considered to be parasites? (p. 146)
28. What are the three traits used to group soft-bodied animals? (p. 149)

Skill Review 

For more help, refer to the Skill Handbook, pages 704-719.

1. **Observe:** Use a hand lens to observe a natural sponge and a sponge made by humans. Predict which one will hold more water.
2. **Infer:** What can you infer about the fact that a tapeworm has no mouth?
3. **Interpret diagrams:** Look at Figure 7-9. Describe how a human might get tapeworms from a pig.
4. **Make and use tables:** Make a table that compares the traits of flatworms, roundworms, and segmented worms.

Finding Out More

Critical Thinking

1. What advantage does a hydra have over a sponge in getting food?
2. Why is a tapeworm able to live without a digestive system?

Applications

1. Make a poster to illustrate and describe the life cycle of the dog heartworm. List the symptoms of a heartworm infection.
2. Write a report on how pearls are made by oysters. Explain the difference between natural pearls and cultured pearls.

CHAPTER PREVIEW

Chapter Content

Review this outline for Chapter 8 before you read the chapter.

8:1 Complex Invertebrates
- Jointed-leg Animals
- Spiny-skin Animals

8:2 Vertebrates
- Chordates
- Characteristics of Fish
- Jawless Fish
- Cartilage Fish
- Bony Fish
- Amphibians
- Reptiles
- Birds
- Mammals

Skills in this Chapter

The skills that you will use in this chapter are listed below.

- In **Lab 8-1,** you will make and use tables, observe, and infer. In **Lab 8-2,** you will interpret data, observe, and infer.
- In the **Skill Checks,** you will understand science words and infer.
- In the **Mini Labs,** you will observe.

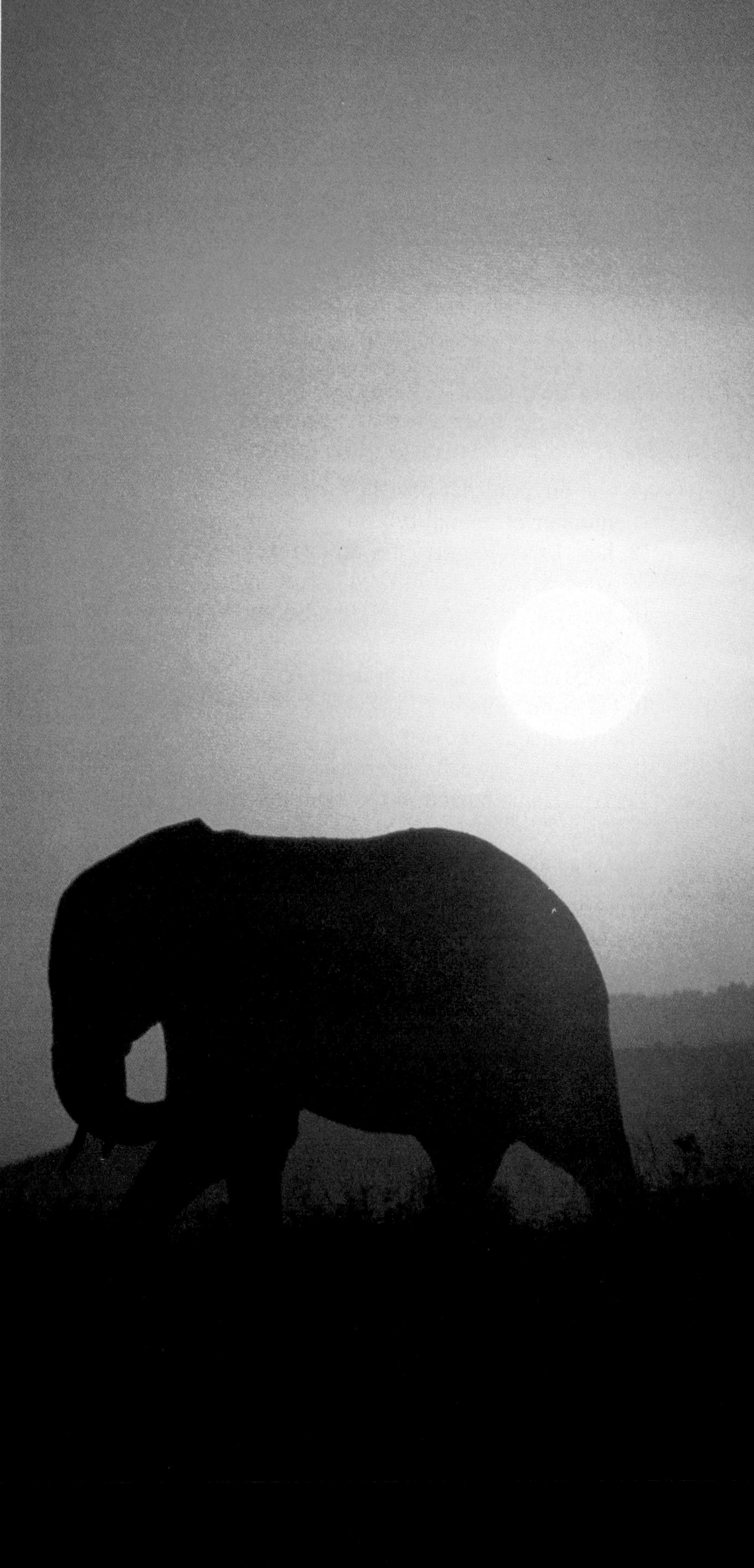

Chapter 8

Complex Animals

The photograph on the left shows one of the largest animals alive today. How can the elephant grow so big? Now look at the smaller photograph on this page. It is a daphnia, or water flea, magnified many times. Daphnia are found in lakes and ponds. Even though a daphnia is only 1 mm long, it is still a very complex animal. Looking into its transparent body, you can see that it is made up of many parts.

How are the daphnia and the elephant alike? In this chapter you will see what these two animals have in common, and how they differ. You will learn the traits of the more complex animals and how they are grouped.

Try This!

What does a fish scale look like? Place a dried fish scale on dark paper. Use a hand lens to count the wide, lighter bands. What do the bands represent?

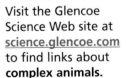

BIOLOGY Online

Visit the Glencoe Science Web site at science.glencoe.com to find links about **complex animals.**

Daphnia

8:1 Complex Invertebrates

Objectives

1. **Identify** the major traits of jointed-leg animals.
2. **Compare** insects with other jointed-leg animals.
3. **Describe** the traits of spiny-skin animals.

Key Science Words

jointed-leg animal
appendage
exoskeleton
molting
antennae
compound eye
spiny-skin animal
tube feet

The animals in the two invertebrate phyla in this chapter are more complex than the animals you studied in Chapter 7. One phylum, containing insects and spiders, is familiar to everyone. Insects are found almost everywhere. It is hard to think of life without insects.

Jointed-leg Animals

Have you ever watched an army of ants moving food back to its nest? Hundreds of these animals work together to carry food that is much larger than they are. Ants crawl along in single file, following a trail that they marked on the way to the food. Back in the nest, hungry ants wait for the hunters to return. Ants are amazing animals!

Ants are jointed-leg animals. A **jointed-leg animal** is an invertebrate with an outside skeleton, bilateral symmetry, and jointed appendages (uh PEN dihj uz). An **appendage** is a structure that grows out of an animal's body. Jointed appendages bend to allow quick movement. Legs, antennae (an TEN ee), and wings are all appendages.

What is an advantage to jointed appendages?

The phylum of jointed-leg animals includes insects, spiders, and crayfish. It has more animal types than any other animal phylum. The graph on page 136 shows that over 80 percent of the known animal types on Earth are jointed-leg animals.

Jointed-leg animals have a segmented body. In many of these animals, the segments are fused into three parts—the head, the thorax (THOR aks), and the abdomen (AB duh mun). Look for these parts in the pictures in this section.

Figure 8-1 Ants work together to carry food back to the nest.

Jointed-leg animals have an exoskeleton. An **exoskeleton** (EK soh skel uht uhn) is a skeleton on the outside of the body. It is made of a hard, waterproof, nonliving substance. It protects the body from drying and injury, and provides a place for muscles to attach.

How does an animal with an exoskeleton grow? The animal must shed its exoskeleton from time to time. Shedding the exoskeleton is called **molting.** After the old skeleton is shed, the body is quite soft. The animal swells by taking in extra water or air while the new skeleton hardens. This swelling gives the animal growing room inside the new skeleton. The animal is not very well protected from its enemies, so it stays hidden until its skeleton hardens.

There are five classes of jointed-leg animals. Crayfish, shrimp, lobsters, crabs, water fleas, and pill bugs are in one class, Figure 8-2. Crayfish and water fleas are found in fresh water. Pill bugs live in damp places on land. The rest of these animals live in the ocean.

Figure 8-3 shows the traits of animals in the first class. Animals in this class have mouthparts that hold, cut, and crush food. They have two pairs of antennae. **Antennae** are appendages of the head that are used for sensing smell and touch. These animals also have compound eyes for seeing. **Compound eyes** are eyes with many lenses. In contrast, you have simple eyes. They each have only one lens.

Most animals in this class have only two body sections. The head and the thorax are fused to make up one section. The abdomen is the second section. These animals usually have five pairs of legs for walking. A clawlike pair of legs at the head grabs and holds food.

The animals in this class are food for many fish and whales. Humans also eat lobsters, shrimp, and crabs.

a

b

Figure 8-2 Crabs (a) and pill bugs (b) belong to the first class of jointed-leg animals.

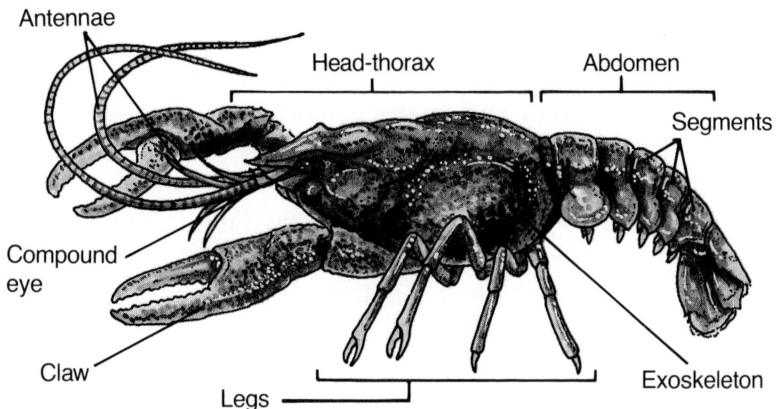

Figure 8-3 Traits of a crayfish

Figure 8-4 Traits of a spider

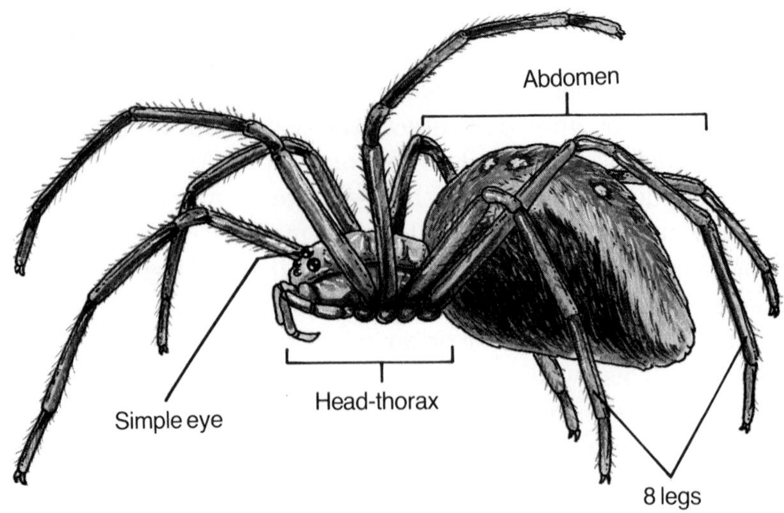

Mini Lab

How Can You Catch a Spider Web?

Observe: Find a spider web covered with dew. Catch the web by placing a piece of black paper against it. Sprinkle some flour over the web on the paper and tap off the excess. What do you see? *For more help, refer to the **Skill Handbook**, pages 704-705.*

The second class of jointed-leg animals includes spiders, scorpions (SKOR pee uhnz), ticks, and harvestmen, also called daddy longlegs. Each has four pairs of walking legs, simple eyes, no antennae, and a body with two sections.

Spiders feed on insects. Most spiders trap insects in webs. The head of a spider has a pair of hollow fangs that connect to poison glands. When a spider traps an insect and bites into it, the poison stuns the insect. You may think that all spiders are dangerous to people. However, only a few spiders, like the black widow, have poison that causes serious sickness in people. Scorpions have a pointed stinger on the end of their abdomen. The stinger contains poison. Scorpions use their stinger for protection against enemies, but they can harm people, too.

The third and fourth classes of jointed-leg animals include the centipedes (SENT uh peedz) and millipedes (MIHL uh peedz). Centipedes and millipedes live on land under rocks or wood. They both have heads, long segmented bodies, and many legs.

Figure 8-5 Centipedes have one pair of legs on each segment.

The name *centipede* suggests that these animals have one hundred legs. Remember that *centi-* means 100. *Pede* means foot. Actually centipedes usually have no more than 30 legs. You can see in Figure 8-5 that each of their segments has one pair of walking legs. The appendages on the first segment are poison claws. They help capture food. Centipedes usually eat insects.

Millipedes are slow-moving animals. They have two pairs of legs on most segments. They do not really have a thousand legs, as the name has you believe. Millipedes do not have poison claws. They usually eat decaying plants.

Insects are the fifth class of jointed-leg animals. There are more kinds of insects than all other animals combined. They live almost everywhere. They live deep in the ocean and high on mountain tops. They live in the air and on the ground, in the tropics and even at the North and South poles. They come in many shapes and colors. They can eat almost anything because they have special mouthparts for chewing, sucking, or lapping.

How can insects eat so many different kinds of food?

How can you tell if an organism is an insect? An insect's body has three main parts as shown in Figure 8-6. The head has two compound eyes and usually three simple eyes, one pair of antennae, and several mouthparts. The thorax has three pairs of walking legs and usually two pairs of wings, so the animal can fly. Insects are the only invertebrates that can fly.

Insects reproduce sexually, with separate sexes producing eggs and sperm. The fertilized eggs are laid in large numbers on leaves or branches of plants.

Insects can be helpful or harmful. Some insects destroy food crops. For example, Mediterranean fruit flies can destroy all of the fruit in an orchard. Moths can eat holes in your clothing when they feed on oils in the fabric or food particles you have left there. Termites destroy wood in buildings and fences. Why do people try to prevent houseflies from touching food? Houseflies carry bacteria that cause diseases.

Some insects help farmers by eating harmful insects. Ladybird beetles, or ladybugs, eat aphids that feed on many types of plants. Bees and other insects carry pollen from flower to flower. Crops must be pollinated in order to reproduce. There are insects that produce food and other useful materials. Honeybees, for example, produce honey.

Bio Tip

Health: Deer ticks carry bacteria that cause Lyme disease. A tick picks up bacteria when it bites a mouse and sucks blood containing the bacteria. Then, when the tick bites a human, the bacteria are transferred. Lyme disease causes rashes and heart damage. It is treated with antibiotics.

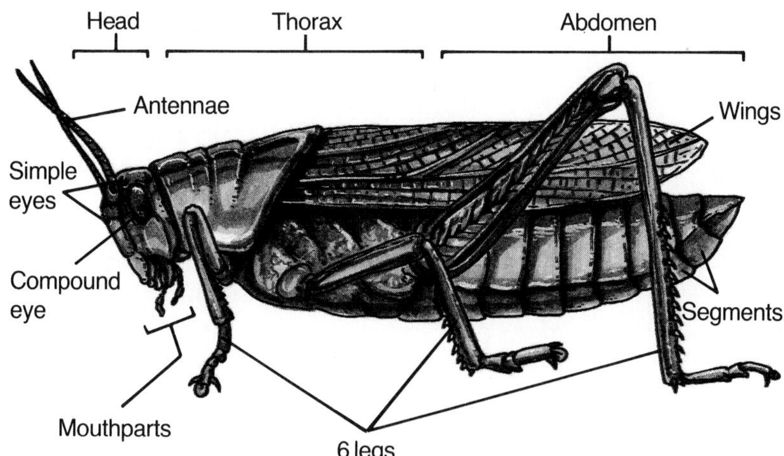

Figure 8-6 Traits of an insect are shown by the grasshopper. How many body sections does an insect have?

Crayfish

Lab 8–1

Problem: What are the traits of a jointed-leg animal?

Skills

make and use tables, observe, infer

Materials

small aquarium with live crayfish

Procedure

1. Make a table similar to the one shown and record your observations as you make them. Use the figure to help you.
2. You will receive a crayfish in a small aquarium from your teacher. **CAUTION:** *Use care when working with live animals.* Leave the crayfish in the aquarium while you make your observations.
3. **Observe:** Touch the crayfish. Observe how it feels. Count the number of body sections.
4. Locate the two pairs of antennae. Are all the antennae the same length?
5. Look closely at the compound eyes. Notice that each eye is located at the top of a short stalk.
6. Locate the first pair of walking legs or claws. On what body section are they found?
7. Observe the other four pairs of walking legs. How are they different from the first pair?
8. Observe the five pairs of swimmerets. On which body section are they found?
9. Find the flipper at the end of the crayfish. How do you think it is used?
10. Return the crayfish to the place your teacher has assigned. Wash your hands.

BODY SECTIONS	PARTS ATTACHED	FUNCTION
Fused head-thorax		
Abdomen		

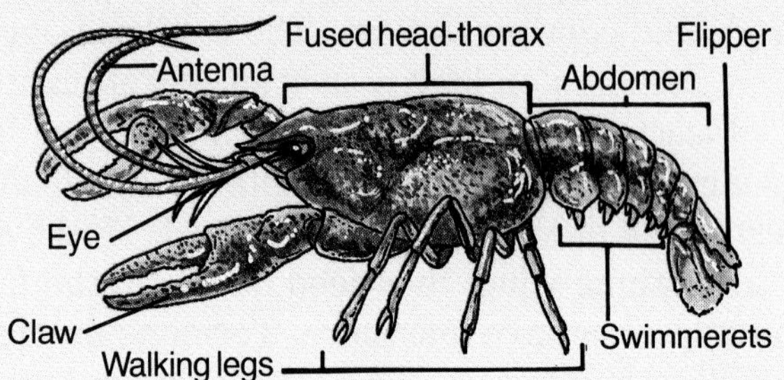

Data and Observations

1. What appendages are attached to the first body section?
2. What appendages are attached to the abdomen?

Analyze and Apply

1. Describe the exoskeleton of the crayfish.
2. What is the function of the exoskeleton?
3. How does the location of the eyes permit the crayfish to look in different directions?
4. **Infer:** How does the structure of the claws aid in getting food?
5. Which parts does the crayfish use to get food?
6. **Apply:** In what way are swimmerets in crayfish and fins in fish similar?

Extension

Observe a spider. Identify the traits that spiders and crayfish have in common.

a

b

c

Figure 8-7 A sea urchin (a), starfish (b), and sand dollar (c) are spiny-skin animals. They all show a five-part body design.

Spiny-skin Animals

The animal in Figure 8-7a looks like a living pincushion. It is a sea urchin. The sea urchin, starfish, and sand dollar belong to the phylum of spiny-skin animals. A **spiny-skin animal** is an invertebrate with a five-part body design, radial symmetry, and spines. Count the arms of the starfish in Figure 8-7b. Sand dollars and sea urchins also show the five-part body design. You may have walked along a beach and seen these animals. They are found only in the oceans. They are very common on rocky shores.

What is the body design of a sand dollar?

What are some other traits of spiny-skin animals? If you look closely at the bottom of a starfish you will see dozens of tube feet, like those in Figure 8-8. **Tube feet** are parts like suction cups that help the starfish move, attach to rocks, and get food. If you ever try to pull a starfish from a rock, you will see how tightly it holds onto the rock with its tube feet. A tube foot works like a medicine dropper. If you squeeze the bulb of a dropper filled with water, you force out the water. If you squeeze the bulb and then place your finger against the open end, you will feel suction when you let go of the bulb. A starfish grips slippery rocks in much the same way. The starfish takes in ocean water and passes it through a series of canals to the tube feet. As this water moves in and out of the tube feet, suction is made and released.

Figure 8-8 Tube feet help a starfish move, attach to rocks, and get food.

Figure 8-9 Starfish can grow back missing arms. What type of reproduction is this?

If a starfish loses an arm it can grow a new one, as seen in Figure 8-9. A whole new animal can grow from one arm if the arm is still attached to part of the central body. This is a form of asexual reproduction. Starfish also reproduce sexually with separate sexes producing eggs and sperm.

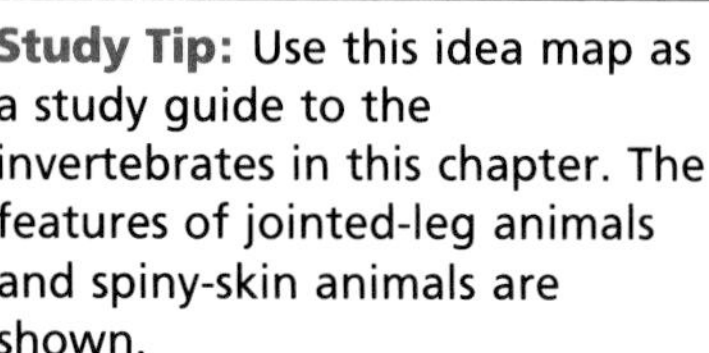

Study Tip: Use this idea map as a study guide to the invertebrates in this chapter. The features of jointed-leg animals and spiny-skin animals are shown.

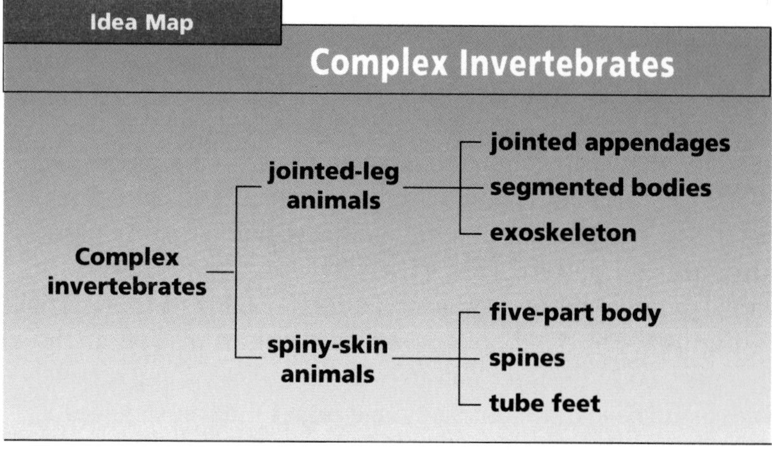

Check Your Understanding

1. List three traits of the jointed-leg animals.
2. How does an animal with an exoskeleton grow?
3. How do starfish use their tube feet for gripping surfaces?
4. **Critical Thinking:** Why are insects able to live in so many more different environments than other animals?
5. **Biology and Reading:** The scientific name for the jointed-leg animals is Arthropoda. The word *arthropoda* is made from two Greek words. The prefix *arthro-* means jointed. The suffix *-poda* means foot. What part of the body do you think the disease arthritis refers to?

8:2 Vertebrates

You have been learning about some of the phyla that make up the invertebrate animals, those with no backbone. Animals that have a backbone are called vertebrates. All vertebrates belong to the chordate phylum. Vertebrates are the most complex organisms in the animal kingdom.

Objectives

4. **Compare** the traits of jawless fish, cartilage fish, and bony fish.
5. **Describe** the major traits of amphibians, reptiles, and birds.
6. **Identify** the characteristics of mammals.

Key Science Words

chordate
endoskeleton
cold-blooded
gill
jawless fish
cartilage
cartilage fish
bony fish
amphibian
hibernation
reptile
warm-blooded
mammal
mammary gland

Chordates

The chordate phylum is probably the most familiar to you. The structures and ways of life of these animals are most like yours. They are your food, your pets, farm animals, the animals you see all around you. Chordates live in water as well as on land. Some chordates, such as bats and birds, are able to fly. You, too, belong to this phylum.

How do you identify a chordate? A **chordate** is an animal that, at some time in its life, has a tough, flexible rod along its back. The chordate phylum is named for this trait.

The chordate phylum contains the largest animals on Earth, one of which is shown in Figure 8-10. How are chordates able to grow so large? They have an endoskeleton (EN doh skel uht uhn). An **endoskeleton** is a skeleton on the inside of the body. An exoskeleton limits growth because it is on the outside of the body. Animals with endoskeletons don't have this problem.

Figure 8-10 A hippopotamus can grow so large because it has an endoskeleton.

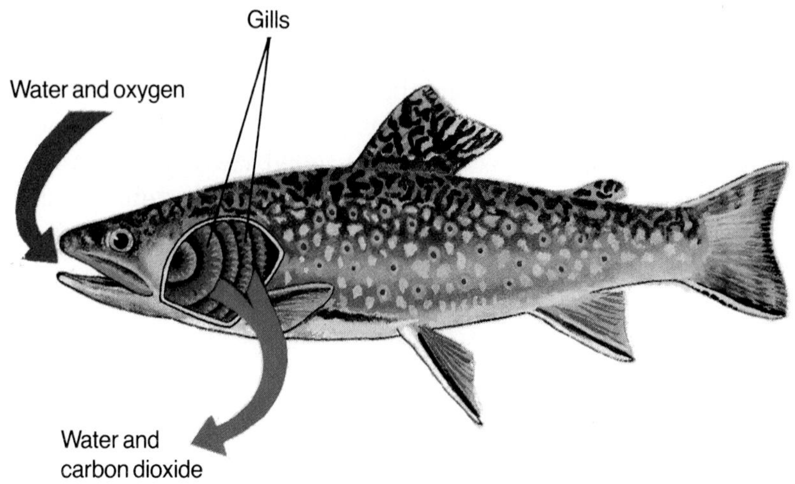

Figure 8-11 Fish have gills that remove oxygen from water. What gas is removed from the fish by the gills?

Vertebrates are chordates. In most vertebrates, the rod along the back is replaced by a backbone. Vertebrates have well-developed body systems. They have a circulatory system with a heart and blood vessels, a digestive system to change food into a useful form, a skeletal system for support, a respiratory system for gas exchange, and a nervous system for control. Vertebrates have large brains, well-developed senses, and are very intelligent animals. The seven vertebrate classes are jawless fish, cartilage fish, bony fish, amphibians, reptiles, birds, and mammals.

Characteristics of Fish

Three of the seven vertebrate classes are fish. All fish have certain traits in common. Fish are cold-blooded vertebrates that live in water and breathe with gills. **Cold-blooded** means having a body temperature that changes with the temperature of the surroundings. Fish have gills on each side of the throat region. A **gill** is a structure used to breathe in water. The fish pumps water into its mouth and out through the gills, Figure 8-11. The gills pick up oxygen from the water as it passes through.

Most fish have scales that cover and protect their bodies. They also have fins that help them swim. The fins steer the fish as it moves its body from side to side through the water. A lateral line runs along each side of the body. The lateral line is an important sense organ for fish because it detects water movement and the presence of objects.

There are three different classes of fish in the chordate phylum. They are jawless fish, cartilage fish, and bony fish.

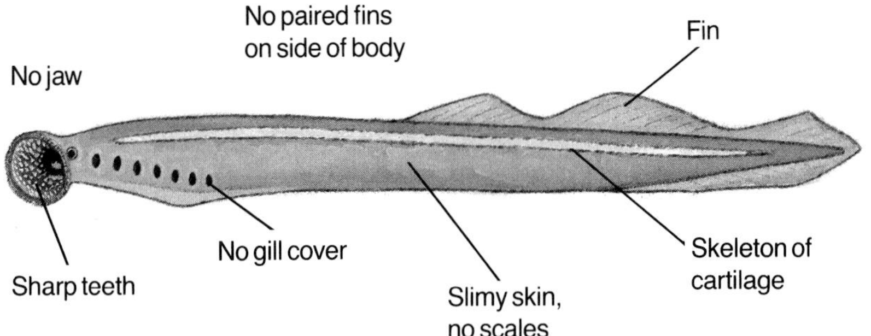

Figure 8-12 Traits of jawless fish are shown by a lamprey.

Jawless Fish

Jawless fish are fish that have no jaws and are not covered with scales. The skeletons of jawless fish are made of cartilage (KART ul ihj). **Cartilage** is a tough, flexible tissue that supports and shapes the body. Your ears and nose tip are made up of cartilage.

The lamprey (LAM pree) shown in Figure 8-12 is a jawless fish. Lampreys have tubelike bodies covered with slime that protects the skin. They don't have paired fins. Lampreys swim by waving their body from side to side.

The lamprey's mouth does not have jaws. It is a round opening lined with toothlike structures. How does the lamprey eat? Many lampreys attach themselves to the sides of other fish with their sharp, toothlike structures. The lamprey cuts a hole through the skin and sucks out the blood and body fluids. Are lampreys parasites?

Cartilage Fish

Cartilage fish are fish in which the entire skeleton is made of cartilage. They have no bone. Unlike the jawless fish, cartilage fish have jaws, toothlike scales, and paired fins. Sharks and rays are cartilage fish. The shark in Figure 8-13 shows these traits.

What kind of skeleton do cartilage fish have?

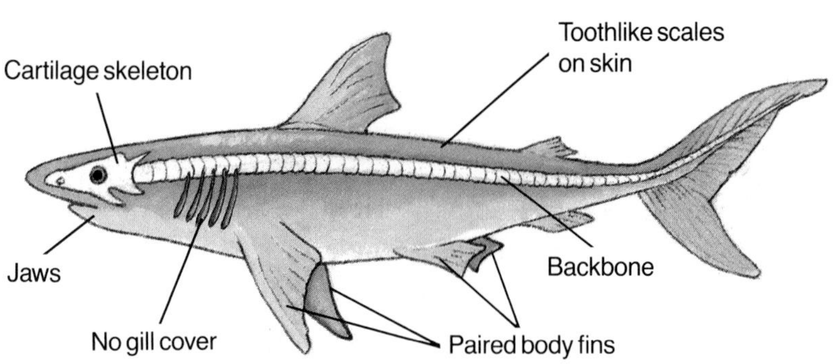

Figure 8-13 Traits of cartilage fish are shown by a shark.

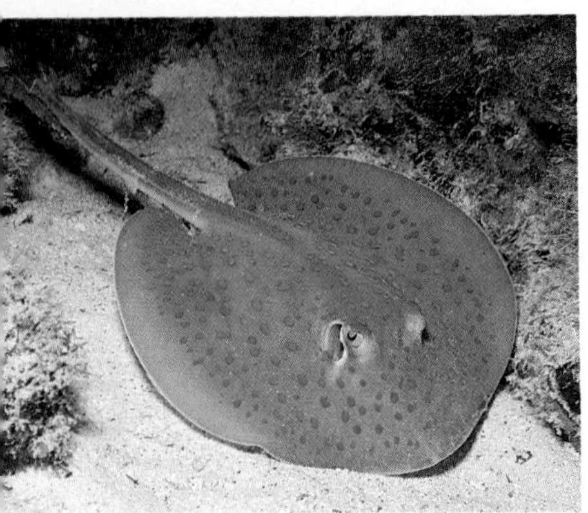

Figure 8-14 A ray has a flat body that is adapted for living on the ocean floor.

Sharks have slim bodies and paired fins that help with movement and balance. Sharks are fast swimmers.

Rows of sharp teeth that slant backward cover the shark's jaws. The teeth help sharks hold and cut up their food. Most sharks eat other animals that live in the ocean. However, the whale shark eats only protists.

Rays are flat and live on the ocean bottom, Figure 8-14. They eat fish and invertebrates in the ocean. Most rays are harmless to humans, but some stingrays have whiplike tails with stingers that can cause a painful wound.

Bony Fish

Most fish known today belong to the class of bony fish. **Bony fish** have skeletons made mostly of bone, not of cartilage. Many fish that you eat, such as perch, bass, flounder, and trout, are in this group. You may have seen some of these fish in a supermarket.

Figure 8-16 shows the important traits of bony fish. Most bony fish have smooth, bony scales that are covered with a slimy coating, but some, like the catfish in Figure 8-15, have slime-covered leathery skin instead. The slime and scales protect the fish from infections and from enemies. The slime also makes it easier for the fish to glide through the water.

Figure 8-15 Catfish have leathery skin instead of scales.

Bony fish have a flap that covers and protects the gills. Jawless fish and cartilage fish do not have this flap. Most bony fish have a swim bladder. A swim bladder is a baglike structure that fills with gases and helps the fish float and go up and down in the water. Fish can change the amount of gas in the swim bladder. As the bladder fills with gas, the fish rises. As the gas is let out, the fish goes deeper.

Most fish need water to reproduce. The female lays large numbers of eggs in the water. The male fish deposits sperm in the water, and some of the eggs are fertilized when the sperm swim to the eggs.

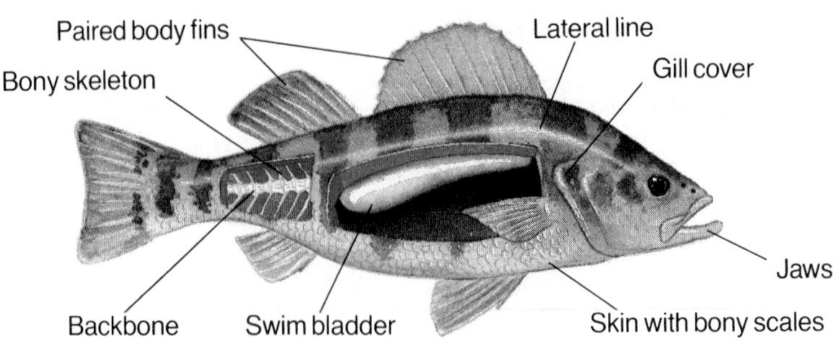

Figure 8-16 Traits of a bony fish. What is the function of the swim bladder?

Amphibians

Amphibians (am FIHB ee uns) include frogs, toads, and salamanders. An **amphibian** is an animal that lives part of its life in water and another part of its life on land. Usually, young amphibians live in water and adult amphibians live mostly on land. Amphibians that live mostly on land still need water. Without water, an amphibian's skin dries out. Amphibian eggs are laid in water so they won't dry out. The female lays large numbers of eggs. The male deposits sperm in the water. Only a few eggs are fertilized.

Amphibian young look very different from the adults. Notice that the tadpole in Figure 8-17b is different from an adult frog in many ways. Tadpoles must live in water and breathe with gills. The adults have lungs and can take in oxygen from the air. They also take in oxygen through their moist skins and the linings of their mouths.

Frogs and toads are probably familiar to you. Figure 8-17a shows some of their traits. Most frogs and toads do not have tails. They have broad mouths with long, sticky tongues for catching insects. They have two pairs of legs. The hind legs are much larger and more powerful than the front legs, so the animal can jump great distances. Frogs and toads have webbed feet for swimming. Their eyes stick out from their heads. This trait allows them to hide under the water with only their eyes showing above the surface. When an insect flies by, the frog flicks out its sticky tongue and catches the insect.

Skill Check

Understand science words: amphibian. The word part *amphi* means both. In your dictionary, find three words with the word part *amphi* in them. *For more help, refer to the **Skill Handbook,** pages 706-711.*

a

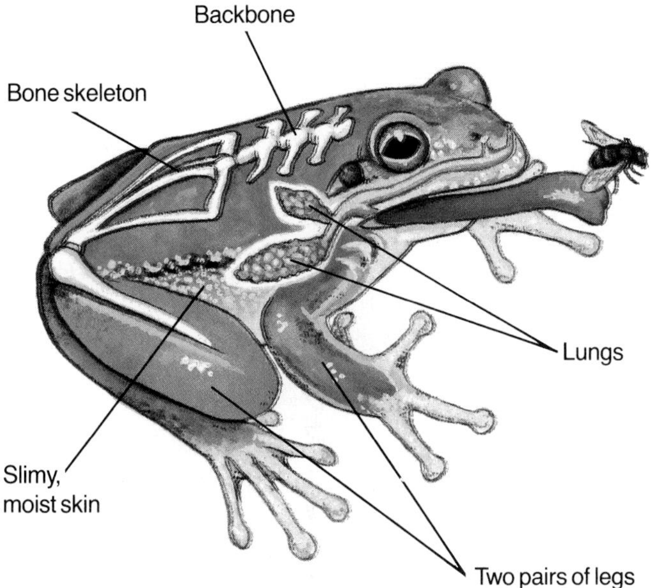

b

Figure 8-17 The traits of an amphibian are shown by a frog (a). A tadpole (b) looks very different from an adult frog and has a different way of life.

a

b

Figure 8-18 A cave salamander (a) and a snake (b) are adapted for living on land.

Salamanders are amphibians with a tail. Their two pairs of legs are about the same size, Figure 8-18a. They live only in moist places. Some salamanders keep their gills throughout their lives and live in water, even as adults.

Amphibians are cold-blooded vertebrates. Their body temperature drops as the temperature of their surroundings drops. Because they are cold-blooded, amphibians become inactive during cold weather. The state of being inactive during cold weather is called **hibernation.** Animals that hibernate eat no food and use only a little oxygen. Their energy needs are met with stored fat.

Amphibians help control insects. Amphibians also are used in life science studies and in medical research. They are eaten by snakes, turtles, birds, and mammals.

Reptiles

You may have seen a snake at one time and thought it looked slimy. If you touched the snake, you probably were surprised to find that it was not slimy at all. Snakes are reptiles. A **reptile** is an animal that has dry, scaly skin and can live on land. Reptiles were the first vertebrates to live and reproduce entirely on land. Snakes, lizards, turtles, crocodiles, and alligators are all reptiles.

What are the traits of reptiles? Look at Figure 8-19. Reptiles are cold-blooded vertebrates with a backbone and an endoskeleton. Their scaly skin protects them, prevents water loss, and keeps them from drying out. Some reptiles are covered by hard plates instead of scales. They have well-developed lungs for breathing air. Most reptiles have two pairs of legs and clawed toes, but snakes and some lizards don't have legs. Most reptiles can move quickly.

Skill Check

Infer: Compare reproduction in reptiles with reproduction in fish and amphibians. Explain why reptiles lay far fewer eggs than amphibians or fish. *For more help, refer to the* ***Skill Handbook,*** *pages 706-711.*

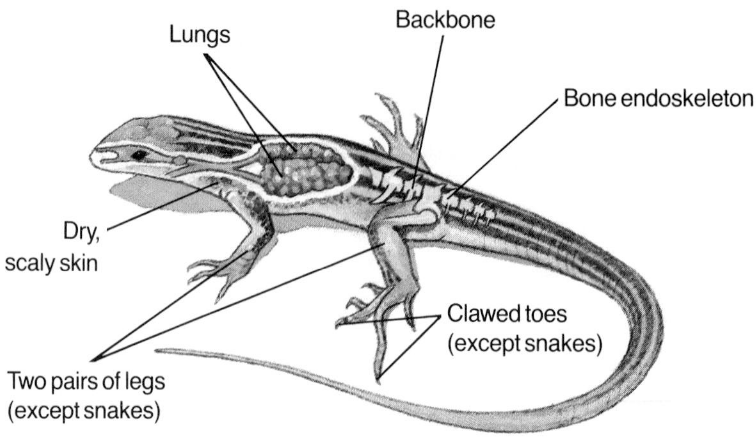

Figure 8-19 A lizard shows the traits of a reptile.

They use their claws for running and climbing, and for digging nests in the soil. A reptile egg has a tough, leathery shell that protects it and keeps it from drying out. Because reptile eggs have shells, they can be laid on land instead of in the water. Even though reptiles do not need to live in water, many still do. Alligators, crocodiles, and turtles, for example, spend a good part of their lives in or near the water.

How are reptile eggs protected?

Reptiles eat insects and pests such as rats and mice. In some areas of the world, people eat reptiles and their eggs.

Mini Lab

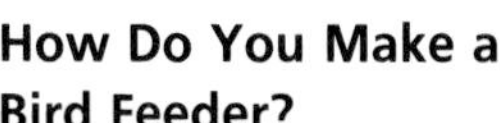

How Do You Make a Bird Feeder?

Observe: Nail bottle caps onto a piece of wood. Fill the caps with peanut butter or birdseed. Hang the feeder from a tree with string. Observe the birds that come to the feeder. *For more help, refer to the **Skill Handbook,** pages 704-705.*

Birds

Birds are vertebrates that have wings, a beak, two legs, and a covering of feathers over most of their bodies. Some of the traits of birds are shown in Figure 8-20. Like reptiles, birds have scales, but the scales are only on their legs. Birds also have claws on their toes. They have well-developed lungs. Female birds, like reptiles, lay eggs with shells from which the young develop.

Most birds are well adapted for flying. They have hollow bones, which makes them light in weight. Birds have powerful muscles to move their wings. Some birds, such as the ostrich, do not fly. Ostriches, however, can run as fast as 48 km an hour, about the speed of a racing bike.

Birds are warm-blooded. **Warm-blooded** means controlling the body temperature so that it stays about the same no matter what the temperature of the surroundings. Feathers help the body keep a constant body temperature.

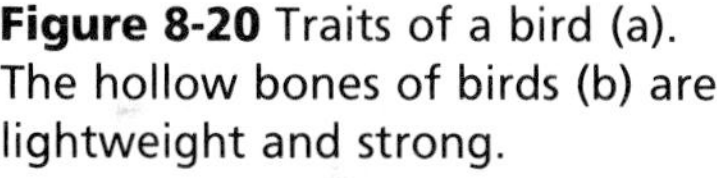

Figure 8-20 Traits of a bird (a). The hollow bones of birds (b) are lightweight and strong.

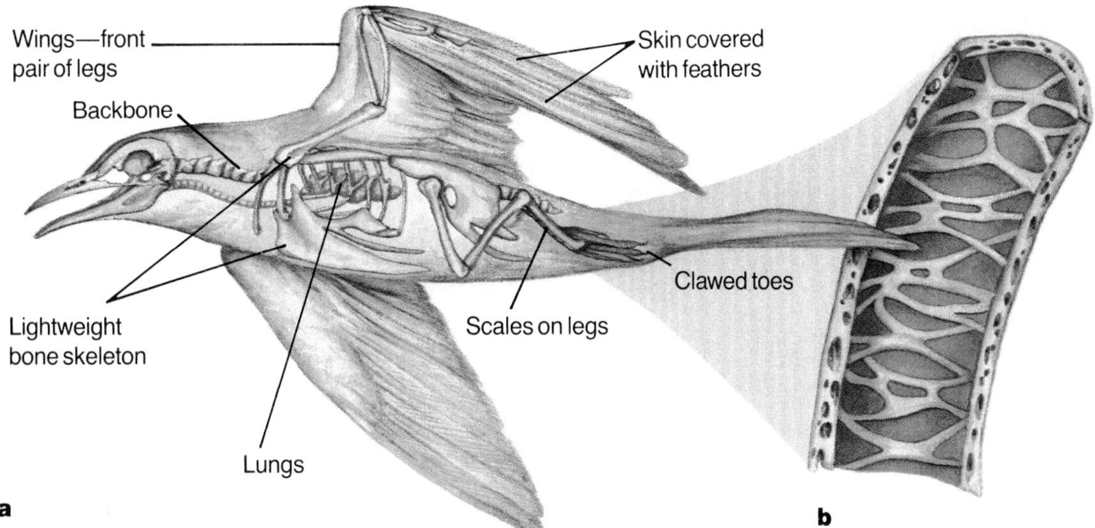

Feathers

Lab 8–2

Problem: What is the structure of feathers?

Skills

interpret data, observe, infer

Materials 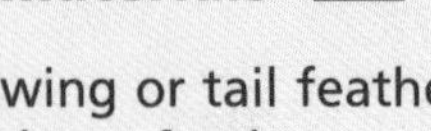

wing or tail feather
down feather
hand lens
scissors
metric ruler

Procedure

1. There are two kinds of feathers. Contour feathers are found on a bird's body, wings, and tail. Down feathers lie under the contour feathers and insulate the body. Look at a contour feather with a hand lens. The hard center tube is the shaft.
2. Copy the data table. Cut 2 cm off the end of the contour feather shaft. **CAUTION:** *Use care when using scissors.* Observe the cut end with the hand lens. Record your observations.

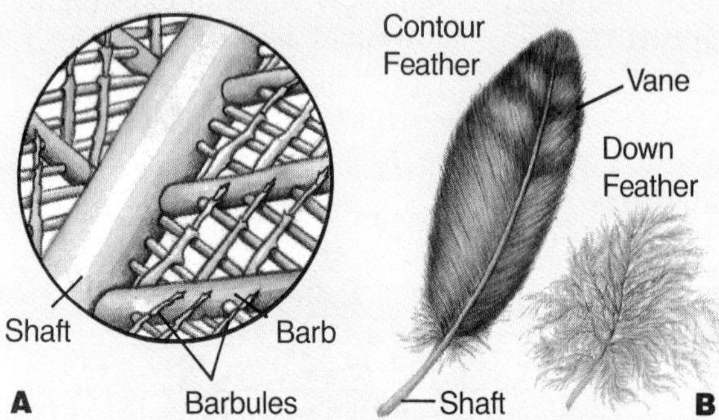

3. **Observe:** Examine the vane with the hand lens. Compare what you see with Figure A. How are the tiny strands of the feather hooked together? This gives the feather strength for flight.
4. Hold the contour feather by the shaft and fan yourself. Hold the down feather by the shaft and fan. Describe what you feel.
5. Observe the shape of the down feather with a hand lens. Describe how it feels. Compare the feather with Figure B.

Data and Observations

1. What connects a contour feather's barbs?
2. Explain any differences observed when you fanned the air with the two feathers.

Analyze and Apply

1. List the parts of a contour feather.
2. How does the shaft of the contour feather aid in flight?
3. **Infer:** How does the structure of a down feather help insulate a bird?
4. **Apply:** What structures in reptiles are like feathers? Are their functions similar?

Extension

Observe pictures of a bird wing and a bat wing. How are they similar and how are they different?

FEATHER	PARTS	OBSERVATIONS
Contour	shaft	
	vane	
	barb	
	barbules	
Down	strand	

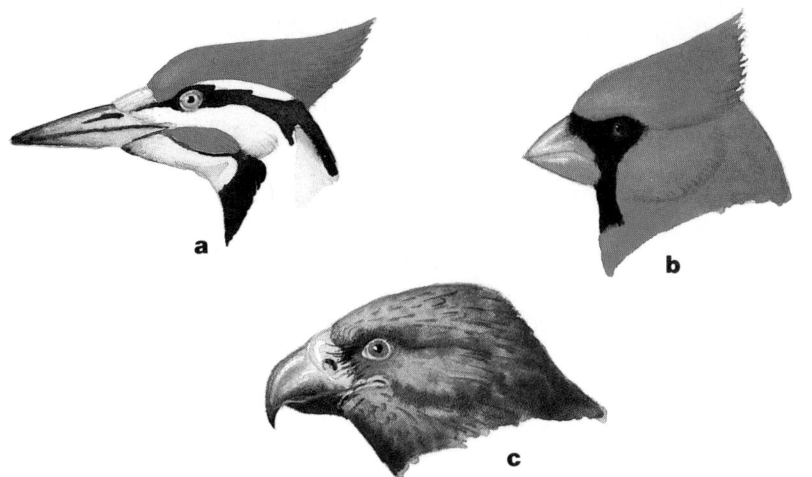

Figure 8-21 The shape of a bird's beak is related to the kind of food it eats. Woodpecker (a), cardinal (b), and hawk (c). What food does each bird eat?

Birds have no teeth. Instead, they have beaks that they use to get food. The kind of beak a bird has is related to the kind of food it eats. Figure 8-21 shows some examples. Woodpeckers have beaks that cut through tree bark to expose insects. Sparrows and cardinals have thick beaks that crack seeds. Hawks have sharp beaks that tear meat.

Birds are important to us in many ways. They help farmers by eating insects and the seeds of weeds. Some birds eat rats and mice. Chickens and ducks provide eggs and meat. Some people have birds as pets. Many people enjoy watching and feeding wild birds. The bald eagle is the national bird in the United States. Do you know the state bird for your state?

Bio Tip

Leisure: Millions of people in the United States enjoy bird watching. Each year, bird lovers spend over $500 million on birdseed, $100 million on bird feeders, boxes, and baths; and $18 million on field guides.

Mammals

What traits are found in the animals in Figure 8-22? First, they have hair and are warm-blooded. The hair helps keep a constant body temperature. Second, the females have milk glands with which they nurse their young. These animals are mammals (MAM ulz). A **mammal** is an animal that has hair and feeds milk to its young. You are a mammal.

Most animals that you have studied hatch from eggs. Mammals usually develop inside the mother's body and are born alive. After birth, young mammals feed on milk produced by the mother's mammary glands. **Mammary** (MAM uh ree) **glands** are body parts that produce milk. Male mammals also have mammary glands, but they don't produce milk. Besides providing milk, all mammals care for their young.

Figure 8-22 Mammals feed their young with milk from their mammary glands.

a

b

Figure 8-23 Kangaroo young (a) finish developing in the mother's pouch. The duck-billed platypus (b) lays eggs.

How do opossum young finish developing?

TABLE 8–1. CLASSIFICATION OF HUMANS
Kingdom: animal
Phylum: chordate
Subphylum: vertebrate
Class: mammal
Order: primate
Family: hominid
Genus: *Homo*
Species: *sapiens*

Two groups of mammals whose young do not develop inside the mother's body are the pouched mammals and the egg-laying mammals. Kangaroos and opossums are pouched mammals. The young of kangaroos and opossums are not fully developed at birth. After birth, they crawl into a pouch outside the mother's body. There they attach themselves to the mother's mammary glands and feed on milk until they are more fully developed. Egg-laying mammals, such as the duck-billed platypus shown in Figure 8-23b, lay eggs like those of reptiles.

Now that you have studied the entire animal kingdom, let's see where you fit in. Remember that you are also a chordate and a mammal. You can see the smaller subgroups in Table 8-1. The name of your genus is *Homo*. Your species name is *sapiens*. These names make up the scientific name of humans, *Homo sapiens*. The Latin word *Homo* means man and *sapiens* means wise. Together *Homo sapiens* means wise man.

Check Your Understanding

6. Compare the scales of jawless fish, cartilage fish, and bony fish.
7. How do birds get food?
8. What does being warm-blooded mean?
9. **Critical Thinking:** Why do mammals have a relatively small number of young?
10. **Biology and Reading:** Why can chordates grow larger than any other animals? What limits the size of animals that are not chordates?

Science and Society

Are Animal Experiments Needed?

Millions of animals are used each year for research conducted in the United States and for dissections in high school classrooms. In most cases, the animals are treated with great care. In other cases, however, the animals are mistreated and many die. Sometimes the research itself is fatal and the animals die. Rats and mice are the animals used most often, but dogs, cats, pigs, rabbits, and frogs are also used. Monkeys and apes are used to study diseases such as cancer and AIDS because these animals are similar to humans. Many high school students dissect frogs, worms, fish, and even mammals. These animals are purchased from companies that usually raise them for the purpose of scientific study, although some animals are captured from the wild.

What Do You Think?

1. Many people in the United States protest the use of animals in experiments and in dissections. These people argue that it is cruel and the animals are made to suffer unnecessarily. They also argue that animals are not enough like humans to give useful results. Should animals be used in laboratory experiments? Should experiments be done if they cause the animals to suffer? Should there be strict laws governing the use of animals in labs?

2. Many people defend the use of animals in lab experiments. Doctors studying living animals say the results of experiments allow many human lives to be saved. Vaccines that protect people from diseases were tested on animals before being given to humans. Animals are used to test the safety of drugs and are being used to develop artificial body parts. Do you think human welfare should be placed over animal welfare? If animals are not used, how should new drugs be tested?

3. Many teachers feel that biology students should dissect animals to learn the structure of body systems and how they work. These teachers say that students cannot learn these things by looking at pictures in textbooks. People who disagree say that this is not a good reason for killing animals. Do you think animal dissections are needed for teaching biology students? Does it make a difference whether the animals used are wild animals captured for scientific use or animals raised for sale to schools?

Conclusion: If animals are not used in laboratories, what are the alternatives?

Animal rights activists

Chapter 8 Review

Summary

8:1 Complex Invertebrates

1. Jointed-leg animals have jointed appendages, segmented bodies, and exoskeletons.
2. Insects are jointed-leg animals with one pair of antennae, special mouthparts, and three pairs of walking legs. There are more kinds of insects than all other animals combined.
3. Spiny-skin animals have tube feet and a body design showing five parts.

8:2 Vertebrates

4. There are three classes of fish. Fish have gills. Most fish have scales. The jawless fish and cartilage fish have skeletons made of cartilage. Bony fish have skeletons made of bone. Most bony fish have a swim bladder that helps them go up and down in the water.
5. Amphibians live both in water and on land. Reptiles are animals with dry, scaly skin that can live on land. Birds are warm-blooded. They have feathers, wings, and light, hollow bones.
6. Mammals are warm-blooded and have hair. The females have milk glands with which they nurse the young.

Key Science Words

amphibian (p. 169)
antennae (p. 159)
appendage (p. 158)
bony fish (p. 168)
cartilage (p. 167)
cartilage fish (p. 167)
chordate (p. 165)
cold-blooded (p. 166)
compound eye (p. 159)
endoskeleton (p. 165)
exoskeleton (p. 159)
gill (p. 166)
hibernation (p. 170)
jawless fish (p. 167)
jointed-leg animal (p. 158)
mammal (p. 173)
mammary gland (p. 173)
molting (p. 159)
reptile (p. 170)
spiny-skin animal (p. 163)
tube feet (p. 163)
warm-blooded (p. 171)

Testing Yourself

Using Words

Choose the word from the list of Key Science Words that best fits the definition.

1. animal that lives part of its life in water and another part on land
2. body temperature that stays the same no matter what the temperature of the surroundings
3. tough, flexible tissue that supports and shapes the body
4. shedding of the exoskeleton
5. structure that produces milk
6. used by fish to breathe
7. skeleton inside the body
8. body temperature that changes with the temperature of the environment
9. animal that, sometime in its life, has a stiff rod along the back
10. has dry, scaly skin and lives on land
11. has many lenses
12. has a body design with five parts

Review

Testing Yourself *continued*

Finding Main Ideas

List the page number where each main idea below is found. Then, explain each main idea.

13. how tube feet are used
14. why frogs have long, sticky tongues
15. what a lamprey is
16. why an animal molts
17. which phylum has the biggest animals
18. what an exoskeleton is
19. what sharks eat
20. why amphibian eggs must be laid in water
21. how feathers help a bird
22. what helps a bony fish float
23. what a mammary gland is
24. what kind of egg a reptile has

Using Main Ideas

Answer the questions by referring to the page number after the question.

25. How do insects differ from spiders? (pp. 160, 161)
26. What are two ways that amphibians, reptiles, birds, and mammals are alike? (pp. 169-174)
27. How does a starfish grip slippery rocks? (p. 163)
28. How does an endoskeleton differ from an exoskeleton? (p. 165)
29. Why are vertebrates believed to be the most intelligent animals on Earth? (p. 166)
30. How do centipedes differ from millipedes? (p. 160)
31. What helps sharks hold their food? (p. 168)
32. Why must tadpoles live in water? (p. 169)
33. How are insects helpful? (p. 161)

Skill Review ✓

*For more help, refer to the **Skill Handbook,** pages 704-719.*

1. **Infer:** Frogs have light coloring on their bottom sides and dark coloring on their top sides. Infer how this coloring can be an advantage to the frog.
2. **Observe:** Examine Figures 8-3 and 8-6. How do crayfish and grasshoppers differ from each other?
3. **Infer:** Animals whose eggs are fertilized outside the body must reproduce in water. Why is this so?
4. **Make and use tables:** Make a table to show the traits of the seven vertebrate classes and give examples of each class.

Finding Out More

TECH PREP

Critical Thinking

1. What problems might occur if all the insects in the world were to die?
2. Considering the advantages and disadvantages of endoskeletons and exoskeletons, why do you think there are more kinds of insects than mammals?

Applications

1. Set up a bird feeder at home. Learn what you can do to attract different kinds of birds to your feeder.
2. Write a report about how honeybees communicate with one another. Discuss how communication helps bees make honey.

CONNECTIONS

Biology in Your World

The Animal World

Classification is used to make sense out of all the many kinds of living things in the world. In this unit, you have seen how organisms are similar. You have also looked at some of the relationships among organisms—from simple, one-celled monerans to complex animals.

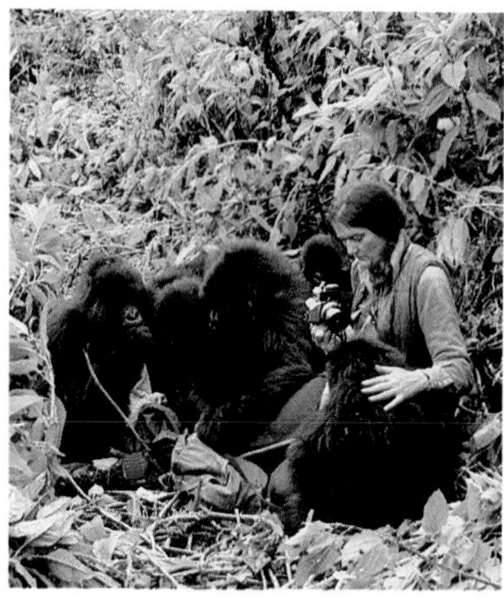

LITERATURE

Primate to Primate

Dian Fossey left her home in Kentucky in 1966 to go to Zaire (ZI uhr) and Rwanda (roo AHN duh) in Africa. There, she studied the endangered mountain gorilla in the Virunga (vuh RUN guh) Mountains.

Fossey learned to accept the animals on their own terms. She was able to learn how gorillas form family groups. She saw behaviors that scientists didn't know existed in gorillas.

During her studies, poachers were a constant threat. A poacher is a person who hunts or takes plants or animals illegally. Fossey fought a never-ending battle to protect the gorillas and their natural habitat.

Fossey studied gorillas for 13 years. Her experiences are discussed in her book, *Gorillas in the Mist.*

GEOGRAPHY

Coral Reefs

Coral reefs are found in tropical oceans. Coral reefs are built by animals called coral polyps. These creatures are related to jellyfish. They live in colonies and make hard limestone skeletons around their soft bodies. As old colonies die, new ones grow on top of the old skeletons.

The largest coral reef is the Great Barrier Reef off the coast of Australia. It is over 1900 kilometers long.

Divers often see crabs, worms, fish, octopuses, and sponges living within a coral reef. Coastline development, oil spills, and pollution threaten these fragile and beautiful communities.

LEISURE

Collecting Fossils

You can have fun searching for fossils of organisms that lived long ago. Fossils are often found in rocks such as shale, clay, sandstone, and limestone that were formed underwater. Since fossils are found in layers, try to find exposed areas such as cliffs, quarries, riverbanks, and road or railway cuts. You will need a shovel, hammer, chisel, notebook, and bags for specimens. You may find fossils of sea lilies, which are related to starfish. Fossil clams, snails, and insects are also common.

Take careful notes, and label and wrap all your specimens. You just might turn up something unusual!

CONSUMER

Tasty Morsels from the Sea

Do you enjoy eating lobster, crab, or shrimp? They are all jointed-leg animals. Scallops, clams, oysters, mussels, and snails are soft-bodied animals. To some people, sea urchins and sea cucumbers, both relatives of the starfish, are delicacies. Fish, such as cod, haddock, and tuna, which are vertebrates, are also enjoyed by many people.

You don't have to live near the ocean or go to a fancy restaurant to enjoy good seafood. Many grocery stores have a seafood department that receives fresh or frozen shipments regularly.

Seafood is best fresh and should be properly cooked. Ask at the store when shipments are received. To be considered fresh, seafood should have been received within the last 24 hours. Keep seafood cold and plan to serve it right away. You can freeze it if it hasn't been previously frozen.

Do you like to help in the kitchen? Try a simple recipe, and surprise your family with a seafood dinner.

Unit 3

CONTENTS

Body Systems—Maintaining Life

What would happen if...

we could live forever? What would it mean? Would we still grow old, or would we stay young? How would everyone's living forever affect the environment? The population would increase at an even greater rate than it does now. We might soon run out of food, space, and resources.

Better medical care and advances in preventing disease have increased the human life span. Further advances may increase it more. However, we can't live forever. In the laboratory, human cells die after a set number of divisions.

Should we think more about improving the quality of life than increasing the length of life?

CHAPTER PREVIEW

Chapter Content

Review this outline for Chapter 9 before you read the chapter.

9:1 What Are the Nutrients in Food?
- Food
- Proteins, Fats, and Carbohydrates
- Vitamins
- Minerals
- Water
- Supplying Nutrients to Your Body

9:2 Calories
- Energy in Food
- Calorie Content of Food
- Using Calories

Skills in this Chapter

The skills that you will use in this chapter are listed below.

- In **Lab 9-1,** you will use numbers, experiment, measure in SI, and make and use tables. In **Lab 9-2,** you will form a hypothesis, experiment, and make and use tables.
- In the **Skill Checks,** you will infer, interpret data, and understand science words.
- In the **Mini Labs,** you will interpret data.

Chapter 9

Nutrition

What influences the types of foods people buy or eat? The photo on the left will give you some clues. It shows foods that can be prepared in a microwave oven. Usually these types of foods can be cooked quickly. Speed of preparation is one thing that influences the kinds of foods we buy. Is price important? Does advertising influence our buying habits? Even ethnic background plays a role in the types of food you eat. You may prefer Greek food as shown in the photo below. Don't stores today have food sections marked ethnic foods?

No matter what influences your choice of food, it's important to your body for several reasons. You will find out how and why food is important as you read this chapter.

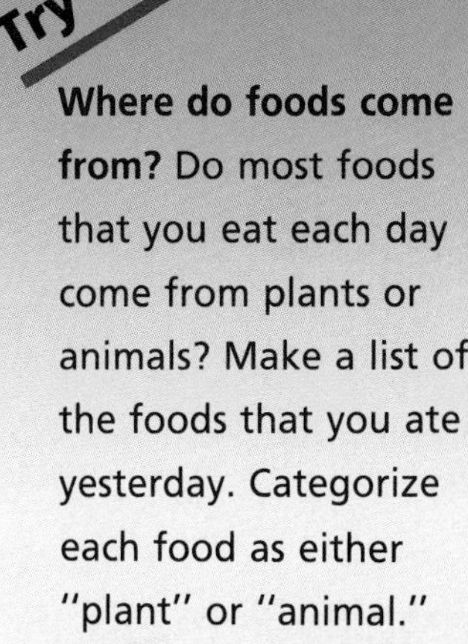

Try This!

Where do foods come from? Do most foods that you eat each day come from plants or animals? Make a list of the foods that you ate yesterday. Categorize each food as either "plant" or "animal." What conclusion can be drawn from your list?

BIOLOGY Online

Visit the Glencoe Science Web site at science.glencoe.com to find links about **nutrition.**

A Greek salad

9:1 What Are the Nutrients in Food?

Objectives

1. **List** six important nutrients that the body needs.
2. **Explain** how each of the six important nutrients is used by the body.
3. **Define** a balanced diet.

Key Science Words

nutrient
nutrition
protein
fat
carbohydrate
vitamin
recommended daily allowance
Percent Daily Value
mineral
balanced diet

If a car runs out of gas, its motor stops. If you don't add fuel to your body, your motor also stops. Food is the fuel for living things.

Food

The cells of your body must be supplied with food or they stop working. Cells use food for growth and repair. Food must be supplied to your body regularly in the right amount and balance. Your body needs fuel to keep working, just as a motor needs fuel to keep running. You and the motor need the right balance of fuel and oxygen to keep working. Without the proper balance of these substances, your body and the motor will slow down or stop.

What are the six nutrients in food?

To keep all your body cells running properly, you must supply them with the right kinds and amounts of nutrients (NEW tree unts). **Nutrients** are the chemicals in food that cells need. The study of nutrients and how your body uses them is called **nutrition** (new TRISH un). The six different nutrients in food are proteins (PROH teenz), fats, carbohydrates (kar boh HI drayts), vitamins, minerals, and water. Each of these six nutrients will be described in the following sections.

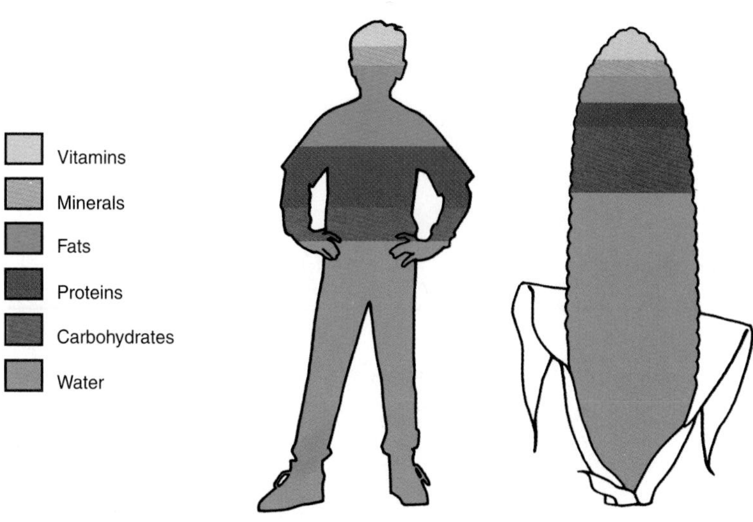

Figure 9-1 The chemicals that make up the human body come from the exact same chemicals found in the foods you eat.

Figure 9-2 These foods are excellent sources of proteins, fats, and carbohydrates.

Proteins, Fats, and Carbohydrates

Proteins are nutrients that are used to build and repair body parts. Proteins make up large parts of tissues such as bone, muscle, and skin. Foods such as meat, eggs, fish, nuts, and chicken supply you with protein.

Fats are nutrients that are used as a source of energy by your body. Fats are compounds that store large amounts of energy. Salad dressing, butter, and cooking oils are foods high in fat.

Carbohydrates are nutrients that also supply you with energy. What then is the difference between fats and carbohydrates? The difference is that your body may store fats. It uses them as an energy source after it first uses up all of your carbohydrate supply. Foods containing starches and sugars, such as bread and fruit, supply you with carbohydrates.

How much of each nutrient is present in your body? Water is an important nutrient that makes up a large percentage of the human body. Figure 9-3 shows that a male's body is 60 percent water. How does this amount compare with what is present in a female? A male has 18 percent fat. Does a female have more or less of this nutrient? Look at Figure 9-3 and compare amounts of carbohydrate for males and females.

How much of each nutrient do you need each day? You might guess that you need more protein than carbohydrate because protein makes up more of your body. You might also guess that you need very little carbohydrate. After all, this nutrient makes up little of your body.

Skill Check

Understand science words: carbohydrate. The word part *carbo* means coal. In your dictionary, find three words that contain the word part *carbo. For more help, refer to the* ***Skill Handbook,*** *pages 706-711.*

Figure 9-3 Compare the average percentages of each nutrient in males and females.

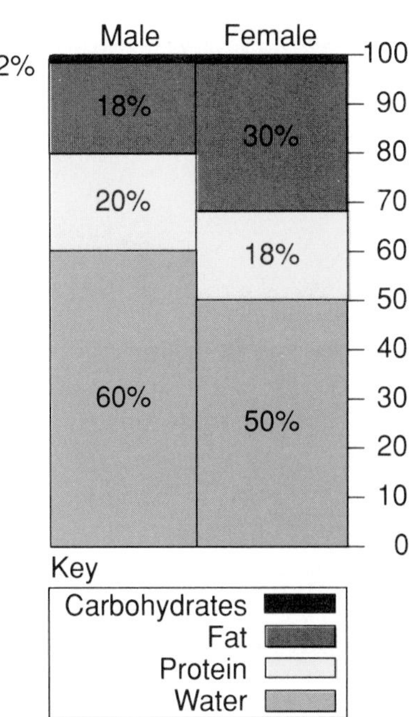

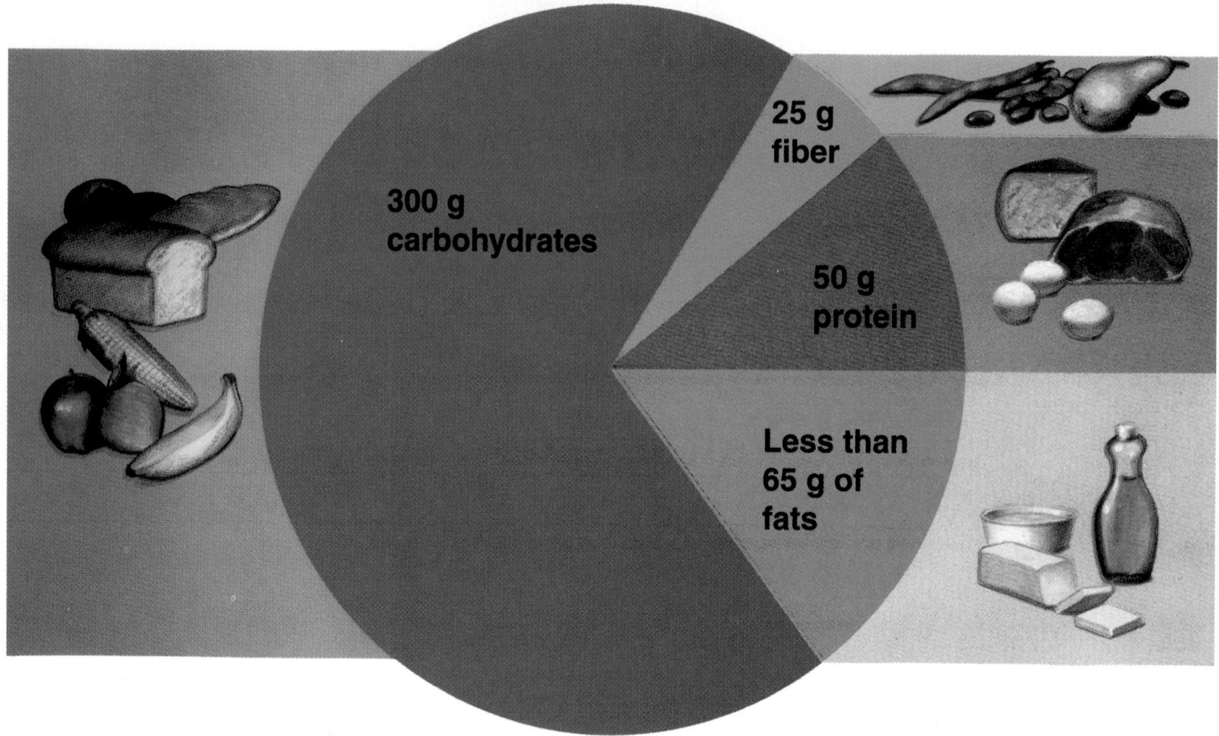

Figure 9-4 This graph shows how much of each nutrient your body needs each day.

Something quite different is true. You need more carbohydrate than protein each day. Why? The body uses carbohydrates quickly. Carbohydrates are the body's main source of energy. They are not stored for a long time. Fats and proteins can be stored for a long time.

Look at Figure 9-4. Which nutrient is needed in the greatest amount each day? Except for fiber, which nutrient is needed in the smallest amount each day? Carbohydrates are needed in the greatest amount and protein is needed in the smallest amount. A person can remain healthy only if he or she takes in the correct amounts of each nutrient. Of course, the amount of each nutrient needed daily may differ slightly for each person. The amount of each nutrient you take in must be balanced with what your body uses. One way to stay healthy is to eat foods that will supply you with the correct amount of each nutrient. Your diet should be from 55 to 65 percent carbohydrates, less than 30 percent fats, and from 10 to 15 percent protein each day for you to stay healthy.

What parts of your body contain or store nutrients? Protein makes up the cytoplasm in your cells. All the body organs are mostly protein. Fats are stored under skin and around body organs. Carbohydrates are stored in the liver and blood.

Mini Lab

Leisure: Planning a meal for a group of friends can be an enjoyable way to learn about nutrients.

Lab 9–1

Milk Nutrients

Problem: What nutrients are present in milk?

Skills

use numbers, experiment, make and use tables, measure in SI

Materials

whole milk
hot plate
medicine dropper
enzyme
stirring rod
250-mL graduated cylinder
200-mL beakers (2)
thermal glove
cheese cloth
funnel

Procedure

1. Copy the data table.
2. **Measure in SI:** Measure 100 mL of whole milk in the graduated cylinder.
3. Pour the milk into a beaker.
4. Place the beaker of milk onto a hot plate and warm the milk. **CAUTION:** *Use a thermal glove when handling hot objects.*
5. Add 20 drops of the enzyme to the milk.
6. Use a stirring rod to mix the milk and enzyme for several minutes.
7. Continue to stir until the milk separates into a solid white part and a liquid part.
8. Line a funnel with a single layer of wet cheese cloth. Position the funnel so that a beaker is below it, as shown in the figure.
9. Pour the milk into the funnel. Wait 2 to 3 minutes until all the liquid has drained through the cheese cloth.

10. The liquid that drains into the beaker is water. Use a graduated cylinder to measure the volume.
11. Record this volume in the data table.
12. The solid material left in the cheese cloth is protein. Record its color and texture in your data table.

Data and Observations

1. What is the percent water in your milk sample?
2. What is the percent protein in your milk sample?

STATE OF MILK	OBSERVATIONS
Original volume of milk used	
Volume of water in milk sample	
Color and texture of protein	
Color of liquid	

Analyze and Apply

1. Why did you add the enzyme to the milk?
2. What two essential nutrients did you study in this lab?
3. Milk can be separated into solid curds and liquid whey. In which step did you do this?
4. **Apply:** If you had tested skim milk instead of whole milk, would nutrients be present in different amounts in the two types of milk?

Extension

Experiment with skim milk and 2-percent milk to measure the same nutrients as those for whole milk.

> **Bio Tip**
>
> **Consumer:** Which has more vitamin C, an orange or a red bell pepper? An orange contains 70 milligrams, while a red bell pepper contains 140 milligrams.

Vitamins

Are you a label reader? Have you ever read a label on a cereal box or bread wrapper? Labels give you useful information about the foods you eat. For example, labels tell you which vitamins are in foods. **Vitamins** are chemical compounds needed in very small amounts for growth and tissue repair of the body. A well-balanced diet will supply you with the vitamins needed each day.

The vitamins have chemical names. For example, ascorbic acid is more commonly known as vitamin C. The chemical name for vitamin A is retinol. When you read a food label, you will see vitamins listed as A, B_1,B_2, B_3, C, and D, or sometimes by their chemical names. Look at Table 9-1 for the names of the vitamins you need each day.

Why are vitamins important?

Certain diseases may appear if too much or not enough of a vitamin is in your diet. Taking too much vitamin A, for example, can cause loss of hair or liver problems. Without enough vitamins, many chemical changes cannot take place within your cells. Table 9-1 lists some vitamins that appear on food labels. The table shows how the body uses the vitamins and describes what happens if you do not have enough of them. For example, if you were to go for many months without fresh fruit and vegetables, you may find your mouth sore, your skin rough, and that you bruise easily.

Study Tip: Use this idea map as a study guide to comparing how the body uses each of the six food nutrients.

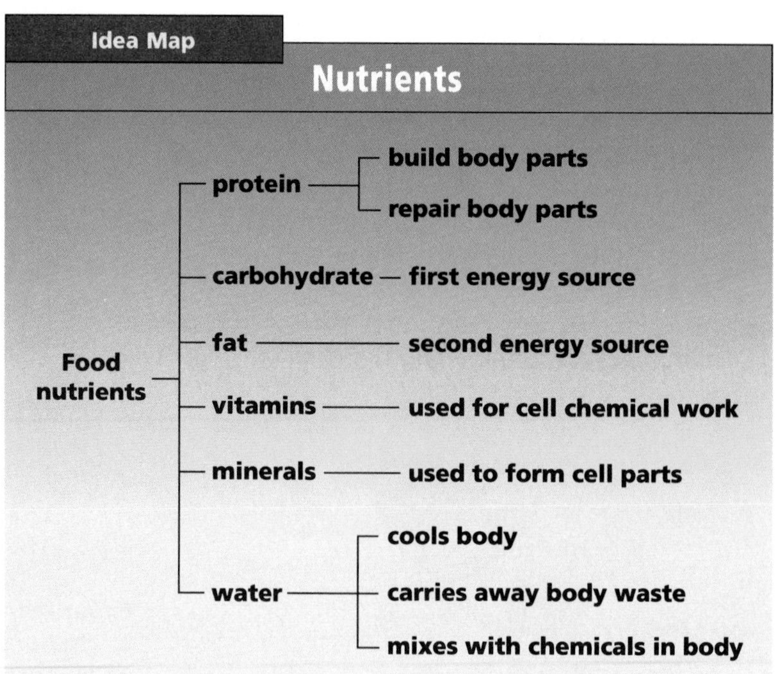

TABLE 9–1. VITAMINS				
Vitamin	**How Used in Body**	**Problems if Not Enough**	**Foods**	**RDA**
A (retinol)	vision, healthy skin	night blindness, rough skin	liver, broccoli, carrots	1000 µg*
B_1 (thiamine)	allows cells to use carbohydrates	digestive problems, muscle paralysis	ham, eggs, raisins	1.5 mg
B_2 (riboflavin)	allows cells to use carbohydrates and proteins	eye problems, cracking skin	milk, yeast, eggs	1.7 mg
B_3 (niacin)	allows cells to carry out respiration	mental problems, skin rash, diarrhea	peanuts, tuna, chicken	20.0 mg
C (ascorbic acid)	healthy membranes, wound healing	sore mouth and bleeding gums, bruises	green peppers oranges, lemons, tomatoes	60.0 mg
D (calciferol)	bone growth	bowed legs, poor teeth	egg yolk, shrimp, milk, yeast	10 µg*

*µg = microgram (1 microgram is 1/1 000 000 of a gram)

Mini Lab

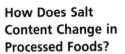

How Does Salt Content Change in Processed Foods?

Interpret data: Find out how much salt is present in a tomato. Compare this with the salt in canned tomato juice. How do these amounts compare? *For more help, refer to the **Skill Handbook**, pages 704-705.*

Notice the column marked RDA in Table 9-1. RDA stands for recommended daily allowance. The **recommended daily allowance** is the amount of each vitamin and mineral a person needs each day to stay in good health. Some of the units in Table 9-1 are listed as mg, or milligrams, and µg, or micrograms. Remember from Chapter 1 that one milligram is one thousandth of a gram. A strand of hair has a mass of more than one milligram. You can see that vitamins are needed in very small amounts.

Let's look at a real food label. Figure 9-5 shows a label from orange juice. The vitamins are highlighted. Each vitamin is listed as a percent. The percent stands for % Daily Value (% DV). **Percent Daily Value** is the percent of nutrient found in one serving of a food compared to 100%. Remember, 100% of each nutrient is needed each day. For example, you eat a cereal that has a % DV of 15% for Vitamin A. This means that you still need 85% more of Vitamin A from other foods to complete the 100% total for that day. % DVs are based on the amount of food that an average person eats in one day.

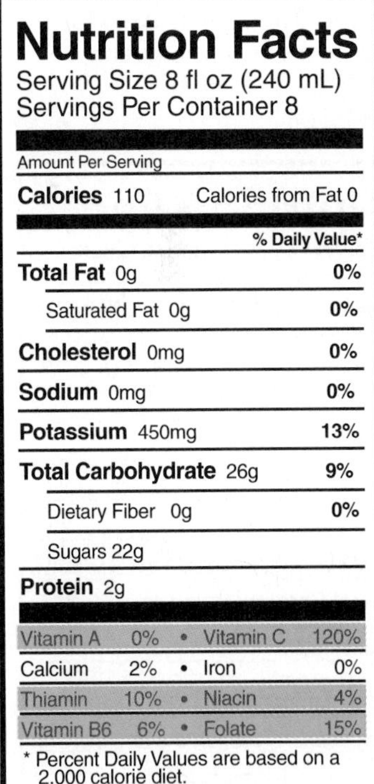

Nutrition Facts
Serving Size 8 fl oz (240 mL)
Servings Per Container 8
Amount Per Serving
Calories 110 Calories from Fat 0
% Daily Value*
Total Fat 0g 0%
Saturated Fat 0g 0%
Cholesterol 0mg 0%
Sodium 0mg 0%
Potassium 450mg 13%
Total Carbohydrate 26g 9%
Dietary Fiber 0g 0%
Sugars 22g
Protein 2g
Vitamin A 0% • Vitamin C 120%
Calcium 2% • Iron 0%
Thiamin 10% • Niacin 4%
Vitamin B6 6% • Folate 15%
* Percent Daily Values are based on a 2,000 calorie diet.

Figure 9-5 Food labels tell which vitamins are present. They are listed in %DV.

TABLE 9–2. MINERALS				
Mineral	**How Used in Body**	**Problems if Not Enough**	**Foods**	**RDA**
Iron	helps form blood cells, helps blood carry oxygen	anemia, feeling tired	liver, egg yolk, peas, enriched cereals, whole grains	10-15 mg
Calcium	helps form bones and teeth	bones and teeth become weak or brittle	milk, cheese, sardines, nuts, whole-grain cereals	800-1200 mg
Magnesium	helps form bones and teeth	muscles twitch	potatoes, fruits, whole-grain cereals	325 mg
Iodine	helps make thyroid gland chemical	causes thyroid gland to enlarge	seafoods, eggs, milk, iodized table salt	150 µg*
Sodium	muscle contractions, nerve messages	dizziness, tired feeling, cramps	bacon, butter, table salt	less than 2400 mg

*µg = microgram (one microgram is 1/1 000 000 of a gram)

Bio Tip

Health: The American Heart Association recommends that you use only 500 mg of salt each day.

Figure 9-6 Food labels tell which minerals are present in one serving. They are listed in %DV and in grams or milligrams.

Minerals

Minerals are nutrients needed to help form different cell parts. Minerals, like vitamins, are chemicals. Your body needs minerals in very small amounts. A diet with a variety of foods provides needed minerals.

Table 9-2 lists five minerals and how they are used by your body.

The table also tells you what can happen when these minerals are missing from your diet. For example, calcium is a mineral needed to form strong bones and teeth. Lack of calcium may cause bones and teeth to become weak and even brittle.

Figure 9-6 shows an actual food label from crackers. The minerals are highlighted. Minerals are shown two ways on labels. First, the quantity of a mineral present in one serving of the food is listed in grams or milligrams, as labeled *A* in Figure 9-6. Second, the % DV is also given for minerals, as labeled *B* in Figure 9-6.

Figure 9-7 You need to replace water that is lost from the body as sweat.

Water

Water is an important nutrient for good health. You learned earlier that the human body is made up of 50 to 60 percent water.

There are three reasons why water is important. First, water cools the body. You see or feel this happening when you sweat, Figure 9-7. Second, many chemicals within the body can combine only with water. Therefore, many chemical changes can take place only in water. Third, water helps carry away body wastes.

The average adult needs about two liters of water each day. Most people don't drink that much water but still get enough of it. Many foods contain water, Figure 9-8. In fact, some foods are over 90 percent water.

Mini Lab

What Are Processed-food Costs?

Interpret data: Find the cost per gram of a raw potato. Compare this with the cost per gram of potato chips. *For more help, refer to the **Skill Handbook,** pages 704-705.*

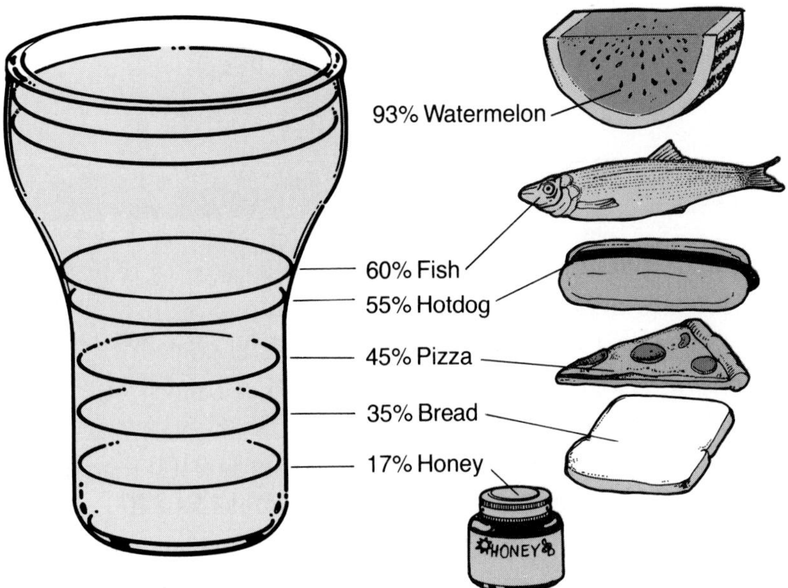

Figure 9-8 Compare the percentage of water in some foods.

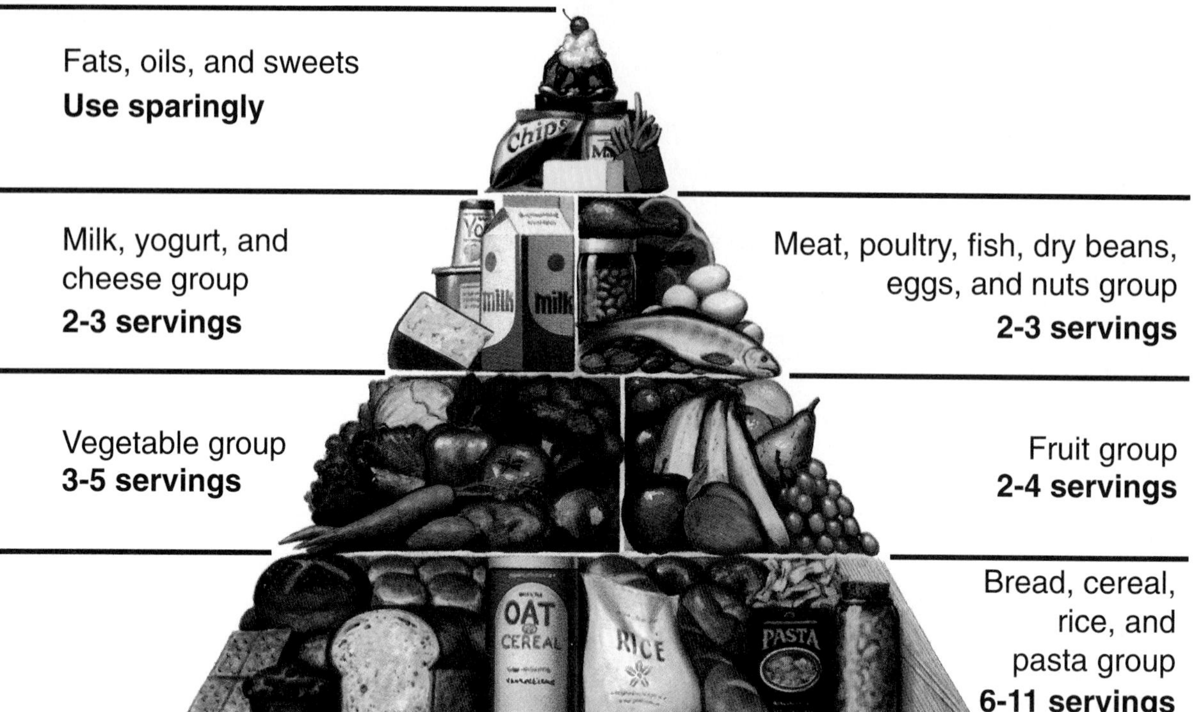

Figure 9-9 A food pyramid shows that most foods each day should come from the base of the pyramid, with very few foods each day coming from the tip of the pyramid.

Supplying Nutrients to Your Body

Most foods contain several nutrients. Whole milk is four percent protein, four percent fat, and five percent carbohydrate. It also has vitamins, minerals, and water.

What is a balanced diet?

How do you know which foods will give you a balanced diet? A **balanced diet** is a diet with the right amount of each nutrient. Figure 9-9 shows foods placed into six groups. Eating the number of servings for each food group every day will help you eat a balanced diet.

Check Your Understanding

1. Name two nutrients that supply the body with energy.
2. How are minerals used by the body?
3. How can you be sure that your diet is balanced?
4. **Critical Thinking:** Why does the body need more carbohydrates than fats or proteins?
5. **Biology and Math:** The %DV of sodium is 2400 mg. Which of the following has about the same mass as 2400 mg: two paper clips, a quarter, a pencil, or a biology book?

9:2 Calories

Did you ever have a soft drink or candy bar when you wanted to eat something that would give you a lift? Besides supplying your body with nutrients, food gives your body energy. You are able to talk, play sports, and read this book because of the energy food gives you. Let's look at how the energy in food is measured and how the body uses food energy.

Objectives

4. **Describe** how a Calorie is used by the body.
5. **Compare** the number of Calories found in different nutrients.
6. **Compare** the number of Calories used in different activities.

Key Science Words

Calorie

Energy in Food

Have you ever noticed how often TV food commercials mention Calories? A **Calorie** is a measure of the energy in food. Foods high in Calories provide a lot of energy. Low-Calorie foods provide less energy. Diet soft drinks usually contain only one Calorie. A candy bar might contain over 200 Calories. Most of those Calories are in the form of fat and sugar.

Food, like wood or coal, can be burned. The heat given off by burning food can change the temperature of water. Scientists describe a Calorie as the amount of heat it takes to raise the temperature of 1000 g of water 1°C, Figure 9-10. If a slice of bread has 70 Calories, what does it mean? It means that if the slice of bread were burned, it would give off enough heat to raise the temperature of 1000 g of water 70°C.

What is a Calorie?

Of course, in your body, food energy is not used to heat water. Food energy is used to keep your body temperature close to 37°C. It is also used to move your muscles, pump your blood, and send messages along your nerves. Food energy is released when your cells carry on respiration.

Figure 9-10 This procedure shows how one Calorie can be measured.

a Start with 1000 g of water at 19°C.

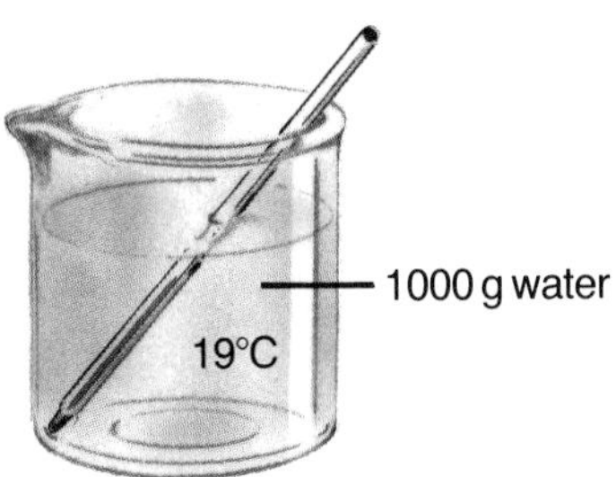

b Burn food to heat the water.

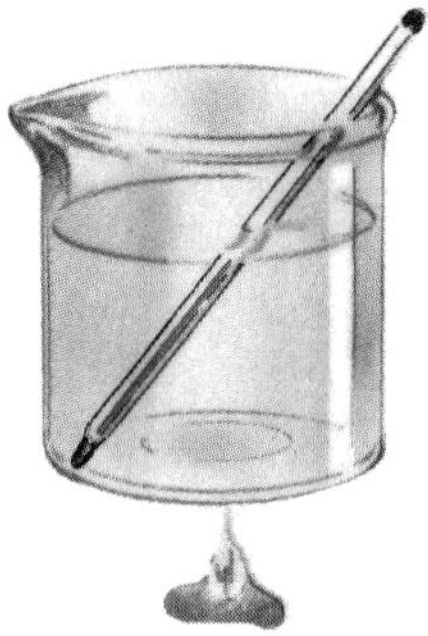

c When one Calorie of food has burned, the temperature of the water is one degree higher.

Energy Foods

Lab 9–2

Problem: Which foods contain sugar?

Skills

form a hypothesis, experiment, make and use tables

Materials

glass slides
razor blade
sugar test paper
food samples for sugar test
droppers
water

Procedure

1. Copy the data table.
2. **Form a hypothesis:** Look at the foods you are to test for sugar. In your notebook, write a hypothesis about which foods contain sugar and which don't.
3. Using a dropper, put a drop of honey on a glass slide. Honey is a food that contains sugar.
4. Touch a piece of sugar test paper to the honey. Wait one minute. When the test paper turns green, this shows that sugar is present.
5. Using a different dropper, put a drop of water on a glass slide. Touch a piece of sugar test paper to the water. Wait one minute. No green color shows that sugar is not present.
6. Using a razor blade, cut a small slice from a sweet potato and touch a piece of sugar test paper to the inside of the slice. **CAUTION:** *Always cut away from yourself when using a razor blade.*
7. Record in your table if a green color appears and if sugar is present.
8. Repeat steps 3 and 4 with maple syrup, molasses, and apple juice. Use a different dropper and piece of sugar test paper for each food tested.
9. Dispose of the food samples according to your teacher's directions.

Data and Observations

1. Which foods contained sugar?
2. Which foods did not contain sugar?

FOOD	GREEN COLOR?	SUGAR PRESENT?
Honey	yes	yes
Water	no	no
Sweet potato		
Maple syrup		
Molasses		
Apple juice		

Analyze and Apply

1. Explain how you can tell if a food
   (a) contains sugar.
   (b) does not contain sugar.
2. What type of nutrient is sugar?
3. Why was it important to test water for sugar?
4. **Check your hypothesis:** Is your hypothesis supported by your data? Why or why not?
5. To which of the six food groups do each of your food samples belong?
6. **Apply:** What is the role of sugar in the diet?

Extension

Experiment using the same foods and iodine solution to test for the presence of starch.

Calorie Content of Food

Foods differ in the amount of energy, or Calories, they contain. How many Calories do you think are in a ham sandwich, a small salad, and a glass of milk?

To find the answer, add the Calories next to the list of foods in Table 9-3.

TABLE 9–3. CALORIES IN CERTAIN FOODS

Foods Eaten	Calories
2 slices of white bread	140 (70 in each slice)
1 spoonful of mayonnaise	70
2 slices of ham	148 (74 in each slice)
1/8 head of lettuce	10
1 slice tomato	6
2 spoonfuls of salad dressing	120 (60 in each spoonful)
1 glass milk	150

This lunch would supply 644 Calories.

You can see from the table that not all foods provide the same number of Calories. However, we have not been comparing equal masses of foods to one another. When we do compare equal masses, we find that fats supply the most Calories. For example, let's compare butter with bread. Look at Figure 9-11b as you read. Butter is mostly fat, and bread is mostly carbohydrate. What would happen if you ate an equal mass of butter and bread? The butter would give your body two and one-half times the number of Calories as you would get from the bread.

Equal masses of protein and carbohydrate are both lower in Calories than fats. Let's compare bread with ham, Figure 9-11a. Remember that bread is mostly carbohydrate. Ham is mostly protein. There are 70 Calories in a slice of bread and 74 Calories in a slice of ham of the same mass. They have almost equal numbers of Calories.

Skill Check

Infer: Cooking oil and lard are both fats. How will the numbers of Calories found in equal masses of these substances compare? *For more help, refer to the **Skill Handbook**, pages 706-711.*

Which foods supply the most Calories?

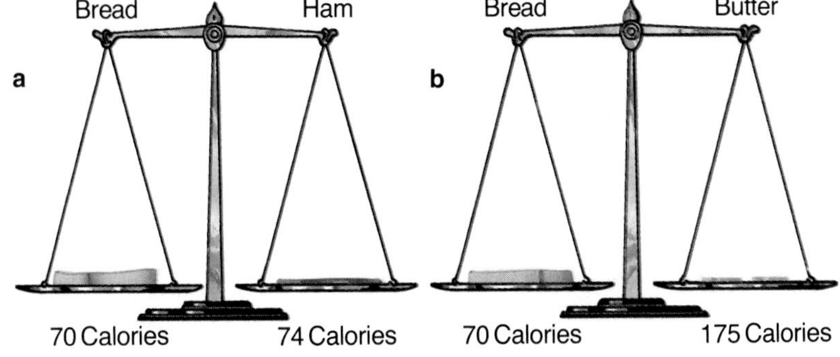

Figure 9-11 Food of equal masses can have very different numbers of Calories.

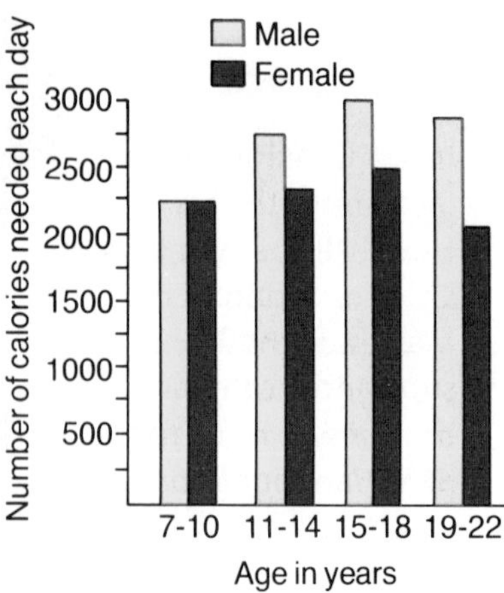

Figure 9-12 Compare the Caloric needs for different ages and sexes. What age group needs the most Calories daily?

Using Calories

Your body needs a certain number of Calories to keep a proper weight. How many Calories does your body need each day? The answer to that question is not simple. Figure 9-12 shows that age and sex of a person are important in figuring Calorie needs. The easiest way to receive the needed number of Calories is to eat different foods from each of the six groups shown on page 192.

How are Calories used by different people? How a person uses Calories depends on his or her size and on how active he or she is. For example, a larger person uses more Calories than a smaller person doing the same activity. Table 9-4 shows that a person who is 73 kg uses 240 Calories walking for one hour. Someone who is 54 kg uses 180 Calories walking for the same amount of time.

The more energy it takes to do an activity, the more Calories a person uses for that activity. For example, a 63-kg person uses 112 Calories standing for one hour. The same person uses 420 Calories playing tennis for one hour. Playing tennis uses more energy than standing.

Taking in too many Calories or too little exercise may result in overweight. If the number of Calories taken in are equal to the number of Calories used, body weight stays about the same. If more Calories are taken in than used, weight is gained. What becomes of the extra Calories? They may be stored in the body as fat.

What happens if more Calories are taken into the body than are used?

What happens if the body takes in less Calories than it uses? The body makes up the difference by using stored Calories, and the person loses weight. Another way of using stored Calories is by exercising.

Figure 9-13 Taking in more or fewer Calories than are needed can result in overweight or underweight. Neither condition is healthy for the body.

TABLE 9–4. CALORIES USED IN 1 HOUR

Type of Activity	Mass of Person		
	54 kg	63 kg	73 kg
Sleeping	48	56	64
Sitting	72	84	96
Eating	84	98	112
Standing	96	112	123
Walking	180	210	240
Playing tennis	380	420	460
Bicycling fast	500	600	700
Running	700	850	1000

Skill Check

Interpret data: Study Table 9-4. What happens to the number of Calories used as the type of activity becomes more difficult? *For more help, refer to the* ***Skill Handbook,*** *pages 704-705.*

Figure 9-14 The number of Calories you will burn during a one-hour walk depends on your body mass.

Check Your Understanding

6. How are Calories important to the body?
7. Which nutrient supplies the body with the most Calories?
8. Who needs more Calories, a 16-year-old boy or a 16-year-old girl?
9. **Critical Thinking:** A 73-kg person and a 54-kg person run at the same speed for one hour. Which person will use more Calories?
10. **Biology and Math:** Tom has a mass of 63 kg. In a typical day, Tom sleeps for 8 hours, sits for 6 hours, plays tennis for 2 hours, eats for 2 hours, walks for 3 hours, and stands for 3 hours. If Tom consumed 2500 Calories on a given day, would he have gained, maintained, or lost weight?

Bio Tip

Health: Are You Overweight? Calculate your Body Mass Index (BMI). Divide your weight in pounds by your height in inches. Divide the answer again by your height in inches and multiply by 705. If your BMI is over 25, you are considered overweight.

TECH PREP **APPLYING TECHNOLOGY**

Get the Fat Out

Fat is a nutrient needed by humans. So why all the fuss about reducing fat intake in the diet? It is known that a diet high in fat contributes to obesity. It also is a major cause of heart disease. How much fat should you include in your diet if you are trying to eat healthy? Check the labels on any food package, and you will see that fat should be limited to less than 65 grams each day. Labels also tell you that you should limit your saturated-fat intake to less than 20 grams of the 65-gram total. Saturated fat is a type of fat that differs chemically from another type, called unsaturated fat. An ideal diet, therefore, should consist of less than 45 grams of unsaturated fat and less than 20 grams of saturated fat each day.

Identifying the Problem

You are a nutritionist. People come to you for advice about the kinds of foods they should and should not be eating. A high school student asks you the following questions. Are "light" fats, such as those found in "light" butter, really light? Do they contain less fat than "regular" fats? Do they really help to cut down on the amount of fat a person eats each day? You want to experiment with "regular" and "light" butter to see if there is any difference between the two product types. This way, as a nutritionist, you will be better able to answer the questions raised earlier by the student.

Nutrition Facts

Serving Size 1 Tbsp. (14g)
Servings Per Container about 32

Amount Per Serving	
Calories 100 Calories from Fat 100	
	% Daily Value*
Total Fat 11g	**17%**
Saturated Fat 8g	**38%**
Cholesterol 30mg	**10%**
Sodium 0mg	**0%**
Total Carbohydrate 0g	**0%**
Protein 0g	
Vitamin A 8%	

Not a significant source of dietary fiber, sugars, vitamin C, calcium, and iron.

*Percent Daily Values are based on a 2,000 calorie diet.

Technology Connection

Nutritionists and scientists are trying to design new artificial foods that look and taste like fats but aren't fats. Their goal is to find a chemical that provides few Calories, but tastes like fat. The newest chemical is a protein product called Simpless. It is being used in "fat-free butter" spreads. Another fat-free chemical currently being used is a product called Olestra. If you were to design a fat substitute, what are some of the qualities that this artificial food would have to show? Explain why these qualities might be needed.

Collecting Information

Visit your local grocery store. Look in the dairy section for "regular" and "light" butter. Butter is a fat. See if the food labels on these products give you a clue as to how the "light" butter differs from the "regular" butter. Also, see if the two butter types differ in the amounts of saturated and unsaturated fats. Make a hypothesis as to how the makers of these products can make butter "light."

Carrying Out an Experiment   

1. Weigh out equal masses of both butter types, about 30 grams of each.
2. Place each sample into a separate labeled test tube.
3. *Put on protective goggles.* Place the tubes into a beaker half-filled with water. Place the beaker and tubes onto a hot plate and heat the water until the butter melts.
4. When each sample is totally melted, turn off the hot plate. Leave the tubes in the water bath and allow the tubes to cool overnight. Examine the tubes the next day in class.
5. Make a prediction as to what the different layers of solids and liquids are that can be seen in the test tubes. Record any measurements, observations, or other data that you feel might be important in explaining how "light" and "regular" butters differ. Diagrams may also be made as part of your data reporting.

Career Connection

- **Food Processing Technician** May work as an inspector for a state or federal agency; helps with quality control by testing for the presence of bacteria, impurities, or toxic substances in foods
- **Food Scientist** Works to improve processes of canning, freezing, storing, packaging, and distributing of foods; may also work in the research and development areas of food processing companies
- **Dietetic Technician** Assists dietitians; work is usually in hospitals, day care centers, nursing homes, or schools; may supervise the ordering, storing, preparing, and serving of food

Assessing Your Results

Compare and contrast the two butter samples. Explain how "light" and "regular" butter may differ. Based on your experimental results, how would you answer the student's original questions? Do "light"-fat foods really contain less fat than "regular"-fat foods? Could "light"-fat foods help to cut down on the amount of fat a person uses each day? Explain how.

Chapter 9

Review

Summary

9:1 What Are the Nutrients in Food?

1. The body needs six different nutrients to stay healthy—proteins, fats, carbohydrates, vitamins, minerals, and water.
2. Fats and carbohydrates are needed for energy. Proteins are needed to build and repair body parts. Vitamins are needed for chemical work in the cell. Minerals are needed to form certain cell parts. Water cools the body. Chemical changes take place in water. Water carries away body wastes.
3. Choosing a variety of foods from the six nutrients will provide you with a balanced diet.

9:2 Calories

4. A Calorie is a measure of the energy in food.
5. When comparing equal masses of food, fats provide more Calories than any other nutrient.
6. How a person uses Calories depends on his or her age, sex, size, and the kind and amount of activity a person does. Using less Calories than the amount taken in may result in weight gain. Using more Calories than the amount taken in may result in weight loss.

Key Science Words

balanced diet (p. 192)
Calorie (p. 193)
carbohydrate (p. 185)
fat (p. 185)
mineral (p. 190)
nutrient (p. 184)
nutrition (p. 184)
Percent Daily Value (p. 189)
protein (p. 185)
recommended daily allowance (p. 189)
vitamin (p. 188)

Testing Yourself

Using Words

Choose the word from the list of Key Science Words that best fits the definition.

1. study of how your body uses food
2. measure of food energy
3. nutrient used to build and repair body parts, such as skin and bone
4. eating the proper amount of all nutrients each day
5. type of chemical in food needed by all living cells
6. amount of a vitamin needed each day
7. energy food, such as oil or butter
8. nutrient, such as niacin, that aids a cell with its chemical work
9. body's main source of energy
10. nutrients, such as calcium or iron

Finding Main Ideas

List the page number where each main idea below is found. Then, explain each main idea.

11. why the number of Calories used during an activity is different for each person

Testing Yourself *continued*

12. how to find the total number of Calories in a meal
13. the roles of vitamins C and D
14. why water is important
15. the nutrients present in dairy foods
16. what happens to unused Calories in the body
17. ways in which vitamins differ from minerals

Using Main Ideas

Answer the questions by referring to the page number after each question.

18. Which food group supplies the body with vitamin C, carbohydrates, water, and magnesium? (pp. 190, 192)
19. How do the numbers of Calories used change as activity increases? (p. 197)
20. Which nutrients are used for growth and repair of skin and bones? (p. 185)
21. What does it mean if a certain food gives you 20 percent of your %DV for niacin? (p. 189)
22. What happens to body weight when
    (a) more Calories are used than taken in? (p. 196)
    (b) more Calories are taken in than used? (p. 196)
23. Which foods would you suggest a person with anemia and muscle twitches eat in an attempt to correct the condition? (p. 190)
24. How many Calories are in equal masses of
    (a) fat and carbohydrate? (p. 195)
    (b) fat and protein? (p. 195)
    (c) carbohydrate and protein? (p. 195)
25. What are the six important nutrients found in food? (p. 184)

Skill Review ✓

*For more help, refer to the **Skill Handbook,** pages 704-719.*

1. **Infer:** Males and females have different amounts of fat in their bodies. What does this tell you about how different sexes store this nutrient?
2. **Use numbers:** Calculate the cost of one gram of canned, cooked kidney beans compared with the cost of an equal amount of plain, uncooked kidney beans. (A can costs $0.85 for 200 gm, while a bag of beans costs $1.25 for 1000 gm.)
3. **Measure in SI:** Use a pan balance to determine the masses of two foods listed in Table 9-3 that have similar Caloric values.
4. **Interpret data:** List the number of Calories that a 63-kg person will use up in one hour of walking and running.

Finding Out More

TECH PREP

Critical Thinking

1. If you were a vegetarian, how would you obtain the proteins necessary for a healthy body? Plan two well-balanced vegetarian dinners.
2. If a person takes vitamin and mineral supplements daily, do they still need a variety of foods? Explain your answer.

Applications

1. Build a tin can calorimeter. Measure the Calories in different foods.
2. Use food charts to plan two different dinners—one low in sodium, and one low in carbohydrates.

CHAPTER PREVIEW

Chapter Content

Review this outline for Chapter 10 before you read the chapter.

Skills in this Chapter

The Skills that you will use in this chapter are listed below.

- In **Lab 10-1,** you will separate and control variables, form a hypothesis, observe, and design an experiment. In **Lab 10-2,** you will classify and interpret diagrams.
- In the **Skill Checks,** you will experiment and understand science words.
- In the **Mini Labs,** you will use numbers and experiment.

Chapter 10

Digestion

Have you ever watered a very dry houseplant? If so, you probably sprinkled water around the base of the plant rather than flooding one spot with water. Look at the photo on the left. By wetting a large area of soil, the amount of water immediately absorbed by the plant is increased.

Your body needs to absorb nutrients from food just as a houseplant needs to absorb water. This is the job of various organs of your digestive system. One such digestive organ is the small intestine. The picture below shows an enlarged view of the lining of the small intestine. How does the irregular surface of this organ aid in the digestion of food? As you read this chapter, you will discover the answer to this question.

Try This!

Does particle size affect the rate at which a substance dissolves? Predict which would dissolve faster, a copper sulfate crystal or an equal mass of powdered copper sulfate. Check your prediction by placing these substances in two beakers, each with 100 mL of warm water.

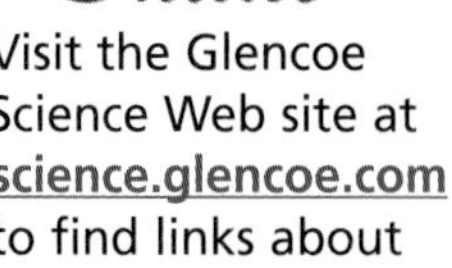

BIOLOGY Online

Visit the Glencoe Science Web site at science.glencoe.com to find links about **digestion.**

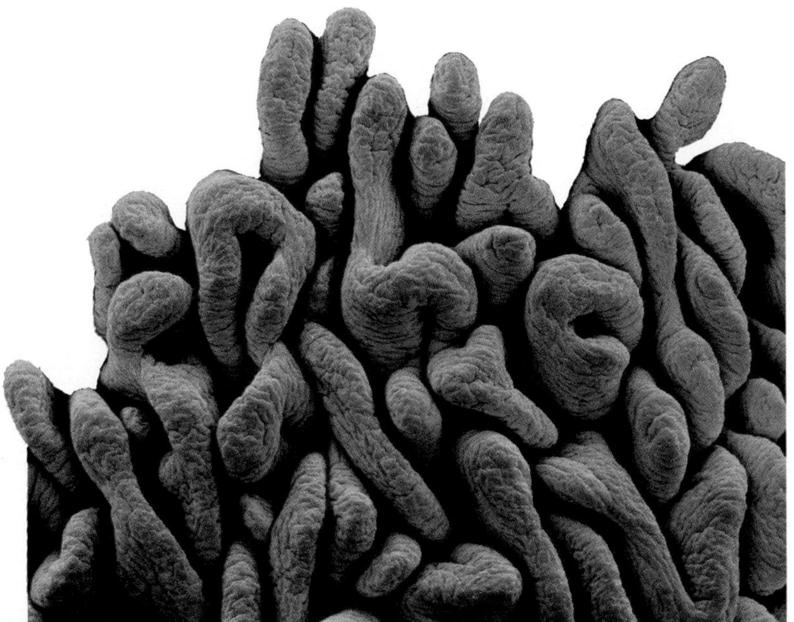

High power view inside the wall of the small intestine

10:1 The Process of Digestion

Objectives

1. **Relate** the importance of the digestive system.
2. **Compare** a physical change and a chemical change in the digestive system.
3. **Explain** the role of enzymes in a chemical change.

Key Science Words

digestive system
digestion
physical change
chemical change
enzyme

When food is first eaten, it is not in a form that can be used by cells in the body. Food must be broken down into a form that cells can use. All cells need food for energy, growth, and repair. How does the body break down food into a form that is usable to cells? The body changes food into a usable form by means of the digestive system. The **digestive system** is a group of organs that take in food and change it into a form the body can use.

Breakdown of Food

Why does food need to be digested?

Much of your digestive system is hollow. Food moves through it just as materials pass through a tube. To understand how the digestive system works, we can compare it to a factory.

Raw materials are brought into a factory. The materials are transported to different locations in the factory. At each location, the materials are changed. The final products that leave the factory are quite different from the original raw materials. For example, iron ore enters a factory as a raw material. As it moves through the factory, the iron ore is changed to steel. The steel is later changed into new products. Figure 10-1 shows these events.

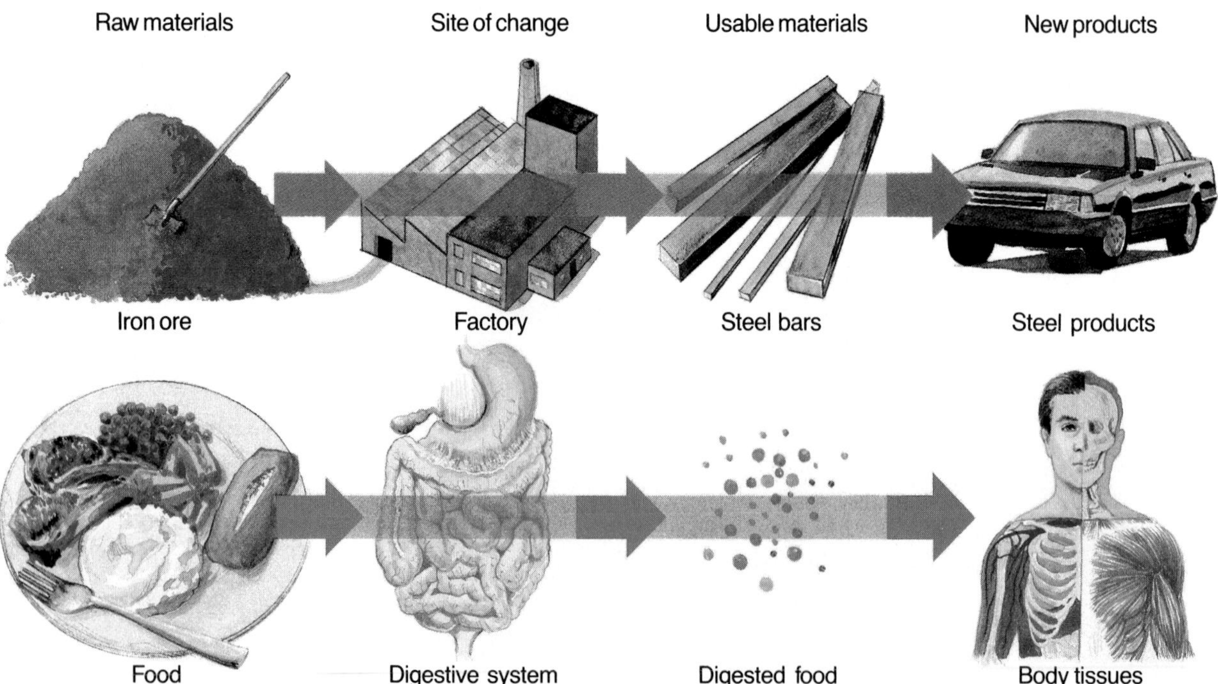

Figure 10-1 The human digestive system is like a factory. Raw materials are used to make products.

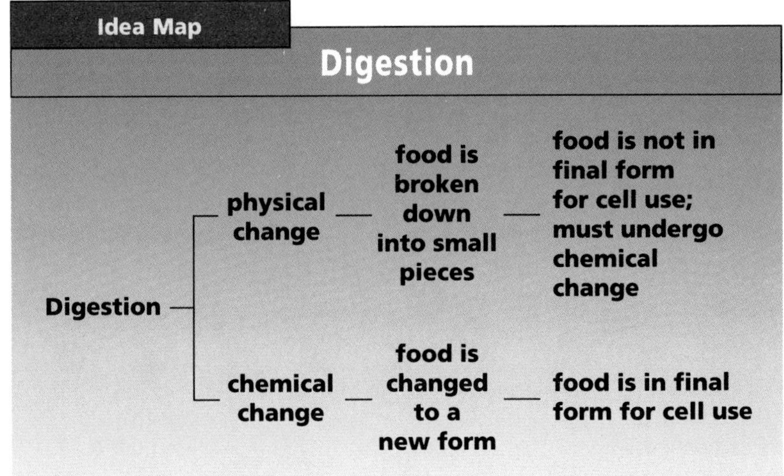
Idea Map
Digestion
Digestion
physical change
food is broken down into small pieces
food is not in final form for cell use; must undergo chemical change
chemical change
food is changed to a new form
food is in final form for cell use

Study Tip: Use this idea map as a study guide to identifying the differences between physical and chemical changes. Which type of change turns food into a form the body can use?

Your digestive system works in a way similar to the factory. Food is the raw material of your digestive "factory." It enters your digestive system through your mouth. As soon as food enters your body, it begins to change form. The changing of food into a usable form is called **digestion.** As food leaves your mouth, it enters the long tube that makes up your digestive system. The food continuously changes form as it passes through the tube. Finally, the food is in a form that can be used to supply the body with energy or to help make bone, skin, or muscle cells. These are the products of your digestive factory.

Skill Check

Experiment: Compare how long it takes for a sugar cube and a crushed sugar cube to dissolve in a glass of warm water, with or without stirring the water. *For more help, refer to the **Skill Handbook,** pages 704-705.*

Physical and Chemical Changes

The digestive system is like a long tube. It is narrow in some places and wide in others. Food is broken down as it moves through this sometimes wide, sometimes narrow tube.

How is food broken down as it passes through the digestive tube? There are two ways. They are physical changes and chemical changes. Physical and chemical changes to food occur at different times and places along the digestive tube.

What causes a physical change in food?

A **physical change** occurs when large food pieces are broken down into smaller pieces. The food is still in the same form. Only the size and shape of food particles are different. Chewing by the teeth causes a physical change in food. Grinding and mixing also cause a physical change. These physical changes occur farther on down the long tube of the digestive system.

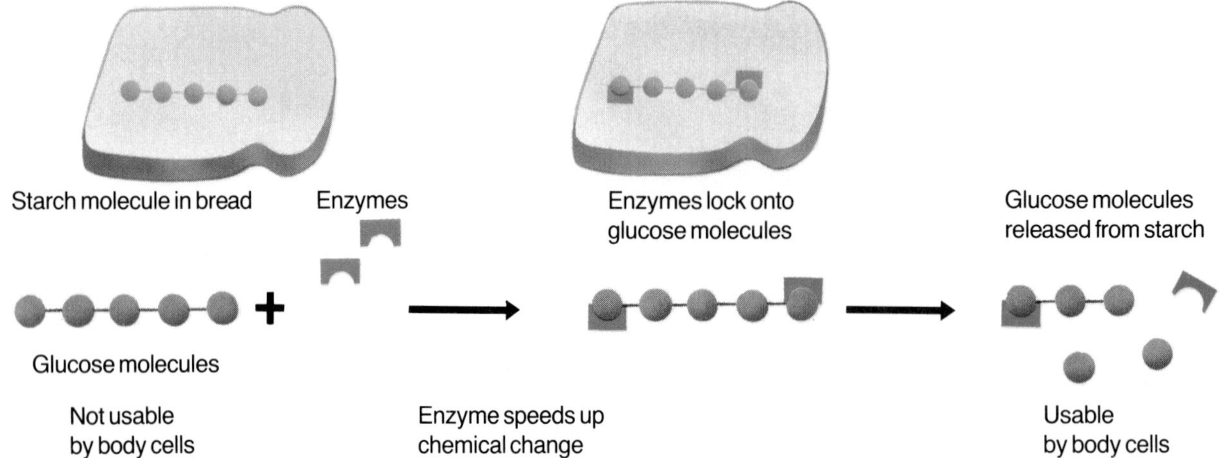

Figure 10-2 Starch is a large molecule made of many glucose molecules strung together. An enzyme breaks apart the starch molecule, releasing glucose molecules that are usable by the body.

How do enzymes help chemical changes?

A chemical change occurs when food changes form. How does a chemical change differ from a physical change? A **chemical change** turns food into a form that cells can use. Your digestive system makes chemicals that help with the chemical changes. These chemicals are added to food as it moves through the organs of your digestive system. The chemicals are called enzymes (EN zimes). **Enzymes** are chemicals that speed up the rate of chemical change. Figure 10-2 shows how enzymes work.

Bread is made of a carbohydrate called starch. Bread itself is not usable to cells. It must be digested first. The starch in bread is made of many molecules of a chemical called glucose (GLEW kohs). Glucose is a type of sugar molecule. An enzyme speeds up the change of starch by removing glucose molecules from the starch. Once the glucose molecules are separated, they are in a form your body cells can use.

Check Your Understanding

1. What is the main job of the digestive system?
2. Identify two ways that food is broken down.
3. What are enzymes?
4. **Critical Thinking:** Explain why bread must be digested.
5. **Biology and Writing:** You have learned that the process of digestion is similar to the work of a factory. Write a paragraph that describes the differences between a real factory and digestion.

Lab 10–1 Digestion

Problem: How are proteins digested?

Skills

separate and control variables, form a hypothesis, observe, design an experiment

Materials

4 paper cups
gelatin
water
4 applicator sticks
refrigerator (optional)
4 plastic spoons
enzyme A
enzyme B

Procedure

1. Copy the data table.
2. Number four paper cups 1 through 4.
3. **Separate and control variables:** Fill each cup as follows:
   Cup 1—2 spoonfuls of gelatin
   This is the control in your experiment.
   Cup 2—2 spoonfuls of gelatin and 1 spoonful of water
   This setup is a variable to show the effect of water on the gelatin.
   Cup 3—2 spoonfuls of gelatin and 1 spoonful of enzyme A
   This setup is a variable to show the effect of enzyme A.
   Cup 4—2 spoonfuls of gelatin and 1 spoonful of enzyme B
   This setup will show the effect of enzyme B.
4. Mix the contents of each cup with a different applicator stick.
5. **Form a hypothesis:** Which cup's contents will show digestion of the gelatin? Write your **hypothesis** in your notebook.
6. Wait 20 minutes. (Or place cups in a refrigerator for 20 minutes.)

CUP	CONTENTS OF CUP	SOLID OR LOOSE?	DIGESTION OCCURRED?
1	Gelatin		
2	Gelatin + water		
3	Gelatin + enzyme A		
4	Gelatin + enzyme B		

7. **Observe:** After waiting, record in your table whether or not the gelatin is solid or loose and whether digestion occurred. (NOTE: If gelatin is *solid,* enzyme *did not* digest gelatin. If gelatin is loose, enzyme *did* digest gelatin.)
8. Dispose of the gelatin as indicated by your teacher.

Data and Observations

1. Which cup showed gelatin digestion?
2. What is your evidence?
3. What was added to this cup?

Analyze and Apply

1. What is an enzyme?
2. Where are digestive enzymes usually made?
3. What is your evidence that water is not an enzyme?
4. **Check your hypothesis:** Is your hypothesis supported by your data? Why or why not?
5. **Apply:** Gelatin is a protein. What body organs form enzymes that help digest this nutrient?

Extension

Design an experiment to determine whether saliva contains an enzyme that can digest gelatin. Try your experiment and report the results.

10:2 The Human Digestive System

Objectives

4. **Trace** the path of food from the mouth to body cells.
5. **Compare** the human digestive system with those of other animals.
6. **Identify** the problems of the digestive system.

Key Science Words

saliva
salivary gland
esophagus
stomach
hydrochloric acid
small intestine
pancreas
liver
bile
gallbladder
large intestine
appendix
villi
mucus

Humans provide a good example of how digestive systems work. If you understand digestion in humans, you should be able to understand it in other animals.

Nutrients Are Digested

Many nutrients must be digested before they can be used, while others are already in a usable form. Table 10-1 shows which nutrients are already usable and which are not.

Water, vitamins, and minerals can move directly from your digestive system into body cells without being changed. Fats, proteins, and carbohydrates must be acted upon by enzymes in your digestive system. There are different kinds of enzymes for each nutrient that must be digested. For example, enzymes that speed up changes in fats cannot change proteins and carbohydrates. Enzymes that speed up changes in proteins cannot help with the breakdown of fats or carbohydrates. The digestive system, therefore, must make different kinds of enzymes for each different nutrient.

TABLE 10–1. NUTRIENTS

Nutrient	Already in Usable Form
Water	yes
Vitamins	yes
Minerals	yes
Fat	no
Protein	no
Carbohydrate	no

A Trip Through the Digestive System

Let's take a trip through the human digestive system to see how it works. To make it a little more interesting, we will look at what happens to a hamburger. Remember that ground meat is mostly protein, mayonnaise is mostly fat, and the bun is mostly carbohydrate. The entire trip of the hamburger through the digestive system takes about 21 hours. Figure 10-3 shows how the entire digestive system looks from one end to the other.

The hamburger enters the system through the mouth. In the mouth, teeth break and grind the food. The breaking and grinding cause the hamburger to change physically. **Saliva** is a liquid that is formed in the mouth and that contains an enzyme. Saliva speeds up chemical changes in the carbohydrates of the bun. It has no chemical effect on proteins or fats. Figure 10-3 shows that saliva is made in the salivary (SAL uh ver ee) glands. **Salivary glands** are three pairs of small glands located under the tongue and behind the jaw. Saliva passes from the glands through small tubes leading to the mouth.

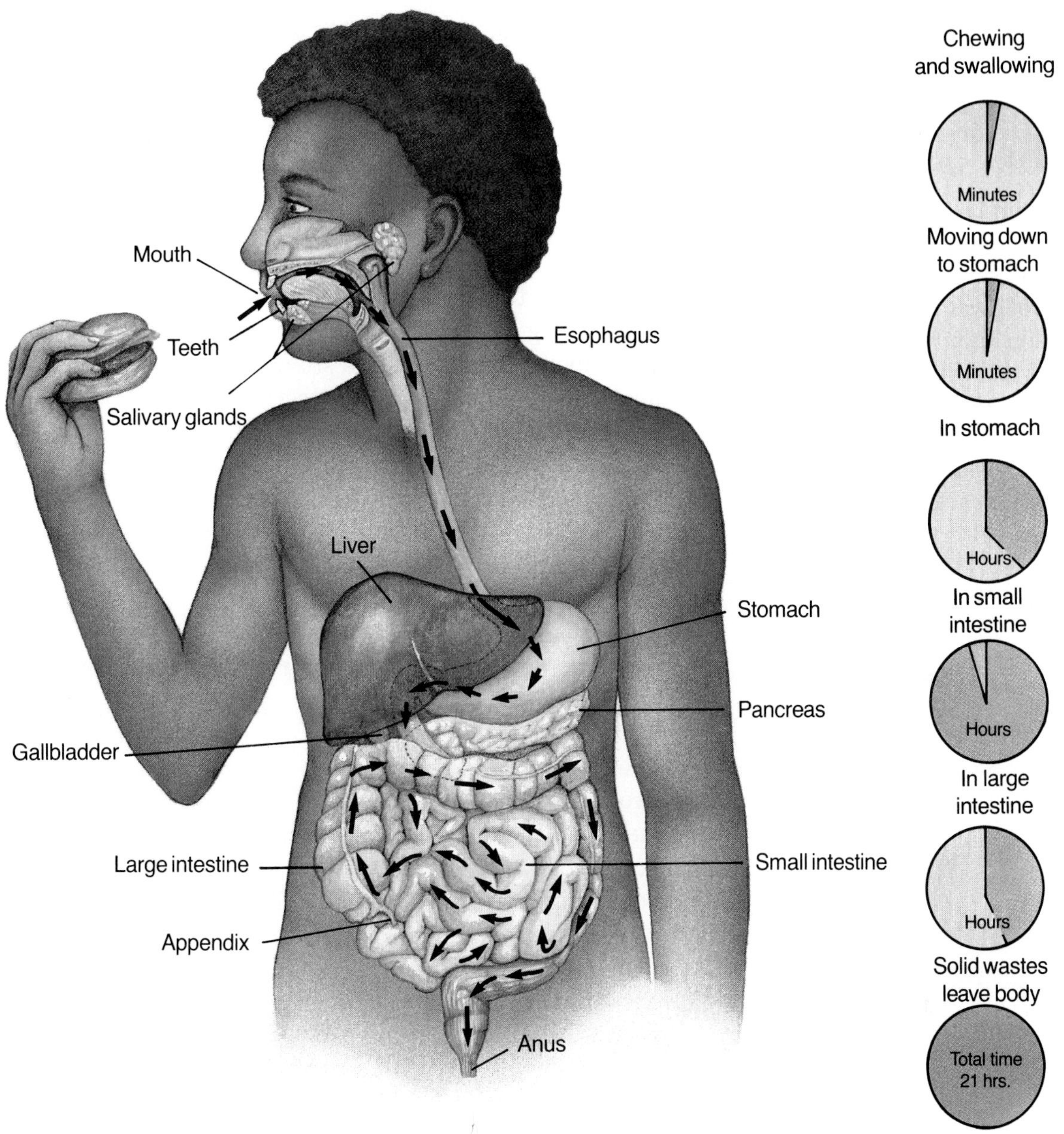

The broken down hamburger remains in the mouth for about one minute. Swallowing moves it from the mouth into the esophagus (ih SAHF uh gus). The **esophagus** is a tube that connects the mouth to the stomach. Muscles in the esophagus push the food toward the stomach. These events are shown in Figure 10-3. This part of the trip takes less than one minute.

Figure 10-3 Follow the path of a meal through the digestive system.

As it leaves the esophagus, the food enters the stomach. The **stomach** is a baglike, muscular organ that mixes and chemically changes protein. It can hold about one liter of liquid and food.

How does food reach your stomach?

Mini Lab

How Long Is the Digestive System?

Use numbers: Diagram and label the lengths and names of the following digestive organs: esophagus (25 cm), stomach (20 cm), small intestine (700 cm), and large intestine (150 cm). *For more help, refer to the **Skill Handbook,** pages 718-719.*

Cells on the inside of the stomach make two chemicals that help in digestion. One is an enzyme. The enzyme speeds up the chemical change of protein or meat in the hamburger. The other chemical made by the stomach is hydrochloric (hi druh KLOR ik) acid. **Hydrochloric acid** is a chemical often called stomach acid.

The muscular walls of the stomach mix and churn the food. Mixing is a physical part of digestion. Look at Figure 10-3 and notice the small clock next to the stomach. It shows how long foods remains in the stomach.

After about four hours, the stomach pushes food into the small intestine. The **small intestine** is a long, hollow, tubelike organ. Most of the chemical digestion of food takes place in the small intestine.

Before describing what the small intestine does, we must take a short detour. Figure 10-4 shows three different organs that are part of the digestive system. These organs are the pancreas (PAN kree us), liver, and gallbladder. Food doesn't pass through these organs.

The **pancreas** makes three different enzymes. One kind of enzyme helps to break down fats, one kind helps to break down proteins, and the third kind helps to break down carbohydrates. The enzymes pass through a small tube from the pancreas to the small intestine.

The **liver,** the largest organ in the body, makes a chemical called bile. **Bile** is a green liquid that breaks large fat droplets into small fat droplets. This change is physical and takes place in the small intestine. Bile is delivered from the liver to the gallbladder. The **gallbladder** is a small, baglike part located under the liver. It stores bile until it is needed by the small intestine.

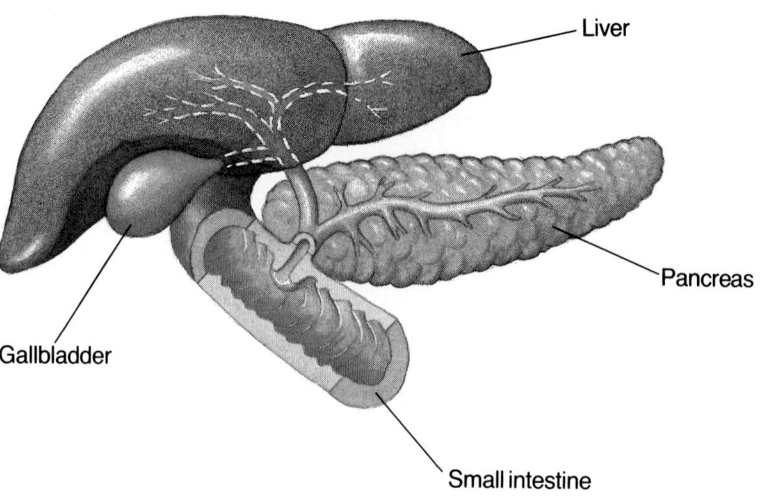

Figure 10-4 Bile from the gallbladder breaks up fat. The pancreas makes three enzymes that help digest fat, protein, and carbohydrates.

Career Close-Up

Dietitian

Steve was interested in a career as a dietitian. As part of his career day project at school, he decided to interview a dietitian to find out more about the job.

Steve: What is a dietitian?

Dietitian: A dietitian is one who understands nutrition. A dietitian plans diets for groups or individuals. These diets may be regular, balanced meals or diets for patients with special needs.

Steve: What type of training is needed?

Dietitian: Usually a four-year college degree is needed. This background will lead to a Registered Dietitian Degree. A six- to eighteen-month internship in a hospital is required.

Steve: Where are you most likely to find a job?

Dietitian: Hospitals are one place. You could also work for a company, school, university, or nursing home that has its own employee or patient dining hall. You might even work for an eating disorder clinic. Large food companies and drug companies hire dietitians.

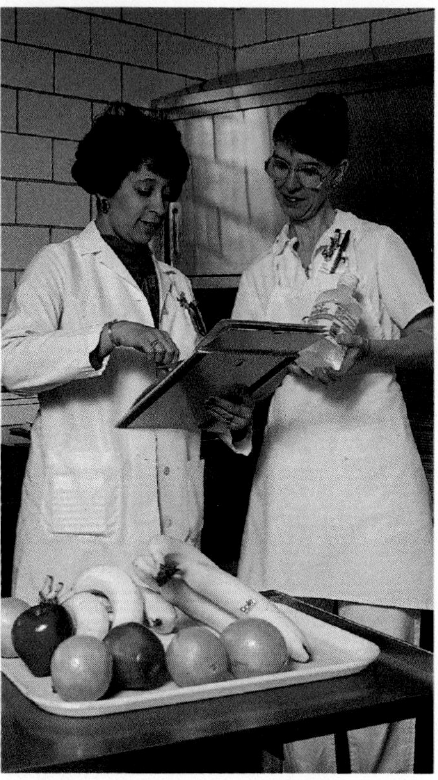

A dietitian approves the meals to be served to hospital patients.

Let's continue our tour and follow the food into the small intestine. Remember, the pancreas and gallbladder have already emptied their chemicals into the small intestine. In addition, the small intestine itself makes several enzymes. Cells that line the small intestine make enzymes that help with digestion of proteins in the hamburger meat and carbohydrates in the bun. All foods that enter the small intestine are finally changed chemically into a form that is usable by the body. The main goal of the digestive system has now been carried out. Figure 10-3 shows these changes. It also shows you that food spends about 12 hours in the small intestine. Why should food spend so much time in this part of the digestive system?

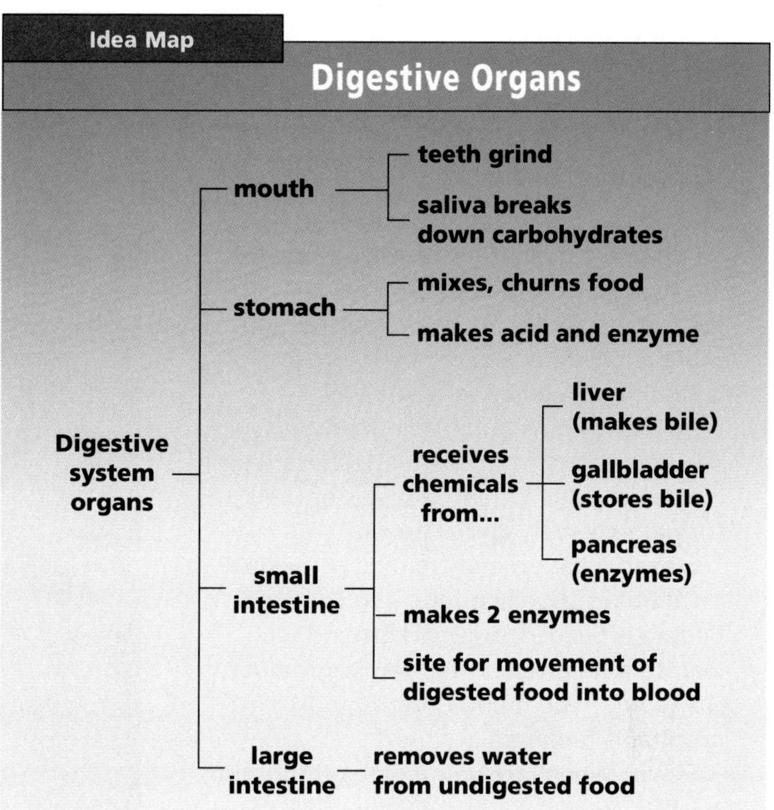

Study Tip: Use this idea map as a study guide to understanding the job of each digestive organ.

After passing through the small intestine, food enters the large intestine. The **large intestine** is a tubelike organ at the end of the digestive tract. It is called the large intestine because of its width. The large intestine is about 5 cm wide, and the small intestine is only 2.5 cm wide. Not much digestion occurs in the large intestine. Its main job is to remove water from undigested food. Water is then returned to the bloodstream. The clock in Figure 10-3 tells you that food spends about five hours in the large intestine. Not all of the food is digested when it reaches the end of the large intestine. It then leaves the body as solid waste through the anus.

One last part of the digestive system should be mentioned. This part is the appendix (uh PEN dihks). The **appendix** is a small fingerlike part found where the small and large intestines meet. The appendix in our body does not take part in the digestion of food.

Altogether your digestive system forms a tube about 900 cm long. The average height of an adult is about 170 cm. Your digestive system is about five times longer than your body! How can such a long tube fit inside of you? Why must it be so long?

Skill Check

Understand science words: appendix. The word part *pend* means to hang. In your dictionary, find three words with the word part *pend* in them. *For more help, refer to the **Skill Handbook,** pages 706-711.*

Lab 10–2 Digestive System

Problem: What are the jobs of the digestive system organs?

Skills

classify, interpret diagrams

Materials

five diagrams of the digestive system
colored pencils: red, blue, green, yellow, purple

Procedure

1. Label your five diagrams A, B, C, D, and E.
2. **Classify:** On diagram A, label the liver, esophagus, large intestine, mouth, small intestine, gallbladder, pancreas, salivary gland, stomach, anus, and appendix.
3. With a regular pencil, shade only the parts through which food actually passes.
4. Label the diagram "Human Digestive System and Food Pathway."
5. On diagram B, label the organs that aid the chemical change of carbohydrates. Color these organs red.
6. Label the diagram "Organs that Help to Digest Carbohydrates."
7. On diagram C, label the organs that aid the chemical change of protein. Color these organs blue.
8. Label the diagram "Organs that Help to Digest Protein."

9. On diagram D, label the organs that help with chemical and physical changes of fat. Color these organs green.
10. Label the diagram "Organs that Help to Digest Fat."
11. On diagram E, label the organs of the digestive system where diffusion of digested food into blood occurs. Color these organs yellow.
12. Using the same diagram, E, label the organs where diffusion of water occurs. Color these organs purple.
13. Label the diagram "Organs that Help Remove Digested Food and Water."

Data and Observations

1. Which organs did you color red?
2. Which organs did you color blue?
3. Which organs did you color green?
4. Which organs did you color yellow?
5. Which organs did you color purple?

Analyze and Apply

1. **Interpret diagrams:** Which two organs are the most important in digestion? Explain your answer.
2. Which organs help digest carbohydrates, protein, and fat?
3. **Apply:** Diagram the digestive path of a slice of cheese and pepperoni pizza. Indicate where the crust, cheese, and pepperoni are digested.

Extension

Draw a diagram of the digestive system. Create a color code to indicate where chemical, physical, both physical and chemical, and no changes occur. Then, color your diagram accordingly.

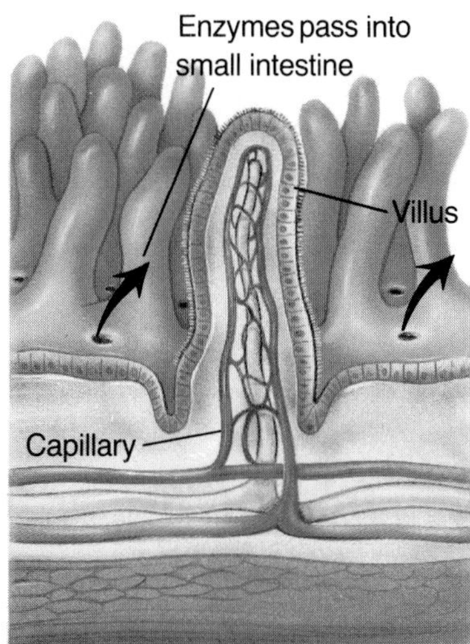

Figure 10-5 The surface area of the small intestine is increased by many hundreds of villi.

Moving Digested Food into Body Cells

Food digested in the small intestine is ready to be used by the body cells. Food cannot stay in the small intestine and do the body any good. It must be carried to all body cells by the blood. How does most food get out of the small intestine and into the blood? Food gets out of the small intestine and into the blood mainly by diffusion. Diffusion is a process that you studied in Chapter 2.

The inside surface of the small intestine helps absorb food molecules. The small intestine is a long, hollow tube much like a garden hose. The main difference between your small intestine and a hose is the inside lining. The lining of your small intestine is not smooth. Figure 10-5 shows that the small intestine has many tiny, fingerlike parts covering its entire inside surface. The fingerlike parts on the lining of the small intestine are called **villi** (VIHL i). Each villus contains blood vessels that carry digested food.

Once inside your blood vessels, digested food is carried by the blood to all body cells. How does digested food get from your blood vessels into your body cells? This happens through a special kind of diffusion. Figure 10-6 shows how food gets into the blood vessels.

The villi allow your small intestine to absorb more digested food than if it were smooth. Why? Villi increase the intestinal surface that comes in contact with digested food. With more intestinal surface, more digested food can pass into the blood.

Mini Lab

How Do Villi Aid Absorption?

Experiment: Compare how one, two, three, and four folded paper towels absorb water. Dip each paper towel into a beaker of water. Record the volume of water left in the beaker. *For more help, refer to the **Skill Handbook,** pages 704-705.*

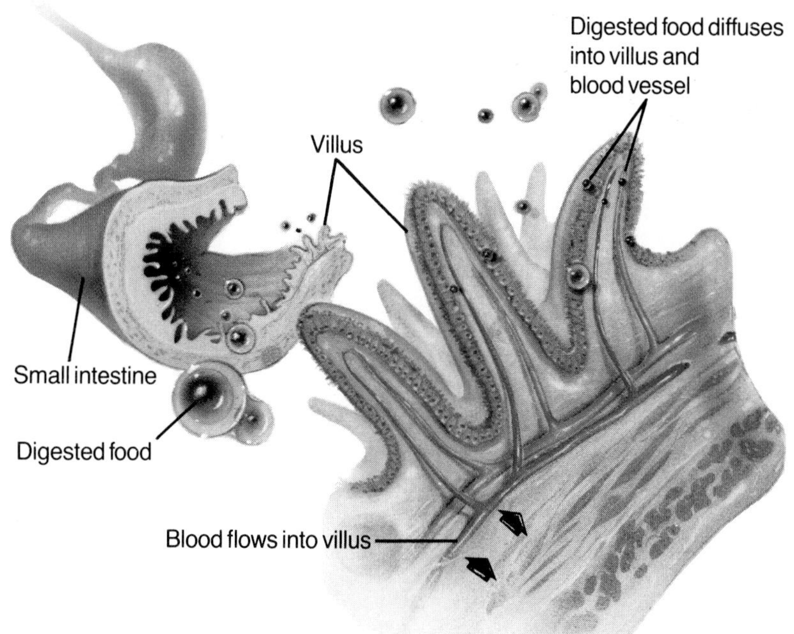

Figure 10-6 Digested food particles pass into the blood through the walls of the villi.

Consider this example. Trapeze artists may use a small net during their act. If they should fall, the chances of landing in this net are small. If they use a larger net, however, they have a better chance of landing in the net. A similar thing happens with the small intestine. The larger the surface of the intestine, the better the chances are that food molecules will come into contact with it. The more food that contacts the surface of the intestine, the more food that passes through the intestine and is picked up by the blood.

Digestion in Other Animals

How does the human digestive system compare with those of other animals? Different kinds of animals eat different kinds of food. The kind of food eaten is related to the animal's digestive system. Animals that eat plants, such as cows and rabbits, usually have long digestive systems. Animals that eat meat, such as cats and wolves, usually have shorter digestive systems. Plant eaters have longer digestive systems because plants are harder to digest than meat. The longer digestive system gives the food more time to change into a usable form.

Why do plant eaters have long digestive systems?

Like humans, many other animals have digestive systems with two openings. For example, the earthworm also has a mouth and an anus, Figure 10-7. Also like humans, the earthworm's digestive system has different organs that have different jobs. Some of the organs are the sites of chemical changes. Other organs produce physical changes in foods. Dogs, birds, snakes, fish, insects, and squid also have digestive systems with two openings.

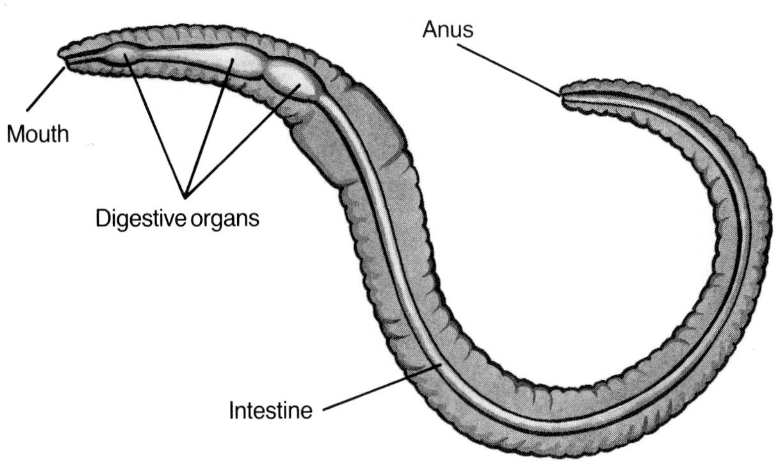

Figure 10-7 There are different organs in the earthworm's digestive system that cause physical or chemical changes in food.

Some animals have digestive systems with only one opening. The hydra is a small, simple animal with only one opening. Its digestive process is simple. The opening serves as both a mouth and an anus as shown in Figure 10-8. Its stomach is just a hollow sac within the animal's body. What other animals you studied in Chapter 7 have one opening to the digestive system?

A few animals have no digestive system. Animals such as tapeworms are examples. Tapeworms live inside the digestive systems of other animals. Digested food passes into their bodies by diffusion. It may not be a very good arrangement for the animal in whose body the tapeworm lives, but it certainly works for the tapeworm.

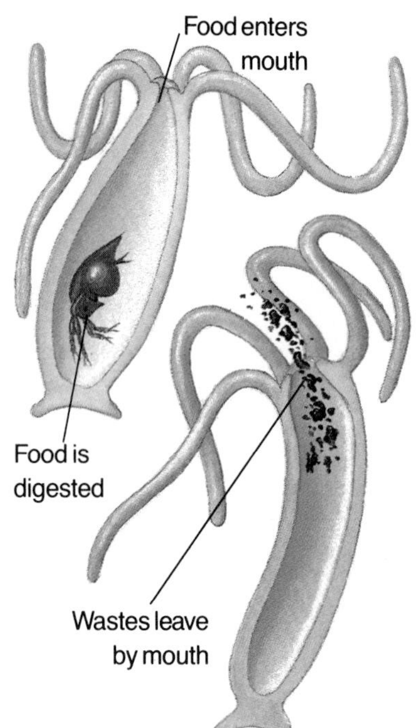

Figure 10-8 The single opening in a hydra acts as both mouth and anus. How are nutrients absorbed by a hydra?

Problems of the Digestive System

The digestive system, like any body system, may have problems. You may have heard of ulcers or heartburn. An ulcer (UL sur) is a sore or hole on the inside of either the stomach or the small intestine. Figure 10-9 shows a cut-away view of the inside of a normal stomach and the same view of a stomach with an ulcer. The ulcer is caused by the stomach lining being digested, or "eaten" away. Remember that enzymes and stomach acids are found in the stomach. These enzymes and acids cause the ulcer.

Normally, however, the stomach or intestine does not digest itself. There is usually a chemical covering on the inside. This covering is called mucus (MYEW kus). **Mucus**

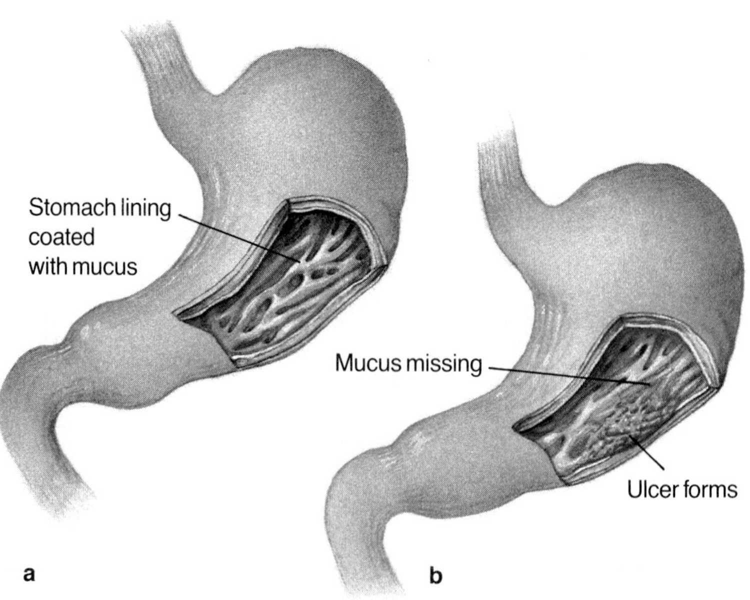

Figure 10-9 A healthy stomach (a) is coated with mucus. A stomach with an ulcer (b) can be very painful.

Figure 10-10 Eating too many rich or acidic foods can cause heartburn. Which food is more acidic, a pickle, or a potato?

is a thick, sticky material that protects the stomach and intestinal linings from enzymes and stomach acid. It's the same sticky, thick material that lines your nose.

Heartburn is a problem caused by stomach acids moving into the esophagus. Most of the time, stomach acids stay in the stomach. They then pass into the small intestine. When stomach acids move into the esophagus, they cause a burning feeling. However, nothing is being "burned." The problem also has nothing to do with the heart. The esophagus lies behind the heart. The pain feels like it is coming from the heart, even though it isn't.

Eating too much at one time may result in heartburn, Figure 10-10. The stomach can hold only so much food. If too much is eaten, some may back up into the esophagus.

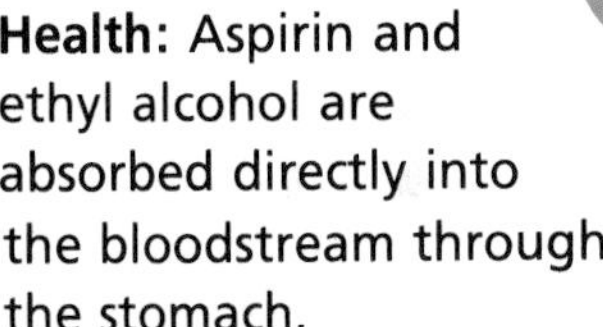

Bio Tip

Health: Aspirin and ethyl alcohol are absorbed directly into the bloodstream through the stomach.

Why does eating too much cause heartburn?

Check Your Understanding

6. Beginning with the mouth, describe the path food takes as it moves through the digestive system.
7. What causes an ulcer?
8. Identify the differences between the digestive process of a human and a hydra.
9. **Critical Thinking:** How do antacids reduce heartburn?
10. **Biology and Reading:** Name three organs that are not part of the digestive tube. How do these organs aid the digestive process?

Chapter 10

Review

Summary

10:1 The Process of Digestion

1. The digestive system—a long, hollow tube—takes in food and changes it into a form usable to body cells.
2. Physical changes grind and break down food. Chemicals called enzymes speed up the changing of food into a usable form.
3. Enzymes help to change fats, proteins, and carbohydrates into usable forms by speeding up chemical changes.

10:2 The Human Digestive System

4. Food passes from mouth to esophagus to stomach to small intestine to large intestine. Villi in the small intestine increase surface area and help with the movement of digested food into body cells.
5. Digestive systems in animals differ in length, number of openings, and specialization.
6. An ulcer is formed by a part of the stomach lining being digested. Mucus protects the stomach and intestine linings from enzymes and acid. Heartburn is caused by stomach acids that enter the esophagus. Eating too much at one time may result in heartburn.

Key Science Words

appendix (p. 212)
bile (p. 210)
chemical change (p. 206)
digestion (p. 205)
digestive system (p. 204)
enzyme (p. 206)
esophagus (p. 209)
gallbladder (p. 210)
hydrochloric acid (p. 210)
large intestine (p. 212)
liver (p. 210)
mucus (p. 217)
pancreas (p. 210)
physical change (p. 205)
saliva (p. 208)
salivary gland (p. 208)
small intestine (p. 210)
stomach (p. 209)
villi (p. 214)

Testing Yourself

Using Words

Choose the word from the list of Key Science Words that best fits the definition.

1. breaking of large food pieces into small pieces
2. connects mouth to stomach
3. process of changing food into a usable form
4. organ that makes three different enzymes
5. changes carbohydrates in the mouth
6. liquid that causes a physical change in fats
7. organ that stores bile
8. fingerlike parts on the lining of the small intestine
9. makes saliva
10. chemical that speeds up the changing of food into a usable form

Review

Testing Yourself *continued*

Finding Main Ideas

List the page number where each main idea below is found. Then, explain each main idea.

11. how much time food spends in each of the following: the stomach, small intestine, and large intestine
12. why so many different enzymes are needed for digestion
13. how the digestive system is like a factory
14. the body parts that help with physical and chemical changes in the mouth
15. how a digestive system with two openings is more specialized than a digestive system with one opening
16. what causes a stomach ulcer

Using Main Ideas

Answer the questions by referring to the page number after each question.

17. What are the organs through which food passes on its trip through the digestive system? (pp. 209, 210)
18. What two chemicals are made in the stomach? (p. 210)
19. What is an example of an animal with no openings to its digestive system? One opening? Two openings? (pp. 215, 216)
20. What is the function of mucus in the digestive system? (p. 217)
21. What is an example of a physical change and a chemical change that takes place in the digestive system? (pp. 205, 206)
22. What causes heartburn? (p. 217)

Skill Review 

*For more help, refer to the **Skill Handbook,** pages 704-719.*

1. **Form a hypothesis:** Which food would enzymes work on more rapidly: a slice of bread or an equal mass of bread crumbs? Explain by using the term *surface area.*
2. **Use numbers:** Determine from Figure 10-3 how long it takes for a meal to pass into the stomach and out into the large intestine.
3. **Interpret diagrams:** Use the information shown in Figure 10-3 to make a bar graph illustrating the amount of time food generally remains in the major digestive organs.
4. **Classify:** Make a list of five foods you have eaten during the past 18 hours. Classify each food according to the main nutrient it contains.

Finding Out More

Critical Thinking

1. Explain how it is possible for a person to live without a stomach.
2. Explain why it is impossible for a person to live without a pancreas.

Applications

1. How do the number and shape of teeth compare among adult animals, such as dogs, cats, cows, and horses? Relate tooth shape with diet.
2. Plants contain the substance cellulose. Report on the relationship between this substance and the length of a plant eater's digestive system.

CHAPTER PREVIEW

Chapter Content

Review this outline for Chapter 11 before you read the chapter.

Skills in this Chapter

The skills that you will use in this chapter are listed below.

- In **Lab 11-1,** you will form a hypothesis, calculate, interpret data, and make and use tables. In **Lab 11-2,** you will formulate models, form a hypothesis, and measure in SI.
- In the **Skill Checks,** you will understand science words.
- In the **Mini Labs,** you will design an experiment.

Chapter **11**

Circulation

Look at the photo of the crowded highway shown on the left. All the cars and trucks may have special pickup or delivery jobs. People in cars may be on their way to work, to school, or even going on vacation. Trucks may be picking up or delivering goods to homes, warehouses, or stores.

The photograph below shows a close-up view of human blood moving through a blood vessel. Blood cells are round. Notice that the walls of a blood vessel are very thin.

How are the jobs of cars and trucks on a highway similar to the jobs of blood cells? How is the job of the highway similar to the job of a blood vessel? Can you think of ways in which they are different?

Try This!

What kinds of sounds does the heart make? Use a stethoscope to listen to the sounds your heart makes. How many different sounds do you hear? Is there a pattern to the sounds? How many sounds do you hear during a one-minute period?

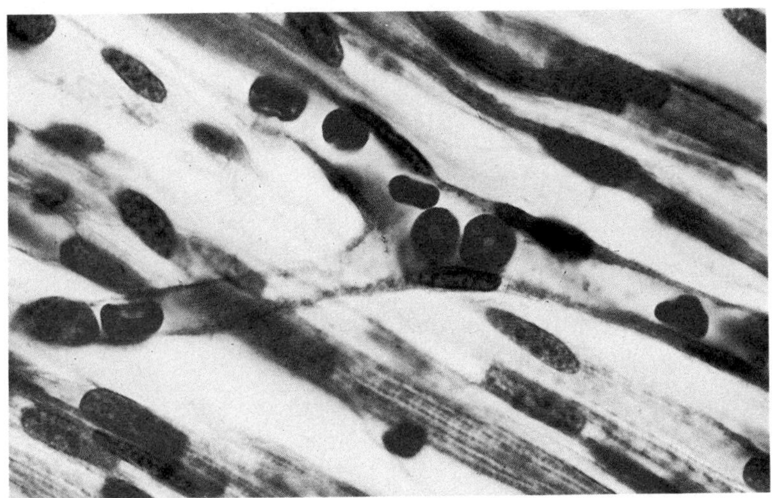

Blood vessel

BIOLOGY Online

Visit the Glencoe Science Web site at science.glencoe.com to find links about **circulation.**

11:1 The Process of Circulation

Objectives

1. **Identify** the circulatory system.
2. **Compare** the circulatory systems of earthworms, insects, and humans.

Key Science Words

circulatory system

Your body is made of billions of cells. Each cell is like a tiny factory that must be supplied with certain chemicals. You have a pickup and delivery system in your body to supply these chemicals.

Pickup and Delivery

Your body's pickup and delivery system is the circulatory system. Your **circulatory system** is made of your blood, blood vessels, and heart. Blood delivers needed materials, such as oxygen, water, and food, to your cells. Blood picks up the cells' waste products, such as carbon dioxide gas.

What makes up the circulatory system?

Blood travels through a series of tubes called blood vessels. Your heart serves as a pump to help move, or circulate, blood through these vessels.

Circulation in Animals

Not all animals have circulatory systems. Simple animals, such as a sponge, sea anemone, or hydra, do not have circulatory systems. The bodies of these animals are just a few cells thick. Water moves freely in and out of their bodies. Nearly every cell in the animal's body comes in contact with the water. Oxygen and nutrients in the water diffuse into the cells. Wastes, such as carbon dioxide, diffuse out of the cells into the water. The water in which these simple animals live acts as a pickup and delivery service for them.

Study Tip: Use this idea map as a study guide to comparing circulation patterns in different animals.

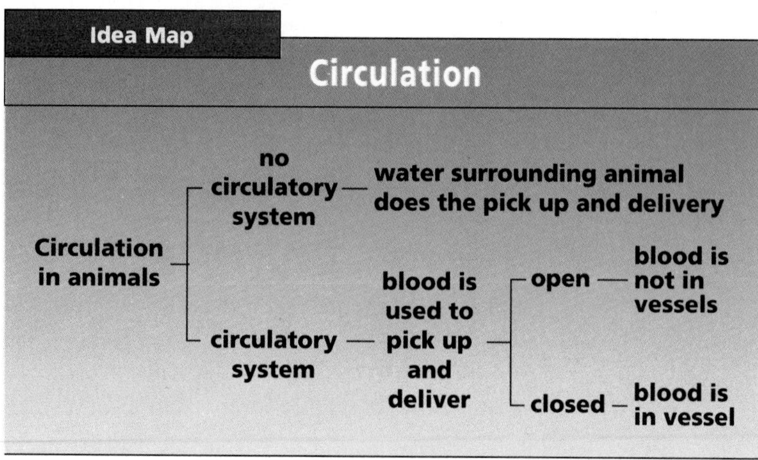

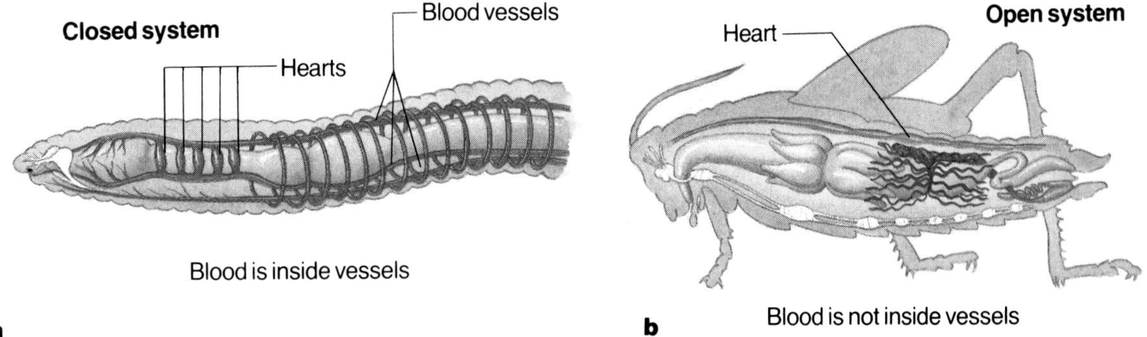

Complex animals, such as earthworms and insects, do have circulatory systems. These animals have many layers of cells and tissues. Oxygen and nutrients would not reach each cell of the animal's body if there were no circulatory system. Look carefully at Figure 11-1 and you will notice an important difference between the circulatory systems of earthworms and insects. Notice that the earthworm has blood vessels and hearts, Figure 11-1b. The circulatory system in the earthworm is said to be closed. Blood is inside vessels. The insect has a heart but no blood vessels contained within. Blood in the insect's body moves about without traveling in vessels. The insect's circulatory system is said to be open. Think about your own body. Your body is made up of many complex systems of organs. You have a closed circulatory system to supply all your cells with the nutrients they need.

Figure 11-1 Earthworms have a closed circulatory system (a). Insects have an open circulatory system (b).

What kind of circulatory system does an insect have?

Check Your Understanding

1. What are the two main jobs of the circulatory system?
2. How do animals that lack a circulatory system get their needed materials?
3. (a) How are the circulatory systems of insects and earthworms alike?
   (b) How are the systems different?
4. **Critical Thinking:** What advantage does a closed circulatory system have over an open circulatory system? (HINT: What would happen if an animal with an open circulatory system got a cut?)
5. **Biology and Reading:** Describe the relationship between the complexity of an animal's body and the type of circulatory system it has.

Pulse Rate

Lab 11-1

Problem: How do a worm's pulse and your pulse compare?

Skills

form a hypothesis, calculate, interpret data, make and use tables

Materials

live earthworm
petri dish with cover
water, room temperature
wall clock

Procedure

1. Copy the data table.
2. Place the live earthworm in a petri dish. **CAUTION:** *Keep the earthworm moist by stroking now and again with wet fingers.*
3. Look for the blood vessel along the top surface of the earthworm, Figure A.
4. Count and record the pulse of the earthworm for exactly 15 seconds.
5. Multiply your answer by 4 to get the worm's pulse for one minute. Record your result in the data table.
6. Repeat steps 4 and 5 three more times.
7. **Calculate** an average pulse for the earthworm for one minute. (Add all 4 pulse rates and then divide the total by 4.)
8. Return the earthworm to your teacher.
9. **Form a hypothesis** about how your average pulse rate will compare with that of the earthworm.
10. Use Figure B as a guide or ask your teacher to help you locate your pulse. Take your own pulse for one minute.
11. Record your results in your data table.
12. Repeat step 10 three more times and calculate an average pulse for yourself.

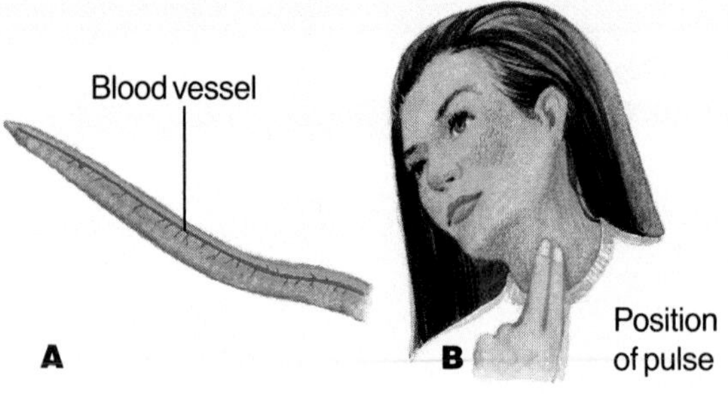

TRIAL	WORM PULSE FOR 15 SECONDS	WORM PULSE FOR 1 MINUTE	YOUR PULSE FOR 1 MINUTE
1			
2			
3			
4			
Total	—		
Average	—		

Data and Observations

1. What was the earthworm's average pulse over one minute?
2. What was your average pulse?

Analyze and Apply

1. **Check your hypothesis:** Is your hypothesis supported by your data? Why or why not?
2. What is a pulse?
3. Why is it important to repeat measurements in an experiment?
4. **Apply:** Your heartbeat rate slows down when you relax. Explain how this change will affect your pulse rate.

Extension

Design an experiment to determine the effect different kinds of exercise have on your pulse rate.

11:2 The Human Heart

Can you believe there is a pump within your chest that started working before you were born? It sounds impossible, but that is what your heart is.

Objectives

3. Describe how blood is pumped through the heart.

4. Explain what causes the sounds the heart makes.

5. Trace the pathway of blood through the left and right sides of the heart.

Key Science Words

atrium
ventricle
artery
vein
valve
tricuspid valve
bicuspid valve
semilunar valve
vena cava
pulmonary artery
pulmonary vein
aorta

Heart Structure

The heart is a muscle that pumps blood through the body. Figure 11-2b shows a photograph of the human heart.

Use Figure 11-2a to learn the parts of the human heart. There are three things to note about this diagram of the heart.

1. It is a cut-away view of the inside of the heart.
2. The colors bright red and blue represent blood within the heart.
3. The uncolored parts are heart muscle.

Because it has two sides, you can think of the heart as two separate pumps, one on the left and one on the right. Figure 11-2a is divided into left and right sides.

You might think that the sides of the heart are backward. The sides are labeled as if the heart were inside someone facing you. So, the heart diagram's right and left sides are opposite to your right and left sides.

Each side of the heart has a small chamber on the top and a large chamber on the bottom. The small, top chambers of the heart are **atria** (AY tree uh). The atria are divided into the right atrium and the left atrium. The large bottom chambers of the heart are called the **ventricles** (VEN trih kulz). The ventricles also are divided into right and left sections.

a

Right atrium
Left atrium
Right ventricle
Left ventricle
Right side
Left side

b

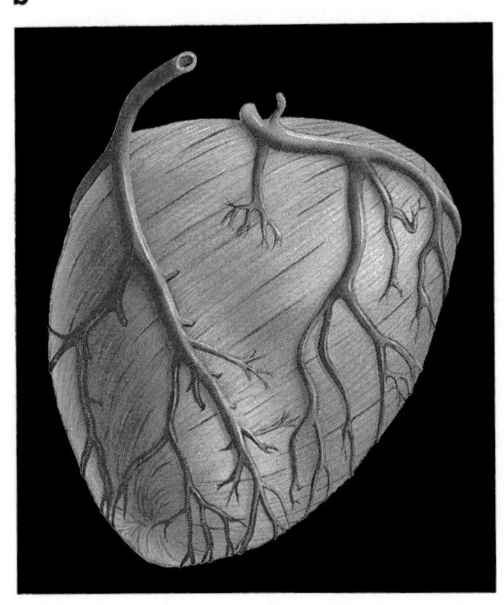

Figure 11-2 The human heart has four compartments. Which compartments of the heart does blood enter from veins?

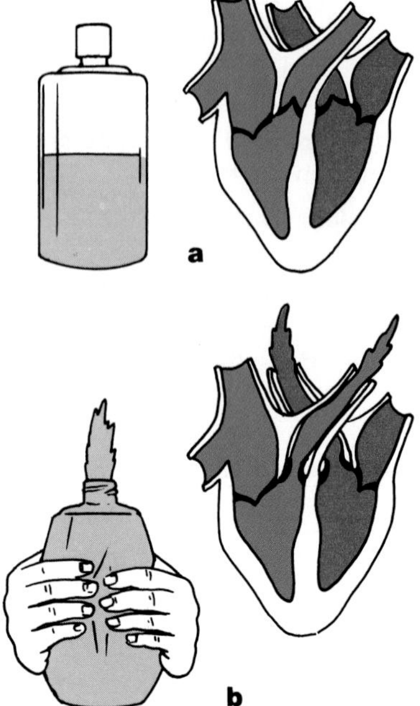

Figure 11-3 When a bottle or heart is not squeezed, no liquid is pushed out. When a bottle or heart is squeezed, liquid is squirted out.

How do the ventricles receive blood?

The Pumping of the Heart

How does the heart work as a pump? Figure 11-3 shows the heart's squeezing action by comparing it to the squeezing of a plastic bottle. The top picture shows the plastic bottle when it is not squeezed. No liquid is pushed out. The same is true for the heart. If the heart muscle does not squeeze, no blood is pumped. The bottom picture shows the plastic bottle being squeezed and liquid squirting out. The bottom diagram of the heart shows that when muscles of the ventricles squeeze, blood is pushed out.

While a person is resting, the heart pumps 60 to 80 times each minute. After running, the heart may pump 150 times in one minute. Each pump of the heart is called a beat. The structure of the heart allows blood to move through the circulatory system in only one direction. Your heart muscle will pump for more than two billion times during your lifetime!

Figure 11-4 shows several large blood vessels called arteries (ART uh reez) and veins. An **artery** is a blood vessel that carries blood away from the heart. A **vein** is a blood vessel that carries blood back to the heart. The arrows in the diagram show the direction of blood flow.

1. Blood from the veins enters the heart's right and left atria, Figure 11-4a. Neither chamber is pumping.
2. Figure 11-4b shows the atria beginning to pump or squeeze. Right and left ventricles are relaxed and are not pumping. These chambers receive blood from the atria.
3. Figure 11-4c shows the right and left ventricles squeezing blood into two large arteries leading to the body and the lungs. When the ventricles are pumping, the atria are relaxed. Blood returning from the body enters the atria and the entire pumping cycle begins again.

Figure 11-4 Follow the steps that show the pumping action of the heart.

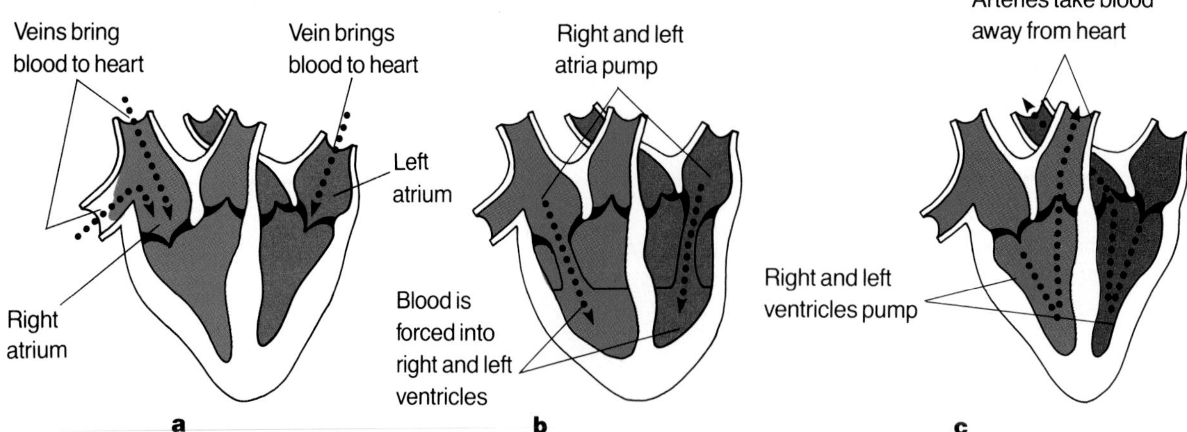

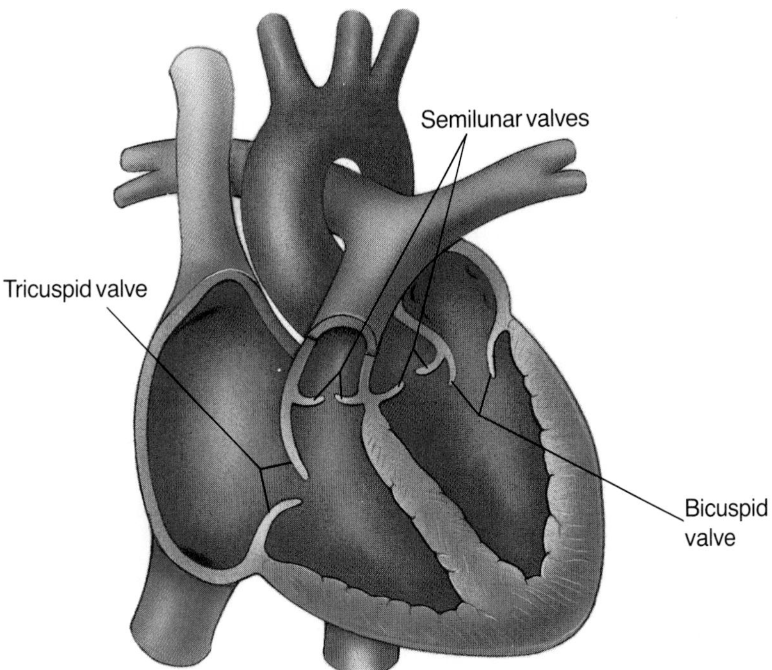

Figure 11-5 Locate the valves of the heart.

Heart Valves

If a pump is to work well, it must keep liquid moving in only one direction. A pump has valves that help with this task. So does the heart. **Valves** are flaps in the heart that keep blood flowing in one direction.

Figure 11-5 shows the location of the valves. Notice that there are two sets of valves. One set is between the atria and ventricles. The valve between the right atrium and right ventricle is called the **tricuspid** (tri KUS pud) **valve.** The valve between the left atrium and the left ventricle is called the **bicuspid** (bi KUS pud) **valve.** Note that these two valves are like one-way doors. They only open downward into the ventricles to keep blood flowing in one direction.

The other valves are the semilunar (sem ih LEW nur) valves. **Semilunar valves** are located between the ventricles and their arteries. Note that these two valves are also like one-way doors. They open only in an upward direction away from the ventricles.

Have you ever heard your heart beat? Do you know what causes the sounds you hear? Some people think they are hearing the heart muscle as it pumps. Actually, the sounds you hear are your heart valves closing. These valves, like doors, make sounds when they close. Doesn't a door make a loud sound when it slams shut?

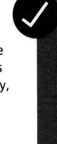

Skill Check

Understand science words: semilunar. The word part *semi* means half. In your dictionary, find three words with the word part *semi* in them. *For more help, refer to the* ***Skill Handbook,*** *pages 706-711.*

Figure 11-6 The lub and dup sounds of the heart are caused by the closing of the heart's valves.

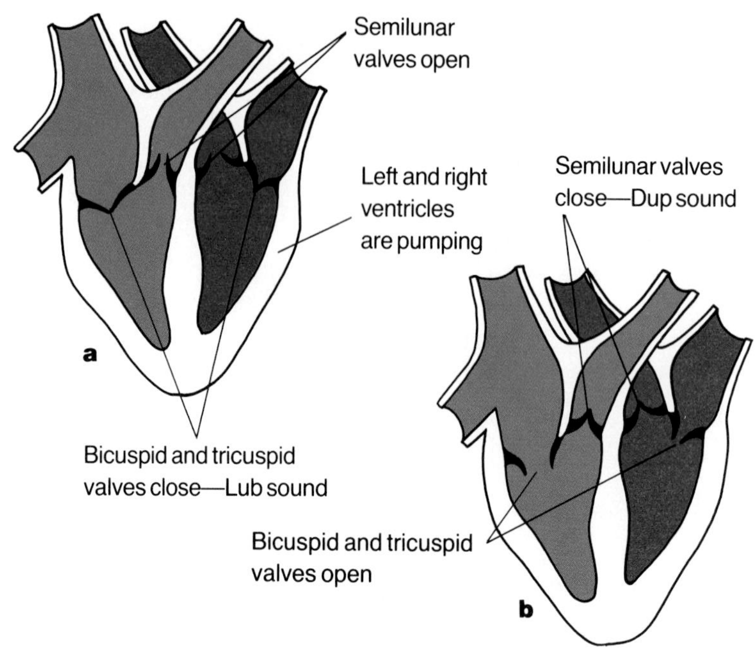

Mini Lab

How Do the Pulses of Males and Females Compare?

Design an experiment: Design an experiment to compare the pulse rates of males and females. Carry out the experiment. Write a report. *For more help, refer to the **Skill Handbook**, pages 704-705.*

Because you have two sets of valves in your heart that close at different times, your heart makes two sounds. The first sound is caused by the closing of your bicuspid and tricuspid valves. Figure 11-6a shows these valves closed. The blood can no longer pass from the atria to the ventricles. When your ventricles squeeze, these valves close and cause a "lub" sound. The second sound occurs when your semilunar valves close. These valves close when your ventricles stop squeezing. A "dup" sound is now heard. Figure 11-6b shows the semilunar valves closed. The blood can no longer pass from the ventricles out of the main arteries.

The two heart sounds heard together make a "lub-dup" sound. The tricuspid and bicuspid valves close harder than the semilunar valves. That is why the sounds differ. A doctor listening to your heart can tell if your heart is working normally. A "lub-dup" sound means your valves are working normally. If the doctor hears a "lub-swish-dup" sound, the bicuspid or tricuspid valves are not closing properly. Blood leaking past the valves causes the "swish" sound. The blood moves in the opposite direction from which it should be going. Blood flowing backward through the heart when valves are not closing tightly is called a heart murmur. Which valves are not closing properly if a doctor hears a "lub-dup-swish" sound? Since the "swish" sound follows the "dup" sound, it must be caused by the semilunar valves not closing properly.

What makes the "lub-dup" sound in your heart?

The Jobs of the Heart

So far we've only seen how your heart pumps. Now let's look at your blood vessels and see how the rest of your circulatory system works. Remember, your heart has a right and left side. Each side has its own job of pumping blood around the body.

Right Side of the Heart

The right side of your heart pumps blood only to your lungs. The path the blood follows can be seen by matching numbers 1 through 5 in Figure 11-7 with statements 1 through 5.

1. Blood enters the heart's right side at the right atrium from a large vein called the vena cava. The **vena cava** (VEE nuh • KAY vuh) is the largest vein in the body and carries blood from the body back to the heart. The blue color is used to show that at this point blood contains a lot of carbon dioxide gas and little oxygen gas.
2. Blood is pumped into the right ventricle.
3. From here, blood is pumped into a large artery. This artery is called the pulmonary (PUL muh ner ee) artery. The **pulmonary artery** carries blood away from the heart to the lungs. Because you have two lungs, the pulmonary artery divides in two. Blood will be pumped to both lungs.
4. Blood enters both lungs. In the lungs, the blood picks up oxygen and loses carbon dioxide. Notice that the color of the blood is now bright red. Blood containing a lot of oxygen is shown in bright red.
5. Blood returns to the heart through the pulmonary veins. **Pulmonary veins** carry blood from the lungs to the left side of the heart. The total time needed for blood to travel from your heart to your lungs and back again is only about 10 seconds.

Figure 11-7 Trace the pathway of blood to and from the lungs.

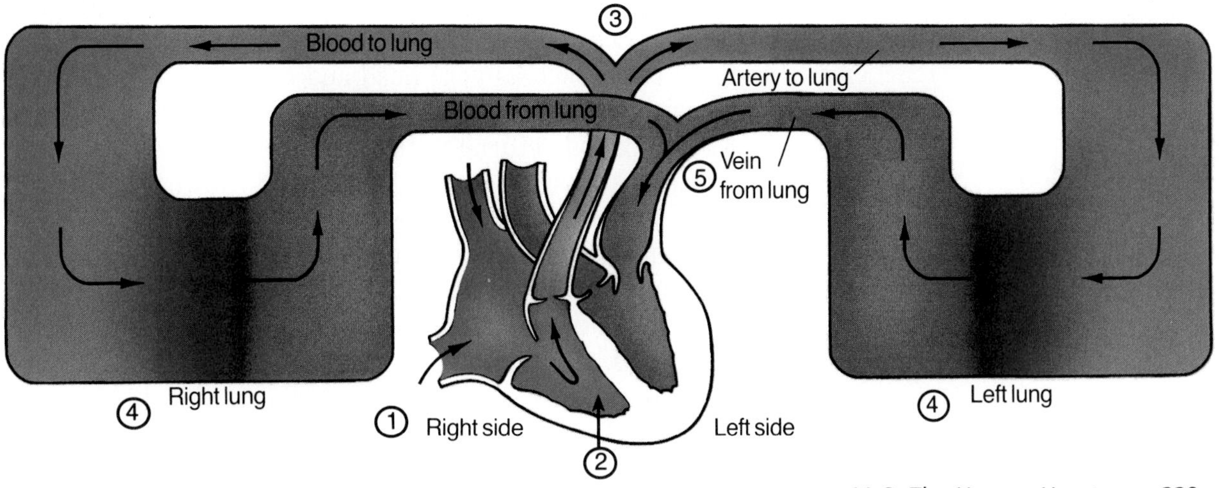

Figure 11-8 Trace the pathway of blood through the body.

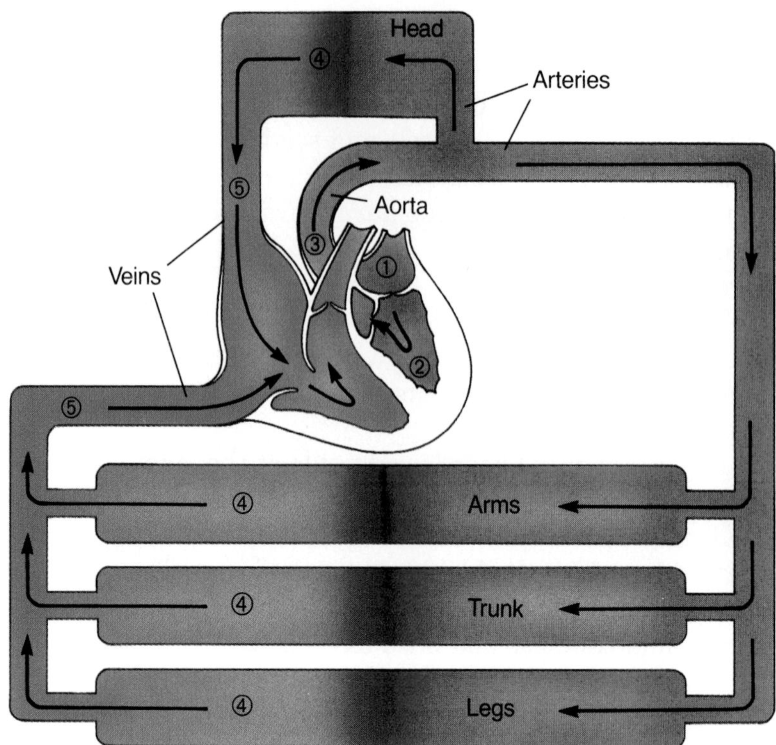

How is oxygen-rich blood supplied to the left side of the heart?

Left Side of the Heart

The left side of the heart pumps blood to all parts of the body. The path the blood follows can be seen by matching numbers 1 through 5 in Figure 11-8 with statements 1 through 5.

1. Blood rich in oxygen has just arrived at the left atrium from the lungs.
2. The blood is pumped from the left atrium into the left ventricle.
3. From the left ventricle, blood is pumped out of the heart into the aorta (ay ORT uh). The **aorta** is the largest artery in the body. It carries blood away from the heart's left side. Then it branches and goes to the body or head.
4. Body parts such as your head and organs receive blood that is rich in oxygen as shown by the bright red color.
5. Finally, blood returns to the heart's right side. Notice that blood has lost its oxygen and is now shown as a blue color. Why does this change occur?

Remember, one main job of blood is to carry two gases, oxygen and carbon dioxide. As blood passes through the lungs, it picks up oxygen. This oxygen is then delivered to all body cells. Carbon dioxide is produced as a waste product by all body cells. The blood picks up this carbon dioxide and drops it off at the lungs.

Bio Tip

Health: A resting heart will pump 5 liters of blood through the body each minute.

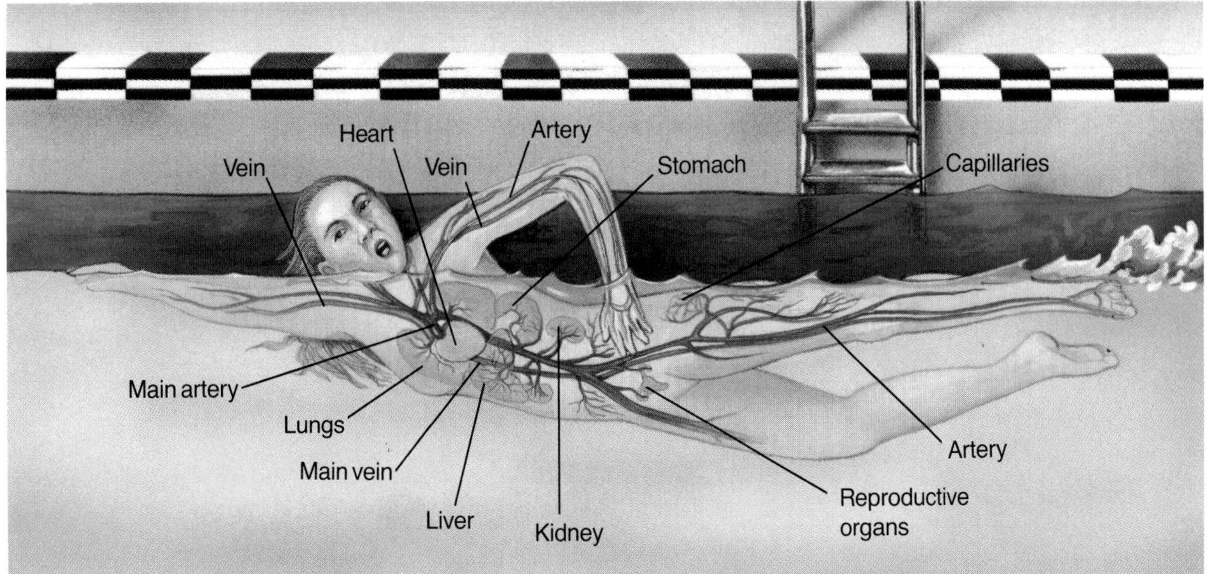

Figure 11-9 Oxygen is delivered to all the organs of your body by arteries. Veins pick up blood that has carbon dioxide in it and return it to the lungs.

There is a very important difference between the left and right sides of your heart. Blood in the right side of your heart contains a lot of carbon dioxide. Blood in the left side of your heart contains a lot of oxygen. Which side of the heart pumps blood only to your lungs? The side with more carbon dioxide. Which side of the heart pumps blood only to your body? The left side. Figure 11-9 helps to show how all the blood vessels supply all the organs of your body.

Check Your Understanding

6. (a) Are ventricles relaxed or pumping when they fill? Describe the condition of the atria as the ventricles fill. (b) Are ventricles relaxed or pumping when they empty? Describe the condition of the atria as the ventricles empty.
7. What causes the sounds of the heart?
8. Trace the pathway of blood once it leaves the (a) left atrium, (b) right atrium, (c) left ventricle, and (d) right ventricle.
9. **Critical Thinking:** An amphibian heart has one ventricle and two atria. How would the blood circulation through an amphibian heart differ from that in a human heart?
10. **Biology and Math:** A person's normal heartbeat rate is 70 beats each minute. About how many times does a person's heart beat in a day?

11:3 Blood Vessels

Objectives

6. **Discuss** the importance of arteries.
7. **List** four traits of veins.
8. **Give the function** of capillaries.

Key Science Words

blood pressure
capillary

Blood vessels are tubelike structures through which blood moves. You have about 96 000 kilometers of blood vessels. If they were all placed end-to-end, they would go around Earth about two and one-half times. The three types of blood vessels in your body are arteries, veins, and capillaries (KAP uh ler eez).

Arteries

Remember that arteries carry blood away from the heart. The aorta and pulmonary artery are two arteries you have studied. In what direction is blood being carried within these vessels?

As you can see in Figure 11-11, arteries are round and have thick walls made of many muscle cells. Arteries, such as the aorta, are quite wide in diameter. The aorta can be as wide as a garden hose. Arteries branch into smaller and narrower vessels as they carry blood away from the heart. Look again at Figure 11-9 on page 231 to see what the entire network of arteries might look like.

How is pressure created within arteries?

One important feature of arteries is blood pressure. What do you think would happen if you squeezed the water-filled plastic bottle shown in Figure 11-10? Water would squirt out of the hollow tube. Why does this happen? When you squeeze the bottle, you are reducing its volume. You give the liquid less room inside the bottle. The liquid has to go somewhere, so it squirts out of the tube. The water in the bottle is put under high pressure because of the squeezing. The water moves into the tube where the pressure is lower.

The heart's pumping action on blood is similar to the plastic bottle example. Blood is squeezed by the heart muscle just as water was squeezed out of the bottle by your hand. Just as water moved into the tube, blood is forced into the arteries. The force created when blood pushes against the walls of vessels is called **blood pressure.** This pressure drives the blood through the blood vessels. When you feel your pulse, you are feeling your blood moving through your arteries. Your pulse rate agrees with your heartbeat rate.

The three important traits of arteries are (1) they carry blood away from the heart (2) they carry blood under high pressure (3) they are round in shape and have thick, muscular walls.

Figure 11-10 If a bottle is squeezed, high pressure forces water to squirt out.

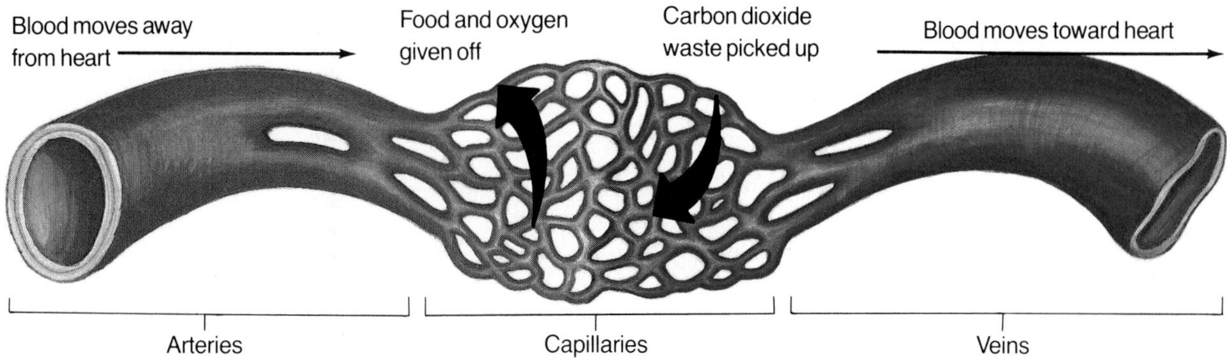

Figure 11-11 The three types of blood vessels are arteries, veins, and capillaries. Where do pickups and deliveries occur?

Veins

Blood must be brought back to the heart in veins. The vena cava and pulmonary vein are two veins you have studied. In what direction is blood being carried within these vessels?

Look again at Figure 11-11. You can see how veins differ from arteries. Veins have less muscle than arteries. Their shape is rather flat, but they may be much wider than arteries. Veins have thinner walls than arteries. Veins in your arms and legs have many one-way valves inside them. These valves keep blood flowing in one direction, toward the heart. Figure 11-12 shows what happens to the valves if blood begins to flow backward away from the heart. The valves are helpful because blood in veins is under low pressure. Without the valves, how would blood in your leg or arm veins return to your heart without flowing back into your feet or hands?

In what direction does blood in the veins travel?

The four important traits of veins are (1) they carry blood to the heart (2) they carry blood under low pressure (3) they are flat in shape and have little muscle (4) they have many one-way valves to keep blood flowing in one direction.

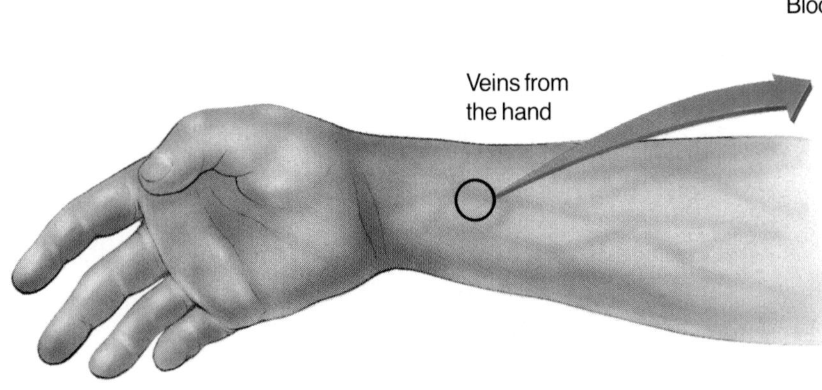

Figure 11-12 One-way valves in veins keep blood flowing in one direction toward the heart.

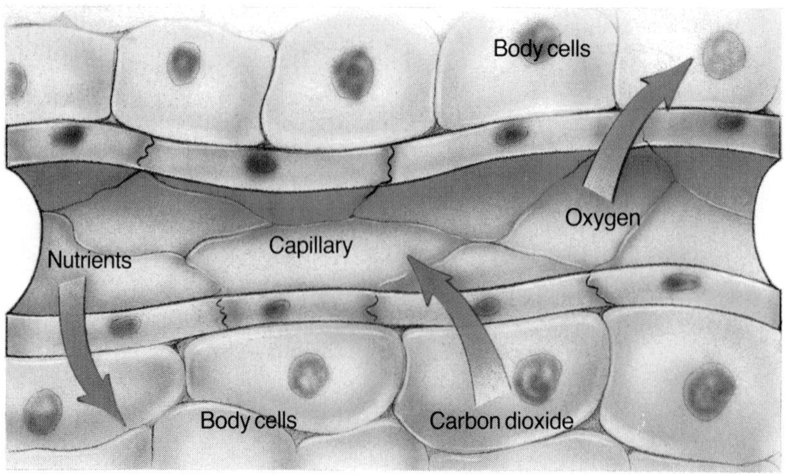

Figure 11-13 Pickups and deliveries between blood and body cells occur in the capillaries.

Capillaries

What occurs in the capillaries?

There are blood vessels in your body that become very narrow until they can no longer be called arteries or veins. A **capillary** is the smallest kind of blood vessel. The wall of a capillary is only one cell thick and may be thinner than a single page in this book. Some blood cells can easily pass between the cells of the capillary wall. You have more capillaries in your body than any other blood vessel type.

Capillaries have a very important job. They bring blood close to every body cell. All the pickups and deliveries occur in the capillaries. Blood in the capillaries can deliver oxygen, food, or other needed materials to all body cells. At the same time, waste chemicals made in body cells, such as carbon dioxide, are picked up by the blood in capillaries. Figure 11-13 shows these events.

Check Your Understanding

11. List three traits of arteries.
12. List four traits of veins.
13. What is the main job of capillaries?
14. **Critical Thinking:** What would be more life-threatening, a cut in an artery or a cut in a vein? Why?
15. **Biology and Math:** You have read that the total length of all your blood vessels is about 96 000 kilometers, enough to go around Earth about two and one-half times. Use this information to calculate the circumference of, or distance around, Earth.

Lab 11-2

Problem: How does blood pressure in arteries and veins compare?

Skills

formulate models, form a hypothesis, measure in SI

Materials

plastic squeeze bottle
pan or sink
rubber stopper and tube assembly
water
meterstick

Procedure

1. Copy the data table.
2. Fill the plastic bottle with water.
3. Insert the rubber stopper and tube assembly into the plastic bottle, Figure A.
4. Position the bottle and meterstick over a pan or sink as shown in Figure B.
5. **Form a hypothesis** about which tube the water will squirt from farther. One tube is glass and the other is plastic.
6. Give the bottle one firm squeeze.
7. Note the distance that water squirts from each tube.
8. **Measure in SI:** Measure in centimeters the distance that water squirts from each tube. Record the distance in your table.
9. Repeat steps 2 through 4 and 6 through 7 two more times.

TRIAL	DISTANCE WATER SQUIRTS (cm)	
	GLASS TUBE	PLASTIC TUBE
1		
2		
3		
Total		
Average		

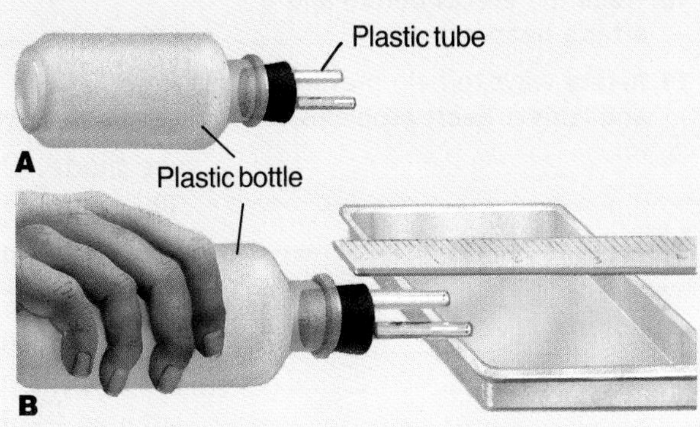

10. Calculate the average distance water squirted for each tube. (Add the 3 distances and divide the total by 3.)

Data and Observations

1. Which tube squirted water farther?
2. Which tube squirted water the shorter distance?

Analyze and Apply

1. (a) What body organ does the plastic bottle represent?
   (b) What body liquid does the water represent?
2. Arteries are firmer than veins.
   (a) Which tube represents an artery?
   (b) Which tube represents a vein?
3. Which tube was under higher pressure?
4. Which tube was under lower pressure?
5. How does blood pressure in arteries and veins compare?
6. **Check your hypothesis:** Is your hypothesis supported by your data? Why or why not?
7. **Apply:** Why are arteries firmer than veins?

Extension

Design an experiment to show the effect of less salt intake on blood pressure.

11:4 Problems of the Circulatory System

Objectives

9. **Explain** how blood pressure is measured.
10. **Trace** the events before and after a heart attack.
11. **Relate** ways to help prevent and correct heart problems.

Key Science Words

hypertension
cholesterol
coronary vessel
heart attack

Like any body part, the circulatory system can develop problems. Many of these problems can be corrected.

High Blood Pressure

When your blood pressure is taken, two measurements are made. One is taken when the ventricles are contracting. This upper pressure is the greater measurement. When the ventricles are not pumping, a lower pressure is taken. The normal range of blood pressure for young adults is 110 to 140 units for the upper pressure and 65 to 90 units for the lower pressure. Blood pressure is recorded as a fraction such as $\frac{120}{80}$. In general, blood pressure increases gradually as a person gets older.

Arteries are partly muscle, which gives them the ability to contract. When the artery contracts, the inside of the artery gets smaller. Blood must now pass through a narrower opening than before. The result is blood pushing through at higher pressure.

Let's use an example with a garden hose representing an artery. Water inside the hose can represent blood.

What happens if you squeeze the hose while the water is on? Water squirts out farther and with more force, Figure 11-14. You have raised the pressure of the water by making it pass through a narrower opening. The same thing happens with arteries.

Hypertension (HI pur ten chun) occurs when blood pressure is extremely high. Hypertension may also refer to a disease caused by high blood pressure. It occurs when arteries are too narrow for easy movement of blood.

Figure 11-14 The water pressure in a hose is higher when it is squeezed.

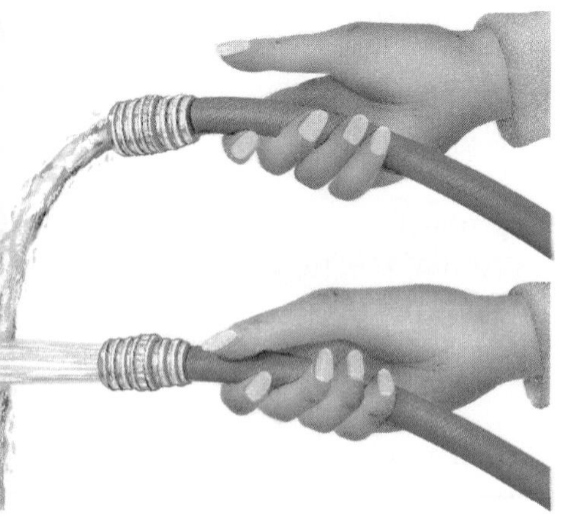

Hypertension affects millions of people. Unlike many diseases, hypertension does not always make a person feel ill. Many people with high blood pressure do not know that they have it.

Hypertension can cause damage to body organs. The high pressure makes the heart work harder, which can cause heart failure. It can also cause a blood vessel to burst, which can cause a stroke. For this reason, many people have their blood pressure checked regularly, Figure 11-15. If the blood pressure is high, a doctor may suggest a change in diet or activities. In some cases, medicines are also given to help reduce high blood pressure.

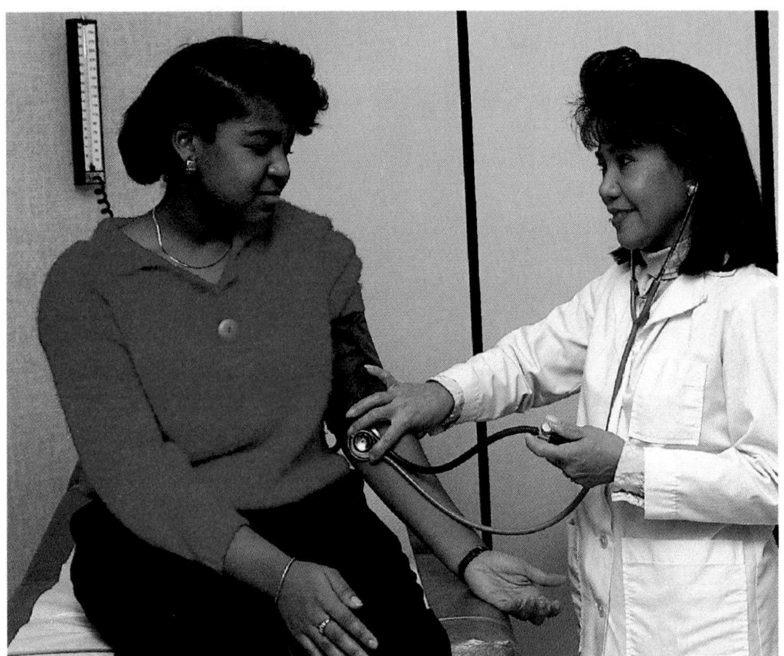

Figure 11-15 Your blood pressure is checked regularly by your doctor.

High blood pressure may be worsened by a person's diet. For example, too much salt in the diet may make hypertension worse. **Cholesterol** (kuh LES tuh ral) is a fatlike chemical found in certain foods. Cholesterol can coat the inside of arteries and cause them to become narrower. This narrowing raises the blood pressure, just as squeezing the hose did.

Bio Tip

Consumer: Foods that come from plants, such as fruit and vegetables, do not contain cholesterol. Natural foods contain much less salt (sodium) than processed foods. The amount of salt in a box of uncooked rice is much less than the amount in a frozen rice dish.

Heart Attack

The heart is made of muscle cells. These cells, just like other cells in the body, must receive oxygen and nutrients. How does the heart receive oxygen and nutrients? The heart has its own supply of blood vessels. The blood vessels that carry blood to and from the heart itself are called **coronary** (KOR uh ner ee) **vessels.** Oxygen and nutrients are carried to the heart by the blood in the coronary vessels.

Coronary vessels, like other blood vessels, can become clogged by cholesterol or a blood clot. When the vessels are clogged, blood cannot reach a part of the heart. This part of the heart then begins to die because of a lack of nutrients and oxygen. The death of a section of heart muscle is called a **heart attack.** If too much heart muscle dies, the heart is unable to pump properly. When blood doesn't deliver enough oxygen to body tissues, the person could die.

What is a heart attack?

TECHNOLOGY

Angioplasty

Angioplasty is a way of opening blocked arteries using a balloon.

1. A deflated balloon is inserted into an artery.
2. The balloon is inflated and pushes cholesterol against the artery wall.
3. The balloon is deflated and removed. The artery now has an open pathway and blood can flow freely in the vessel.

This type of operation is easier than removing a blocked blood vessel. Patients who have angioplasty may return to normal activities in a few days.

Not all patients can be treated this way. For many patients, replacing clogged vessels is the best form of treatment. A problem with angioplasty is that the cholesterol that caused the problem is still present.

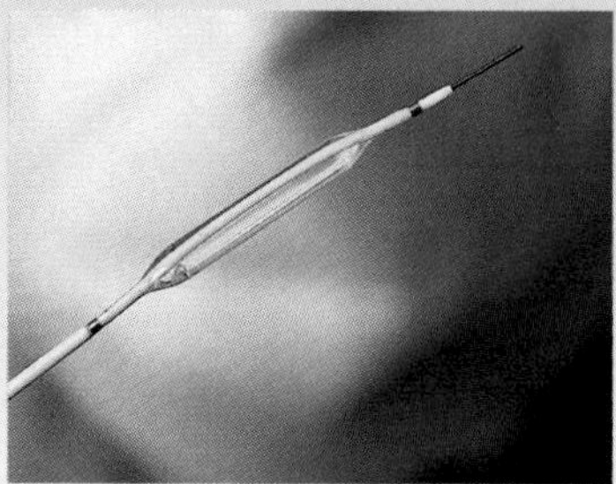

A deflated balloon is prepared for angioplasty.

Preventing Heart Problems

What can a person do to try to prevent heart problems? Scientific research has shown three things that may help. They are exercise, proper diet, and not smoking.

Why is smoking harmful to the heart?

Why is exercise important? Just as leg or arm muscles need exercise, the heart muscle also needs exercise. Physical exercise makes your heart muscle stronger because your heart beats more often when you exercise. Increasing the heartbeat rate regularly is good exercise for your heart.

Diet is also important. Eating a balanced diet can help a person avoid being overweight. Being overweight puts a strain on the heart because fat forms around the heart. The heart has to work harder to pump blood throughout the body. Eating a balanced diet also may help to reduce your cholesterol intake. Table 11-1 shows the cholesterol content of popular foods. How many foods that are high in cholesterol do you eat?

Studies show that a person who smokes is more likely to have heart problems than someone who doesn't smoke. Nicotine, a chemical in tobacco smoke, causes blood vessels to narrow in size. The heart must work harder to pump blood through these narrow vessels.

TABLE 11–1. CHOLESTEROL VALUES FOR CERTAIN FOODS

Food	Cholesterol in One Normal Serving (in milligrams)
Chicken liver	630
Egg yolk	252
Pork spareribs	120
Ham	100
Chicken	90
Flounder	70
Pizza	60
Milk (whole)	34
Butter (1 tbsp.)	20
Milk (skim)	5
Egg white	0

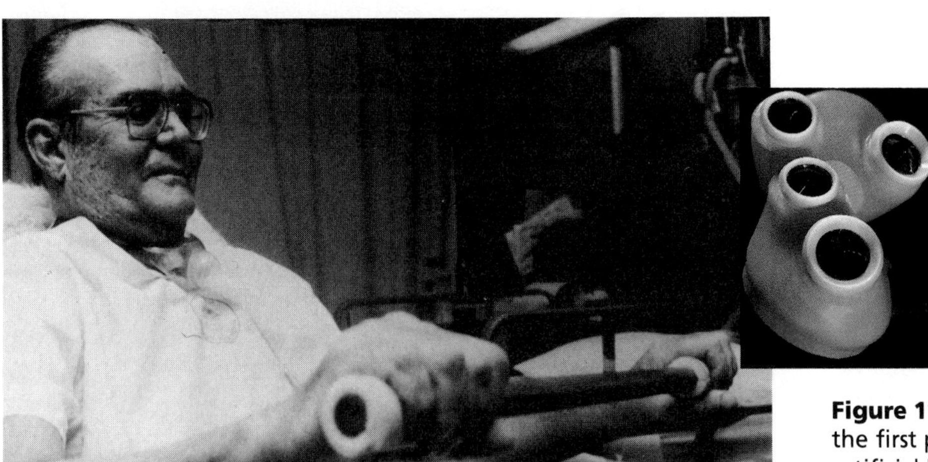

Figure 11-16 Barney Clark was the first person to receive an artificial heart like the one shown (inset). He died in 1983 after living 112 days with the plastic and metal heart.

What can be done for the person who can't be helped with exercise or diet? If blocked coronary vessels are the problem, unclogged vessels from another part of the patient's body can replace those in the heart. There are drugs that dissolve blood clots inside blood vessels. Heart transplant operations are also possible. Hearts are donated by families of persons who die.

An artificial heart made of plastic and metal has been used for heart patients in an emergency until a human heart is donated. There were 3698 people on the waiting list for a donor heart at the end of 1996. Figure 11-16 shows what this heart looks like. This type of heart has kept a person alive for over 600 days.

Check Your Understanding

16. What two measurements are made when a person's blood pressure is taken?
17. What causes a section of heart muscle to die?
18. What are three ways of preventing heart problems?
19. **Critical Thinking:** An overweight person's blood pressure is usually higher than the blood pressure of a person of average weight. Why does blood pressure increase with body weight?
20. **Biology and Reading:** A heart attack occurs when blood vessels inside the heart muscle are damaged. The scientific name for a heart attack comes from the name for these blood vessels. What is this name?

Chapter 11 Review

Summary

11:1 The Process of Circulation

1. The circulatory system works as a pickup and delivery system for the body.
2. Insects have an open circulatory system. Earthworms and humans have closed systems.

11:2 The Human Heart

3. Your heart is a muscle with four chambers. Blood is pumped from top to bottom chambers.
4. The closing of heart valves produces heart sounds.
5. The heart's right side pumps blood only to the lungs. The heart's left side supplies blood to the rest of the body.

11:3 Blood Vessels

6. Arteries have thick walls and carry blood away from the heart. Blood in arteries is under high pressure.
7. Veins have valves and carry blood to the heart.
8. Capillaries are very small blood vessels.

11:4 Problems of the Circulatory System

9. Blood pressure is caused by blood pushing against the walls of the blood vessels.
10. A heart attack is caused by blockage of the coronary blood vessels.
11. Many heart problems can be avoided by exercise, proper diet, and not smoking.

Key Science Words

aorta (p. 230)
artery (p. 226)
atrium (p. 225)
bicuspid valve (p. 227)
blood pressure (p. 232)
capillary (p. 234)
cholesterol (p. 237)
circulatory system (p. 222)
coronary vessel (p. 237)
heart attack (p. 237)
hypertension (p. 236)
pulmonary artery (p. 229)
pulmonary vein (p. 229)
semilunar valve (p. 227)
tricuspid valve (p. 227)
valve (p. 227)
vein (p. 226)
vena cava (p. 229)
ventricle (p. 225)

Testing Yourself

Using Words

Choose the word from the list of Key Science Words that best fits the definition.

1. flap in a vein that keeps blood flow going in one direction
2. smallest blood vessel
3. largest artery in body
4. valve between left atrium and ventricle
5. carries blood away from heart to lungs
6. death of a section of heart muscle
7. your heart, blood vessels, and blood
8. bottom heart chamber
9. disease caused by high blood pressure

Testing Yourself *continued*

Finding Main Ideas

List the page number where each main idea below is found. Then, explain each main idea.

10. what the "lub-dup" sounds are
11. how the traits of arteries aid blood movement
12. the condition of blood leaving the lungs
13. how simple animals can live without a circulatory system
14. what causes a heart murmur
15. what causes high blood pressure
16. what special jobs the heart's right and left sides have
17. what the main parts of the human heart are

Using Main Ideas

Answer the questions by referring to the page number after each question.

18. What is a heartbeat? (p. 227)
19. What are four traits of veins? (p. 233)
20. What is the cause of most heart attacks? (p. 237)
21. What does your circulatory system pick up and deliver? (p. 222)
22. How can people prevent heart problems? (p. 238)
23. What are the locations and names of all heart valves? (p. 227)
24. What occurs in the heart when one measures the upper and lower blood pressure? (p. 236)
25. How is the pumping action of the heart similar to the squeezing of a plastic bottle? (p. 232)
26. What is the pathway that blood takes as it leaves your heart's right and left ventricles? (p. 226)

Skill Review ✓

For more help, refer to the ***Skill Handbook,*** *pages 704-719.*

1. **Design an experiment:** Design an experiment to find out whether the average pulse rate of a young child is different from that of a teenager.
2. **Understand science words:** What do the word parts *bi* and *tri* mean?
3. **Formulate a model:** Construct and label a three-dimensional model of the heart and the blood vessels that lead to and from it.
4. **Calculate:** The total sodium (salt) content of a bag of popcorn is 680 mg. The total sodium content of a bag of potato chips is 1140 mg. The popcorn bag contains 4 servings while the bag of potato chips contains 6 servings. Determine the sodium content of a single serving of each snack. Which snack has the lower sodium content?

Finding Out More

Critical Thinking

1. Some babies are born with a hole between the right and left ventricles. Why is this a problem?
2. Suppose you are asked to design an artificial heart. What technical problems might you need to solve?

Applications

1. Design meals for three days that would be suitable for a person who wishes to reduce his cholesterol.
2. Explain why your heartbeat rate speeds up during exercise.

CHAPTER PREVIEW

Chapter Content

Review this outline for Chapter 12 before you read the chapter.

Skills in this Chapter

The skills that you will use in this chapter are listed below.

- In **Lab 12-1,** you will use a microscope, measure in SI, and compare. In **Lab 12-2,** you will use a microscope, compare, and classify.
- In the **Skill Checks,** you will infer and understand science words.
- In the **Mini Lab,** you will formulate a model.

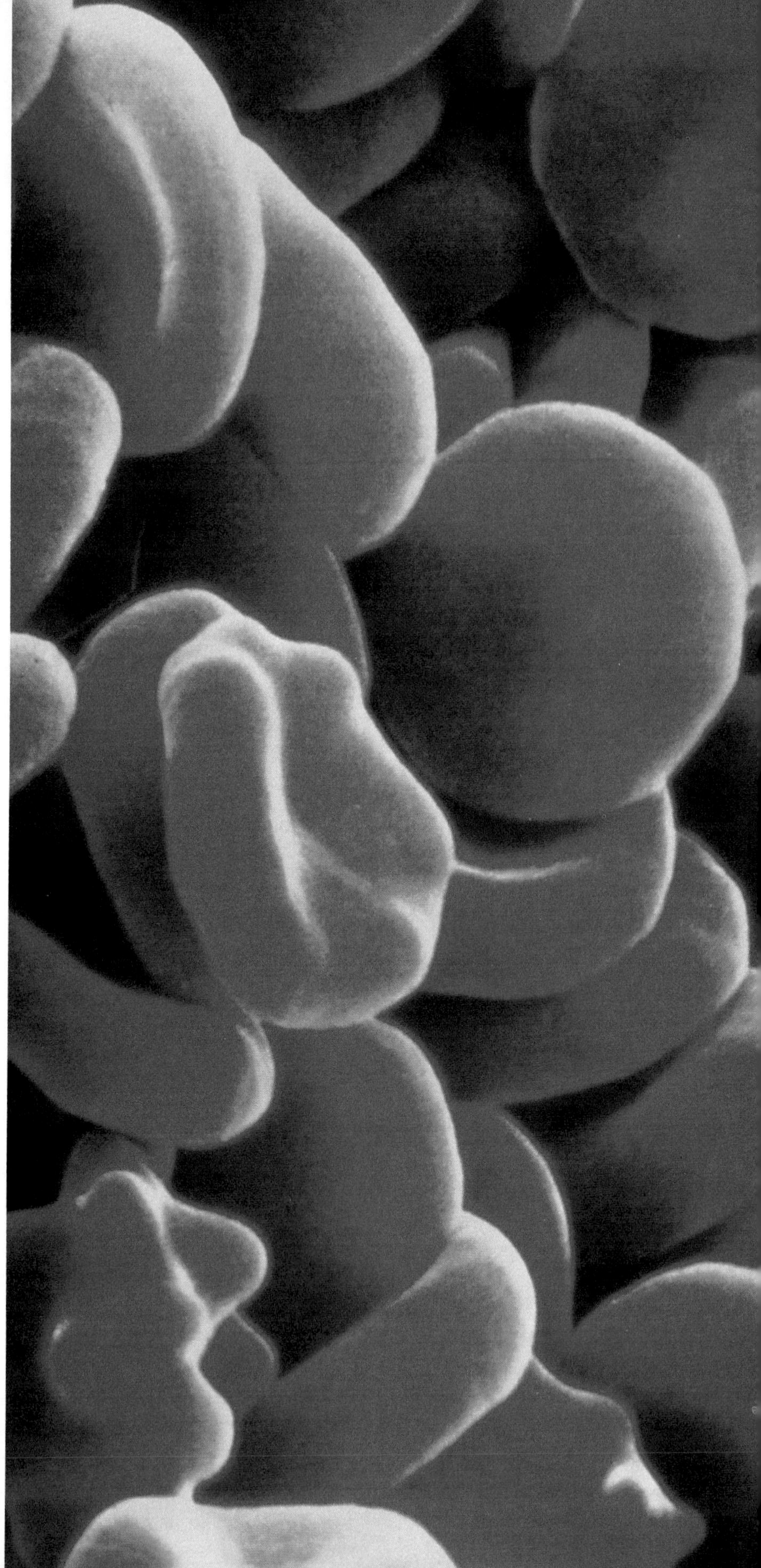

Chapter 12

Blood

Suppose you were asked to classify blood as either a solid or a liquid. If you have ever had a bloody nose or deep cut, you would probably classify blood as a liquid. But if you viewed blood through a microscope, you might classify it as a solid. The large photograph on the left shows what your blood looks like magnified by an electron microscope. Notice the solid particles. These round, red parts are cells. The small photograph below shows bottles of fruit juices with about the same volume of liquid as you have blood in your body, or about five liters. You may be asking yourself: if blood is made of cells, how can it be liquid in form? After reading this chapter, you will know the answer to this puzzling question.

Try This!

How much blood does the body contain? The body of an adult human contains about 5 liters of blood. List three ways of showing this quantity. For example, your body contains as much blood as 5 quarts of milk. (1 liter is a little more than 1 quart.)

BIOLOGY Online

Visit the Glencoe Science Web site at science.glencoe.com to find links about **blood.**

Your body holds five liters of blood.

12:1 The Role of Blood

Objectives

1. **Identify** the pickup and delivery jobs of the blood.
2. **Explain** the pickup and delivery system in animals without blood.

In Chapter 11 you learned that blood is part of your circulatory system. Blood has many important jobs. You will learn about the roles of blood as you read this chapter.

Blood Functions

Blood is a body tissue that is part liquid and part cells. What are the functions of blood? Because it's part of the circulatory system, the blood's functions are to pick up and deliver nutrients and wastes.

First, let's list some of the delivery jobs of blood.

1. Blood delivers digested nutrients from fats, proteins, and carbohydrates to all body cells.
2. Blood delivers oxygen to all body cells.
3. Blood delivers chemical messengers to some body cells.
4. Blood delivers water, minerals, and vitamins to cells.

Next, let's list the pickup jobs of blood.

1. Blood picks up carbon dioxide waste from cells and carries it to the lungs.
2. Blood picks up chemical waste from cells and carries it to the kidneys.
3. Blood moves excess body heat into the skin.

Where does blood carry the wastes that it picks up from body cells?

You may have noticed the red flush that appears on the skin of a person after heavy exercise. Exercise results in the buildup of excess body heat. The red color is due to the widening of blood capillaries in the skin, Figure 12-1.

a

b

Figure 12-1 As you begin to exercise, the body heats up (a). The blood vessels in the skin get wider and excess heat is given off through the skin (b).

Figure 12-2 Many sea animals, such as these sea anemones, have no blood or blood vessels. Water brings oxygen and nutrients to their cells.

Other functions of blood include helping to fight diseases and helping to stop bleeding.

Not all animals have blood. Animals such as sponges, jellyfish, and flatworms have no blood or blood vessels. How do the cells of these animals get oxygen or food? How are they able to get rid of waste chemicals? Most animals without blood live in water. Water serves as their "blood." It brings needed oxygen and food to all cells. It also carries away any waste chemicals from the body.

How is blood to a human like seawater to a jellyfish?

Check Your Understanding

1. What are six substances blood delivers to body cells?
2. What three substances are picked up by blood?
3. How do animals without blood get oxygen or food?
4. **Critical Thinking:** How does blood help a person maintain normal body temperature?
5. **Biology and Reading:** One of the following sentences is a fact. The other sentence can be made into a fact by changing the boldfaced word. Change the word and then write both facts on your paper.

The average adult body contains about five **liters** of blood.
Blood is a body **system** made up of liquid and cells.

Red Blood Cells

Lab 12–1

Problem: How do red blood cells of different animals compare?

Skills

use a microscope, measure in SI, compare

Materials

microscope
petri dish cover
metric ruler
prepared slides of human and frog blood

Procedure

1. Copy the data table.
2. Use the cover of a petri dish to draw a circle on a piece of paper.
3. **Use a microscope:** Look at a prepared slide of human red blood cells under low power on the microscope. Change to high power. Most of the cells you see are red blood cells. Cells that appear blue are not red blood cells.
4. Draw a red blood cell in your circle. Assume your circle is the same as your field of view. Draw the cell to scale.
5. **Measure in SI:** Use a ruler to measure the size in millimeters of the blood cell in your drawing. Record your answer in your data table.
6. Repeat steps 3 through 5 with a prepared slide of frog blood.
7. On your diagrams, check for the presence of a nucleus and a cell membrane. Check these parts in your table if present.
8. Describe and record the shapes of the blood cells in your table. Use the following choices—round or oval.
9. Complete the rest of the data table using the photo of bird blood.

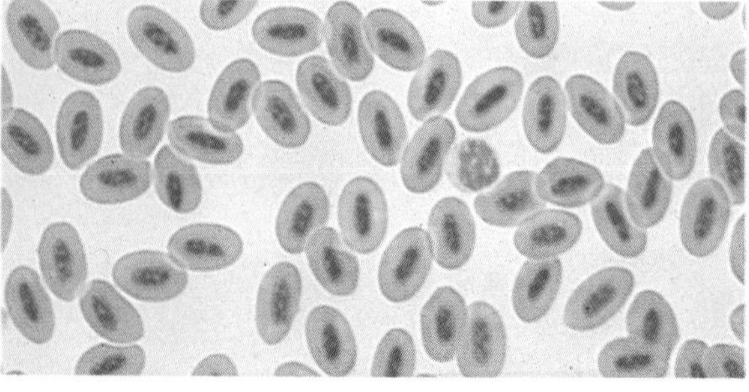

BLOOD FROM	SIZE	SHAPE	PARTS PRESENT?	
			NUCLEUS	CELL MEMBRANE
Human				
Frog				
Bird	—			

Data and Observations

1. Which animals have red blood cells:
   (a) with a nucleus?
   (b) that are round?
   (c) that are oval?
2. Which animal has the larger red blood cells—frog or human?

Analyze and Apply

1. What is the main job of red blood cells?
2. In what blood vessels do red blood cells transport oxygen?
3. **Apply:** Based upon the appearance of red blood cells, which two animals are most closely related?

Extension

Calculate: The photo of bird red blood cells is magnified 1000 times. What is the actual measurement of the diameter of one blood cell?

12:2 Parts of Human Blood

Blood is different from most body tissues. Not only is it able to flow, but it's also part living and part nonliving.

Objectives

3. **Describe** the living and nonliving parts of blood.
4. **Give the functions** of the blood cells and cell-like parts.
5. **Explain** the problems that may occur with blood cells and platelets.

Key Science Words

plasma
red blood cell
bone marrow
hemoglobin
anemia
white blood cell
leukemia
platelet
hemophilia

Plasma

Let's look at two tubes of blood. The tube on the left in Figure 12-3a shows blood just removed from a person's vein. The blood is a deep red color. No special parts can be seen. Now look at the tube on the right. This tube has been sitting for about an hour. Notice how the blood has separated into two parts. Blood cells have settled to the bottom. A yellow liquid appears on top.

Plasma (PLAZ muh) is the nonliving, yellow liquid part of blood. Blood plasma is 92 percent water. The remaining eight percent includes blood proteins, nutrients, salts, and waste chemicals.

Most of the pickup and delivery jobs of blood are carried out by plasma.

Red Blood Cells

Red blood cells are cells in the blood that carry oxygen to the body tissues. Red blood cells make up most of the blood. There are about five million red blood cells in a small drop of blood. Red blood cells are part of the living portion of blood. They give blood its red color.

Which substance makes up the nonliving part of blood?

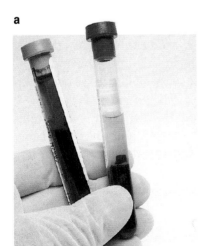

Figure 12-3b shows red blood cells as seen under a light microscope. Look at this photograph as you read about red blood cells.

a

b

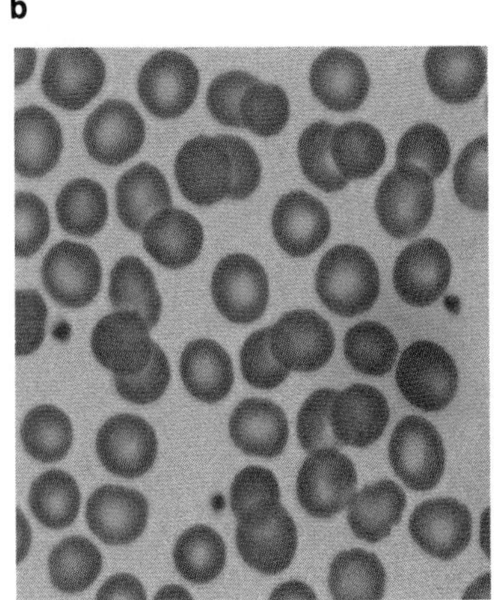

Figure 12-3 When blood is left to stand (left tube) the red cells and plasma separate out (right tube) (a). Blood cells and plasma can be examined under a light microscope (b).

Study Tip: Use this idea map as a study guide for comparing the living and nonliving parts of blood.

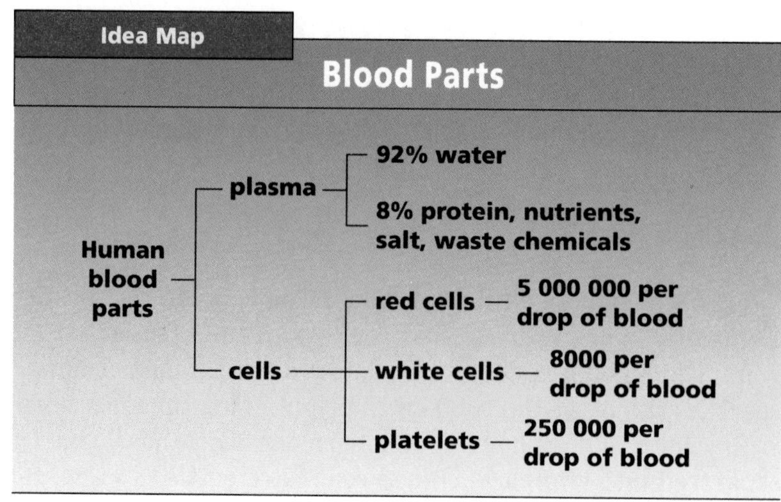

Red blood cells are round. They have thin centers and thick edges. They look like doughnuts without holes. Red blood cells have a nucleus only when they are first formed. The nucleus is lost as the cell matures and begins to move through the blood vessels.

Red blood cells have a life span of about 120 days. The body must constantly make new red cells. Red blood cells are made primarily in your bone marrow. **Bone marrow** is the soft center part of the bone.

Mini Lab

How Is Blood Made Up?

Formulate a model: Add 20 mL of cooking oil to 20 mL of water and a drop of red food coloring. Shake the tube. Allow the liquids to separate. What do the layers represent? *For more help, refer to the* ***Skill Handbook,*** *pages 706-711.*

Roles of Red Blood Cells

Red blood cells are the main part of blood that helps with the blood's job of pickup and delivery. They deliver oxygen to all body cells. Oxygen coming into the lungs is picked up by red blood cells as the cells pass through capillaries in the lungs. Red blood cells are then carried by the plasma to all body parts. Once these cells reach the cells of the body, oxygen is given up by the red blood cells.

How can red blood cells perform this job of oxygen pickup and delivery so well? The answer lies in a special chemical found in red blood cells. This chemical is called hemoglobin (HEE muh gloh bun). **Hemoglobin** is a protein in red blood cells that joins with oxygen and gives the red cells their color. Hemoglobin in blood cells carries oxygen to other body cells.

What is anemia? How is it treated?

Hemoglobin contains the mineral iron. If a person's diet is low in iron, the total number of red blood cells and hemoglobin in the blood decreases. This decrease in red blood cells and hemoglobin can lead to anemia. **Anemia** is a

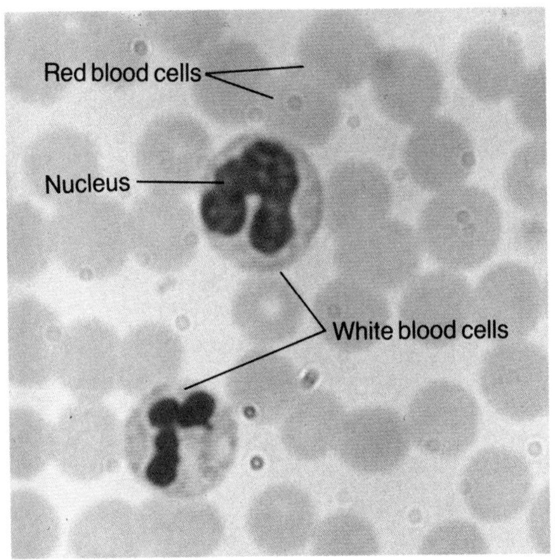

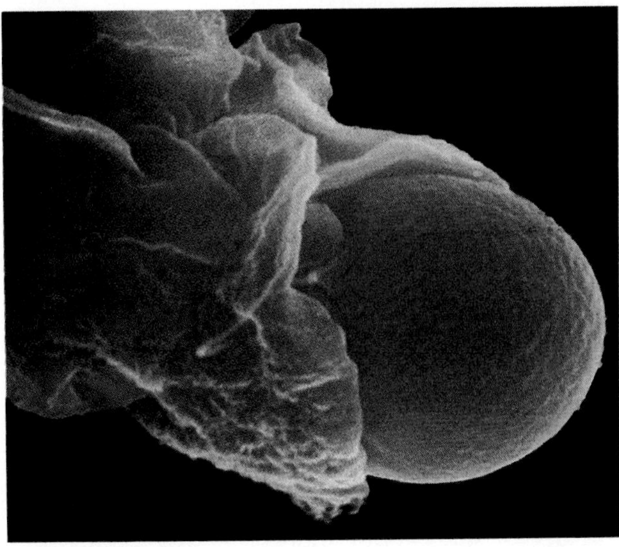

condition in which there are too few red blood cells in the blood. A person with anemia will usually feel weak, tired, and short of breath. These symptoms are caused by too little oxygen reaching body cells. Extra iron in the diet may help solve the problem. How much iron is needed each day? Look back at Table 9-2 on page 190 for the answer.

Figure 12-4 White blood cells as seen under a light microscope (left) are larger than red blood cells. In fact, when a red blood cell dies, white cells surround and destroy it, as shown under an electron microscope (right).

White Blood Cells

The second type of blood cell is the white blood cell. **White blood cells** are the cells in the blood that destroy harmful microbes, remove dead cells, and make proteins that help prevent disease.

White blood cells are part of the living portion of blood. Figure 12-4 shows white blood cells magnified under a light microscope and an electron microscope.

A white blood cell has a nucleus. The nucleus appears dark blue in both cells in Figure 12-4 (left). Also notice that white blood cells are larger than red blood cells. Certain white blood cells may live for months or years, but most have a life span of about 10 days. A healthy person has about 8000 white cells in a very small drop of blood.

Like red cells, white cells are made in bone marrow. Unlike red cells, however, white cells are also made in organs such as the spleen, thymus gland, and tonsils.

Some kinds of white blood cells can move out of capillaries and travel among the body cells. Figure 12-5 shows how white blood cells move like amoebas and at times even look like them.

Figure 12-5 An important trait of white blood cells is their ability to move into body tissues from the capillaries.

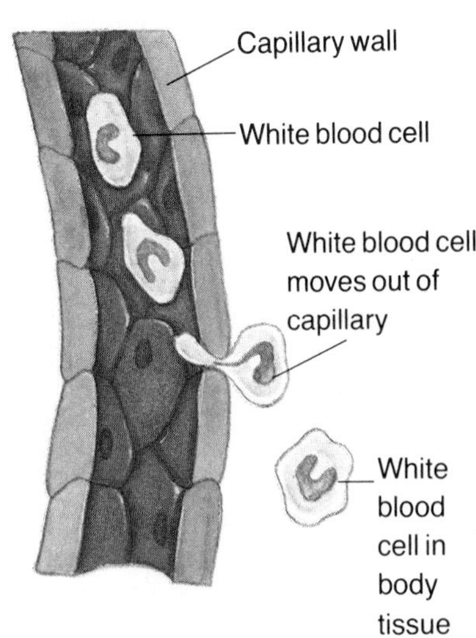

TECHNOLOGY

Eye "Fingerprints"

Blood vessels can be seen clearly in the human eye. These vessels run along the back surface of the eye. That is why a doctor sometimes shines a flashlight into your eyes. The condition of these blood vessels gives the doctor clues about your general health.

The pattern of blood vessels in the eye is different for every person. It's like having a "fingerprint" of your eye.

Using a computer scanner, scientists can identify people based on blood vessel patterns in their eyes. This technology can be used to screen applicants for drivers' licenses. It prevents people who already have licenses from obtaining duplicates. The states of Wisconsin and California are already planning to use this new technology in the issuing of drivers' licenses.

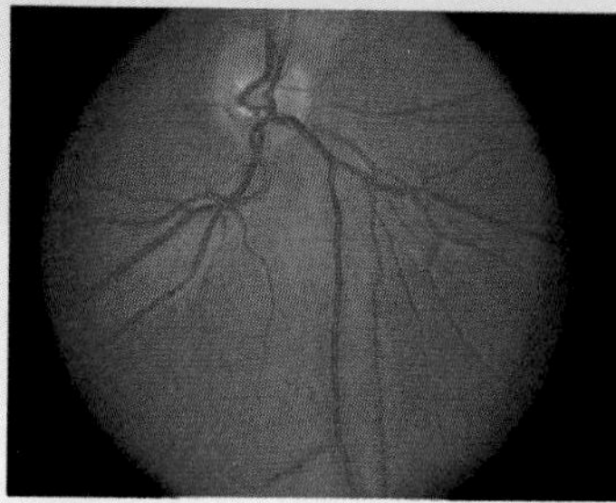

A view of the back of your eye

Roles of White Blood Cells

Have you ever had a cut that became infected? An infection is usually caused by an attack on your body cells by bacteria.

White blood cells move to an infection and destroy the bacteria causing it. There's a rapid increase in white cells at the time of an infection. The added numbers of white blood cells help to destroy more bacteria at a faster rate. After an infection is over, the number of white cells returns to normal. Another job of white blood cells is to rid the body of dead cells. Certain white cells can move about the body and "eat" dead cells just as they do bacteria, Figure 12-6.

Figure 12-6 A white blood cell detects a bacterium and moves toward this foreign body to destroy it.

Increased amounts of white blood cells can sometimes cause problems such as leukemia (lew KEE mee uh). **Leukemia** is a blood cancer in which the number of white blood cells increases at an abnormally fast rate.

There are three main differences between the rise in white blood cell numbers during an infection and during leukemia.

1. During leukemia, the number of white cells in each drop of blood may reach 100 000 or more. During an infection, the number rarely goes above 30 000.

2. With leukemia, the number of white cells does not return to normal.
3. The cells formed in such large numbers are not normal cells. They can't do the job performed by normal white blood cells.

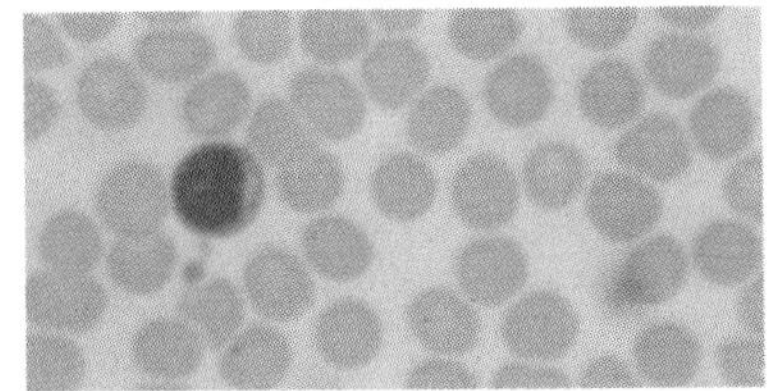

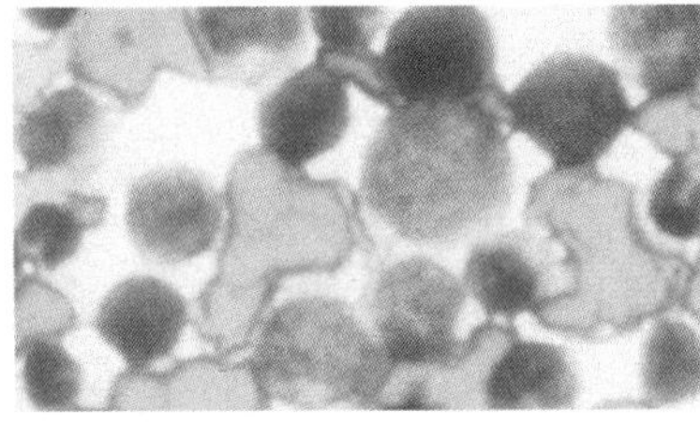

Figure 12-7 Which photo (top or bottom) is from a person with leukemia?

Platelets

Cuts and scrapes are common events in our lives. We usually don't worry about these injuries because we know that the flow of blood will stop quickly. A clot forms, and in a few days a scab appears. A blood clot is blood that has formed a plug to stop the bleeding. Figure 12-8 (left) shows a clot starting to form.

The forming of blood clots leads us to the last living part of blood, platelets (PLAYT lutz). **Platelets** are cell parts that aid in forming blood clots. Platelets are not complete cells. However, platelets do come from cells that break apart, so they are still a living part of blood. Notice in Figure 12-8 (right) that the platelets are much smaller than red cells.

What is the main job of platelets?

When an injury occurs in the body, platelets break apart and release a chemical. This chemical starts the formation of a clot. Several chemicals are involved in clot formation.

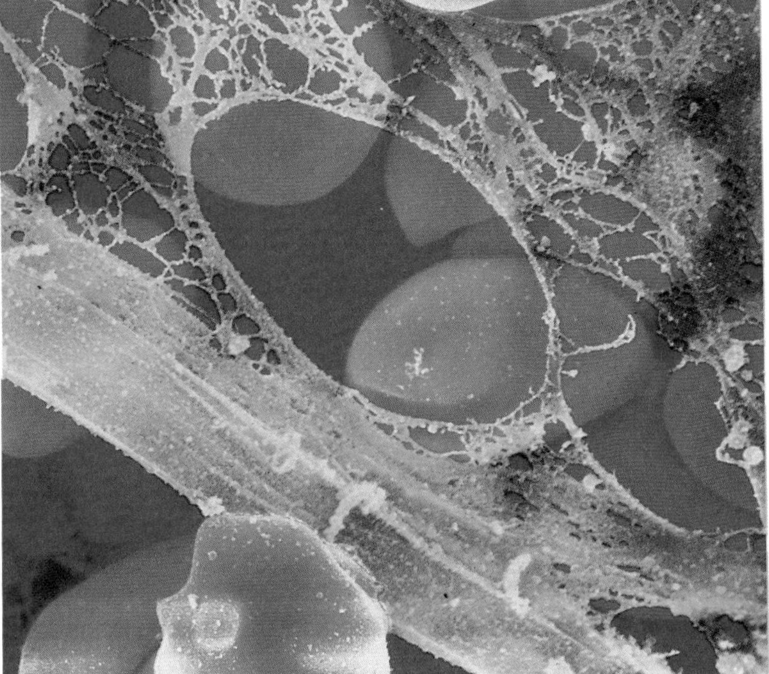

Figure 12-8 Red blood cells caught in fibers form the beginning of a blood clot, as seen under an electron microscope (left). Platelets, as seen under a light microscope (right), produce a chemical that binds the fibers together.

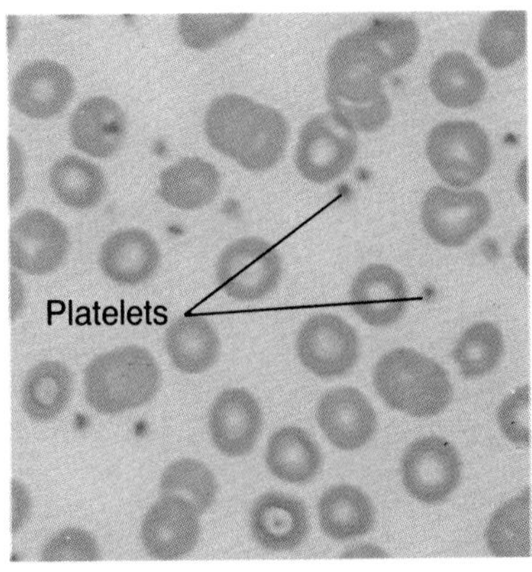

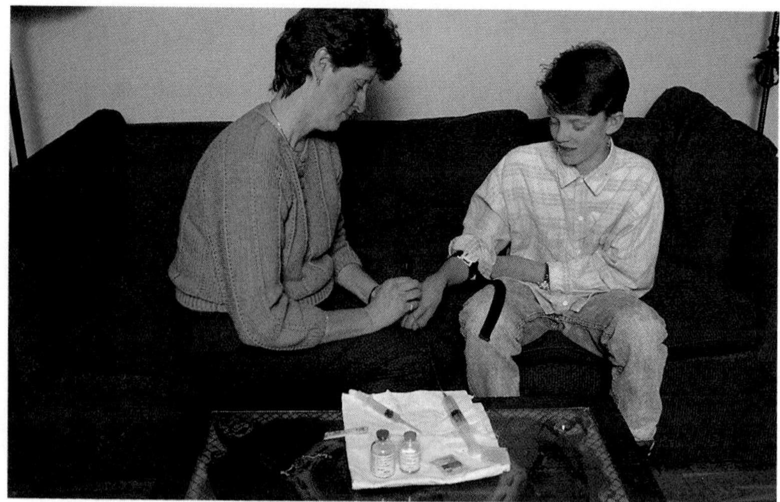

Figure 12-9 A person with hemophilia needs a transfusion of blood clotting factor, even after a minor accident.

> ### Skill Check
> **Understand science words: hemophilia.** The word part *hemo* means blood. In your dictionary, find three words that contain the word part *hemo. For more help, refer to the* ***Skill Handbook,*** *pages 706-711.*

A small drop of blood contains about 250 000 platelets. The life span of platelets is very short, only about five days. Like red cells and some white cells, platelets are made in the bone marrow.

Two problems can happen with platelets. First if the number of platelets is too low, it's hard for blood clots to form. Second, if platelets lack the clotting chemical, this causes hemophilia (hee muh FIHL ee uh). **Hemophilia** is a disease in which a person's blood won't clot. A minor cut or bruise can be very dangerous to a person with hemophilia because of the large amount of blood that might be lost. Hemophilia is a genetic disease. This means that a person inherits the disease from his or her parents.

Check Your Understanding

6. What substances make up the living part of blood?
7. What are three jobs of white blood cells?
8. What are three ways the rise in white blood cell numbers during leukemia differs from the rise in white blood cell numbers during infection?
9. **Critical Thinking:** You have learned that increasing the amount of iron in one's diet is a way of treating anemia. What does this indicate about the relationship between iron and red blood cells?
10. **Biology and Reading:** What is the maximum life span of a white blood cell? What is the average life span of most white blood cells?

Lab 12–2

Human Blood Cells

Problem: How can human blood cells be counted, identified, and compared?

Skills

use a microscope, compare, classify

Materials

sheet of clear plastic
marking pencil
prepared slide of human blood
microscope

Procedure

1. Copy the data table.
2. Figures A and B show what blood cells would look like under the microscope. Each figure represents blood from a healthy person. Place a piece of clear plastic over Figure A.
3. Count and record the number of red blood cells. Use the marking pencil to check the cells counted so that you do not count them twice.
4. Count and record the number of white blood cells.
5. Count and record the number of platelets.
6. Repeat steps 2 through 5 for Figure B.
7. **Use a microscope:** Look at a prepared slide of human blood under low power on the microscope. Change to high power.
8. Draw and label a red blood cell, white blood cell, and platelet.

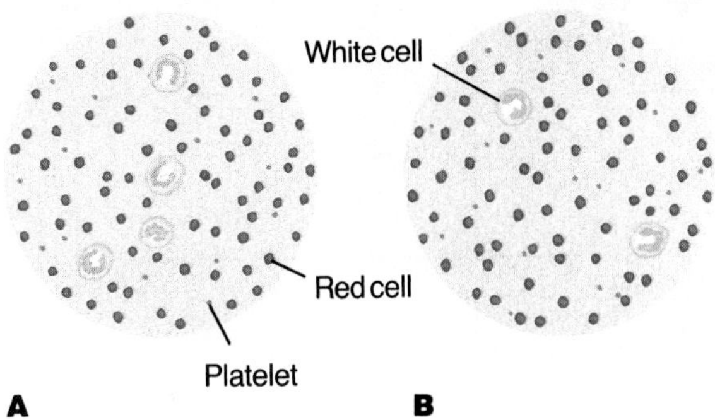

NUMBER OF CELLS	FIGURE A	FIGURE B	PREPARED SLIDE
Red cells			
White cells			
Platelets			

9. Record the numbers of red blood cells, white blood cells, and platelets in an area similar to that in the figures.

Data and Observations

1. According to your results with Figures A and B:
   (a) which blood cell type is most common?
   (b) which type is second-most common?
   (c) which type is least common?
2. According to your results with the prepared slide:
   (a) which blood cell type is most common?
   (b) which type is second-most common?
   (c) which type is least common?

Analyze and Apply

1. **Classify:** List three ways that:
   (a) red cells differ from white cells.
   (b) red cells differ from platelets.
2. Which count of cells, from figures or from a prepared slide, is more accurate?
3. **Apply:** How is a cell count important when diagnosing an illness?

Extension

Compare: Examine a prepared slide of frog blood under high power. Count the numbers of red blood cells and white blood cells. Compare frog and human blood.

12:3 Blood Types

Objectives

6. **Compare** the red cell and plasma proteins in the four main blood types.
7. **Explain** why blood types can't be mixed.

There are different kinds or types of blood. If you were to donate blood, your blood usually would be given to someone with the same type. If the wrong blood type were given, the person receiving the blood could have serious problems or even die.

Blood Types Differ

There are four main blood types: A, B, AB, and O. What makes each blood type different from the others? The difference is due to proteins found on red blood cells and in blood plasma. Look at Figure 12-10 and notice the different model shapes used to show red cell proteins and plasma proteins. If the protein is on a red blood cell, we shall call it a red cell protein. If it is in the plasma, we shall call it a plasma protein.

How is the plasma protein of type A blood like the red cell protein of type B blood?

Note that the red cell protein shapes on types A and B blood cells are different. Also note that the red cell protein shapes on type AB are the same as both types A and B. Type O red blood cells have no red cell proteins.

Now notice the shapes of the plasma proteins. Type A plasma proteins are different from those in type B. Type AB has no plasma proteins, and type O has plasma proteins like types A and B.

Figure 12-10 Blood types differ in their proteins. Each blood type has red cell and plasma proteins that don't fit each other.

Blood type	Red cell protein present	Plasma protein present	Plasma protein and cell protein
A	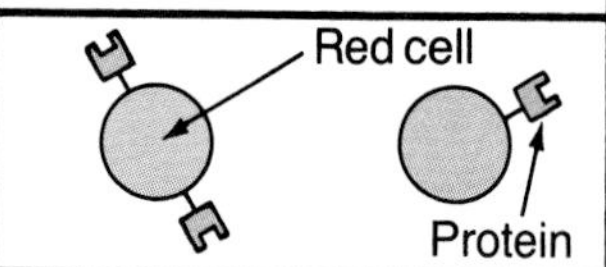 Red cell Protein		 No fit
B	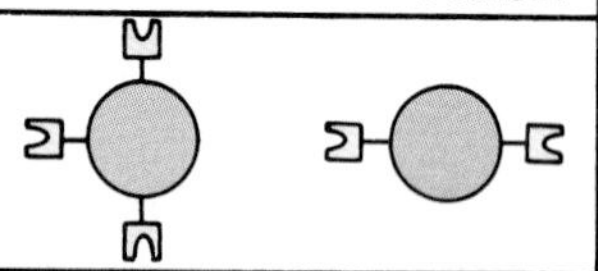		 No fit
AB		None	 No fit
O	 None		 No fit

Mixing Blood Types

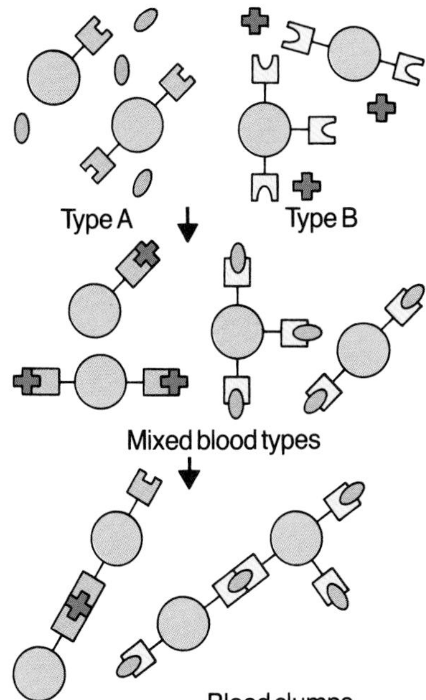

Figure 12-11 Blood clots form when blood types A and B are mixed.

Why is it safe to mix certain blood types and not others? The answer lies in the shapes of the red cell proteins and plasma proteins. Figure 12-10 is a model that shows how the shape of the red cell protein in type A blood can't fit into the shape of its own plasma protein. The red cell and plasma proteins have different shapes. The same is true for B, AB, and O types as well.

What happens when different blood types are mixed? Our models can help explain the result. Let's see what happens when a person with type A blood receives type B blood. Proteins in the plasma of type A blood fit like puzzle pieces into the red cell protein shape of type B blood. Proteins in the plasma of type B blood also fit like puzzle pieces into the red cell protein shape of type A blood. Note this happening in Figure 12-11.

Many type A and B blood cells now join together like all the completed pieces in a puzzle and large clumps of blood cells form. These clumps are too large to pass through capillaries. They plug these blood vessels and prevent the normal flow of blood. If these clumps block vessels in the heart, brain, or lungs, death can occur.

Why can't blood types be mixed?

Why wouldn't it be safe for a person with type A blood to receive type AB blood? Why wouldn't it be safe for a person with type O blood to receive blood types A or B? Now you know why most people receive blood that is the same type as their own.

Check Your Understanding

11. What are the four main blood types?
12. Describe the model shapes of the red cell proteins and plasma proteins found in type O blood.
13. What happens to blood proteins and plasma when blood types A and B are mixed?
14. **Critical Thinking:** Sometimes AB plasma is transfused from one person to another. What trait of this plasma makes it safer to use in a transfusion than whole blood?
15. **Biology and Reading:** Look at Figure 12-10. Notice the shapes of plasma proteins for type O blood. Now, look at the shapes of the red cell proteins for type AB blood. What would happen if you mixed type O blood with type AB blood?

12:4 Immunity

Objectives

8. Describe how the immune system works.

9. Explain how shots prevent disease.

10. Discuss the AIDS virus.

Key Science Words

immune system
antibody
antigen
immunity
AIDS

Another job of your blood helps you stay healthy. Certain blood cells play an important role in helping your body get rid of disease-causing viruses and bacteria.

Your Immune System

Your immune system cures you of the flu, measles, mumps, or even boils. The **immune system** is made of proteins, cells, and tissues that identify and defend the body against foreign chemicals and organisms. It helps to keep you free of disease. Actually, the word *immune* means "to free."

Teardrops, mucus, and skin are part of your immune system. They prevent disease-causing organisms from entering your body. Other parts of the immune system are shown in Figure 12-12. Locate the parts of the immune system and examine their roles.

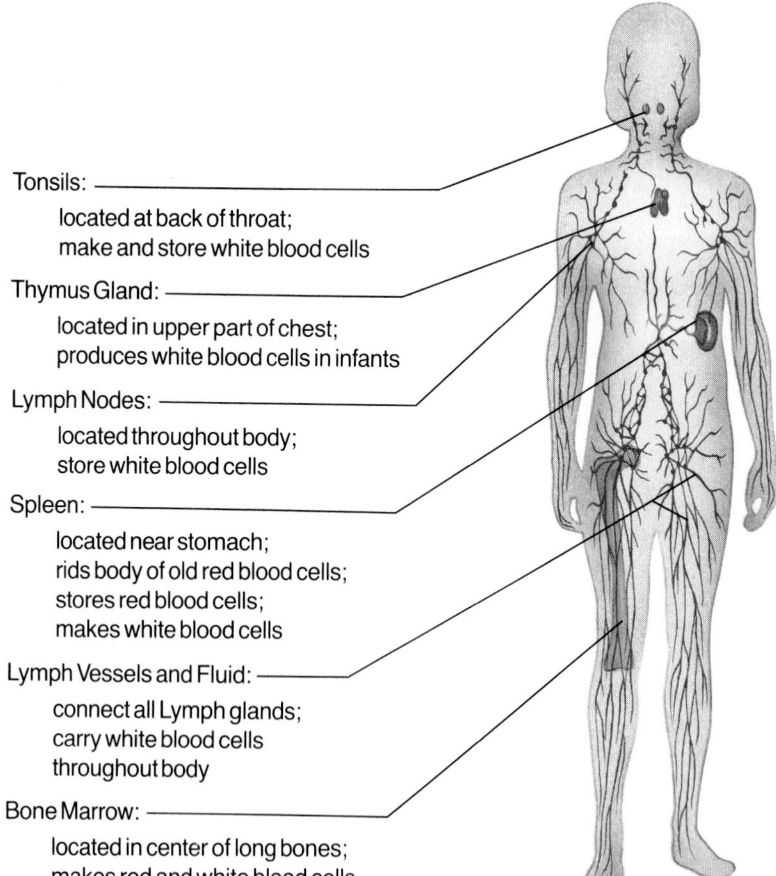
Tonsils:
located at back of throat;
make and store white blood cells
Thymus Gland:
located in upper part of chest;
produces white blood cells in infants
Lymph Nodes:
located throughout body;
store white blood cells
Spleen:
located near stomach;
rids body of old red blood cells;
stores red blood cells;
makes white blood cells
Lymph Vessels and Fluid:
connect all Lymph glands;
carry white blood cells
throughout body
Bone Marrow:
located in center of long bones;
makes red and white blood cells

Figure 12-12 In which organs of the immune system are white cells made?

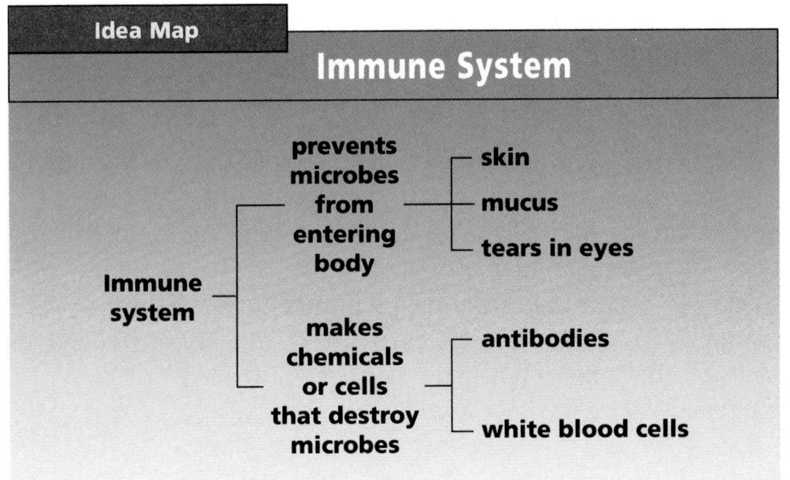

Study Tip: Use this idea map as a study guide for identifying the functions of the immune system.

How the Immune System Works

There are many kinds of white blood cells. Each kind has its own job in the protection of the body. Some white blood cells have the job of making special chemicals called antibodies (ANT ih bohd eez). **Antibodies** are chemicals that help destroy bacteria or viruses. Bacteria, viruses, and other foreign substances are examples of antigens (ANT ih junz) that enter our bodies. **Antigens** are foreign substances, usually proteins, that invade the body and cause diseases.

Bio Tip

Health: Have you had your shots? Check Table 12-2 on page 258 to see which shots you might need.

What are antigens?

Models can show how antibodies made by white cells get rid of antigens such as bacteria. Figure 12-13 shows a white blood cell with a model of an antibody on its cell membrane. The antibody looks like the letter *M*. Certain antibodies stay attached to the cell membrane. Other antibodies are released into the blood plasma. These released antibodies match a bacterium that has an exactly opposite "shape" on its surface as shown in Figure 12-13. Note that the antibody and antigen shapes fit together like a lock and key.

When the antigen and antibody on the surface of a white cell fit together correctly, the bacterium may be easily destroyed. First, the cell membrane of the bacterium may break open. When the cell membrane is broken, the bacterium will die. Second, with antibodies stuck to it, the bacterium can now be destroyed by other white blood cells. Each different type of virus or bacterium that enters the body has a different antigen shape on its surface. The white blood cells must have a similar number of different antibody shapes.

Figure 12-13 An antigen is destroyed when a white cell produces a matching antibody.

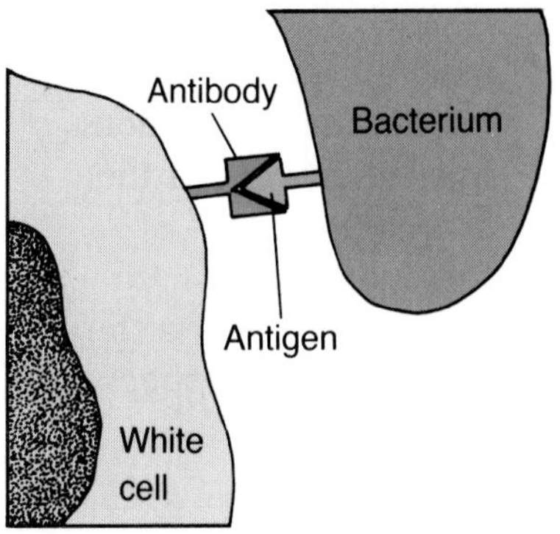

TABLE 12–1. CAUSES OF SOME DISEASES	
Disease	**Cause**
Diphtheria	Bacteria
Tetanus (Lock Jaw)	Bacteria
Pertussis (Whooping cough)	Bacteria
Polio (Infantile paralysis)	Virus
Measles	Virus
Rubella (German measles)	Virus
Mumps	Virus

The immune system now does a remarkable thing. It begins to make many new white blood cells, each with the shape of the antibody just used to get rid of the antigen. It's as if the immune system has a memory. White blood cells will reproduce antibodies in the bloodstream for many months or years. Their job is to prevent future disease caused by the same type of bacteria. If the same type of bacteria gets into your body five years later, the white blood cells with the antibodies will quickly destroy the bacteria. This time, the bacteria are so quickly destroyed that you may not even know you were invaded by the bacteria.

How Shots Keep You Healthy

What occurs in a person's body after receiving a DPT shot?

When you receive shots to prevent a disease, your immune system's memory often is being refreshed. DPT shots against diphtheria (dif THIHR ee uh), pertussis (pur TUH suhs), and tetanus (TET nus) are good examples. The DPT shot is made up of proteins from these three bacterial diseases.

When you receive a DPT shot, it's as if you are getting antigens of these three diseases. Your immune system begins to make antibodies against these antigens. Or, if you have already come into contact with the disease, your body may already have a memory of the antigens. Later on, if the actual bacteria causing these diseases were to enter your body, your immune system would be ready and waiting. You will have gained immunity. Table 12-1 lists other diseases for which you may have received shots. **Immunity** is the ability of a person who once had a disease to be protected from getting the same disease again. Table 12-2 shows the schedule for receiving shots.

Skill Check

Infer: It's common for a person to get the measles only once during his or her lifetime. Yet, a person can get the flu many times during his or her lifetime. What does this indicate about the body's "memory" of these immunities? *For more help, refer to the* ***Skill Handbook,*** *pages 706-711.*

TABLE 12–2. IMMUNIZATION SCHEDULE FOR PERSONS AGE 7 AND OLDER*

Spacing	Vaccines	Explanation
1st visit	Td (adult) TOPV M/M/R	Td—tetanus and diphtheria adult vaccine TOPV—trivalent oral polio vaccine M/M/R—one dose of measles/mumps/rubella combined vaccine
2 months after first	Td (Adult) TOPV	
6—12 months after 2nd	Td (Adult) TOPV	

*who did not begin their immunization series before age 15 months

AIDS and the Immune System

Acquired Immune Deficiency Syndrome, or **AIDS,** is a disease of the immune system. AIDS is caused by a virus that reproduces only inside one kind of white blood cell.

What happens when this type of white blood cell is totally destroyed? Other diseases can then easily invade the body. Diseases that the body's immune system would normally overcome can now cause death.

How can one be infected by the AIDS virus, Figure 12-14? The virus is present in body fluids, such as blood and semen. There are four known ways the virus can be passed from one person to another. One is during sexual intercourse, when body fluids containing the virus may be passed along through broken body tissues. Second, another way is by sharing needles during drug use. The needle may have a small amount of blood on it. If the blood has the AIDS virus in it, the virus can be injected into the next person's blood. Third, the AIDS virus may be passed from a pregnant woman to her unborn child. Fourth, it's possible to get AIDS from a blood transfusion if the blood is infected. Modern technology has made it possible to test and reject blood that has the AIDS virus.

AIDS is spreading rapidly and there is no known cure for it. Through mid-1996, there have been more than 510 000 recorded cases of AIDS in the United States. Of this number, more than 315 000, or over 60 percent, have died of AIDS.

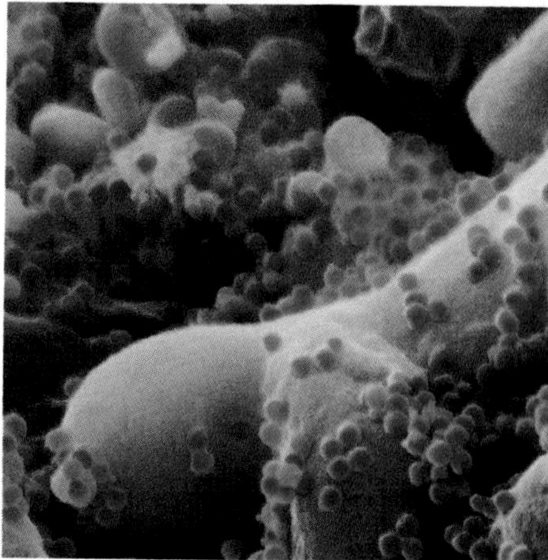

Figure 12-14 AIDS virus particles attack white blood cells and destroy the immune system.

What effect does the AIDS virus have on a person's immune system?

Check Your Understanding

16. Compare antigens and antibodies.
17. What is a DPT shot made up of?
18. What are four ways the AIDS virus can be passed from one person to another?
19. **Critical Thinking:** Why do you think there's not a shot against AIDS?
20. **Biology and Writing:** When a person has an organ transplant, the transplanted organ has antigens different from those of the person receiving the organ. Usually, the person receiving the organ is given drugs that lower the body's immune response. Explain how this treatment might help the transplant procedure.

Chapter 12 Review

Summary

12:1 The Role of Blood

1. Blood is a body tissue that helps with delivery and pickup of chemicals, nutrients, gases, and wastes.
2. Animals without blood usually live in water. Water delivers oxygen and food to the cells and picks up wastes.

12:2 Parts of Human Blood

3. Blood plasma is a nonliving, yellow liquid that makes up 55 percent of blood. Red blood cells, white blood cells, and platelets are the living parts of blood.
4. Red blood cells carry oxygen in the blood. White blood cells destroy harmful microbes and remove dead cells. Platelets aid in blood clotting.
5. Health problems can occur in people who lack the proper amount of red blood cells, white blood cells, or platelets.

12:3 Blood Types

6. The four main blood types are A, B, AB, O.
7. Different blood types cannot be mixed.

12:4 Immunity

8. The immune system keeps the body free of most diseases. Antibodies destroy antigens that enter the body.
9. Shots help prevent disease.
10. AIDS is a serious disease caused by a virus.

Key Science Words

AIDS (p. 259)
anemia (p. 248)
antibody (p. 257)
antigen (p. 257)
bone marrow (p. 248)
hemoglobin (p. 248)
hemophilia (p. 252)
immune system (p. 256)
immunity (p. 258)
leukemia (p. 250)
plasma (p. 247)
platelet (p. 251)
red blood cell (p. 247)
white blood cell (p. 249)

Testing Yourself

Using Words

Choose the word from the list of Key Science Words that best fits the definition.

1. helps form blood clots
2. gives red blood cells their color
3. part of blood that is nonliving, yellow liquid
4. disease in which white cell numbers increase

Review

Testing Yourself *continued*

5. a problem resulting from too little hemoglobin in red cells
6. disease caused by a virus that destroys the immune system
7. chemicals that defend the body against foreign chemicals and organisms
8. a disease in which blood does not clot
9. cell with the job of carrying oxygen

Finding Main Ideas

List the page number where each main idea below is found. Then, explain each main idea.

10. why blood types differ
11. what the pickup jobs of blood are
12. several delivery jobs of blood
13. the main parts of your immune system
14. which animal groups do not have blood
15. what happens when blood type A is mixed with type B
16. three ways of getting AIDS
17. how antibodies and antigens differ

Using Main Ideas

Answer the questions by referring to the page number after each question.

18. What are the nutrients that make up blood plasma? (p. 247)
19. How do antibodies help get rid of bacteria? (p. 257)
20. When do white blood cells act as if they have a memory? (p. 258)
21. How do red cell proteins and plasma proteins differ in the four blood types? (p. 254)
22. How do animals with no blood get nutrients and oxygen? (p. 245)
23. How do shots help to prevent diseases? (p. 258)

Skill Review ✔

*For more help, refer to the **Skill Handbook,** pages 704-719.*

1. **Infer:** The common cold is a viral disease that most people survive easily. Explain why a cold can be life-threatening to a person infected with the AIDS virus.
2. **Measure in SI:** The red blood cells in the large photo on page 242 are magnified about 6000 times. Measure the diameter of one cell and determine its actual size in micrometers. (1 mm = 1000 micrometers)
3. **Use a microscope:** Examine a prepared slide of human white blood cells. Compare the sizes of the different kinds of white cells and the different shapes of their nuclei.
4. **Understand science words:** The word part *anti* means against. Define *antibody* and *antigen* using the term *against* in your definitions.

Finding Out More

TECH PREP

Critical Thinking

1. Trace the events that occur in your immune system after a mumps shot.
2. A person seriously needs a blood transfusion but none is available. What could be given temporarily to help?

Applications

1. Research the progress of scientists' search for a cure for AIDS.
2. Interview a local dentist to determine how AIDS has altered even the most routine dental procedures.

CHAPTER PREVIEW

Chapter Content

Review this outline for Chapter 13 before you read the chapter.

Skills in this Chapter

The skills that you will use in this chapter are listed below.

- In **Lab 13-1,** you will form a hypothesis, measure in SI, and interpret data. In **Lab 13-2,** you will observe, interpret data, make and use tables, and infer.
- In the **Skill Checks,** you will understand science words, interpret diagrams, and define words in context.
- In the **Mini Labs,** you will experiment, formulate a model, and design an experiment.

Chapter 13

Respiration and Excretion

If you were asked, "How much air can your lungs hold?" could you come up with the correct answer? The person in the photo on the left is snorkeling. Different people can stay under water for different lengths of time. This time can vary from a few seconds to several minutes. The small photo below shows one way to measure your own lung volume. What gases enter the lungs when you breathe in? What gases leave the lungs when you breathe out? Why is breathing important? What are some of the effects of breathing in polluted air? This chapter will help you answer these questions.

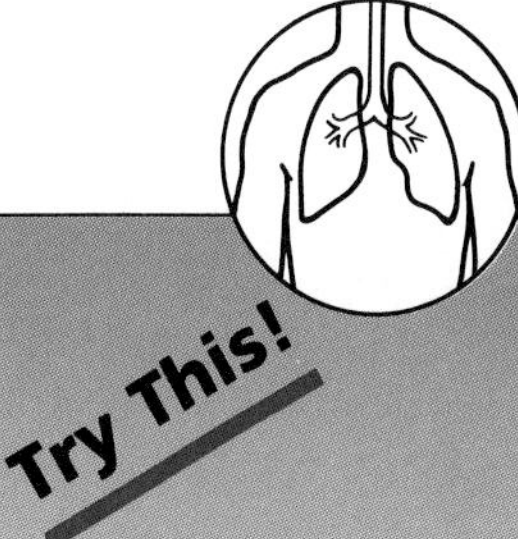

Try This!

How many breaths do you take? Count the number of times you breathe in and out for exactly one minute. What is your breathing rate?

BIOLOGY Online

Visit the Glencoe Science Web site at science.glencoe.com to find links about **respiration and excretion.**

Measuring the volume of air you breathe out

13:1 The Role of Respiration

Objectives

1. **Define** the role of the respiratory system.
2. **Explain** why cells need oxygen and give off carbon dioxide.
3. **Compare** the respiratory systems of different animals.

Key Science Words

respiratory system

When you talk of respiration, many people think of breathing. Is there more to respiration than this? Yes, respiration involves more body parts besides the nose and mouth.

Gases of Respiration

Chapter 12 described how your blood delivers oxygen to all your body cells. It also explained how blood removes carbon dioxide from your cells.

What is the role of the respiratory system?

Oxygen is brought into your body by the respiratory (RES pruh tor ee) system. Carbon dioxide is carried out of your body by the respiratory system. Your **respiratory system** is made of body parts that help with the exchange of gases. It brings in oxygen and removes carbon dioxide.

What waste gas is given off by all cells?

What happens to oxygen in cells? Oxygen is a gas used by cells in cellular respiration. During this chemical process, sugar reacts with oxygen, and energy is released for all of the cells' activities. At the same time, carbon dioxide and water are formed. Carbon dioxide is a waste gas given off by all cells. Water is an important nutrient for all cell processes. Remember that cellular respiration occurs in cells of all organisms, including plants. Cellular respiration can be shown as an equation:

$$\underset{\text{Oxygen}}{O_2} + \underset{\text{sugar}}{C_6H_{12}O_6} \xrightarrow{\text{Energy}} \underset{\text{carbon dioxide}}{CO_2} + \underset{\text{water}}{H_2O}$$

Study Tip: Use this idea map as a study guide to the role of the respiratory system. What is the process that releases energy from food?

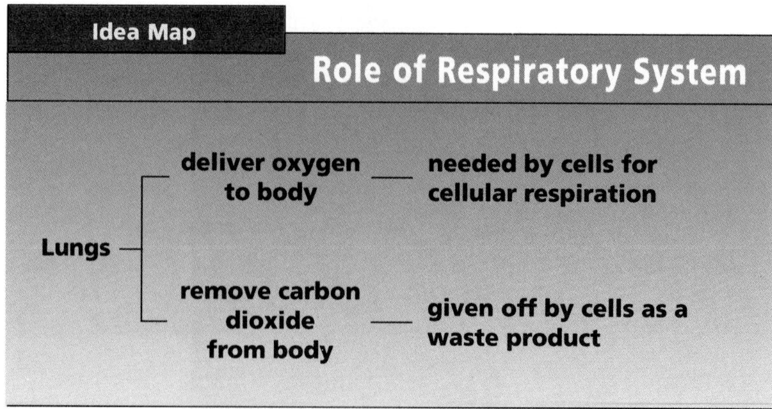

Respiration in Animals

Most animals have respiratory systems to help with the exchange of gases. In Figure 13-1, the blue shading shows the respiratory systems of some animals. Earthworms exchange gases through their skins. Insects have a large network of small tubes that carry air to all body parts. Fish carry on gas exchange through organs called gills. Frogs exchange gases through two types of organs—lungs and skin. Humans, as you know, also have lungs.

How do fish exchange gases?

The systems may look different in different animals, but all these systems are adapted for gas exchange. One important trait of all respiratory systems is that there is a large surface area over which air passes. This large area provides a surface for gases to diffuse into or out of the animals' bodies. In the earthworm, for example, the entire body surface acts as an area for gas exchange.

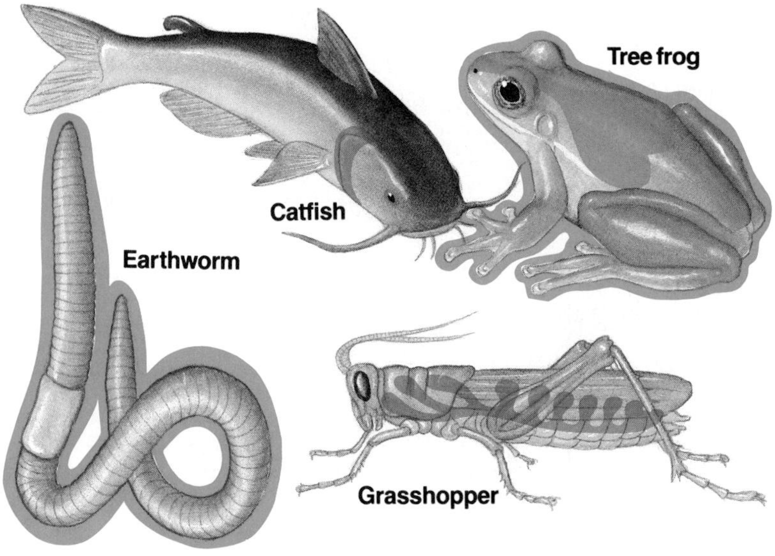

Tree frog
Catfish
Earthworm
Grasshopper

Figure 13-1 Compare the respiratory systems of some animals. The shaded areas in blue show where animals exchange gases with air.

Check Your Understanding

1. Which gases are exchanged by the respiratory system?
2. Why is oxygen needed by our cells?
3. Why do respiratory systems have a large surface area?
4. **Critical Thinking:** Why do birds need a more efficient respiratory system than humans?
5. **Biology and Writing:** Write a four to ten line paragraph that describes one kind of respiratory system.

Breathing Rate

Lab 13–1

Problem: Does temperature affect the breathing rate of fish?

Skills

form a hypothesis, measure in SI, interpret data

Materials

250-mL beaker
net
ice
goldfish in aquarium tank
thermometer
watch or wall clock

Procedure

1. Copy the data table.
2. Fill the beaker with water from the aquarium.
3. Use a net to transfer one goldfish from the tank to the beaker. **CAUTION:** *Handle animals with care.*
4. Locate the fish's gill cover. Note that the gill cover opens and closes as the fish breathes in and out.
5. **Measure in SI:** Use a thermometer to measure the water temperature in the beaker. Record this temperature in your table.
6. Count the number of times the fish opens its gill cover during one minute. Record this number in your data table.
7. Repeat step 6 two more times, recording your results in the table as trials 2 and 3.
8. Add a small piece of ice to the beaker to slowly drop the water temperature. Continue to add ice until the temperature of the water is 10 degrees lower than at the start. Wait at least 5 minutes before going on to step 9. As you wait, write a **hypothesis** as to whether you think the fish's breathing rate will speed up, slow down, or remain the same with the cooler water.
9. Repeat steps 5 through 7.
10. Complete the data table.
11. Allow the water temperature to rise to room temperature. Then, use the net to return your fish to the aquarium.

Data and Observations

1. What were you measuring when you counted gill cover openings?
2. **Interpret data:** Were there more or fewer gill openings with cold water when compared with warm water?

WATER TEMPERATURE:		WATER TEMPERATURE:	
TRIAL	NUMBER OF GILL OPENINGS	TRIAL	NUMBER OF GILL OPENINGS
1		1	
2		2	
3		3	
Total		Total	
Average		Average	

Analyze and Apply

1. Cold water contains more oxygen than warm water. Explain why a fish in cold water breathes more slowly than a fish in warm water.
2. **Check your hypothesis:** Is your hypothesis supported by your data? Explain.
3. **Apply:** Why might a fish die if the water temperature gets too warm?

Extension

Predict the number of gill openings at 5 degrees lower than the water was at the start. Experiment to test your prediction.

13:2 Human Respiratory System

What are the parts of the human respiratory system and how do they work? When you breathe, all you may feel is the slight movement of your chest. What else is involved in breathing?

Objectives

4. **Give the function** of different parts of the respiratory system.
5. **Compare** the pathways of gases as they enter and leave the body.
6. **Sequence** the movement of the diaphragm and rib cage during the breathing process.

Key Science Words

epiglottis
trachea
bronchi
alveoli
diaphragm

Respiratory Organs

Use Figure 13-2 to tour the respiratory system and study its parts.

1. Our tour starts with the nose. As you breathe in, air enters the nose. Hairs inside the nose trap dust particles and prevent them from entering the lungs.
2. The nasal chamber is a large space above the roof of your mouth. It warms or cools the entering air. Mucus is formed inside the nose. Mucus keeps the air moist as it passes through the nose.
3. Air enters the nose and passes to the windpipe. The windpipe is covered by the epiglottis (ep uh GLAHT uhs). The **epiglottis** is a small flap that closes over the windpipe when you swallow. This closing prevents food or liquid from entering the windpipe when you are eating. If food passed into your windpipe, you would choke.
4. The air passes into your windpipe, which is called the trachea (TRAY kee uh). The **trachea** is a tube about 15 centimeters long that carries air to two shorter tubes that lead into your lungs.
5. **Bronchi** (BRAUN ki) are two short tubes that carry air from the trachea to the left and right lung.

Skill Check

Understand science words: epiglottis. The word part *glott* means language. In your dictionary, find three words with the word part *glott* in them. *For more help, refer to the **Skill Handbook**, pages 706-711.*

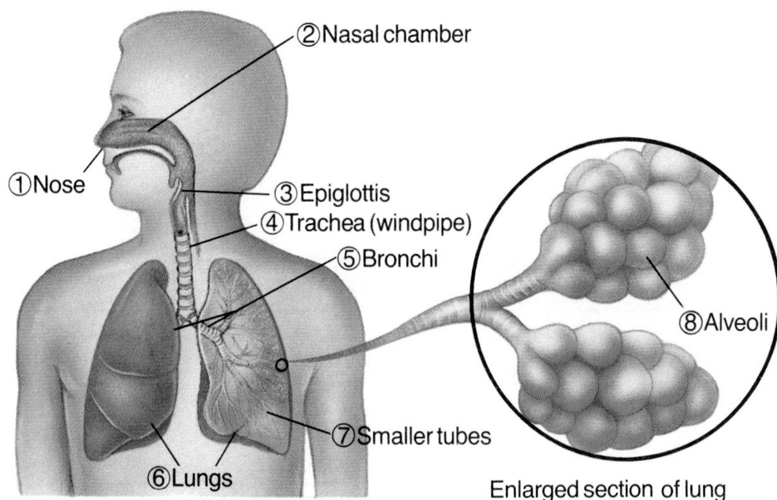

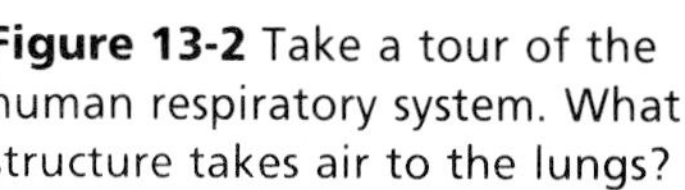

Figure 13-2 Take a tour of the human respiratory system. What structure takes air to the lungs?

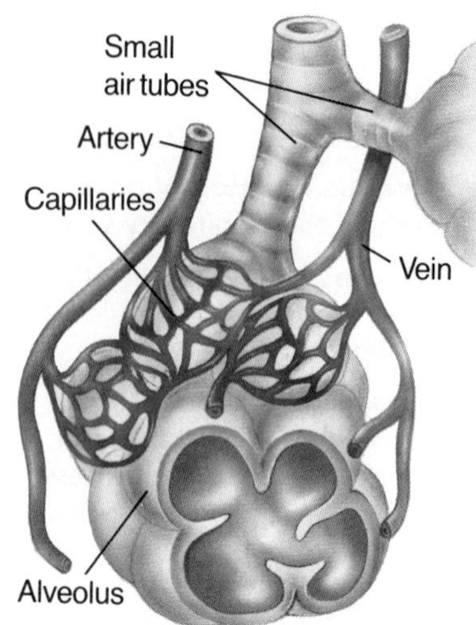

Figure 13-3 In the alveoli, gas exchange takes place in less than one-tenth of a second.

6. The two lungs are major organs in your respiratory system. Lungs are large, soft organs in which oxygen and carbon dioxide are exchanged.
7. As the bronchi enter each lung, they branch into thousands of smaller and smaller tubes. Each tiny tube finally leads into tiny air sacs.
8. There are about 300 million alveoli (al VEE uh li) found in each lung. **Alveoli** are the tiny air sacs of the lungs. The alveoli look like clusters of grapes.

Each air sac, or alveolus, is surrounded by tiny blood vessels called capillaries, Figure 13-3. Oxygen in the air sacs moves through the cell membranes into the blood. Carbon dioxide moves out of the blood into the air sacs. If all the air sacs in a person's lungs were opened and spread out, they would cover an area equal to about 1/5 of the area of a basketball court, Figure 13-4. The air sacs of your lungs provide a large surface area for gases to pass into and out of your blood.

Why are alveoli important?

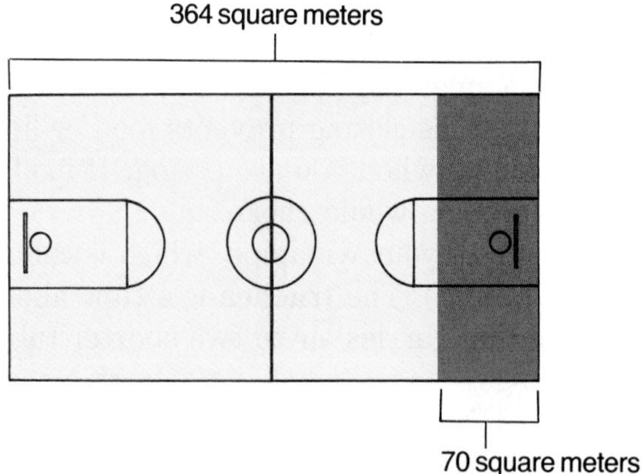

Figure 13-4 The area covered by a person's air sacs equals about one-fifth the area of a basketball court.

Gas Exchange

The function of your lungs is to exchange gases. Table 13-1 shows how much oxygen (O_2), carbon dioxide (CO_2), and nitrogen (N_2) gas we breathe in and out.

TABLE 13–1. MAKEUP OF AIR WE BREATHE

Gas	Chemical Symbol	Air Entering Lungs	Air Leaving Lungs
Oxygen	O_2	19.97%	16.00%
Carbon Dioxide	CO_2	.03%	4.00%
Nitrogen	N_2	80.00%	80.00%

Blood in the capillaries around the alveoli is low in oxygen. As you breathe in, air and the oxygen in it fill each alveolus. Oxygen passes out of the air sacs into the blood of the capillaries by diffusion, Figure 13-5.

Red blood cells carry oxygen to all body cells. The oxygen diffuses into the body cells, Figure 13-6. Why, then, does less oxygen move out of the lungs than goes in as shown in Table 13-1? Less moves out because some oxygen is transported to all your body cells.

Now, why is more carbon dioxide given off than is taken in by the lungs? Carbon dioxide is formed in all body cells as a waste gas during cellular respiration. The gas moves by diffusion from body cells into nearby capillaries, Figure 13-6. In time, blood loaded with carbon dioxide will travel back to the lungs. When blood arrives at the lungs, carbon dioxide diffuses out of the blood in the capillaries and into the alveoli, Figure 13-5. All the carbon dioxide gas given off by your body cells during cellular respiration is carried by your blood to your lungs. Once there, it is given off in the air you breathe out. Do the data in Table 13-1 show this? Why do you think the amount of nitrogen that enters and leaves the lungs stays the same? It's because nitrogen gas is not used by body cells.

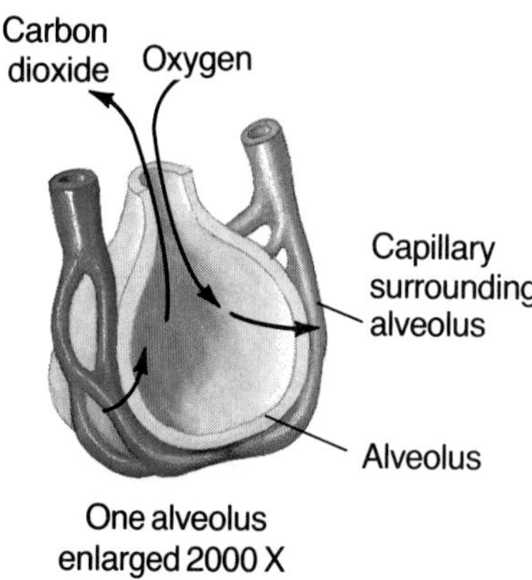

Figure 13-5 The pathways of oxygen and carbon dioxide between the alveolus and its blood capillaries

Why is more carbon dioxide given off than is taken in?

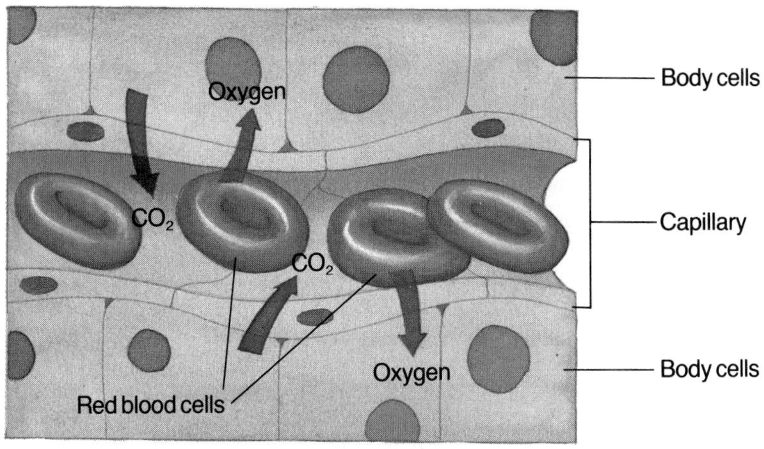

Figure 13-6 Oxygen moves by diffusion from red blood cells into body cells. Carbon dioxide (CO_2) diffuses into capillaries from the cells.

Breathing

Usually, you don't think about your breathing. If you did think about it, however, you would probably notice that you breathe in and out about 12 to 16 times in one minute. How does your breathing work to bring air in and push air out of your lungs? Your rib cage and a muscle at the bottom of your chest help you to breathe.

Mini Lab

Is Carbon Dioxide Present?

Experiment: You can detect carbon dioxide with bromthymol blue. Exhale 40 times through a straw into a beaker filled with the chemical. What color does the chemical turn? *For more help, refer to the* ***Skill Handbook,*** *pages 704-705.*

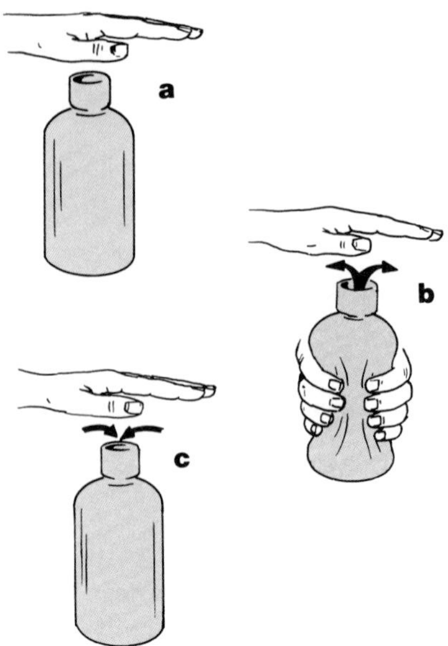
a
b
c

Figure 13-7 No movement of the bottle's sides results in no air moving in or out (a). Pushing the bottle's sides in causes air to move out (b). The bottle sides snap back and air is pulled in (c).

Hold your hand over a plastic bottle such as the one shown in Figure 13-7. You will feel no air moving in or out of the bottle opening. Squeeze the bottle as shown in Figure 13-7. You will feel air rushing out. Why? As the sides of the bottle are pushed in, air is pushed out. Now allow the sides of the bottle to snap back to their original shape, as in Figure 13-7. You can feel air rushing back into the bottle.

Your chest is like the plastic bottle. Its shape and size change, allowing you to breathe in and out. The size of your chest changes due to the action of a muscle located along the bottom of your chest. This muscle is your diaphragm (DI uh fram). The **diaphragm** is a sheetlike muscle that separates the inside of your chest from the intestines and other organs of your abdomen. Read the following statements and match them with parts of the diagram in Figure 13-8.

What happens when the diaphragm flattens and moves downward?

1. The diaphragm is relaxed. When it relaxes, it is domelike and pushes upward.
2. There is space between the lungs and the inside of the chest wall. The diaphragm pushes up against the lungs, and the space gets smaller.
3. Your lungs are soft. They are squeezed as the space grows smaller.
4. Air is pushed out of the alveoli as the lungs are squeezed. You breathe out.
5. The diaphragm begins to contract. As it tightens up, it flattens and moves downward.
6. The space between the lungs and the inside of the chest wall becomes larger.

Figure 13-8 Notice the changes in the diaphragm, lungs, rib cage, and chest cavity during breathing out (a) and breathing in (b).

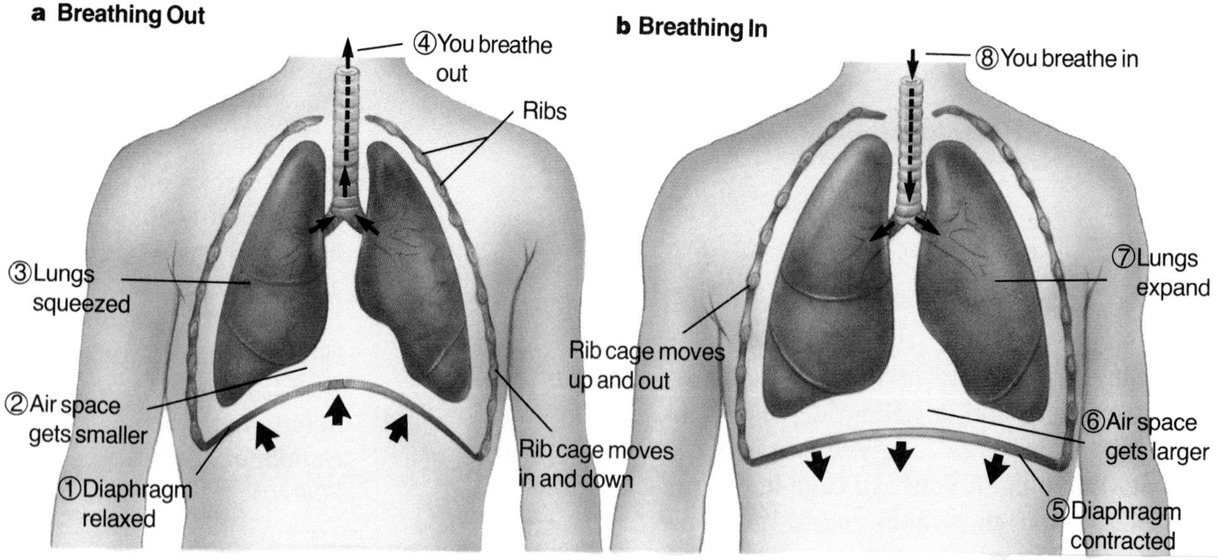
a Breathing Out
④You breathe out
Ribs
③Lungs squeezed
②Air space gets smaller
①Diaphragm relaxed
Rib cage moves in and down
b Breathing In
⑧You breathe in
⑦Lungs expand
Rib cage moves up and out
⑥Air space gets larger
⑤Diaphragm contracted

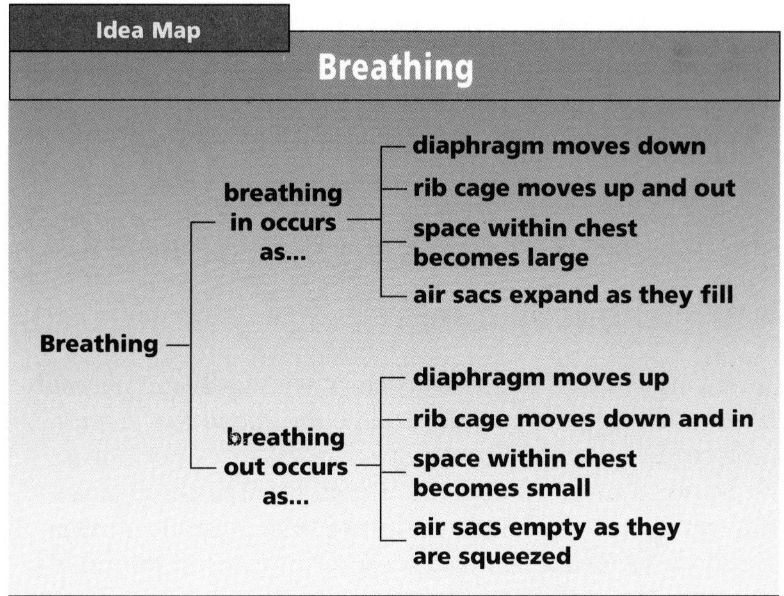
Idea Map
Breathing
Breathing
breathing in occurs as...
diaphragm moves down
rib cage moves up and out
space within chest becomes large
air sacs expand as they fill
breathing out occurs as...
diaphragm moves up
rib cage moves down and in
space within chest becomes small
air sacs empty as they are squeezed

Study Tip: Use this idea map as a study guide to breathing. Does breathing occur in your chest or abdomen?

7. Your lungs expand because the space surrounding the lungs has increased.
8. Air is pulled into the alveoli as the lungs expand. This causes you to breathe in.

Your rib cage also helps with breathing. The rib cage moves in and out as well as up and down, Figure 13-8. The muscles between each rib help with this movement. The ribs can be pulled closer together or moved apart. When your rib cage moves in and slightly down, the ribs close up slightly and the space within your chest gets smaller. Air is squeezed out and you breathe out. As your rib cage moves out and slightly up, the ribs move apart and the space within your chest expands. Air rushes in from the outside and you breathe in.

Mini Lab

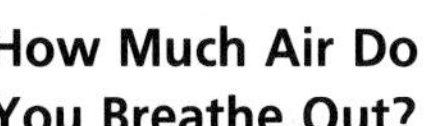

How Much Air Do You Breathe Out?

Design an experiment: Use a 1-L graduated cylinder, rubber tube, pan, and water to measure how much air you breathe out. Check with your teacher before doing it. *For more help, refer to the **Skill Handbook**, pages 704-705.*

Check Your Understanding

6. What are the functions of the trachea, bronchi, lungs, and alveoli?
7. How is carbon dioxide formed in your cells, and how does it pass out of your body?
8. In which directions do your rib cage and diaphragm move as you breathe in?
9. **Critical Thinking:** What causes a person to choke?
10. **Biology and Reading:** Where does air go after it leaves the smallest tubes of the bronchi?

13:3 Problems of the Respiratory System

Objectives

7. **Relate** the effects of carbon monoxide poisoning and how to avoid the gas.
8. **Explain** the cause of pneumonia.
9. **Describe** emphysema.

Key Science Words

carbon monoxide
pneumonia
emphysema

Problems of the respiratory system can be caused by diseases of the respiratory organs. Some problems, however, are caused by the quality of the air we breathe.

A Breathing Problem

A person can live for many days without food. Without oxygen, however, a person can live for only a few minutes. Cells can't live without oxygen.

Some gases are poisonous. If you breathe these gases, they act by taking the place of oxygen in your blood. One poisonous gas is carbon monoxide (CO). **Carbon monoxide** is an odorless, colorless gas sometimes found in the air. If carbon monoxide is inhaled, it is more easily picked up by your red blood cells than oxygen.

Why are exhaust fumes dangerous?

Where does carbon monoxide come from? You know that when fuels are burned, oxygen is used up and carbon dioxide is released—just as in cellular respiration. Carbon monoxide is also given off in small amounts when fuels are burned. For example, carbon monoxide is given off in exhaust fumes when gasoline is burned in a car. The small amount of gas mixes with the surrounding air and is not a health problem. It becomes a serious problem, however, if the exhaust is given off in a closed area, such as a garage. A common cause of carbon monoxide pollution is from the smoke of cigarettes, Figure 13-9.

Bio Tip

Leisure: Swimming is an excellent sport for people with asthma. It provides exercise for the body at the same time as helping a person relax.

Figure 13-9 A person who breathes too much carbon monoxide can't deliver enough oxygen to his or her cells.

Respiratory Diseases

Pneumonia (noo MOH nyuh) is a lung disease caused by bacteria, a virus, or both. These microbes invade the alveoli and multiply. Their presence causes the lung tissue to form extra mucus. Mucus and pus from microbes fill the alveoli and cause difficulty in breathing. The blocked lungs prevent oxygen from getting into the blood. Because pneumonia is caused by bacteria or viruses, it is communicable. You read in Chapter 4 that a communicable disease is passed from one person to another.

Why is pneumonia communicable?

Some diseases are not passed on from person to person. Emphysema (em fuh SEE muh) is a respiratory disease that is not communicable. **Emphysema** is a lung disease that results in the breakdown of alveoli. A person with emphysema always seems to be out of breath because there are not enough alveoli to get oxygen into the blood. A common cause of emphysema is the breathing in of chemicals while smoking or from air pollution, Figure 13-10.

What happens to a person's lungs with emphysema?

Skill Check

Define words in context: If you say that you are "catching a cold," would this describe a communicable disease? *For more help, refer to the **Skill Handbook,** pages 706-711.*

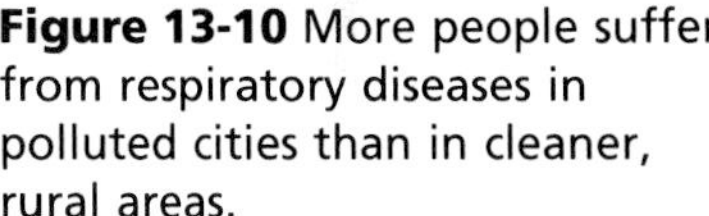

Figure 13-10 More people suffer from respiratory diseases in polluted cities than in cleaner, rural areas.

Check Your Understanding

11. Why is carbon monoxide poisonous?
12. What causes pneumonia?
13. What is emphysema?
14. **Critical Thinking:** A spirometer measures lung capacity and volume. Why are these measurements important?
15. **Biology and Reading:** What happens to the surface area of the lung when alveoli are lost as a result of emphysema?

13:4 The Role of Excretion

Objectives

10. **Discuss** the importance of the excretory system.
11. **Sequence** the steps for filtering of blood in a kidney.
12. **Describe** the skin as part of the excretory system.

Key Science Words

excretory system
urea
ureter
urinary bladder
urethra
nephron
urine

Chemical wastes are formed by living cells. Carbon dioxide is removed from your body by your lungs. Other organs, such as your skin and kidneys, help in getting rid of other waste materials.

Waste Removal

Your blood contains many different chemical wastes. These wastes are chemicals that are not needed by the body and may be harmful. If wastes weren't removed from the body, the tissues would fill with poisonous waste products. The wastes would destroy cells and tissues. Fever, poisoning, or even death can result from a buildup of wastes in the tissues.

Wastes are either made by your body cells or are the remains of undigested food in your diet. Getting rid of wastes is called excretion. In Chapter 10, you read that solid wastes are removed by the digestive system. Liquid wastes are removed by the excretory (EK skruh tor ee) system. The **excretory system** is made up of those organs that rid the body of liquid wastes.

Why does urea need to be removed from the body?

Urea (yoo REE uh) is a waste that results from the breakdown of body protein. It's poisonous and must be removed from the body. Urea is picked up from the cells by the blood and carried to the kidneys for removal. The kidneys are the most important organs of the excretory system.

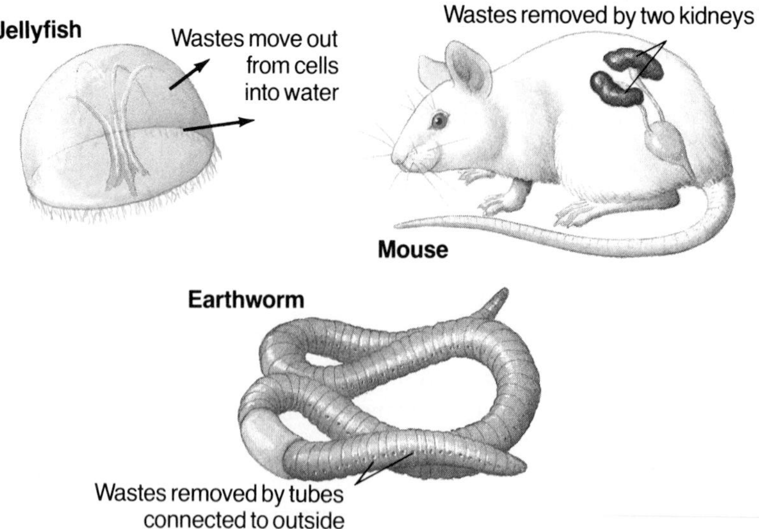

Jellyfish
Wastes move out from cells into water
Wastes removed by two kidneys
Mouse
Earthworm
Wastes removed by tubes connected to outside

Figure 13-11 Compare the excretory organs in a jellyfish, earthworm, and mouse.

Do all other animals have kidneys as we do? Many animals, such as mice, fish, and frogs, have kidneys. Simpler animals without kidneys, however, must still get rid of wastes. Simple animals such as sponges and jellyfish don't need kidneys. Wastes made by their cells pass out of their cells directly into the water in which they live. Earthworms excrete their wastes through a pair of tubes in each body segment. These tubes connect the inside of the animal to the outside. Compare the organs of excretion in the animals in Figure 13-11.

Human Kidneys

The human body has two kidneys, each about as big as a fist. Kidneys could be described as blood filters. To understand the job of your kidneys, let's look at how a filter works, Figure 13-12. Let's assume you have some muddy water and you want to separate the mud from the water. The muddy water can be poured through filter paper. The mud will be caught by the paper, while the water will pass through. Your kidneys work in a way similar to the filter paper. They filter wastes from your blood. During one day, your kidneys filter up to 200 liters of blood.

If wastes were not removed from your body, they would build up in your blood and act as a poison. In order to carry out their job of removing wastes, your kidneys are hooked up to your blood vessels. Match the following numbered steps to parts of the human excretory system shown in Figure 13-13.

Mini Lab

What Does a Kidney Do?

Formulate a model: Filter some muddy water to model how the kidneys work. *For more help, refer to the* ***Skill Handbook,*** *pages 706-711.*

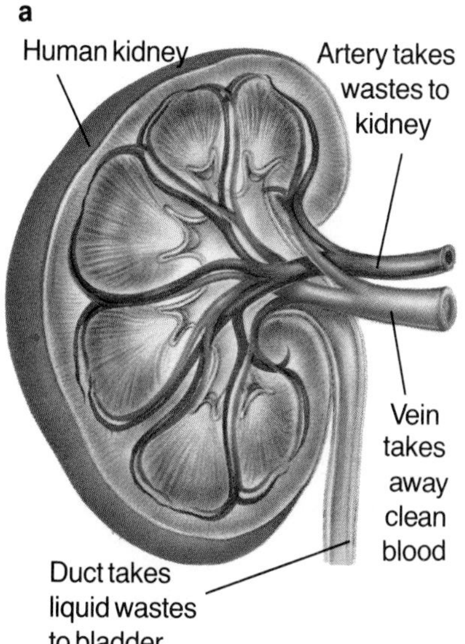

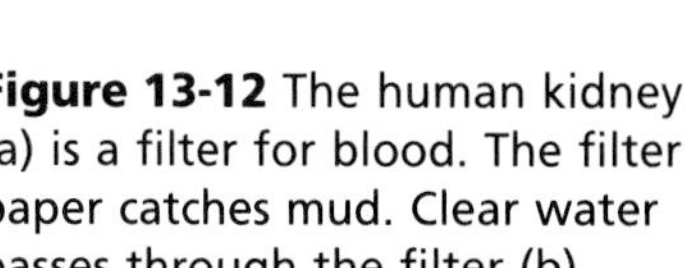
Figure 13-12 The human kidney (a) is a filter for blood. The filter paper catches mud. Clear water passes through the filter (b).

Figure 13-13 Follow the path of blood as it is cleaned by the human excretory system.

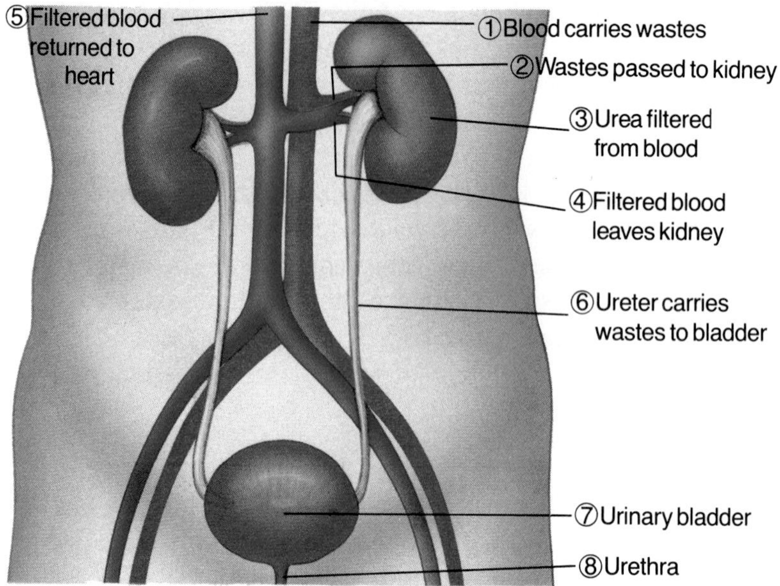

⑤Filtered blood returned to heart
①Blood carries wastes
②Wastes passed to kidney
③Urea filtered from blood
④Filtered blood leaves kidney
⑥Ureter carries wastes to bladder
⑦Urinary bladder
⑧Urethra

1. Blood carrying wastes moves through the body's arteries.
2. Small arteries bring blood to be filtered into each kidney.
3. The kidneys filter urea from the blood.
4. The filtered blood leaves the kidney through a vein. The blood is now free of wastes.
5. The blood in the veins is taken to the heart and other parts of the body.
6. Wastes from the blood leave each kidney through a ureter (YOOR ut ur). A **ureter** is a tube that carries wastes from a kidney to the urinary bladder.
7. The **urinary bladder** is a sac that stores liquid wastes removed from the kidneys.
8. The **urethra** (yoo REE thruh) is a tube that carries liquid wastes from the urinary bladder to outside the body.

What is the path of wastes from the blood through the body?

How do the kidneys do their job of filtering blood? Look at Figure 13-14. It shows an enlarged view of a nephron (NEF rahn), which is a small part of a kidney. A **nephron** is a tiny filter unit of the kidney. Each kidney has about one million nephrons that allow the kidneys to filter your blood every day.

Use the figure and the numbered steps to follow the pathway of blood through a nephron.
1. Blood with blood cells, salts, sugar, urea, and water enters each nephron.
2. The blood first passes into a tightly coiled capillary inside a cuplike part of the nephron.

3. Water, salts, sugar, and urea pass into the cuplike part of the nephron. Everything but blood cells is forced out of the blood.
4. Follow the flow shown by the dashed black arrows that indicate the movement of these materials into a long tube.
5. Notice that the capillary with the original blood cells from step 2 now twists itself around the long tube. The movement of blood cells is shown by the dashed red arrows.
6. All materials except urea, water, and salts diffuses back into the blood. This movement is shown by the solid black arrows. Water passes back into the blood by osmosis.
7. The cleaned blood now leaves the kidney and returns to the heart.
8. The filtered wastes stay in the long tube and are carried out of the kidney through the ureter to the bladder. Waste liquid that reaches the ureter is called **urine.** Urine passes out of the body.

Taking in more water or salt means more of these nutrients will be excreted. Less water or salt taken in means less will be excreted. Kidneys regulate the amount of water and salts kept within the body. A healthy person can excrete about one liter of urine each day.

Skill Check

Interpret diagrams: What feature of Figure 13-14 makes it easier for you to follow the pathways of blood, wastes, and water? *For more help, refer to the* ***Skill Handbook,*** *pages 706-711.*

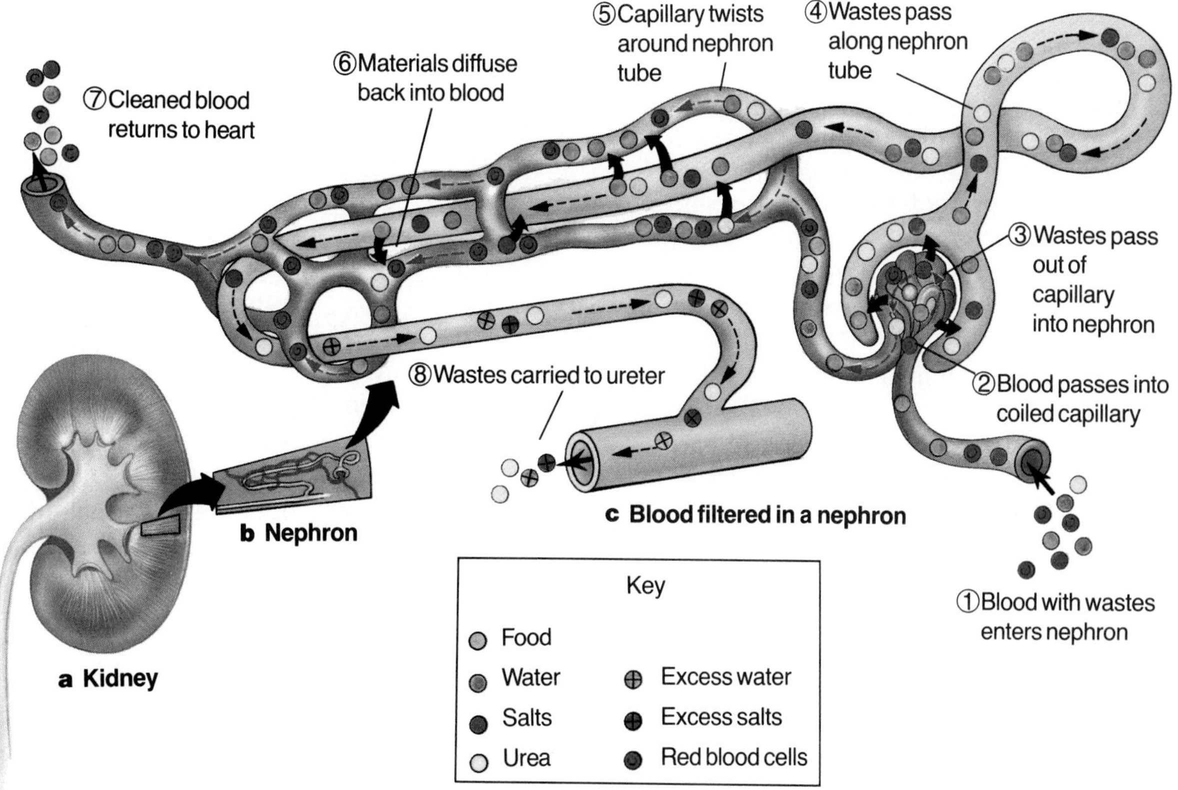

⑤Capillary twists around nephron tube
④Wastes pass along nephron tube
⑥Materials diffuse back into blood
⑦Cleaned blood returns to heart
③Wastes pass out of capillary into nephron
②Blood passes into coiled capillary
⑧Wastes carried to ureter
①Blood with wastes enters nephron
a Kidney
b Nephron
c Blood filtered in a nephron
Key
Food
Water
Salts
Urea
Excess water
Excess salts
Red blood cells

Figure 13-14 A section of a kidney (a) shows the position of a nephron (b). Follow the pathway of materials through a nephron (c).

Urine

Lab 13–2

Problem: What chemicals can be detected in urine?

Skills

observe, interpret data, make and use tables, infer

Materials

4 test tubes
test-tube rack
distilled water
salt water
normal artificial urine
abnormal artificial urine
silver nitrate in bottle with dropper
labels
glucose
glucose test tape
droppers
4 glass slides

Procedure

1. Copy the data table.
2. Use labels to number four test tubes 1, 2, 3, and 4 and place them in a rack.
3. Half fill each tube with the following– 1: water, 2: salt water, 3: normal artificial urine, 4: abnormal artificial urine.
4. Add five drops of silver nitrate to each tube. If a white haze appears, this means that salt is present. **CAUTION:** *Do not spill silver nitrate on skin or clothing. Rinse with water if spillage occurs.*
5. **Make and use tables:** Record your before and after results in the table.
6. Use labels to number four slides 1, 2, 3, and 4.
7. Add one drop of the following to each slide– 1: water, 2: glucose, 3: normal artificial urine, 4: abnormal artificial urine.
8. Touch a small piece of glucose test tape to each drop. If a green color appears on the tape, it means that the sugar glucose is present.
9. Record your before and after results in the table.
10. Dispose of your samples as indicated by your teacher.

Data and Observations

1. How can you tell if a liquid contains salt?
2. How can you tell if a liquid contains glucose?

TUBE	TUBE CONTENTS	HAZE PRESENT BEFORE?	HAZE PRESENT AFTER?	SALT PRESENT?
1				
2				
3				
4				

SLIDE	SLIDE CONTENTS	COLOR OF PAPER BEFORE	COLOR OF PAPER AFTER	GLUCOSE PRESENT?
1				
2				
3				
4				

Analyze and Apply

1. Using your data, does the normal artificial urine contain glucose or salt?
2. Using your knowledge, would urine from your body contain glucose or salt? Why?
3. Using your data, does the abnormal urine contain glucose?
4. **Apply:** Why is it useful to know if urine contains salt or glucose?

Extension

Experiment: Add table sugar to urine and test for glucose.

Excretion and the Skin

Skin has three main jobs. First, it protects the body from infection. Second, it is a sense organ. Third, it helps in the excretion of water and salts.

Figure 13-15 shows what a slice through skin looks like. The top layer of this organ is made of a layer of living cells covered with dead cells that are constantly flaking off. A lower layer of skin has capillaries and sweat glands. The sweat glands open onto the surface of the skin. Hair grows from this bottom layer. This layer also has many different types of nerve cells. You will learn in Chapter 16 how nerves in the skin make it an important sense organ.

Your body has two to five million sweat glands like those shown in Figure 13-15. Each gland gives off water and salts much as the kidneys do. Unlike kidneys, however, your skin can't control water or salt loss. It's possible for you to lose half a liter of water each day through the skin without noticing it, unless you sweat heavily.

Does loss of water through the skin help the body? This water loss is one way the body cools itself. You usually sweat more when the weather is warm or when you exercise. Water moving onto the skin from the sweat glands evaporates. The evaporation cools the body.

Bio Tip

Consumer: Sweating helps cool the body. Antiperspirants block the sweat glands and stop sweating. Deodorants, however, just mask the odor of sweat.

How does sweat help the body?

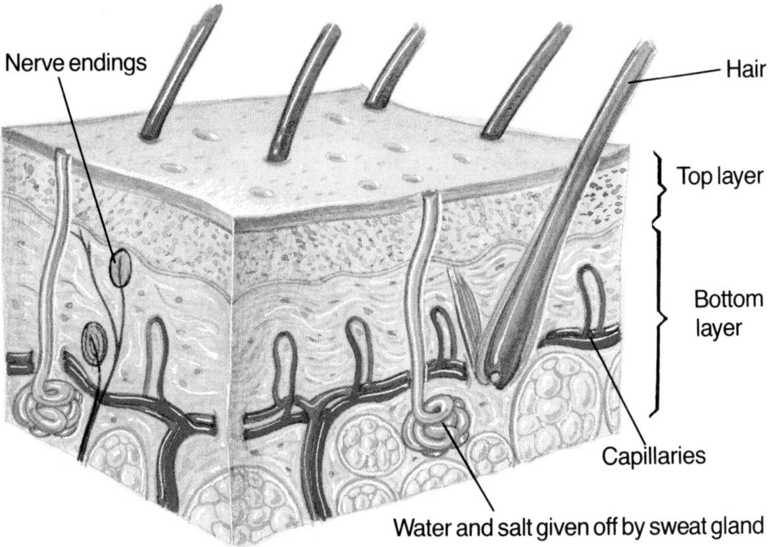

Figure 13-15 The sweat glands are a part of the excretory system. Find these glands in this cross section of skin.

Problems of the Excretory System

Like other organs, the kidneys may become diseased. The kidneys can be damaged over time if a person has high blood pressure. Another cause of kidney damage is infection caused by bacteria.

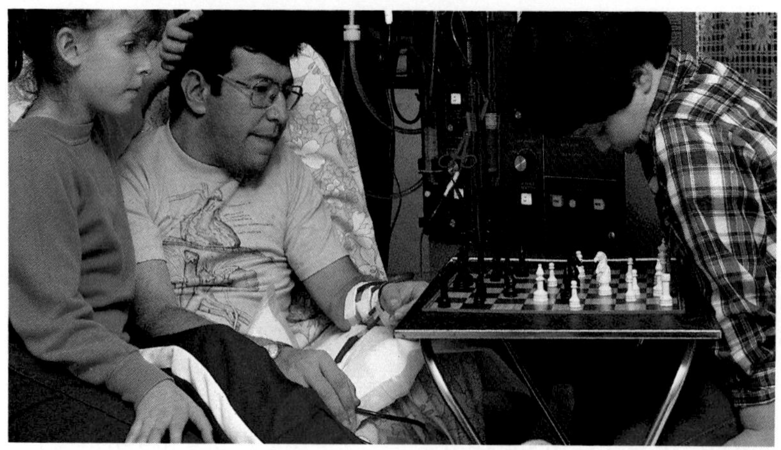

Figure 13-16 Each session on an artificial kidney machine can take up to five hours.

Humans are lucky in that they can live with only one kidney. Thus, if one kidney is damaged, we still have the other one to help with excretion. But, what happens when both kidneys are damaged? In such cases, either a kidney machine must be used, or a kidney transplant operation should be performed.

How does a kidney machine compare with a kidney?

An artificial kidney machine, Figure 13-16, removes urea from the blood when the kidneys can't function. The artificial kidney machine works by passing a person's blood through a very long tube made from a thin membrane. This membrane is surrounded by liquid. Urea in the blood diffuses through the tube membrane and into the liquid. Fresh liquid is constantly added and the used liquid is constantly removed. Blood without urea is then returned to the person. The kidney machine is designed to act just like your nephrons. A person may need up to three sessions a week on the kidney machine.

Check Your Understanding

16. What are the jobs of your ureters and urethra?
17. Name three things that kidneys filter out of the blood.
18. What are the three main jobs of the skin?
19. **Critical Thinking:** How does the process of diffusion allow materials in a nephron to be returned to the blood?
20. **Biology and Reading:** There's a pair of tubes in each body segment of an earthworm that removes wastes made by the cells. What organ in humans has the same function as these tubes in earthworms?

Science and Society

Organ Transplants

After a kidney transplant operation, a person has a new chance to live a long and healthy life.

The first successful kidney transplant operation was performed in the 1960s. A kidney was donated from one identical twin to the other. Up to that time, the major problem with organ transplants was that tissues from another person's body were usually rejected. Today, we use drugs that reduce rejection. Kidney transplants are now rather common. Why is the kidney such a good organ for transplants? Remember, each person has two of these organs and could survive with only one. It is not unusual for a healthy person to donate a kidney to another member of his or her family or to an unrelated person. Healthy kidneys are also available from people who have agreed to donate their organs when they die.

What Do You Think?

1. There are a greater number of people needing kidney transplants than there are available kidneys. Thousands of patients are now waiting for a healthy kidney donor. The demand is much greater than the supply. Where should the supply of organs come from? Could they be taken from animals such as baboons, terminally ill patients, or from prisoners on death row? At the present time, the most common source is from the bodies of people in fatal auto accidents. Should everyone be expected to carry around a permission card to donate their kidneys?

2. Kidney and other transplant operations are very expensive. Should only those who can afford to pay for the operations be able to have them? If the government or insurance companies pay for the transplant operations, who will decide who gets an organ? Is the life of a four-year-old child worth more than that of a sixty-year-old adult? How should these choices be controlled?

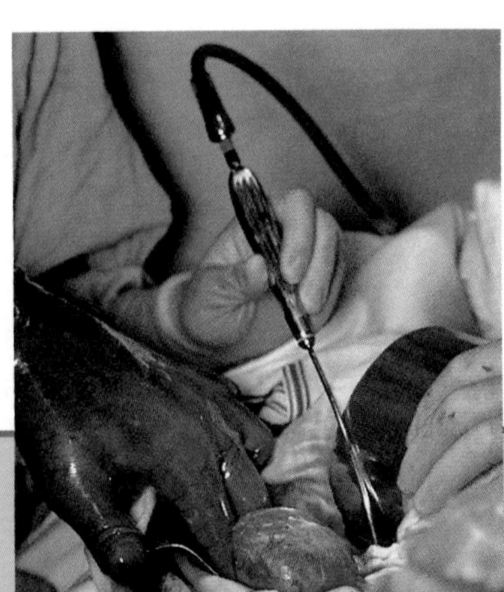

A kidney operation

3. An illegal market exists that encourages healthy people to sell one of their kidneys to the highest bidder. Prices as high as $13 000 for a kidney have been reported. Many of these kidneys come to the United States from foreign countries. These organs may not be in a state suitable for transplanting. A person may spend a lot of money and then discover the organ to be useless. Should the United States pass laws that prevent the importing of human organs?

Conclusion: What can be done to assure that anybody who needs a transplant can have one?

Chapter 13

Review

Summary

13:1 The Role of Respiration

1. Oxygen is taken in and carbon dioxide is given off in the respiratory system.
2. Oxygen is used by cells to convert food to energy. Carbon dioxide is given off as waste.
3. Some animals have no respiratory system. Animals with a respiratory system use skin, gills, lungs, or pairs of tubes for gas exchange.

13:2 Human Respiratory System

4. The respiratory system is made up of tubes that supply air to the alveoli in the lungs.
5. Oxygen moves from lungs to blood to body cells. Carbon dioxide moves from body cells to the lungs.
6. Your diaphragm relaxes and the rib cage moves in when you breathe out. Your diaphragm contracts and the rib cage moves out when you breathe in.

13:3 Problems of the Respiratory System

7. Carbon monoxide is a poisonous gas.
8. Pneumonia is caused by bacteria and viruses.
9. Emphysema causes breakdown of lung tissues.

13:4 The Role of Excretion

10. Kidneys filter wastes such as urea, excess water, and salts from the blood.
11. Blood, carrying wastes from body cells, is filtered in nephrons in the kidney.
12. Your skin excretes water and cools the body.

Key Science Words

alveoli (p. 268)
bronchi (p. 267)
carbon monoxide (p. 272)
diaphragm (p. 270)
emphysema (p. 273)
epiglottis (p. 267)
excretory system (p. 274)
nephron (p. 276)
pneumonia (p. 273)
respiratory system (p. 264)
trachea (p. 267)
urea (p. 274)
ureter (p. 276)
urethra (p. 276)
urinary bladder (p. 276)
urine (p. 277)

Testing Yourself

Using Words

Choose the word from the list of Key Science Words that best fits the definition.

1. tiny filter unit of kidney
2. communicable disease of lungs
3. flap that keeps food out of trachea
4. tube that carries urine from kidney to bladder
5. tiny air sacs of lungs
6. sheetlike muscle that separates chest from rest of body

Review

Testing Yourself *continued*

7. tubes that carry air from trachea to both lungs
8. tube that carries urine to the outside of your body
9. poisonous gas in exhaust fumes

Finding Main Ideas

List the page number where each main idea below is found. Then, explain each main idea.

10. why more carbon dioxide is given off than taken in
11. which chemicals are forced out of the capillaries of the nephrons
12. which two gases are exchanged during cellular respiration
13. what makes the space around your lungs get smaller as you breathe out
14. why water loss through the skin can be helpful
15. why a person with emphysema is always short of breath
16. the events that take place in the alveoli

Using Main Ideas

Answer the questions by referring to the page number after each question.

17. Why do your lungs expand when your diaphragm contracts? (p. 271)
18. How do the respiratory systems of earthworms and fish compare? (p. 265)
19. What is the sequence of organs through which air passes, starting with the nose? (pp. 267, 268)
20. How do the following pairs of words differ–urea and urine, ureter and urethra, and carbon dioxide and carbon monoxide? (pp. 268, 272, 274, 276, 277)
21. What can be done to lower the chances of getting emphysema? (p. 273)

Skill Review ✓

*For more help, refer to the **Skill Handbook**, pages 704-719.*

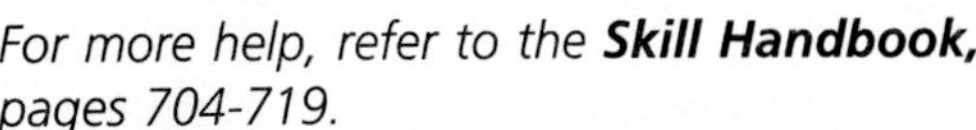

1. **Experiment:** Blow up a round balloon for 15 seconds. Tie it closed. Run on the spot for one minute, then blow up a second round balloon for 15 seconds. Tie it closed. Compare the volume of each balloon by measuring their circumferences. How did exercise affect your breathing efficiency?
2. **Formulate a model:** Make a model of your lungs using everyday household materials.
3. **Understand science words:** Use the dictionary to find two words closely related to trachea that are used in biology. Give their meanings.
4. **Infer:** Why can you see your breath on a cold morning? What is being excreted?

Finding Out More

Critical Thinking

1. People with high blood pressure are advised to reduce salt intake. What reasons might be given for this?
2. Compare the chemicals in blood entering and leaving the kidney.

Applications

1. Place your hand just below your ribs and feel the movement of your diaphragm and rib cage as you breathe in and out.
2. Which organs of the respiratory system are used when you cough, cry, hiccup, laugh, yawn, and sing?

CHAPTER PREVIEW

Chapter Content

Review this outline for Chapter 14 before you read the chapter.

Skills in this Chapter

The skills that you will use in this chapter are listed below.

- In **Lab 14-1,** you will interpret data, form a hypothesis, experiment, and measure in SI. In **Lab 14-2,** you will formulate a model, experiment, and infer.
- In the **Skill Checks,** you will infer, classify, and understand science words.
- In the **Mini Labs,** you will formulate a model, observe, and experiment.

Chapter 14

Support and Movement

All your body systems tackle mighty jobs each day. They help you walk, sit, write, tie your shoelaces, and even watch a TV show. But what about the extra work that you give your body? Do you jog, dance, do judo, lift heavy weights, swim, cycle, or play baseball? These are not activities you need to do to survive. So, how can your body put up with these extra demands?

Your bones and muscles give your body support and strength, just as the steel frame supports the bridge shown in the photo on the left. The strength of your skeleton and muscles can be relied on in most exercises, such as the one shown in the photo below. In this chapter, you will see how the pieces and parts of the skeleton and muscles work together.

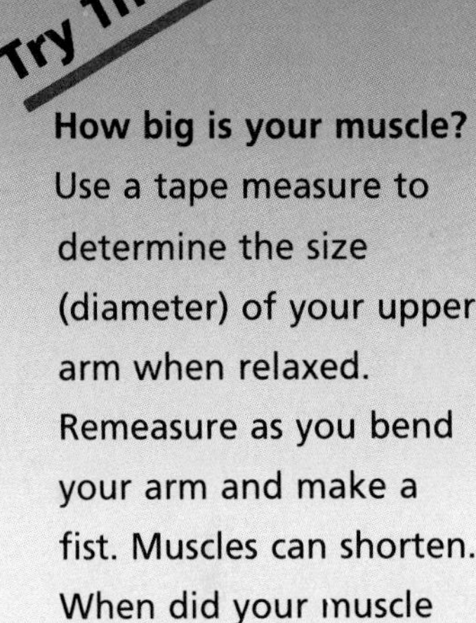

Try This!

How big is your muscle? Use a tape measure to determine the size (diameter) of your upper arm when relaxed. Remeasure as you bend your arm and make a fist. Muscles can shorten. When did your muscle shorten?

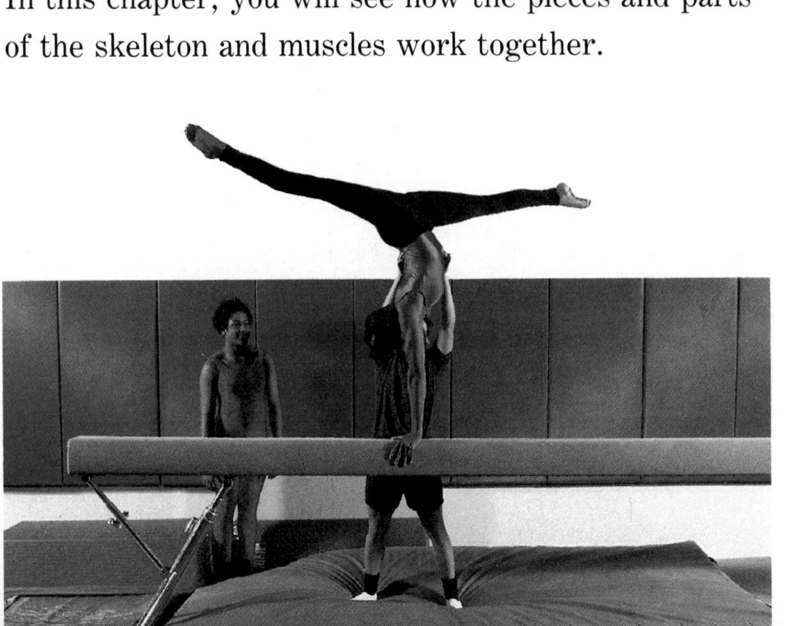

Visit the Glencoe Science Web site at science.glencoe.com to find links about **support and movement.**

14:1 The Role of the Skeleton

Objectives

1. **Explain** the functions and growth of the skeleton.
2. **Define** the six tissues of bones.
3. **Identify** three types of joints.

Key Science Words

skeletal system
solid bone
spongy bone
ligament
hinge joint
ball-and-socket joint
fixed joint

All vertebrates, from fish to mammals, have a skeleton. The skeleton provides support to their bodies. It also has other jobs such as protection, storage of materials, production of blood cells, and helping with movement.

Functions of the Skeleton

You know that all body systems are made of organs. Bones are the organs in the skeletal system. The **skeletal system** is the framework of bones in your body. As a framework, the skeleton helps to support the entire body. A skeleton also has other jobs. The skeleton helps to protect certain body organs. Your brain, heart, and lungs are organs with soft tissues. They could easily be damaged. Your skull protects your brain against injury. Ribs form a cage that protects your heart and lungs.

What part of bone produces blood cells?

Many bones in your body make blood cells. You learned in Chapter 12 that bone marrow is the soft center part of bone. Blood cells are made within the bone marrow. You may have eaten a ham steak. A ham steak is the muscle of a pig's thigh, and the bone in the steak is a section of the pig's thigh bone. At the center of the bone is the marrow.

Another job of bone is to store calcium (Ca). Remember that calcium is a mineral used by the body and is part of all bones. Calcium gives bones their strength. Without calcium, your bones would become weak or brittle.

The last job of your skeleton is to provide a place for muscles to attach. If muscles were not attached to your bones, you would not be able to move.

Study Tip: Use this idea map as a study guide to the functions of the skeleton. What bones allow you to write the answer to this question?

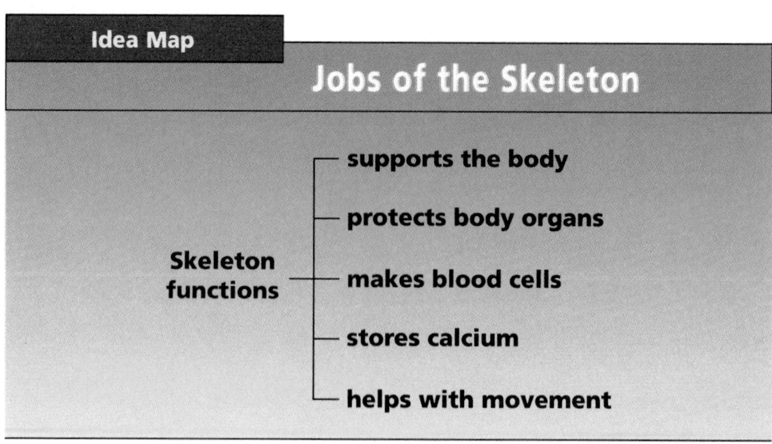

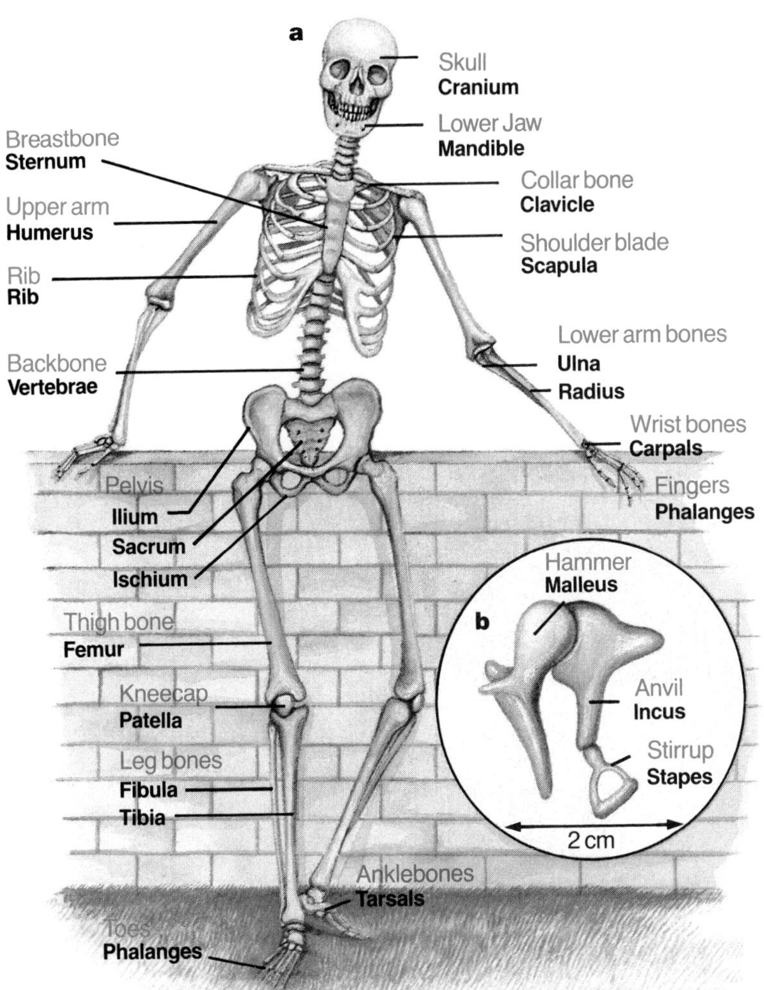

Figure 14-1 The skeleton is made of flat bones in the skull and ribs, irregular bones in the backbone, long bones in the arms and legs, and short bones in the hands and feet (a). The smallest bones are in the ear (b).

Bones in the Skeleton

If you counted all the bones in an adult human skeleton, you would find 206 of them. Examine the human skeleton in Figure 14-1. All the bones can't be counted in this diagram because not all of them are shown. Also, some bones are joined with other bones, making them look like one. For example, there are 22 different bones in the human skull. You can see only a few of these in the diagram.

Notice that the common name of a bone is usually different from the medical name. The common names are printed in blue. The medical names are printed in black. The three small ear bones shown to the bottom right are the body's smallest bones. Together, they measure only two centimeters long. They look so much like a hammer, an anvil, and a stirrup that these are their common names.

Mini Lab

What Happens to Bone When Calcium Is Removed?

Experiment: Place a chicken bone into a beaker with dilute HCl. Leave the bone overnight. Rinse carefully to remove any acid. Now, describe the bone. *For more help, refer to the* ***Skill Handbook,*** *pages 704-705.*

Figure 14-2 The hand and wrist bones of an infant (a) and an adult (b) indicate an increase in numbers of bones with age.

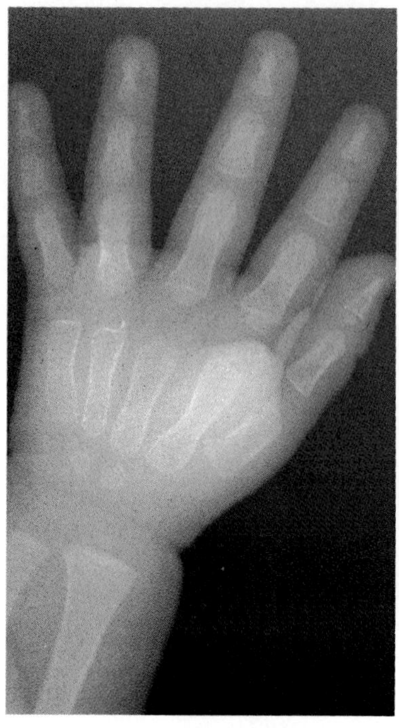

a

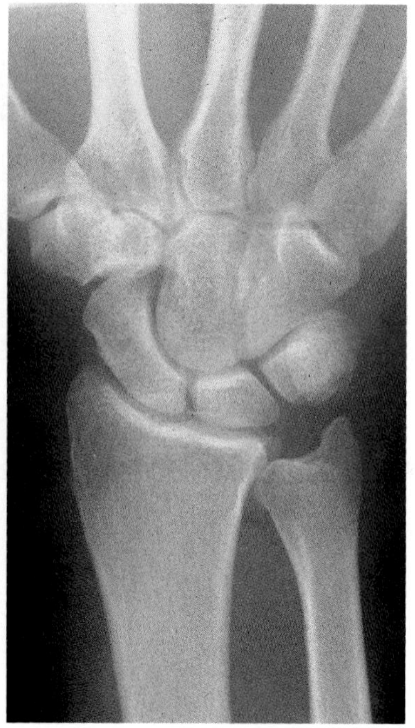

b

Bone Growth

Bone tissue is alive and is made of cells, just as in other organs. Because bone cells are living, they can reproduce and make more bone. This results in bone growth. You know that bones grow because you are taller now than when you were younger.

Not only does bone size change as you grow, the number of your bones increases. Figure 14-2a shows an X ray of the hand and wrist bones of a three-year-old child. The X-ray photo on the right shows the hand and wrist bones of an adult. It is clear that the size of the bones has increased during growth. Has the number of bones increased? Count the number of bones shown in the wrist of each X ray. An infant's wrist has five bones. An adult's wrist has eight bones.

Bio Tip

Consumer: Many foods provide you with a natural supply of calcium. They include milk, milk products, green leafy vegetables, oranges, dried peas and beans, and shellfish.

Bone Structure

What are six tissues of bone?

Most bones are made of six kinds of tissues. Figure 14-3 shows the positions of these six tissues in a bone.

You read in Chapter 8 that cartilage is a tough, flexible tissue that supports and shapes the body. This tissue acts as a cushion where bones come together. Cartilage doesn't store calcium, so it is much softer than bone. Your ears and the tip of your nose can be bent from side to side. They are shaped by cartilage. When eating chicken, you may have noticed cartilage on the ends of the bones. Cartilage in foods such as chicken and beef is often called gristle.

Bone is covered by a thin sheet of tissue that forms an outer membrane. This membrane has many nerves and blood vessels that supply the bone with blood. This is the part that hurts when a bone gets bumped or bruised. It is also the growth area for new bone.

The inside of a bone shows other tissues. **Solid bone** is the very compact or hard part of a bone. It is usually found along the outer edges of bones. This part of the bone is very strong. Calcium is stored in solid bone and makes it strong.

Spongy bone is the part of a bone that has many empty spaces, much like those in a sponge. Spongy bone is usually found toward the ends of bones. It is strong like solid bone and gives the bone strength. Spongy bone also stores calcium but is lightweight because of the many empty spaces. Bird skeletons are lightweight because much of their bone tissue is spongy bone. How is this feature helpful to birds?

Remember that marrow is the soft center in bone. Blood cells and platelets are made in the marrow.

Ligaments (LIGH uh munts) are tough fibers that hold one bone to another. They are usually found at bone joints where two bones come together.

Mini Lab

How Strong Can a Hollow Bone Be?

Formulate a model: Curl a sheet of paper into a tube about 3 cm in diameter. Tape it so it does not unwind. How many books can be supported on top of the cylinder? *For more help, refer to the **Skill Handbook**, pages 706-711.*

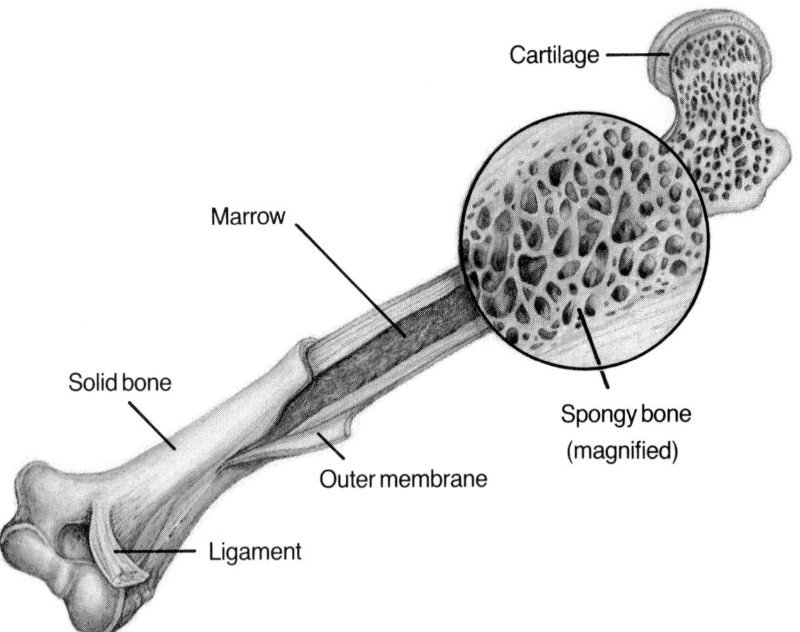

Figure 14-3 The six tissues of bone are shown in this cut-away diagram.

Bone Density

Lab 14–1

Problem: How do bird and mammal bone densities compare?

Skills

interpret data, form a hypothesis, experiment, measure in SI

Materials

100-mL graduated cylinder
water
balance
bones of cow, pig, chicken, turkey

Procedure

1. Copy the data table.
2. **Measure in SI:** Use a balance to determine the mass in grams of a cow bone.
3. Record the mass of the bone.
4. Measure 50 mL of water in the graduated cylinder.
5. **Experiment:** Place the bone into the cylinder. Read the new volume of water. Remember to read the volume at eye level.
6. Subtract 50 from the new volume reading. This will give you the volume of the bone in millimeters.
7. Record this volume of the bone in your data table.
8. To calculate the density of the bone, divide the mass of the bone by its volume. (NOTE: Density is a measure of how compact a material is. The more compact it is, the higher its density. The less compact it is, the lower the density.)
9. Write a **hypothesis** in your notebook about which type of bone you think will have higher density, bird or mammal. Justify your hypothesis. (Hint: Read Section 14:1 again.)
10. Repeat steps 2 to 8 using a pig bone, a chicken bone, and a turkey bone.
11. Wash your hands after handling the animal bones.

	MAMMAL BONE		BIRD BONE	
	COW	PIG	CHICKEN	TURKEY
Mass				
Volume				
Density				

Data and Observations

1. Which kind of bone, mammal or bird, has the higher density?
2. Which kind of bone, mammal or bird, has the lower density?

Analyze and Apply

1. **Check your hypothesis:** Is your hypothesis supported by your data? Explain.
2. How would you describe a mammal bone in terms of its density? How would you describe a bird bone's density?
3. (a) Which material would be more dense, a piece of steel or a piece of wood?
   (b) What kind of experiment could you do to prove your answer correct?
4. **Interpret your data** to explain if mammal bone is mainly solid or spongy.
5. Using your data, is bird bone mainly solid or spongy? Explain.
6. **Apply:** Using your data, explain how the type of bone might help birds and mammals survive.

Extension

Design an experiment that would help you measure the density of each of the three tissues of bone: spongy bone, solid bone, and marrow.

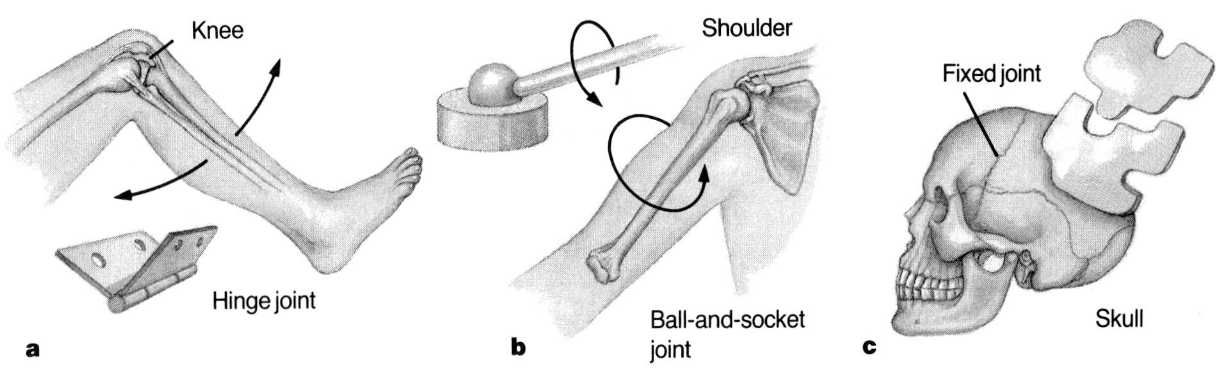

Figure 14-4 Hinge joints allow only back and forth movements, as in the hinge of a door (a). Ball-and-socket joints allow twisting and turning movements, as in a radio antenna (b). Fixed joints in the skull allow no movement, as in a jigsaw puzzle (c).

Joints

The place where bones come together is called a joint. There are several types of joints in the body. Most joints allow for different kinds of bone movement. Another type of joint allows no movement at all.

Hinge joints allow bones to move only back and forth. Your knee and elbow joints are examples of hinge joints, Figure 14-4a. Compare a knee and an elbow to a real hinge. Why do you think you can open your mouth? Your lower jaw is attached to the skull by a hinge joint.

What joint allows only back and forth movement?

Your upper arm bone meets the shoulder, and your upper leg bone meets the pelvis in ball-and-socket joints. A **ball-and-socket joint** allows you to twist and turn the bones in a circle where they meet. Figure 14-4b compares the movement of a ball-and-socket joint to an antenna.

The skull has joints where the bones come together. These joints don't move. Joints that don't move are called **fixed joints.** In Figure 14-4c, the zigzag lines on the skull are the fixed joints.

Check Your Understanding

1. What mineral supplies strength to bone?
2. Describe the jobs of cartilage, ligaments, and marrow.
3. Describe the movements of a hinge joint and a ball-and-socket joint. Give examples of each.
4. **Critical Thinking:** What are some sports you couldn't do if your shoulder and upper arm were joined by a hinge joint? Explain your answer.
5. **Biology and Math:** Is the smallest bone in the body less than two centimeters long? How do you know?

14:2 The Role of Muscles

Objectives

4. **Compare** three kinds of muscles.
5. **Explain** how muscles work to move body parts.
6. **Relate** how muscles work in pairs.

Key Science Words

muscular system
skeletal muscle
voluntary muscle
cardiac muscle
involuntary muscle
smooth muscle
tendon

The ball flies in your direction and you leap into the air to meet it. Your sister's pass from the other side of the volleyball net was a clever one, but you manage to return the ball. At the same time, you notice your shoelace is undone and you kneel to tie it. In these few seconds, you have used hundreds of different muscles in your body.

Movement

Muscle is a body tissue that can change its shape and length and thus cause movement. The kinds of movements muscles carry out depend on where the muscles are attached. For example, arm and leg muscles allow you to move the bones of your arms and legs, such as when you leap to punch a volley ball, Figure 14-5. Your heart is an organ with very strong muscle tissue. Its main job is to help pump or move blood. Your stomach and intestines also have muscle tissues. The movement of muscles in these organs helps to push food along.

The **muscular system** is all the muscles in your body. Almost all animal groups have a muscular system, which allows them to move about.

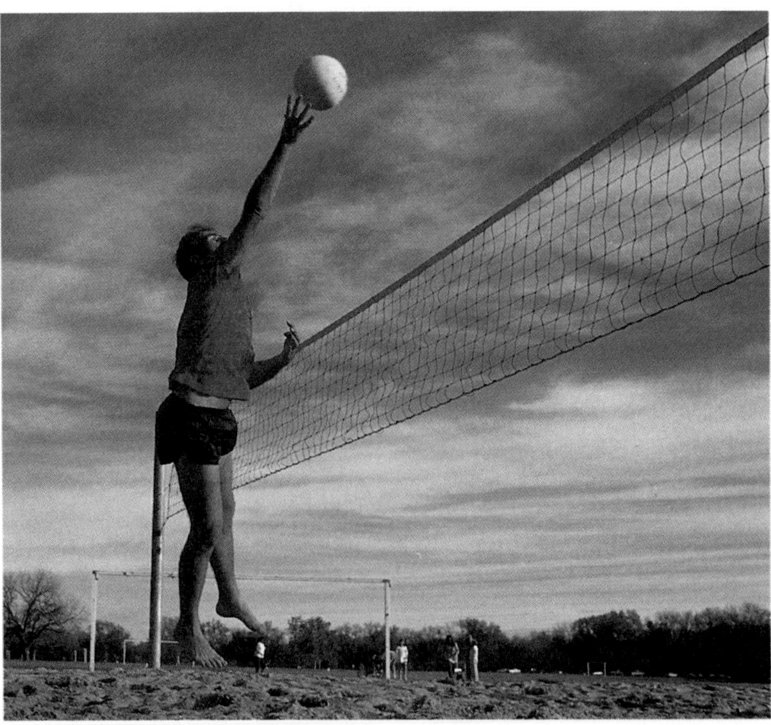

Figure 14-5 Muscles allow your body to perform a variety of different movements.

Career Close-Up

Athletic Trainer

Are you interested in sports? You might want a job working with athletes. You might want to work with students in high school, but you could also train athletes in college. Maybe one day you will be training a professional team. To become an athletic trainer you will need to understand how muscles and bones make up the human body. Take courses in biology, health, human physiology, chemistry, and physics in order to start your career. Other useful courses include nutrition, drugs, first aid, health, and human anatomy.

In a typical day you may be asked to work directly with athletes. As the trainer, you will supervise exercises and you might be required to treat sprains, muscle cramps, or other minor body injuries. Other activities for an athletic trainer include meal planning, repairing of sports equipment, and ordering of training room supplies.

As an athletic trainer, your future is rewarding. With a college degree, the starting salary can be up to $26 000 per year.

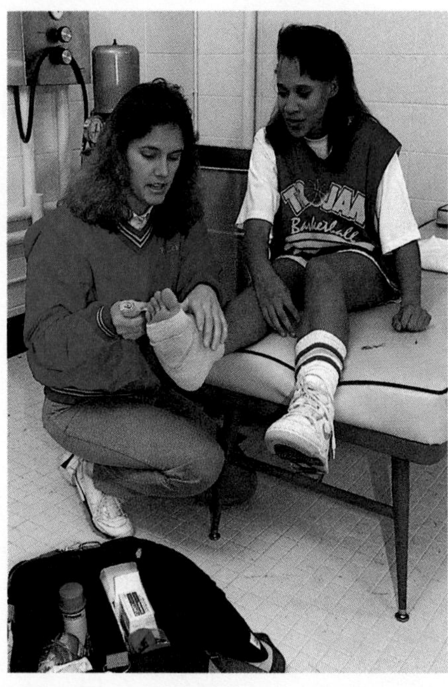

An athletic trainer understands the importance of warm-up exercises to avoid muscle damage.

Human Muscle Types

The human body has three different kinds of muscle. They are skeletal, smooth, and cardiac (KAR dee ac) muscle. Each muscle type has a different structure and pattern that make it different from the other. Each type of muscle has a different job, and each has a different location in the body.

Skeletal muscles are muscles that move the bones of the skeleton. Your arms and legs are moved by this muscle type. Skeletal muscles make up the bulk of your body. Important skeletal muscles for athletic activity include those in the arms, the legs, the abdomen, the chest, and the shoulders.

What are the three different kinds of muscles?

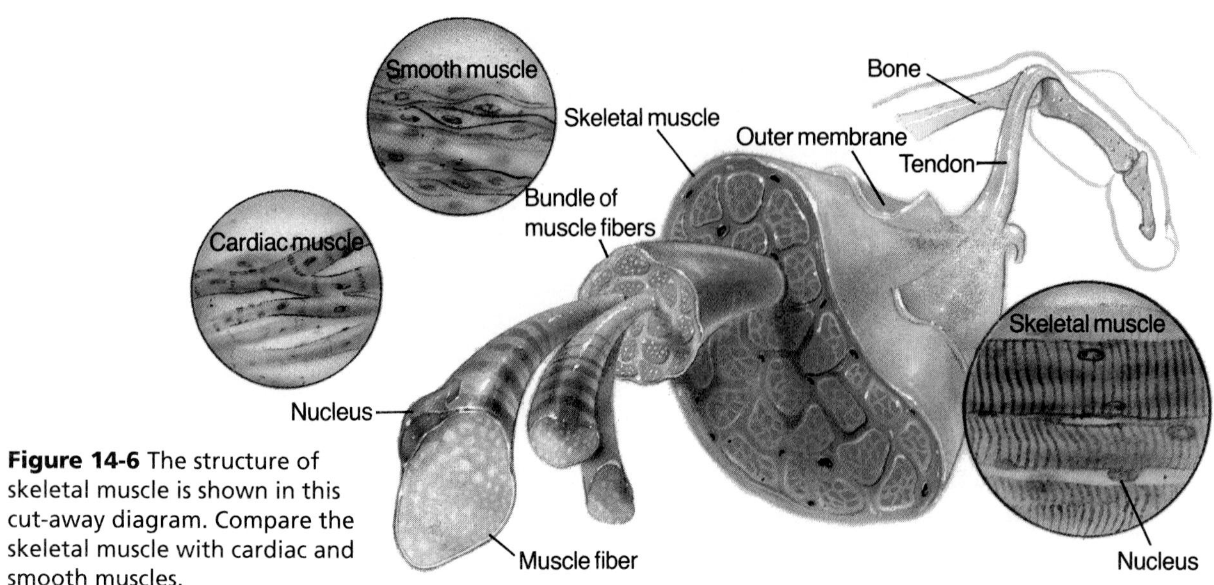

Figure 14-6 The structure of skeletal muscle is shown in this cut-away diagram. Compare the skeletal muscle with cardiac and smooth muscles.

Figure 14-6 shows how skeletal muscle is made of bundles of fibers. Each fiber is a muscle cell. Notice that skeletal muscle looks striped. The stripes are caused by two kinds of proteins that make up the muscle fibers. The whole muscle is surrounded by a layer of fibrous tissue.

Skeletal muscles are voluntary. **Voluntary muscles** are muscles you can control. If you decide to lift something, you control the muscles needed for lifting. If you pull a rope, jump, look out the window, or point, you decide what you want your muscles to do. You control your arm, leg, head, and finger movements.

Most of the animal tissue that people eat is skeletal muscle. If you eat meat, you are usually eating muscle. For example, the parts of a fish that you eat are its muscles. You eat mostly muscle when you eat a chicken drumstick. Hamburger is the ground-up muscle of beef cows.

Cardiac muscle is the muscle that makes up the heart. The heart is the only body organ made of this muscle type. Cardiac muscle is not connected to any bones, Figure 14-7.

Cardiac muscle is made of bundles of fibers, just as in skeletal muscle. Again, this muscle type looks striped. The main difference between cardiac and skeletal muscles is that the bundles of cardiac muscle fibers form a tight weave. The tight weave makes cardiac muscle very strong.

Cardiac muscle is involuntary. **Involuntary muscles** are muscles you can't control. You don't think whether your heart needs to pump slow or fast. Your brain controls this movement as needed.

Skill Check

Classify: In what ways are skeletal, cardiac, and smooth muscle alike? How do they differ? *For more help, refer to the **Skill Handbook** pages 715-717.*

Figure 14-7 The heart is the only organ with cardiac muscle.

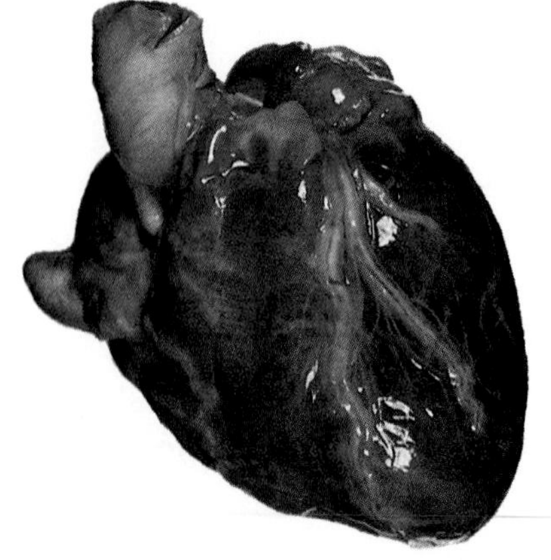

Smooth muscle is involuntary muscle that makes up the intestines, arteries, and many other body organs. You have no control over the working of your intestines or blood vessels. The cells of smooth muscle are long and spindly, but much shorter than the fibers of skeletal muscles. Smooth muscle does not have stripes. Like cardiac muscle, smooth muscle is not connected to any bones.

How Muscles Work

A muscle works by changing its length. For a muscle to do its job, its fibers must shorten. The entire muscle shortens when the muscle fibers contract. The contraction of muscles allows you to move your body parts. Muscles don't lengthen to move body parts, they only shorten.

Let's look at the working of a skeletal muscle. Skeletal muscles are connected to bones by tendons (TEN duhns). A **tendon** is a tough, fibrous tissue that connects muscle to bone. The tendons cause the bone to be pulled when the muscle contracts.

How would your muscles work if you were to pull a boat in to dock with a rope? Look at Figure 14-8a to see the shape of your upper arm muscles before you begin to pull. They are relaxed. The muscle fibers have not contracted or shortened. Figure 14-8b shows how your arm muscles would look while you were pulling on the rope. Your muscle fibers are working. When they work, they contract or shorten. The bulging you see when a muscle is working is due to the thickening of muscle fibers as they shorten.

Mini Lab

What is the Structure of a Chicken Leg?

Observe: Examine a cooked chicken drumstick. Locate the muscles, tendons, cartilage, and bone. Wash your hands after handling the drumstick. Why are so many tendons needed? Where do tendons lead to? *For more help, refer to the **Skill Handbook,** pages 704-705.*

How do skeletal muscles move body parts?

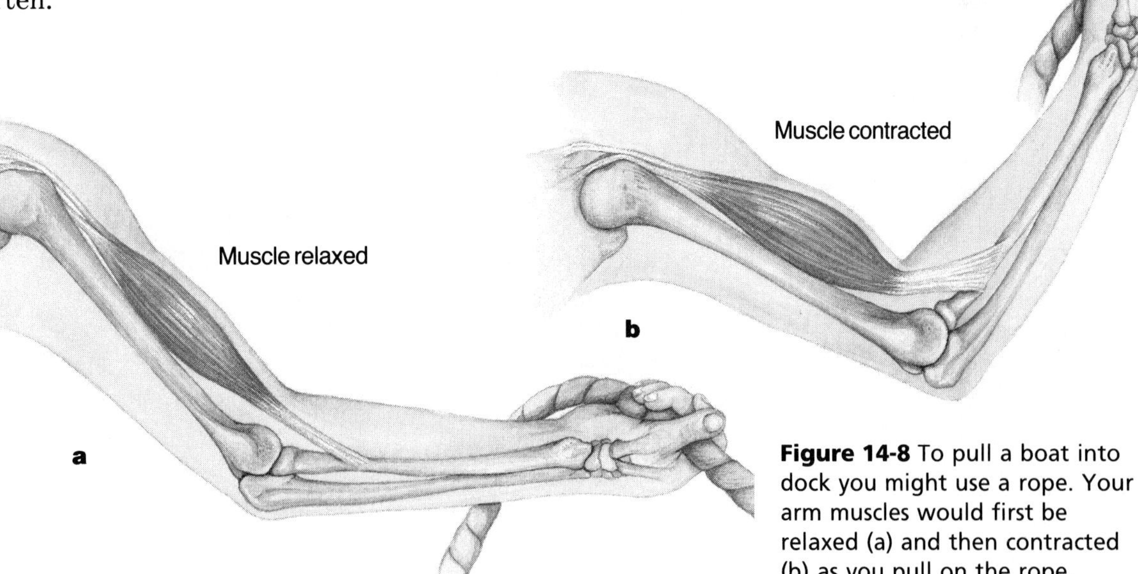

Figure 14-8 To pull a boat into dock you might use a rope. Your arm muscles would first be relaxed (a) and then contracted (b) as you pull on the rope.

Skill Check

Infer: If muscles shorten when they contract, where does the excess muscle tissue go? *For more help, refer to the* ***Skill Handbook,*** *pages 706-711.*

Bones with skeletal muscles also allow you to walk. Walking is an exercise that strengthens muscles because of the continual contracting and stretching of pairs of muscles. Let's look at this movement.

Figure 14-9a shows the muscles that are used to raise or lower your foot. All the muscles in this diagram are relaxed. This means that the foot is not moving up or down. Notice that one end of muscle A is attached to the top of the tibia (TIHB ee uh), your leg bone. This end won't move during contraction. The lower end of the tibia is attached by a tendon to the tops of the tarsals, the bones of your foot. This end moves during contraction. One end of muscle B is attached to the bottom of the femur, your thighbone. This end won't move during contraction. The other end is attached by a tendon to the back of your heel bone. This end moves.

Figure 14-9b shows that muscle A is relaxed. You can tell this because the muscle is the same length as muscle A shown in Figure 14-9a. Muscle B, however, is contracted. You can tell this because it is much shorter than it was in Figure 14-9a. As muscle B contracts, it pulls up the bottom of your heel. A hinge joint in the ankle allows for this movement.

How is your foot pulled up? A different muscle must be used this time. Muscle B is not used. Muscle B in Figure 14-9c is now relaxed. Muscle A is contracted. When it shortens, it pulls up the foot. You can now move forward.

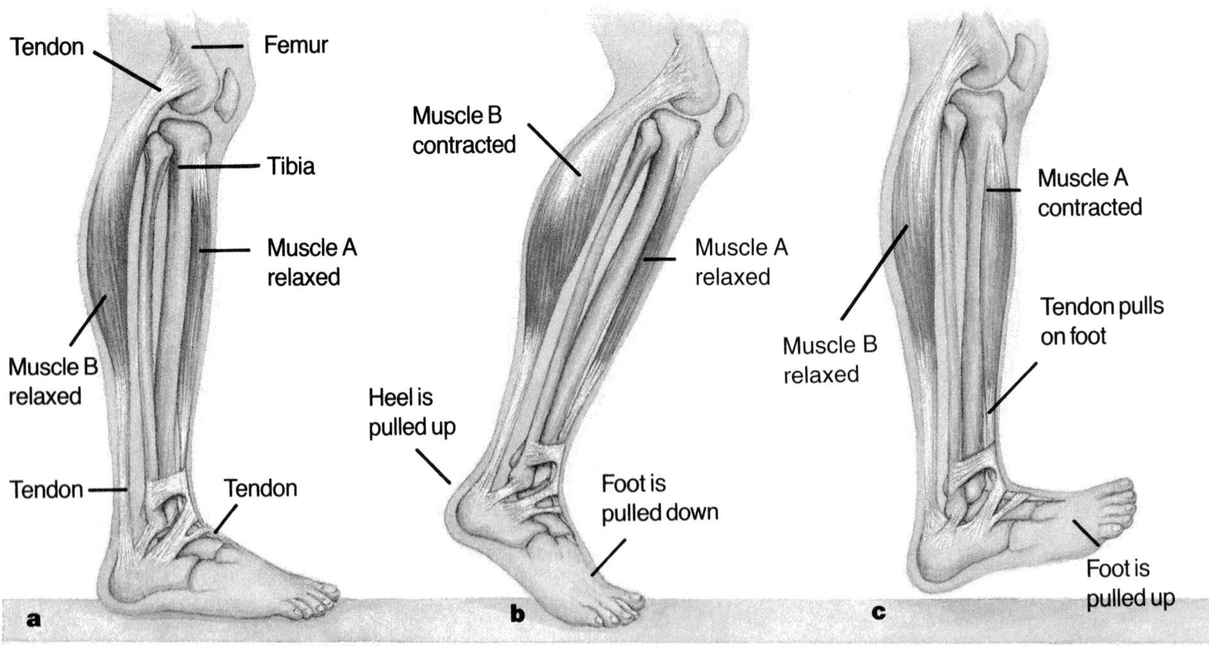

Figure 14-9 When you walk, you lift your feet. At first, your muscles are relaxed (a). Then, the heel is pulled up (b), and the foot is lifted off the ground (c). Why is cycling a good exercise for leg muscles?

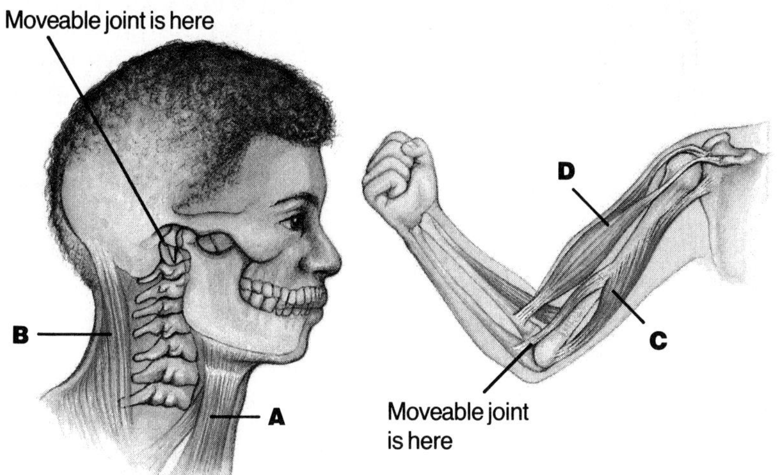

Figure 14-10 Which way will the head and arm move as different muscles contract and relax?

Muscles Work in Pairs

How do skeletal muscles work in pairs?

Because muscles can only shorten, it takes a pair of skeletal muscles to move bones back and forth. When one muscle contracts, the opposite muscle must relax. In the examples just used, one muscle was needed to pull your foot up. A different muscle was needed to lower your foot.

Look at these other examples in Figure 14-10. See if you can figure out which muscle has which job. Remember, muscles move body parts only when they shorten. The body part moves because muscles pull on a bone.

How does the head move when muscle A relaxes and muscle B contracts?

How does the arm move when muscle C contracts and muscle D relaxes?

Check Your Understanding

6. Give examples of the three types of muscles and where they are found in the body.
7. What is the job of tendons?
8. How does the length of a relaxed muscle compare with that of a contracted muscle?
9. **Critical Thinking:** Muscle cells have more mitochondria than other cells. Why?
10. **Biology and Reading:** You have some muscles in your eye that you can control. Other muscles in your eye work automatically. What two kinds of muscles must control movements in your eye?

Muscles

Lab 14–2

Problem: Do muscles work in pairs?

Skills

formulate models, experiment, infer

Materials

index card
scissors
string
metal fastener
tape
paper punch

Procedure

1. Copy the data table.
2. Cut one index card in half lengthwise. Attach one half of the card to the other half using the metal fastener as shown in Figure A. One half card is the foot, and the other half card is the leg.
3. Punch two holes in the top of the leg bone card, and thread a piece of string, 15 cm long, through each hole. Attach the strings to the base of the foot card as shown in Figure B.
4. **Experiment:** Pull up on string A. Record in your table how the foot part of the model moves. Also note if string B gets shorter or longer between the hole on the leg card and the tape on the foot card.
5. Record your findings in the table.
6. Pull up on string B. Record in your table how the foot part of the model moves. Also note if string A gets shorter or longer between the hole on the leg card and the tape on the foot card.

Data and Observations

1. In your model:
   (a) what does the metal fastener represent?
   (b) what do strings A and B represent?
   (c) what do the pieces of tape represent?
2. When the foot moved down, which string got shorter? Which got longer?

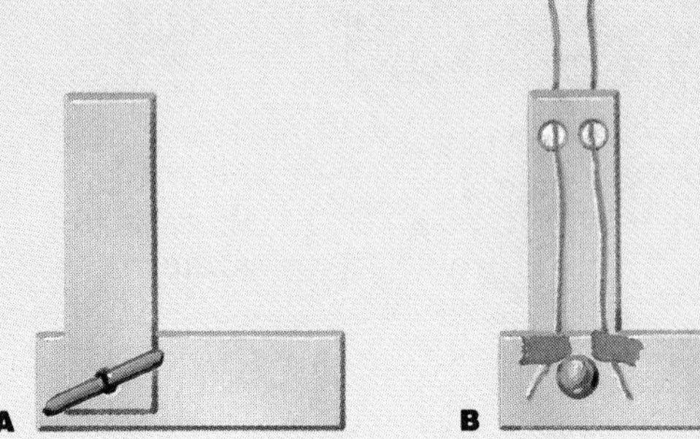

3. When the foot moved up, which string got shorter? Which got longer?

	FOOT MOVES UP OR DOWN?	STRING A LENGTHENS OR SHORTENS	STRING B LENGTHENS OR SHORTENS
String A pulled up			
String B pulled up			

Analyze and Apply

1. (a) Was a short string supposed to show a contracted or a relaxed muscle?
   (b) Was a long string supposed to show a contracted or a relaxed muscle?
2. Can the same string (muscle) move the foot up and down? Explain.
3. **Apply:** Which other sets of muscles in your body act the same way as those in your model?

Extension

Formulate a model of a weight-lifting machine that would act to strengthen the thigh muscles while moving the leg forward and backward.

14:3 Bone and Muscle Problems

From studying this chapter, you know that the skeletal and muscular systems work together. So, a problem in one system may cause a problem with the other. Modern science and technology is used to solve some of the problems that affect the skeletal and muscular systems.

Objectives

7. **Discuss** the problems of the skeletal system.
8. **Give examples** of problems of the muscular system.
9. **Explain** the purpose of new designs for products.

Problems of the Skeletal System

Key Science Words

arthritis
sprain
muscular dystrophy

Some problems with the skeletal system are caused by diseases of the joints. **Arthritis** (ar THRIT us) is a disease of bone joints. One type of arthritis results in breakdown of the cartilage at the joints. The result is pain and swelling at the diseased joint. In time, a person may not be able to bend or move the affected part of the body, Figure 14-11.

What are the effects of arthritis?

Today, there is some help for the problems of arthritis. Artificial joints made of plastic or metal are sometimes used to replace diseased joints. Hip, ankle, and knee joints have been replaced by this technology. Figure 14-11 shows an artificial hip joint. It is made of metal and replaces the ball part on the femur, which is attached to the ball-and-socket joint in the hip.

Bones are strong because they store calcium. If the level of calcium is low, bones will break easily. Bones tend to become brittle as a person ages. Exercise and eating dairy foods helps to keep bones strong.

Skill Check

Understand science words: arthritis. The word part *arthr* means joint. In your dictionary, find three words with the word part *arthr* in them. *For more help, refer to the* ***Skill Handbook,*** *pages 706-711.*

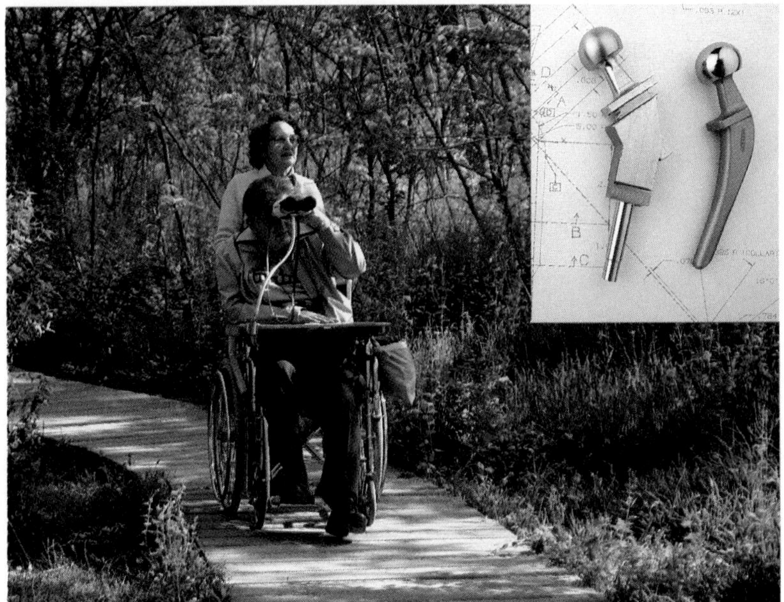

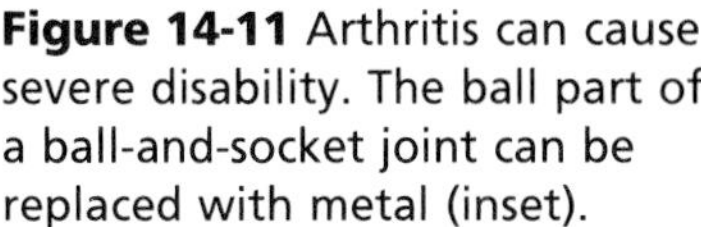

Figure 14-11 Arthritis can cause severe disability. The ball part of a ball-and-socket joint can be replaced with metal (inset).

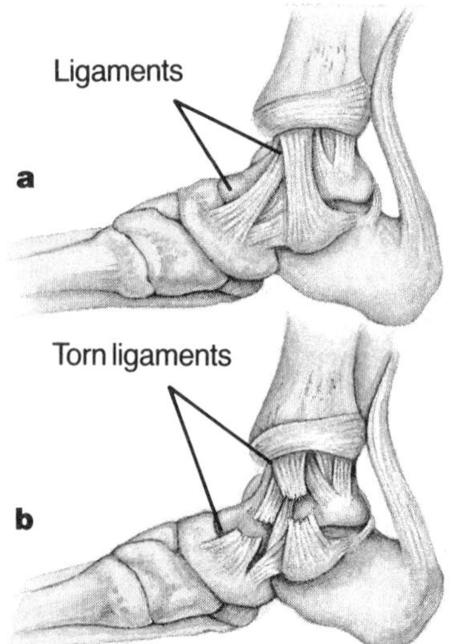

Figure 14-12 Normal (a) and torn (b) ligaments of an ankle with a sprain

Most bones in the body are connected by ligaments. Have you ever twisted your ankle? If you have, the pain you felt was caused by injury to your ankle ligaments, Figure 14-12. **Sprains** are injuries that occur to your ligaments at a joint. A sprain results when the ligaments are torn and blood vessels are damaged.

Problems of the Muscular System

Have you ever strained a muscle while lifting or after a sudden move? Have you ever had muscle cramps while swimming or during a long-distance run? Strains and cramps are two quite different problems.

What is a cramp?

You might strain a muscle if you haven't exercised for a while. A sudden use of a poorly exercised muscle may tear the fibers. Regular exercise will help you avoid straining your muscles. A muscle cramp is when the muscle contracts strongly and then can't relax. Muscle cramps result from a poor supply of oxygen. If you exercise one set of muscles too long, the oxygen supply may run low. When oxygen returns, the muscles will recover.

Skeletal muscles are controlled by nerve cells. A disease called muscular dystrophy (MUS kyuh lur • DIHS truh fee) blocks the nerve messages to muscles. **Muscular dystrophy** is a disease that causes the slow wasting away of skeletal muscle tissue. Muscular dystrophy can be inherited and is most common in males, Figure 14-13.

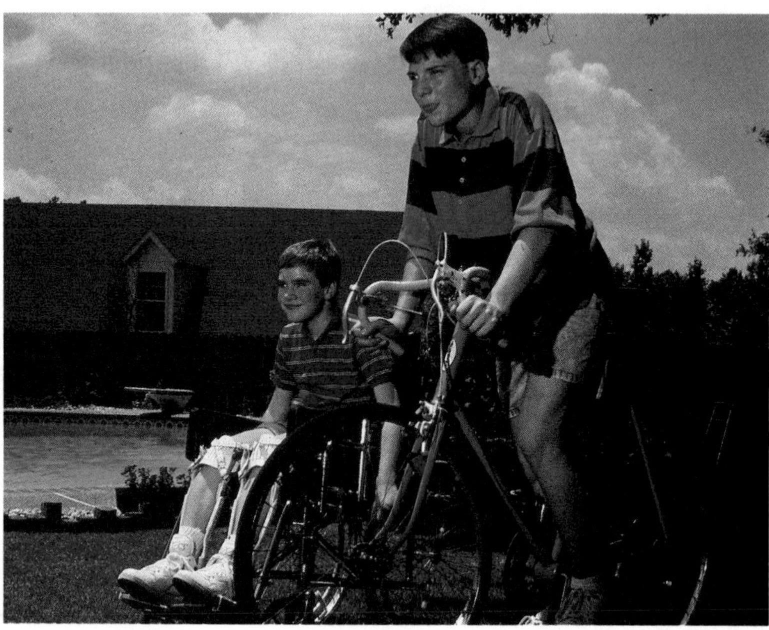

Figure 14-13 By transplanting healthy muscle cells into a person with muscular dystrophy, scientists have given new hope to many people with this disease.

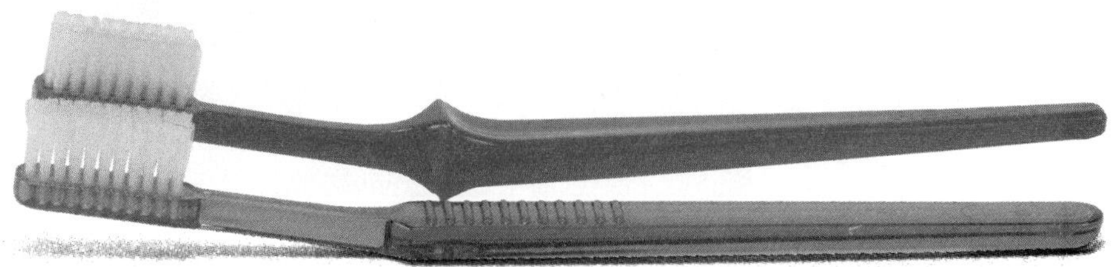

Figure 14-14 How are these toothbrushes designed to fit your body?

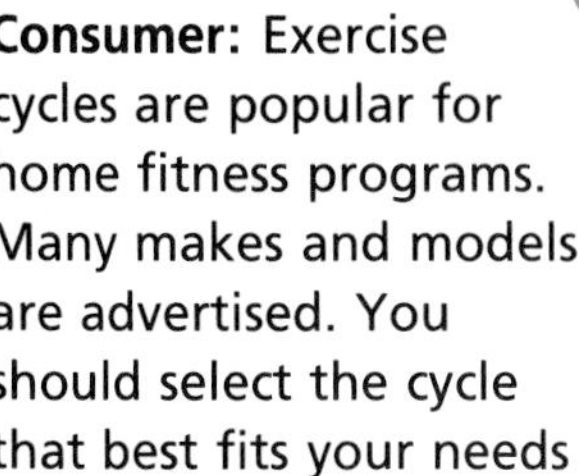

Bio Tip

Consumer: Exercise cycles are popular for home fitness programs. Many makes and models are advertised. You should select the cycle that best fits your needs and body shape.

Making Products Fit People

How often have you sat in class and complained that the desk or chair was very uncomfortable? Backaches, muscle fatigue, and general discomfort are often the result of poor product design. Why? In the past, the most important thing for a new product was that it should look good. It wasn't important that a product fit the shape of the human body.

All of this is changing, however. New product design that fits the user's body shape is now becoming important. Automakers have put lumbar supports in their car seats. The lumbar region is the "small of the back" that gets tired after long periods of sitting.

What kind of product design is becoming popular?

Look at the toothbrushes in Figure 14-14. Will their handles reduce the chances of dropping them? Will they feel more comfortable in your hands? Will the bristles reach all the surfaces of your teeth and gums? Their design helps solve these kinds of problems.

Check Your Understanding

11. What is arthritis?
12. What happens to ligaments when the ankle is sprained?
13. What new design in autos makes drivers more comfortable?
14. **Critical Thinking:** Low back pain and neck pains are often due to muscle spasms related to stress. What might be some ways to avoid these problems?
15. **Biology and Math:** There are several kinds of muscular dystrophy. In the kind of muscular dystrophy that occurs most often in the U.S., 3 out of every 10 000 boys will be expected to show symptoms of the disease. How many boys would be expected to show symptoms of muscular dystrophy in a city of 60 000?

Chapter 14 Review

Summary

14:1 The Role of the Skeleton

1. The skeleton serves as a framework, protects body organs, makes blood cells, stores calcium, and provides a place for muscle attachment. Bone is alive and grows.
2. The six tissues of bones include a fibrous membrane around solid bone, spongy bone, and marrow, as well as cartilage and ligaments.
3. Three types of joints are hinge, ball-and-socket, and fixed.

14:2 The Role of Muscles

4. Muscles bring about body movement. The three types of muscle are skeletal, cardiac, and smooth.
5. Muscles contract in order to work. They are connected to bone by tendons.
6. Muscles must work in pairs in order to bring about movement of body parts.

14:3 Bone and Muscle Problems

7. Arthritis, lack of calcium, and sprains are three common problems of the skeletal system.
8. Strains and cramps are two common muscle problems. Muscular dystrophy is a more serious and often inherited muscle problem.
9. New products are being designed to fit the shape of the human body.

Key Science Words

arthritis (p. 299)
ball-and-socket joint (p. 291)
cardiac muscle (p. 294)
fixed joint (p. 291)
hinge joint (p. 291)
involuntary muscle (p. 294)
ligament (p. 289)
muscular dystrophy (p. 300)
muscular system (p. 292)
skeletal muscle (p. 293)
skeletal system (p. 286)
smooth muscle (p. 295)
solid bone (p. 289)
spongy bone (p. 289)
sprain (p. 300)
tendon (p. 295)
voluntary muscle (p. 294)

Testing Yourself

Using Words

Choose the word from the list of Key Science Words that best fits the definition.

1. allows arm to move in a full circle
2. involuntary muscle that makes up intestines and blood vessels
3. muscle that makes up the heart
4. a disease of bone joints
5. compact or hard part of bone
6. attaches muscle to bone
7. holds two bones together
8. type of muscle that works in pairs
9. framework of all body bones

Review

Testing Yourself *continued*

Finding Main Ideas

List the page number where each main idea below is found. Then, explain each main idea.

10. examples of the three types of joints in the human body
11. how the problem of arthritis is treated
12. five main jobs of the skeletal system
13. what a sprain is
14. a disease that causes the blocking of nerve messages to muscle
15. the six tissues of bone
16. why two different muscles are needed to move bones back and forth
17. the medical names of the bones of the leg, arm, collar, and shoulder blade
18. how products can be designed to fit the shape of the human body

Using Main Ideas

Answer the questions by referring to the page number after each question.

19. What type of joint is found in each of the following: lower jaw, upper leg and pelvis, and shoulder? (p. 291)
20. What are two examples of bones protecting body organs? (p. 286)
21. Why do you feel pain when you sprain your ankle? (p. 300)
22. What type of muscle is found in or as part of each of the following: stomach, blood vessels, heart, and arms? (pp. 294, 295)
23. How do you move your foot up and down? (p. 296)
24. What is the difference between voluntary and involuntary muscles? (p. 294)
25. What happens to muscle when the oxygen supply runs low? (p. 300)

Skill Review ✓

*For more help, refer to the **Skill Handbook,** pages 704-719.*

1. **Infer:** When weight lifters flex their muscles, are the muscles shortening, relaxing, or stretching?
2. **Understand science words:** The word *involuntary* means without choice. The word part *in* means without. In your dictionary, find another meaning of the word part *in.*
3. **Classify:** Which words or statements are not correct for all muscles: voluntary, living, contract, made from cells, and form a tight weave?
4. **Infer:** A muscle that moves your leg up or down depends on what direction of action at your knee joint?

Finding Out More

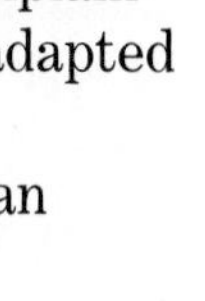

Critical Thinking

1. Draw an imaginary animal. Build your new animal's skeletal system from balsa wood or paper. Hinge bones together with wire or thread. Explain to the class how your animal is adapted to survive.
2. How does the skeleton of a human compare with the skeleton of an alligator? Explain how each skeleton helps the organism survive.

Applications

1. Find out the causes and problems of rheumatoid arthritis, myasthenia gravis, and gout.
2. Make a list of the different types of bone fractures that can occur.

Biology in Your World

An Amazing Machine

Even as you sleep, your digestive system is breaking down food. Your circulatory system is delivering oxygen and nutrients to cells. Eating well and getting plenty of exercise will keep your body systems running smoothly as they do their tasks.

LITERATURE

Think Positive

Your heart really pumps life into your body. It is a remarkable, complex machine.

Norman Cousins was an editor and author. For many years he was editor of the magazine *Saturday Review.* He had a massive heart attack in 1980. Instead of giving up, he looked at his illness as a challenge. He began to take part in medical decisions concerning his health. Cousins knew that panic and depression could make his heart condition worse. He used positive thought, humor, and relaxation to avoid panic and depression so his heart could recover. In *The Healing Heart,* Cousins tells about his illness and recovery. He stresses that a person's attitude has a lot to do with recovering from an illness. Cousins died in 1990 at the age of 78.

ART

The Inside Story

One of the ways medical students learn about the inside parts of the human body is through dissection. In the 1600s, the bodies of executed criminals were used. In 1632, the Dutch painter Rembrandt van Rijn was asked to paint a group portrait of the Amsterdam Guild of Surgeons. His painting, *The Anatomy Lecture of Dr. Nicolaes Tulp,* records a public dissection. In the painting, Dr. Tulp is showing the tendons of the arm. This painting was one of the first group portraits Rembrandt did in Amsterdam. It is considered one of the greatest dramatic group portraits.

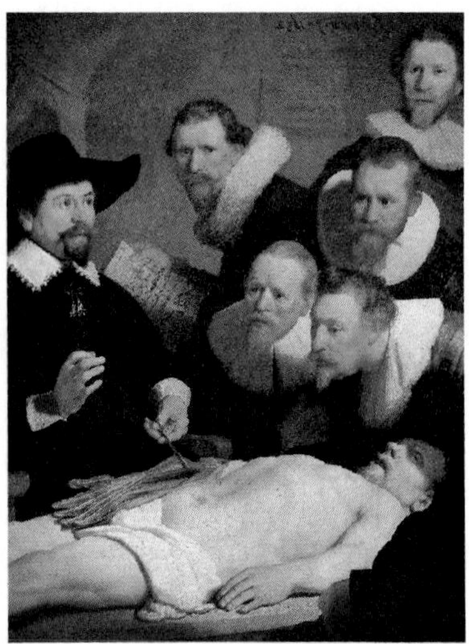

LEISURE

How Fit Are You?

Exercise can be an enjoyable way to spend leisure time. The benefits of exercise are long-lasting. Endurance exercises done on a regular basis will help your heart, lungs, and circulatory system work better. These exercises include walking, jogging, bicycling, and swimming. Lifting weights will build muscle strength, and stretching will increase flexibility.

To improve your overall fitness, plan an exercise program. Before you begin one, get advice from a doctor. Your program need not take large amounts of time or money. You may want to exercise with a friend.

Besides an exercise program, there may be other ways to get exercise. Are there daily activities you could change? Maybe you could walk to the store rather than ride in a car.

CONSUMER

Healthy Dieting

Many people diet to lose weight. What is a safe diet? A safe diet provides the nutrients the body needs every day, but with fewer calories. An effective diet includes an exercise program. Many doctors think a person should lose no more than two pounds per week. Before beginning a diet and exercise program, a person should see a doctor.

How safe are fad diets? Some fad diets promise weight loss without exercise. Others stress eating foods with a lot of a certain nutrient. Some diets promise a loss of several pounds a week. How safe and effective do you think these diets are? How well do they meet safe diet requirements?

Unit 4

CONTENTS

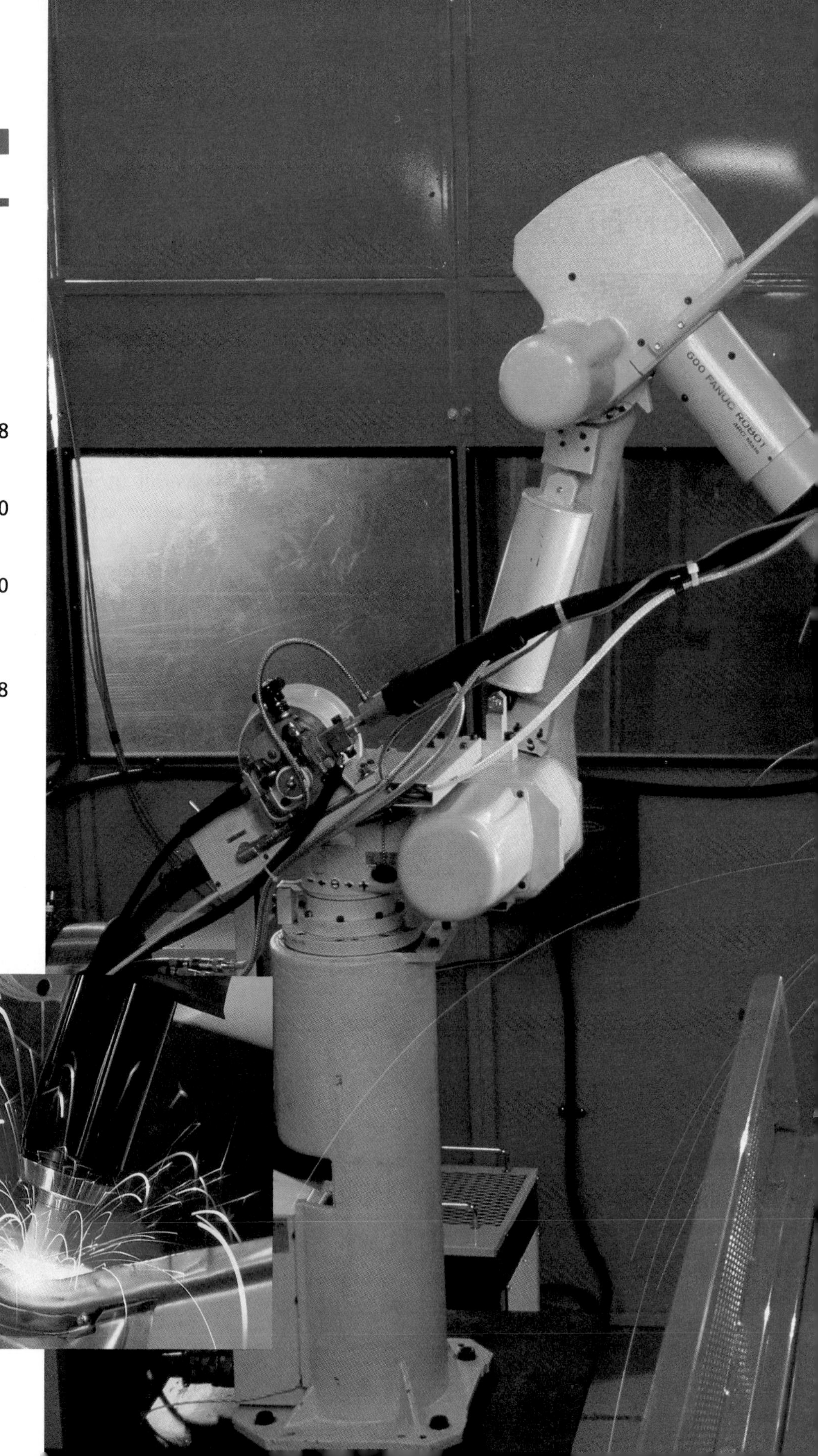

Body Systems—Controlling Life

What would happen if...

robots could think and make decisions? So far, robots can't think, guess, imagine, create, or make decisions. But researchers hope to produce computer programs that mimic human thought.

With such computer programs, the possible uses of robots would increase greatly. More and more jobs could be taken over by robots.

Being able to run a factory or business without people has many advantages. Robots are not paid, and they could work 24 hours a day. Dirty, boring, and dangerous jobs could be done by robots. But, would people who lose their jobs to robots find other jobs or would they become unemployed? What would be the impact on society of increased unemployment?

CHAPTER PREVIEW

Chapter Content

Review this outline for Chapter 15 before you read the chapter.

15:1 The Role of the Nervous System
- Response
- Nervous Systems of Animals

15:2 Human Nervous System
- Nerve Cells
- Pathways of Messages
- Nervous System Parts
- Jobs of the Brain
- Reflexes

15:3 The Role of the Endocrine System
- Chemical Control
- The Pituitary Gland
- The Thyroid Gland

15:4 Nervous and Endocrine System Problems
- Nervous System Problems
- Endocrine System Problems

Skills in this Chapter

The skills that you will use in this chapter are listed below.

- In **Lab 15-1,** you will form a hypothesis, calculate, and infer. In **Lab 15-2,** you will experiment and interpret data.
- In the **Skill Checks,** you will interpret diagrams and understand science words.
- In the **Mini Lab,** you will observe.

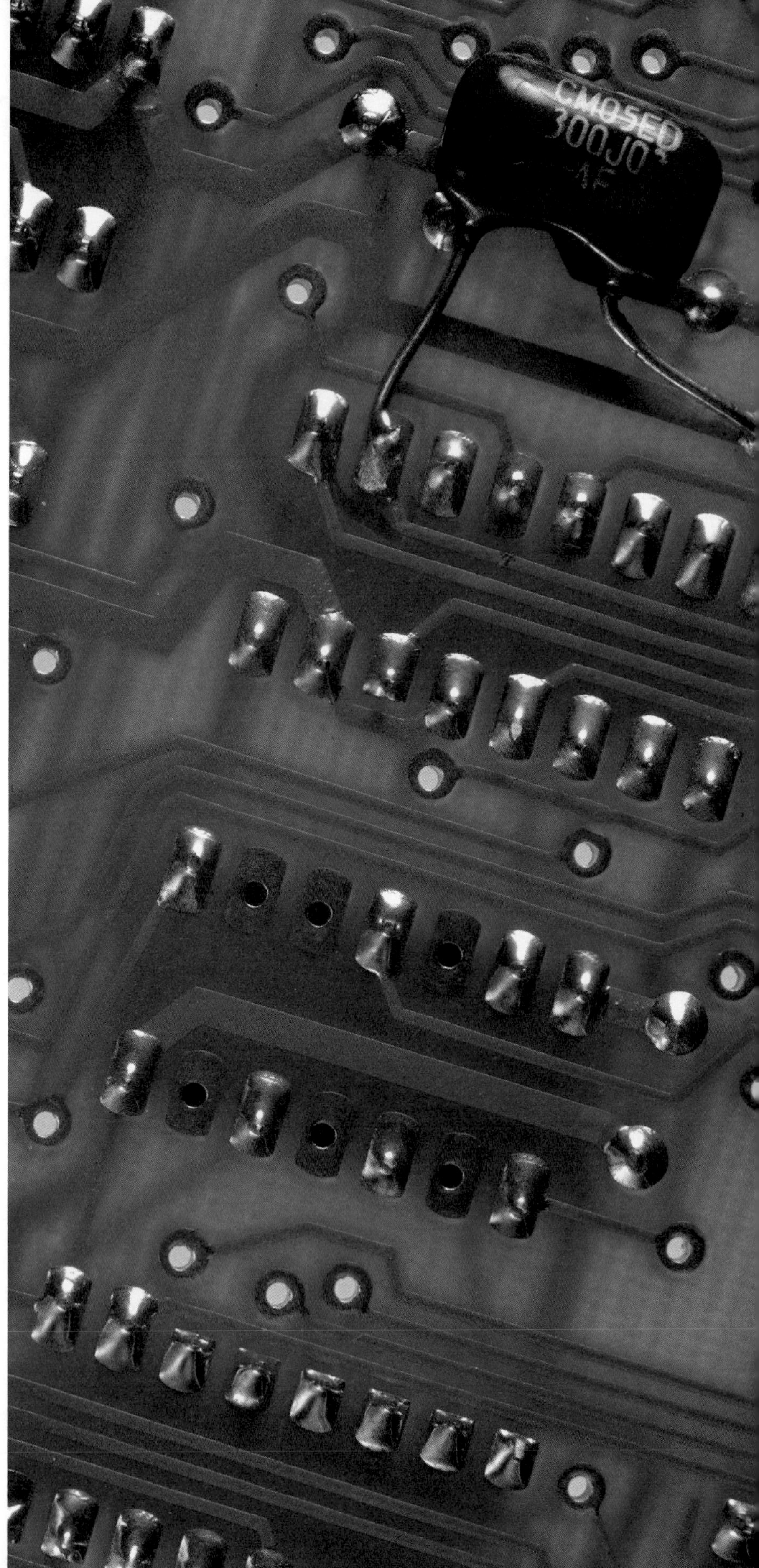

Chapter 15

Nervous and Chemical Control

The larger photo on the left is a photograph of a circuit board from a computer. The smaller photo below shows the human brain. It is often said that a computer and the human brain are very much alike. The brain is made of living cells. There are billions of cells in the human brain. Each cell sends and receives messages from nearby cells. These messages are interpreted as sights, sounds, smells, or thoughts. A computer has thousands of computer chips. Each chip sends and receives messages from the other chips in the computer. Both brains and computers process information. How is the processing of information by the brain important? In what other ways are messages carried through the body?

Try This!

Can you control your reflexes? Put on a pair of safety goggles. Have someone toss a paper wad toward your eyes. Do you blink? Can you keep yourself from blinking? What is this involuntary action called?

BIOLOGY Online

Visit the Glencoe Science Web site at science.glencoe.com to find links about **nervous and chemical control.**

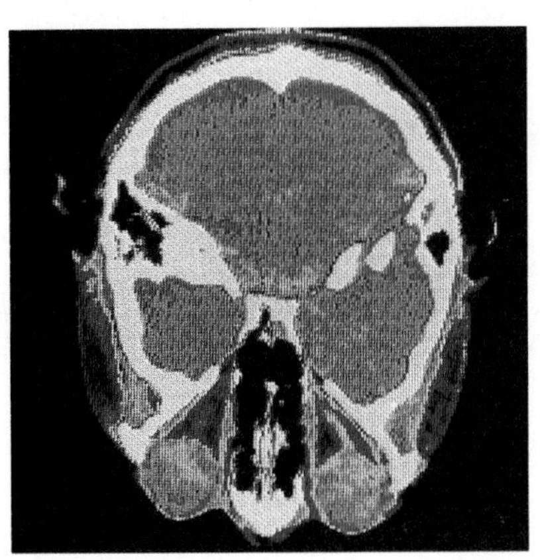

CAT scan of the human brain

15:1 The Role of the Nervous System

Objectives

1. **Explain** how animals keep in touch with their body parts and their surroundings.
2. **Compare** the nervous systems of common animals.

Key Science Words

nervous system

Most animals have special cells and organs that help them keep in touch with parts of their own bodies. They are able to send messages to all body parts and receive messages from those body parts. These special cells also allow an animal to receive messages from its surroundings.

Response

Most animals can quickly detect changes that take place around them. Usually, they respond quickly to these changes. A response is the action of an organism because of a change in its environment.

Let's use an example to show how quickly animals respond to changes around them. Have you ever seen a dog chase a cat? It seems impossible for the dog to catch the cat. Why? First, the cat usually sees the dog. The cat's brain gets the message that something nearby is moving toward it. The brain responds by sending messages to the cat's leg muscles. The cat runs up a tree to protect itself.

The cat sees and responds to the dog because it has a nervous system. A **nervous system** is made of cells and organs that let an animal detect changes and respond to them. This system is made of nerve cells, sense organs, and usually a brain.

Nervous Systems of Animals

The simplest animals to have a nervous system are the stinging-cell animals. Figure 15-1a shows what this system looks like in the hydra. Hydras have a very simple nervous

Figure 15-1 Hydras (a) and planarians (b) have simple nervous systems.

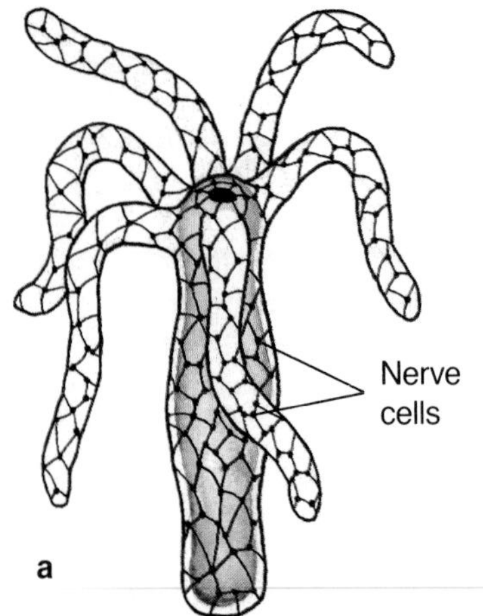

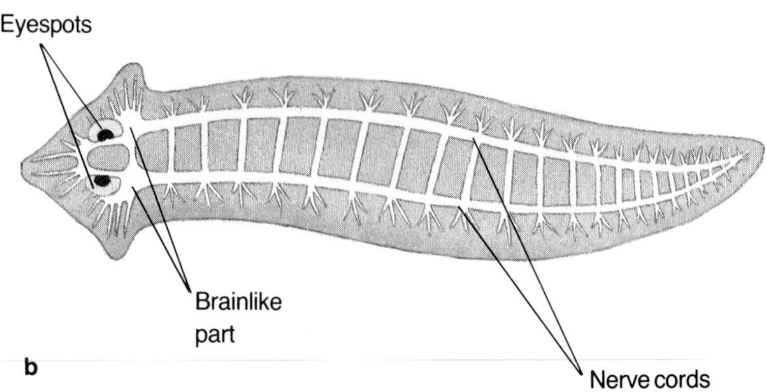

system made up of nerve cells that form a net throughout the body. This simple grouping of nerve cells lets these animals detect and respond to changes around them.

What kind of nervous system does a hydra have?

A flatworm has a nervous system that is more complex than that of the hydra. The nervous system of a flatworm has several parts. You can see in Figure 15-1b that a planarian has three parts that a hydra doesn't have. The planarian has a brainlike part, eyespots, and two nerve cords. Planarians can respond to changes in light with their eyespots. Because they have a brainlike part and nerve cords, they also can respond more quickly than a hydra.

How are animals able to respond to changes around them?

Jointed-leg animals have a more complex nervous system than flatworms. Figure 15-2 shows the inside of a spider. A spider has four pairs of eyes. It has a brain, nerve cord, and many nerves going to its body parts. Spiders can respond even more quickly than planarians.

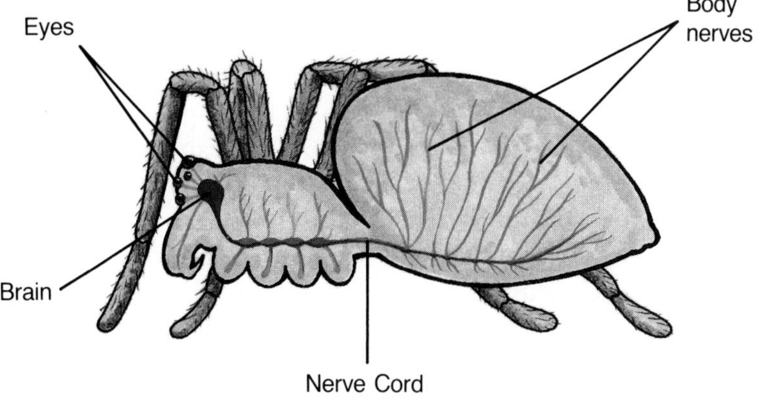

Figure 15-2 Nervous system of a spider

Check Your Understanding

1. What is meant by *response?* Give an example not found in this chapter.
2. What is a nervous system? What parts form a nervous system?
3. Compare the nervous systems of a hydra, planarian, and spider.
4. **Critical Thinking:** What cell parts do protists use to respond to changes in the environment?
5. **Biology and Writing:** A planarian has a nervous system. Do you think a planarian can learn? Write a paragraph of at least three sentences to explain your answer.

15:2 Human Nervous System

Objectives

3. **Explain** how nerve cells carry messages through the body.
4. **State** the functions of the three major organs of the nervous system.
5. **Describe** a reflex action.

Key Science Words

neuron
nerve
dendrite
axon
synapse
brain
spinal cord
cerebrum
cerebellum
medulla
reflex

Humans have one of the most complex nervous systems. If you understand how the human nervous system works, then you will have an idea of how nervous systems work in many other animals. Let's look at parts of the human nervous system in more detail.

Nerve Cells

Nerve cells are the main part of any nervous system. They carry messages through the nervous system. Your body contains billions of nerve cells. Nerve cells are called **neurons** (NOO rahnz).

Neurons are often compared to electrical wires. Wires are very thin and long. Wires carry messages and have a covering of insulation around them. Wires usually connect two things, such as an electric outlet and a lamp. Neurons are very thin and can be very long. They are as long as one meter in some parts of your body. Neurons also have a covering. Neurons carry messages through your body. They make it possible for body parts to keep in touch with each other. Figure 15-3 shows how a wire and a neuron compare.

We can also compare a cable with a nerve. A cable is made of many wires bunched together. Cables are thick. A **nerve** is many neurons bunched together. Nerves can be thick, too. Figure 15-3 compares a nerve with a cable. Notice how thick the nerve is compared to the single neuron.

Figure 15-3 Neurons and nerves can be compared to wires and cables.

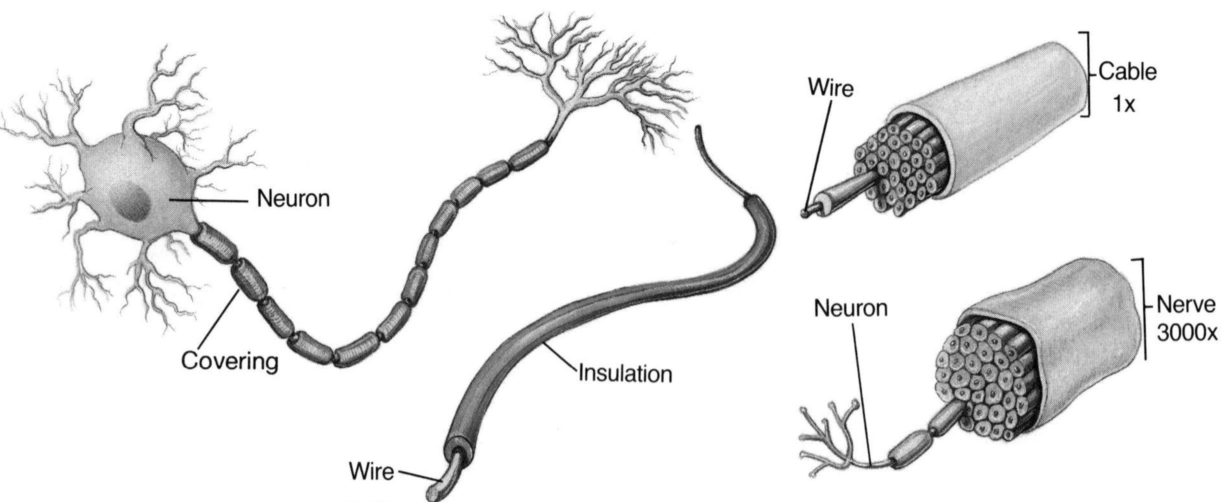

A neuron, shown in Figure 15-4, has many of the same parts that other cells have. There are a few differences, however, between neurons and other cells. Two differences are the length and the shape of the neuron. The long, thin shape helps the neuron do its job well. The third difference between neurons and other cells is in the shape of the ends. Notice the branching shape of both ends of the neuron. One end is called the dendrite. **Dendrites** are parts of the neuron that receive messages from nearby neurons. The other end, usually longer, is called the axon (AK sahn). The **axon** is the part of the neuron that sends messages to surrounding neurons or body organs.

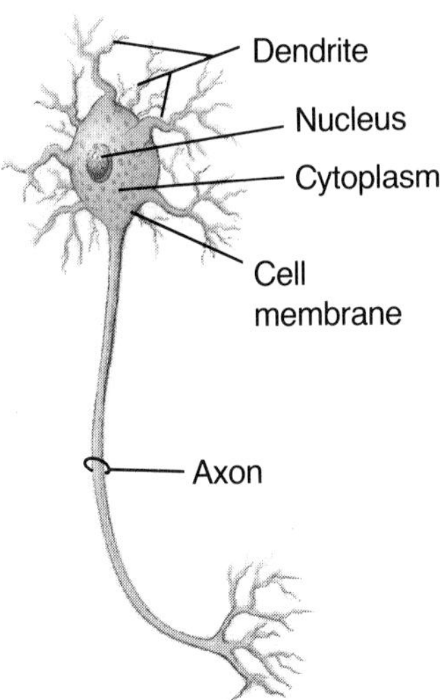

Figure 15-4 The long, thin shape of a neuron helps it do its job of carrying messages.

Pathways of Messages

How do nerves carry messages from one part of your body to another? Figure 15-5 shows what the path of nerves would look like. There are three important things that you should see in the diagram.

1. The pathway that carries messages from brain to hand is not the same pathway that carries messages from hand to brain. The different colored arrows in this diagram show the directions in which messages move along the pathways.
2. The neurons that make up a long pathway do not touch each other. There is a very small space called a synapse (SIN aps) between one neuron and the next. A **synapse** is a small space between the axon of one neuron and the dendrite of a nearby neuron.
3. The dendrites of one neuron are always next to the axon of another neuron.

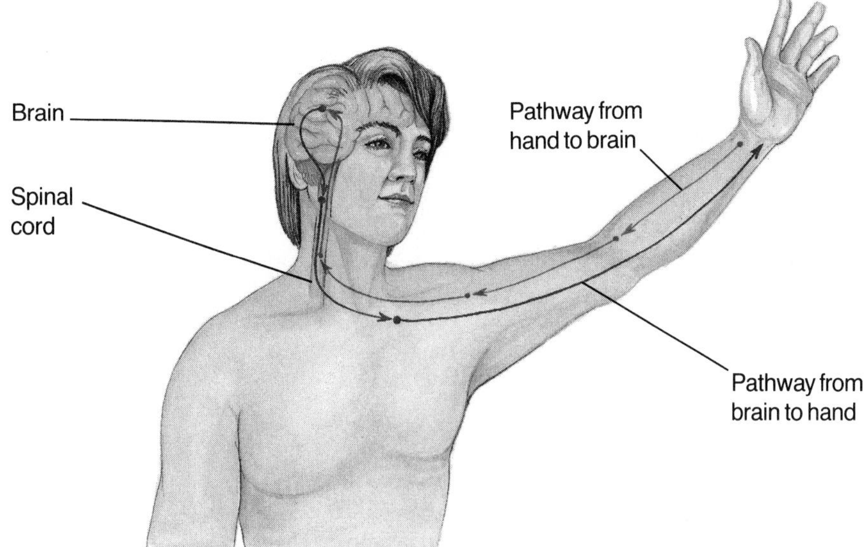

Figure 15-5 The nerve pathways that carry messages away from the brain are different from the nerve pathways that carry messages to the brain.

Messages move along a neuron from one end to the other. The message is an electrical charge that moves along the axon just as electricity moves along a wire. It flows along the neuron from dendrite end to axon end. How can this message travel across the synapse between one neuron and the next?

How do messages move across a synapse?

Figure 15-6 shows how messages move across a synapse. First, the message moves along a neuron from the dendrite to the axon, Figure 15-6a. Next, the message reaches the tip of the axon, Figure 15-6b. Notice that a chemical is given off by the axon when the message arrives there. This chemical then passes across the synapse and reaches the dendrite of the next neuron as shown in Figure 15-6c. The chemical restarts the message and the message continues along the new neuron.

Now you know why different pathways are needed to carry messages from the brain to the hand and from the hand to the brain. Messages do not travel in both directions along the same neuron. Only the axon of the neuron gives off the chemical that crosses the synapse.

Skill Check

Interpret diagrams: Look at Figure 15-6. How does the message travel from one neuron to the next? *For more help, refer to the* ***Skill Handbook,*** *pages 706-711.*

Nervous System Parts

Nerve cells make up three important parts or areas of your nervous system. These parts are the brain, spinal cord, and body nerves.

The first major organ of the human nervous system is the brain. The **brain** is the organ that sends and receives messages to and from all body parts. It also records and interprets these messages. It is made of billions of neurons.

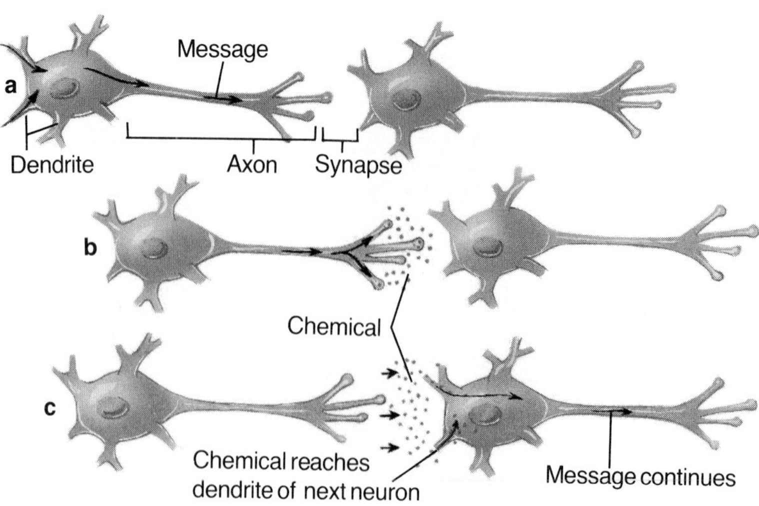

Figure 15-6 Messages move from one neuron to the next with the help of a chemical messenger that crosses the synapse.

What are the three major parts of the nervous system?

The second major part of the human nervous system is the spinal cord. Your **spinal cord** is the part that carries messages from the brain to body nerves or from body nerves to the brain. Messages travel up the spinal cord on the way to your brain and down the spinal cord to body nerves. The spinal cord is made of millions of neurons. Figure 15-7 shows the location of your spinal cord. It runs down the center of your back. The spinal cord can be compared to the power lines that enter or leave a power plant. The power plant could be compared to the brain.

Figure 15-8 shows how well the brain and spinal cord are protected against injury. Both are covered by bone and membranes. Both are also surrounded by fluid that cushions them. The brain is covered by your skull. The spinal cord is covered by bones called vertebrae. The word *vertebrate,* used in Chapter 7, sounds very similar. What do *vertebrae* and *vertebrate* mean?

The third part of your nervous system is the group of nerves that enter and leave the spinal cord. Body nerves connect organs, muscles, and skin to the spinal cord. Messages move along body nerves from organs or muscles to the spinal cord and then to the brain. Messages also move from the brain to the spinal cord and then along body nerves to organs or muscles. Body nerves are like the wires that lead from the power lines to your house.

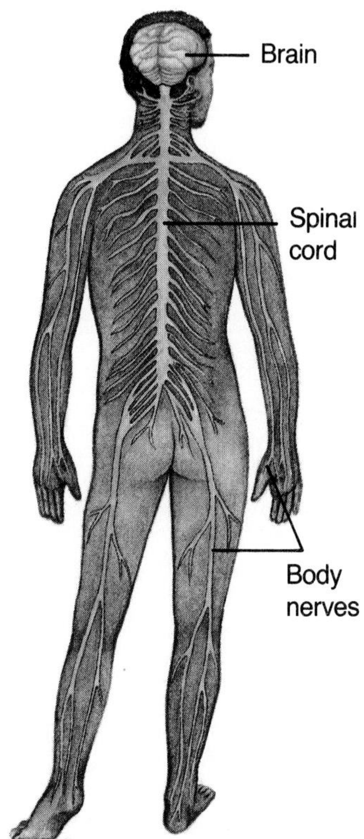

Figure 15-7 The parts of the nervous system include the brain, spinal cord, and body nerves.

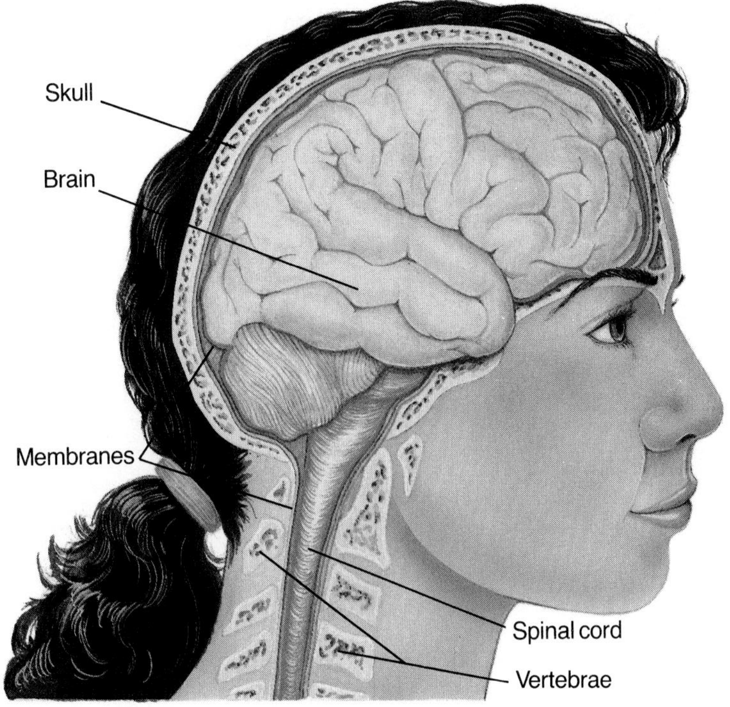

Figure 15-8 The brain and spinal cord are well protected.

Jobs of the Brain

The brain, like other body organs, has several different parts. Each part has a different job. The human brain has three main parts. Let's look at each part. They are the cerebrum (suh REE brum), cerebellum (ser uh BEL uhm), and medulla (muh DUL uh). The **cerebrum** is the brain part that controls thought, reason, and the senses. Look at Figure 15-9. Most of what you see is the cerebrum. It is the largest part of the brain. It looks like a huge walnut because of its folds.

The cerebrum has many jobs. One of its jobs is to store messages. We call stored messages memory. The cerebrum receives messages from all the sense organs. For example, sounds may be interpreted as music, laughter, or a whistle. Sights may be interpreted as brightly colored flowers or dark thunder clouds. The cerebrum is also the center for muscle control. Messages for moving the arms and legs start in the cerebrum. Messages about pain or touch end up in the cerebrum. The cerebrum also controls personality.

Many jobs of the cerebrum are voluntary. Remember from Chapter 14 that voluntary means you have control. For example, you decide if you want to move a toe or foot.

Figure 15-9 also shows a top view of the cerebrum. It is divided into left and right sides. The left side of the cerebrum controls the right side of the body. The left side also receives messages coming from the right side of the body. The right side of the cerebrum controls and receives messages from the left side of the body.

Skill Check

Understand science words: cerebrum. The word part *cerebr* means brain. In your dictionary, find three words with the word part *cerebr* in them. *For more help, refer to the* ***Skill Handbook,*** *pages 706-711.*

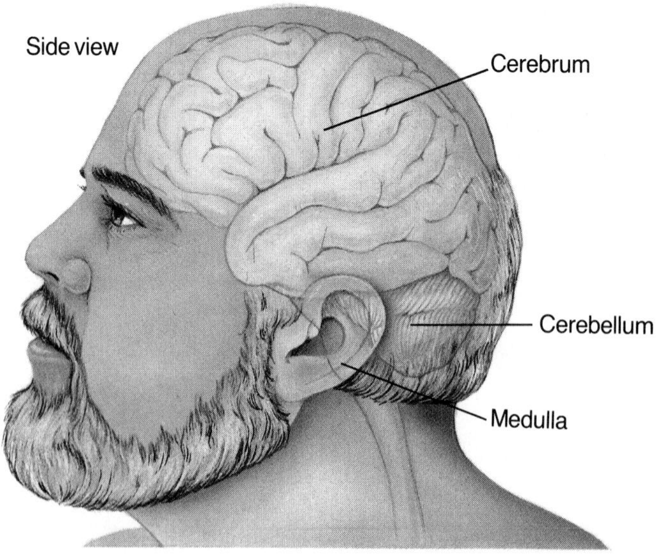

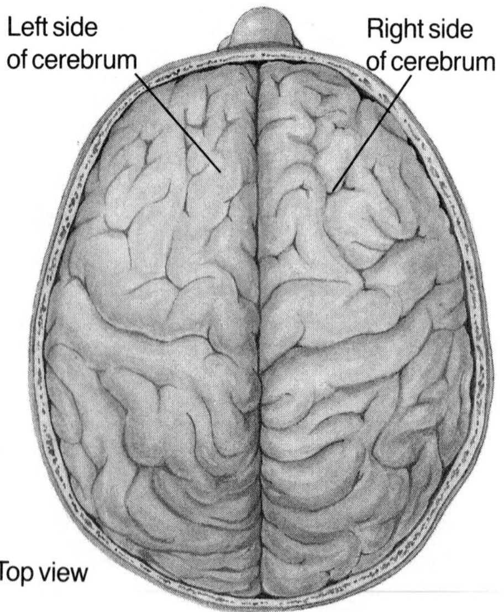

Figure 15-9 The three main parts of the brain are the cerebrum, cerebellum, and medulla. Which part controls involuntary jobs only?

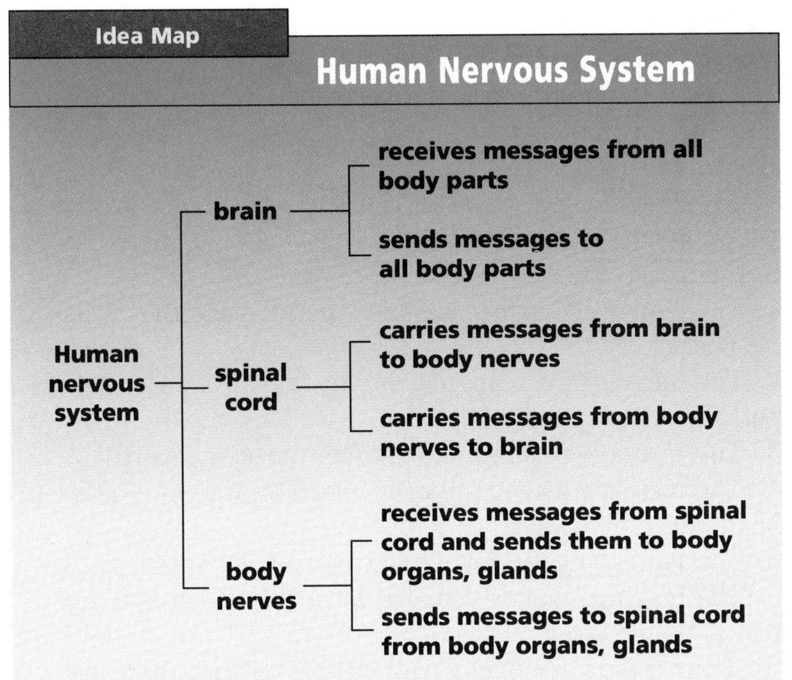

Study Tip: Use this idea map as you study the three main parts of the human nervous system. What are the main jobs of the brain, spinal cord, and body nerves?

Locate the cerebellum in Figure 15-9. The **cerebellum** is the brain part that helps make your movements smooth and graceful, rather than robotlike. How does the size of the cerebellum compare to that of the cerebrum? Where is the cerebellum located in relation to the front and back of your head? All nerves that enter or leave the brain on their way to and from the muscles deliver messages to the cerebellum. The cerebellum helps you keep your balance. The cerebellum's actions are involuntary. This means that you can't control them.

The third part of the brain is the medulla. The medulla may look as if it were part of the spinal cord, but its job is very different. The **medulla** is the brain part that controls heartbeat, breathing, and blood pressure. All jobs handled by the medulla are involuntary. The medulla works without your thinking about it. When was the last time you thought about keeping your heart beating?

Mini Lab

What Is an Eye Reflex?

Observe: Have a partner cover one eye. Shine a small flashlight into the other eye. Uncover the first eye. Are the pupils the same size? *For more help, refer to the **Skill Handbook,** pages 704-705.*

Reflexes

In most cases, any message received by your body must get to the brain before you can react to it. Think of what happens when you see a ball flying toward you. You raise your arms or duck after your brain gets the message that the ball is coming toward you.

What is a reflex?

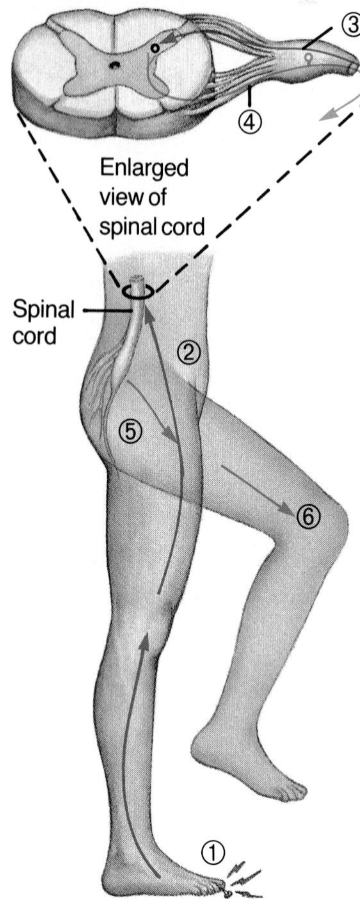

Figure 15-10 In a reflex, the message moves from the body part to the spinal cord and back to the body part.

Some messages do not make it to the brain. They go into the spinal cord and quickly back out to the muscles. The body is able to react in a very short time. Quick, protective reactions that occur within the nervous system are called **reflexes.**

How does a reflex work? Think of what happens when you accidentally step on a tack. Follow the numbered steps in Figure 15-10.

1. The skin on your foot receives the message that it has been stuck.
2. The message goes up your leg by way of a body nerve pathway.
3. The message reaches and enters your spinal cord.
4. The message leaves the spinal cord by way of a different body nerve pathway.
5. The message goes down your leg to your muscles.
6. The message makes your leg muscle contract and your foot is pulled away.

Four things are true for all reflexes. First, they are involuntary. Second, they happen very quickly. The events in Figure 15-10 took a fraction of a second. Third, reflexes may or may not involve the brain. In the tack example, your brain may have received a pain message. But, the message reached the brain *after* you pulled your foot away. If your brain had to receive the pain message before you could react, it would take more time. Injury could result or be more severe. This brings us to the fourth point about reflexes. Most reflexes are helpful. They help protect you from further harm. Coughing, blinking, and swallowing are reflexes. How does coughing or blinking protect you?

Check Your Understanding

6. Compare the jobs of dendrites and axons.
7. Describe the locations and jobs of the brain's three parts.
8. How do reflexes help? Give two examples.
9. **Critical Thinking:** In humans, the cerebrum is very large and folded. How does having a large cerebrum help humans?
10. **Biology and Reading:** You have learned that neurons are like electrical wires. You also learned that neurons have coverings, just as electrical wires do. From this analogy, what are the coverings on neurons for?

Lab 15–1 Reaction Time

Problem: What can change reaction time?

Skills

form a hypothesis, calculate, infer

Materials

metric ruler

Procedure

Reaction time is how long it takes for a message to travel along your nerve pathways.

1. Copy the data table.
2. Have your partner hold a metric ruler at the end with the highest number.
3. Place the thumb and first finger of your left hand close to, but not touching, the end with the lowest number.
4. When your partner drops the ruler, try to catch it between your thumb and finger.
5. Record where the top of your thumb is when you catch the ruler.
6. Repeat steps 2 to 5 three more times.
7. State if you think the ruler will fall farther if you catch it with your right hand. Write your **hypothesis** in your notebook.
8. Repeat steps 2 to 5 four more times using your right hand to catch the ruler.
9. Switch roles and drop the ruler for your partner.
10. **Calculate:** To complete your data table, calculate the time in seconds needed for the ruler to fall. Multiply the distance by 0.01. Use a calculator if you have one.
11. Find the average for each column.

Data and Observations

1. Did you catch the ruler faster with your left hand or your right hand?
2. Which hand is your writing hand?

Analyze and Apply

1. **Check your hypothesis:** Is your hypothesis supported by your data? Why or why not?
2. **Infer:** Compare your results with several classmates. Which hand was faster at catching the ruler? Why? (HINT: Which hand do most people use to write?)
3. Why was it a good idea to run several trials?
4. **Apply:** Explain why a message moving along nerve pathways takes time.

Extension

Design an experiment to show how age affects reaction time.

TRIAL	LEFT HAND		RIGHT HAND	
	DISTANCE RULER FALLS (cm)	TIME IN SECONDS	DISTANCE RULER FALLS	TIME IN SECONDS
1				
2				
3				
4				
Total				
Average				

15:3 The Role of the Endocrine System

Objectives

6. **Explain** the function of the endocrine system.
7. **Relate** the importance of the pituitary and thyroid glands.

Key Science Words

endocrine system
hormone
pituitary gland
thyroid gland
thyroxine

Many animals have an additional system for sending messages through their bodies. This system does not use nerve cells. It uses chemicals formed in special glands.

Chemical Control

What are hormones?

The second system that allows different parts of your body to keep in touch is called the endocrine (EN duh krin) system. The **endocrine system** is made of small glands that make chemicals for carrying messages through the body. Endocrine glands are found throughout the body. The chemicals made by endocrine glands are called hormones (HOR mohnz). **Hormones** are chemicals made in one part of an organism that affect other parts of the organism. Hormones are released into the blood. Once in the blood, hormones travel to different organs of the body. Changes take place in the organs when they receive the chemical messages that hormones carry.

Figure 15-11 shows the main endocrine glands in the human body. Note that their jobs, or functions, are also given. Some of these glands will be studied in greater detail in the next few sections.

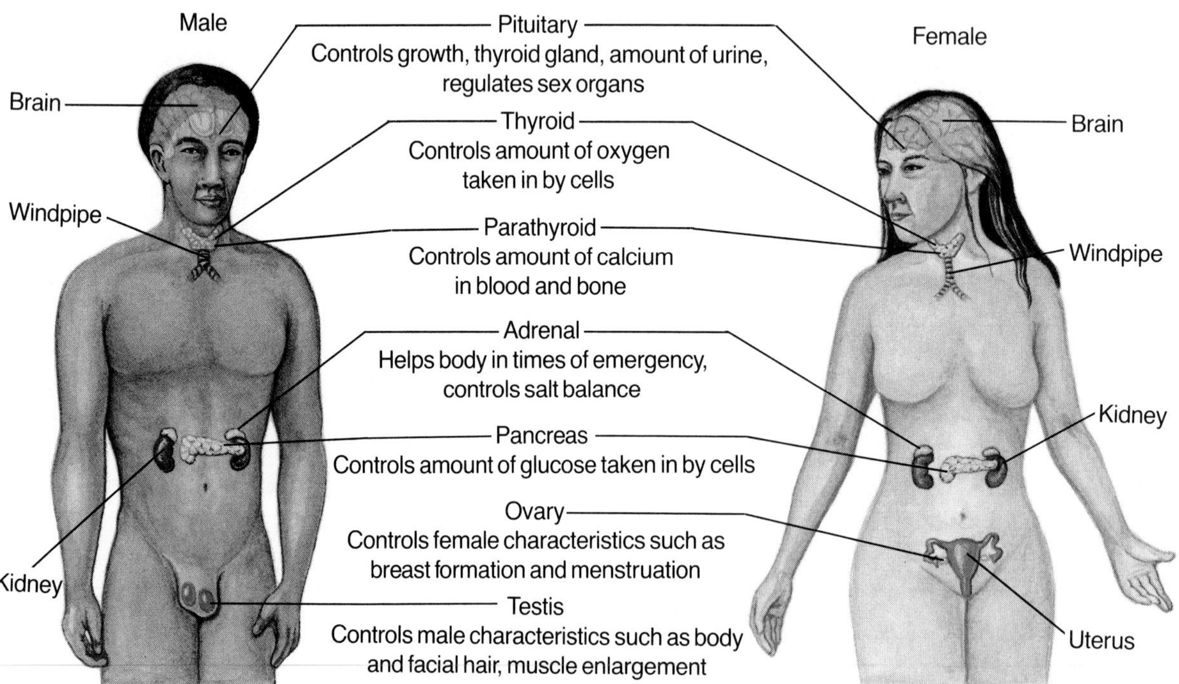

Figure 15-11 The endocrine system is made up of several glands and organs that control different body functions.

The Pituitary Gland

The **pituitary** (puh TEW uh ter ee) **gland** is an endocrine gland that forms many different hormones. It is often called the master gland because it makes hormones that regulate other endocrine glands. The pituitary forms more hormones than any other endocrine gland in the body. These hormones control many organs, other endocrine glands, and different body functions. Figure 15-12 shows where the pituitary gland is located in the human body. It is located just below the brain.

As a master gland, the pituitary has many jobs. For example, it makes hormones that control body growth. It causes a person to reach sexual maturity, also called puberty. This gland helps to control the amount of urine that you form.

Let's take a closer look at one hormone made by the pituitary gland. This hormone controls body growth and is called growth hormone. How did biologists find out that the pituitary controls body growth? They used the scientific method. Their experiments are shown in Figure 15-13. Look over their work.

Rats with their pituitary glands removed did not grow. Those with a pituitary gland did grow. The rats that had their pituitary glands removed and received injections of pituitary gland hormone grew like normal rats.

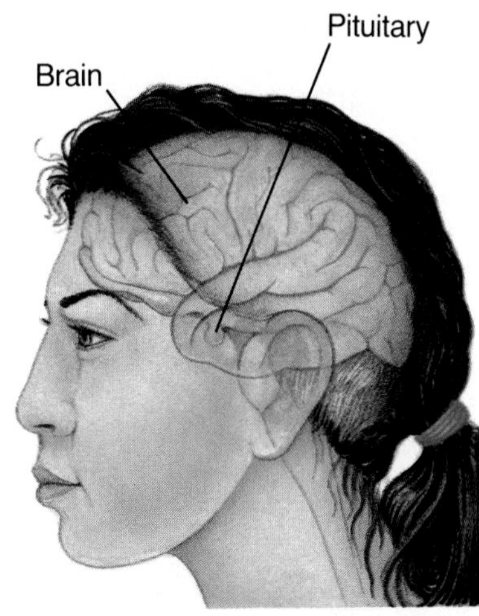

Figure 15-12 The pituitary gland is located below the brain.

How is body growth controlled?

Figure 15-13 The experiment pictured here showed that growth hormone from the pituitary gland controlled growth.

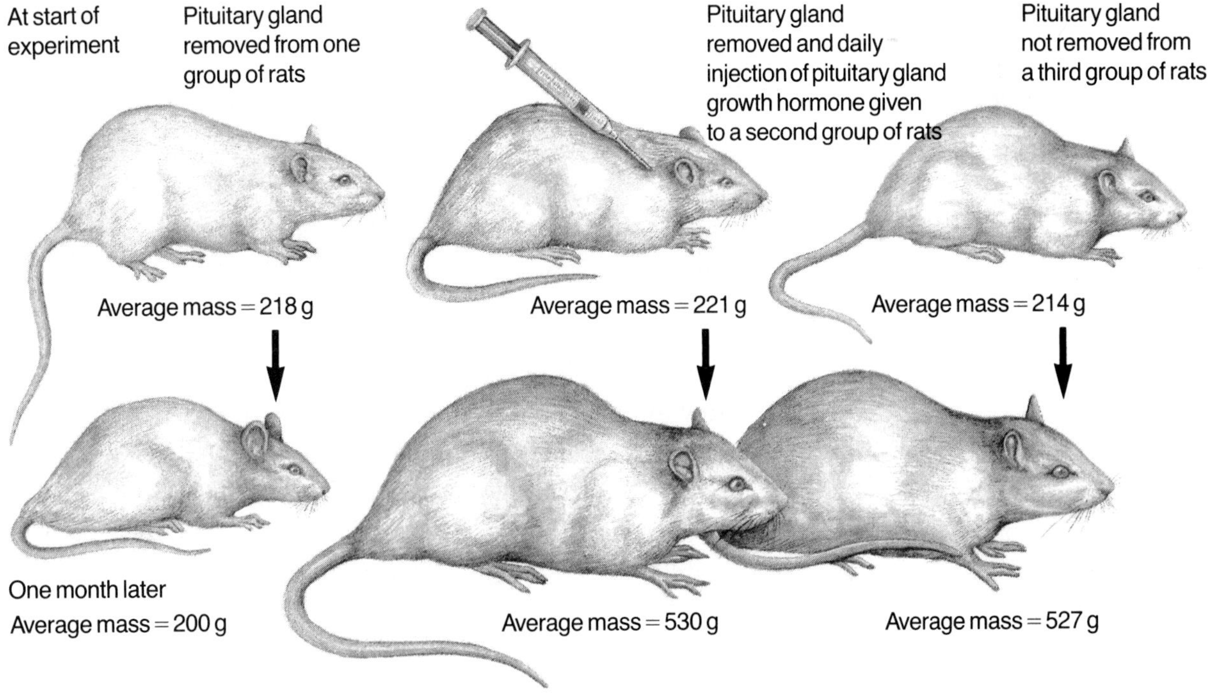

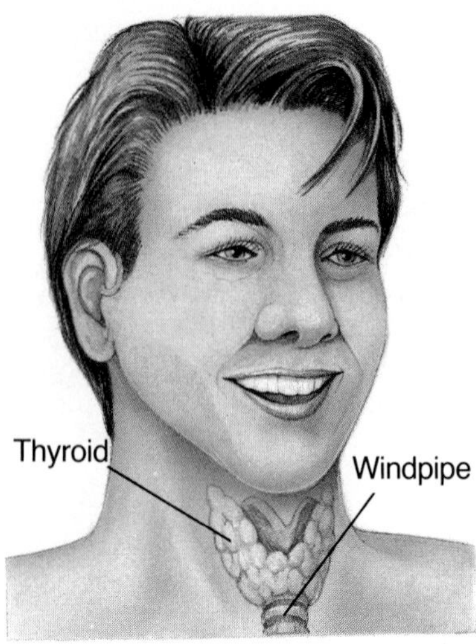

Figure 15-14 The thyroid gland makes a hormone that controls how fast your cells release energy from food.

The Thyroid Gland

The **thyroid** (THI royd) **gland** is an important endocrine gland found near the lower part of your neck. It lies in front of the windpipe and is about the size of your ear. Figure 15-14 shows its location. The thyroid's job is to make thyroxine (thi RAHK sun). **Thyroxine** is the hormone that controls how fast your cells release energy from food.

Sometimes the body may form too little or too much of a hormone. If the thyroid makes too little thyroxine, a person may gain weight and feel tired. How is this problem related to what thyroxine does? Could it be that food does not release enough energy? If the thyroid makes too much thyroxine, a person may lose weight and feel nervous. How could this problem be related to thyroxine's job?

Study Tip: Use this idea map as you study the human endocrine system. What are the jobs of the endocrine system glands?

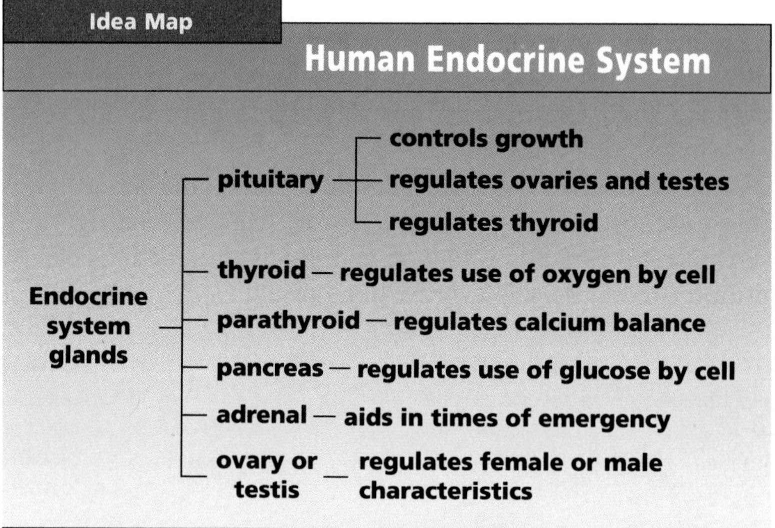

Check Your Understanding

11. What is the job of your endocrine system?
12. How is the pituitary gland important for growth?
13. How are the thyroid gland and thyroxine related?
14. **Critical Thinking:** If a person is overweight and can't lose weight through diet and exercise, what endocrine gland should he or she have checked? Why?
15. **Biology and Reading:** You learned that the pituitary is sometimes called the master gland. What do you think the word *master* means when used in this way?

15:4 Nervous and Endocrine System Problems

Many health problems are caused by diseases of the nervous and endocrine systems. Luckily, some of these diseases can be treated or cured.

Objectives

8. Identify problems that damage the brain.

9. Explain the importance of insulin.

Key Science Words

insulin
diabetes mellitus

Nervous System Problems

The brain has many blood vessels covering its surface, Figure 15-15a. The brain receives food and oxygen from these blood vessels. These blood vessels may become weak and break. When this happens, a person is said to have a stroke.

What is a stroke?

What happens when blood vessels of the brain break? It depends on where the blood vessel is and how much blood is lost. Usually, a person loses the use of a part of the brain because the brain cells die when they no longer receive food or oxygen. Losing the use of part of the brain causes the body part that it controls not to work. Figure 15-15b shows what happens when certain parts of the brain lose their oxygen supply.

Endocrine System Problems

The pancreas is a familiar gland. You studied it in Chapter 10 with the digestive system. Refer back to Figure 15-11 to see the location of the pancreas. This gland is also part of the endocrine system. It makes a hormone called insulin (IHN suh lun). **Insulin** is a hormone that lets your body cells take in glucose, a sugar, from your blood.

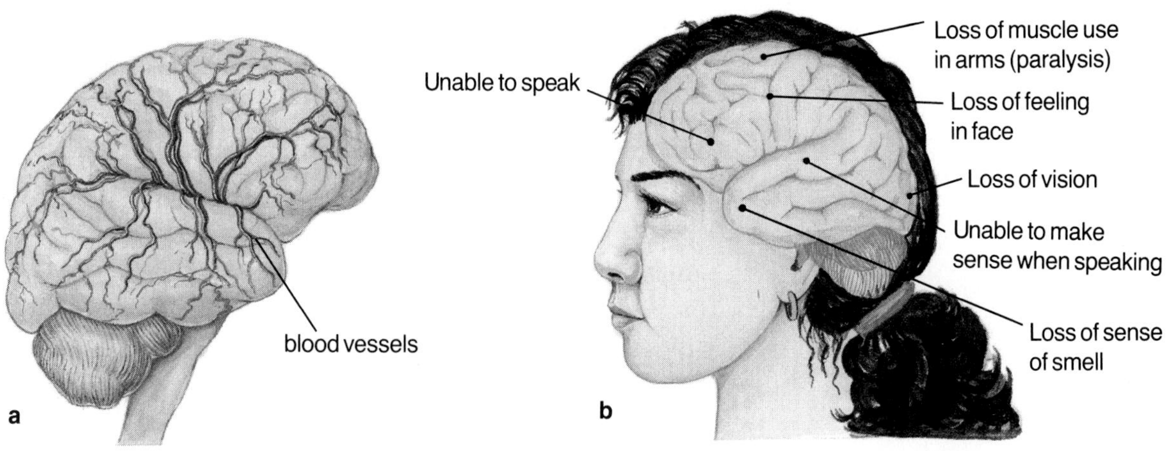

Figure 15-15 A stroke is the breaking of blood vessels that serve the brain and can result in loss of body functions.

Figure 15-16 Insulin helps cells take in glucose from the blood.

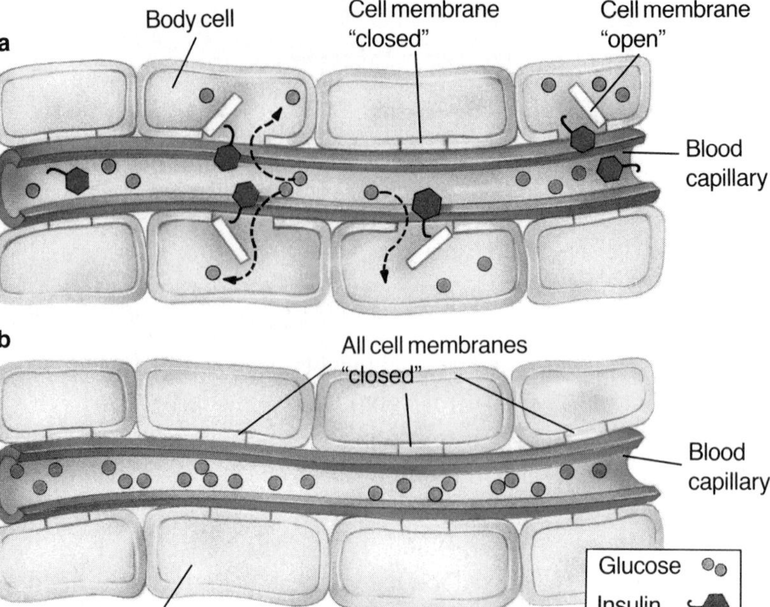

What is insulin?

> **Bio Tip**
>
> **Health:** Scientists can grow microbes that make human insulin. The scientists insert the chromosome material that controls the making of human insulin into the microbes. The microbes then multiply and produce large amounts of human insulin.

Figure 15-16a shows how insulin works. The blue circles stand for glucose. The red figures stand for insulin. Insulin opens the cell membrane, allowing glucose to enter the cells. Once inside your cells, glucose can be used as food.

What if the pancreas stops making insulin? If insulin is missing, glucose can't get into the cells. It remains in the blood. As a result, a person may lose weight. Not treating this problem may lead to blindness, heart disease, or death.

The problem just described has a name. **Diabetes mellitus** (di uh BEET us • MEL uht us) is a disease that results when the pancreas doesn't make enough insulin. People with this disease have too much glucose in their blood and not enough glucose in their cells. The excess glucose that is in the blood leaves the body in the urine.

Check Your Understanding

16. What happens when the brain loses its oxygen supply?
17. How does insulin do its job?
18. What is the cause of diabetes mellitus?
19. **Critical Thinking:** If you were a doctor, how would you test someone for diabetes?
20. **Biology and Reading:** Why does the part of the brain damaged by a stroke no longer receive food or oxygen?

Lab 15–2 Diabetes

Problem: Is glucose found in the urine of a person with diabetes?

Skills

experiment, interpret data

Materials

glucose test paper
glass slides
normal urine sample (A)
urine sample from diabetic (B)
urine samples in test tubes marked C-E
test-tube rack
wax pencil
8 droppers

Procedure

1. Copy the data table.
2. Use a wax pencil to draw two circles on a glass slide. Mark the circles A and B.
3. Add two drops of normal urine to the circle marked A. Add two drops of urine from a diabetic to the circle marked B.
4. Touch a small piece of glucose test paper to each urine sample. A green color means glucose is present. A yellow color means that no glucose is present. Record the color of the paper in your data table.
5. Draw one circle on each of three slides. Label the slides C to E.
6. Using a dropper, put two drops of urine from test tube C into the circle on slide C. Do the same with test tubes D and E.
7. Test each urine sample with glucose test paper. Record the color of the paper in your table. Complete your data table.
8. Dispose of your samples as directed by your teacher.

Data and Observations

1. In which samples was glucose present?
2. In which samples was glucose not present?

Analyze and Apply

1. **Interpret data:** Which samples tested could be from a normal person? A person with diabetes? Explain.
2. Explain why glucose is present in the urine of a person with diabetes.
3. Name the endocrine gland not working if a person has diabetes.
4. **Apply:** What kinds of foods should a person with diabetes avoid? Why?

Extension

Experiment to test if glucose test paper will detect different amounts of glucose.

SAMPLE	COLOR OF TEST PAPER	GLUCOSE PRESENT?	DIABETES PRESENT?
Normal urine (A)			
Urine from diabetic (B)			
C			
D			
E			

Get a Kick Out of This

Reflexes are usually protective. This means that the body is somehow protected as a result of an occurring reflex. Think about coughing. Coughing is a reflex that you don't have to learn or think about. It occurs when food or liquid gets into your lungs. You start to cough in an attempt to bring the food or liquid up and out of the lungs. Did you have to think about starting the process of coughing? Can you stop yourself from coughing once it starts?

Identifying the Problem

You may have heard of a reflex called the knee-jerk reflex. If lightly struck with a soft object just below the knee, your leg kicks upward. The question you are trying to answer is as follows: Can you prevent this reflex from happening if you think about it and concentrate on it not happening? Form a hypothesis that will guide you in your answering of this question.

Collecting Information

Review the meaning of a reflex. Check on the pathway that a reflex follows through the nervous system. Be sure that you understand the role of the spinal cord in a reflex. Use a reference to check on the meaning of the terms *motor neuron* and *sensory neuron.* Study the figure on this page, and locate the motor and sensory neurons that are

Technology Connection

The spinal cord carries messages from the brain to all body parts and from all body parts back to the brain. When cut or damaged, neurons in the spinal cord do not repair themselves, and a person is left paralyzed. There is some new hope for people with this type of injury. Researchers have found that if spinal cord neuron cells from embryo rats are placed into the spinal cord of adult rats with spinal cord damage, the adult rats show improvement in movement. This supports the idea that some repair of spinal cord neurons has occurred.

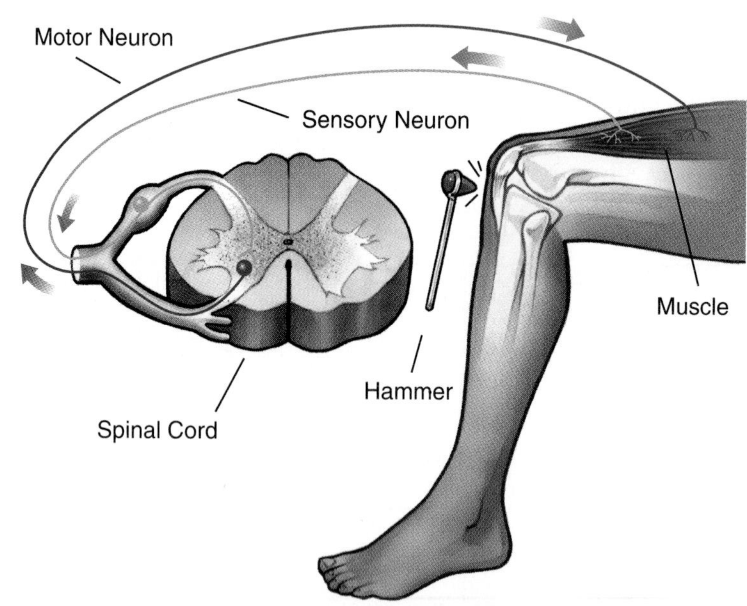

used in a knee-jerk reflex. Describe the job of the motor and sensory neurons in this figure.

Carrying Out an Experiment

You will need to work in a group of three to carry out this activity. Student A will record results. Student B will be the subject, and Student C will start the reflex by using the soft hammer.

1. Student B is to sit on a desk or lab table with his or her feet dangling, that is, not touching the floor.
2. Student C *lightly* taps the area just below the kneecap of Student B, using a soft hammer. See the diagram below.
3. Student A notes the reaction of Student B's leg and records the observation. It may be necessary to tap Student B several times before the reflex is seen. Record each tap as a separate trial.
4. To test your hypothesis, Student B should concentrate on not allowing the leg to jerk up when struck with the hammer. Again, Student A should record what happens. Note: use several trials rather than just one.
5. Change roles so that all students in your group have a chance to try steps 1-4.

Assessing Your Results

Describe the knee-jerk reflex. Explain the role of the following in the knee-jerk reflex: soft hammer, sensory neuron, motor neuron, spinal cord. Did a knee jerk occur with each and every trial? Offer an explanation for why this may have happened. Were you or your teammates able to prevent the knee-jerk reflex from occurring? Offer a reason why this might be so. Was your hypothesis supported or not? Trace the pathway of the message that passes through your body during a knee-jerk reflex.

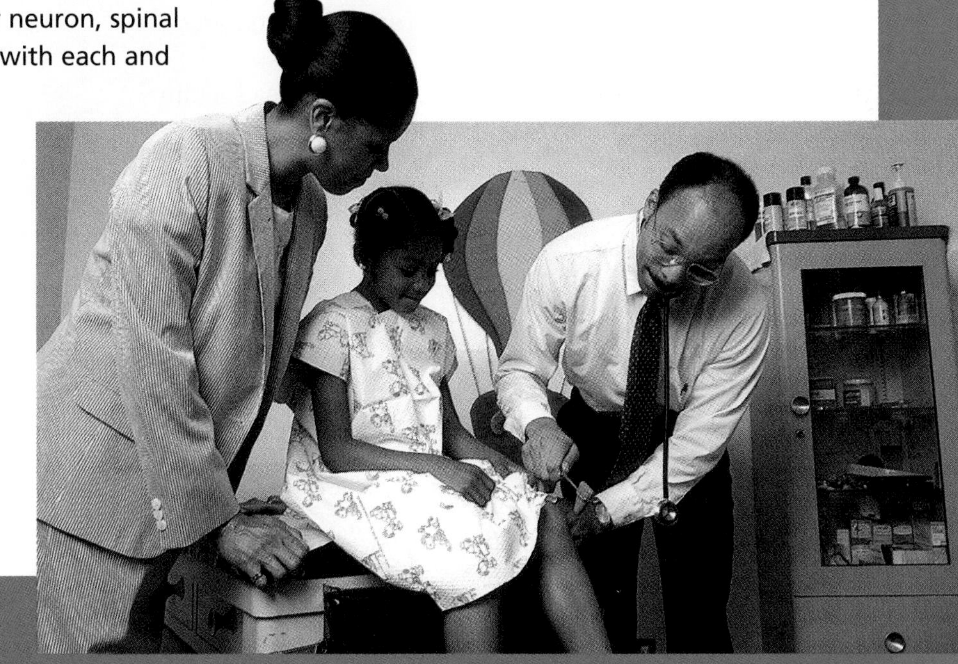

Career Connection

- **Home Health Aide** Helps people recovering at home after a hospital stay or aids the handicapped; gives baths, massages, changes bandages and bedding; may give medicines and help with household chores
- **Nurse's Aide** Cares for patients in a hospital or nursing home; takes blood pressure, temperature, and pulse of patients
- **Radiologic Technician** Works in a hospital, doctor's office, or private laboratory; takes X rays, MRIs, and CAT scans; develops X-ray film

Chapter 15 Review

Summary

15:1 The Role of the Nervous System

1. The nervous system allows animals to receive messages from their surroundings.
2. Nervous systems become more complex as animal groups become more complex.

15:2 Human Nervous System

3. Neurons are nerve cells. They have ends called dendrites and axons. Neurons make up nerve pathways.
4. The brain, spinal cord, and body nerves form the human nervous system.
5. Reflexes are involuntary, quick reactions of the nervous system that protect the body.

15:3 The Role of the Endocrine System

6. The endocrine system consists of glands that send hormones throughout the body.
7. The pituitary gland makes many hormones that control different body functions. Thyroxine is the hormone made by the thyroid gland.

15:4 Nervous and Endocrine System Problems

8. Blood vessels that supply the brain with food and oxygen may break, resulting in a stroke.
9. Insulin is a hormone made by the pancreas. Insulin controls glucose movement into your cells.

Key Science Words

axon (p. 313)
brain (p. 314)
cerebellum (p. 317)
cerebrum (p. 316)
dendrite (p. 313)
diabetes mellitus (p. 324)
endocrine system (p. 320)
hormone (p. 320)
insulin (p. 323)
medulla (p. 317)
nerve (p. 312)
nervous system (p. 310)
neuron (p. 312)
pituitary gland (p. 321)
reflex (p. 318)
spinal cord (p. 315)
synapse (p. 313)
thyroid gland (p. 322)
thyroxine (p. 322)

Testing Yourself

Using Words

Choose the word from the list of Key Science Words that best fits the definition.

1. quick reaction that is protective
2. disease caused by lack of insulin
3. special name for a nerve cell
4. brain area that controls heartbeat
5. chemical made by the thyroid gland
6. neuron end that receives messages
7. hormone made by the pancreas
8. gland found in the neck
9. system that sends chemical messages through the body

Testing Yourself *continued*

Finding Main Ideas

List the page number where each main idea below is found. Then, explain each main idea.

10. the four main features of a reflex
11. the functions of the adrenal, pituitary, and parathyroid glands
12. the brain regions that are voluntary
13. how insulin works
14. how to prove with experiments that the pituitary gland controls growth
15. the pathway that a reflex follows

Using Main Ideas

Answer these questions by referring to the page number after each question.

16. How do the nervous systems of simple and more complex animal groups differ? (p. 311)
17. Which side of the cerebrum receives and controls messages from the left side of your body? (p. 316)
18. How does a message move from one neuron to the next? (p. 313)
19. Name the locations of the pituitary, thyroid, parathyroid, pancreas, adrenal, and ovary. (p. 320)
20. How do animals keep in touch with their surroundings? (p. 310)
21. What can happen when someone has a stroke? (p. 323)
22. How do the brain and spinal cord work together? (p. 315)
23. What happens if the pancreas does not produce enough insulin? (p. 324)
24. What happens if the thyroid produces too much thyroxine? (p. 322)
25. Why are different pathways needed to carry messages from hand to brain and brain to hand? (p. 313)

Skill Review ✓

*For more help, refer to the **Skill Handbook,** pages 704-719.*

1. **Interpret data:** Which rat in the experiment in Figure 15-13 was the control? Why?
2. **Interpret diagrams:** Study the diagrams of neurons in this chapter. How does the shape of a neuron help it do its job?
3. **Infer:** What would happen to messages traveling along the spinal cord if the spinal cord were cut? How could this type of injury affect body parts?
4. **Observe:** Examine slides of thyroid tissue. Diagram what you see.

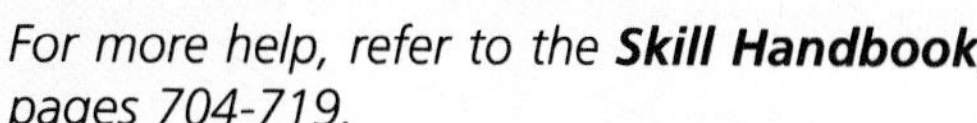

Finding Out More

Critical Thinking

1. Find out what the following diseases are and what their causes are: multiple sclerosis, cerebral palsy, epilepsy, polio.
2. What are the symptoms of diabetes mellitus?

Applications

1. Design and build a model of a neuron, a neuron pathway of at least three neurons, and a nerve. (HINT: Spaghetti might work well.)
2. Prepare a large diagram that shows the locations and jobs of the different areas of the cerebrum.

CHAPTER PREVIEW

Chapter Content

Review this outline for Chapter 16 before you read the chapter.

Skills in this Chapter

The skills that you will use in this chapter are listed below.

- In **Lab 16-1,** you will formulate a model, observe, and measure in SI. In **Lab 16-2,** you will interpret data and make and use tables.
- In the **Skill Checks,** you will interpret diagrams and understand science words.
- In the **Mini Lab,** you will observe.

Chapter **16**

Senses

Think about all the things you can enjoy because of your senses. You can enjoy the aroma of baking pizza because of your sense of smell. You can savor the rich sweetness of a chocolate chip cookie because of your sense of taste. Your senses let you experience the richness of the world around you.

But, did you realize there are things your senses can't detect? What do you think an insect can see? Insects can see ultraviolet light from the sun that you can't see. The flower on the left is what you see when you look at a flower. The flower on the right is what an insect sees. This flower was photographed in ultraviolet light. The insect's ability to see the middle of the flower as a darkened area helps guide the insect to the flower. Like you, the insect's senses allow it to keep in touch with the world around it.

Try This!

Did you hear that? Listen to the sounds you make as you speak, breathe, and chew. Plug your ears with your fingers and listen to these sounds again. How do the sounds differ? Can you explain why?

Visit the Glencoe Science Web site at science.glencoe.com to find links about **senses.**

Flowers photographed under natural light (left) and ultraviolet light (right)

16:1 Observing the Environment

Objectives

1. **Relate** ways a planarian and an earthworm sense light.
2. **Compare** the sense organs of a cricket and a snake.

Key Science Words

sense organ

Living things must know what is going on around them. They may use eyes. What if they do not have eyes? They will have other ways to keep in contact with their surroundings.

Senses Aid Survival

The human body has many sense organs. **Sense organs** are parts of the nervous system that tell an animal what is going on around it. Eyes, ears, nose, tongue, and even our skin are sense organs.

Could you stay alive for very long without sense organs? It would be very hard. What about other animals? How well would they survive without any sense organs?

Sense Organs of Animals

The sense organs of four different animals will be studied in this section. These animals are a flatworm, a segmented worm, an insect, and a snake.

How does a planarian sense light?

A planarian is a simple, small animal that lives in ponds or streams. Figure 16-1a shows two of its sense organs. Eyespots of a planarian are simple. They can tell only light from dark. A planarian has knobs on its front end that detect the direction of water currents and food.

Earthworms are a little more complex than planarians. If you look at an earthworm's body closely, you cannot see anything that looks like an eye. Yet, the animal can detect light. How? Earthworms have nerve cells just below the skin that detect light.

Figure 16-1b shows three different sense organs of a cricket. The cricket has parts for detecting sound on the

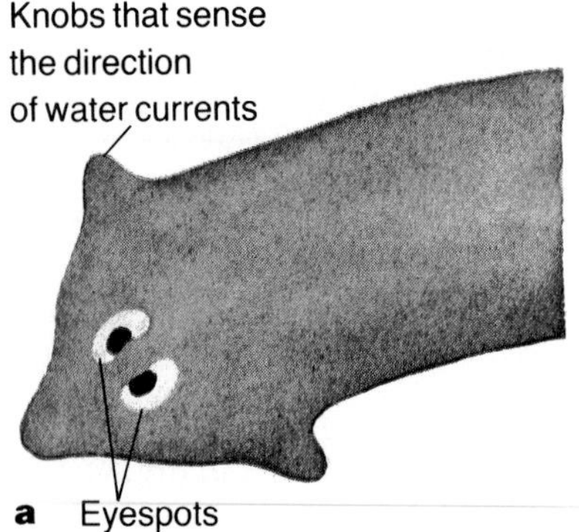

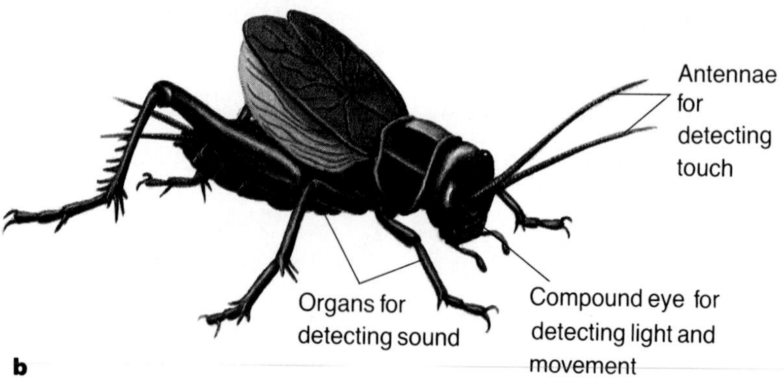

Figure 16-1 A planarian (a) and a cricket (b) have sense organs that help them keep in touch with their surroundings.

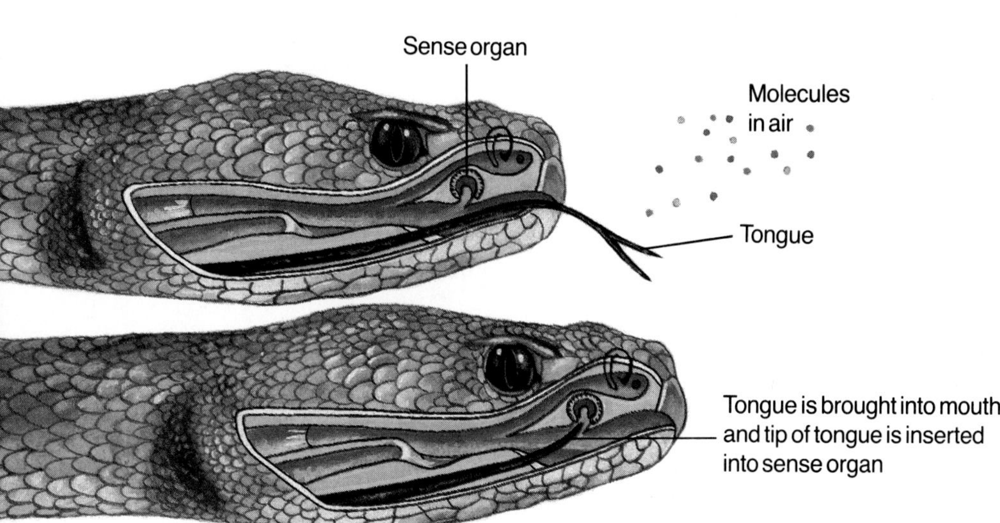

front legs. The antennae detect touch and chemicals. The eyes detect movement and light and dark. Insects have compound eyes. In this case, the word *compound* means many. A compound eye is made of hundreds or even thousands of very small eyelike parts. These eyelike parts are grouped together to form the compound eye.

Figure 16-2 A snake has a sense organ in the roof of its mouth for sensing odors.

The sense organs of snakes and other reptiles are more complex than those of insects. Snakes have eyes similar to human eyes. Their eyes have many of the same parts that human eyes have.

How does a snake smell?

Snakes have a very good sense of smell. Have you ever noticed how a snake darts its tongue in and out? A snake picks up odor molecules in the air with its tongue, then brings it back into its mouth. The snake then sticks the tip of its tongue into special sense organs found on the roof of its mouth. These sense organs look like two curled grooves as shown in Figure 16-2. Nerve cells line these grooves. The nerve cells pick up the molecule messages brought in by the tongue. Messages are then sent to the snake's brain.

Check Your Understanding

1. Name five different sense organs in the human body.
2. Explain how an insect can detect light.
3. How does a snake use its tongue as a sense organ?
4. **Critical Thinking:** Since earthworms live underground, why do they need to detect light?
5. **Biology and Reading:** How do senses aid survival?

16:2 Human Sense Organs

Objectives

3. State the functions of the parts of the eye.

4. Explain the jobs of the nose, tongue, and skin as sense organs.

5. Describe the parts of the outer, middle, and inner ear.

Key Science Words

sclera	cone
iris	optic nerve
pupil	taste bud
cornea	olfactory nerve
lens	eardrum
lens muscle	cochlea
retina	auditory nerve
vitreous humor	semicircular canal
rod	epidermis
	dermis

The human body has a variety of sense organs. Each sense organ is specialized as to what it can detect. Eyes detect light. The nose and tongue detect chemical molecules. Ears detect moving air molecules. Skin detects heat, cold, pain, touch, and pressure.

The Eye

What are the parts of the eye?

How do our eyes give us information about our surroundings? What are the parts of the human eye? Figure 16-3 shows a front view and a side view of the eye. Let's look at what the parts of the eye do.

The eyelid protects and moistens the outside of the eye. Each time you blink, liquid spreads over the front of your eye.

The **sclera** (SKLER uh) is the tough, white outer covering of the eye. The sclera protects the eye.

The **iris** is a muscle. The iris controls the amount of light entering the eye. It is also the part that gives the eye its color. When you say that someone has blue or brown eyes, you are describing the color of the iris.

The **pupil** is an opening in the center of the iris. Light enters the eye through the pupil. As the amount of light changes, the size of both the iris and the pupil changes. The pupil becomes larger in dim light and smaller in bright light. How does a larger pupil help you see in the dark?

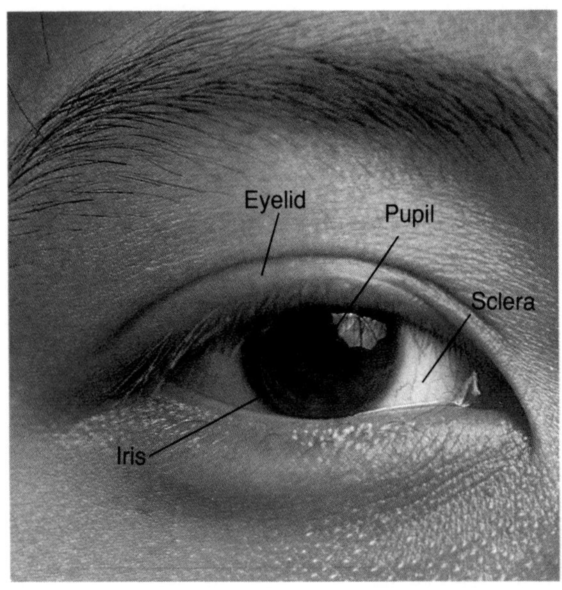

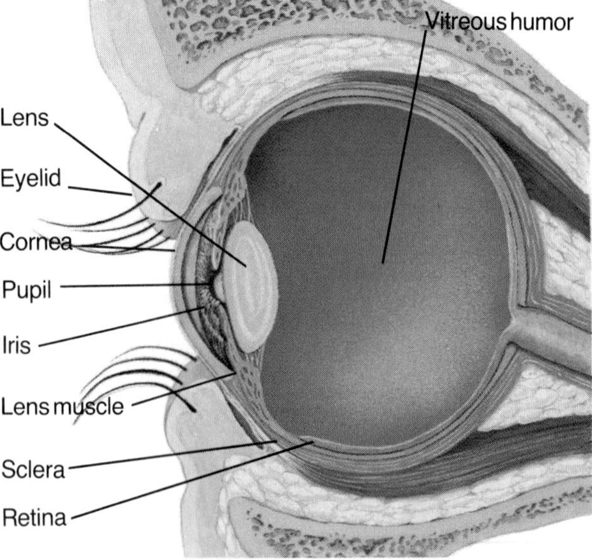

Figure 16-3 The parts of the human eye are shown here.

The clear, outer covering at the front of the eye is the **cornea** (KOR nee uh). It protects the eye and allows light to enter the eye through the pupil. The cornea also helps to focus the light by bending the light rays as they enter the eye.

The **lens** is a clear part of the eye that changes shape as you view things at different distances. The lens muscle attaches to the lens. The **lens muscle** pulls on the lens and changes its shape. This helps to focus objects that are close up or far away.

The **retina** (RET nuh) is a structure at the back of the eye made of light-detecting nerve cells. The retina is often compared to film in a camera.

The **vitreous** (VI tree us) **humor** is a jelly-like material inside the eye. It pushes out on the eye parts. Its main job is to keep the eye round in shape.

Now that we have looked at the parts of the eye, we can follow the path that light takes through the eye. Follow the numbered steps in Figure 16-4.

1. Light rays enter the eye through the cornea. Notice that the cornea slightly bends the light rays. The light rays are bent toward each other.
2. Light passes through the pupil and then through the lens. Light is bent again as it passes through the lens. Again, the light rays are bent toward each other.
3. Because the vitreous humor is clear, light continues to pass through the eye.

Mini Lab

What Is an Afterimage?

Observe: Stare out of the classroom window for at least 30 seconds. Then, close your eyes for 15 seconds. What you see is called an afterimage. What does the afterimage look like? *For more help, refer to the **Skill Handbook**, pages 704-705.*

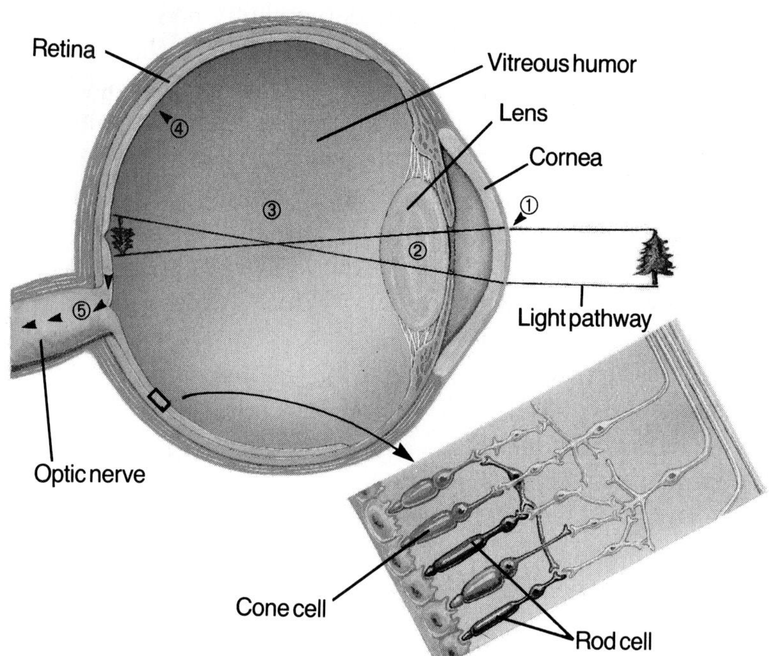

Figure 16-4 Light passes through the cornea, lens, and vitreous humor before it strikes the retina.

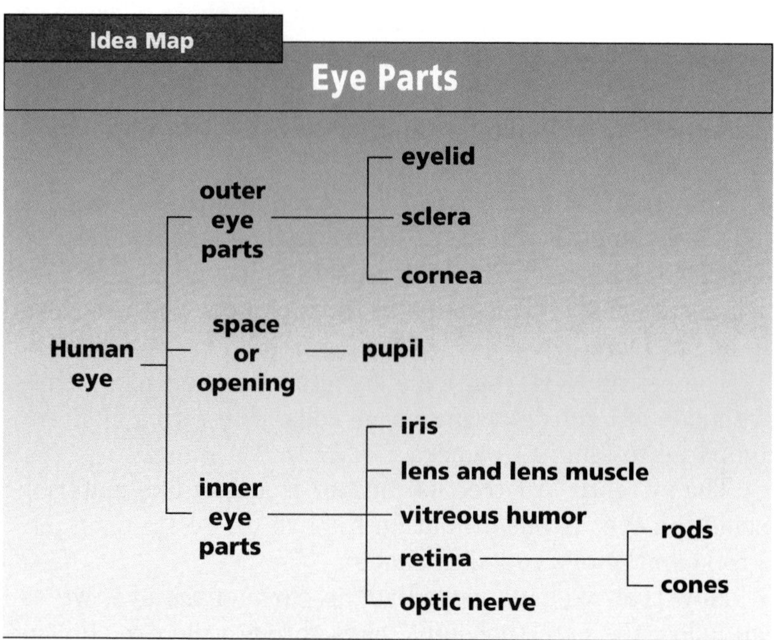

Study Tip: Use this idea map as you study the parts of the human eye and their functions. Write down the function of each eye part.

4. Finally, the light rays strike the retina. Notice that the retina is made of two types of nerve cells. These cells are called rods and cones because of their shapes. **Rods** are nerve cells that detect motion and help us to tell if an object is light or dark. They also help us to detect the shapes of objects. We use our rods when looking at something in dim light.

Cones are nerve cells that can detect color. There are three types of cones. Each type detects only one color: red, green, or blue. All other colors that we see are the result of two or three of the different types of cones acting together. Figure 16-5 shows a section of the retina enlarged 1000 times. Notice the many rods and cones.

Look closely again at the retina in Figure 16-4. When the image of the tree forms on the retina, it is upside down. It appears upside down because the light rays crossed over one another when they were bent. Which two eye parts bend light?

5. The message sent by the light leaves the retina and enters the optic (AHP tihk) nerve. The **optic nerve** is a nerve that carries messages from the retina to the brain. Once a message arrives at the vision center of the cerebrum, two things happen. One, the message is interpreted. In our example, the brain decodes the message and we see a tree. Two, the brain interprets the message as an object right side up. We see the tree in its normal position.

Figure 16-5 The retina is made up of thousands of rods and cones.

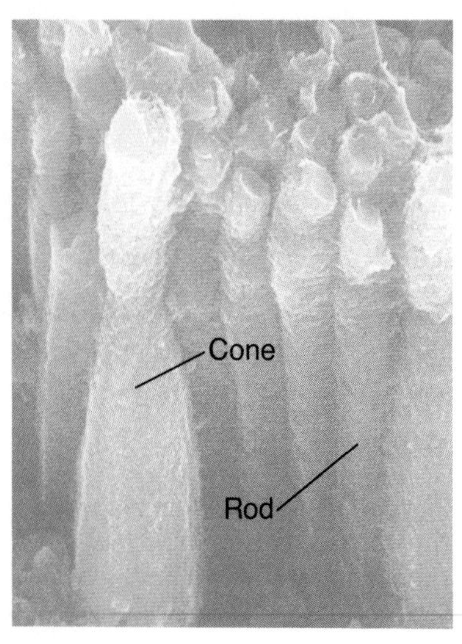

Lab 16-1 The Eye

Problem: How does the eye work?

Skills

observe, measure in SI, formulate a model

Materials

black paper
ruler
clear tape
white paper
scissors
top part of a hand lens

Procedure

1. Make a data table in which to diagram your results.
2. Use a ruler to trace a letter *L* onto black paper. Make the letter 15 cm high and 8 cm long. Make each leg of the *L* 3 cm wide.
3. Cut out the letter and tape it onto the classroom window right side up.
4. Hold a piece of white paper as shown in the figure. The paper should be 40 to 60 cm away from the window.
5. With your back toward the window, look at the white paper.
6. Hold the top of the hand lens about 4 cm from the paper. Move the lens slowly toward and away from the paper until you see the letter *L* clearly on the paper.

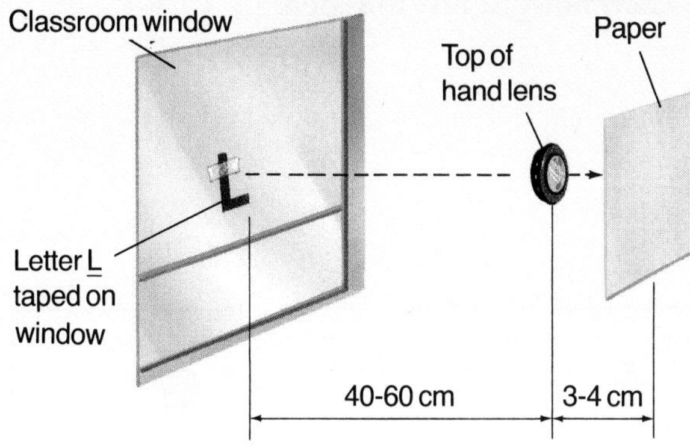

7. Complete the data table by making a diagram of the letter *L* as it appears on the window and on the paper.
8. **Measure in SI:** Measure the distance between the lens and paper in cm. Record this distance in your data table.
9. Determine the shape of the lens by rubbing your fingers over both sides of it. Diagram the shape of the lens in your data table.

Data and Observations

1. What was the appearance of the letter *L* on the window?
2. What was the appearance of the letter *L* on the paper?
3. Which of the following better describes the shape of the lens: (a) thicker toward the edges and thinner toward the center or (b) thinner toward the edges and thicker toward the center?

Analyze and Apply

1. Which part of the human eye is represented by the:
   (a) lens?
   (b) white paper?
2. (a) In a real eye, what structures would make up the paper?
   (b) In a real eye, what shape would the lens have?
3. **Apply:** In this lab, you moved the lens back and forth to obtain a clear view of the letter *L* on the paper. What does the action of moving the lens back and forth represent in the real eye?

Extension

Formulate a model to show what causes nearsightedness. How does the distance between the lens and white paper change?

The Senses

Lab 16-2

Problem: How reliable is your sense of vision?

Skills

interpret data, make and use tables

Materials

white paper
green and red pencils

Procedure

1. Copy the data table.
2. Close your left eye. With your right eye, stare at the cross in the figure for about 10 seconds.
3. Very slowly move the page toward you. At a certain distance, the dot in the figure will disappear. When it disappears, light from the dot is falling on your blind spot. The blind spot is an area on the retina with no rods or cones.
4. Close your right eye. Repeat steps 2 and 3, only this time stare at the circle.
5. Record in your table whether each eye has a blind spot.
6. Stare at the triangles for 30 seconds. Then, look at a blank piece of paper. You will see a similar figure on the blank paper. Record in your table what colors you see in the outer triangle and the inner triangle in the second figure.

TEST		OBSERVATIONS
Blind spot	Right eye	
	Left eye	
Colors of triangles	Outer triangle	
	Inner triangle	

Data and Observations

1. Is your blind spot present in both eyes?
2. What happened to the colors of the triangles when you looked at the white paper?

Analyze and Apply

1. Why aren't you usually aware of your blind spot?
2. Which parts of the eye detect the colors of the triangles?
3. **Apply:** Was the change in color you observed in the triangles taking place in your eyes or in your brain? Explain your answer.

Extension

The size of an object determines how far away the object must be before the light from it falls on your blind spot. **Design an experiment** to test this idea.

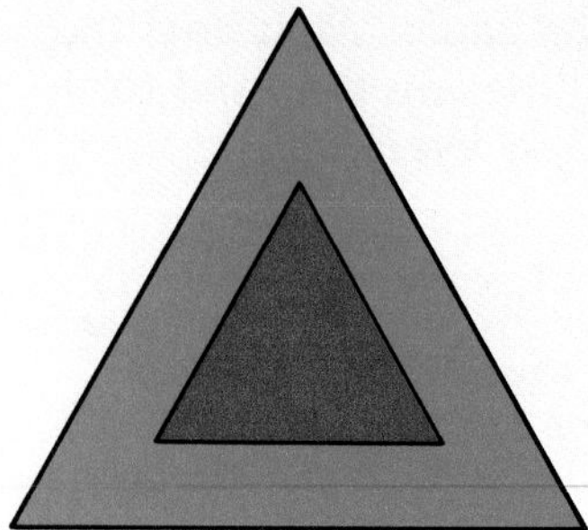

The Tongue and the Nose

The tongue is the major sense organ for taste. The tongue detects molecules dissolved in water. The tongue senses different molecules and you taste the sweetness of a candy bar or the sourness of a lemon.

Notice in Figure 16-6a that the surface of the tongue has many small bumps. On each of these bumps are tiny pits that contain taste buds. **Taste buds** are nerve cells in the tongue that detect chemical molecules. You have about 10 000 taste buds. Figure 16-6a shows a greatly magnified view of several taste buds.

There are four different kinds of taste buds on the tongue. Each kind is located in a different area of the tongue. Each area of the tongue detects a certain kind of molecule, or taste. The four different tastes are sour, salty, sweet, and bitter. You know that there are more than just four tastes. What you taste are different combinations of the four basic tastes.

Have you ever had trouble tasting things when you had a cold? That's because the senses of taste and smell work together. They both detect molecules. Your nose detects gas molecules in the air. These gas molecules are then interpreted by the brain as odors.

Figure 16-6b shows how your nose works. Molecules in the form of a gas are detected by neurons that line the top of the nasal chamber. There are seven different types of nerve cells in the nose. Each type detects one kind of odor. The **olfactory** (ohl FAK tree) **nerve** then carries the message to your brain. The cerebrum interprets the message as smoke or perfume or some other odor.

Bio Tip

Science: Your sense of smell is 1000 times more sensitive than your sense of taste.

Figure 16-6 Taste buds (a) detect four kinds of tastes. The photograph shows the surface of the tongue magnified 1000 times. The large, round structures are tastebuds. The nose (b) detects odors in the form of gases.

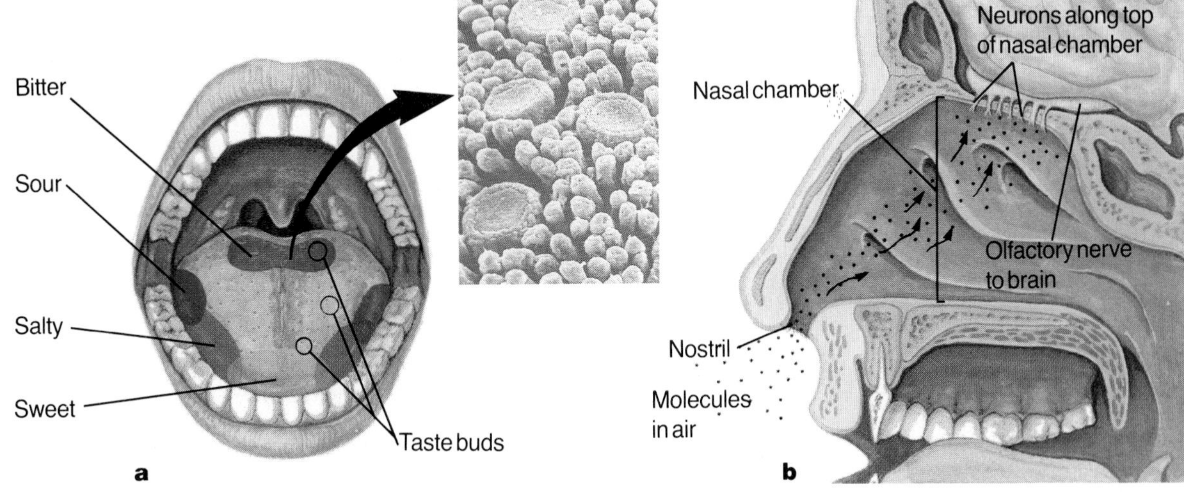

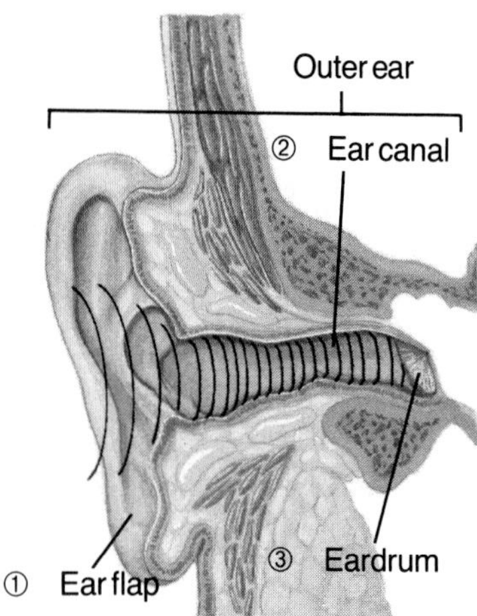

Figure 16-7 Sound waves are directed into the ear by the ear flap and ear canal.

The Ear

Ears are sense organs for detecting sound waves. Sound waves are air molecules in motion. How does the ear detect moving air molecules and allow you to hear?

The human ear is divided into three areas. These areas are called the outer, middle, and inner ear. The following description tells how these parts work together and enable you to hear.

Figure 16-7 shows the three parts of the outer ear. Look at this figure as you read steps 1 through 3.

1. The only part of your ear that you can see is the ear flap. The ear flap helps direct sound waves into a narrow tube leading into the ear. The ear flap is made of cartilage and connective tissue. It is soft and easily bent.
2. The narrow tube leading into the ear is called the ear canal. The ear canal carries sound waves to the middle ear.
3. Sound waves bump against a thin membrane called the eardrum. The **eardrum** is the membrane that vibrates at the end of the ear canal. The vibrations, or back-and-forth movements, are caused by sound waves striking the eardrum. A fatlike chemical, earwax, is made by cells in the ear canal. Earwax helps to keep insects and other foreign matter out of the ear and keeps the eardrum soft.

Figure 16-8 shows the middle ear. It has two main parts. Look at this figure as you read steps 4 and 5.

4. Three small bones make up one of the main parts of the

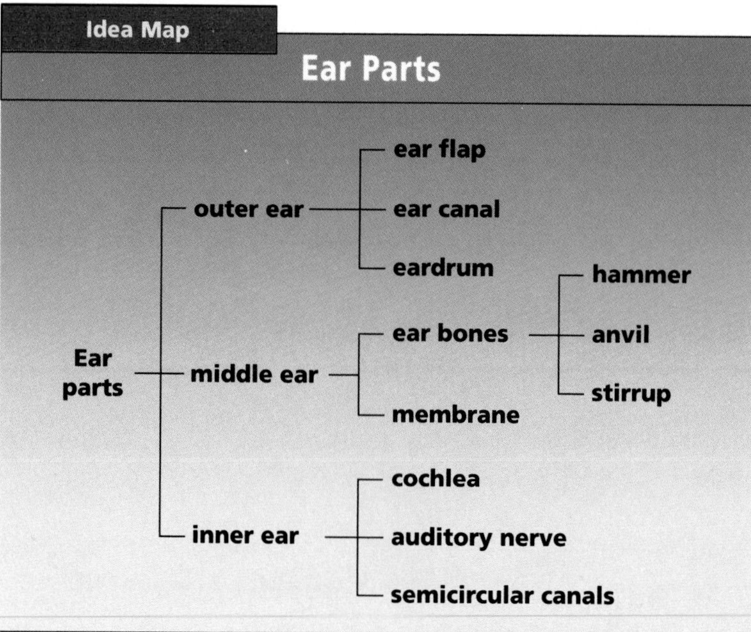

Study Tip: Use this idea map as you study the parts of the human ear and their functions. Write down the function of each ear part.

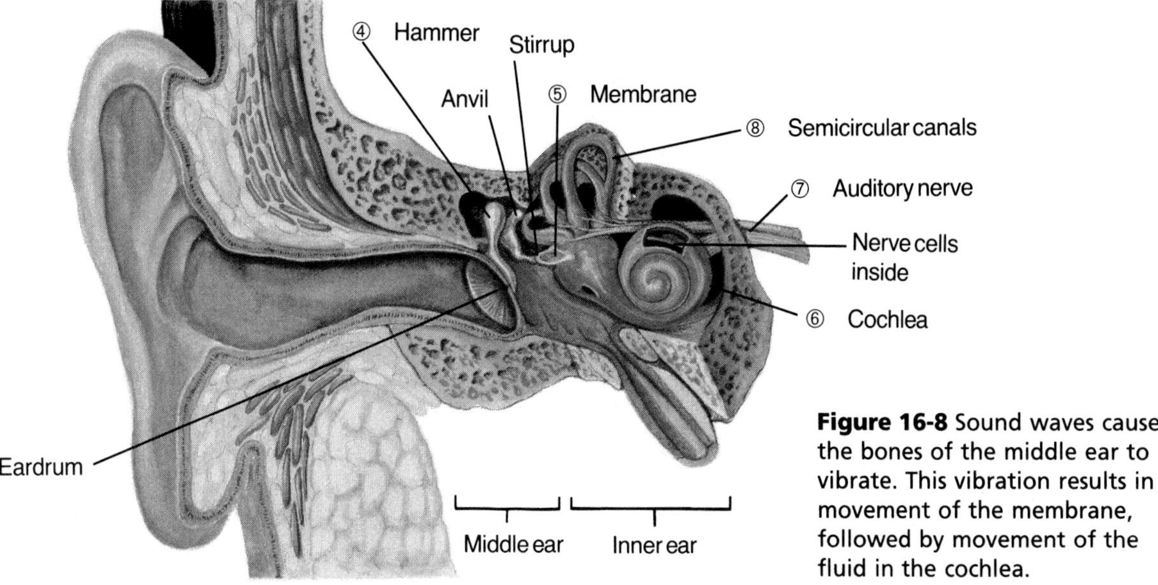

Figure 16-8 Sound waves cause the bones of the middle ear to vibrate. This vibration results in movement of the membrane, followed by movement of the fluid in the cochlea.

middle ear. Because of their shapes, the bones are called the hammer, anvil, and stirrup. These bones are quite small and are connected to one another. The hammer is also connected to the eardrum. When the eardrum vibrates, the hammer moves. The movement of the hammer causes the other two bones to move.

5. The stirrup is connected to a membrane in the middle ear that vibrates with the motion of the ear bones.

The inner ear also has three main parts. Look at Figure 16-8 and read steps 6, 7, and 8 to see what they do.

How does the cochlea work?

6. This part is called the cochlea (KAHK lee uh). *Cochlea* is Latin for "snail shell." The **cochlea** is a liquid-filled, coiled chamber in the ear that contains nerve cells. When the middle ear membrane vibrates, it makes the liquid in the cochlea move. The nerve cells in the cochlea detect this movement.

7. Each nerve cell in the cochlea is connected to a large nerve, the auditory (AHD uh tor ee) nerve. The **auditory nerve** carries messages of sound to the brain.

Once messages reach the cerebrum of your brain, they are interpreted as particular sounds. You hear a whistle, a barking dog, or some other sound.

8. The semicircular canal has a very special job not related to hearing. The **semicircular canals** are inner ear parts that help us keep our balance. Nerve messages pass from these ear parts to your brain's cerebellum. When these messages arrive in the brain, they send messages to body muscles to contract or relax as needed to maintain balance.

Skill Check

Interpret diagrams: Study Figure 16-7. Which of the ear parts shown is the first to respond to sound? *For more help, refer to the* ***Skill Handbook****, pages 706-711.*

Career Close-Up

Licensed Practical Nurse

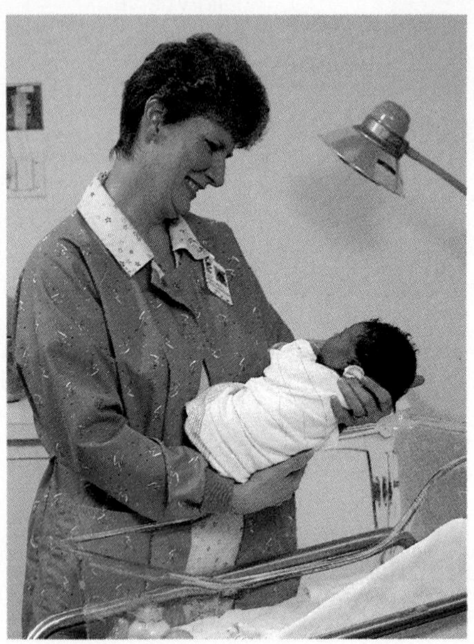

An LPN cares for patients.

Did you know that you can be a nurse without going to college for four years? You can if you become a licensed practical nurse, or LPN. An LPN must graduate from high school and take a year of special training at a junior college or hospital. When the LPN finishes training, he or she can go right to work helping doctors and registered nurses do their jobs.

A practical nurse works directly with patients. The LPN takes temperatures, changes bandages, and feeds and bathes patients. Practical nurses are also licensed to give drugs to patients. They learn about how drugs affect patients and the proper use of drugs.

Practical nurses may work in hospitals, nursing homes, and homes for the elderly. Practical nurses are licensed by the state in which they work.

Have you ever experienced motion sickness? It's a terrible feeling! Motion sickness results when the eyes and semicircular canals receive conflicting messages. For example, when you are in a car, your inner ear will be telling you that you are moving. If for some reason you can't look out of the car to see where you are going, your eyes will be telling you that you aren't moving. The messages are mixed up and you feel ill. One way to prevent motion sickness is to look out the front window of the car. Most drivers don't experience motion sickness because they are always looking out the front window.

The Skin

How does the skin function as a sense organ?

You may not think of your skin as a sense organ. It does, however, have many different kinds of nerve cells that detect changes around the body. Before we describe these nerve cells, let's look at the skin itself. Figure 16-9

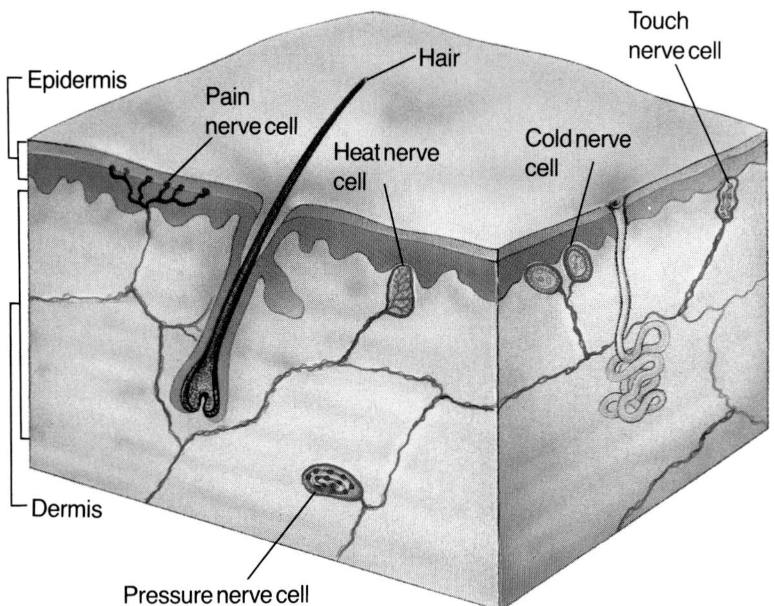

Figure 16-9 The skin contains five kinds of nerve cells.

shows a closeup of a small section of skin. Two different layers form the skin. The **epidermis** (ep uh DUR mus) is the outside layer of cells of an organism. The **dermis** is a thick layer of cells that form in the inner part of the skin.

Notice in Figure 16-9 that five nerve cell types are shown. Each nerve cell detects a different condition. The nerve cells detect pain, pressure, touch, heat, and cold. Note that most nerve cells are found in the dermis. Only nerve cells that detect pain are found in both the epidermis and the dermis.

Like messages from other sense organs, messages from the nerve cells in the skin also travel to the cerebrum. There they are interpreted as hot, cold, pain, pressure, or touch messages.

Skill Check

Understand science words: epidermis. The word part *epi* means outer. In your dictionary, find three words with the word part *epi* in them. *For more help, refer to the* ***Skill Handbook,*** *pages 706-711.*

Check Your Understanding

6. What is the relationship between the iris and the pupil?
7. How do the bones of the middle ear enable you to hear?
8. How are taste and smell related? How do they differ?
9. **Critical Thinking:** Many animals are active only at night. Would these night creatures have larger or smaller pupils than humans? Why?
10. **Biology and Reading:** Where are the nerve cells of the eye found? What are they called?

16:3 Problems with Sense Organs

Objectives

6. **Describe** how common vision problems are corrected.
7. **Discuss** reasons for protecting your ears against loud noises.
8. **Explain** ways of correcting hearing problems.

Key Science Words

nearsighted
farsighted

Many people have lost their sense of hearing or vision completely. There are many more people with hearing or vision problems. What causes some of these problems? What can be done to correct them?

Correcting Vision Problems

Have you ever wondered what is meant by 20/20 vision? A person with 20/20 vision is said to have normal vision. Figure 16-10a tells you what the numbers mean. Notice that the units used by eye doctors are not in metric or SI. A person with 20/20 vision sees certain things clearly at 20 feet. A person with 20/60 vision sees clearly at 20 feet what a person with 20/20, or normal, vision would see clearly at 60 feet.

Not everyone has 20/20 vision. Prove it to yourself by looking around the classroom. How many of your classmates are wearing glasses or contact lenses?

Why do some people have normal vision and others have problems seeing clearly? Figures 16-10b-d show you. The lines from the letter *T* to the retina show the path of light into the eye. Notice that the lines in Figure 16-10b meet on the retina. This eye has 20/20 vision. The eye in Figure 16-10c is nearsighted. **Nearsighted** means being able to see clearly close up but not far away. The path of light meets in front of the retina, instead of on it. This is because the nearsighted eye is longer from front to back than the normal eye.

Figure 16-10 A person with 20/20 vision can see a 1 cm high letter clearly at 20 feet (a). In a normal eye (b), light rays meet on the retina. In a nearsighted eye (c), light rays meet in front of the retina. In a farsighted eye (d), light rays meet behind the retina.

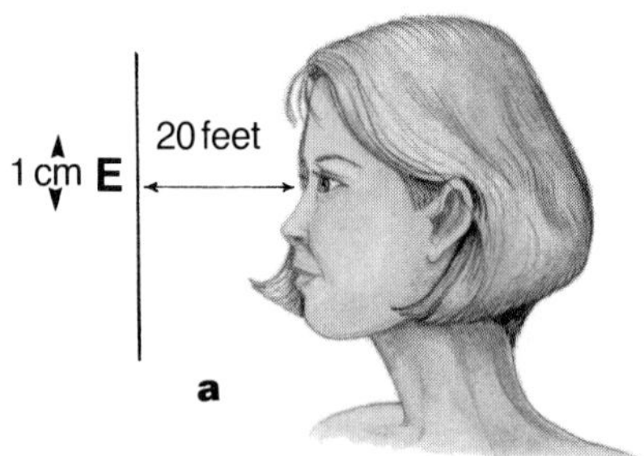

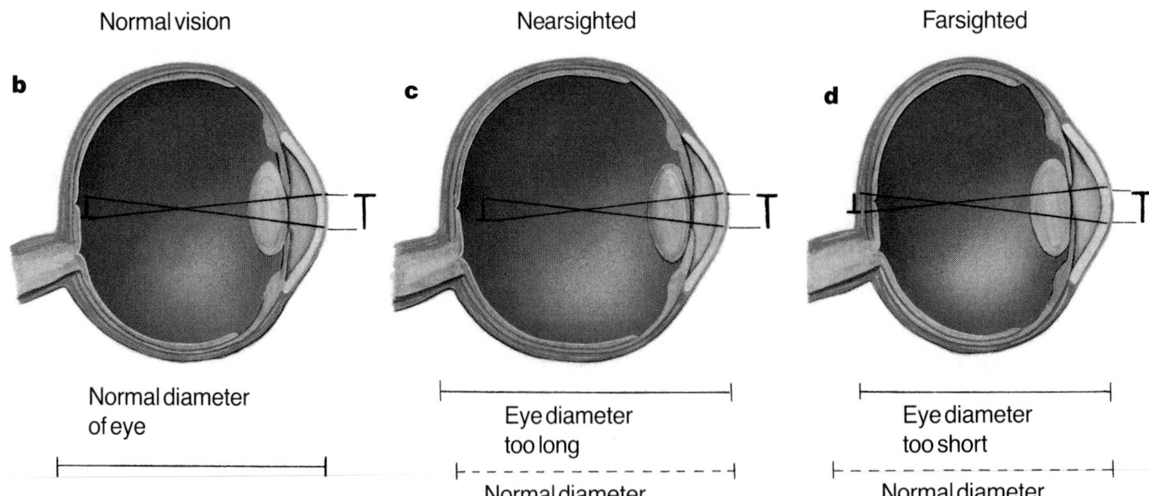

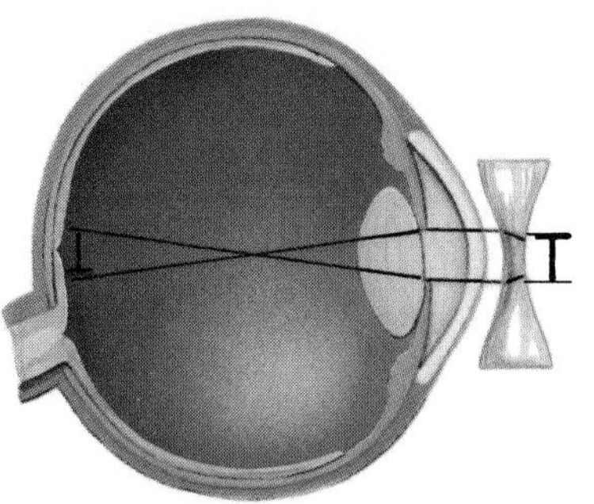

Lens to correct for nearsighted vision

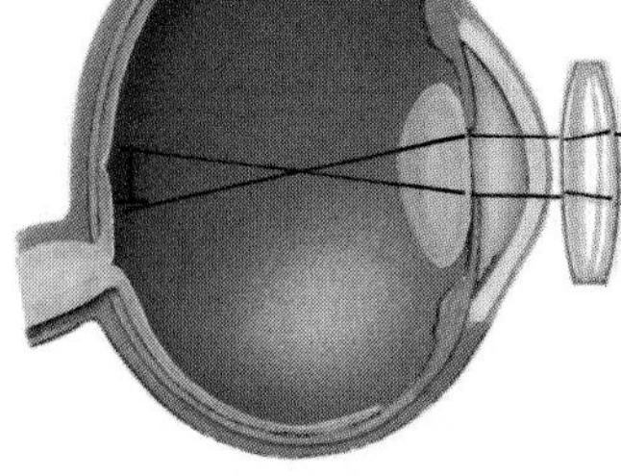

Lens to correct for farsighted vision

Figure 16-11 Lenses bend the light entering the eye and cause it to fall on the retina.

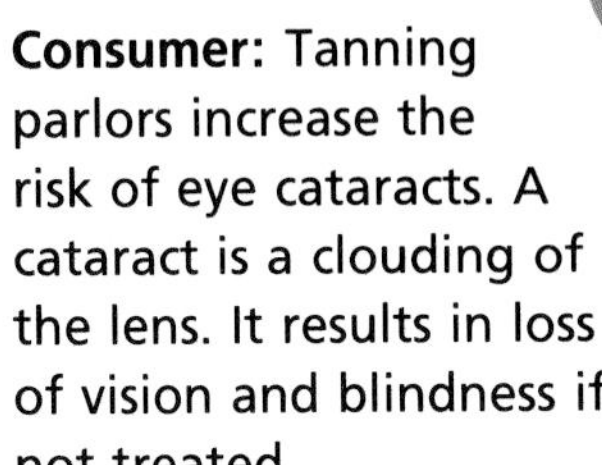

Bio Tip

Consumer: Tanning parlors increase the risk of eye cataracts. A cataract is a clouding of the lens. It results in loss of vision and blindness if not treated.

Figure 16-10d shows what happens within the eye when a person is farsighted. **Farsighted** means being able to see objects clearly far away but not close up. When the path of light enters the eye, it meets in back of the retina, instead of on it. Notice in the figure that the farsighted eye is shaped differently than the normal eye. The farsighted eye is shorter from front to back.

How are these kinds of vision problems corrected? They are corrected with lenses people can wear. Figure 16-11 shows how the lenses of glasses or contacts help correct vision problems. The lenses bend the light before it enters the eye. This extra bending allows the light rays to meet on the retina as they do in a normal eye. As a result, objects appear clear and sharp instead of fuzzy or blurry.

How does wearing glasses correct nearsightedness?

Protecting Against Hearing Loss

Can listening to loud music damage your hearing? The answer to this question is yes. Listening to any kind of loud noise over a long period of time can cause permanent damage to your hearing.

How does listening to loud noises damage hearing? Loud noises damage the inside of the cochlea. The cochlea contains fluid and thousands of hairlike cells. The hairlike cells vibrate as the fluid within the cochlea is set into motion by sound waves. The movement of the hairlike cells starts a message in the auditory nerve leading to the brain. Whenever the ear is exposed to loud noises, some of the hairlike cells become flattened. Once the hairlike cells are flattened, they can no longer send messages to the auditory nerve. If enough hairlike cells are damaged, hearing loss results. Also, the hairlike cells can't repair themselves. Hearing that is damaged this way can't be restored.

How loud a sound can damage your hearing?

How loud must a sound be before it damages your hearing? Table 16-1 shows the loudness of different sounds that you hear frequently. The loudness of sound is measured in units called decibels (DEH suh bulz). The softer a sound is, the lower its decibel value will be. The louder a sound is, the higher its decibel value will be. Listening to sounds that have decibel values over 85 can damage your hearing. Notice in the table that the decibel values for a motorcycle, jet, and rock concert are over 85.

What can be done to prevent hearing loss? Avoiding the sources of loud noises is one way. If loud noises can't be avoided, then hearing protection can be used. People who work at airports wear earmuffs to protect their hearing. People who work in factories often use earplugs made of plastic or foam rubber. It is important to prevent hearing loss. Once the hearing has been damaged by loud noises, it can't be restored.

TABLE 16–1. DECIBEL VALUES OF COMMON SOUNDS

Decibels	Sound	
20	rustling of leaves	
30	whisper	
50	light traffic	
60	air conditioner	
70	loud talk	
	television	
	vacuum cleaner	
85	food blender	
90	motorcycle at 10 meters	hearing damage may result
	shouting	hearing damage may result
	power mower	hearing damage may result
100	snowmobile	hearing damage may result
	jet overhead	hearing damage may result
110	portable radio with headphones	hearing damage may result
120	thunderclap	hearing damage may result; painful to the ear
	rock concert	hearing damage may result; painful to the ear
140	jet nearby	hearing damage may result; painful to the ear
160	shotgun	hearing damage may result; painful to the ear

Correcting Hearing Problems

There are many causes of deafness or hearing loss. Sometimes the problem lies with the auditory nerve or cochlea. Hearing problems can happen if the middle ear bones don't move smoothly. Let's look at some new ways of treating these problems.

A person may become deaf if the nerve cells in the cochlea don't work. An electronic ear can help, Figure 16-12. The electronic ear has a long wire that is placed inside the cochlea. The wire carries messages from outside

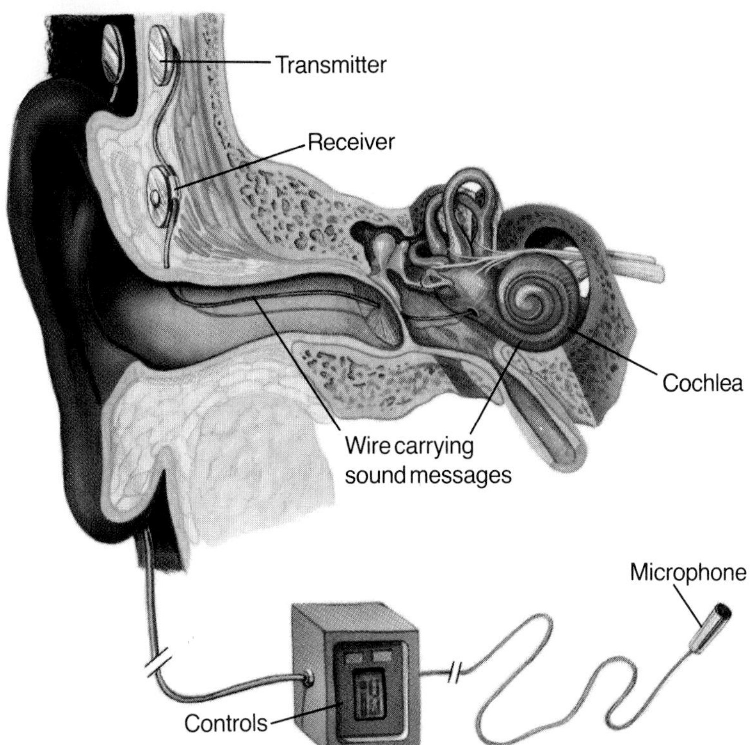

Figure 16-12 An electronic ear helps to restore certain kinds of hearing loss.

the ear to the cochlea. The electronic ear lets people hear sounds that they couldn't hear without it.

Remember that the job of middle ear bones is to move back and forth. Deafness may result if the ear bones can't move. The ear bone that most often gets stuck is the stirrup. When it does, the middle ear membrane can't move, and messages no longer reach the cochlea. To treat this problem, the stirrup is removed during an operation. It is replaced with a small piece of plastic. The plastic stirrup vibrates properly and moves the membrane.

How does an electronic ear work?

Check Your Understanding

11. Explain what causes a person to be nearsighted or farsighted.
12. How does loud noise cause hearing loss?
13. Describe two ways of correcting hearing problems.
14. **Critical Thinking:** How would damage to the optic nerve affect vision?
15. **Biology and Reading:** What system of measurement do eye doctors use?

Chapter 16

Review

Summary

16:1 Observing the Environment

1. Sense organs differ among animals. Planaria have eyespots. Earthworms have light-sensitive skin.
2. Crickets detect sound with their front legs, touch and chemicals with antennae, and light with compound eyes. Snakes smell with their tongues.

16:2 Human Sense Organs

3. The human eye has parts for reducing or increasing the amount of light entering the eye, focusing, and for seeing color or black and white.
4. The tongue detects molecules dissolved in water. It senses sour, salty, sweet, and bitter. The human nose detects molecules or odors in the form of gases. Skin contains nerves that detect pain, pressure, touch, heat, and cold.
5. The human ear detects air molecules in motion. It is divided into an outer, middle, and inner area.

16:3 Problems with Sense Organs

6. Nearsighted and farsighted vision are corrected with glasses or contact lenses to provide 20/20 vision.
7. Loud noise causes hearing loss by damaging the nerve cells in the cochlea.
8. Certain types of deafness can be helped with a plastic stirrup or electronic ear.

Key Science Words

auditory nerve (p. 341)
cochlea (p. 341)
cone (p. 336)
cornea (p. 335)
dermis (p. 343)
eardrum (p. 340)
epidermis (p. 343)
farsighted (p. 345)
iris (p. 334)
lens (p. 335)
lens muscle (p. 335)
nearsighted (p. 344)
olfactory nerve (p. 339)
optic nerve (p. 336)
pupil (p. 334)
retina (p. 335)
rod (p. 336)
sclera (p. 334)
semicircular canals (p. 341)
sense organ (p. 332)
taste bud (p. 339)
vitreous humor (p. 335)

Testing Yourself

Using Words

Choose the word from the list of Key Science Words that best fits the definition.

1. liquid inside the eye
2. membrane at the end of the ear canal
3. liquid-filled chamber of the ear
4. carries messages from the nose to the brain
5. inner layer of the skin
6. detects molecules on the tongue
7. can see clearly close up but not far away
8. adjusts amount of light entering eye
9. nerve cell that detects color
10. helps us keep our balance
11. changes shape for viewing distant and nearby objects

Testing Yourself *continued*

Finding Main Ideas

List the page number where each main idea below is found. Then, explain each main idea.

12. how the eyes of planaria, insects, and humans differ
13. the eye parts that protect the eye and the eye parts that adjust light
14. two causes of deafness
15. layer of skin in which you find many pressure, touch, and pain nerves
16. how you are able to taste more than just four types of taste
17. how the cochlea aids hearing
18. how nearsighted and farsighted vision differ

Using Main Ideas

Answer the questions by referring to the page number after the question.

19. What effect do sound waves have on the eardrum, ear bones, and middle ear membrane? (p. 341)
20. Which eye parts (a) are muscles, (b) protect the eye, (c) give the eye shape, (d) are made of nerves, and (e) bend light? (pp. 334-335)
21. What is meant by 20/60 vision? (p. 344)
22. How does a cricket detect sound, touch, and movement? (pp. 332-333)
23. What is the job and the sense organ associated with the following nerves?
    (a) olfactory (p. 339)
    (b) auditory (p. 341)
    (c) optic (p. 336)
24. How does a compound eye differ from a human eye? (p. 333, pp. 334-336)
25. Describe two jobs of the cerebrum when messages from the eye reach it. (p. 336)

Skill Review ✔

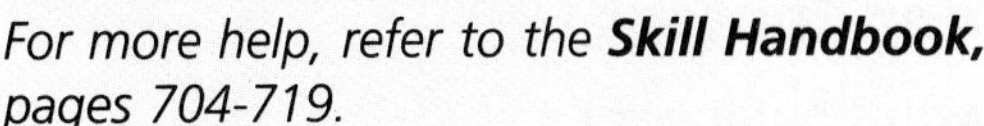

For more help, refer to the ***Skill Handbook,*** *pages 704-719.*

1. **Interpret diagrams:** Study Figure 16-6. Where on the tongue are the four tastes detected?
2. **Predict:** How would puncturing the eardrum affect the ability to hear?
3. **Observe:** Look at your pupils in dim light, then in bright light. How does the size of the pupil change?
4. **Formulate a model:** Strike a tuning fork with a rubber mallet. Note the sound it makes. Place the tuning fork into a glass of water. What happens to the water? How does this model show what happens within the cochlea?

Finding Out More

TECH PREP

Critical Thinking

1. Why is it impossible to transplant the retina or optic nerve? Why is it possible to transplant the cornea?
2. Why is the pupil black?

Applications

1. Design an experiment to test which sense is better developed in most people, taste or smell.
2. Using a large ball and strips of cloth, construct a model of the human eye. Use the cloth to represent the muscles located on the outside of the eye. Determine the job of each muscle pair. Indicate these jobs on your model.

CHAPTER PREVIEW

Chapter Content

Review this outline for Chapter 17 before you read the chapter.

17:1 Behavior
- Steps of Behavior
- Innate Behavior
- Learned Behavior

17:2 Special Behaviors
- Behaviors for Reproduction
- Behaviors for Finding Food
- Protective Behaviors
- Migration
- Parental Behavior

Skills in this Chapter

The skills that you will use in this chapter are listed below.

- In **Lab 17-1,** you will predict, separate and control variables, observe, and design an experiment. In **Lab 17-2,** you will form hypotheses, observe, and make and use tables.
- In the **Skill Checks,** you will understand science words and form hypotheses.
- In the **Mini Labs,** you will infer and observe.

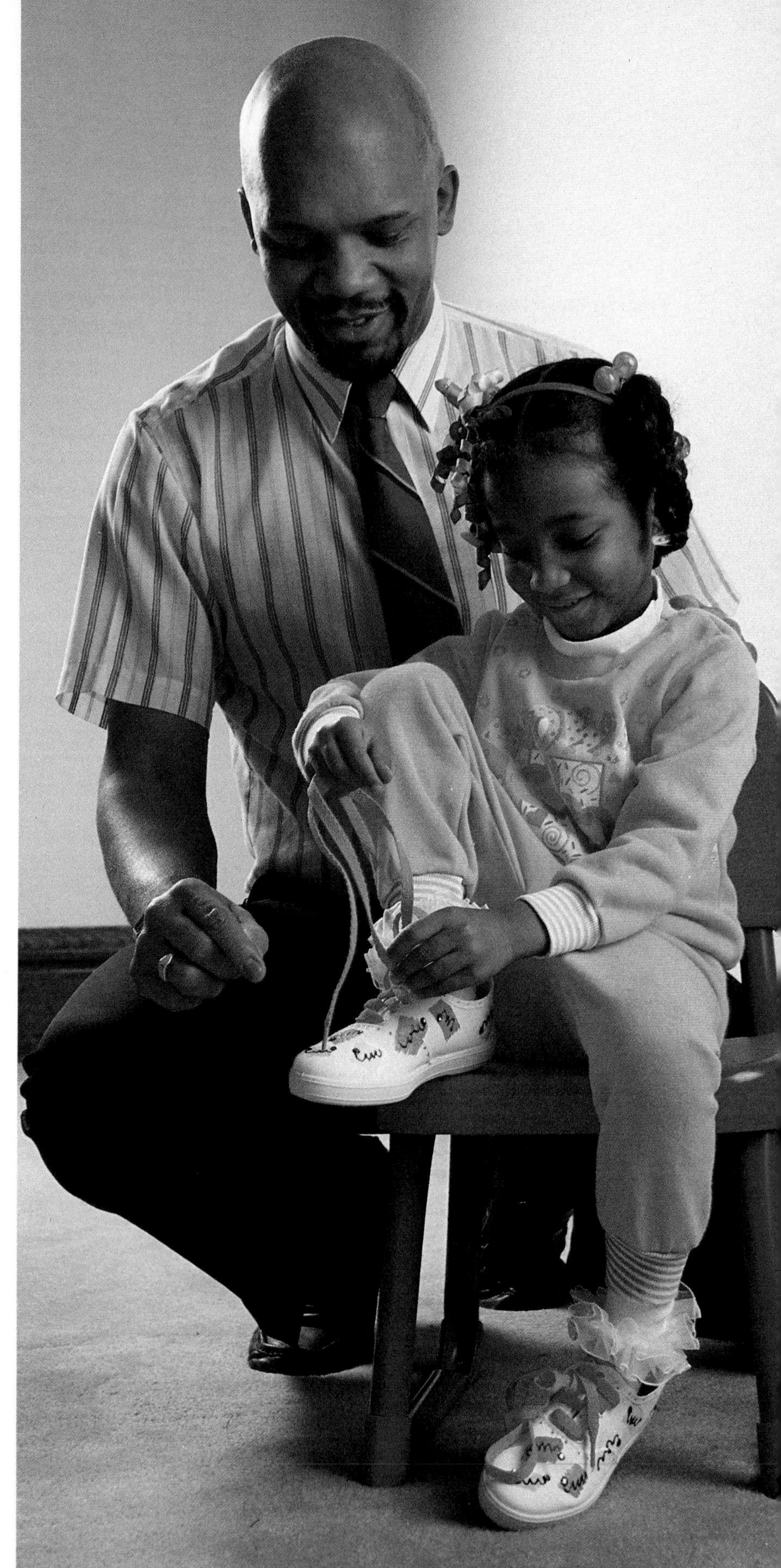

Chapter 17

Animal Behavior

Children must learn how to tie their shoelaces. This skill is learned at home or in school. The child in the photo on the opposite page is being taught to tie her shoelaces. After a lot of practice, she will have learned this skill.

A bird building a nest has never gone to school to learn nest-building. How does the bird shown in the photograph below know how to build a nest without learning about it first? How does learning how to tie shoelaces differ from nest-building? These are some of the questions you will explore in this chapter.

Try This!

How do you learn? Find a book on origami, or paper folding. Follow directions on folding an animal, while you time how long it takes you to do this. Repeat the folding of the same animal several more times. Note how long each new trial takes. Does the length of time change as you practice?

BIOLOGY Online

Visit the Glencoe Science Web site at science.glencoe.com to find links about **animal behavior.**

Birds don't have to learn how to build nests.

17:1 Behavior

Objectives

1. **Explain** the difference between a stimulus and a response.
2. **Explain** innate behavior and give examples.
3. **Describe** how behavior is learned and give examples of learned behavior.

Key Science Words

behavior
stimulus
innate behavior
instinct
learned behavior

How do animals know what is happening around them? How do they react to events occurring in their environment? Animals react to changes around them in many different ways. However, they all use their sense organs, nervous systems, and muscles when they react.

Steps of Behavior

Everything that an animal does is part of its behavior. **Behavior** is the way an animal acts. The way an animal acts depends on what is going on around it.

Let's use an example to see how behavior works. The feeding behavior of an owl can be broken down into steps. Follow the steps in Figure 17-1 as you read.

1. The owl sees a mouse on the ground.
2. Messages pass along nerves from the bird's eyes to its brain.
3. The brain sends out new messages along nerve pathways to muscles.
4. Messages arriving at muscles allow the owl to grab the mouse.

Figure 17-1 Steps in the behavior of an owl catching a mouse. What is the stimulus?

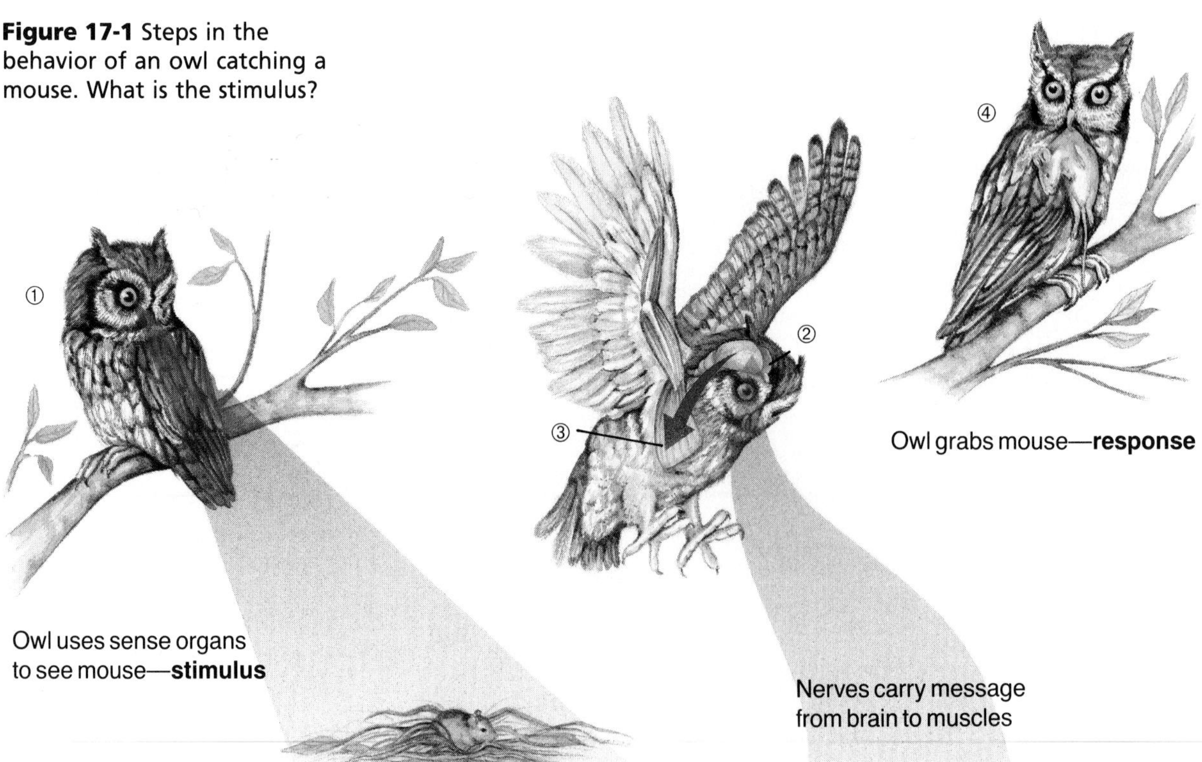
①
②
③
④
Owl grabs mouse—**response**
Owl uses sense organs to see mouse—**stimulus**
Nerves carry message from brain to muscles

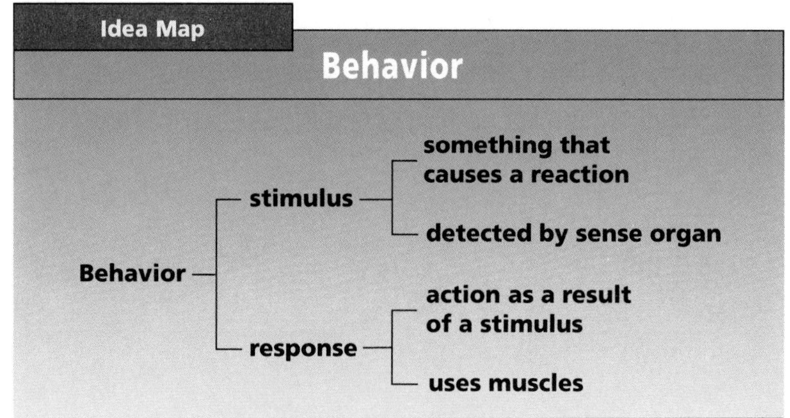

Study Tip: Use this idea map as a study guide to stimulus and response. Can you give an example of a stimulus and its response?

There are two important parts to all behavior. They are stimulus and response. A **stimulus** is something that causes a reaction in an organism. In the owl example, the mouse was the stimulus. Seeing it led to the owl's behavior.

What are the two parts of behavior?

Moving to get the mouse was the owl's response. You read in Chapter 15 that a response is the action of an organism as a result of a change, in this case a stimulus. Notice that the bird's response used muscles. How were they used?

Here is a different example. You are standing in a dark room and you think you are alone. You don't see anyone, but someone else is in the room with you and taps you on the arm. You jump and scream. What was the stimulus in this example? Which sense organs detected the stimulus? What were the responses? Which muscles were used to respond?

Innate Behavior

There are two main types of behavior in animals—innate and learned. When a robin builds a nest, it is using innate behavior. **Innate behavior** is a way of responding that does not require learning. You did not learn to grab for objects, yawn, smile at your mother, or cry as a baby. All of these behaviors are innate. One type of innate behavior is an instinct. An **instinct** is a complex pattern of behavior that an animal is born with. Nest-building and mouse-catching are instincts. They are more complex than simple innate behavior, such as a yawn. In nest-building, for example, the bird must search for the proper material, gather it, and build the nest. The behavior may take several days to complete.

Skill Check

Understand science words: innate. The word part *in* means inside. In your dictionary, find three words with the word part *in* in them. *For more help, refer to the **Skill Handbook,** pages 706-711.*

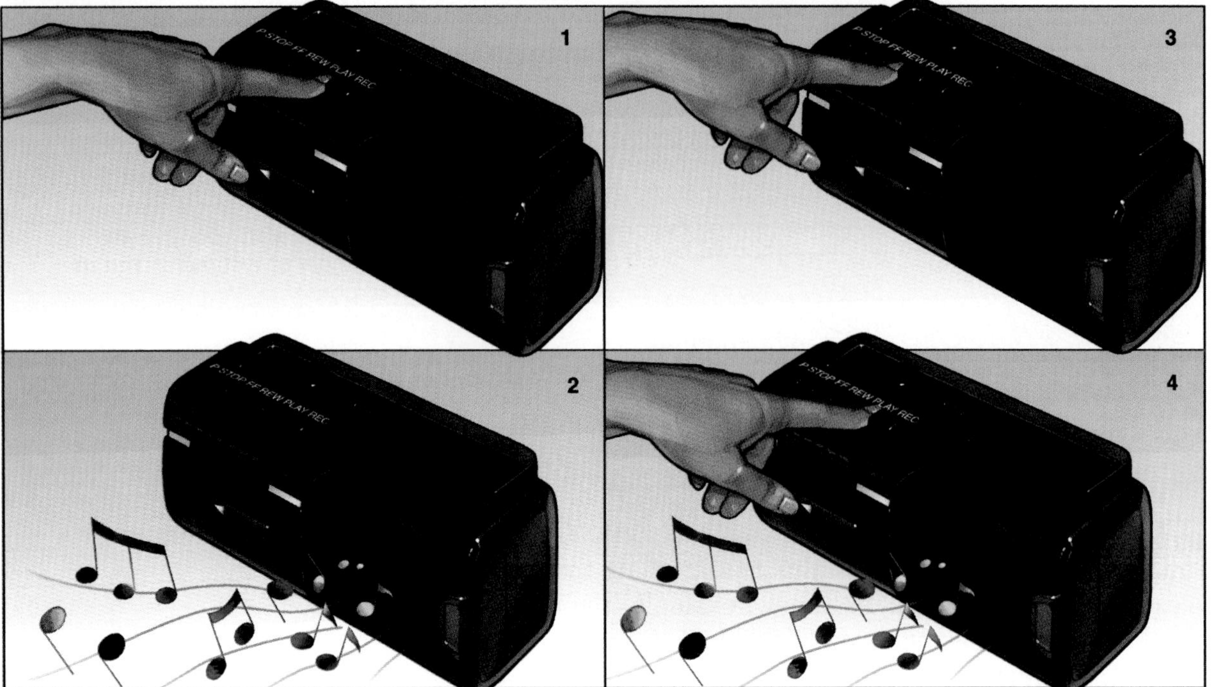

Figure 17-2 Innate behavior can be compared to the playing of a prerecorded tape.

Let's compare innate behavior to running a tape recorder. Follow the steps in Figure 17-2 as you read.

1. You put a tape of your favorite song into a tape recorder. Then you press the play button.
2. The recorder plays the song.
3. You rewind the tape and press play again.
4. The recorder plays the same song as long as you continue to play the same tape.

Innate behavior is similar to playing this pre-recorded tape. Putting the tape into the recorder is like putting the stimulus into the nervous system. The playing song is like the response to the stimulus.

Can innate behavior be changed?

Each time a certain stimulus for an innate behavior occurs, the animal has the same response. Innate behaviors cannot be changed. Wasn't this also the case with our tape recorder? Let's look at a behavior that you have experienced but never had to learn. You swallow, and food goes down your trachea, or windpipe, instead of your esophagus. Your response is to cough. Coughing brings the food up and out of your windpipe. Why is coughing innate behavior? The same thing happens each time. You don't have to think about it and you can't change it.

Instincts, too, are automatic. A robin does not have to think about what kind of nest to build. All robins build the same shape nests, made out of the same materials. They can't change the kind of nest they build.

Bio Tip

Health: An innate behavior can be seen in infants when they grasp a rattle. The middle finger always closes on the handle first.

Lab 17–1 Innate Behavior

Problem: How do vinegar eels respond to gravity?

Skills

predict, separate and control variables, observe, design an experiment

Materials

microscope	2 test tubes, small
dropper	2 stoppers to fit tubes
glass slide	2 labels
coverslip	test-tube rack
vinegar eels	

Procedure

1. Copy the data table.
2. Use a dropper to place a drop of liquid containing vinegar eels onto a glass slide. Gently add a coverslip.
3. **Observe:** Observe the eels under low power of your microscope. Note their movement.
4. Draw several eels in the circle in the table.
5. **Predict:** Certain animals move toward gravity, while others move against gravity. Predict whether the vinegar eels will move toward gravity, against gravity, or will show no response to gravity.
6. Fill a small test tube with the liquid containing vinegar eels. Stopper the tube. If there is any air in the tube, remove the stopper and add a few more drops of liquid. Replace the stopper.
7. Fill a second tube half full with the liquid containing vinegar eels. Stopper the tube.
8. Place both tubes in the test-tube rack and label them with your name. Let the tubes stand overnight.
9. The following day, note carefully the location of the vinegar eels in each tube. **NOTE:** *Do not move or touch the tubes.*
10. Record on the tube outlines in the table where most of the eels are found.
11. Return the eels to their original container.

OBSERVATIONS OF VINEGAR EELS

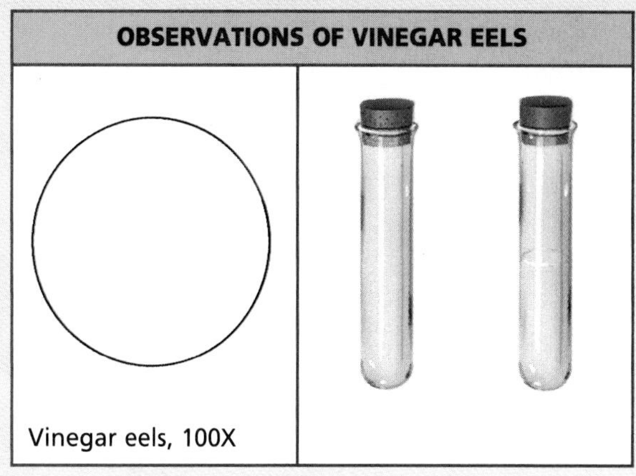

Data and Observations

1. Describe the appearance and movement of vinegar eels under the microscope.
2. If vinegar eels move toward gravity, they will be found near the bottom of the test tube. If they move away from gravity, they will be found near the top of the test tube. If they do not show any behavior in response to gravity, they will be found evenly throughout the tube. What type of behavior do vinegar eels show toward gravity?

Analyze and Apply

1. What evidence do you have that the animals are responding to gravity and not to the air at the top of the tube?
2. Is the behavior of vinegar eels in response to gravity innate or learned? Explain.
3. **Apply:** How might vinegar eels benefit from the type of behavior that they show in response to gravity?

Extension

Design an experiment to show the type of behavior vinegar eels might show in response to light and dark.

TECHNOLOGY

Blinking Reveals Behavior Secrets

Blinking your eyes is an innate behavior. When you blink your eyes, you are protecting them from dust or smoke, and you are providing moisture for the cornea to prevent it from drying out. You don't have to think about blinking. It happens by itself.

Normally, people blink about 10 to 20 times each minute. Scientists have found that if a person is lying, he or she blinks more often.

Scientists also found that a person who is tired will blink about 30 or 40 times per minute. This information would be helpful in alerting pilots to the fact that they are too tired to be flying airplanes. Scientists could design an instrument that would tell the pilot if his or her blinking has increased beyond the normal rate. Thanks to people who study blinking, flying may be safer in the future.

Airplane pilot

Learned Behavior

The second main type of behavior is learned behavior. **Learned behaviors** are behaviors that must be taught. These behaviors must be practiced until a certain stimulus results in a certain response. Behaviors such as tying your shoelaces or writing your name are learned behaviors.

Let's look at how learned behavior compares to recording a tape. Follow the steps in Figure 17-3 as you read.

1. You put a blank tape into a tape recorder. Then you press the play button. You hear nothing.
2. You rewind the tape and record a song.
3. You play back the tape. This time the recorder plays the song you just recorded.

You can continue to record the song until it sounds good to you. You can correct mistakes and record again.

Learned behavior is like recording a tape. There may be no response the first time a stimulus is received. The same kind of thing happens with a blank tape. It plays nothing.

Responding over and over to a stimulus can lead to learned behavior. Recording over and over can lead to a song that sounds good to you. Just as you recorded on the blank tape, the nervous system can "record" a behavior.

Mini Lab

Do Onions Make You Cry?

Infer: Cut an onion. Hold the cut edge close to your face. Describe what happens. Could you have prevented this behavior? Is this behavior innate or learned? Why? *For more help, refer to the* ***Skill Handbook,*** *pages 706-711.*

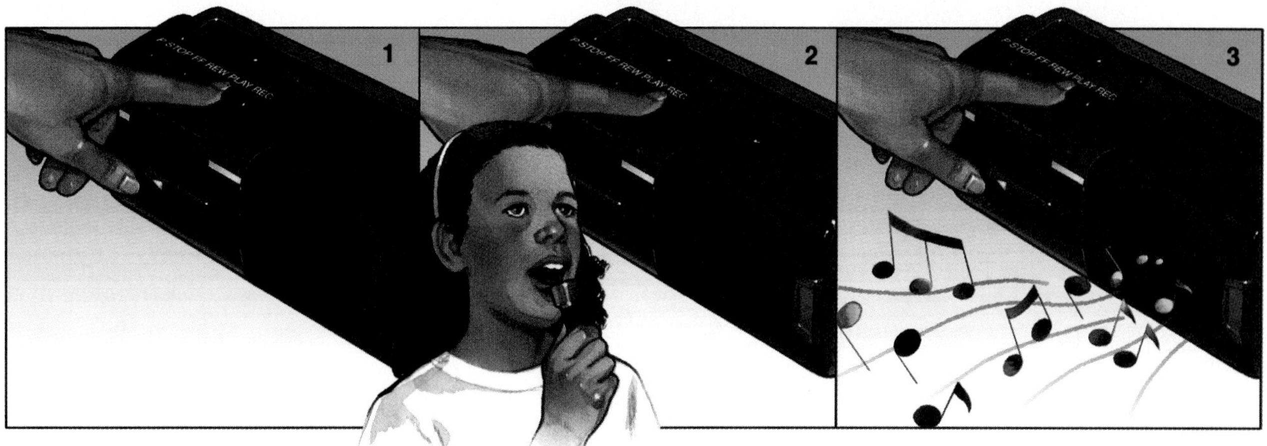

Suppose you want to teach a dog to raise its paw. When you first tell the dog to raise its paw, nothing happens. Repeat the message over and over and pick up the dog's paw each time. The dog will learn what to do. Then, each time the dog hears the stimulus word, "paw," it will respond the same way. It lifts its paw.

Figure 17-3 Learned behavior can be compared to the recording of a tape.

Can learned behavior change? Yes, you can teach a dog to lift its paw and bark at the same time. You can even teach the dog to lift its paw, bark, and then roll over.

Most animals learn new behaviors faster if there is a reward. Animal trainers often use food as a reward when an animal shows the correct behavior. The next time you see dolphins jump through a hoop at an aquarium, look for the reward of fish they receive from their trainer. What do dog trainers give dogs as a reward for correct behaviors?

What helps animals learn faster?

Check Your Understanding

1. What is the difference between stimulus and response? Give an example of each.
2. How does an instinct differ from simple innate behavior?
3. What type of behavior must be practiced?
4. **Critical Thinking:** Why would you expect the behavior of a honeybee to be more complex than that of a sponge?
5. **Biology and Reading:** Read the following sentences. Write the one that summarizes the information in this section.
   a. Behavior is how an animal knows what is happening.
   b. Behavior is everything that an animal does and is made up of unlearned and learned responses.
   c. Behavior is all of the complex learned responses to stimuli by animals that have brains.

Behavior

Lab 17-2

Problem: How do innate and learned behavior differ?

Skills

form hypotheses, observe, make and use tables

Materials

watch with second hand
mirror
pencil and paper

Procedure

Part A

1. Copy the data table.
2. **Observe:** Look at your eyes in the mirror and note the size of your pupils with the lights on and the lights off.
3. The teacher will turn off the lights. When the lights are turned on, look at the watch. Time how long it takes your pupils to change size. Record your data.
4. Write a **hypothesis** to tell whether the length of time it takes your pupils to change size will increase, decrease, or stay the same with repeated trials.
5. Repeat step 3 four more times.

Part B

1. Write your name on a piece of paper.
2. Position a mirror as shown in Figure B.
3. While looking *only* into the mirror, try to trace your name. Have a partner time you. Record the data in the table.

TIME IN SECONDS		
Trial	Pupil Change	Trace Name
1		
2		
3		
4		
5		

A

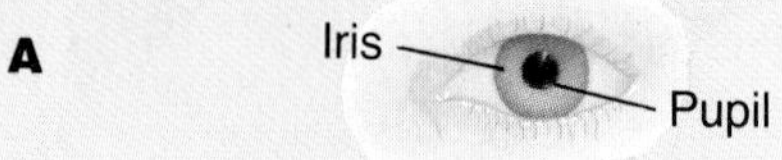

B

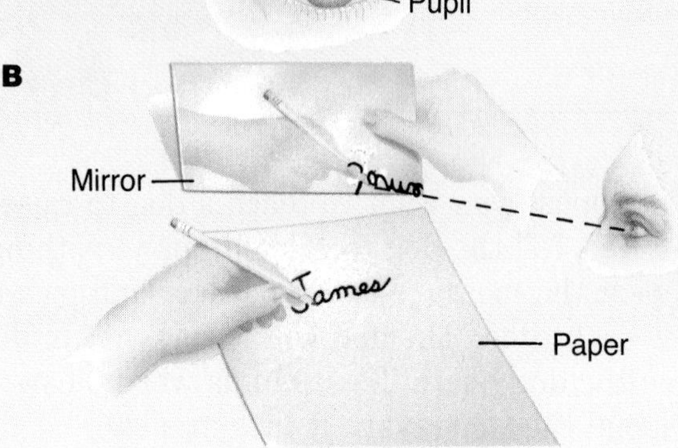

4. Write a **hypothesis** to tell whether the length of time it takes you to trace your name will change with repeated trials.
5. Repeat steps 2 and 3 four more times.

Data and Observations

1. Did the pupils take the same amount of time to change with repeated trials?
2. Did the time needed to trace your name change with repeated trials?

Analyze and Apply

1. **Check your hypotheses:** Are your **hypotheses** supported by your data? Why or why not?
2. Was the behavior in Part A innate or learned? Was the behavior in Part B innate or learned? How do you know?
3. Compare innate and learned behaviors.
4. **Apply:** How does the behavior in Part A help humans with their vision?

Extension

Design an experiment to show if learned behavior is forgotten if it is not practiced.

17:2 Special Behaviors

How do certain behaviors help animals? Some types of behavior help animals find food. Other types may help an animal find a mate. Certain behaviors protect animals from their enemies. Finding a place to live while having young is another type of special behavior.

Objectives

4. **Explain** how courting behaviors help an animal species reproduce.
5. **Describe** how behavior helps an animal find food, shelter, and protection from enemies.
6. **Compare** reproduction in animals that give parental care with reproduction in animals that give no care.

Key Science Words

courting behavior
pheromone
social insects
migration
parental care

Behaviors for Reproduction

Male frogs and crickets make sounds. You may have heard these sounds on a summer night. The males use these sounds to attract females. This behavior brings males and females together for mating. The sounds are the stimulus, and the female responds by finding the male. Does this type of behavior use sense organs or a nervous system? The answer to this question is that both are used. The female must hear the sounds and her brain must send the proper signals to her muscles so she can respond. Is a frog's croaking or a cricket's chirping innate or learned behavior? Why do you think so?

Some animals, such as birds and fish, have complex courting behaviors when they reproduce. **Courting behaviors** are behaviors used by males and females to attract one another for mating. Figure 17-4 shows a peacock with its tail feathers spread. The feather display is a courting behavior. The male uses it to attract the female. The colorful neck flap of certain lizards helps attract females and threaten other males. Figure 17-4 shows the neck flap on an anole lizard.

Bio Tip

Leisure: People enjoy watching fireflies. Males of different species make special flash patterns when they look for mates. A female of the same species flashes back the same signal. He flies down to mate with her. If all species had the same flash, males would mate with the wrong females.

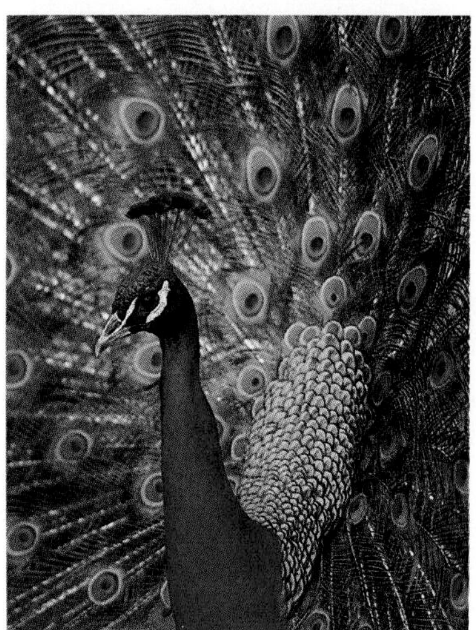

Figure 17-4 Peacocks and anole lizards have colorful courting behavior.

What is a pheromone?

Not all male animals make sounds or have displays to attract their mates. Some animals give off chemicals with distinct odors that do the same job. Chemicals that affect the behavior of members of the same species are called **pheromones** (FER uh mohnz).

Many female insects use pheromones to attract mates. One example is the silk moth. The female silk moth attracts the male by giving off a pheromone that is carried by wind currents. Males can detect the odor of the chemical with their antennae from many kilometers away. They respond by flying toward the odor. When the males reach the female, she chooses one male and allows him to mate with her.

How did scientists prove that a chemical attracts the male moth? Find the answer in Figure 17-5. Notice that it doesn't matter whether the males can see the females. They aren't attracted to the females by sight. Males are attracted to the odor of the pheromone, which is given off by the females.

How could you prove that the male senses the chemical with its antennae? You could remove the antennae from the males and note their responses. Scientists did this, and the male moths flew in all different directions. They could not smell the stimulus and find the female.

Skill Check

Form hypotheses: What would happen to the behavior of the male moths in Figure 17-5 if the pheromone were added to the paper before the female moth was placed in the airtight cage? *For more help, refer to the* ***Skill Handbook,*** *pages 704-705.*

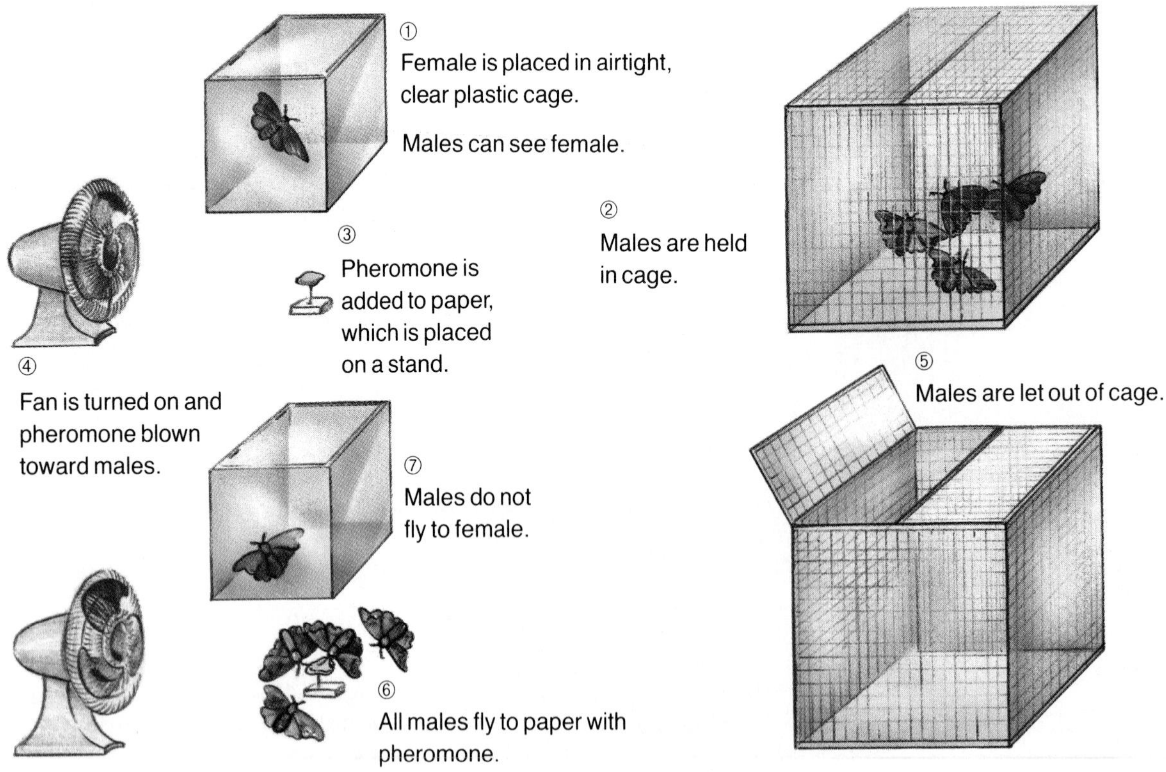

①
Female is placed in airtight, clear plastic cage.
Males can see female.
②
Males are held in cage.
③
Pheromone is added to paper, which is placed on a stand.
④
Fan is turned on and pheromone blown toward males.
⑤
Males are let out of cage.
⑥
All males fly to paper with pheromone.
⑦
Males do not fly to female.

Figure 17-5 Male moths are attracted to the female's pheromone, as shown in this experiment.

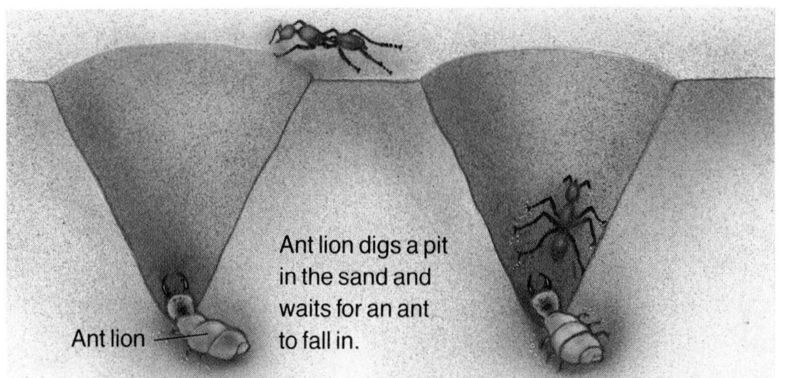

Ant lion digs a pit
in the sand and
waits for an ant
to fall in.
Ant lion

Figure 17-6 An ant lion digs a pit to help in getting food.

Behaviors for Finding Food

How would you like to walk along a beach and fall into a pit? Falling into the pit might not be too bad in itself. What if a hungry animal were waiting for you at the bottom? Things could go downhill very quickly.

Ant lions catch food by building a pit. Figure 17-6 shows how this behavior works. They wait for animals to fall into their pits. When an animal such as an ant falls in, it becomes a meal for the ant lion. Do you think that pit–building is innate or learned? Why do you think so?

Honeybees have complex behavior that helps them find food. They "talk" to one another and tell each other where food is located. The way bees "talk" is explained in Figure 17-7. When bees "talk," they move in a figure eight pattern. The "talking" bee faces the direction of the food when it is in the middle part of the figure eight. How do bees tell one another how far the food is? The "talking" bee wags its abdomen. The number of wags tells the other bees how far the food is from the hive.

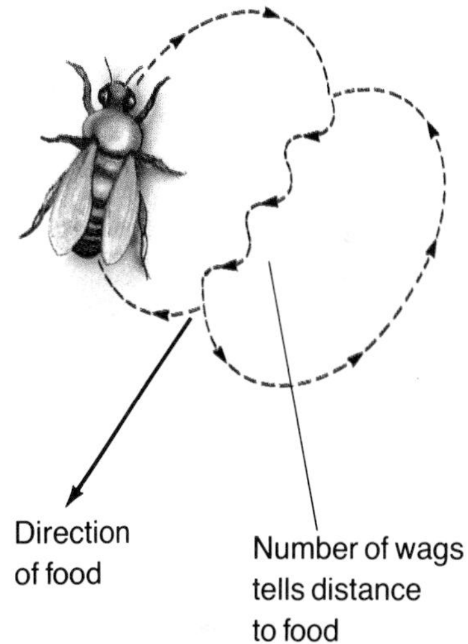

Figure 17-7 Honeybees use special behavior in "talking" to other bees.

Some spiders build webs to trap food. Each species of spider makes a web with a special design, as shown in Figure 17-8. Do you think this behavior is innate or learned?

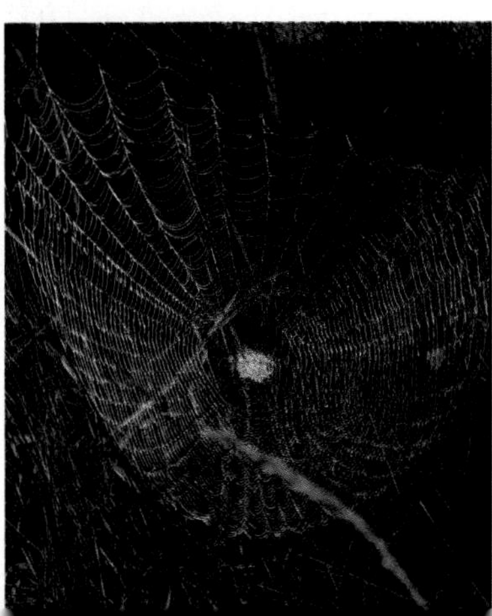

Figure 17-8 Spiders use their webs to trap food. Each species of spider builds a web of a unique design.

Mini Lab

How Do Earthworms Behave?

Observe: Place an earthworm in a box. Cover half the box with black paper. Note if the worm moves toward or away from light. What behavior is this? How does it help a worm? *For more help, refer to the* ***Skill Handbook,*** *pages 704-705.*

Why do some animals live in groups?

Protective Behaviors

Musk oxen, such as in Figure 17-9a, are mammals that live in the far north of Alaska and Canada. They are never found living alone. They are always found in groups. Living in a group gives them protection.

Imagine a wolf pack moving toward a herd of musk oxen. These wolves hunt the oxen for food. The oxen see, smell, or hear the wolves. They set up a defense against the wolves by forming a ring, as shown in Figure 17-9b. The adults stand on the outside. The young stay inside the ring. If a battle takes place, the older and larger oxen are in a good position to fight. The wolves can't get to the young inside the ring. The wolves also have a harder time fighting a group than they would a single animal. This type of innate group behavior helps to protect all the animals in the herd.

Many social insects have innate protective behaviors. **Social insects** are insects that live in groups, with each individual doing a certain job. Bees, wasps, ants, and termites are social insects. Bees have workers, drones, and a queen. The queen is the only female in the hive to reproduce. Drones are males. They mate with the queen. Workers are females that can't reproduce. Workers gather food, keep the hive clean, and protect it and the queen.

Ants and termites have an additional job. They have soldiers that help to protect the entire group. If an enemy enters an ant or a termite nest, the soldiers respond. They destroy the enemy by cutting it apart with their sharp jaws. A social insect lives in a group for protection.

Figure 17-9 Musk oxen (a) form a ring (b) to defend their young against wolves.

a

b

a

b

Figure 17-10 The fur seal (a) and the Atlantic golden plover (b) migrate to find a better place to live in winter.

Migration

Many animals move from one place to another. **Migration** is a kind of behavior in which animals move from place to place in response to the season of the year. Many birds, such as the Atlantic golden plover in Figure 17-10b, migrate. The plovers' path is shown in Figure 17-11. Plovers spend spring, summer, and fall in northern Canada. As winter comes, the plovers migrate to South America. They fly a total distance of 10 000 kilometers, more than twice the distance across the United States.

When plovers migrate, they find more food. Food is plentiful during the spring through fall in Canada. As winter arrives, food supplies disappear. The plovers migrate to where there is more food.

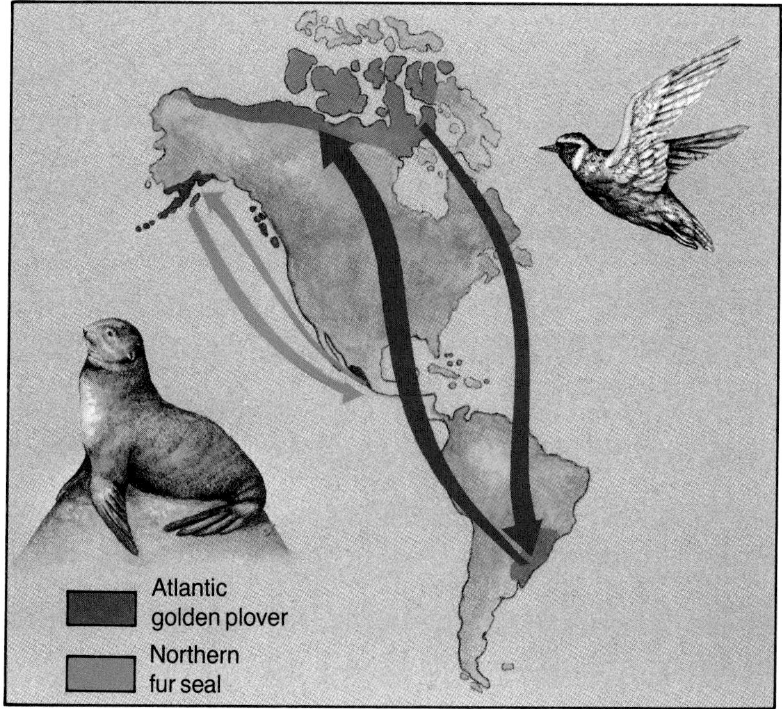

Figure 17-11 Migration routes of the Atlantic golden plover and the Northern fur seal. Why do these animals migrate?

Idea Map

Special Behaviors

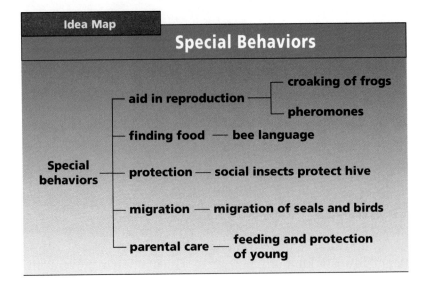

Special behaviors		
	aid in reproduction	croaking of frogs
		pheromones
	finding food	bee language
	protection	social insects protect hive
	migration	migration of seals and birds
	parental care	feeding and protection of young

Study Tip: Use this idea map as a study guide to special behaviors. How many other examples can you give?

Fur seals, shown in Figure 17-10a, migrate every year. They swim from the coast of Alaska to the coast of Mexico and back again. That's a total distance of about 7000 kilometers each year, the distance between Chicago and England! Their path of migration is also shown in Figure 17-11.

The fur seals' migration helps the animals to reproduce. By moving south, seals find better places to reproduce and care for their young. Seals mate and raise their young on land. It is easier to raise young where it is warm.

How do animals that migrate find their way? This question is of great interest to scientists. Some of the answers can be found in studies made with penguins. Adelie (uh DAY lee) penguins, shown in Figure 17-12, live in the Antarctic. They have nests along the coast, where they live and reproduce. Could these animals find their way back to their nests from miles away?

Figure 17-12 Adelie penguins use the sun's position to find their way.

Scientists moved the penguins far away from their nests and followed them. They found that the penguins used the sun for direction. On sunny days, the penguins moved in a straight line toward their original nests. On cloudy days, the penguins did not move in the correct direction.

Is this type of behavior innate or learned? Here is a clue. Scientists moved baby penguins to a new location. The babies also used the sun to find their way back. They did not have the adults to follow.

How do birds that migrate at night find their way? Scientists found that some birds use the stars for migration. Other birds seem to have built-in "compasses" that tell them the direction of the magnetic north pole.

Parental Behavior

Figure 17-13 Many animals, such as birds, feed and protect their young. What is this kind of behavior called?

After you were born, many things had to be done for you. You could not get your own food. You could not find a safe place to live. You could not protect yourself in any way. You would have died if your parents or other adults hadn't taken care of you. Are humans the only animals that care for their young? Most other mammals and birds care for their young. Some fish, amphibians, and reptiles provide protection for their eggs until they hatch.

Parental care is a behavior in which adults give food, protection, and warmth to eggs or young, Figure 17-13. The female parent is often the one to give parental care. The male parent also may protect the eggs or young. Male penguins often care for eggs while the female is feeding.

Parental care becomes more important in animals that produce fewer young. Most fish produce hundreds or thousands of eggs at a time, but do not provide any care at all. Some eggs are eaten, some never hatch, and some survive and grow to adulthood. Large numbers of eggs must be produced to be sure that some will survive.

Why is parental care important in animals that produce few young?

Animals that provide better parental care can produce fewer young and have them survive. Wolves, squirrels, and birds give lots of parental care to their young. These animals usually have no more than six young at a time. Most of the young live because of the care they receive from their parents.

Check Your Understanding

6. How do crickets and silk moths differ in courting behavior?
7. Describe the path of migration for Atlantic golden plovers, and tell why this behavior helps the animals.
8. Why must an animal that has few young give parental care?
9. **Critical Thinking:** What advantages might a blackbird have by living in a flock of blackbirds rather than by living alone?
10. **Biology and Reading:** In this section you learned that social insects live in groups and have protective behaviors. What word means the opposite of *social* as used in the sentence above? Name an insect that can be described using this word.

Chapter 17 Review

Summary

17:1 Behavior

1. Behavior is the way an animal acts. All behavior is a response to a stimulus.
2. Innate behavior does not require learning. An animal is born with it. Instincts are complex patterns of innate behavior.
3. Learned behavior must be taught. Animals are not born with it. Learned behavior must be practiced, can be changed, and can be learned more quickly if a reward is given.

17:2 Special Behaviors

4. Courting behavior and response to pheromones are innate behaviors that help reproduction.
5. Many animals use special innate behaviors for catching or finding food. Some animals protect themselves from attack by staying in large groups. Social insects live in groups and have specialized behavior. Animals migrate in response to seasonal changes.
6. Certain animals care for their young. Young animals that receive parental care have a better chance of surviving than those that do not.

Key Science Words

behavior (p. 352)
courting behavior (p. 359)
innate behavior (p. 353)
instinct (p. 353)
learned behavior (p. 356)
migration (p. 363)
parental care (p. 365)
pheromone (p. 360)
social insects (p. 362)
stimulus (p. 353)

Testing Yourself

Using Words

Choose the word from the list of Key Science Words that best fits the definition.

1. behavior that must be taught
2. behavior that does not have to be learned
3. the way an animal acts
4. a complex pattern of responding to a stimulus that is not learned
5. behavior used to attract males or females for mating
6. adults giving food or protection to young
7. moving from place to place in response to the season
8. insects that may have workers, drones, soldiers, and a queen
9. something that causes a reaction
10. chemical that affects the behavior of other members of the same species

Testing Yourself *continued*

Finding Main Ideas

List the page number where each main idea below is found. Then, explain each main idea.

11. how the number of young produced and the number that live compare between animals that give parental care and those that do not
12. how a bee tells others where food is
13. why learned behavior may be compared to recording a tape
14. why animals migrate great distances each year
15. why innate behavior may be compared to playing a prerecorded tape
16. how sense organs and muscles are related to stimulus and response
17. the experimental evidence used to show that penguins use innate behavior to find their nests
18. why a reward might be given when teaching a new behavior

Using Main Ideas

Answer the questions by referring to the page number after the question.

19. What experimental evidence shows that male silk moths use their antennae to detect pheromones? (p. 360)
20. What is the sequence of stimulus and response when a bird sees and catches a mouse? (p. 352)
21. What are two examples of parental care? (p. 365)
22. How do birds find their way while migrating at night? (p. 364)
23. What are three examples of courting behavior in animals? (p. 359)
24. Is courting behavior an instinct? Explain your answer. (p. 353)

Skill Review ✔

*For more help, refer to the **Skill Handbook,** pages 704-719.*

1. **Infer:** Why was pheromone used to attract the male moths in Figure 17-5?
2. **Infer:** How do you know that Adelie penguins use the sun to find the way?
3. **Form hypotheses:** You train goldfish to come to the top of a fish tank to eat when you turn on the light. The next time you feed them, you tap the tank instead of turning on the light. Hypothesize what the fish will do.
4. **Make and use tables:** Make a table with five examples each of human innate and learned behaviors.

Finding Out More

TECH PREP

Critical Thinking

1. Decide if the following human baby behaviors are innate or learned: crying, suckling, grasping, crawling, toilet training, yawning, putting on clothes. Explain your answers.
2. Male killer bees are attracted by a pheromone given off by a female. How might the number of bees be reduced?

Applications

1. Design a plan to teach a dog to fetch the newspaper. Include the following terms: stimulus, nervous system, brain, response, sense organs, reward, practice.
2. Make a list of animals that give parental care and those that don't. Find out how many offspring each produces. Relate amount of care to the number and survival rate of offspring.

CHAPTER PREVIEW

Chapter Content

Review this outline for Chapter 18 before you read the chapter.

Skills in this Chapter

The skills that you will use in this chapter are listed below.

- In **Labs 18-1** and **18-2,** you will observe, interpret data, and experiment.
- In the **Skill Checks,** you will sequence, understand science words, and make and use tables.
- In the **Mini Labs,** you will experiment, observe, and infer.

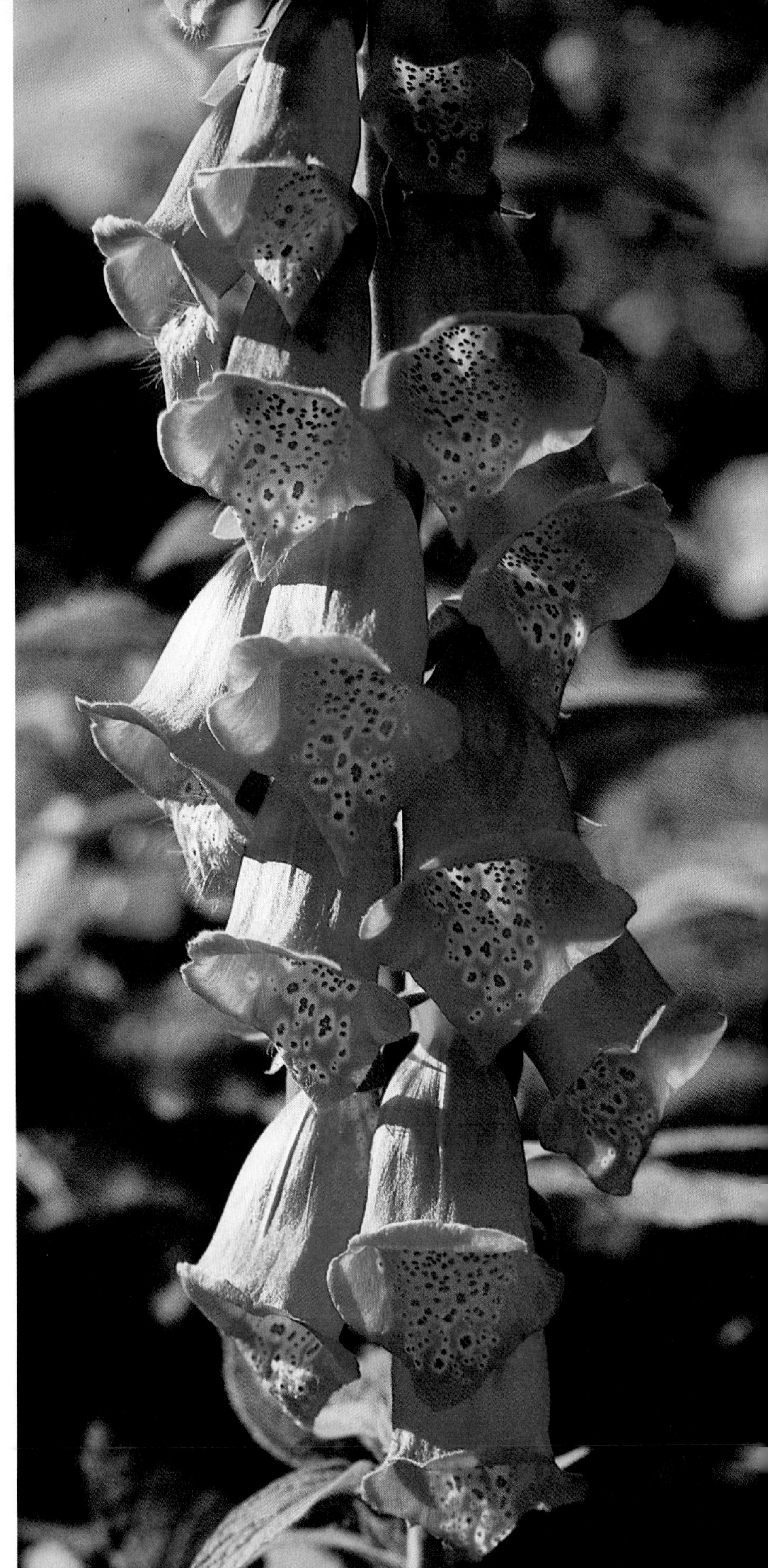

Chapter 18

Drugs and Behavior

The plant in the photo on the opposite page is foxglove, a member of the genus *Digitalis*. The small photo on this page shows several pills. What do the plant and the pills have in common? Foxglove leaves are used to produce a drug called digoxin. Digoxin is used to treat heart failure, a disease that is caused by damaged heart muscle. The digoxin pills in the small photo help to increase the force of the heart contractions so the weakened heart can pump blood to all parts of the body.

Try This!

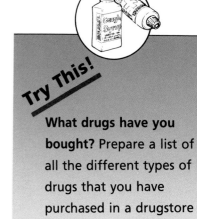

What drugs have you bought? Prepare a list of all the different types of drugs that you have purchased in a drugstore this past year. Describe why you bought each drug.

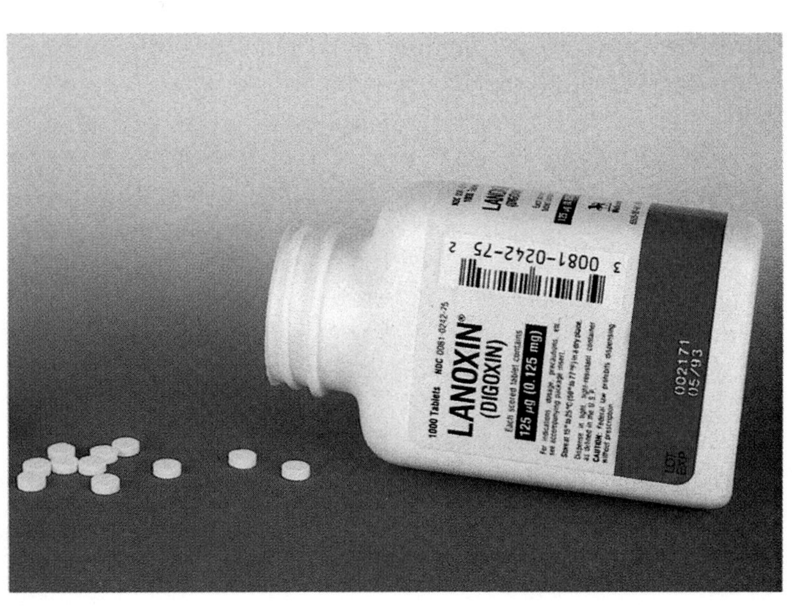

BIOLOGY Online

Visit the Glencoe Science Web site at science.glencoe.com to find links about **drugs and behavior.**

Digoxin tablets

18:1 An Introduction to Drugs

Objectives

1. **Compare** prescription and over-the-counter drugs.
2. **Relate** the importance of information listed on drug labels.
3. **Describe** how drugs stop pain.

Key Science Words

drug
symptom
controlled drug
prescription drug
over-the-counter drug
side effect
dosage
overdose

Where do most drugs come from? The answer may surprise you. Most drugs come from plants. Aspirin comes from the willow tree. Castor oil comes from the seeds of the castor bean. Marijuana comes from the hemp plant. Some drugs come from fungi. In Chapter 5 you learned that the drug penicillin comes from a fungus called *Penicillium.* Other drugs come from bacteria, and a few come from animals.

What Is a Drug?

What is a drug? A **drug** is a chemical that changes the way a living thing functions when it is taken into the body. Drugs are used for the treatment, cure, and prevention of disease. Drugs may be familiar chemicals such as those found in tea, coffee, or cigarettes. Did you brush your teeth today? If you did, you probably used a drug called fluoride in your toothpaste.

Why do people use drugs? Most drugs are used for medical reasons. These drugs are often called medicines. Doctors use certain drugs to kill organisms that cause diseases. Have you ever had strep throat? This disease is caused by bacteria. Doctors prescribe drugs like penicillin to cure strep throat.

People use other drugs to treat disease symptoms. **Symptoms** are changes that occur in your body as a result of a disease. You may know you have strep throat when your throat is so sore that it hurts to swallow. A sore throat is a symptom. Coughing and rashes are other examples of symptoms. These symptoms can be treated with cough medicines and ointments.

Most drugs are used to help people. If drugs are used incorrectly, however, they can harm people. For this reason, the manufacture and sale of many drugs that might be dangerous if used incorrectly are controlled by law. Drugs that are controlled by law are called **controlled drugs.** Marijuana, cocaine, and tranquilizers are all controlled drugs. Many people use controlled drugs illegally to change their behavior.

If you need a controlled drug for medical reasons, a doctor will direct you to take it, and you will be able to buy it or get it legally from the doctor. If you use a controlled drug without getting it from a doctor or drugstore, you are using the drug illegally.

Figure 18-1 Medicines, and the fluoride in toothpaste, are drugs.

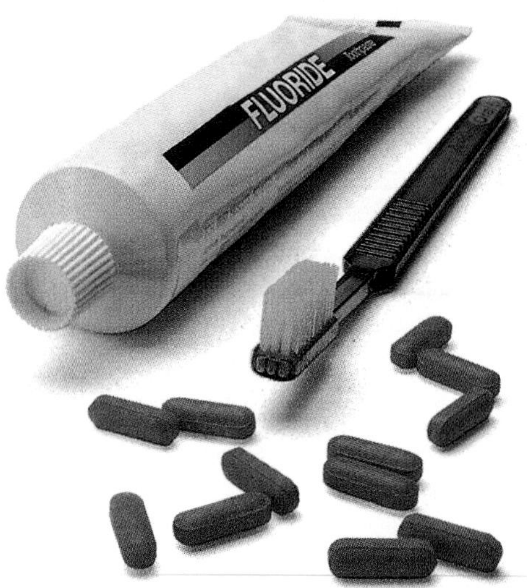

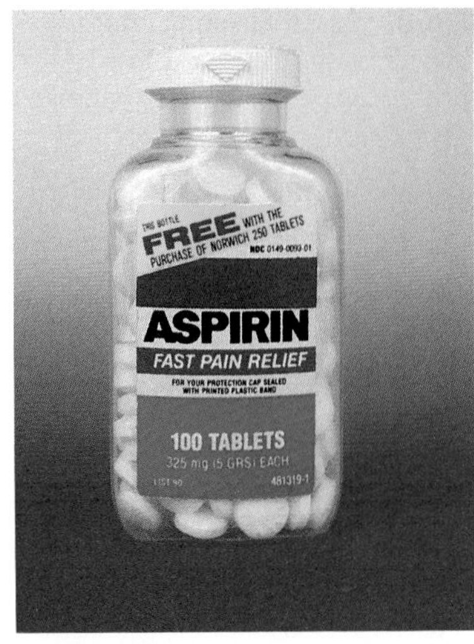

Figure 18-2 Of all the over-the-counter drugs, aspirin is bought most often.

Obtaining Drugs Legally

A legal drug is a drug used legally to treat a disease or its symptoms. There are two ways to obtain drugs legally. One way is with a prescription, and the other way is by buying them over the counter without a prescription.

What is a prescription drug?

A **prescription drug** is one that a doctor must tell you to take. The word *prescription* means direction. The doctor directs you to take a certain drug. Drugs obtained this way are legal.

Why do you need a doctor to tell you which drug to take? Do you know which drug will help your specific illness? Do you know how much of the drug to take? How would you know of any problems that could result from using the drug? You can see why a doctor must prescribe certain drugs for you. Some drugs are harmful if used incorrectly.

Over-the-counter drugs are those that you can buy legally without a prescription. They are also called nonprescription drugs. Figure 18-2 shows the huge selection of over-the-counter drugs in a supermarket. It includes everything from cough medicine to sunburn ointment to aspirin.

Mini Lab

Are Garlic and Onions Antibiotics?

Experiment: Apply bacteria to a petri dish of agar. Add paper disks treated with garlic or onion juice. Use a plain disk as a control. Observe growth after two days. What do you see? *For more help, refer to the **Skill Handbook,** pages 704-705.*

Reading Drug Labels

What can you learn from a drug label? Most drug labels tell you what the drug can be used for. The label also may have special instructions on what to do before using the drug.

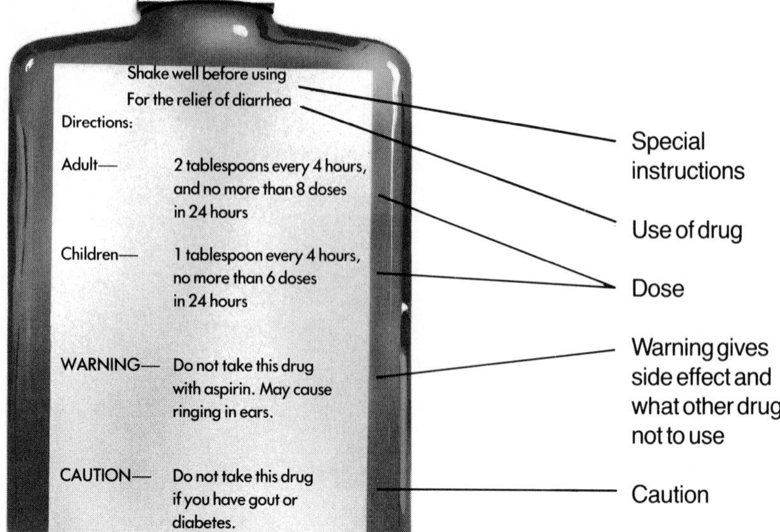
Shake well before using
For the relief of diarrhea
Directions:
Adult— 2 tablespoons every 4 hours, and no more than 8 doses in 24 hours
Children— 1 tablespoon every 4 hours, no more than 6 doses in 24 hours
WARNING— Do not take this drug with aspirin. May cause ringing in ears.
CAUTION— Do not take this drug if you have gout or diabetes.
Special instructions
Use of drug
Dose
Warning gives side effect and what other drug not to use
Caution

Figure 18-3 Warnings, special instructions, and side effects are present on all legal drug labels.

Mini Lab

Can a Drug Package Be Made Childproof?

Observe: Survey the methods used to make over-the-counter drugs tamperproof and childproof. Design a childproof package for a product that you now use. *For more help, refer to the* ***Skill Handbook,*** *pages 704-705.*

Look at the drug label in Figure 18-3. One of the most important parts of a label is the warning. Warnings tell about possible side effects. A **side effect** is a change other than the expected change caused by a drug. This change often is not wanted or desirable. What is the side effect that may take place with use of the drug in Figure 18-3?

Warnings also tell you which foods or other drugs must not be taken with the drug. Mixing drugs with certain other drugs or foods can be dangerous. Mixing drugs can cause an unexpected and unwanted change in the body. For example, some cold medicines help get rid of cold symptoms. If used with certain diet pills, a steep rise in blood pressure, confusion, and nervousness can result.

Drug labels also include cautions. A caution may tell you to keep the drug away from children.

What do drug labels tell you?

Another very important thing a drug label tells you is the dosage. The **dosage** is how much and how often to take a drug. "Take one pill every four hours" is an example of a drug dosage.

Age influences drug dosage. Very young children must receive smaller amounts of a drug than adults. It is important to make sure that a child is not given too much of a drug. Drug dosage also is related to body size. Most drugs taken into the body leave it quickly or are changed into a different form by the body. The kidney, lungs, and liver are three organs that help the body get rid of drugs. For a drug to do its job, a certain amount of it must be in the body for a certain length of time. There must be enough drug taken in to make up for the amount leaving. This amount varies with body size.

Lab 18–1 Aspirin

Problem: Which medicines contain aspirin?

Skills

observe, interpret data, experiment

Materials

4 glass slides
wax marking pencil
aspirin-testing chemical, in dropper bottle
6 test solutions of painkillers
8 droppers
water
aspirin solution

Procedure

1. Copy the data table.
2. Use a marking pencil to draw two circles on a glass slide. Label the circles 1 and 2.
3. Add one drop of water to circle 1. Use a clean dropper to add one drop of aspirin solution to circle 2.
4. Add one drop of aspirin-testing chemical to each circle. **CAUTION:** *Aspirin-testing chemical is poisonous and can burn the skin.*
5. **Observe:** Wait one minute. Note the color in each circle. A violet color means that aspirin is present. Record your results.
6. Draw two circles on each of three more glass slides. Label the circles A through F.
7. Record in your data table the name of each medicine being tested.
8. Add a drop of a different test solution to each circle. Use a clean dropper for each solution.
9. Add one drop of aspirin-testing chemical to each circle.
10. Wait one minute. Record the color that appears in each circle. Complete the last column of your data table.
11. Dispose of your solutions and wash your glassware.

Data and Observations

1. Which circles showed a violet color?
2. Which circles did not show a violet color?

CIRCLE	CONTENTS	COLOR WITH TEST CHEMICAL	ASPIRIN PRESENT?
1	water		
2	aspirin		
A			
B			
C			
D			
E			
F			

Analyze and Apply

1. What was the purpose of the slide labeled circle 1 and circle 2?
2. Where should aspirin be listed if it is present in a medicine?
3. Ask your teacher to show you the labels from the medicines tested. Which medicines contain aspirin? Does this agree with your data?
4. **Apply:** Why should aspirin be listed as an ingredient if it is present in a medicine?

Extension

Caffeine is often added to aspirin products because the aspirin works more rapidly if taken with caffeine. Check the labels of aspirin products available at a drugstore to see which products contain caffeine. **Make a table** of the products that contain both drugs and those that contain only aspirin.

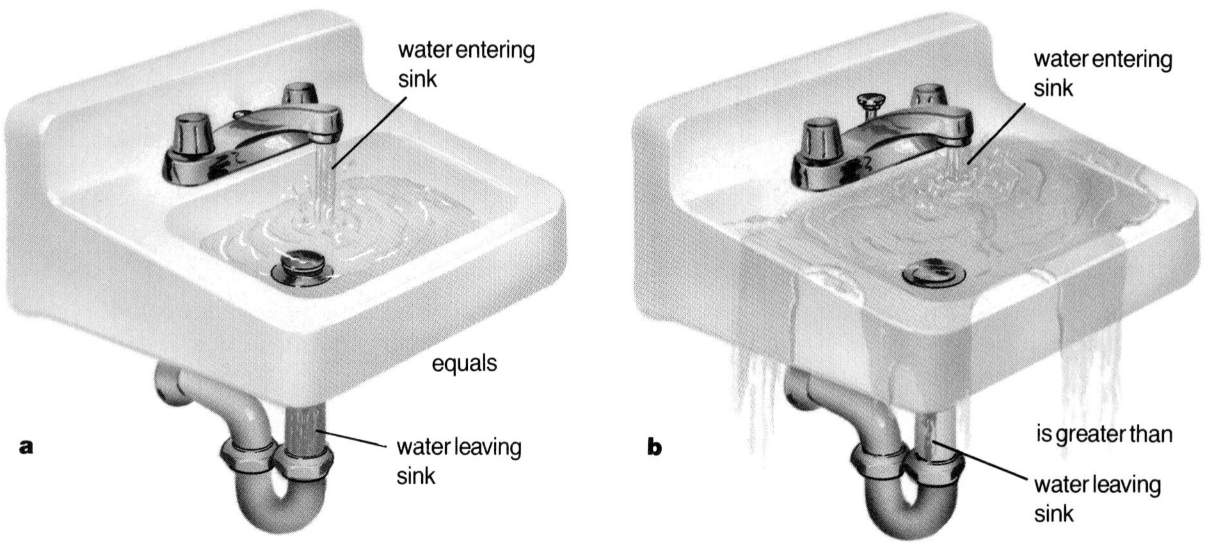

Figure 18-4 Normal drug dose (a) – the amount of drug entering the body equals the amount leaving the body. Overdose (b) – the amount of drug entering the body is greater than the amount leaving the body.

Figure 18-4a uses a leaking sink to help explain how the body balances the amount of drug entering and leaving it. Think of the water going into the sink as a drug entering the body. The leaking stopper stands for the organ removing the drug. If the water entering the sink is equal to the water leaving it, the water will not overflow. Nor will the sink become empty. The same type of thing happens with the proper drug dosage. The amount of drug entering the body equals the amount leaving it.

What happens if a person does not take the correct drug dose? Not taking enough of a drug may result in the drug not doing its job. Taking too much of a drug can result in a drug overdose. An **overdose** is the result of too much of a drug in the body. Let's look at the sink example again. Figure 18-4b shows what might lead to a drug overdose. Too much of a drug is added to the body. The body can't get rid of the drug fast enough, and a drug overdose results. Drug overdoses are dangerous because they prevent the body from working properly.

Bio Tip

Health: Do not store any medicine in the glove compartment of a car. High temperatures may alter the chemicals in the drug.

Speed of Drug Action

There are two ways that people commonly take drugs. These ways are swallowing or getting a shot. Suppose a pain comes from your toe. How can swallowing a pill or getting a shot stop that pain? Look at Figure 18-5 as you read the following steps.

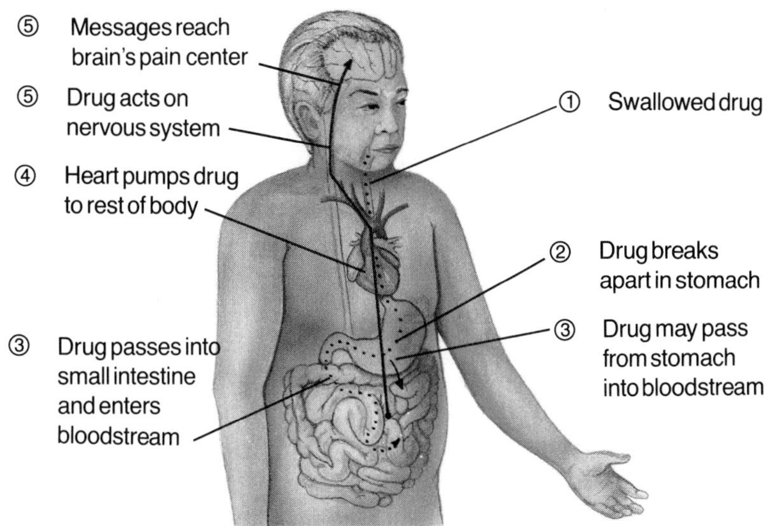
⑤ Messages reach brain's pain center
⑤ Drug acts on nervous system
④ Heart pumps drug to rest of body
③ Drug passes into small intestine and enters bloodstream
① Swallowed drug
② Drug breaks apart in stomach
③ Drug may pass from stomach into bloodstream

Figure 18-5 Path followed by a swallowed drug. What body system is responsible for pain relief?

1. A person swallows a drug.
2. The drug reaches the stomach and begins to break up.
3. Some of the drug may pass from the stomach directly into the bloodstream. The rest will continue into the small intestine and then into the bloodstream.
4. Once the drug is in the blood, the heart pumps it to all parts of the body.
5. The drug acts on the nervous system. The nervous system sends messages to the brain and shuts off the brain's pain center.

How does a painkiller stop pain after it is swallowed?

Drugs taken by mouth don't work as quickly as injected drugs. Injected drugs go directly into blood capillaries. They don't have to spend any time passing through the digestive system first.

Check Your Understanding

1. How is buying a prescription drug different from buying an over-the-counter drug?
2. Name four things that a drug label tells you.
3. Describe how a drug stops pain in the body.
4. **Critical Thinking:** If you are very sick, the doctor may give you a shot of antibiotic in the office and then give you pills to take later. Why do you think he or she gave you both a shot and pills? Why not just pills?
5. **Biology and Reading:** Look at the drug label in Figure 18-3. Why should this drug be taken? How often should the person take this drug?

18:2 How Drugs Affect Behavior

Objectives

4. **Describe** the effects of stimulants and depressants on the body.
5. **Discuss** how psychedelic drugs and inhalants affect the body.

Key Science Words

stimulant
depressant
psychedelic drug
inhalant

Three of the ways in which drugs affect the body are to speed up the body's activities, slow down the body's activities, or affect a person's senses or way of thinking. These changes in the body cause changes in behavior.

Stimulants

A **stimulant** (STIHM yuh lunt) is a drug that speeds up body activities that are controlled by the nervous system. Many stimulants are controlled drugs. Examples of stimulants that are controlled drugs are cocaine and amphetamines (am FEHT uh meenz). Caffeine and nicotine are also stimulants, but they are not controlled drugs.

How does a stimulant speed up the body's activities? To answer this question, we must look at nerve cells in the brain.

In Chapter 15 you learned how neurons carry messages. Chemical messengers given off by the axon end of one neuron move across the synapse and are picked up by the dendrite end of the next neuron. Usually these chemical messengers are destroyed after crossing the synapse. This prevents the original message from going on continuously. Figure 18-6 shows how this works.

How do stimulants change this pattern? They may cause the axon of a neuron to give off more of the chemical

Bio Tip

Consumer: An average of four prescription drugs per year, at a yearly cost of $22, are used by people age 16 or younger. Eighteen prescription drugs per year, at a yearly cost of $138, are used by those 65 or older.

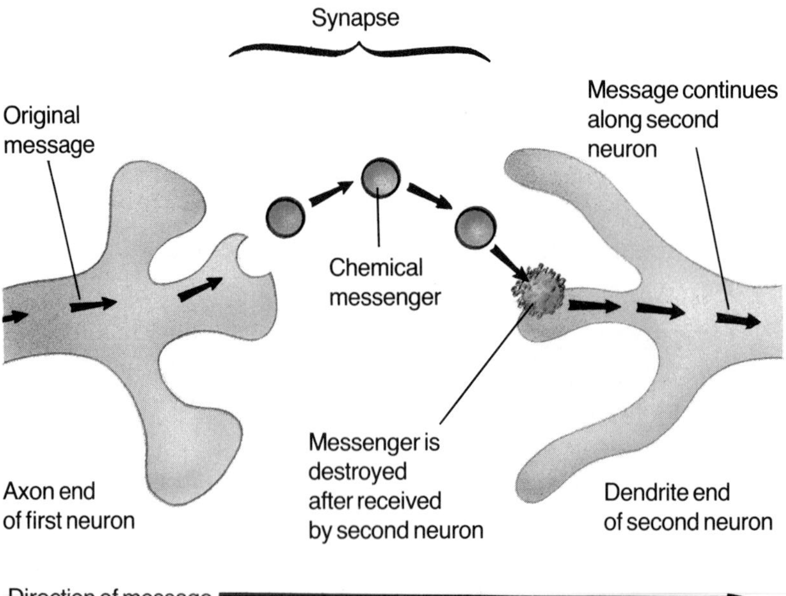

Figure 18-6 Messages move from one neuron to the next across a synapse using a chemical messenger.

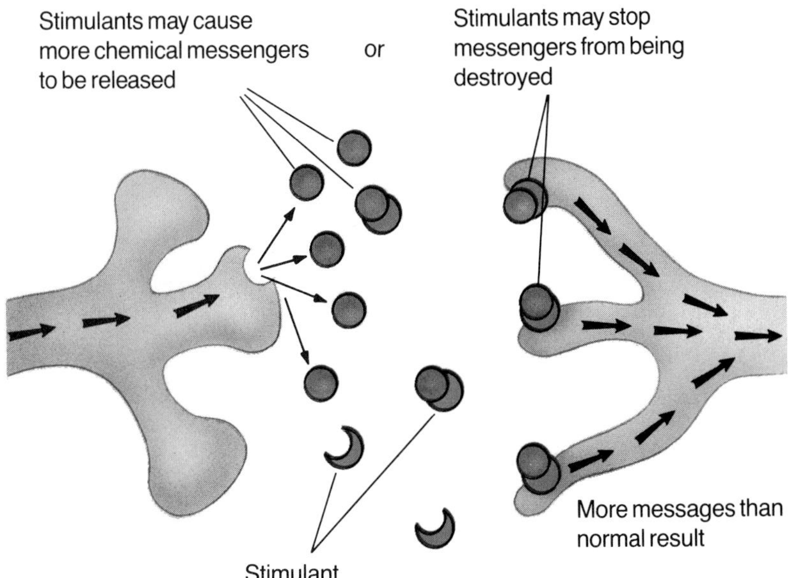
Stimulants may cause more chemical messengers to be released
or
Stimulants may stop messengers from being destroyed
Stimulant
More messages than normal result

Figure 18-7 There are two ways that stimulants can change the chemical messengers in a synapse.

messenger than normal. Or, stimulants may prevent the chemical messenger from being destroyed once it reaches the dendrite of the next neuron. In both cases, the second neuron keeps receiving chemical messengers. Figure 18-7 shows changes caused by stimulants. With stimulants, messages move from one neuron to the next for a longer time.

How does a stimulant affect the nervous system?

What are some effects of stimulants on the body? Stimulants speed up the heart rate. They increase blood pressure. There is a decrease in appetite. A person taking stimulants has a speeded-up feeling. He or she may feel very alert. Over time, these changes may harm the body because they use up the body's supply of nutrients and energy too quickly.

Depressants

The word *depress* means to push down. A **depressant** (dih PRES unt) is a drug that slows down messages in the nervous system. Depressants slow down the body the way brakes slow down a car. Depressants are controlled drugs.

The main roles of depressants are to calm behavior, reduce pain, and help people sleep. Many mild sleep aids are sold over the counter, even though they are controlled drugs. Codeine (KOH deen), morphine, and barbiturates (bahr BICH uh ruhts) are other depressants. These are legal drugs that may be prescribed by doctors.

Skill Check

Sequence: Draw the path that a message takes to go from one neuron to another. Then draw the same path to show what happens if a depressant is present. *For more help, refer to the* ***Skill Handbook,*** *pages 706-711.*

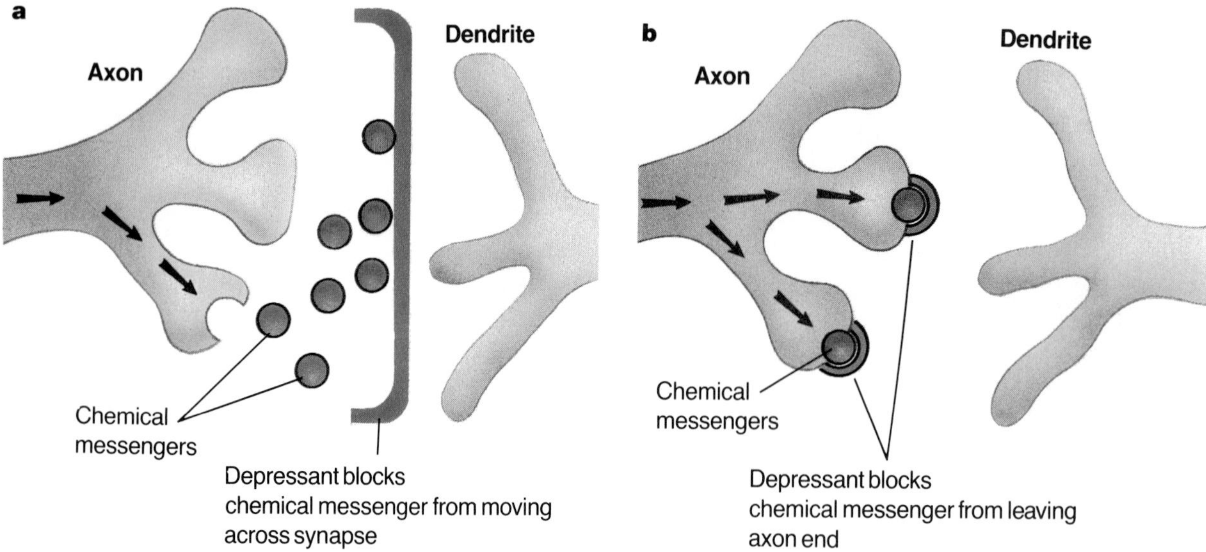

Figure 18-8 Depressants change chemical messengers at a synapse in two ways.

Let's see how a depressant stops pain. Suppose you had a bad toothache. Neurons carry the pain message to your brain. At the brain, the message enters the brain's pain center. Then you feel the discomfort called pain.

Painkillers work in your brain at the synapse, just as stimulants do. Certain depressants may block the movement of chemical messengers across the synapse. This is shown in Figure 18-8a. Some depressants block the axon of a neuron from giving off the chemical messenger. This is shown in Figure 18-8b. Painkillers don't change what causes the pain. They simply keep you from feeling pain by stopping the chemical messenger from doing its job.

Figure 18-9 The leaves, stems, flowers, and pollen of marijuana contain several drugs.

Psychedelic Drugs

A third group of drugs that change behavior is the psychedelic (si kuh DEL ihk) drugs. A **psychedelic drug** is one that alters the way the mind works and changes the signals we receive from our sense organs. Hearing, seeing, and thinking are changed. For example, someone may say that an object appears brighter than it really is. Sounds become louder. Senses blend together. Someone may report "tasting colors" or "seeing music."

There are two general groups of psychedelic drugs. One group is found in nature. Natural psychedelic drugs are those found in certain kinds of mushrooms, cactus plants, marijuana, and the leaves of other desert and jungle plants. The second group of psychedelic drugs is synthetic, or made by humans. Drugs in both groups are controlled.

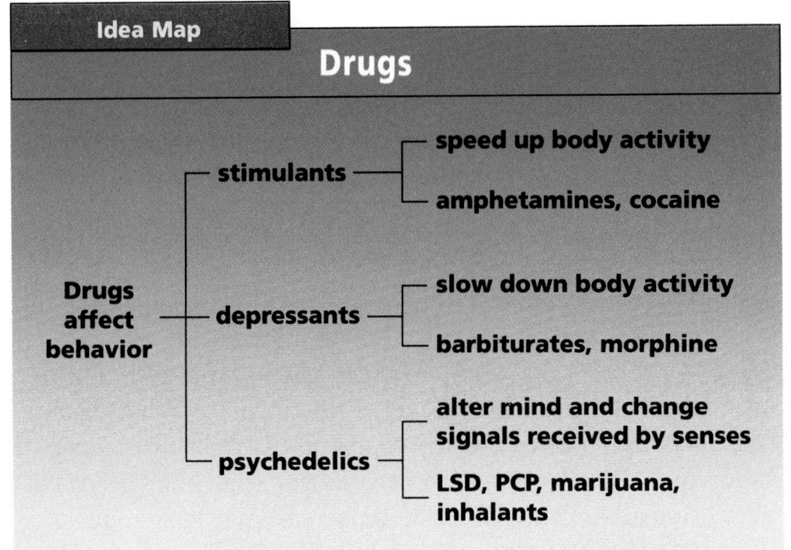

Study Tip: Use this idea map as a study guide to drugs that affect behavior. How many other examples can you give?

What problems can psychedelic drugs cause?

Synthetic drugs that alter the mind include PCP and LSD. PCP is also known as "angel dust." Use of PCP causes a variety of physical changes. These include high blood pressure, difficulties with walking or standing, and numbness. PCP also causes changes in behavior. Users of PCP may become very violent. They may also have a loss in memory or try to harm themselves.

Related to the psychedelics are inhalants. An **inhalant** is a drug breathed in through the lungs in order to cause a behavior change. The chemicals in glues, paints, and typewriter correction fluid are inhalants. Their use can cause irregular heartbeat and liver damage. The heart may stop beating, and death may occur.

Check Your Understanding

6. Describe the action of a stimulant on the nervous system.
7. What behavior changes are caused by psychedelic drugs?
8. What are the main roles of depressants in the body?
9. **Critical Thinking:** How do diet pills work at the nerve level? Are they stimulants or depressants?
10. **Biology and Reading:** According to what you learned in this section, what word means the opposite of depressant?

18:3 Uses of Over-the-Counter Drugs

Objectives

6. **Give the function** of antihistamines and cough suppressants.
7. **Explain** the role of antacids in the body.

Key Science Words

antihistamine
antacid

There are thousands of over-the-counter drugs available. They relieve minor symptoms of illnesses, but they do not cure diseases. How do they work?

Antihistamines

You probably have experienced the symptom of a stuffy nose when you have had a cold, Figure 18-10. What causes this? When you have a cold, your body is fighting a virus of some kind. Blood goes to the capillaries that line the nasal membranes. A clear fluid, the blood plasma, leaks out of the capillaries into the surrounding tissue. This plasma causes the tissues in your nose to swell. The swelling is what you feel when you have a stuffy nose.

Do you use a nasal spray or nose drops when your nose is stuffy? Nasal sprays and nose drops contain chemicals called antihistamines (ANT i HIHS tuh meenz). An **antihistamine** is a drug that reduces swelling of the tissues by stopping the leaking of blood plasma from capillaries. Most nasal sprays cause the leaking and swelling to stop. Then the stuffy feeling goes away. Antihistamines are also found in pills. They are used to reduce symptoms of hay fever, bee stings, and colds.

Skill Check

Understand science words: antihistamine. The word part *anti* means against. In your dictionary, find three words with the word part *anti* in them. *For more help, refer to the* ***Skill Handbook,*** *pages 706-711.*

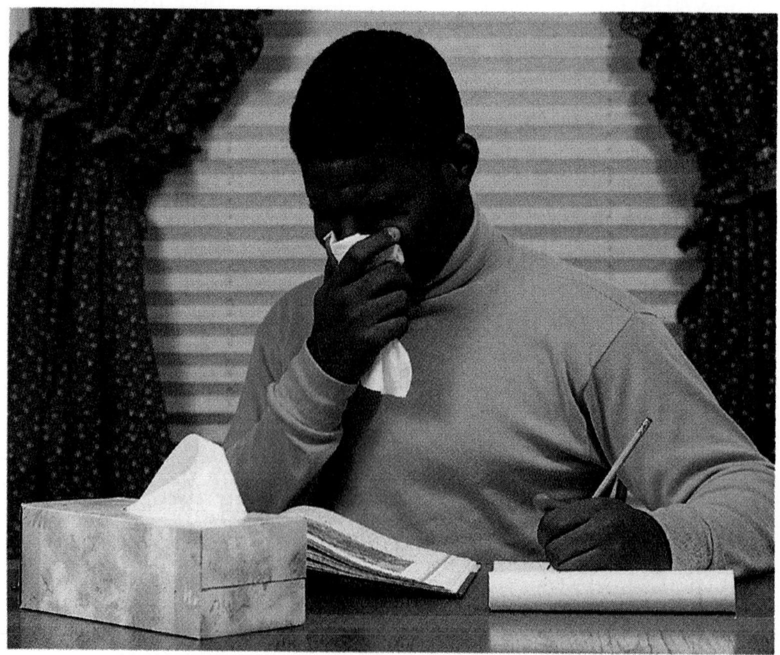

Figure 18-10 Antihistamines relieve the stuffy feeling of a cold. How do antihistamines work?

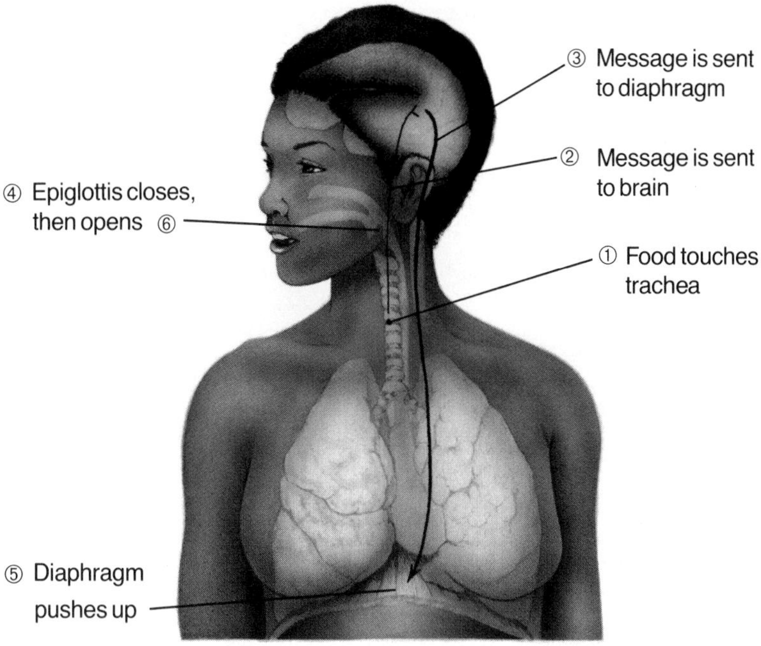

③ Message is sent to diaphragm

② Message is sent to brain

④ Epiglottis closes, then opens ⑥

① Food touches trachea

⑤ Diaphragm pushes up

Figure 18-11 How and why you cough

Cough Suppressants

Coughing is your body's way of getting rid of whatever blocks your windpipe or lungs. Figure 18-11 shows why you cough. Something, such as food, may be touching the sides of your trachea. Nerves in the trachea sense the food and send a message to the part of the brain that controls coughing. The brain then sends the message to your diaphragm and your epiglottis, the flap that covers the opening of your windpipe. The epiglottis closes. Your diaphragm pushes up into your chest. Pressure builds up within your chest and the epiglottis opens. A burst of air under high pressure is pushed from your lungs, and you cough.

What is the role of cough suppressants in coughing? Suppressants are medicines that slow down or suppress the part of the brain that controls coughing. Cough medicines usually contain depressants. These drugs work on the chemical messenger of the synapse in the same way as other depressants.

Mini Lab

What Determines the Price of a Drug?

Infer: Compare the price of a "name brand" drug with a less known brand. Determine the cost per dose for each. Name some factors that might affect the price. *For more help, refer to the* ***Skill Handbook****, pages 706-711.*

What kind of drug do cough suppressants contain?

Figure 18-12 Cough suppressants are available in many different forms.

+

Figure 18-13 Baking soda can change an acid to water and a salt.

Antacids

Sodium bicarbonate (bi KAR buh nate) is the chemical name for baking soda. Many people have it in their homes. It is used to make cakes rise during baking. It is also useful as a drug. Let's see how. The stomach normally makes acid for digestion. When the stomach makes too much acid, you might get heartburn. Baking soda gets rid of the extra acid by causing a chemical change. You see this change in Figure 18-13. In the figure, baking soda has changed vinegar, an acid, to water and a salt. It does the same thing to stomach acid. The chemical change releases carbon dioxide gas. You burp the gas. The pain goes away because water and salt do not irritate the stomach as the acid did.

What does an antacid do to an acid?

Drugs like baking soda are called antacids (ant AS udz). An **antacid** is a drug that changes acid into water and a salt. The word *antacid* means against acid.

Check Your Understanding

11. Describe the changes that take place in your nose when you use an antihistamine during a cold.
12. What type of drug is present in cough medicines?
13. How does an antacid help relieve heartburn?
14. **Critical Thinking:** Is coughing voluntary or involuntary? What nerve pathway is involved in coughing?
15. **Biology and Reading:** What class of drug that you read about in Section 18:2 is similar to the suppressants that you read about in this section?

18:4 Careless Drug Use

Most people use drugs in a wise and careful way. When drug use becomes careless, serious problems can result.

Objectives

8. List reasons for not using other people's drugs.

9. Discuss the problems caused by using caffeine, nicotine, cocaine, and alcohol.

Key Science Words

drug abuse
dependence
cocaine
caffeine
nicotine
ethyl alcohol

Drug Misuse and Abuse

"Not feeling well? Why don't you try some of my pills?" Have you ever heard someone say that? Should you use drugs that a doctor has told someone else to use? The answer is *no*. Using someone else's drugs is dangerous.

Why is using someone else's drugs dangerous? First, a doctor prescribed the drug for a health problem someone else had. You may not have the same problem.

Second, the dose prescribed for someone else may not be correct for you. Dosage varies with a person's weight and age.

Third, you may be allergic (uh LUR jik) to the drug. If you have a drug allergy (AL ur gee), you are sensitive to a particular drug. You could end up with a rash, itchy eyes, or a runny nose. Sometimes the results are much more serious. An allergic reaction to a drug can make you unable to breathe, or it might cause a drop in blood pressure. These kinds of allergic reactions may cause death.

What are three reasons for not taking another person's drugs?

Using a drug for a health purpose, but using it in the wrong way or amount, is called drug misuse. Using drugs when they are not needed at all is another problem. For example, cocaine and morphine are painkillers. They often end up being abused. **Drug abuse** is the incorrect or improper use of a drug. Taking a controlled drug illegally is drug abuse.

Drug abuse can lead to dependence. **Dependence** means needing a certain drug in order to carry out normal daily activities. Many drugs cause dependence. Morphine, heroin, codeine, alcohol, and even the drugs in coffee, tobacco, and some soft drinks are drugs that people may become dependent upon.

If a person becomes dependent on a drug and stops using it, that person will suffer from withdrawal. Withdrawal sickness causes loss of appetite, vomiting, stomach pains, and other symptoms. These symptoms continue until the person's nervous system has recovered from the effects of the drug. Withdrawal symptoms can be so severe that sudden withdrawal from certain drugs can cause death.

Figure 18-14 The flower of the opium poppy is the source of morphine and codeine.

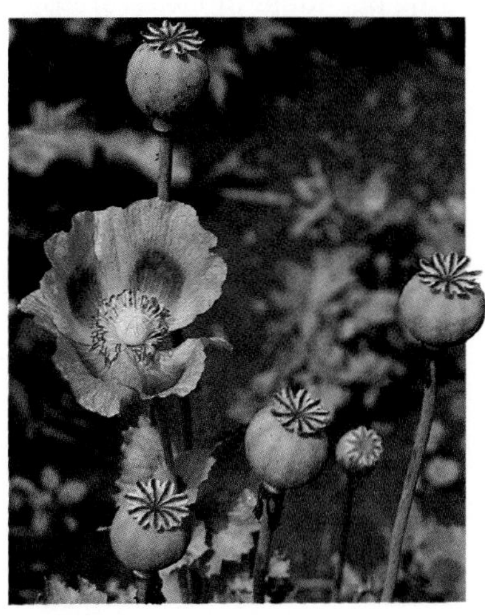

Cocaine

The leaves of a coca bush are used to make a drug called cocaine. **Cocaine** is a controlled drug used for its stimulant effects. Cocaine is a major cause of drug-related deaths in the United States. It may be injected with a needle, inhaled through the nose, or smoked. When inhaled, the drug enters capillaries in the nose and then goes through the circulatory system to the brain.

Upon first taking cocaine, a person may become more active. Blood pressure rises, and breathing and heart rate speed up. If too much of the drug is taken, the reverse may occur. The activity of the medulla slows down greatly. Remember from Section 15:2 that the medulla controls involuntary actions, such as breathing and heart rate. The slowing of the activity of the medulla can cause breathing or heart rate to slow or stop suddenly. It is not known how large a dose is needed for this to happen, and the drug acts differently in different people.

What is crack?

Crack is a very strong form of cocaine. Crack is made from cocaine that has been formed into a paste and then allowed to dry and harden. When smoked, the drug enters the blood by way of the lungs. In this way, it acts on the body in a matter of seconds. Use of crack can cause heart attacks and lung damage. Crack is so dangerous that it can cause death, even after the first use.

Crank, also called "ice," is another drug that is smoked. Crank is even more dangerous than crack because it causes quicker dependence than crack. Crank often causes death.

Figure 18-15 Police often find large quantities of illegal drugs (a) when they make a raid (b).

a

b

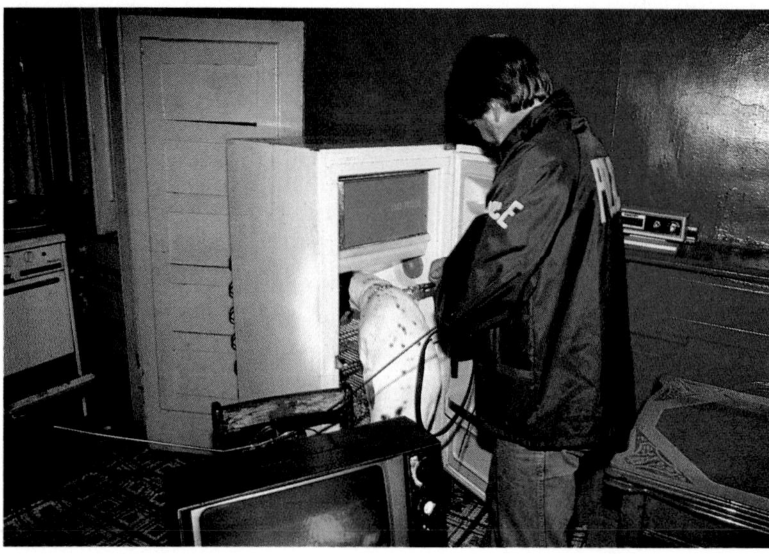

Figure 18-16 Coffee beans are the seeds of the coffee tree and the source of the stimulant caffeine. People in the United States drink about 400 million cups of coffee each day.

Caffeine and Nicotine

Caffeine is a drug found in coffee and tea. It also is found in cocoa, chocolate, and some soft drinks. Some over-the-counter drugs, such as cold remedies and diet pills, have caffeine.

How does caffeine affect a person? You can almost guess by what people say about coffee. "It helps me wake up." "It gives me a lift." Caffeine is a stimulant.

How much caffeine is found in certain foods? Table 18-1 shows you.

TABLE 18–1. CAFFEINE CONTENT OF SOME FOODS

Food or drug	Milligrams of caffeine
cup of coffee	80
cup of tea	40
regular cola soft drink, 1 can	40
chocolate bar	10
chocolate chip cookie	4

How much caffeine causes a change in behavior? As little as 100 milligrams may cause a change. A person may feel more awake and alert. He or she will notice that the heart rate speeds up. Caffeine may make it hard to sleep. Caffeine also causes dependence.

Nicotine is a stimulant found in tobacco. Nicotine speeds up the heart and increases blood pressure. Cigarettes, cigars, snuff, and chewing tobacco have nicotine.

Bio Tip

Health: In the United States, there are 359 000 tobacco-related deaths a year. This is equal to 1000 deaths each day.

Figure 18-17 Tobacco products all contain nicotine.

What problems are caused by using tobacco? First, the nicotine in tobacco causes dependence. Second, the other chemicals in tobacco cause lung, throat, and mouth cancer. Third, smoking is strongly linked with heart disease. Warning labels appear on all cigarette packages. Fourth, second-hand smoke can harm people who are not smoking but who breathe other people's smoke. Tobacco smoke contains poisonous gases.

Alcohol

Beer, wine, and whiskey contain a drug called ethyl (ETH ul) alcohol. **Ethyl alcohol** is a drug formed from sugars by yeast and is found in all alcoholic drinks. Some over-the-counter drugs, such as cough medicines, contain ethyl alcohol. There are other types of alcohol, such as rubbing alcohol, but they are not used in drugs because they are poisonous.

Figure 18-18 shows the path of swallowed alcohol in the body. Alcohol enters the blood from the stomach. Once in the blood, alcohol acts on the nervous system.

Many people think that alcohol is a stimulant because it speeds up heart and breathing rates. Alcohol is actually a depressant because it slows down the nervous system, leading to dangerous changes.

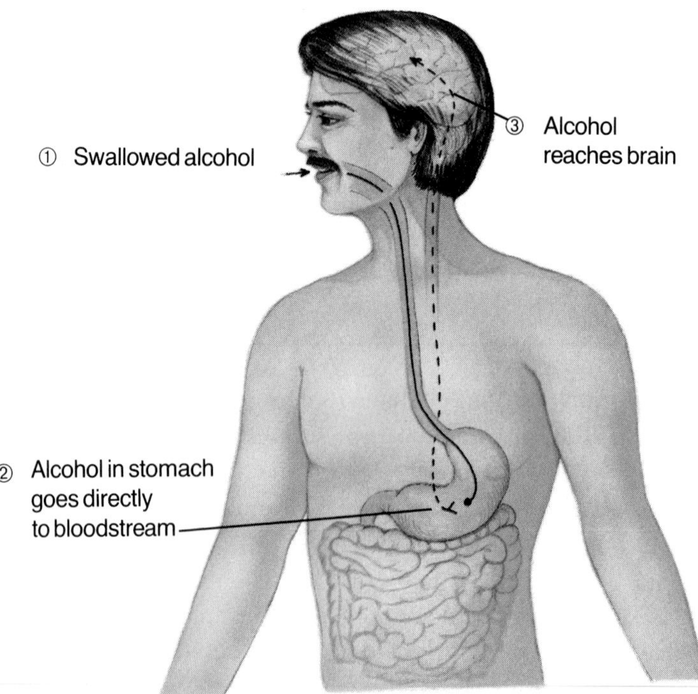
① Swallowed alcohol
③ Alcohol reaches brain
② Alcohol in stomach goes directly to bloodstream
Figure 18-18 Swallowed alcohol goes quickly into the bloodstream and reaches the brain. What kind of drug is alcohol?

Lab 18–2 Alcohol

Problem: Is alcohol present in over-the-counter drugs and household items?

Skills

observe, interpret data, experiment

Materials

4 glass slides
wax marking pencil
water
alcohol
8 droppers
alcohol-testing chemical, in dropper bottle
6 test solutions of over-the-counter drugs and household items

Procedure

1. Copy the data table.
2. Use a marking pencil to draw two circles on a slide. Label the circles 1 and 2.
3. Add one drop of water to circle 1. Use a clean dropper to add one drop of alcohol to circle 2.
4. Add one drop of alcohol-testing chemical to each circle. **CAUTION:** *Rinse immediately with water if alcohol-testing chemical is spilled on skin or clothing.*
5. **Observe:** Wait five minutes. Look for a color change. A yellow to orange color means alcohol *is not present.* A green, deep green, or blue color means that alcohol *is present.* Record your results.
6. Draw two circles on each of three glass slides. Label the circles A through F.
7. Record in your data table the name of each drug or household item being tested.
8. Add a drop of a different test solution to each circle. Use a clean dropper for each.
9. Add one drop of alcohol-testing chemical to each circle.
10. Wait five minutes. Record any color changes. Complete the last column of your data table.
11. Dispose of your solutions and wash your glassware.

CIRCLE	CONTENTS	COLOR WITH TEST CHEMICAL	ALCOHOL PRESENT?
1	water		
2	alcohol		
A			
B			
C			
D			
E			
F			

Data and Observations

1. Which products showed a color change to blue or green?
2. Which products did not show a color change to blue or green?

Analyze and Apply

1. What was the purpose of circles 1 and 2?
2. Ask your teacher to show you the labels from the products tested. Do your data agree with the labels?
3. Did all the products that contained alcohol also show alcohol listed on their labels?
4. What does your answer to question 3 tell you about the accuracy of these labels?
5. **Apply:** Why is it important that alcohol be listed if it is present in a product?

Extension

There are many different kinds of alcohol (rubbing, grain, wood, etc.). Will the alcohol-testing chemical detect all types of alcohol? **Design an experiment** to test this.

Figure 18-19 About one half of all car accident deaths involve the use of alcohol.

Table 18-2 shows what happens to the body as the amount of alcohol in the blood increases. The letters *BAC* mean *b*lood *a*lcohol *c*oncentration. The BAC is a way of describing the amount of alcohol in 100 mL of blood.

TABLE 18–2. EFFECTS OF ALCOHOL ON THE BODY	
% BAC	**Effect**
0.01 – 0.05	heart and breathing rates increase, judgment decreases
0.06 – 0.10	alertness, coordination, and judgment decrease
0.11 – 0.15	reaction time is slower, speech is slurred, balance is upset
0.16 – 0.29	frequent staggering or falling, loss of awareness
0.30 – 0.39	stupor, as if under an anesthetic
0.40 and up	unconsciousness, breathing stops, death

Skill Check

Make and use tables: Alcohol is a depressant. How does Table 18-2 show that drinking alcohol slows down body activities? *For more help, refer to the **Skill Handbook,** pages 715-717.*

Another problem is the effect of alcohol on unborn babies. Alcohol passes from mother to baby during pregnancy. The baby can be born with very serious health problems such as fetal alcohol syndrome. These babies are smaller than normal, may be mentally retarded, and have heart defects. Many are born dependent on alcohol.

Check Your Understanding

16. Give three reasons a person should not use someone else's drugs.
17. Describe problems caused by using cocaine and crack.
18. Name four foods that contain caffeine.
19. **Critical Thinking:** Think of a way that might help people stop abusing alcohol or tobacco.
20. **Biology and Reading:** From what you learned in this section, what do you believe was the author's purpose for writing about drugs?

Science and Society

Steroids

Testes in the male have two jobs. Not only do they make sperm cells, but they also make a hormone called testosterone (teh STAHS tuh rohn). Testosterone causes the male voice to deepen, hair to grow on the face, and muscles to increase in size. All of this is normal when it occurs at puberty. It is now possible to increase the amount of this hormone in the body by taking testosterone-like drugs. These drugs are called anabolic steroids (an uh BAHL ik • STIHR oydz). They cause a person's muscles to become larger and stronger, and they increase endurance. These changes occur faster than through normal exercise. Some athletes see these drugs as helpful.

What Do You Think?

1. Athletes who use steroids are finding that the drugs can make them the strongest, fastest, or biggest person around. Many of these athletes train for years, and sports are their lives. Taking steroids is as much a part of their training routine as lifting weights or running. Taking the drugs is helping them to win medals. Many athletes, however, have been kicked off teams or stripped of their medals when it was discovered they were taking steroids. Should an athlete be allowed to use drugs and then compete? Would it be fair to you if you were in a contest with a person who was using anabolic steroids?

2. Many sports organizations, high schools, and colleges test all participating athletes for the presence of drugs in their urine. If drugs are found, the athlete is usually suspended for a length of time and prevented from competing. These drug tests are required of all professional athletes. The United States Constitution guarantees all citizens the right to privacy. Do you think these drug tests violate the right to privacy? What would happen if an athlete refused to be tested?

3. Steroids cause major health problems. Their use causes liver damage, heart disease, high blood pressure, and sterility. They can cause baldness, growth of breasts, and personality changes in males. Females may develop facial hair, a deep voice, stopping of the menstrual cycle, or a decrease in breast size. Many of these changes are not reversed when the athlete stops taking the drug. Who will pay the athletes' medical bills?

Conclusion: Should laws be passed that prevent or limit the use of steroids by athletes? Is it a government's business what a person does with his or her body?

Bodybuilders may be tested for drugs before they can compete.

Chapter 18

Review

Summary

18:1 An Introduction to Drugs

1. Prescription drugs are obtained through a doctor. Over-the-counter drugs are obtained without a prescription.
2. Drug labels give the use, dosage, and warnings against side effects.
3. Drugs that are injected work faster in the body than those that are swallowed.

18:2 How Drugs Affect Behavior

4. Stimulants increase the amount of chemical messenger given off into a synapse. Depressants keep messages from moving across a synapse.
5. Psychedelic drugs and inhalants alter the mind by changing messages received by sense organs.

18:3 Uses of Over-the-Counter Drugs

6. Antihistamines stop plasma loss from capillaries. Cough suppressants slow the brain's cough center.
7. Antacids change acid into water and a salt.

18:4 Careless Drug Use

8. Using someone else's drugs can be dangerous.
9. Cocaine is a controlled stimulant. Caffeine and nicotine are both stimulants. Alcohol is a depressant. All of these drugs cause dependence.

Key Science Words

antacid (p. 382)
antihistamine (p. 380)
caffeine (p. 385)
cocaine (p. 384)
controlled drug (p. 370)
dependence (p. 383)
depressant (p. 377)
dosage (p. 372)
drug (p. 370)
drug abuse (p. 383)
ethyl alcohol (p. 386)
inhalant (p. 379)
nicotine (p. 385)
overdose (p. 374)
over-the-counter drug (p. 371)
prescription drug (p. 371)
psychedelic drug (p. 378)
side effect (p. 372)
stimulant (p. 376)
symptom (p. 370)

Testing Yourself

Using Words

Choose the word from the list of Key Science Words that best fits the definition.

1. using drugs incorrectly or improperly
2. drug bought legally without a prescription
3. unexpected change while using a drug
4. having too much of a drug in the body
5. how much and how often a drug should be used
6. drug whose manufacture and sale are controlled by law
7. drug that speeds up the nervous system
8. drug found in coffee or tea

Review

Testing Yourself *continued*

9. drug that slows the nervous system
10. drug that a doctor must prescribe

Finding Main Ideas

List the page number where each main idea below is found. Then, explain each main idea.

11. how baking soda works in the stomach
12. how an antihistamine works
13. problems that result as blood alcohol concentration increases
14. the difference between prescription and over-the-counter drugs
15. changes that a psychedelic drug, such as PCP, may cause
16. the path that a swallowed drug takes to reach the brain's pain center
17. what causes withdrawal sickness
18. how the body gets rid of drugs
19. how a stimulant speeds up the body

Using Main Ideas

Answer these questions by referring to the page number after each question.

20. What happens when a person has an allergy to a drug? (p. 383)
21. What two factors are used to determine a drug dose? (p. 372)
22. Is each of the following drugs a stimulant or a depressant? caffeine, nicotine, morphine, cocaine, amphetamine (pp. 376-377)
23. What is dependence and what drugs can you become dependent on? (p. 383)
24. Why and how do you cough? (p. 381)
25. How do each of the following drugs work? antacids (p. 382); antihistamines (p. 380); cough suppressants (p. 381)
26. What is a controlled drug and why must it be controlled? (p. 370)

Skill Review ✓

*For more help, refer to the **Skill Handbook**, pages 704-719.*

1. **Make and use tables:** Using Table 18-1, estimate the average amount of caffeine you consume in a day. How can you decrease this amount?
2. **Infer:** Use the sink and running water idea to infer what happens when a child takes an adult dose of a drug rather than the correct child dose.
3. **Interpret data:** If you do the Mini Lab on page 371, how do you know that the paper disk itself does not prevent bacteria from growing?
4. **Sequence:** List the steps that lead to a cough. What step is blocked by a cough suppressant?

Finding Out More

Critical Thinking

1. Drivers suspected of drinking while intoxicated blow up a balloon, and the police test the air in the balloon for alcohol. How does alcohol get into the suspect's breath?
2. Explain why urine can be tested to determine if a person has used drugs.

Applications

1. Examine the labels of two over-the-counter drugs. List and compare the following for each drug: use, dosage, how often it should be used, when to use, warnings.
2. Make a report on fetal alcohol syndrome or the effect of crack on an unborn baby.

Biology in Your World

Appreciating Our Senses

In this unit, you learned how body systems are controlled. Our bodies can receive information from outside our bodies. Our ears, eyes, nose, mouth, and skin collect this outside information. Factors such as disease and injury can cause damage to any of our sense organs.

LITERATURE

A Silent and Dark World?

In 1881, illness left 18-month-old Helen Keller unable to see, hear, or talk. Most people thought she would not be able to learn. For the next six years, Keller lived in a world she described as a "no-world."

When Keller was 7 years old, a young teacher named Anne Sullivan came to live with her. Sullivan opened up the world for Keller. Using the manual alphabet and braille, Keller was able to communicate. In the manual alphabet, finger positions stand for letters. Braille is a method of reading and writing using raised dots.

At age 22, Keller wrote *The Story of My Life.* Her writings have helped educators design training programs for blind and deaf people. They have also inspired many blind and deaf people to overcome their handicaps.

ART

Optical Painting

Must an artist mix paints to get certain shades? No, mixing of colors can take place in the brain. George Seurat (suh ROH) was a French painter who wanted his paintings to show the colors of nature in a scientific way. He covered his canvases with tiny dots of pure color. Because the dots are so small and regular, the eye does the mixing. This technique, called pointillism, is seen in Seurat's 1885 painting, *Le Bec du Hoc, Grandcamp.* Look closely at the colors to see for yourself.

LEISURE

World Without Sound

All during your waking hours, and sometimes in your sleep, sounds surround you. How much do you depend on sound? How would a hearing loss affect your activities? To find out, wear earplugs while at home some evening or on a Saturday. Be sure to insert the earplugs carefully. Do not participate in an activity that would be unsafe without your hearing. As you participate in an indoor hobby, sport, or relaxation activity, make a list of the ways your "hearing loss" affects the activity. You may find that you can't do some of your usual activities. You may find that you don't enjoy the activity as much with reduced hearing. Did you also find that you could do some things better?

You may notice that your "hearing loss" affects other people. Did you have to turn the television volume up so high that it became too loud for others? Did others have to speak more loudly to get your attention?

CONSUMER

Protect Your Sight

With many types of sunglasses available, you can be sure to make a fashion statement. But you also need to make choices when it comes to the level of protection sunglasses offer. It is important to protect your eyes from the effects of harmful ultraviolet (UV) radiation. Scientists think that too much UV radiation can lead to a clouding of the lens of the eye.

Sunglasses do not have to be expensive to be effective. Some inexpensive sunglasses provide good UV protection. Several grades of UV blockage are available. Look for a stamp on the lens that tells you the grade.

The next time you consider buying sunglasses, remember that your eyes are worth protecting—you only have one pair.

Unit 5

CONTENTS

Plant Systems and Functions

What would happen if...

there were no forests? Without forests there would be more land available for farming. Then, perhaps enough food could be produced to feed the world's population. Also, more land would be usable for housing, factories, and other types of development.

Without forests, though, there would be no shelter or food for many animals. There would not be enough lumber and other wood products for our needs. Plants use carbon dioxide and give off oxygen during photosynthesis. Without forests, the amount of carbon dioxide might increase and there might not be enough oxygen.

Can we keep our forests and still meet our food and space needs? Scientists are looking for ways to better manage our resources so we can meet our needs without upsetting the environment.

CHAPTER PREVIEW

Chapter Content

Review this outline for Chapter 19 before you read the chapter.

Skills in this Chapter

The skills that you will use in this chapter are listed below.

- In **Lab 19-1,** you will measure in SI, interpret diagrams, and classify. In **Lab 19-2,** you will observe, infer, and experiment.
- In the **Skill Checks,** you will understand science words and infer.
- In the **Mini Labs,** you will observe and experiment.

Chapter 19

The Importance of Leaves

Have you ever wondered where your energy to run, cycle, or play sports comes from? How were you able to grow from the size of a book bag as a baby to the size you are today? The answer to both of these questions is food. The baker in the photo to the left and the leaves in the photo below are both making food. Energy is needed to bake the bread and for the leaves to make food. Where does the energy come from to bake the bread? Where do the leaves get the energy to make food?

The baker is making food for himself and other humans. The leaves are making food for the leaf cells and other plant parts. Why do you need food? Why do plants need food? They don't move about as you do, but they do grow.

Try This!

How do leaves vary? Search around your area for the largest and the smallest leaf. Compare the colors, shapes, and sizes of the two leaves.

BIOLOGY Online

Visit the Glencoe Science Web site at science.glencoe.com to find links about **the importance of leaves.**

19:1 The Structure of Leaves

Objectives

1. **Examine and compare** the parts of a leaf in different plants.
2. **Give the function** of the cells in a leaf.
3. **Explain** the process of transpiration.

Key Science Words

blade
midrib
epidermis
palisade layer
spongy layer
stoma
guard cell
transpiration
wilting

Look around your neighborhood for plants. Apart from flowers, the plant parts that most catch your eye are probably leaves. Leaves are usually green but may show other colors. A plant often can be named just from looking at one of its leaves. Each species has a distinct set of leaf traits.

Leaf Traits

Most leaves are flat and green. The thin, flat part of a leaf is the **blade.** The blades of leaves can be round, heart-shaped, long and narrow, or short and broad. There are leaf blades that don't even look like leaves. The leaf blades of pine trees look like bundles of needles. The leaf blades of cedar trees look like the scales on a fish. Some leaves are adapted to one type of environment. The fleshy leaves of a jade plant are adapted for life in a dry, hot desert. The leaf blade is the main part of a plant that makes food.

A thin stalk usually joins the leaf to the stem. Stalks can be many different lengths and thicknesses. The part of the celery plant that you eat is the leaf stalk. It may be four to eight centimeters wide at the bottom. Many leaf stalks are no larger around than your pencil lead. Some plants, such as corn and grass, have leaves that are not attached to the stem by stalks. The leaf blades are joined directly to the stem. Figure 19-1 shows the stalk and the blade of a leaf.

Remember from Chapter 6 that there are two kinds of cells in the vascular tissue of plants. The stalk of a leaf contains veins with phloem and xylem vessels. These

Bio Tip

Health: Humans get poison ivy rashes when their skin comes in contact with the leaves of the poison ivy plant. You can identify poison ivy by its leaves that have three leaflets.

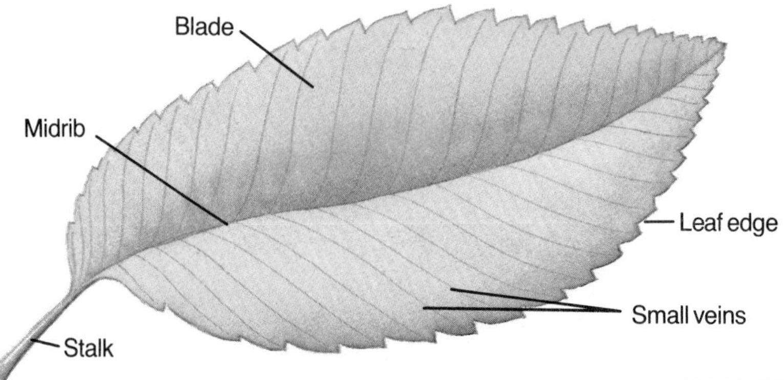

Figure 19-1 A leaf has two main parts—a stalk and a blade.

a

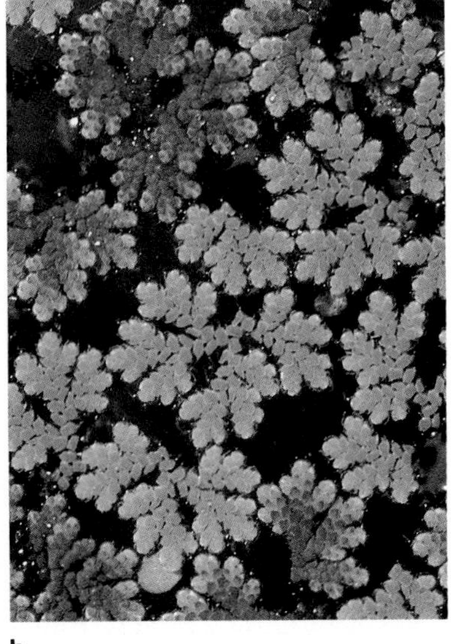

b

Figure 19-2 A royal water lily leaf (a) can be large enough to support a baby, and a water fern (b) has leaves smaller than your fingernail.

vessels, just like the blood vessels in your body, transport nutrients around the tissues of the plant. In the leaf stalk, vessels carry food and water between stem and leaf.

The stalk of a leaf appears to run into the blade and continues to the tip of the leaf. This part of the leaf blade is the midrib. The **midrib** is the main vein of the leaf. Notice in Figure 19-1 the smaller veins branching from the midrib. The midrib and smaller veins are made of phloem and xylem vessels that carry food and water in the leaf. The midribs and veins form different patterns in different leaves.

What leaf traits do scientists use to identify plants?

Scientists use leaf traits to identify and group plants. They use leaf shape, kind of edge, vein pattern, and the arrangement on the stem. It would be difficult to show all the different sizes and shapes of leaves. Look at Figure 19-2. Compare the size of the royal water lily leaf from South America with that of a water fern leaf from North America. Some tropical palm trees have leaves that are as long as a three-story building is high!

How would you describe the shape of grass leaves? Bladelike? Think of some other plants that have leaves shaped like grass leaves. Look at the different leaf shapes in Figure 19-3. Which are shaped like grass leaves?

The leaves in Figure 19-3 also show different kinds of edges. Leaves may have smooth edges or a toothed edge. There may be hundreds of teeth along the edge of a leaf. Notice also in Figure 19-3 that the edges of some leaves curve in and out again to form "fingers" or lobes. The numbers and shapes of lobes are used to identify some plants.

Figure 19-3 Leaves have many shapes and vein patterns. Which of these leaves have parallel veins?

Leaves

Lab 19–1

Problem: What are some traits of leaves?

Skills

classify, interpret diagrams, measure in SI

Materials

15 leaves of different plants
metric ruler

Procedure

1. Copy the data table and make a row for each one of your 15 leaves.
2. **Measure in SI:** Choose a leaf. Measure the length of the leaf blade in centimeters from the tip of the leaf blade to its base. For help, see Figure A.
3. Compare the leaf with the shapes given in Figure B. Record the leaf shape.
4. Compare the leaf edge with those in Figure C. Record the leaf edge.

TRAITS OF LEAVES				
Leaf	Blade Length (cm)	Shape	Edge	Veins
1				
2				

5. Compare and record the vein pattern on your leaf with the vein patterns in Figure D.
6. Repeat steps 2 through 5 with the other 14 leaves.

Data and Observations

1. Which leaves were the longest?
2. How many of your leaves had parallel veins? How many had palm veins?

Analyze and Apply

1. Which type of leaf edge is most common in your leaf samples?
2. These are the traits of two leaves: (a) oval, seven pointed lobes, small toothed edge, feather veins; (b) star shape, small teeth, palm veins. Are these two leaves the same species? Explain your answer.
3. **Apply:** How are leaf traits important?

Extension

Classify the 15 leaves by using a field guide.

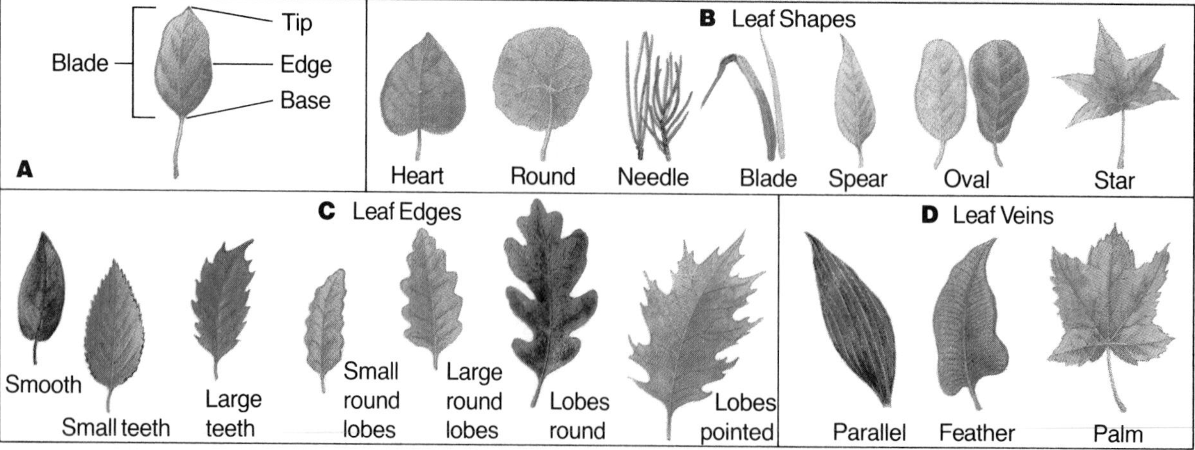

Leaves are arranged in different ways on the stems. Figure 19-4 shows three main leaf arrangements along a stem. Some plants have pairs of leaves opposite each other along the stem. Some plants have one leaf at each point along the stem. They alternate from one side to the other up the stem. In some plants, there are three or more leaves attached around one point on the stem to form a circle or whorled pattern. No matter how they are arranged, the leaves seldom overlap. Each leaf is arranged on the stem to catch the most sunlight it can.

Figure 19-4 Leaves are arranged in an opposite, alternate, or whorled pattern on the stem.

Cells of the Leaf

If you cut a leaf blade in half, you would see its cells. The cells inside a leaf blade are arranged in layers as shown in Figure 19-5. Most leaves are covered with a waxy layer. The waxy layer protects the leaf from water loss and from feeding insects. This layer is not made of cells. Did you ever notice how water forms beads on a newly waxed car? This layer of wax forms a protective coat for the paint on a car. Water cannot pass through the waxy layer on a car, nor can water leak through the waxy layer of a plant. The layer of wax on leaves makes them appear shiny, just as the wax makes a car shiny.

Beneath the waxy layer of a leaf is a layer of cells called the epidermis (ep uh DUR mus). The **epidermis** is the outer layer of cells of a plant. The epidermis is like a skin and is usually only one cell thick.

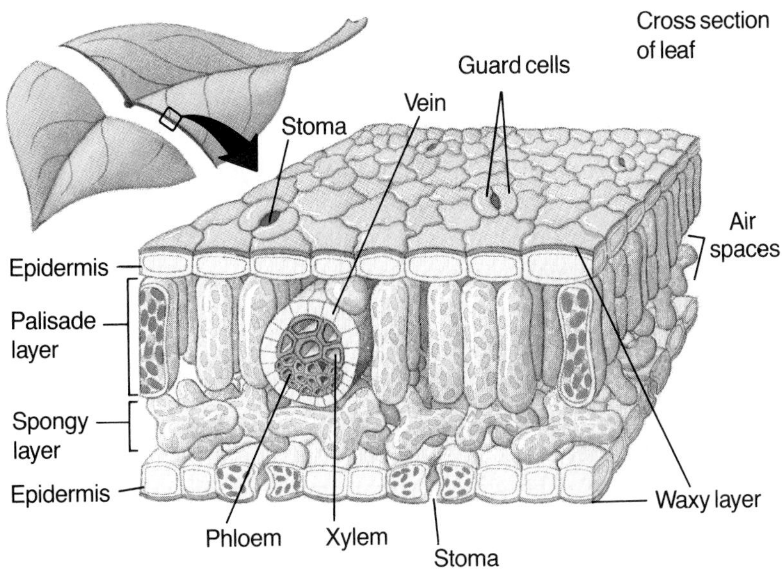

Figure 19-5 The cells of a leaf are arranged in layers. In which layer are there air spaces?

Figure 19-6 Water enters the guard cells and causes them to swell. The stoma opens (a) because the guard cells push apart. The stoma closes (b) when water leaves the guard cells. Below is a cross section through a stoma, as seen from the edge of the leaf.

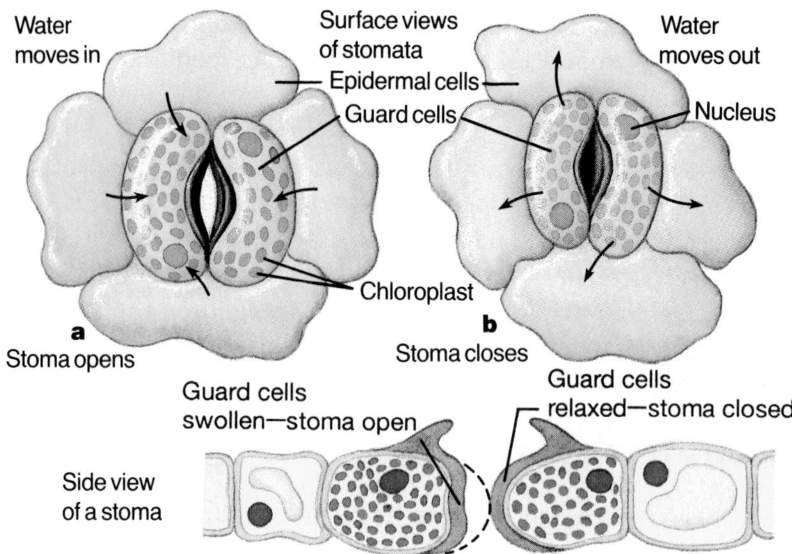

What is the palisade layer?

Examine Figure 19-5 again. Look at the layer of cells below the epidermis. The layer of long, green cells below the upper epidermis of a leaf is the **palisade layer.** Cells of the palisade layer make most of the food for the plant. These cells contain a lot of chloroplasts. Remember that chlorophyll is the green pigment in chloroplasts that helps the plant make food.

Directly below the palisade layer of a leaf is a layer of round, green cells called the **spongy layer.** These cells are loosely arranged and have spaces between filled with water vapor and air. The cells of the spongy layer also make food for the leaf. Leaf veins are often found among the palisade and spongy layers of cells.

Below the spongy layer is another layer of epidermis with its waxy layer. Find a stoma (STOH muh) in the lower epidermis of Figure 19-5. A **stoma** is a small pore or opening in the epidermis of a leaf. The plural of stoma is stomata. Leaves have stomata on both the upper and lower epidermis. The stoma acts as a doorway for gases, including water vapor, to enter and leave the leaf. Stomata are usually open during the day to take in carbon dioxide for photosynthesis. They are closed at night when carbon dioxide isn't needed.

The size of a stoma changes. How does this happen? Around the stoma are two bean-shaped cells called guard cells. **Guard cells** are green cells that change the size of the stoma in a leaf. The size of the stoma changes as the guard cells swell or shrink when they take in or let out water by osmosis, Figure 19-6.

Mini Lab

How Does a Leaf Epidermis Do Its Work?

Experiment: Put two drops of water on waxed paper, on a paper towel, and on a leaf from a potted plant. Compare what happened to the drops of water. *For more help, refer to the **Skill Handbook,** pages 704-705.*

Figure 19-7 Transpiration is most obvious over the treetops of a tropical forest.

Water Loss in Plants

What is transpiration?

Plants lose water daily through their stomata. The process of water passing out through the stomata of leaves is **transpiration** (trans puh RAY shun). Each day, a plant may lose up to 90 percent of the water it takes up through its roots.

Plant cells are mostly water. Water in a plant keeps the cells firm. Sometimes, a plant loses too much water and wilts. **Wilting** is when a plant loses water faster than it can be replaced. A punctured tire losing air is similar to a plant losing water. The tire goes flat if the escaping air is not replaced at the same rate. The plant shown in Figure 19-8 wilted when it didn't get watered for several days.

During hot, dry days, water loss is greater than usual and the stomata close. On cooler, damp days, water loss slows down and the stomata remain open.

Skill Check

Understand science words: transpiration. The word part *trans* means through. In your dictionary, find three words with the word part *trans* in them. *For more help, refer to the* ***Skill Handbook,*** *pages 706-711.*

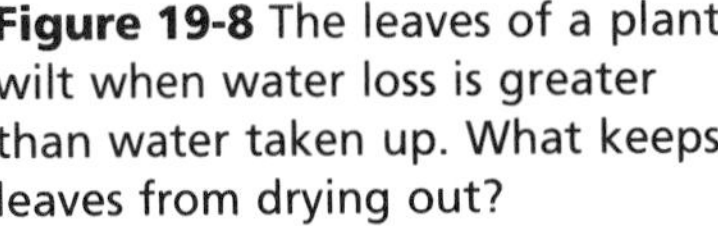

Figure 19-8 The leaves of a plant wilt when water loss is greater than water taken up. What keeps leaves from drying out?

Figure 19-9 A greenhouse provides a warm, wet environment for plants.

Mini Lab

Why Do Leaves Wilt?

Observe: Remove a leaf from a potted plant and place it on a paper towel at the start of class. What is the texture of the leaf at the start and end of the period? Explain what happened to the leaf. *For more help, refer to the* ***Skill Handbook,*** *pages 704-705.*

Do you have a houseplant at home? If you placed this plant outdoors, would it survive? It's unlikely the houseplant would be adapted to your local climate. For example, many ferns couldn't grow where it is hot and dry. Houseplants are often grown first in greenhouses before they are sold to you, Figure 19-9. If you have ever been inside a greenhouse, you may have noticed how humid and warm it is. The temperature is kept the same at all times. The air is kept moist so plants lose less water through their stomata.

Enough water passes through the plants in a small field of corn in 100 days to fill a bathtub at least 10 000 times. Each corn plant may lose up to 500 liters of water in one season. Now you may understand why farmers need to be concerned about the weather. Without rain, their corn crops would suffer.

Check Your Understanding

1. What three parts of most leaves can you see without a microscope?
2. What cells of the epidermis control sizes of stomata?
3. Explain what causes a plant to wilt.
4. **Critical Thinking:** Would a desert plant have as many stomata in its leaves as a forest plant? Explain.
5. **Biology and Math:** Plants normally lose about 99 percent of the water they absorb. What part of the water they absorb is used for growth?

19:2 Leaves Make Food

Objectives

4. **Describe** the process of photosynthesis.
5. **Explain** the chemical equation for photosynthesis.
6. **Compare** the importance of sugar in photosynthesis and respiration.

Plants use materials from their environment to make their own food. The food made by plants is also used by other living things. In meadows, deserts, forests, oceans, and rivers, animals depend on green plants for food.

Building from Raw Materials

At one time, people thought that plants got their food from the soil. Experiments have since shown that as a plant grows taller and heavier, the amount of soil around its roots hardly changes. Today, we know that plants make their own food. The leaf is the main part of a plant that makes food.

Figure 19-10 shows the frame of a house. The way a green leaf makes food by photosynthesis can be compared to the way a carpenter builds a house. Both processes build a product from raw materials.

How is building a house like photosynthesis?

A carpenter starts out by using raw materials such as lumber, nails, and shingles. As the carpenter builds, these raw materials begin to take on the shape of a building. Energy is needed to build a house. Human energy is used to carry, arrange, and join materials, and to drive nails into wood. Electrical energy is used to run power saws or drills.

Figure 19-10 The raw materials for building a house include wood and nails.

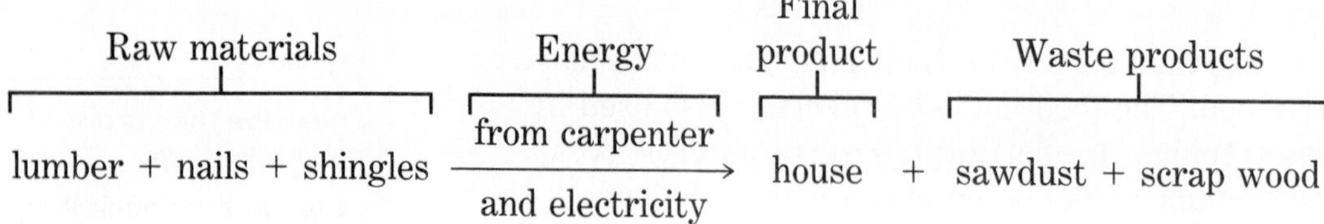

> **Bio Tip**
>
> **Consumer:** Fertilizer for houseplants is often called plant food. Look at a label on a packet of fertilizer to find out what it really contains.

When a house is built, some waste products are formed. Sawdust and scrap wood are waste products. If collected, the sawdust or wood scraps may be used for some other purpose. The equation above shows a formula for building a house. Notice how the raw materials are combined using energy to give a product and waste materials.

Keeping the example of house building in mind, let's think about the raw materials needed for photosynthesis. This process of food-making uses water and carbon dioxide gas as raw materials. Use Figure 19-11 to trace the path of carbon dioxide and water as they enter a plant. Plant roots take water from the soil. Water passes from the roots, up the stem, and into the leaves. Carbon dioxide in the air enters the leaves through the stomata.

Changing Raw Materials into Sugar

The product of photosynthesis is food in the form of sugar. Just as in house building, plants also have waste products from building food.

How does a plant make sugar? Remember that the chemical formula for carbon dioxide is CO_2. The letter C stands for the element carbon. The letter O stands for the element oxygen. The number 2 means there are two atoms of oxygen in each molecule of carbon dioxide.

Figure 19-11 The raw materials for photosynthesis are the waste products of cellular respiration.

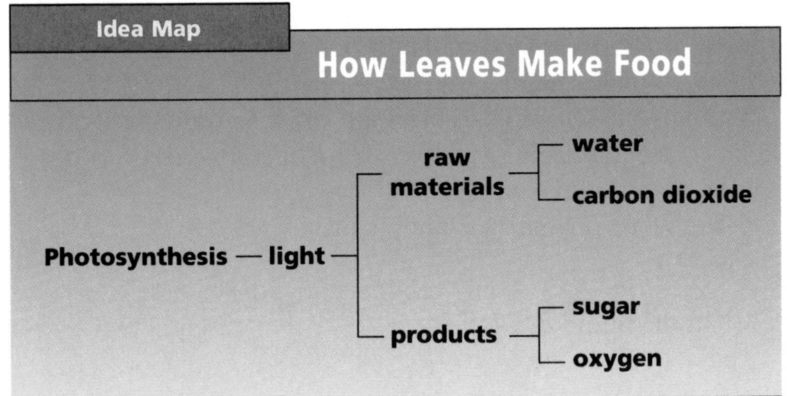

Study Tip: Use this idea map as a study guide to how leaves make food. From where does a plant get its raw materials?

The chemical formula for water is H_2O. The letter H stands for hydrogen. The formula tells us that a water molecule is made of two atoms of hydrogen and one atom of oxygen. The three elements found in carbon dioxide and water—carbon, oxygen, and hydrogen—are the same raw materials that plants need to make sugar, Figure 19-11.

What three elements make up sugar?

The general formula for sugar is $C_6H_{12}O_6$. This is a simple sugar called glucose. The three elements taken in by the plant are arranged in different ways in the leaf to make sugars. All sugars are made of carbon, hydrogen, and oxygen. During photosynthesis, plants first make glucose. This simple sugar is used by the plant to make other more complex carbohydrates such as starch. When food is moved around the plant, another sugar called sucrose is made. Sucrose is the same as your table sugar.

Leaves use six molecules of water and six molecules of carbon dioxide to make one molecule of sugar. How can a leaf cause molecules to change?

Remember how the carpenter used human and electric energy to build the house. Plants also use energy to make sugar. This energy comes from light, Figure 19-12.

The chemical equation for photosynthesis is:

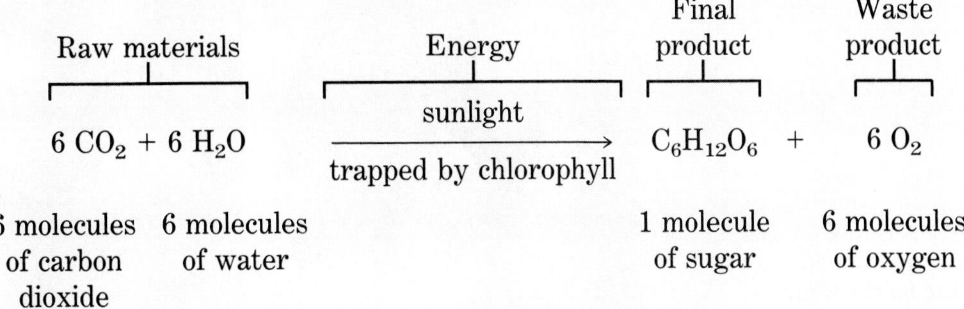

$$6\,CO_2 + 6\,H_2O \xrightarrow[\text{trapped by chlorophyll}]{\text{sunlight}} C_6H_{12}O_6 + 6\,O_2$$

> **Skill Check**
>
> **Infer:** Plants growing beneath other plants sometimes turn yellow. Infer what will happen to these plants if they continue to be covered. *For more help, refer to the* ***Skill Handbook,*** *pages 706-711.*

Light from the sun or even light from a light bulb is the source of energy used by plants. How does a plant use light energy to make sugar? You have read in Chapter 6 that plants have chlorophyll. The chlorophyll in the leaves is able to trap light energy. Once trapped, the energy is used by the plant to make sugar.

The word formula for photosynthesis is:

$$\text{Carbon dioxide} + \text{Water} \xrightarrow[\text{by chlorophyll}]{\text{Light trapped}} \text{Sugar} + \text{Oxygen}$$

Oxygen is a waste product of photosynthesis. Just as the carpenter can recycle the waste products sawdust and wood scraps after building a house, a plant can recycle the oxygen from photosynthesis. Plants and other living things use the oxygen for cellular respiration.

How Is Sugar Used?

Suppose you wanted to tear down an old house. Some of the materials in the old house could be reused to build other things such as a shed or garage. The sugar made during photosynthesis is also broken down and used to build other molecules for growth in the plant. When sugar is broken down, energy is released. Remember from Chapter 2 that this process is called cellular respiration. The energy released when sugar is broken down is used for life processes by all living things.

You have read that plants use carbon dioxide and water to make sugar, with oxygen as a waste product. In

a

b

Figure 19-12 A carpenter uses electrical energy to build a house (a). Plants use light energy to make sugars (b).

Figure 19-13 If the energy stored in this field of corn could be converted to electrical energy, you and two of your neighbors would have enough electricity for your homes for one year.

respiration, plants and animals use oxygen to break down sugar. Carbon dioxide and water are the waste products. Notice in Figure 19-11 that the raw materials of photosynthesis are the waste products in cellular respiration. The raw materials for respiration are the products in photosynthesis.

How is energy stored in plants?

Plants store large amounts of energy in the sugar they make during photosynthesis. Suppose plants could change the chemical energy in the food they make into electrical energy. In 100 days, a cornfield the size of a basketball court could make enough electricity in three homes to run their lights, appliances, and heating systems for one year! Plants are a major source of energy for all living things.

Check Your Understanding

6. What is the energy source for photosynthesis?
7. What chemical elements are found in sugar made by plants? What chemical elements are used by a plant for photosynthesis?
8. How is the sugar made during photosynthesis important to the plant and to animals?
9. **Critical Thinking:** Why can't animals use the energy from sunlight to make sugar?
10. **Biology and Math:** The formula for sugar is $C_6H_{12}O_6$. How many atoms of each kind are in the formula?

19:3 Leaves for Food

Objectives

7. **Determine** why plants are important to other living things.
8. **List** different uses of leaves.
9. **Explain** why leaves change color.

You live on a planet that is sometimes called the green planet. Why? Flying a few hundred miles above Earth, you would see it as green. Much of the green color comes from plant leaves. The gases exchanged by all these leaves make Earth the perfect environment for humans and all other living things that need oxygen for respiration.

Animals Depend on Plants

What do leaves provide to most living things?

Without photosynthesis, most living things would disappear from Earth. Mice, rabbits, owls, and fish as well as humans would die. Why? Leaves are the major food-producing organs of plants. Food produced by leaves and stored throughout a plant is eaten by animals. Plant eaters digest the plant food and use the released energy for all their life processes.

Many insects are consumers of plant tissues. Figure 19-14a shows insects called leaf miners eating their way through the inside of a leaf. Note the odd-shaped tunnels they make. The tunnels show where they have eaten through the inside of the leaf. Which cells of the leaf have they eaten? All the layers of cells with green chlorophyll have been eaten. These are the cells that produce sugar.

In Figure 19-14b notice the caterpillar eating a leaf. Does a caterpillar eat a leaf in the same way as a leaf miner? What cells are being eaten by the caterpillar?

Animals that eat plants are a source of energy for animals that feed only on other animals. Figure 19-14c shows this kind of consumer eating another animal. The

Study Tip: Use this idea map as a study guide to the importance of leaves. What are two ways that both humans and other animals depend on plants?

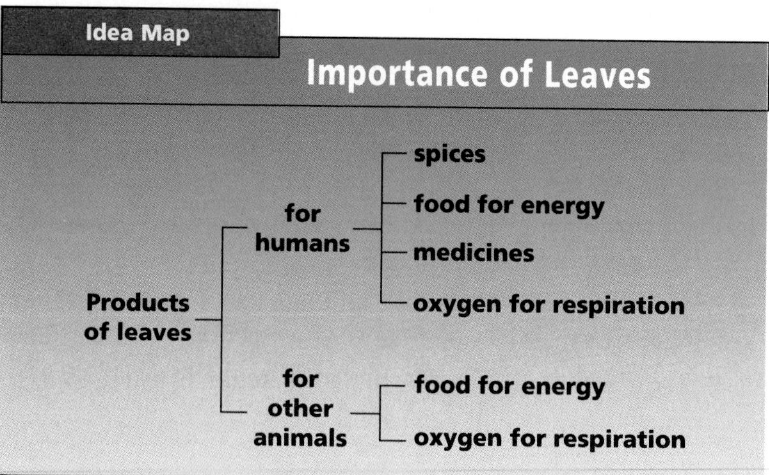

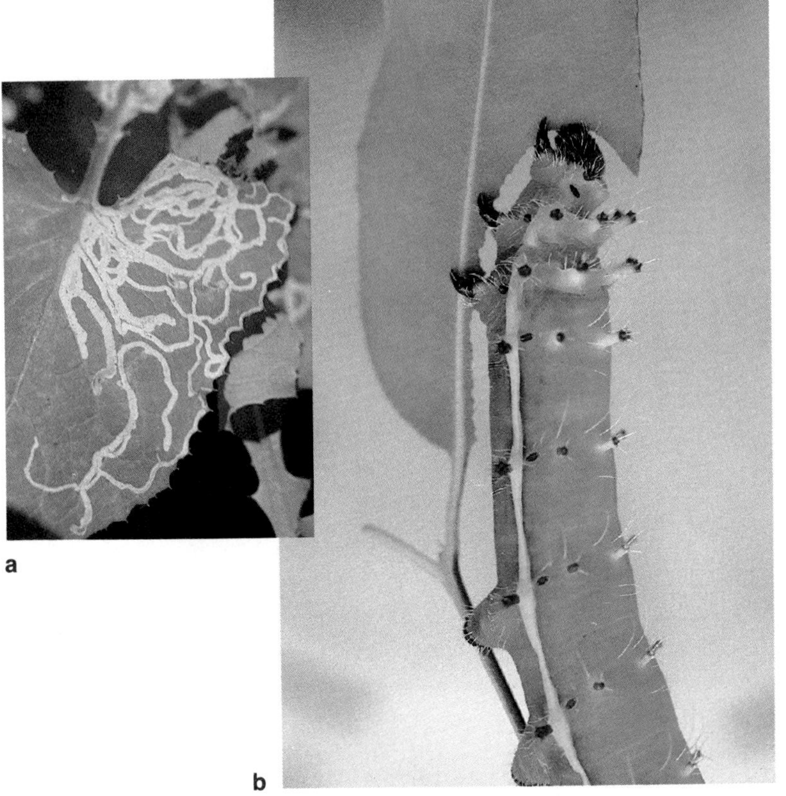

Figure 19-14 Leaf miners tunnel through leaves (a), caterpillars eat whole leaves (b), and birds eat insects that feed on leaves (c).

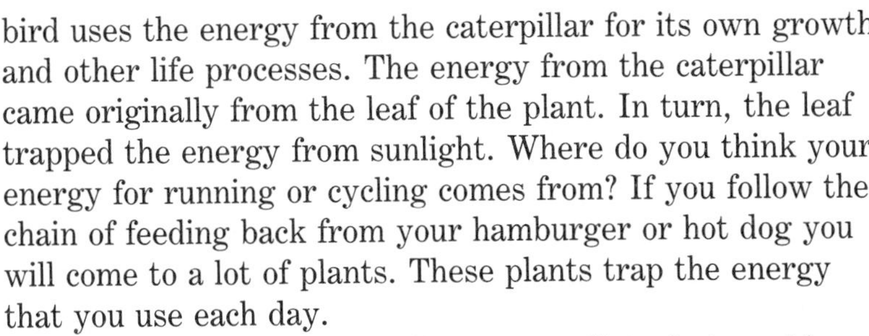

bird uses the energy from the caterpillar for its own growth and other life processes. The energy from the caterpillar came originally from the leaf of the plant. In turn, the leaf trapped the energy from sunlight. Where do you think your energy for running or cycling comes from? If you follow the chain of feeding back from your hamburger or hot dog you will come to a lot of plants. These plants trap the energy that you use each day.

What if all plants died? Can you predict what would happen to all the plant-eating consumers? They would die because the plants that they depend on for food would be dead. If there were no insects or other plant eaters, such as mice and rabbits, other consumers that eat animals would die. If there were no leaves, there would be less oxygen. Most living things depend on oxygen for respiration.

Practical Uses of Leaves

Many animals, including humans, use leaves for food. You know that cattle, sheep, and other grazing animals eat the leaves of grass. What leaves have you eaten? You may have tried the leaves of cabbage, lettuce, spinach, or onions. In the produce section of a grocery store, you could make a list of many of the leaves people eat.

Bio Tip

Leisure: Develop a vegetable garden. You can grow tomatoes, lettuce, carrots, and radishes in an area less than ten square feet and with very little effort.

Figure 19-15 Leaves have many uses.

Some leaves are used for flavorings or for spices. If you drink tea, you are drinking boiled water in which tea leaves have been soaked. Peppermint and spearmint are two other flavorings from leaves. Sage, bay, and parsley are leaves used as spices in cooking. Looking at the spice shelves of a grocery store would help you think of other leaves used as spices. Which of the products from plant leaves in Figure 19-15 have you used?

What are foxglove leaves used for?

Many plant leaves contain drugs that are useful in treating diseases. The foxglove is a flowering plant that produces the chemical digitalis (dihj uh TAL us) used as a drug to treat heart disease. As you can see, leaves are very useful to humans.

Changes in Leaves

Have you ever seen grass that has been growing under a board for several days? The grass may have changed color from green to yellow. What happened to the grass under the board? Grass leaves and many other plant leaves contain pigments other than chlorophyll. Two other common pigments are yellow and red in color. Usually leaves are green because the green pigments in chlorophyll cover up any other pigments present. Chlorophyll forms only when there is a supply of light. When the grass leaves were covered by the board, no chlorophyll was made and the green color was lost.

Science and Society

Farms of the Future

Growing plants without soil

Farmland that once was used to grow crops for food is now being sold for housing or industry. Less land for farming and growing crops could mean less food for humans. Scientists are now growing plants in the laboratory without soil. They found that as long as light was available, plants could grow while supported only in water that contained minerals. Can this technique of growing plants without soil be used to grow food crops for humans? One plant factory in Illinois grows plants such as lettuce and spinach in four weeks. Containers of sprouted seeds are placed on a slow-moving conveyer belt and given a certain amount of light, water, and minerals. Minerals are added as they are used by the plants. A sprout of lettuce uses 2.5 liters of water and the water is recycled. The same lettuce planted in the field would take six to nine weeks to grow and would use 25 liters of water.

What Do You Think?

1. In an artificial environment, there are no pests, such as weeds or insects, to harm the crops. If the plants were outside, the farmer would have to use pesticides. Inside the factory, the farmer needs to supply the plants with heating, cooling, and lighting. All of these things need electricity. In the fields outside, the farmer would use the energy from the sun. Is it more economical to grow plants in factories or in fields?

Which apple would you eat?

2. If we can't afford to build more plant factories, will we be able to grow enough food? If you look around where you live, you might notice that there are more houses, hotels, and offices than there were ten years ago. Did this land once have farm crops? As the human population expands each year, we will build more houses and use up more crop land. What will we all eat? Will we eat more meat?

3. When you buy a head of lettuce, do you put it back if you see a bruise or leaves with holes? If you bought factory grown plants, they wouldn't have holes from insects. If you bought plants grown in the fields and not sprayed with pesticides, you would probably see some insect damage. How important is it for your vegetables to be perfect? Would you put up with insect-damaged fruits if you knew you were helping to reduce use of chemicals in the environment?

Conclusion: Could plant factories completely change farming all over the world?

a

b

Figure 19-16 Some plants have pigments that make them show colors other than green (a). Maple trees in the northeastern parts of the United States give a spectacular show of colors in the autumn (b).

What happens to chlorophyll in the autumn?

Some leaves are always a color other than green. Examine the *Coleus* plant in Figure 19-16. How do you explain the colors of the *Coleus* leaves? Have you noticed other plants with colored leaves?

In the area where you live, the leaves of trees may change colors in autumn. When chlorophyll breaks down in autumn, the other pigments in the leaves can be seen. If a leaf contains red and yellow pigments, it will turn red or orange when the chlorophyll is gone.

Why do leaves change their colors in autumn? In the fall, the flow of sap in the stem slows down. The temperature drops and there are fewer hours of sunlight in the autumn. With fewer nutrients and less light, chlorophyll breaks down and the other colored pigments in the leaves show through.

Check Your Understanding

11. Why would most living things disappear from Earth if plants didn't undergo photosynthesis?
12. In what ways are leaves useful to people?
13. Why do some leaves change color in autumn?
14. **Critical Thinking:** If Earth's climate changed so that there was always a thick cover of clouds, what might happen to plants and other living things?
15. **Biology and Reading:** List three leaves that you read about that are used as spices.

Lab 19–2 Pigments

Problem: What pigments are found in a leaf?

Skills

experiment, observe, infer

Materials   

plant leaf
coin
small jar
metric ruler
filter paper strip (13 x 2 cm)
liquid solvent
applicator stick
masking tape

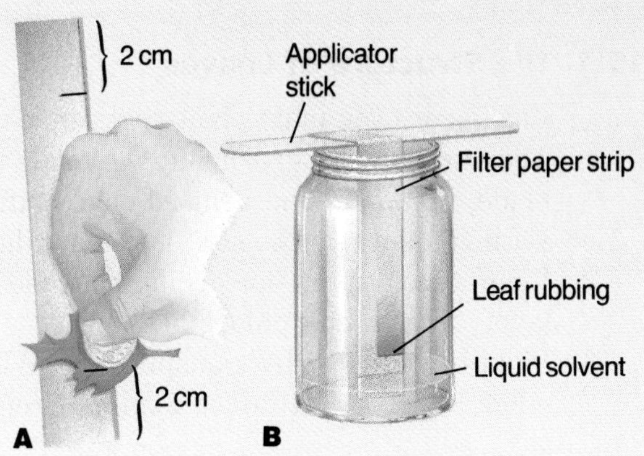

Procedure

1. Lay the leaf across the filter paper, 2 cm from one end. Rub the coin back and forth over the leaf as shown in Figure A. This action will make a dark green line at the 2-cm mark.
2. Pour the liquid solvent into the jar up to a depth of 1 cm. **CAUTION:** *Fumes from liquid solvent are dangerous. Do not breathe in the fumes. Do not use near a flame. Use a fume hood if you have one.*
3. Attach the strip of filter paper to the stick with masking tape, as shown in Figure B. The end with the leaf rubbing should hang down.
4. Hang the end of the paper strip with the leaf-rubbing line on it down into the solvent as shown in Figure B. Make sure the solvent level does not come above the green line. The paper strip may touch the bottom of the jar as long as the solvent level does not cover the leaf-rubbing line.
5. **Experiment:** Allow solvent to move up the paper until it is 2 cm from the top. Watch it carefully so that it doesn't run off the top of the strip.
6. Remove the paper and allow it to dry. Use a fume hood if you have one.
7. Dispose of the solvent as indicated by your teacher.

Data and Observation

1. What color was the paper strip at the leaf-rubbing line before you put the strip in the solvent?
2. What happened to the color of the paper strip at the leaf-rubbing line when the solvent reached it?
3. What colors showed up on the paper strip between the starting and ending lines?

Analyze and Apply

1. Did you see all of the colors the leaf contained before the paper strip was put in the solvent? Explain your answer.
2. **Infer:** Is there more than one shade of any color? If so, what might this mean?
3. Do you think chlorophyll was present? How do you know?
4. **Apply:** How could you find out whether other leaves have the same colors as the one you used in this activity?

Extension

Experiment with a multi-colored leaf and compare the results with those here.

Chapter 19 Review

Summary

19:1 The Structure of Leaves

1. Leaves are the main plant parts that make food. They are usually green and are made up of a stalk, a blade, a midrib, and veins. Leaves differ in arrangement, shape, size, and leaf edges.
2. Leaf cells form an epidermis, palisade layer, spongy layer, xylem, and phloem.
3. Plants lose water by transpiration. Wilting occurs when water is lost faster than it is replaced.

19:2 Leaves Make Food

4. In photosynthesis, leaves with chlorophyll use sunlight energy, carbon dioxide, and water to make sugar. This sugar is used as food by living things.
5. The raw materials for photosynthesis are carbon dioxide and water. The products are sugar and oxygen, a waste product. The energy source is sunlight, which is trapped by chlorophyll.
6. Sugar is broken down by cellular respiration into carbon dioxide and water. A large amount of energy is released during respiration.

19:3 Leaves for Food

7. Plants are necessary to supply animals, including humans, with food and oxygen.
8. Leaves are used by people for flavorings, spices, foods, and medicines.
9. Less light and cooler temperatures in autumn cause less chlorophyll to be made in leaves.

Key Science Words

blade (p. 398)
epidermis (p. 401)
guard cell (p. 402)
midrib (p. 399)
palisade layer (p. 402)
spongy layer (p. 402)
stoma (p. 402)
transpiration (p. 403)
wilting (p. 403)

Testing Yourself

Using Words

Choose the word from the list of Key Science Words that best fits the definition.

1. layer of long, thin cells in a leaf
2. small pore in a leaf's epidermis
3. thin, flat part of a leaf
4. process in which plants lose water faster than it can be replaced
5. process in which plants lose water

Review

Testing Yourself *continued*

6. main vein of a leaf
7. cells on the surface of a leaf blade

Finding Main Ideas

List the page number where each main idea below is found. Then, explain each main idea.

8. what plants need to make sugar
9. what raw materials are changed into sugar in a leaf
10. the result of plants losing water too fast
11. the products of photosynthesis
12. what happens to the sugar made during photosynthesis
13. the uses people have for leaves
14. why animals depend on plants

Using Main Ideas

Answer the questions by referring to the page number after each question.

15. What happens to the sugar that a plant makes? (p. 408)
16. What two layers of leaf cells can carry on photosynthesis? (p. 402)
17. What is the relationship between wilting and transpiration? (p. 403)
18. What is the drug from foxglove leaves? (p. 412)
19. What features can you use to identify different kinds of leaves? (p. 398)
20. What elements make up sugar? (p. 407)
21. How are plants, insects, and birds connected by the process of photosynthesis? (p. 410)
22. What are the raw materials needed for photosynthesis? (p. 407)
23. Other than raw materials, what else do leaves need for photosynthesis? (p. 408)
24. Why do leaves change color? (p. 414)

Skill Review ✓

For more help, refer to the ***Skill Handbook,*** *pages 704-719.*

1. **Classify:** Classify the leaves of plants near your home by shape, edge, and vein pattern.
2. **Measure in SI:** Grow a bean plant from a seed. Measure the volume of water that you use to water the plant each week. After four weeks, find the total volume of water the plant was given. Be careful not to overwater the plant.
3. **Understand science words:** Use a dictionary to find the meaning of the word part *derm*. Find two other words that have this word part.
4. **Interpret diagrams:** Look at Figure 19-6. Which of the drawings shows the stoma in cross section?

Finding Out More

Critical Thinking

1. If plants require water for photosynthesis, how is it possible for desert plants to make food during the dry season?
2. Explain why the number of humans that can live on Earth is related to the rate of photosynthesis in plants.

Applications

1. Make a collection of leaves. Press the leaves in a catalog for two weeks, then use a field guide to identify them.
2. Visit the produce and spice areas of a grocery store. List the leaves or leaf parts that are for sale as food.

CHAPTER PREVIEW

Chapter Content

Review this outline for Chapter 20 before you read the chapter.

Skills in this Chapter

The skills that you will use in this chapter are listed below.

- In **Lab 20-1,** you will observe, formulate a model, infer, and design an experiment. In **Lab 20-2,** you will observe, classify, form hypotheses, and design an experiment.
- In the **Skill Checks,** you will sequence, make and use tables, and understand science words.
- In the **Mini Labs,** you will use numbers.

Chapter **20**

Plant Support and Transport

Plants, like animals, support themselves. Unlike animals, plants have no skeleton. How do you suppose the plants in the photo on the left stay upright without a skeleton? Which plant part supports them?

You may have noticed how pipes supply your home with water and gas. Look at the water pipes in the photo below. Like these pipes, the cells of a plant carry needed materials to all parts of the plant. Which parts of a plant supply the plant with these materials? Look again at the picture on the facing page. How do you suppose the plants get some of their needed materials? Can you tell where the root stops and the stem begins? How is a root different from a stem?

Try This!

Do roots of plants grow only in soil? Find a plant that grows up the sides of buildings. Some of these plants are called vines. Find out which part of the vine attaches to the building.

BIOLOGY Online

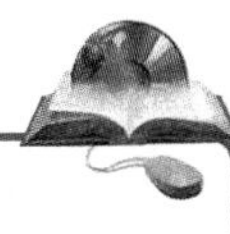

Visit the Glencoe Science Web site at science.glencoe.com to find links about **plant support and transport.**

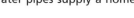

Water pipes supply a home.

20:1 Stem Structure

Objectives

1. **Discuss** the traits of herbaceous stems.
2. **Describe** the five layers of woody stems.
3. **Explain** how stems grow.

Key Science Words

herbaceous stem
cortex
woody stem
cork
cambium
terminal bud
lateral bud
annual ring

A stem is a plant part that supports the plant and transports materials. Roots are on one end of a stem and leaves are attached along the stem's length. The external parts of stems have features that help identify plants. Think how the stem of a daisy differs from the stem of an oak tree.

Herbaceous Stems

Why aren't herbaceous stems tall?

There are two stem types found in plants. They are either woody or herbaceous (hur BAY shus). **Herbaceous stems** are soft, green stems. Plants with herbaceous stems usually grow no taller than two meters. If they grew taller, they would probably fall over. Why? Herbaceous stems don't have many xylem cells. Xylem cells help support a plant stem. Recall from Chapter 6 that xylem tissue has tubelike cells that carry water and minerals throughout a plant.

Bean plants have herbaceous stems. If you slice the stem of a bean plant and look at a section under a microscope, it looks like Figure 20-1a. Many herbaceous vegetables have this pattern. Notice how the bundles of xylem and phloem cells are arranged in a circle. Other herbaceous stems, such as corn, lily, and wheat, have scattered bundles of xylem and phloem, as in Figure 20-1b.

Notice the cortex in the two herbaceous stems. The **cortex** is a food storage tissue in plants. Food is usually stored in plants in the form of starch. The epidermis of a herbaceous stem makes up its outer covering.

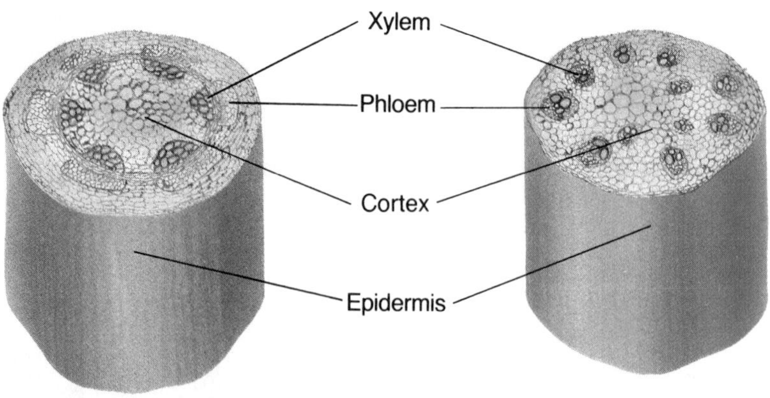

Figure 20-1 There are two patterns of xylem and phloem bundles in herbaceous stems.

Woody Stems

Figure 20-2 Beech trees can be identified by their smooth bark.

Trees and shrubs have woody stems. A **woody stem** is a nongreen stem that grows to be thick and hard. The rough outer covering of a woody stem is called the bark. Bark protects a woody stem better than epidermis protects a herbaceous stem. If you ever look at a tree stump, look for the dark outer covering. This is the bark. Many trees can be identified by the traits of their bark. The bark of some trees is very smooth, Figure 20-2. Other trees have rough bark. The bark of some trees is scaly.

A woody stem is made up of five different cell layers, or tissues. Each layer has a certain job. By following the numbers in Figure 20-3, you will be able to see where these layers are found.

What is the outer layer of a woody stem?

1. The outer layer of a woody stem is the **cork.** Cork is made of dead cells and protects the stem from insects, disease, and water loss.
2. Inside the cork layer is the layer of cells called the cortex. Remember that cortex cells store food.
3. Inside the cortex is a ring of phloem cells. Recall that phloem carries food from leaves to all parts of the plant.
4. The next layer is the cambium. **Cambium** is a thin layer of cells that divides to form new phloem on the outside and new xylem on the inside. Each year, the cambium produces new xylem and phloem cells that make the stem thicker. Thus, woody stems grow wider with age. The four layers—cork, cortex, phloem, and cambium—together make up the bark of a woody stem.

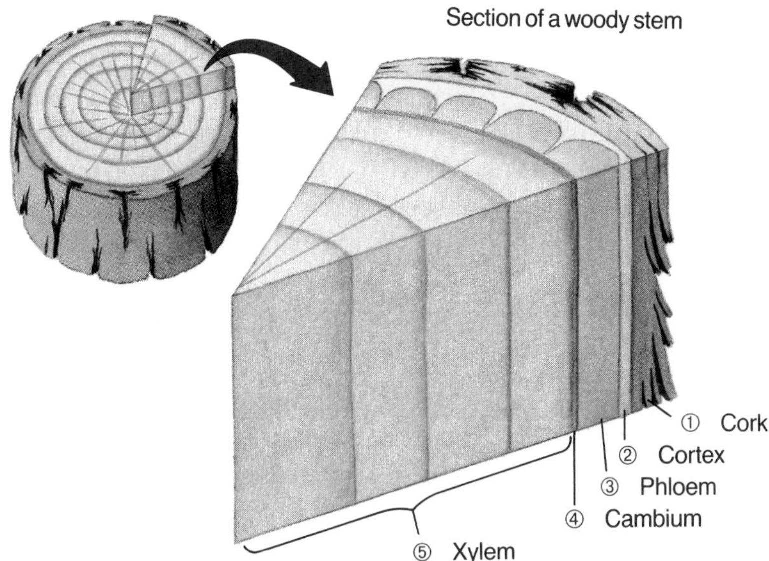

Figure 20-3 A woody stem is made of five cell layers. How many layers make up bark?

Career Close-Up

Nursery Worker

A nursery worker must know how to select the best quality plants.

Many people visit a nursery for gardening advice and to buy plants for their garden. In a large nursery, there are often several nursery workers who are there to help you.

The job of nursery worker includes planting a variety of plants, either at the nursery or a customer's home. Being able to understand the measurements in a landscape plan is an important skill for this job. The worker is responsible for preparing the soil and then setting out the plants.

When a nursery worker is not tending plants, he or she may spend some time repairing equipment. During colder months, he or she may help care for houseplants.

A high school education is important for a career as a nursery worker. You should have a general interest in caring for plants. As a nursery worker, you will need to learn about pruning, plant diseases, and insect pests. A class in horticulture (HOR tuh kul chur) will provide you with this knowledge. Most plant nurseries offer excellent on-the-job training.

5. The innermost layer of a woody stem is xylem. The xylem cells have very thick cell walls that help support the plant. Xylem cells take water up through the stem. Dead xylem cells make up the wood of woody stems. This means that wood used for building homes or furniture came from the xylem of a tree.

Stem Growth

Plants with herbaceous stems usually grow for only one year or one season. Plants with woody stems grow for more than one season. How much do plant stems grow in one year? Some, such as an oak tree, grow less than 15 cm while others, such as bamboo, grow nearly six meters.

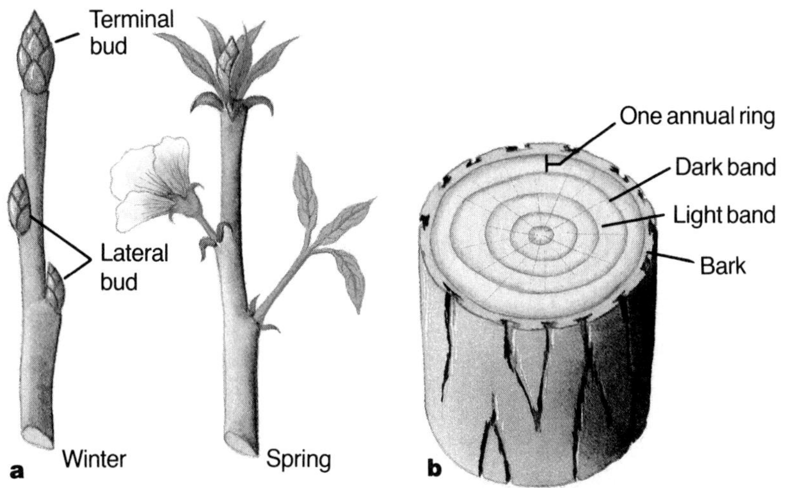

Figure 20-4 Plants grow in length from terminal buds. New branches, leaves, or flowers grow from lateral buds (a). A tree trunk is made of annual rings of wood (b).

What is a terminal bud?

How do stems grow in length? Look at Figure 20-4a. Notice that there are two kinds of buds on the stem. The **terminal bud** is the bud at the tip of a stem. This bud is responsible for the plant's growth in length. Along the sides of a stem are **lateral buds** that give rise to new branches, leaves, or flowers.

A cross section of a tree trunk shows the trunk is made up of a series of rings, Figure 20-4b. Remember that the cambium forms a ring of new xylem cells each year. Each ring of xylem is called an **annual ring.** Each annual ring is often made up of a dark band and a light band of cells. The lighter band forms in the spring when there is plenty of rain and xylem growth is more rapid. The band looks lighter because the xylem cells are larger. Growth is slower in the summer when there is less rain. The band looks darker because the xylem cells are smaller.

Mini Lab

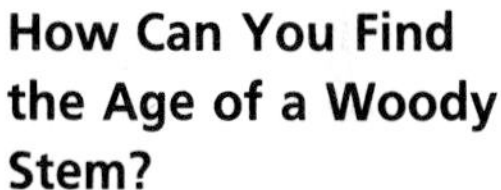

How Can You Find the Age of a Woody Stem?

Use numbers: Find a tree stump or fireplace log and look at the rings. Calculate the age of the woody stem. *For more help, refer to the* ***Skill Handbook,*** *pages 718-719.*

Check Your Understanding

1. List two common plants with herbaceous stems.
2. How can woody stems be used to help identify trees?
3. How does the cambium cause a woody stem to grow?
4. **Critical Thinking:** If you drive a nail into a young tree trunk two feet from the ground, how high up the trunk do you think the nail will be in twenty years?
5. **Biology and Writing:** Plants have two ____ of stems. They are either woody or soft and green. Which word best fits in the blank? kinds, lengths, colors

20:2 The Jobs of Stems

Objectives

4. **Explain** how a stem serves as a transport system.
5. **Discuss** the storage function of stems.
6. **Identify** how stems are useful to humans.

The stem is an important link between the roots and the leaves of a plant. Roots take up and store materials that leaves need. Leaves produce materials that roots and stems need. The stem is the part of the plant through which materials travel between the roots and leaves.

Transport

Water and minerals are needed by plant parts for growth. Leaves, as you read in Chapter 19, make food for the plant by photosynthesis. Water enters the leaves from the stems. How does the stem get this water? The water is taken up by the roots of the plant.

Water taken in by the roots is transported up through the stem. Minerals from the soil are dissolved in this water and are also carried up the stem. All water and minerals are carried up in the long, tubelike cells called xylem.

What cells in a stem carry water and minerals?

What causes water to move through a plant from its roots to its leaves? Biologists have developed a theory that explains how water moves upward in a plant. One part of the theory is that water moves through a plant much as water moves into a paper towel, Figure 20-5. If you put a dry paper towel into water, the water is absorbed by the paper towel.

The second part of the theory explains how water passes up and out of a plant. Water moves up the plant through xylem cells and is lost from the leaves through the stomata. You read in Chapter 19 that transpiration is the loss of water through the stomata of leaves. This movement of water through a plant is like a thread being pulled through a straw, Figure 20-6a. The molecules of water stick together in a threadlike stream through a plant.

Figure 20-6b shows how important the stem is as a link between the leaves and roots. As water moves out of the leaves, more water is pulled into the leaves from the stem. The water in the stem is pulled up from the roots. New water enters the roots by osmosis.

Food made in leaves is also transported through stems. Food in the form of sugars moves down the stem through the tubelike cells called phloem. Why does food move downward? Root cells are alive and also need energy in order to stay alive. They would die without food. The sugars made in the leaf are used by the roots and all the other cells of the plant.

Figure 20-5 Water enters a plant in a way that can be compared to how a paper towel absorbs water.

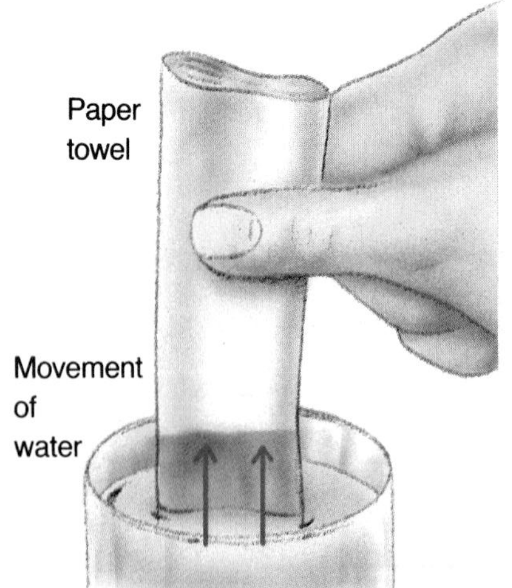

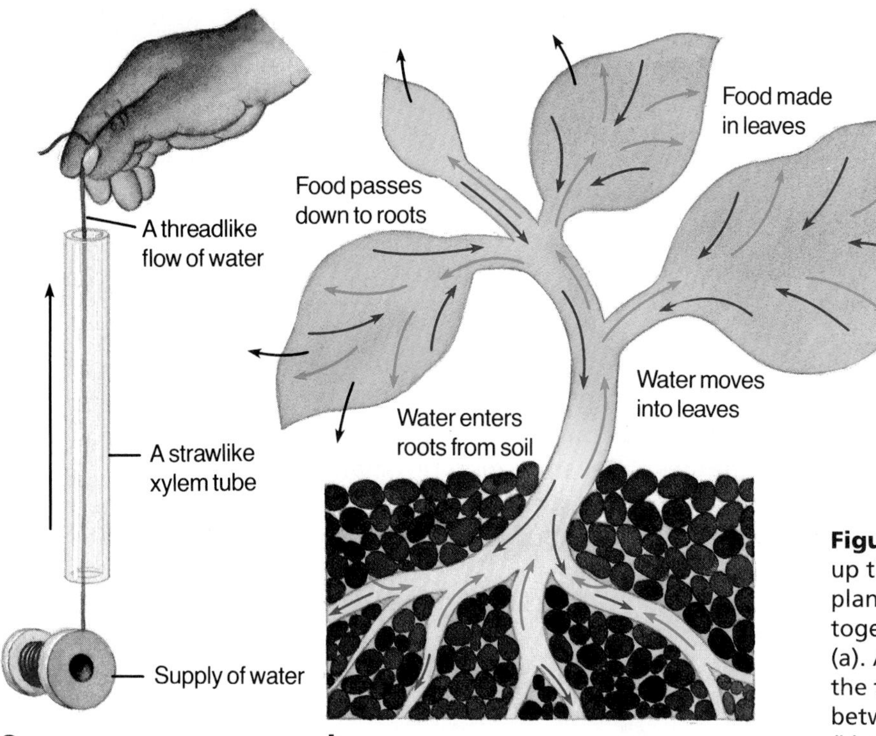

Figure 20-6 When water moves up through xylem tubes in a plant, the molecules stick together in a continuous stream (a). A stem is the plant's organ for the transport of water and food between the roots and the leaves (b). The red arrows show the pathway of food. The blue arrows show the path of water, and the black arrows show water loss from leaves.

Storage

How does water storage help a plant?

Look at the food products stored in containers on the pantry shelf. A container is opened when food is needed. Stems also store materials that are used when needed. What types of materials are stored by the stem? Stems store water, and food in the form of starch. Water comes from the roots and the starch is made from sugars that are carried from the leaves.

One benefit of water storage in herbaceous stems is that it prevents wilting. Water helps keep the cells of the stems stiff. You might say that these stiff cells support a plant like the skeleton supports an animal's body. Plants that live in dry areas have stems that store large amounts of water. A cactus stores a lot of water in the stem. This water is used for photosynthesis during the long, dry season.

Some stems store large amounts of starch. Starch is stored in stems until needed by a plant for its new growth. This food is a reserve supply for other plant parts. When needed, the starch is changed back into sugar. The sugar moves through the phloem cells from the stem to the root or the leaves. Many plants use much of their stored food for new growth each spring.

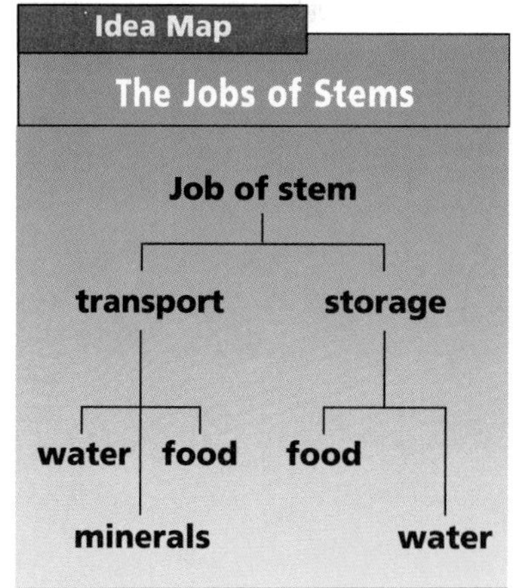

Study Tip: Use this idea map as a study guide to the jobs of stems. What cells transport water in stems?

TECHNOLOGY

Newer Medicines from Plants

Several centuries ago, explorers brought plants from the tropical forests of the New World back to Europe. The explorers had discovered that these plants were being used for medicines by native peoples.

For example, quinine was found to be used as a cure for malaria. The natives of South America had made an extract of quinine by soaking the bark of the *Cinchona* (sihn CHOH nuh) tree in water. Drinking the water that contained the bark extract helped cure malaria.

Plants use the sugar from photosynthesis and minerals from the soil to make many different compounds that have medicinal properties. Many of the original plant medicines have since been made from synthetic chemicals. Now, researchers are going back to the tropical rain forests in search of new plant medicines. AIDS and cancer are two diseases for which we need new medicines.

Why is the tropical rain forest a good place to hunt for medicinal plants? There are more species of plants in the tropical rain forests than anywhere else in the world. Scientists believe that with all these undiscovered species, there must be some with new medicinal value.

Many plant scientists are involved in the search for new plants in the world's rain forests. The work is very time consuming. Will all of this effort help society? Imagine the benefits of discovering a plant that produces a chemical that can cure AIDS.

A chemical from the bark of the Peruvian Cinchona tree helps cure malaria all over the world.

How Are Stems Used?

The most important product we get from stems is wood. When you write with a wooden pencil, you are using a product from a stem. Look around your home or classroom and you will find many places where wood is used. You read in Chapter 6 how conifers are important for paper products. Paper products you use come from the wood of these tree stems. This book, dollar bills, food packets, and many other paper products are part of your daily life, Figure 20-7a. Your life would be very different without stems.

Does paper come from woody or herbaceous stems?

a

b

Figure 20-7 Plants provide us with paper products (a) and sap or liquid products, such as maple syrup and chewing gum (b).

Stems are useful also as food. When you eat asparagus, cauliflower, and broccoli, you are eating the flower stems of these plants. People in some countries grow bamboo plants and eat the young, soft stems. Food seasonings such as cinnamon and dill are from stems. Some foods are made by taking sap from trees. Maple sugar is made from the sap of sugar maple trees. Sap is the liquid that flows up the stem in the xylem cells. In spring, the sap contains up to four percent sugar, which was stored in the roots over the winter.

Rubber, turpentine, and materials for making chewing gum are made from other liquids in the stems of different trees. After the liquids are collected from the trees, they are processed for our use, Figure 20-7b.

Skill Check

Make and use tables: Use reference materials to make a table of plant products. Examples include corn syrup and corn starch. *For more help, refer to the **Skill Handbook**, pages 715-717.*

Check Your Understanding

6. What happens to minerals in the soil before they can get into roots?
7. How does water storage help herbaceous stems?
8. What are five products we get from stems?
9. **Critical Thinking:** How would the rate of water movement through a plant on a rainy day compare with the rate on a hot, sunny day?
10. **Biology and Writing:** Make a complete sentence from each of the three sentence fragments.
    1. A stem supports a plant and transports
    2. are dissolved in water and carried through roots.
    3. One benefit of water storage in herbaceous stems

20:3 Root Structure

Objectives

7. Compare taproots and fibrous roots.

8. Sequence the layers of root cells.

9. Describe how roots grow.

Key Science Words

taproot
fibrous root
root hair
endodermis

A root is the plant part that takes in water and minerals for the plant. The following sections will describe the structures of roots and how they grow.

Taproots and Fibrous Roots

What is a taproot?

There are two basic types of root systems—taproots and fibrous roots. The part of a carrot that we eat is the root. In plants like the carrot, there is one main root called a taproot. A **taproot** is a large, single root with smaller side roots. The thick taproots of some plants store water and food during dry or cold periods. In the vegetable section of a grocery store, you can see many different kinds of taproots. Beets, carrots, and turnips are common vegetables that are taproots.

If you weed a garden or lawn, you know that some weeds such as dandelions are very hard to pull out of the soil, Figure 20-8a. Their long, thick taproots will usually break off before you can pull the plant from the soil.

Not all plants have taproots. Some plants have fibrous (FI brus) roots, Figure 20-8b. **Fibrous roots** are many-branched roots that grow in clusters. Unlike the taproot, there is no main, large root. As the name suggests, the roots are fiberlike. Fibrous roots spread out over a large area. Thus, fibrous roots collect water from a larger area than reached by taproots.

Oak trees and hickory trees have long, thick taproots. Some maple trees and beech trees have fibrous roots. Which kind of tree is more likely to blow over in a harsh windstorm?

a

b

Figure 20-8 A taproot has stored food (a), and a fibrous root has many branches (b).

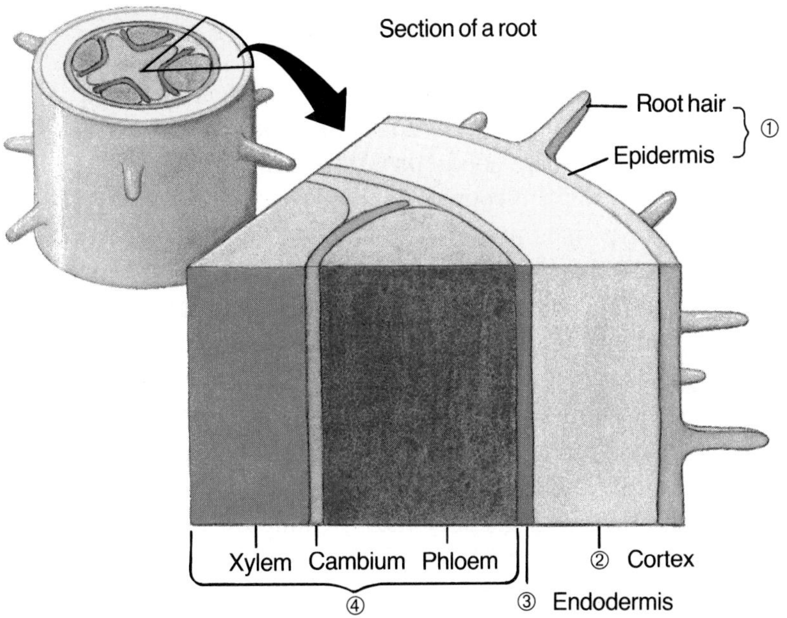

Figure 20-9 A root is made of four main layers of cells. From which layer do root hairs grow?

Cells of a Root

The root is made of several layers of cells. The different layers have different jobs. Using Figure 20-9, let's examine the cell layers of a root.

What is the outer layer of a root?

1. The outside layer of a root is the epidermis, which protects the root. Growing from the epidermis are the root hairs. **Root hairs** are threadlike cells of the epidermis that absorb water and minerals for a plant.
2. Inside the epidermis is the layer of cells called the cortex. Just as in a stem, the cells of the cortex store food.
3. The next layer is the endodermis. The **endodermis** is a ring of waxy cells that surrounds the xylem in roots. This layer of cells helps the plant retain the water taken in by the xylem.
4. Inside the endodermis, notice the bundles of xylem. Between these xylem cells are phloem cells. You know that xylem and phloem carry materials in the plant. Xylem cells in the roots carry water and minerals up to the stem. Phloem cells bring food from the leaves.

If you eat a raw or cooked carrot, you might notice a difference in the feel and taste between the outer and inner layers of the carrot root. The outer ring is softer and sweeter. This is the cortex with its stored food. The inner part is tougher and not as sweet. What makes this layer taste different? This layer is made of the xylem and phloem. There are also more fibers in this ring.

Skill Check

Understand science words: fibrous root. The word part *fibr* means threadlike. In your dictionary, find three words with the word part *fibr* in them. *For more help, refer to the **Skill Handbook**, pages 706-711.*

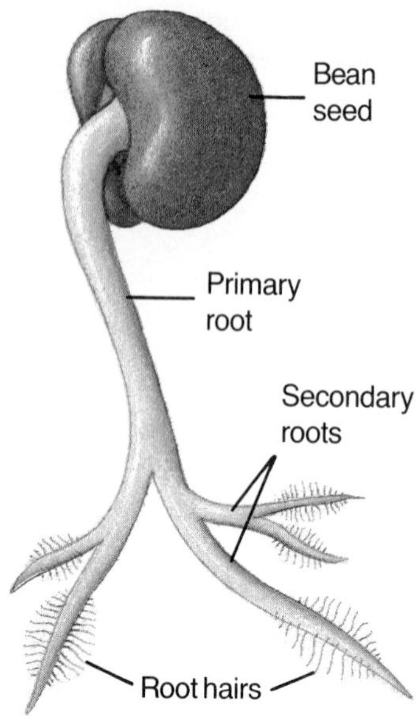

Figure 20-10 The first root of a new plant is its primary root. Other roots are secondary roots.

How does a primary root grow?

Root Growth

If you split open a bean seed, you will see the plant embryo inside. This new, young plant has leaves, a stem, and a root. When the seed sprouts, the first part to grow out of the seed is the root.

Only one root is present in the seed. When it grows out of the seed, it is the primary root. The word *primary* means first. The primary root is the first and largest root to form in a plant. It usually grows straight down, toward the pull of gravity.

Later, secondary roots form. Secondary roots are smaller roots that branch from the primary root. The thinner, shorter, threadlike parts near the tips of the secondary roots are the root hairs. Compare the primary roots and secondary roots in Figure 20-10.

Most roots keep growing as long as the plant lives. When roots grow, new cells are added to the tips of the roots. This is how the root grows down into the soil.

How large can a root system grow? In 1926, an American scientist asked the same question. To find the answer, he collected a lot of data. First he carefully removed all the fibrous roots of a rye plant, a grasslike cereal plant. He then measured the lengths of each of the roots and added all the lengths together. The results showed that if the roots were placed end to end, they could form a line 612 km long, as long as the state of Tennessee!

Next, the scientist measured the root hairs. The lengths of all the root hairs were added together. If every root hair were put end to end, the line would go from the east to the west coast of the United States and back again! You can see that the rye plant has a very large root system.

Check Your Understanding

11. How does a taproot differ from a fibrous root?
12. List the layers of cells found in a root and describe the job of each.
13. How do secondary roots form?
14. **Critical Thinking:** Which layers of cells do roots and stems have in common?
15. **Biology and Reading:** Which would be easier to pull up—a dandelion that has a taproot, or a tomato plant that has a fibrous root system?

20:4 The Jobs of Roots

Is the root vital to a plant? The plant would die if its roots were removed. Roots anchor the plant in one place and supply a plant with nutrients.

Objectives

10. **Describe** root hairs and their function.
11. **Relate** how roots anchor the plant and store nutrients.
12. **Discuss** the benefits and problems of roots.

Absorption

Plants get needed materials for growth through their roots. Roots absorb water from their surroundings by osmosis. In Chapter 2, you read that osmosis is the movement of water molecules across a cell membrane. Roots also take in minerals from their surroundings with the water they absorb.

How does water enter a root?

Water enters the root through the root hairs. Root hairs increase the surface area of the root and so increase the amount of water that can be absorbed or taken in by the plant. When a gardener moves a plant from one part of a garden to another, the plant's root hairs may be torn off. The amount of water these plants can absorb is reduced. To protect the plant, the gardener digs up a ball of soil around the plant's roots and moves this soil along with the plant, Figure 20-11a. This protects the root hairs.

The root supplies the plant with raw materials. Water is a raw material for photosynthesis. Minerals are raw materials for the plant to make proteins and other chemicals such as chlorophyll.

Like all living cells, root cells must be supplied with oxygen. Root cells get oxygen from air in the spaces between soil particles. Gardeners and landscapers break up the soil around plants to ensure that air gets into the soil, Figure 20-11b. Sandy soils or soils rich in organic material have more air spaces than clay soils.

Skill Check

Sequence: List the parts of a plant in the order in which water passes in and out of the plant. *For more help, refer to the **Skill Handbook**, page 706.*

Figure 20-11 Roots are fragile and easily broken when a plant is moved (a). When the soil is broken up, air can circulate around plant roots (b).

a

b

Absorption

Lab 20–1

Problem: How does a root hair work?

Skills

observe, formulate a model, infer, design an experiment

Materials

paper towel strips:
- 10 X 15 cm (1 strip)
- 1 X 10 cm (4 strips)

stapler
large plastic cup
pencil

Procedure

1. Staple the four small paper strips to the larger strip, Figure A. Bend the smaller strips so they are at right angles from the larger strip.
2. Fill the cup to within 2.5 cm of the top with water.
3. Lay a pencil across the rim of the cup. Lay the root model over the pencil. See Figure B. The smaller towel strips will hang down into the water.
4. **Observe** the root model for 20 minutes.
5. Remove the root model from the cup.

Data and Observations

1. What happened to the smaller strips when they were put in water?
2. Did the large strip absorb any water? If so, where did this water come from?

Analyze and Apply

1. Which part of your model represented the root hairs?
2. Which part of your model represented the main root from which root hairs grow?
3. In what way did the small paper strips act like root hairs?
4. Why are there hundreds of root hairs on a single root?
5. **Infer:** What would happen to a plant if its root hairs were damaged?
6. **Apply:** Would you expect desert plants or tropical rain forest plants to have more root hairs? Why?

Extension

Design an experiment that shows the rate of water absorption in a root. Use a carrot, water, and food coloring.

A

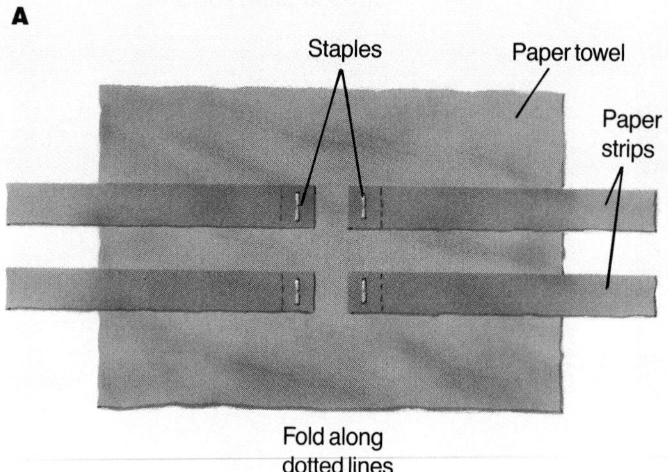

B

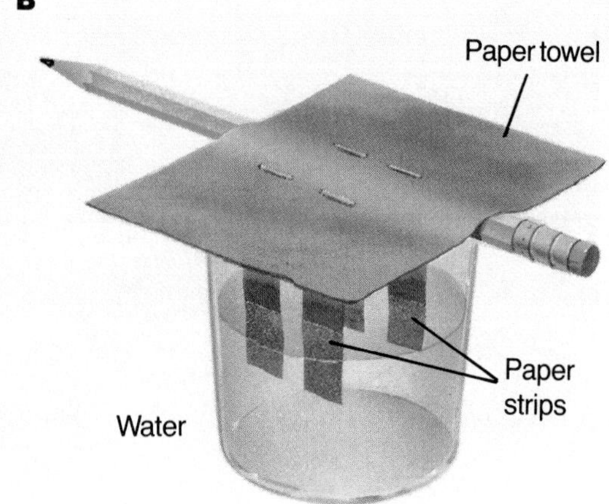

Anchorage and Storage

A major job of a root is to anchor the plant. A root is similar to a boat anchor. An anchor catches on the rocks or soil under the water and holds the boat in one place. The roots of most plants spread through the particles of soil and hold the plant in one place, Figure 20-12a.

Many roots have large cells that store food. The usual food stored is starch. Compare the size of the storage cells with the other cells in Figure 20-12b. You have read that these storage cells make up the cortex. The food in these cells is used later by other cells in the plant. Large amounts of food are stored in taproots. Carrot and turnip roots are good examples of plants that have many food storage cells. The taproot of a dandelion stores food for the plant. You may have noticed that a dandelion will soon grow back if you don't pull up its whole root. It uses its stored food to grow new leaves, stems, and flowers.

Roots store food made in one growing season to use at the start of the next growing season. Many herbaceous plants depend on the stored food in the roots to survive the winter. If food were not stored in the roots for the winter, the plant could not grow new parts the next spring.

Idea Map

The Job of Roots

- **Job of roots**
  - **absorption**
    - **water enters**
    - **minerals enters**
  - **storage**
    - **stores food**
    - **new growing season**

Study Tip: Use this idea map as a study guide to the jobs of roots. What two things do roots absorb?

Why do roots store food?

How Are Roots Used?

You have already learned that people eat many kinds of roots such as beets, carrots, and sweet potatoes. Products made from the roots are also eaten. Sassafras (SAS uh fras) roots are used to make tea. Horseradish roots are ground and used as a sauce to season other foods. Roots are also used to make products such as perfumes and medicines.

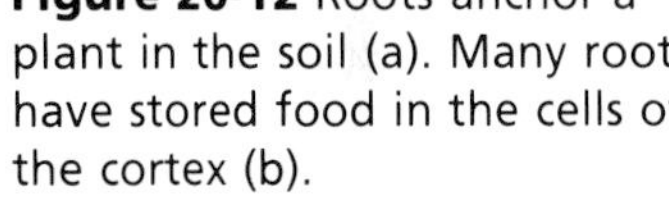

Figure 20-12 Roots anchor a plant in the soil (a). Many roots have stored food in the cells of the cortex (b).

a

b

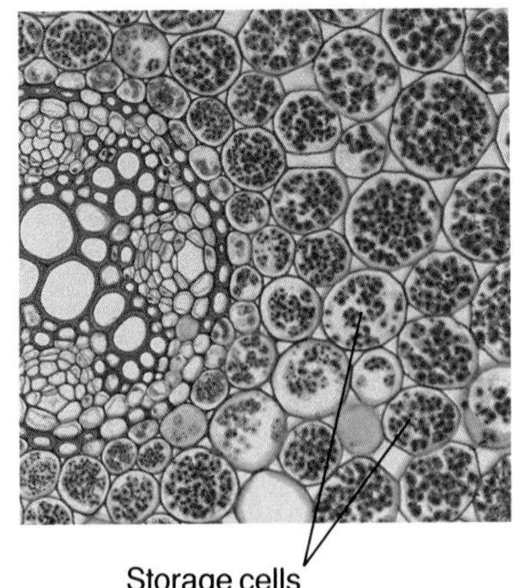

Storage

Lab 20–2

Problem: What parts of roots store starch?

Skills

observe, classify, form hypotheses, design an experiment

Materials

carrot root
yam root
radish root
razor blade
forceps
3 petri dish halves
iodine solution
dropper
hand lens

Procedure

1. Copy the data table.
2. Using the razor blade, cut a very thin slice of each root as shown in the figure. **CAUTION:** *Use extreme care with the razor blade.*
3. Place each slice in a separate petri dish.
4. Write a **hypothesis** about which part of the carrot root will show the presence of starch. Will the other roots show the presence of starch? (HINT: Read pages 429 and 433 again.)
5. Add one dropperful of iodine to each root slice as shown in the figure. Make sure each slice has iodine on it at all times. **CAUTION:** *Iodine is poisonous and can stain and burn the skin.*

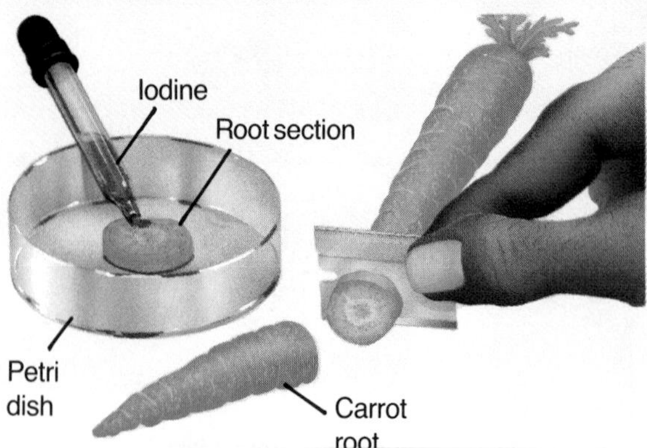

6. **Observe:** During the next 15 minutes, examine the root slices with the hand lens. Look for blue-black areas. A blue-black color means starch is present. Note in your data table in which areas starch is present.
7. Draw each root slice in your notebook. Label the cortex and xylem in the carrot drawing.

Data and Observations

1. Which roots showed starch was present? How could you tell?
2. Which part of each root slice showed more starch present?

TYPE OF ROOT SLICE	BLUE-BLACK AREAS AFTER ADDING IODINE
Carrot	
Yam	
Radish	

Analyze and Apply

1. What cells make up the center of the root? What do they do?
2. What cells make up the outer part of the root? What do they do?
3. How are the starch storage areas of the three roots alike?
4. **Classify:** Are the roots used taproots or fibrous roots? Explain your answer.
5. **Check your hypothesis:** Is your hypothesis supported by your data? Why or why not?
6. **Apply:** How is starch made by a plant?

Extension

Design an experiment to test different stems for stored food.

a

b

Figure 20-13 These root nodules contain bacteria that fix nitrogen for the plant (a). Growth of tree roots can sometimes damage sidewalks or sewage pipes (b).

Roots are beneficial to humans in other ways. If you were to pull up a small clump of grass growing in a yard, you would see soil sticking to the roots. Roots of plants help hold soil together. The soil is not easily carried away by rain, melting snow, or wind.

How do roots help soil?

The roots of many trees and plants, such as beans and clover, help make the soil rich in nitrogen. Some bacteria can change nitrogen in air to a form that plants can use. These bacteria live on plant roots, Figure 20-13a. Soil rich in nitrogen is good for growing corn and other food crops.

Sometimes roots are a nuisance, Figure 20-13b. The roots of some trees, like the willow and maple, grow into underground pipes. Roots can block the pipes that carry water and wastes away from houses. Root growth is so strong it can break pipes. It's a good idea to find out where pipes are laid before planting trees. Sometimes sidewalks or roads are also broken by root growth.

Bio Tip

Biology: Many cities don't plant silver maples and willow trees because their roots cause damage to sewers.

Check Your Understanding

16. Why are root hairs important?
17. What are three jobs of a root?
18. What are three benefits of roots?
19. **Critical Thinking:** You want to plant a shade tree in your yard. What things do you need to find out about the tree and your yard before you plant?
20. **Biology and Reading:** You have read that roots are sometimes a nuisance. What does it mean for a plant's roots to be a nuisance?

Chapter 20

Review

Summary

20:1 Stem Structure

1. Herbaceous stems are soft and green.
2. Woody stems are thick and hard. From outside to inside, they are made up of cork, cortex, phloem, cambium, and xylem.
3. Stem growth in length occurs in terminal and lateral buds.

20:2 The Jobs of Stems

4. A stem transports materials from the roots to the leaves.
5. Stems store food and water in the cortex.
6. Stems provide humans with wood, food, seasonings, rubber, and turpentine.

20:3 Root Structure

7. A taproot is a large single root with smaller side roots. Fibrous roots are many-branched roots.
8. Roots are made up of an outer epidermis, cortex, xylem, and phloem.
9. The first root to grow out of a seed is the primary root. Secondary roots grow out from the primary root.

20:4 The Jobs of Roots

10. Root hairs absorb water and minerals by osmosis.
11. Roots anchor the plant and store starch.
12. Roots are used for food and to make medicines. Roots also help hold the soil together.

Key Science Words

annual ring (p. 423)
cambium (p. 421)
cork (p. 421)
cortex (p. 420)
endodermis (p. 429)
fibrous root (p. 428)
herbaceous stem (p. 420)
lateral bud (p. 423)
root hair (p. 429)
taproot (p. 428)
terminal bud (p. 423)
woody stem (p. 421)

Testing Yourself

Using Words

Choose the word from the list of Key Science Words that best fits the definition.

1. a ring of xylem in a woody stem
2. root with a single, large root
3. threadlike cell of the root epidermis
4. dead cells on the outer part of a tree that partly make up bark
5. stem part that causes growth in length
6. gives rise to new branches, leaves, or flowers

Review

Testing Yourself *continued*

7. cells that form new xylem and phloem
8. soft, green stem
9. a nongreen, hard stem
10. ring of waxy cells around the xylem and phloem in the root
11. plant tissue that stores food

Finding Main Ideas

List the page number where each main idea below is found. Then, explain each main idea.

12. stems storing food for the next growing season
13. how water lost in leaves is replaced
14. how stems are useful for paper products
15. how a root takes in water
16. how a primary root develops

Using Main Ideas

Answer the questions by referring to the page number after each question.

17. How are primary and secondary roots different? (p. 430)
18. How does an annual ring form? (p. 423)
19. Why is transporting water to leaves an important function of stems? (p. 424)
20. Starting with the center, what is the order of these cell types as they would occur in a root: cortex, phloem, endodermis, and epidermis? (p. 429)
21. How can roots be a nuisance to humans? (p. 435)
22. How is the arrangement of xylem and phloem different in bean and corn stems? (p. 420)
23. Why are roots a good source of food for animals? (p. 433)
24. Why do plants in dry areas store much water in stems? (p. 425)

Skill Review

For more help, refer to the ***Skill Handbook,*** *pages 704-719.*

1. **Sequence:** What are the root cells and stem cells in order from the outside to the inside of a root and stem?
2. **Observe:** Observe the plants in the produce section of a grocery store. Make a list of those that are roots.
3. **Classify:** Make a collection of ten weeds from your yard or street. Use a field guide to name them. Group them as either woody or herbaceous.
4. **Use numbers:** Calculate the age of a tree that is 20 cm in diameter if each annual ring is 5 mm thick.

Finding Out More

Critical Thinking

1. What causes knots in wooden boards?
2. What makes a tree ooze sap where limbs are removed? Explain why this happens more often during spring and summer.

Applications

1. Find out which plant parts of onions and white potatoes are the stems.
2. Visit a furniture store and find out what kinds of wood are used in making furniture and why.

CHAPTER PREVIEW

Chapter Content

Review this outline for Chapter 21 before you read the chapter.

21:1 Plant Responses
- Growth and Flowering
- Tropisms
- Other Plant Responses

21:2 Growth Requirements
- Seasonal Growth
- Light
- Water
- Minerals
- Soil
- Temperature

21:3 Plant Diseases and Pests
- Bacteria and Viruses
- Fungi
- Insect Pests

Skills in this Chapter

The skills that you will use in this chapter are listed below.

- In **Lab 21-1,** you will measure in SI, form hypotheses, interpret data, and experiment. In **Lab 21-2,** you will observe, form hypotheses, and interpret data.
- In the **Skill Checks,** you will observe and understand science words.
- In the **Mini Labs,** you will experiment.

Chapter **21**

Plant Growth and Disease

Plants, like animals, respond to changes in their environments. Plants don't have sense organs, nor do they move about. However, plants can still move in response to changes in their environments.

One well known response in plants can be seen in the photo to the left. If you live in the middle or northern states of the United States, you may have seen the new growth of plants in the spring. What causes a plant to grow in the spring, make flowers, or drop its leaves in the fall?

Another change in plants is caused by disease. Plants, like animals, can become sick as shown in the photo below. This chapter will explain how a plant responds to its environment, what it needs to grow, and some of the causes of plant diseases.

Try This!

How can you see a plant move? Place a houseplant in a window. Notice that the stems are straight. Look at the plant after two days. Has the plant moved? In what direction have the stems or leaves moved?

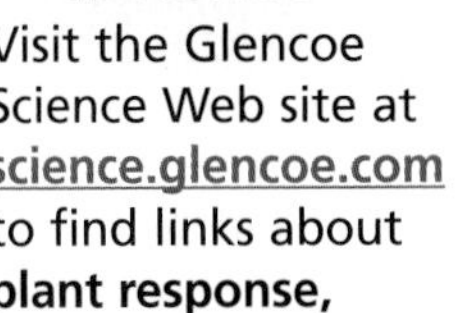

BIOLOGY Online

Visit the Glencoe Science Web site at science.glencoe.com to find links about **plant response, growth, and disease.**

Diseased fruit is not edible.

21:1 Plant Responses

Objectives

1. **Explain** how hormones affect plant growth.
2. **Compare** short-day, long-day, and day-neutral plants.
3. **Distinguish** among tropisms and other plant movements.

Key Science Words

short-day plant
long-day plant
day-neutral plant
tropism
phototropism
gravitropism
thigmotropism

If you took a picture of a bean plant early in the morning and then again later in the day, you would see that some parts of the plant had moved. Early in the morning, bean leaves droop down close to the stem. The photograph taken later in the day would show the leaves held out and away from the stem. Bean leaves respond to a change between night and day. Is the movement of bean leaves a growth response, or do they move by some other means? The following sections describe and explain various plant responses.

Growth and Flowering

The photos in Figure 21-1 are of the same plant. The photos were taken in spring, just a week apart. In just a week, the plant has grown. Growth, flowering, and branching are all plant processes that are controlled by plant growth hormones. In Chapter 15, you read that hormones are chemicals made in one part of an organism that then affect another part of the organism.

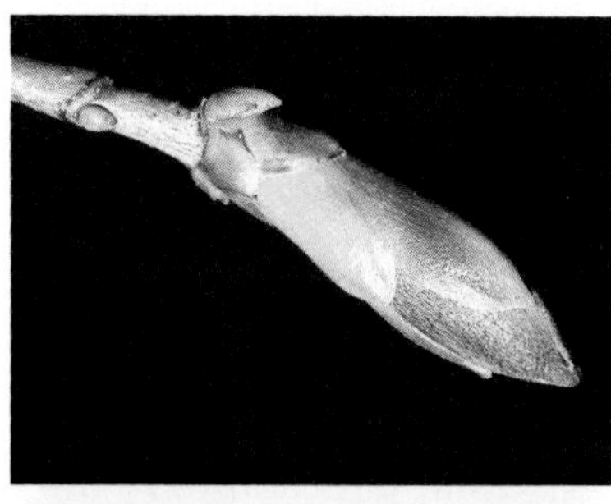

Figure 21-1 A maple tree before (top) and after (bottom) its buds burst open in the spring

Hormones that affect growth are made in the cells of the growing regions of plants. Locate the growth regions of a stem tip and root tip in Figure 21-2a. Growth in the stem tip causes a plant to grow taller and branches to grow larger. Growth in the root tip allows the roots to grow longer. Notice that the cells producing the growth hormones make up only a small area in the growth regions at the very tips of both shoots and roots.

Just what happens when a plant grows? Look at Figure 21-2a. Growth hormones near the root and stem tips cause cells to increase in number. These cell divisions make the root and shoot grow. Farther back from the growing tip, the hormone causes each cell to get longer. The increase in length of the cells pushes the root downward into the soil or the stem higher into the air.

What happens when you remove terminal buds?

The hormones in the terminal buds allow the terminal bud to grow while preventing growth of the lateral bud. The side branches that usually develop from lateral buds grow more slowly, Figure 21-2b. If you want a tree or shrub to grow more dense and bushy, you could trim off the tips of a few of branches. The lateral buds on these branches will then take over the role of the terminal bud and grow faster. The plant will get bushy because of the new growth of lateral branches.

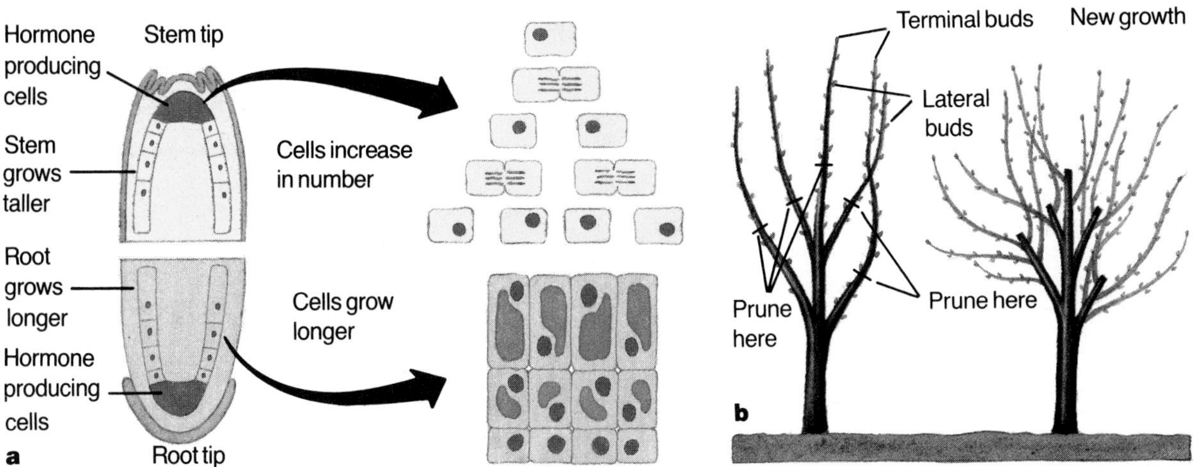

Figure 21-2 Cells in the growth regions of a stem and a root first increase in number at the tips and then get longer farther back (a). If terminal buds are removed, plants grow bushy (b).

You may have noticed that plants don't flower all at the same time. Flowering is controlled by hormones and by the length of time a plant receives light and dark, Figure 21-3. Some plants bloom in the autumn when the days are getting shorter and nights are getting longer. As the days get shorter, there comes a time when a flowering hormone is produced. Plants that flower when the day length falls below 12 to 14 hours are called **short-day plants.** Ragweed, chrysanthemum, and poinsettia are well know short-day plants. Many spring wildflowers are short-day plants. The flower buds form in the fall as the days get shorter. They don't open up until the next spring.

What plants bloom only in summer?

Other plants bloom in the summer as the days get longer. **Long-day plants** are plants that flower when the day length rises above 12 to 14 hours. Long-day plants include lettuce, clover, and gladiolus. Some plants, such as roses and dandelions, can produce flowers at most times of the year. They don't depend on any particular number of hours of light to produce flowering hormones. Once they start flowering, they bloom until frost stops further growth. These plants are day-neutral. **Day-neutral plants** are plants in which flowering doesn't depend on day length.

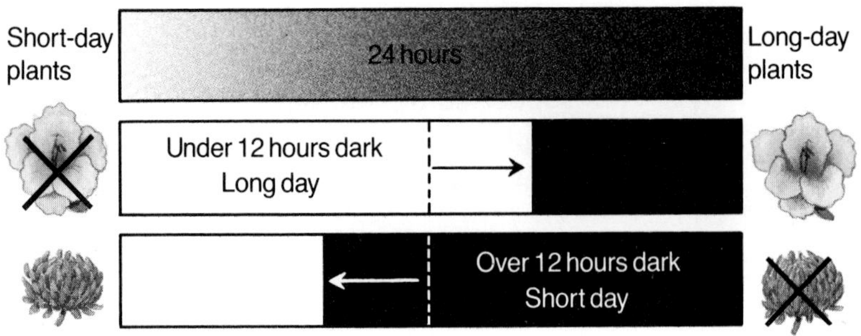

Figure 21-3 Flowering is often controlled by the number of hours of light and dark that a plant receives.

Root Growth

Lab 21–1

Problem: Where is the growth region in a root?

Skills

measure in SI, form hypotheses, interpret data, experiment

Materials

2 bean seedlings
black India ink
paper towel
1 mm graph paper
plastic bag
razor blade
toothpick

Procedure

1. Copy the data table.
2. Place a bean seedling on the graph paper. The graph paper will serve as a ruler.
3. **Measure in SI:** Dip a toothpick in the ink. Starting from the tip of the root, draw three lines 2 mm apart, Figure A.
4. Repeat steps 2 and 3 with one more bean.
5. Choose one bean seedling and use a razor blade to remove 2 mm from the root tip. **CAUTION:** *Be careful when using a razor blade.*
6. Label two areas of a paper towel Bean 1 and Bean 2. Bean 2 will be the seedling with no root tip.

DAY	DAY 1		DAY 2		DAY 3	
BEAN NO.	1	2	1	2	1	2
Tip to first mark		—		—		—
First to second mark						
Second to third mark						

7. **Experiment:** Wet the paper towel. Place the two seedlings in place. Put the towel in a plastic bag.
8. Write a **hypothesis** in your notebook about which part of the root will grow the most in three days.
9. **Measure in SI:** Examine the seedlings for three days. Each day, use the graph paper to measure the distances between the ink marks. Record these data in your table.

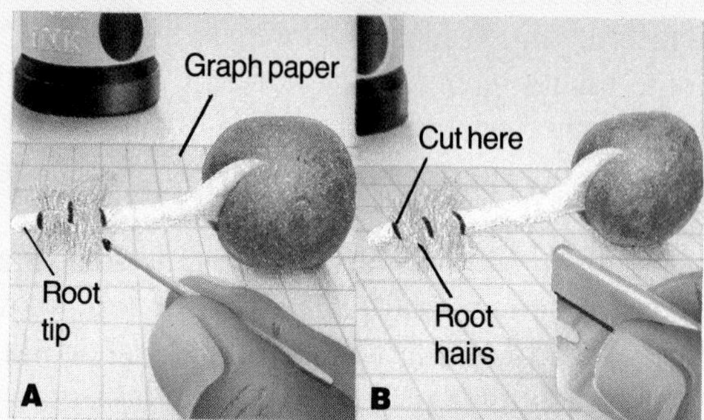

Data and Observations

1. Which bean had the most root growth?
2. In which part of the root tip of Bean 1 was there the most root growth?

Analyze and Apply

1. **Interpret data:** Why did the distance between some lines increase or stay the same?
2. Did growth take place in the bean with no root tip? Explain your answer.
3. **Check your hypothesis:** Is your hypothesis supported by your data? Why or why not?
4. **Apply:** What substance was present in the areas that caused the roots to grow?

Extension

Compare how roots of other seedlings grow.

Tropisms

Plants respond to stimuli such as light and gravity. Animals respond to stimuli because they have sense organs. How do plants respond to stimuli if they don't have a nervous system? Some plant responses, such as flowering and movement towards light, are caused by changes in growth patterns of cells.

A **tropism** (TROH pihz um) is a movement of a plant caused by a change in growth as a response to a stimulus. You may not have seen a tropism because most of the movements are so slow. When a root grows down or when a stem bends over, the plant is showing a tropism.

One tropism that is easy to recognize is phototropism (foh toh TROH pihz um). **Phototropism** is the growth of a plant in response to light. A plant showing phototropism bends toward the light. When leaf stalks bend toward the light, more light falls on the leaves. How is this useful to the plant? It's clear in Figure 21-4 that light is very important to the growth of sunflowers.

Hormones play a role in tropisms. The growth hormone in a stem tip is affected by light. Light causes the growth hormone to move to the dark side of the stem where it causes the cells to lengthen, Figure 21-5. Because the cells on the dark side of the stem become longer, the stem bends toward the light.

Figure 21-4 Sunflowers were probably named for their obvious response to sunlight.

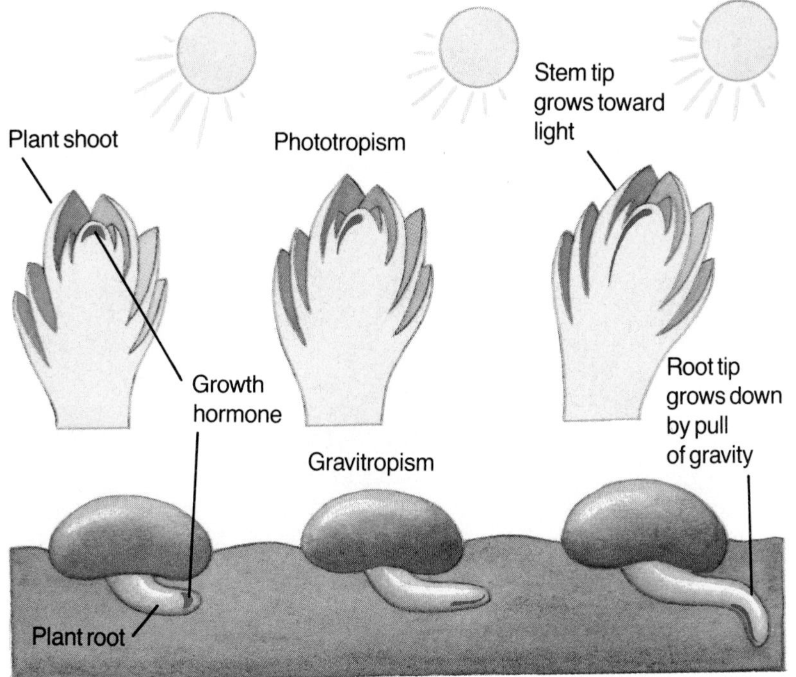

Figure 21-5 Phototropism causes a stem to grow towards light, and gravitropism causes roots to grow down.

Skill Check

Observe: Observe a climbing plant. Plant a pea seed. Place a pencil in the pot. Observe thigmotropism as the plant grows. *For more help, refer to the* ***Skill Handbook,*** *pages 704-705.*

Roots grow downward into the soil as a result of gravitropism (grav uh TROH pihz um). *Gravi* means having weight. **Gravitropism** is the response of a plant to gravity. As in the stem, the growth region of the root produces a growth hormone.

If you put a sprouting bean on its side on top of some potting soil, the root will grow downward into the soil within a few days. Large amounts of growth hormone in the root move through the cells to the lower side of the root, as if pulled down by their weight, Figure 21-5. This diffusion is in response to gravity. The hormone slows down the growth in length of the lower cells. The cells above continue to grow longer and the root grows down into the soil.

What is thigmotropism?

Grape vines, honeysuckle, and greenbriar are very common climbing plants in the forests of the United States. These climbing plants hold onto trees or fences by thigmotropism. *Thigma* means touch. **Thigmotropism** is a plant growth response to contact. Many climbing plants have twining parts called tendrils that respond to contact with a likely support by coiling around this object. The start of coiling can happen within ten minutes. Peas, beans, and squash are examples of food plants with tendrils that show thigmotropism, Figure 21-6.

Figure 21-6 Thigmotropism helps a plant climb a fence.

Lab 21–2 Plant Responses

Problem: How does gravitropism affect the growth of roots?

Skills

observe, form hypotheses, interpret data

Materials 

petri dish
2 paper towels
marking pen
transparent tape
4 presoaked corn seeds

Procedure

1. Copy the data table.
2. Soak two paper towels in water.
3. Wrinkle the towels and place them in the bottom half of the petri dish.
4. Place four presoaked seeds on top of the wet paper towels as shown in Figure A. Place the seeds so the narrow end is pointing toward the center of the dish.
5. Put the top on the petri dish. The lid will press the seeds into the wet towels.
6. Seal the lid with transparent tape.
7. Draw an arrow on the top of the petri dish with a marking pen as shown in Figure A. This will show the direction of the force of gravity.

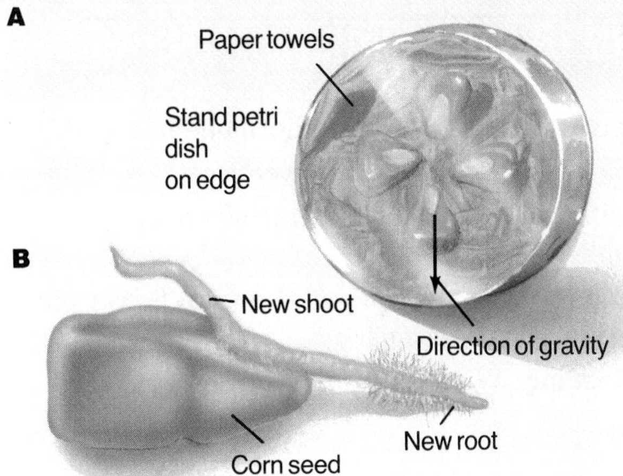

8. Stand the petri dish on edge so that the arrow is pointing straight down.
9. Write a **hypothesis** in your notebook about which direction you think the roots will grow.
10. **Observe** the seeds for the next four days. Each day, fill in the data table with arrows to show the direction of root growth for each of the four seeds.

Data and Observations

1. In what direction were the roots growing on Day 1?
2. What is the direction of root growth at the end of four days?

SEED	DIRECTION OF GROWTH OF ROOT			
	Day 1	Day 2	Day 3	Day 4

Analyze and Apply

1. **Interpret data:** To what factor of the environment are the roots responding?
2. What is this response called?
3. **Check your hypothesis:** Is your hypothesis supported by your data? Why or why not?
4. **Apply:** If this same experiment were carried out in space, would the results be the same? Explain your answer.

Extension

Design an experiment that shows the effect of phototropism.

Figure 21-7 Leaf and flower movements in these bloodroots (left) and movements of hairs on a sundew leaf (right) are controlled by changes in cell pressures. Where are the cells that control these movements?

> **Skill Check**
>
> **Understand science words: biennial.** The word part *bi* means two. In your dictionary, find three words with the word part *bi* in them. *For more help, refer to the* ***Skill Handbook,*** *pages 706-711.*

Other Plant Responses

Not all responses of plants are caused by hormones. Some plant parts move because of changes in pressure inside the cells. You read in Chapter 19 that stomata in the leaves open and close for gas exchange. Stomata open and close when the water pressure inside the guard cells changes. A stoma opens when the guard cells are full with water. When water is low in the cell, the stoma closes.

There are many other plant movements that are caused by changes in cell pressure. For example, the movement of flowers and leaves shown in Figure 21-7 and described earlier for bean plant leaves are caused by changes in cell pressure. At night, the cells at the bases of petals and leaf stalks in some plants lose water and these parts fold up. The cells regain water in the daytime, and the leaves and flowers open out.

Some plants have leaves that trap insects, Figure 21-7. When the hairs on the leaves are touched, such as when an insect lands on them, the pressure in the cells at the base of hairs changes, causing the hairs to fold over rapidly and trap the insect.

Check Your Understanding

1. What is the effect of a high level of hormone in a bud?
2. What causes a short-day plant to flower?
3. What is the difference between phototropism and thigmotropism?
4. **Critical Thinking:** How could you get plants to flower out of their usual flowering season?
5. **Biology and Reading:** Why is it difficult to observe a plant tropism?

21:2 Growth Requirements

Objectives

4. Compare annual, biennial, and perennial plants.

5. List the growth requirements of plants.

Key Science Words

annual
biennial
perennial
fertilizer

Plants can live from a few months to thousands of years. As a plant grows, the numbers of cells increase. Plants use the energy from the sun and water, minerals, and carbon dioxide from the environment to build these new cells. Factors needed by a plant for proper growth are called growth requirements. Growth requirements for plants include light, air, water, minerals, and the right temperature. Water and minerals are usually taken from the soil, so the condition of the soil is often another growth requirement. Growth may be slowed if any of these factors is not present in the right amount.

Seasonal Growth

What are five growth requirements of plants?

Most of the plants you know, such as oak trees, pine trees, poison ivy, daisies, lettuce, and onions reproduce by seeds. Seed-producing plants can grow from one to many years. Some herbaceous plants complete their growth, produce seeds, and then die within one year. Plants that complete their life cycle within one year are called **annual** (AN yul) plants. Garden plants like corn, peas, and lettuce are annual plants, Figure 21-8a.

Some herbaceous plants need two years to complete growth and produce seeds. These are called biennials, Figure 21-8b. **Biennial** (bi EN ee ul) plants produce seeds at the end of the second year of growth and then die. Cabbages and turnips are biennial plants. This is why you don't see flowers or seeds on a head of cabbage.

When do biennial plants produce seeds?

a

b

Figure 21-8 Pea plants (a) are annuals, hollyhocks (b) are biennials. What are dandelions?

A plant that lives longer than two years is a perennial (puh REN ee ul). A **perennial** plant is one that doesn't die at the end of one or two years of growth. Perennials usually produce seeds year after year. Many woody plants, such as shrubs and trees, are perennials. Their woody aboveground parts don't die at the end of each growing season. Some herbaceous plants, such as daffodils, onions, and many spring wildflowers, are also perennials. The aboveground parts of herbaceous perennials do die at the end of each growing season. Herbaceous perennials have underground parts with stored food that is used to produce new growth in the spring. Daffodils, onions, and many spring wildflowers are herbaceous perennials.

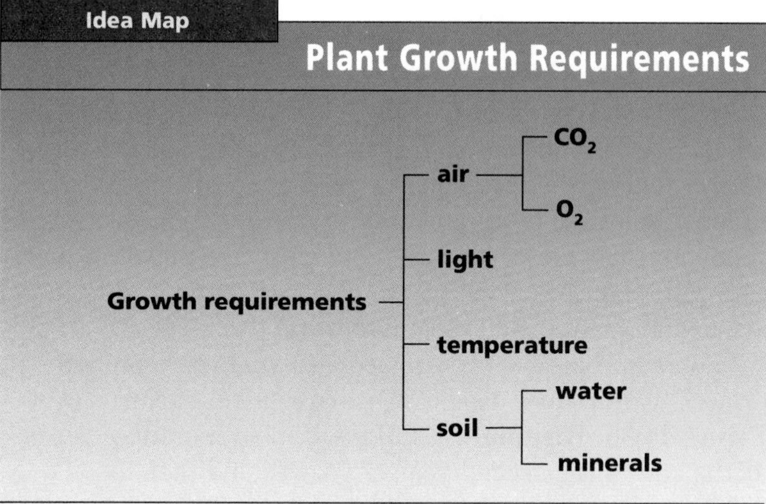

Study Tip: Use this idea map as a study guide to plant growth requirements. Which two requirements are found in soil?

Mini Lab

Do Plants Need Light?

Experiment: Plant 12 radish seeds in moist soil in each of two paper cups. Place one cup in a well-lit area and the other in a cupboard. Compare the growth of the plants for a week. *For more help, refer to the **Skill Handbook,** pages 704-705.*

Light

Plants need light for proper growth. The amount of light a plant gets controls much of its development. Production of chlorophyll, growth of buds, time of flowering, and ability to make food are all plant processes that need light.

Some plants need direct sunlight in order to grow well, others don't. Impatiens, popular bedding plants with brightly-colored flowers, grow best if they aren't planted in direct sunlight. Tomatoes and corn grow best in direct sunlight. Corn grown in the shaded part of a garden will produce short stalks and small ears. Tomatoes will take longer to ripen if grown in the shade. Plants grown in areas that are too shady will grow tall, be lighter green in color, and have fewer, but larger, leaves.

a

b

Figure 21-9 The environments of a cactus (a) and a fern (b) are different in rainfall levels.

Water

The two plants shown in Figure 21-9 have different water needs. The cactus can grow in a dry, sandy soil. The fern needs a moist, rich soil to grow well. A desert cactus takes up water from rain that comes only a few times each year. The cactus stores the water for use between rainfalls. The fern isn't adapted to store water. It needs to live in an area where the soil is damp most of the time.

How can some plants live in a desert with little rainfall?

Water has many important uses in plants. Water makes up 30 to 90 percent of a plant's mass. Plants use some of this water for photosynthesis. Minerals taken up by the roots are dissolved in water. Once inside a plant, this water carries the minerals throughout the plant. A farmer or gardener needs to water crops if there isn't enough rain to supply the needs of the plants.

Minerals

Most plants grow in soil. Soil contains many minerals important to plant growth. Plants also use minerals for making chlorophyll and cell walls. Three important minerals used by plants are nitrogen, phosphorus, and potassium.

Plants may absorb more of one type of mineral than another. Why? Since plants need some minerals such as nitrogen, in larger amounts than others, the soil can become drained of that mineral. If the soil of a garden or field is low in one kind of mineral, fertilizer can be added. A **fertilizer** is a substance made of minerals that improves soil for plant growth. When you want to help your flowering plants bloom or your shrubs grow faster, you might apply different types of fertilizer. The labels on fertilizer packets give the percentages of nitrogen, phosphorus, and potassium that the fertilizers contain, Figure 21-10.

Figure 21-10 Fertilizer can supply missing minerals to soil.

Bio Tip

Leisure: Because flowers bloom at different times of the year, you can design a garden so that you will have flowers all year round.

Soil

You have learned that soil supplies water and minerals to plants. Soil also gives the plant a place to set down its roots. Soil is a mixture that contains minerals, living organisms, dead or decaying matter, air, and water. The minerals come from rocks that have been worn down by wind and rain.

There are three main types of soil, Figure 21-11. Clay soil is hard because the soil particles are small and packed closely together. There are fewer air spaces. Other soil is crumbly because the soil particles are larger. There are more air spaces between the soil particles. Crumbly soil is better for plant growth. The air spaces in this type of soil give roots room to grow. The air spaces also keep roots dry. If roots are soaked too long, they may rot.

If the soil in your garden were like clay, you might first loosen the soil with a hoe. Also, you could add some sand and peat moss. Adding sand and dead organic matter to loosened clay soil increases the number of air spaces and adds nutrients.

Some soils are poor for plant growth. Pure sandy soil is an example of poor soil because it allows water to drain through too fast. The air spaces can't trap and hold water. Plants in desert environments often have enormous root systems to collect as much water as possible.

Temperature

Which plant in Figure 21-12 will be most affected by colder temperatures? Some plants can live in cold climates, while others die when the temperature falls below freezing.

Figure 21-11 Clay soil (a) is very fine; crumbly soil (b) is rich in organic matter; and sandy soil (c) is coarse and can't hold water.

a

b

c

a

b

Figure 21-12 Some plants live only in cold climates (a), whereas others, such as these palm trees, couldn't survive low temperatures (b).

Temperature affects the rate at which a plant will take up water, and the rates of photosynthesis and growth.

Farmers and gardeners know that they can plant most seeds or plants only after all danger of frost has passed. If they do their planting before this time, their plants will freeze or their seeds won't grow. Some seeds, however, need cold temperatures. For example, one kind of wheat called winter wheat is planted in northern states in the fall. It starts to grow until the temperature gets near freezing. The wheat is then protected by snow cover throughout the winter and begins to grow again in the spring. Some seeds, such as those of apples, will not grow unless they have first been frozen.

Check Your Understanding

6. What are two ways in which annual and perennial plants differ?
7. What are three minerals important to the growth of plants?
8. Compare the light requirements of two different plants.
9. **Critical Thinking:** If you collected plants that were native to Texas and brought them home with you to Maine, what would you need to do to keep your plants alive?
10. **Biology and Reading:** Why don't you see flowers or seeds on a head of cabbage?

21:3 Plant Diseases and Pests

Objectives

6. **Explain** how bacteria and viruses affect plants.
7. **Describe** how a fungus can kill a plant.
8. **Relate** two ways that insects damage plants.

Diseases and insect pests can slow down or stop plant growth. Diseases affect all plants.

Bacteria and Viruses

Bacteria cause many plant diseases. They can enter a plant through the stomata or small cuts. Once inside, the bacteria destroy plant cells when they invade the cytoplasm. If the bacteria spread throughout the plant, the plant might die. Bacteria are a common cause of blister spots on fruit and leaves, Figure 21-13a.

What causes yellow spotting on leaves?

Viruses also cause plant diseases. Garden plants and houseplants can be affected by a plant virus that causes small yellow spots on leaves. Eventually the yellow spots darken as the tissue dies. This kind of disease is called a mosaic disease, Figure 21-13b.

Viruses often cause the growth of tumors in leaves. The leaves become deformed. Plant viruses don't infect animals so it doesn't hurt you to touch these leaves.

Fungi

How does a fungus destroy an elm tree?

Many plant diseases are caused by fungi, Figure 21-13c. A plant disease caused by a fungus can spread rapidly through entire fields of crops.

Dutch elm disease is caused by a fungus that has killed hundreds of thousands of American elms. A small bark beetle carries the spores of the fungus from infected to healthy trees. The fungus enters the stem, begins to grow, and eventually plugs up the xylem tissue.

Figure 21-13 Plants can be infected by bacteria (a), viruses (b), or fungi (c).

a

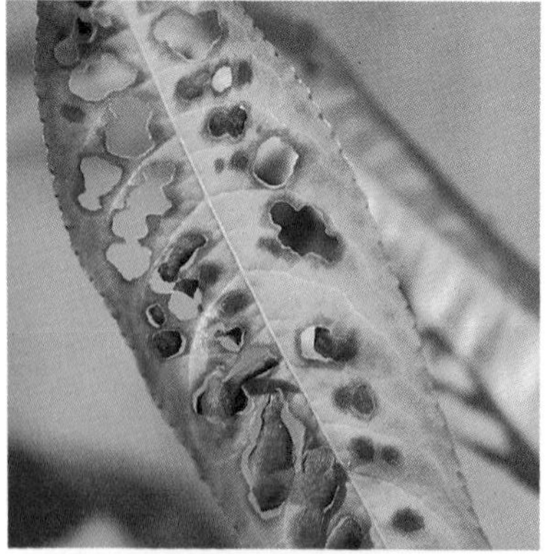

b

c

a

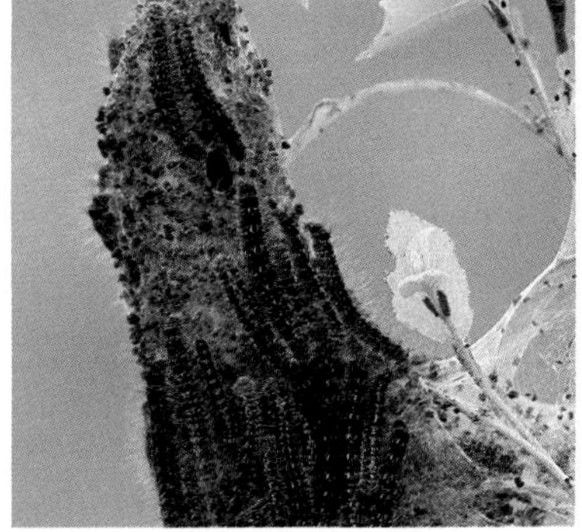
b

Figure 21-14 An insect can transfer viruses from its mouthparts to uninfected plants (a). Tent caterpillars can eat all the leaves on a tree in a few days (b).

Other crop diseases caused by fungi include wheat rusts and corn smuts. There are also a wide variety of fungi that cause fruit and vegetable rot. Recall that spores are the reproductive structures of fungi. Many fungi are spread by the wind and rain that carry the spores.

Insect Pests

How do insects transfer viruses to plants?

In general, insects are a great help to plants. They carry pollen from flower to flower. Sometimes, however, insects cause plant diseases. Suppose an insect visits a plant that has a virus, bacterium, or fungus in its cells. The insect takes in these microbes when it eats tissues of the plant, Figure 21-14a. When the insect visits the next plant, it transfers the disease microbe from its mouthparts to the uninfected plant. The disease soon develops in that plant.

Insects also can damage or even kill plants by eating too many leaves. A plant with damaged leaves can't make as much food. For example, the tent caterpillar, Figure 21-14b, builds a web in a tree and can eat all of the leaves on that tree in a few days.

Bio Tip

Ecology: Insects like praying mantis and lady bugs are helpful to gardeners because they eat harmful insects. Take care not to spray these insects with pesticides.

Check Your Understanding

11. What is mosaic disease?
12. How do spores of Dutch elm disease enter a tree?
13. How can an insect reduce photosynthesis in a plant?
14. **Critical Thinking:** What is a way to control insect pests in your garden without spraying pesticides?
15. **Biology and Reading:** What two kingdoms contain organisms that cause plant diseases?

How Much Nitrogen?

One of the most important nutrients needed by plants is nitrogen, a substance that plants use to make proteins such as chlorophyll. Air is mostly nitrogen, but most plants can't make proteins from the nitrogen in the air. To solve this problem, nitrogen can be added to soil as fertilizer. Most fertilizers have varying amounts of nitrates, chemicals that contain nitrogen. Farmers, gardeners, and landscapers test for nutrients in the soil and add fertilizers to help their plants grow the best.

Identifying the Problem

Your school is planning to plant new grass on its athletic fields. You are a landscaper who has been hired to do the work. As the landscaper, part of your job is to decide what changes should be made to the soil to make it best for growing grass. You will need to choose a type of grass that grows well in your climate and find out what soil conditions are best for that type of grass. After testing the soil on the playing field and experimenting with a couple of new grass seed mixtures, you'll be ready to suggest what type of fertilizer would help make the new grass grow the healthiest.

Collecting Information

Go to a local lawn and garden center or farm supply store, and find out about different types of grass. Collect information about the growing conditions and nutritional needs of each type. Decide which type of grass you want to use in your experiment, one that would grow best in the conditions around your school. Think about ways you can find the best soil conditions for the type of grass you have chosen.

Technology Connection

As plant experts learn more about the nutritional needs of plants, fertilizer companies must design new fertilizers and other products to meet those needs. If you were to design a new fertilizer, what are some of the qualities it would have to show? Explain why these qualities might be needed.

Carrying Out an Experiment

1. Collect several plastic trays, a bucket of soil from the area around your school, several sheets of newspaper, grass seed, fertilizer, and a soil test kit. Soil test kits are found in your local lawn and garden center. They can be used to test soil pH, nutrient content, and drainage.
2. Examine the soil from your school. What materials can you see in the soil? Is it sandy, or does it have clumps of clay? Is it wet or dry? Is it packed tightly or loosely?
3. Test the soil with the soil test kit. Record the results.
4. Predict how much fertilizer you should add to the soil to make the grass grow the best. See if the fertilizer label has information that is helpful to you.
5. Mix different amounts of fertilizer into samples of the school soil, and place them in the plastic trays. As a control, use a sample of the school soil with no added fertilizer. Make a recipe for each of the soils you mix. Record what and how much of each ingredient you add.
6. Test each of the soils with the soil test kit.
7. Plant the same amount of grass seed in each sample.
8. Place the trays near a window and keep the soil samples moist. Allow the grass to grow for two to three weeks. Record your results.

Assessing Your Results

How well did your plan work? What would you change if you were to do the experiment again? Compare and contrast the soils you mixed. Which of the soil samples had the best plant growth? Based on your results, what soil recipe would you suggest using for the athletic fields?

Career Connection

- **Nursery Worker** Sells and takes care of nursery plants; gives advice to customers about plant care and diseases; delivers and plants trees and bushes
- **Lawn Care Technician** Applies fertilizers, herbicides, and insecticides to lawns; analyzes lawn problems and decides on treatments
- **Soil Specialist** Tests soil chemistry and recommends treatments to improve soil

Chapter 21

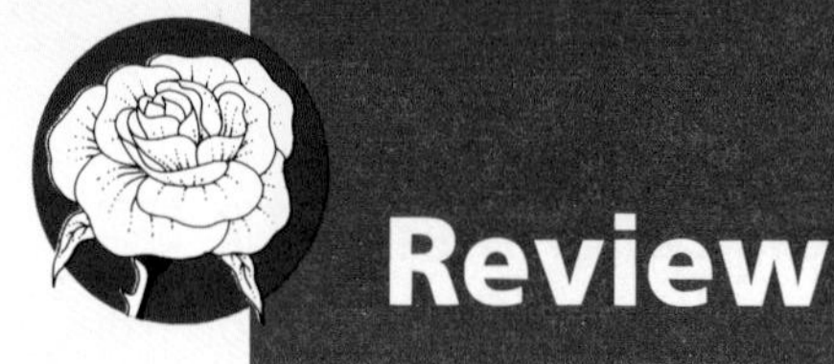

Review

Summary

21:1 Plant Responses

1. Hormones in plants control growth and flowering.
2. Short-day plants flower when day length shortens. Long-day plants flower when there are longer periods of daylight. The flowering of day-neutral plants does not depend on the length of daylight.
3. Tropisms are growth responses to stimuli such as light, gravity, and touch. Other plant movements are caused by changes in cell pressure.

21:2 Growth Requirements

4. Annuals complete their life cycle in one year. Biennials produce seeds at the end of the second year. Perennial plants can live for many years.
5. Growth requirements for plants include varied amounts of light, water, minerals, air, and temperature. The type of soil is often important.
6. Different plants in different environments need different levels of light, temperature, water, and minerals.

21:3 Plant Diseases and Insects

7. Bacteria and viruses cause many plant diseases. They enter the plant through cuts, stomata, or roots.
8. Fungal diseases spread from plant to plant by spores.
9. Insects eat plants and damage them by carrying bacteria, viruses, and fungi from plant to plant.

Key Science Words

annual (p. 447)
biennial (p. 447)
day-neutral plant (p. 441)
fertilizer (p. 449)
gravitropism (p. 444)
long-day plant (p. 441)
perennial (p. 448)
phototropism (p. 443)
short-day plant (p. 441)
thigmotropism (p. 444)
tropism (p. 443)

Testing Yourself

Using Words

Choose the word from the list of Key Science Words that best fits the definition.

1. plant that blooms in the autumn
2. plant response to gravity
3. substance made of nutrients that improves soil for plant growth
4. plant response to a stimulus
5. the response of a plant to light
6. plant that blooms in the summer
7. plant growth in response to touch
8. plant that dies at end of one growing season

Testing Yourself *continued*

9. plant that blooms from early summer until frost
10. plant that produces seed year after year

Finding Main Ideas

List the page number where each main idea below is found. Then, explain each main idea.

11. how insects spread plant diseases
12. minerals used by a plant to make cell walls and chlorophyll
13. what causes stems to bend toward light
14. how a person could cause bushy growth in plants
15. how plant diseases are caused by bacteria
16. how climbing vines hold onto trees and fences
17. what causes Dutch elm disease
18. when biennial plants produce seeds

Using Main Ideas

Answer the questions by referring to the page number after each question.

19. What are the growth requirements for plants? (p. 447)
20. Where are growth hormones found in plants? (p. 440)
21. How do bacteria invade a plant and cause a disease? (p. 452)
22. What are three kinds of tropisms in plants? (pp. 443, 444)
23. What keeps lateral buds from growing faster than the terminal buds? (p. 440)
24. What effect does a tent caterpillar have on a tree? (p. 453)
25. What are three main minerals used by plants? (p. 449)

Skill Review ✓

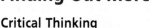

*For more help, refer to the **Skill Handbook,** pages 704-719.*

1. **Observe:** Observe plants that are growing in the shade of other plants. Explain how they differ from similar plants growing in the sun.
2. **Experiment:** Germinate two bean seeds with a moist paper towel. Experiment to find the effects of different stimuli on root and shoot growth.
3. **Measure in SI:** For five days, measure the height of a bean plant that you planted.
4. **Interpret data:** You have a garden of roses. One day, you notice some of the leaves have black spots. A week after a heavy rain, you notice that most of the leaves have black spots. The leaves then begin to drop off the plants. What's wrong with your roses?

Finding Out More

Critical Thinking

1. Explain why cutting grass with a mower doesn't stop grass growing.
2. How would a scientist decide whether or not soil is missing some minerals needed for plant growth?

Applications

1. Find out how the amounts of light and water provided for plants are controlled in a greenhouse.
2. Find out the common names of some hormones made by plants. Visit the garden section of a department store or hardware store for help.

CONNECTIONS

Biology in Your World

Healthy Plants, Healthy People, Healthy Environment

In this unit, you learned about plants and their importance. Everyone depends on plants as a food source. Safe farming practices and pest control are needed for healthy food crops. A healthy environment and successful farming are goals of scientists and farmers.

LITERATURE

Deep Woods

John Burroughs was a writer who loved nature and loved to write about it. Many of his essays and books are about his own experiences exploring the Catskill Mountains of New York, where he spent much of his life.

Burroughs brings the out-of-doors to life for his readers. His vivid, interesting writing style makes you feel as if you are with Burroughs in the woods or on a mountaintop.

Deep Woods is a collection of essays from several of Burroughs' books. Included are essays about the Catskills, Maine, Alaska, and Yosemite. Sit back and enjoy Burroughs' descriptions of the beauties and wonders of nature!

HISTORY

The Potato Famine

For more than 100 years, potatoes had been the main part of the diet in Ireland. But in the fall of 1845, the potato crop rotted in about half of Ireland. Starvation and disease followed as crop after crop rotted over the next three years. We know now that a parasitic mold caused crops to rot. The mold had spread rapidly from North America to England, and then to Ireland.

Today, chemicals can be used to control the growth of molds. Use of these chemicals can reduce the loss of crops. The chemicals may also improve the quality of crops. Maybe famines like the one in Ireland will not occur again.

LEISURE

Do You Have a Green Thumb?

You are never too old to enjoy getting dirty! Ask permission to cultivate a small, sunny area of your yard. You will need a spade, a rake, a hoe, and some seed catalogs. Choose several types of fruits and vegetables. All plants have slightly different requirements. You may have better luck with some than with others.

When you prepare your garden, make sure the soil is broken up well for proper drainage. If the soil is mostly clay, add peat moss, sand, or decayed leaves.

At harvest time, notice which fruits and vegetables did well. How can you improve your harvest next year? For each crop you grew, which part do you eat—leaf, stem, fruit, or root?

CONSUMER

Using Chemicals Wisely

Are the fruits and vegetables you buy in grocery stores safe to eat? Aflatoxin is a chemical produced by a fungus. The fungus can grow on grains and other crops. Aflatoxin is a natural cancer-causing agent. It can affect peanuts, corn, and the milk of corn-fed cows.

In 1989, scientists tested corn for the presence of aflatoxin. They found unacceptable levels of the chemical in 6 percent of the tested samples.

A certain chemical works against aflatoxin. The chemical reduces the amount of aflatoxin present in the milk of cows that eat the treated grain.

Should farmers spray their crops with chemicals to kill fungi? Not all fungi are harmful to plants. One soil-living fungus helps plants remove needed nutrients from the soil. Thus, farmers should use caution when applying chemicals so that the chemicals don't enter the soil and kill the fungi there.

Unit 6

CONTENTS

Reproduction and Development

What would happen if...

the present rate of population growth continued? The population of North America would increase to about 487 000 000 by the year 2000.

If a population of 487 000 000 people were spread out evenly over all of North America, there would be about 20 people per square kilometer. But the population is not spread out evenly. Few people live in desert and mountain areas. Most people are in cities along rivers, lakes, or oceans. The population density in cities can be hundreds of people per square kilometer.

The world population was about 5 880 000 000 in 1997. At a growth rate of 1.4%, the world population would be about 6 147 000 000 in the year 2000. That's about 41 people per square kilometer. How might this many people affect life on Earth?

CHAPTER PREVIEW

Chapter Content

Review this outline for Chapter 22 before you read the chapter.

Skills in this Chapter

The skills that you will use in this chapter are listed below.

- In **Lab 22-1,** you will observe, formulate a model, and sequence. In **Lab 22-2,** you will use numbers, make and use tables, form hypotheses, and interpret data.
- In the **Skill Checks,** you will classify and understand science words.
- In the **Mini Labs,** you will measure in SI.

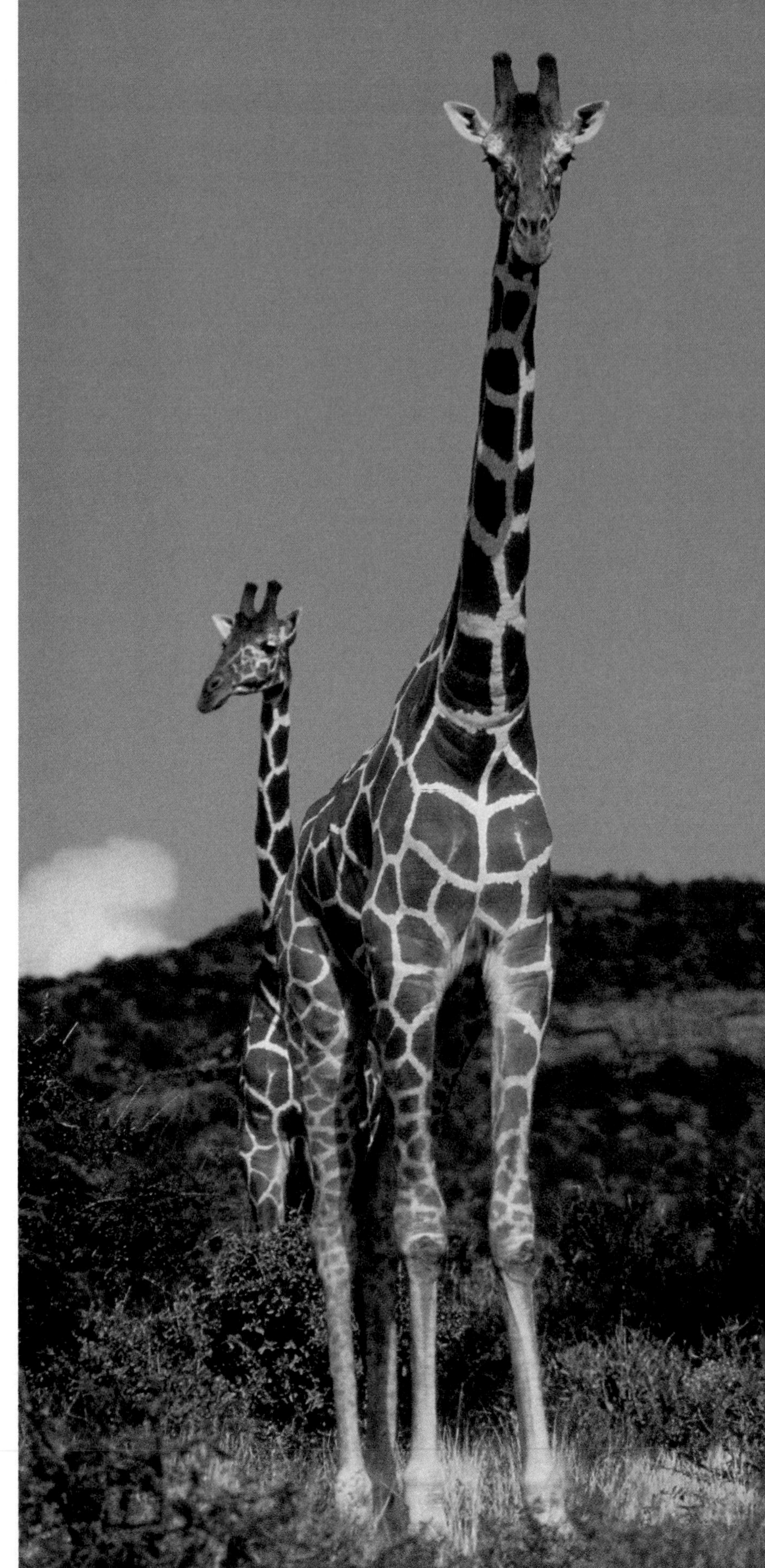

Chapter 22

Cell Reproduction

Every part of your body is made up of cells. You and all other living things started out as single cells. In some plants and animals, that cell came from a part of one parent. In most plants and animals and in humans, the cell came from the joining of a cell from a male parent and a cell from a female parent. You have learned that these two origins of living things are the difference between asexual and sexual reproduction. Look at the difference in size between the parent and its young in the photo on the left. The larger animal has millions more cells than its offspring. Look at the tree seedling below. You know that the parent tree is much larger. How does one cell become millions of cells? Where do these new cells come from?

Try This!

How many chromosomes do onion cells have? Use a microscope to examine a prepared slide of onion root tip cells under high power. Do all the cells look like they are dividing?

BIOLOGY Online

Visit the Glencoe Science Web site at science.glencoe.com to find links about **cell reproduction.**

A young tree

22:1 Mitosis

Objectives

1. **Relate** some benefits of mitosis.
2. **Identify** cell parts involved in mitosis.
3. **Trace** the steps of mitosis.

Key Science Words

mitosis
body cell
sister chromatids

You started life as a single cell. You now have millions of cells in your body. Somehow, you have grown. The new cells in your body came from cell reproduction.

Body Growth and Repair

Living things grow. You were growing even before you were born. Living things also repair themselves when they are injured. As a child, you probably cut or scraped your skin often. You must have seen new skin grow back on your hands and knees many times. Your body must make new cells to grow and to repair itself. New cells are made by the process of cell reproduction.

One kind of cell reproduction in organisms is called mitosis (mi TOH sus). **Mitosis** is cell reproduction in which two identical cells are made from one cell. Each new cell grows in size until it too is ready to reproduce by mitosis. All body cells in humans are formed by mitosis. **Body cells** are cells that make up most of the body, such as the skin, blood, bones, and stomach. Figure 22-1 shows that all body cells don't live for the same length of time. Cells carry on mitosis at different rates in different organs to replace cells that are worn out. In which body organs do cells carry on mitosis most often? In which organs do cells carry on mitosis least often? In some body cells, such as muscle cells, mitosis never occurs after birth. You are born with all the muscle cells you will ever have.

Why are new cells important to your body?

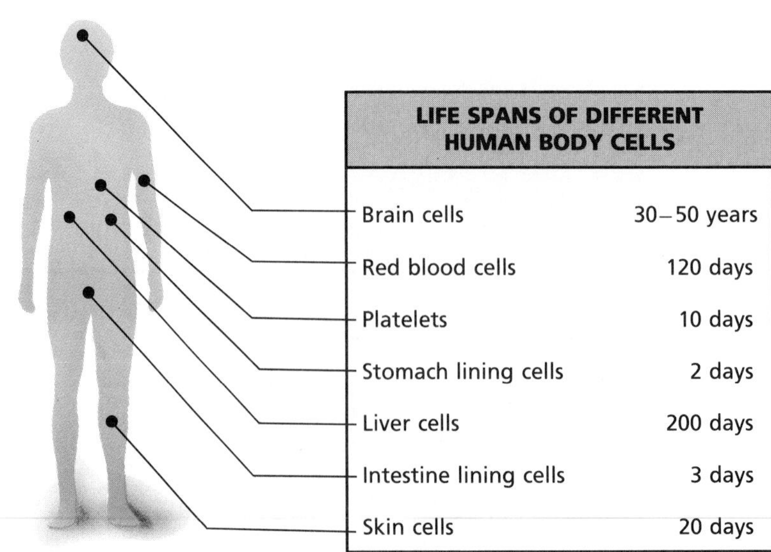

LIFE SPANS OF DIFFERENT HUMAN BODY CELLS	
Brain cells	30–50 years
Red blood cells	120 days
Platelets	10 days
Stomach lining cells	2 days
Liver cells	200 days
Intestine lining cells	3 days
Skin cells	20 days

Figure 22-1 The length of time that human cells live in the body depends on the type of cell.

TECHNOLOGY

Hope for Spinal Cord Injury Patients

It has long been thought that nerve cells don't divide by mitosis after you are born. This used to explain why a person with a spinal injury was often permanently disabled. The nerve pathways from brain to body muscles were cut and would never heal.

Recent research shows that nerve cells can regrow and repair themselves. Scientists carried out the following experiment. The optic nerves leading from the eyes to the brains of several hamsters were cut and sections were removed. Normally, this would have meant the loss of sight for the animals. The removed section was then replaced with a piece of nerve taken from another part of the body of each animal. After seven weeks, nerve endings had regrown. The piece of body nerve that was added had somehow acted as a path for the original nerve to follow as it grew. Were the animals able to see again? The scientists have already found that the hamsters could detect changes in light intensity. This means that the optic nerve is working to some extent.

Will this type of scientific breakthrough lead to the healing of spinal cord injuries? As with most scientific research, a long time is needed to solve medical problems.

Hamsters

Mitosis goes on during a person's entire life. You are constantly forming new cells to replace those worn out or lost. Mitosis begins before you are born and continues up to the time you die.

An Introduction to Mitosis

How does mitosis work? We could compare it to how a photocopy machine works. Look at Figure 22-2. Suppose you have a sketch of your pet. Your friend wants a copy. You copy the sketch on a photocopy machine. You now have two identical sketches of your pet.

The process of mitosis is similar. You start with one cell. The end result is two identical cells. The photocopy machine makes a copy in about five seconds. Mitosis may take several hours.

Figure 22-2 How are photocopying and mitosis similar?

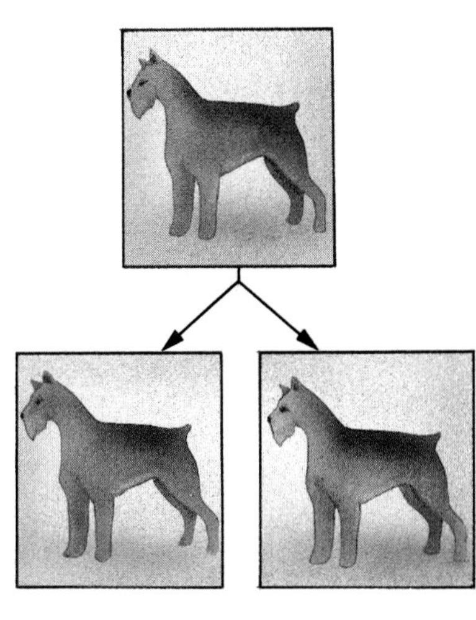

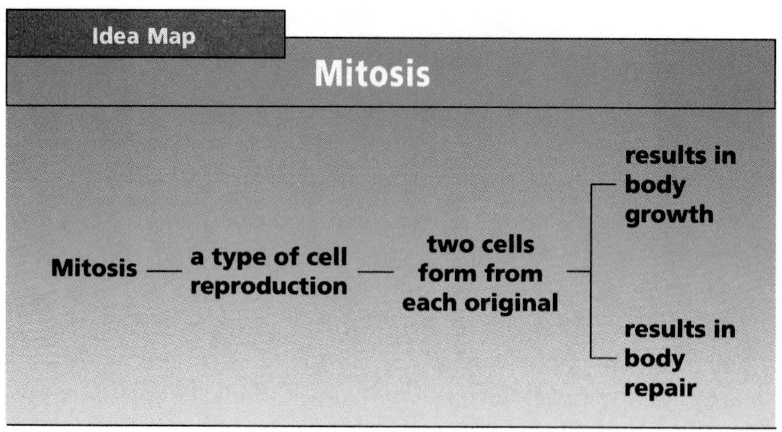

Study Tip: Use this idea map as a study guide to mitosis. If you cut your finger, how does mitosis help you?

Figure 22-3 Study the parts of an animal cell.

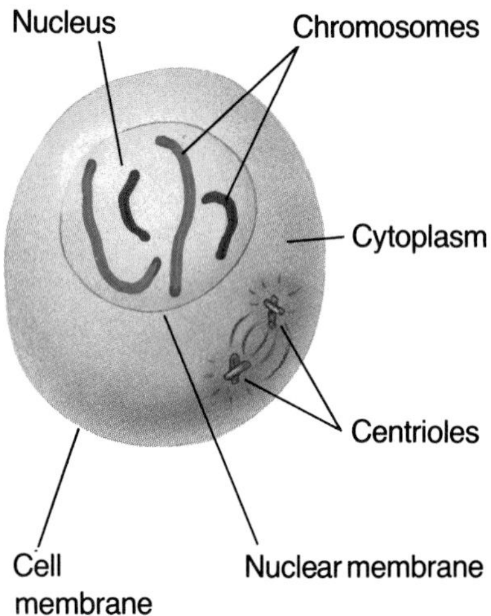

What cell parts are used in mitosis? Figure 22-3 shows what a typical animal cell might look like. Remember from Chapter 2 that the entire cell is surrounded by a cell membrane. Most of the material inside the cell is called cytoplasm. A cell nucleus is present in the cytoplasm. A nuclear membrane surrounds the nucleus. You have also read that inside the nucleus there are chromosomes. They may not be easy to see at this time, but they are there. Notice that there are also centrioles in the cytoplasm.

What does each chromosome do before mitosis begins?

Most cells contain many chromosomes. To explain how a cell goes through mitosis, let's imagine a cell with just four chromosomes. An important step takes place in the nucleus of every cell before mitosis begins. Each chromosome becomes doubled. Figure 22-4 shows how this works. Notice how each chromosome looks after it has doubled. The two strands of each doubled chromosome are held together at one point. The two strands of a chromosome after it becomes doubled are called **sister chromatids** (KROH muh tidz). Each chromatid is an exact copy of the original chromosome. The centrioles also double just before mitosis begins.

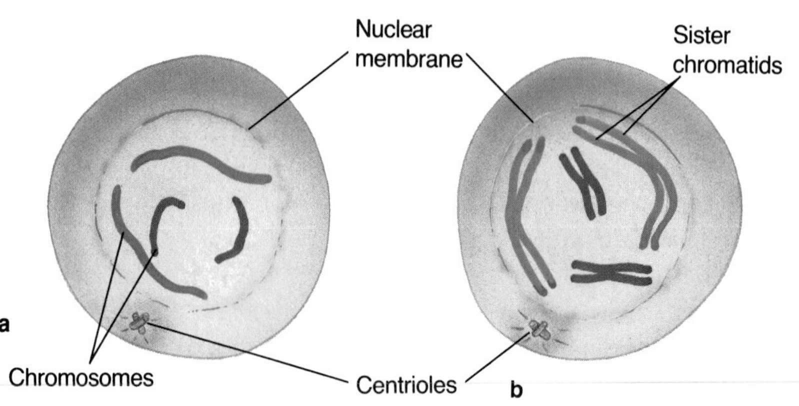

Figure 22-4 Chromosomes become doubled before mitosis begins.

Lab 22–1

Steps of Mitosis

Problem: What do the steps of mitosis look like?

Skills

observe, formulate models, sequence

Materials

paper
pencil

Procedure

1. Look at Figure 22-4 on page 466 showing the stages of mitosis. Notice in Figure 22-4a that the chromosomes have not yet formed sister chromatids. In other words, they're not yet doubled.
2. **Formulate a model:** Trace the cell diagram in Figure 22-4a onto a piece of paper.
3. Draw a new cell that shows the next change before mitosis begins. Trace the diagram in Figure 22-4b.
4. **Sequence:** Use Figure 22-5 on pages 468 and 469 to help you draw the next four steps that would take place in this cell during mitosis.
5. In your diagrams, use the following words to label the parts:
   Original cell—cell membrane, nucleus, nuclear membrane, cytoplasm
   Change before mitosis—sister chromatids
   Step 1—sister chromatids, spindle fibers
   Step 2—fibers, sister chromatids at cell center
   Step 3—separation of sister chromatids, fiber, chromosomes
   Step 4—fibers disappearing, building a new cell wall, nuclear membrane reforms, chromosomes
   Body cells—chromosomes, nucleus, cell membrane, cytoplasm, body cells

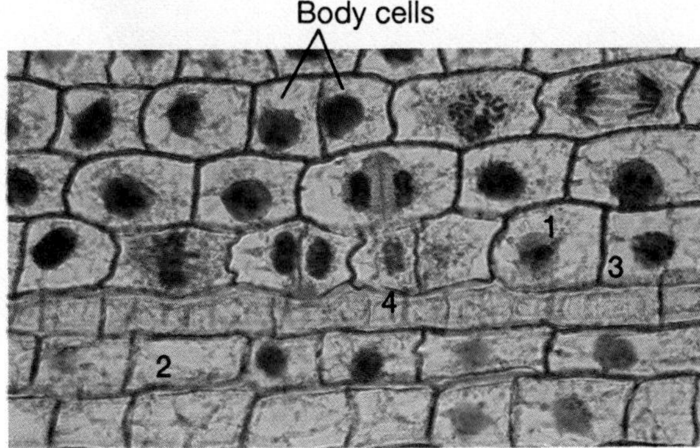

Data and Observations

1. At what step in mitosis do the sister chromatids separate?
2. Would your diagrams differ if two chromosomes had been used? Explain.

Analyze and Apply

1. What is the difference between chromosomes and sister chromatids?
2. How would your diagrams differ if this were meiosis in terms of:
   (a) the final number of cells?
   (b) the final number of chromosomes in each cell if you had started with 46 chromosomes?
   (c) the number of steps needed?
   (d) the labels on the last diagram?
3. **Apply:** Give two examples of human body organs where you would find:
   (a) mitosis taking place.
   (b) meiosis taking place.
4. How are fibers important to cell division?

Extension

Formulate a model: Yarn and paper clips may be used to model the series of changes that occur to two chromosomes before and during mitosis.

Steps of Mitosis

Follow the numbers in Figure 22-5 as you read here about the four steps of mitosis.

Step One

1. Sister chromatids begin to shorten and thicken. If you were to look at a cell in this step of mitosis under a light microscope, you would be able to see the chromosomes. However, you might not be able to see the strands of each sister chromatid.
2. The nuclear membrane begins to break down. The pairs of sister chromatids now look like they are floating in the cytoplasm.
3. The centrioles move away from each other.
4. Fibers form between the centrioles. These fibers are strands of protein that form between the two ends of a cell.

In which step of mitosis are the chromosomes pulled to the center of the cell?

Step Two

1. The centrioles move apart to opposite ends of the cell. The fibers look like they are stretched between the two ends of the cell.
2. Sister chromatids become attached to the fibers at the point where the two strands are joined to each other. The

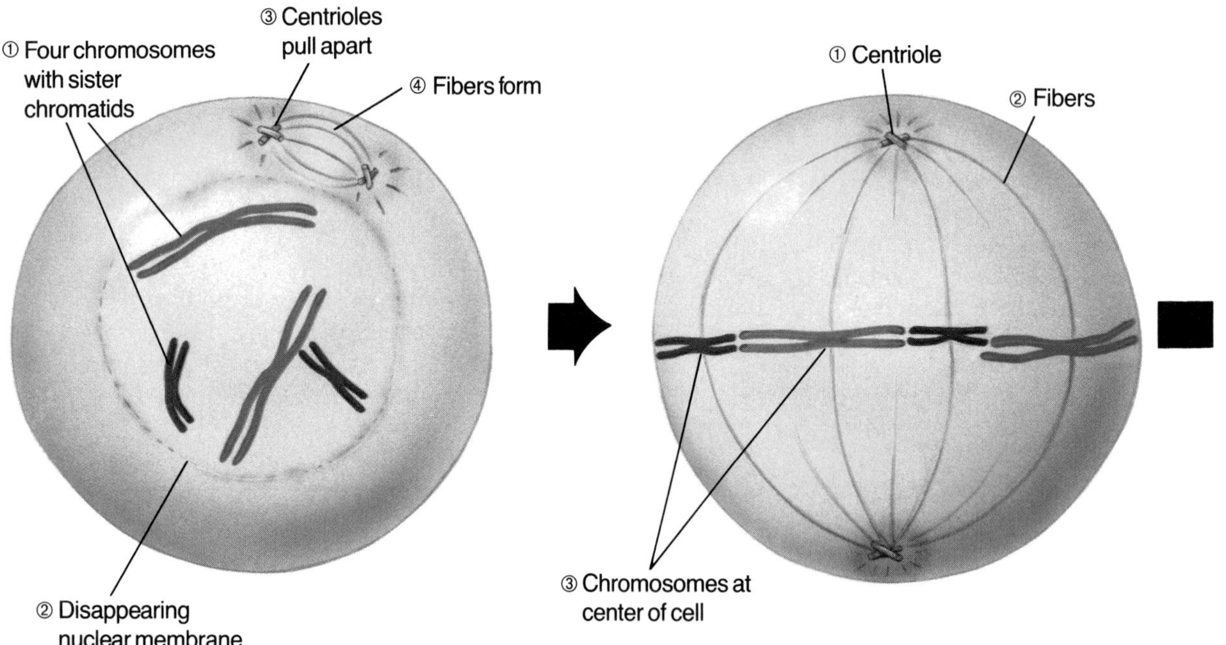

Figure 22-5 Use these diagrams to help you follow the steps of mitosis as you read the text. How many times does the cell divide?

pairs of sister chromatids are pulled toward the center of the cell.

3. The four pairs of sister chromatids are now lined up at the center of the cell.

Step Three

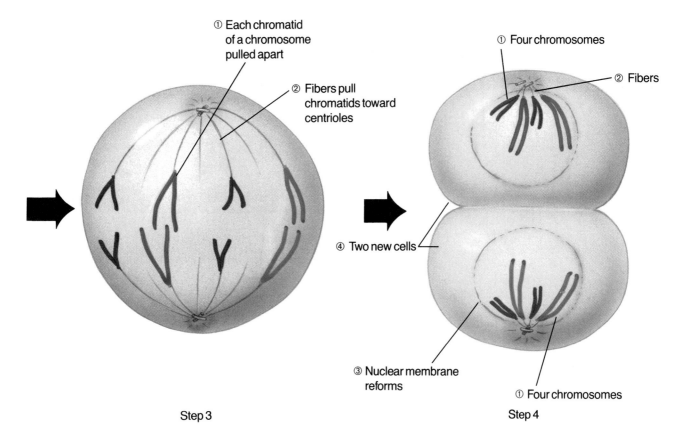

1. The sister chromatids are pulled apart by the fibers. Each chromatid is now separate from its identical partner.
2. The fibers pull each chromatid strand toward the centrioles at opposite ends of the cell. Remember that each chromatid is an exact copy of one original chromosome.

Step Four

1. Each end of the cell now has a complete set of chromosomes. In this example, a complete set is four chromosomes. Notice that this is the same as the number we started with in Step One.
2. The fibers begin to disappear.
3. The nuclear membrane begins to reform.
4. The cell membrane begins to pinch in until the cytoplasm is divided in half. Two new cells have formed. They each have identical chromosomes. They also have the same number of chromosomes as the cell they originally came from.

Bio Tip

Consumer: You may have seen advertisements that claim their creams will make your skin soft and as new as a baby's. Many of these creams work by adding moisture to your skin, not by causing new cells to grow.

Figure 22-6 Mitosis in plant cells: Step One (a), Step Two (b), Step Three (c), Step Four (d)

a

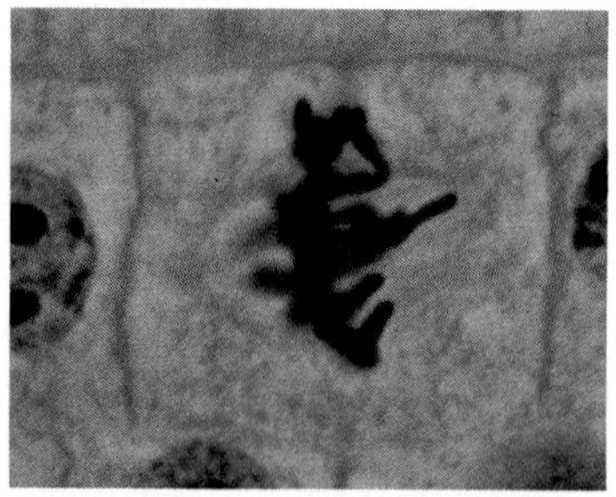
b

c

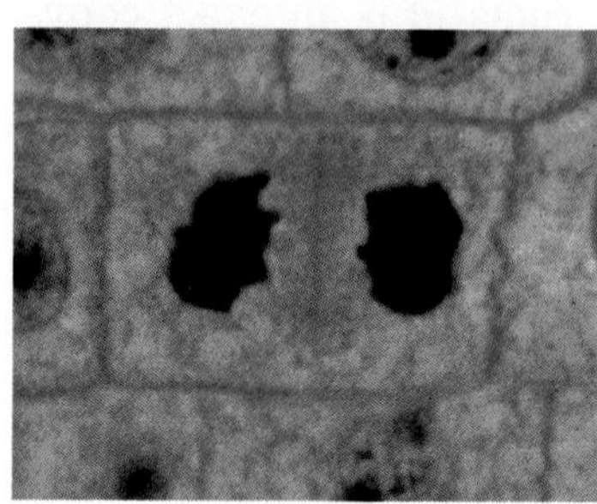
d

Mini Lab

How Rapid Is Mitosis?

Measure in SI: Measure the amount of growth in the root and shoot of a radish seedling over a 24 hour period. *For more help, refer to the* ***Skill Handbook****, pages 718-719.*

The two new cells are smaller in size than the original. However, each new cell now begins to grow.

What are the benefits of mitosis? First, mitosis helps us grow by producing new cells. Second, mitosis replaces cells lost through cell death and injury, such as when you cut your finger.

Figure 22-6 shows photographs of plant cells in the four steps of mitosis. As you can see, mitosis in plant and animal cells is very similar. Mitosis in plant cells shows two differences from mitosis in animal cells. Plant cells lack centrioles, and at the end of cell division a cell wall is laid down.

Check Your Understanding

1. Why is mitosis important to your body?
2. What are the jobs of protein fibers during mitosis?
3. What happens to chromosomes before the nuclear membrane breaks down in mitosis?
4. **Critical Thinking:** How does mitosis help a plant stem grow in width?
5. **Biology and Math:** Some cells divide every two hours. How many cells would result from mitosis in twenty-four hours if you started with one cell?

22:2 Meiosis

Not all cells reproduce by mitosis. A different type of cell reproduction is important to the passing on of traits to offspring.

Objectives

4. **Explain** the results of meiosis.
5. **Trace** the steps of meiosis.
6. **Compare** the sex cells of males and females.

An Introduction to Meiosis

Key Science Words

sex cell
meiosis
puberty
testes
ovary
polar body

Besides body cells, most living things also have sex cells. **Sex cells** are reproductive cells produced in sex organs. Sperm are sex cells made by the male. An egg is a sex cell made by the female. Sex cells are made during meiosis (mi OH sus). **Meiosis** is a kind of cell reproduction that forms eggs and sperm.

To explain how a cell goes through meiosis, let's look again at our cell with four chromosomes, Figure 22-7. The four chromosomes make up two matching pairs.

In meiosis, a cell divides twice. When the cell divides for the first time, each chromosome in a pair moves away from its partner. Each chromosome of a pair goes to a different cell. The sister chromatids stay joined together. The two cells then divide again.

Look at Figure 22-7. How many chromosomes are in each of the four final cells? The number of chromosomes in each cell is one-half the original number. The original cell started with four chromosomes. Each new cell now has only two, or exactly one-half of the original number. Let's examine the process of meiosis in more detail to see how the chromosome number was halved.

After meiosis, how does the number of chromosomes in a new cell compare with those in the original cell?

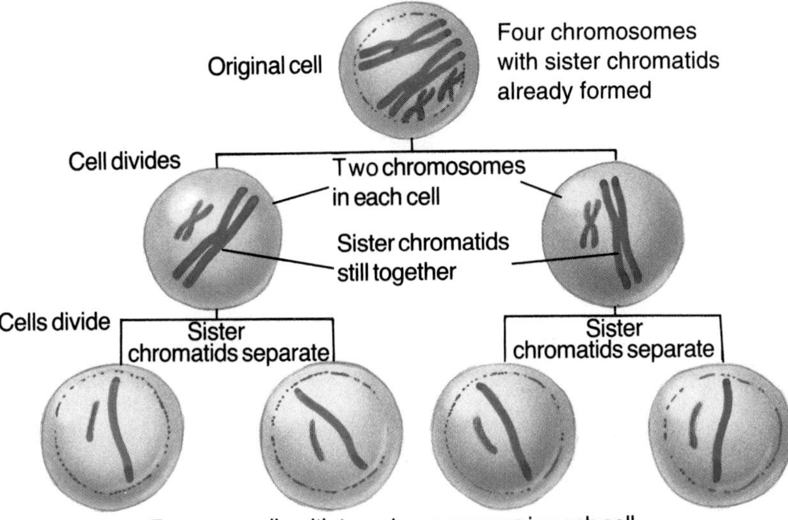

Figure 22-7 Meiosis begins with one cell and results in four cells. Each new cell has one-half the original chromosome number.

Steps of Meiosis

Follow the numbers in Figure 22-8 as you read here about the six steps of meiosis. Just as in mitosis, each chromosome forms sister chromatids before meiosis begins.

Step One

1. The sister chromatids shorten and thicken.
2. The nuclear membrane begins to break down.
3. The centrioles begin to move away from one another and fibers form.
4. The matching chromosomes now come together to form pairs. Remember that each chromosome has two strands of sister chromatids. The pair of chromosomes now looks like a set of four strands and can be seen with a light microscope. Remember that in mitosis this pairing step doesn't take place.

Step Two

1. The centrioles have moved to opposite ends of the cell.
2. The sister chromatids become attached to the fibers.
3. Fibers move the two pairs of matching chromosomes to the center of the cell. Note that each chromosome is made up of sister chromatids that are still attached to one another. However, the paired chromosomes are not joined.

Figure 22-8 Follow the steps of meiosis as you read the text. How many times do the cells divide?

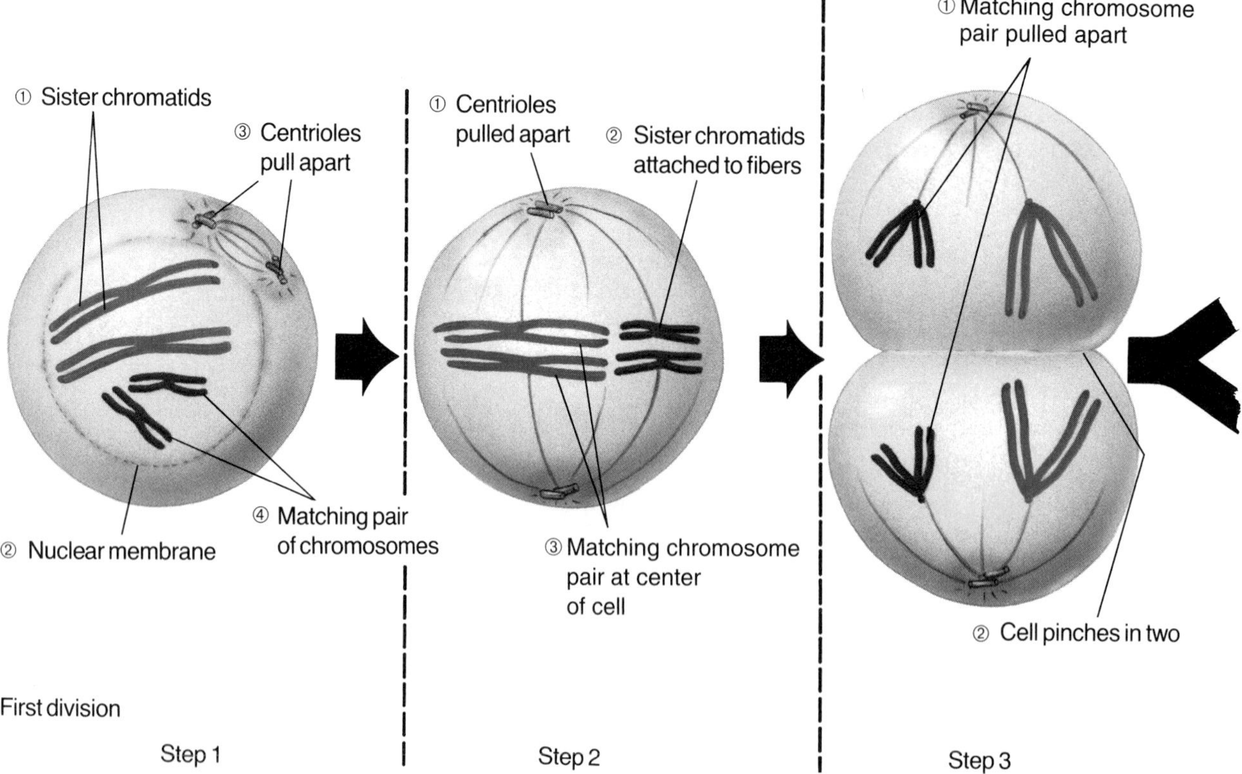

Step Three

1. Fibers move the matching chromosomes apart. Remember that in mitosis the sister chromatids came apart. In meiosis, the sister chromatids remain joined, but each matching pair of chromosomes separates.
2. The cell membrane begins to pinch the cell into two and divides the cytoplasm in half.

In which step of meiosis do the pairs of sister chromatids separate?

Step Four

1. Two new cells have now formed. Each cell has two chromosomes, each with sister chromatids.
2. The centrioles double and fibers form again.
3. A new nuclear membrane doesn't form at this time.

Step Five

The second division of meiosis now begins. Steps five and six now occur in both of the new cells. The movement of chromatids is similar to that in steps one to four of mitosis.

1. The centrioles move apart, and the fibers are formed between them.
2. The fibers connect to the sister chromatids at the point where the chromatids are joined together.
3. The sister chromatids are pulled to the center of each cell.

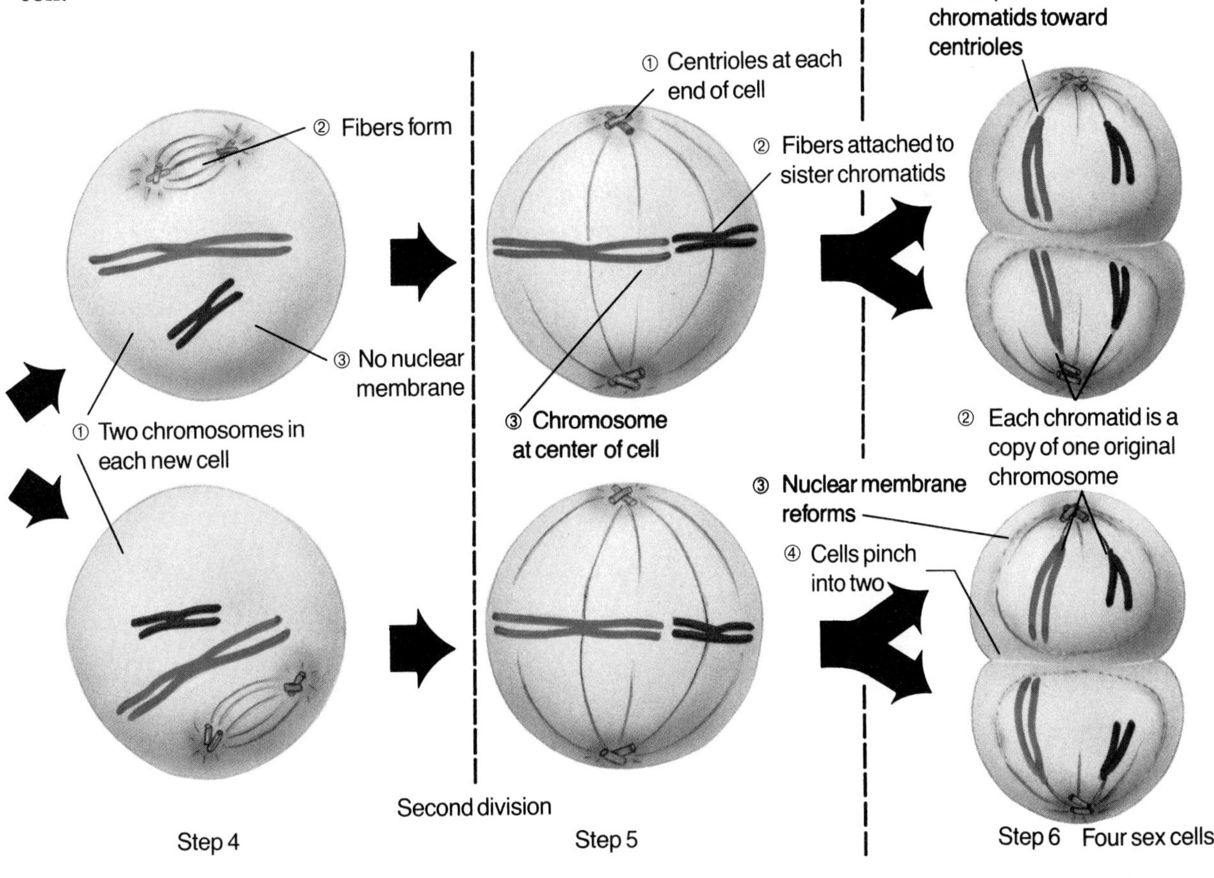

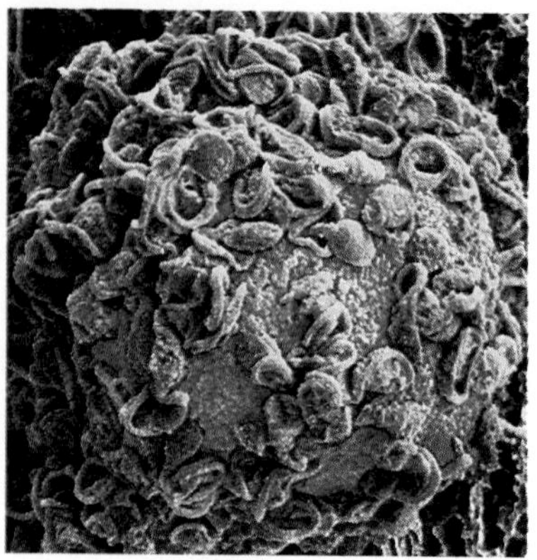

Figure 22-9 Human egg and sperm cells

Step Six

1. The fibers pull each strand of the sister chromatids apart and to opposite ends of the cell.
2. Each sister chromatid is an exact copy of just one of the original chromosomes.
3. The nuclear membrane begins to reform around each new set of chromosomes.
4. Cell membranes begin to pinch each cell in two along the center. The cytoplasm is divided between the new cells. The four new cells are sex cells.

Remember, that in our cell we started with four chromosomes. We now have two chromosomes in each of four cells. In a normal cell, there may be many more chromosomes. But, at the end of meiosis there will always be half the original number of chromosomes in each of four sex cells.

Skill Check

Understand science words: ovary. The word part *ovi* means egg. In your dictionary, find three words with the word part *ovi* in them. *For more help, refer to the* ***Skill Handbook,*** *pages 706-711.*

Sperm, Eggs, and Fertilization

Figure 22-9 shows a photograph of human egg and sperm cells. How are the cells alike? How are they different? The cells are alike in four ways. One, both are sex cells. Two, both formed during meiosis. Three, each has half the number of chromosomes found in body cells. Four, in humans, both cells begin to develop by meiosis at puberty (PEW bur tee). **Puberty** is the stage in life when a person begins to develop sex cells. It takes place between the ages of 10 and 15.

In which organs are sperm formed?

How are sperm and eggs different? Eggs are much larger than sperm. Each sperm has a tail. The tail helps it move. Sperm form in the testes (TES teez). In animals, **testes** are the male sex organs that produce sperm. Eggs form in the ovaries (OHV uh reez). In animals and some plants, **ovaries** are the female sex organs.

Study Tip: Use this idea map as a study guide for comparing eggs and sperm. Which sex cells are smaller?

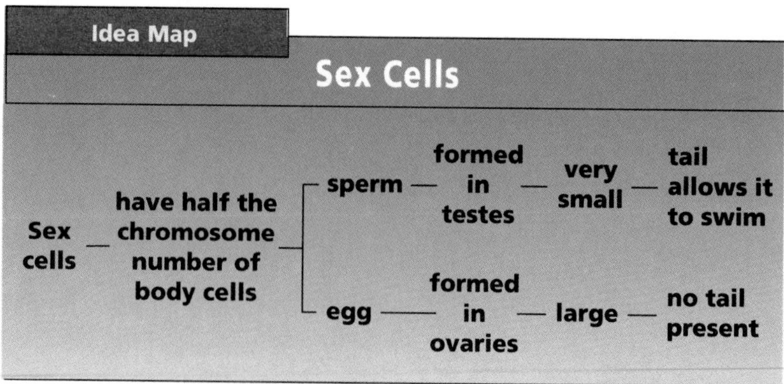

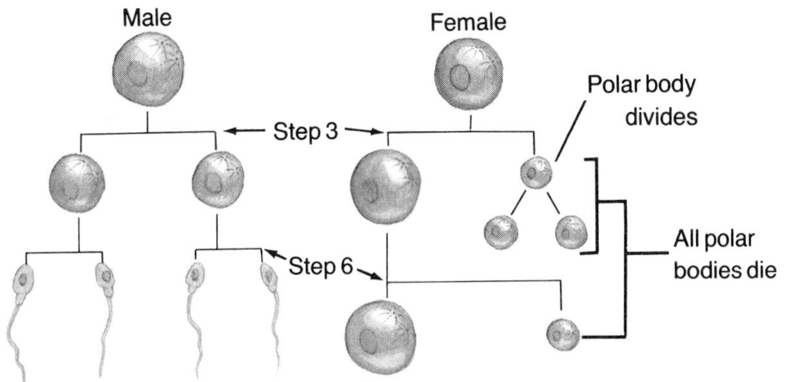

Figure 22-10 Compare the products of meiosis between human males and females. Step 3 and Step 6 refer to the steps of meiosis shown on pages 472 and 473.

During meiosis, each original male cell becomes four sperm, Figure 22-10. Meiosis in males occurs all the time from the beginning of puberty.

Now, look at how eggs are formed, Figure 22-10. When the cell first pinches in half at the end of the first division, one cell is smaller than the other. The smaller cell is called a polar body. A **polar body** is a small cell formed during meiosis in a female. The polar body divides and then dies. The large cell that remains forms another polar body when the cell pinches in half again. This third polar body also dies. The large cell that remains becomes the egg. An egg is formed once a month from the onset of puberty.

When sperm and egg join, the chromosomes from each cell also come together. The new organism has a complete set of chromosomes in each one of its body cells. Half the chromosomes in the organism come from the father, and half come from the mother.

Check Your Understanding

6. How do the number of chromosomes at the beginning and end of meiosis compare?
7. Describe changes in chromosomes during meiosis.
8. Compare the results of meiosis between human male and female sex cells.
9. **Critical Thinking:** What is the advantage of producing more sperm than eggs?
10. **Biology and Math:** If the body cells of an animal have 50 chromosomes, how many chromosomes would its sex cells have after meiosis?

22:3 Changes in the Rate of Mitosis

Objectives

7. Explain some effects of aging.

8. Describe the causes and effects of cancer.

Key Science Words

cancer

Changes often occur in the growth of cells. The rate of mitosis can speed up or slow down. Let's look at two effects caused by a change in the rate of mitosis.

Aging

Aging is the process of becoming older. All living things age. Loss of hair, wrinkled skin, and loss of calcium in bones are some of the common signs of aging in humans.

How does aging affect fingernail growth?

Many changes that result from aging have something to do with mitosis. Let's take fingernail growth for example. Your fingernails grow by the process of mitosis. The new cells push out the old cells. Old cells harden, die, and form the fingernail. As a person ages, mitosis in cells of the fingernails slows down. Fingernail growth then slows down.

Changes in heart and body muscle also occur as you age. Muscle cells do not undergo mitosis. You are born with a certain number of muscle cells. As you grow, the muscle cells get bigger, but no new cells are made. As you become older, the muscle cells wear out, and no new ones replace them. This is why your heart weakens and cannot pump blood as well as you get older. Remember that your heart is made of muscle.

Table 22-1 shows a comparison between body functions at age 20 and age 70. Each difference may be the result of a slowing in the rate of cell mitosis. Mitosis is not a very well understood process. Scientists don't know why mitosis slows down with age.

TABLE 22–1. CHANGES THAT OCCUR IN THE HUMAN BODY WITH AGE

Body System	Trait	20-Year-Old	70-Year-Old
Skin/Nails	Rate of fingernail growth	1 mm/week	.6 mm/week
Nervous	Reaction time	.8 seconds	.95 seconds
Circulatory	Pumping action of heart	3.7 liters/minute	2.9 liters/minute
Nervous	Memory	14 of 24 words recalled	7 of 24 words recalled
Respiratory	Lung volume with deep breath	5.5 liters/ inhalation	3 liters/ inhalation
Muscular	% of body fat (male)	15%	30%

Lab 22–2 Rate of Mitosis

Problem: Does mitosis occur in all cells all the time?

Skills

use numbers, make and use tables, form hypotheses, interpret data

Materials

clear plastic sheet
marking pencil
facial tissue
microscope

Procedure

1. Copy the data table.
2. Lay a plastic sheet over the diagrams of tissue cells in the figure.
3. **Use numbers:** Count the total number of cells in each diagram. Use a marking pencil to check off each cell as you count. Record the number of cells.
4. Wipe off the plastic with the facial tissue. Place it over the diagrams again.
5. **Form a hypothesis:** Read the table in Figure 22-1 on page 464 again. State which of the three tissues you think will show the most cells in mitosis. Write your **hypothesis** in your notebook.
6. Repeat steps 1 and 2, only this time count the numbers of cells reproducing and the numbers of cells not reproducing by mitosis. Record these numbers.

Data and Observations

1. **Interpret data:** Which tissue in the diagrams showed the most cells in mitosis?
2. Which tissue type in the diagrams showed the least number of cells in mitosis?

TISSUE	TOTAL NUMBER OF CELLS	NUMBER OF CELLS NOT IN MITOSIS	NUMBER OF CELLS IN MITOSIS
Liver			
Stomach Lining			
Skin			

Analyze and Apply

1. Which tissue type would:
   (a) repair itself the fastest?
   (b) repair itself the slowest?
2. **Check your hypothesis:** Is your hypothesis supported by your data? Why or why not?
3. **Apply:** Root tip cells reproduce faster than skin cells, but slower than liver cells. Draw 50 root tip cells. Show how many you think would be reproducing.

Extension

Use numbers: Calculate the average number of dividing cells in one field of view of a prepared slide of onion root tip cells.

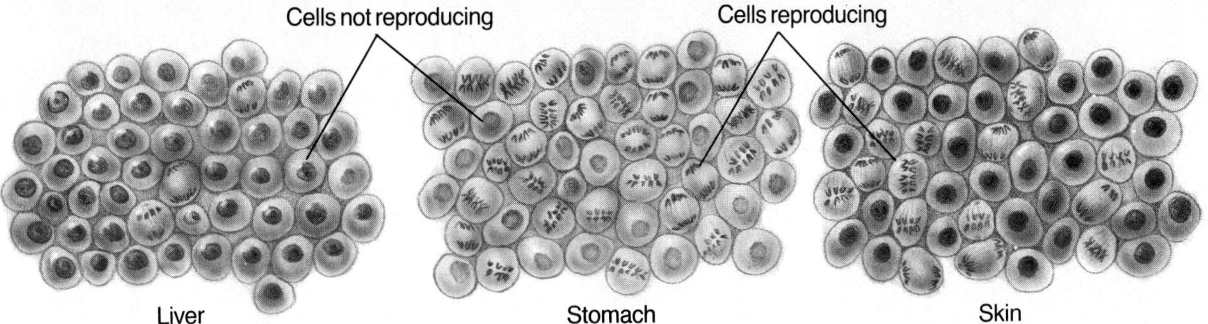

Cancer

Bio Tip

Health: Exposure to radiation can affect the cell divisions in a person's sex cells. When you receive X rays at a hospital or dentist you should wear an apron made of lead for protection. X rays can't pass through lead.

Healthy cells have regular rates of reproducing. For example, a skin cell undergoes mitosis once every 20 days. During the other 19 days, it doesn't reproduce. A liver cell may undergo mitosis once every 200 days.

Cancer is a disease in which body cells reproduce at an abnormally fast rate. Follow the steps in Figure 22-11 that show cancer forming in the lungs. The cells are shown about 250 times their real size. Figure 22-11a shows normal lung cells. A thin layer of cells covers the lungs. Notice the changes in Figure 22-11b. The outer layer of cells has changed from one cell layer to many cell layers.

How does cancer in the lungs affect lung tissue?

In Figure 22-11c, the outer layer of cells is no longer thin and even. Rapid mitosis has increased the number of cells. The shapes of the cells and their nuclei have also changed. The abnormal cells begin to crowd the lung tissues inside. The outer cells have become cancer cells. They will continue to increase in number until they crowd out all normal lung tissue.

Why do some cells start reproducing abnormally and become cancer cells? Three well-known causes include chemicals, radiation, and viruses. Cells may become cancer cells if they are in contact with poisonous chemicals for a long time. For example, it has been shown that chewing tobacco may cause cancer of the mouth. Skin cells may become cancer cells if a person spends too much time in the sun. High levels of X rays may cause cancer in bones. Even viruses cause certain cancers.

Figure 22-11 Follow the changes in lung tissue as the covering cells become cancer cells.

a

b

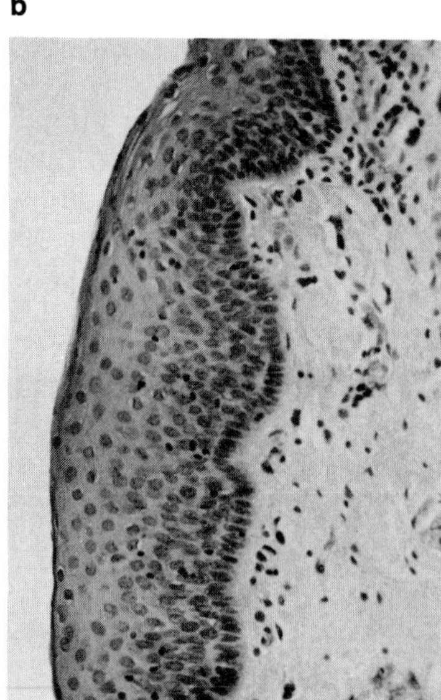

c

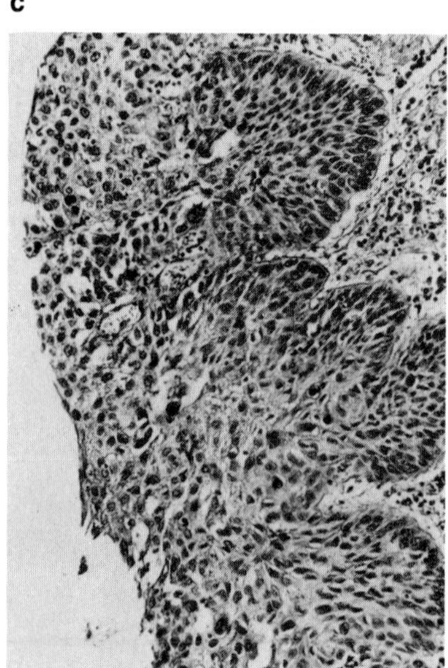

Figure 22-12 Chewing tobacco is a habit that may cause mouth cancer.

Skill Check

Classify: The following are possible causes of cancer. Classify them as chemical, radiation, or other: smoking, radon, asbestos, smog, UV rays, and paint thinner. *For more help, refer to the* ***Skill Handbook,*** *pages 715-717.*

Cancer is caused by many things. There are probably many causes that scientists have not yet discovered. For these reasons, the problem of how to treat cancer is a very difficult one to solve. Over the past 30 years, regular medical treatment that includes doses of radiation has resulted in a doubling of the survival rates of cancer patients.

Check Your Understanding

11. What are some of the changes that take place as a person ages?
12. What is the effect of cancer on mitosis?
13. List four possible causes of cancer.
14. **Critical Thinking:** Your skin produces a dark protective pigment when exposed to the sun's burning rays. There are a variety of suntan lotions. Some help you tan faster, some block the rays of the sun. Which suntan lotion should you choose to protect you from developing skin cancer?
15. **Biology and Reading:** Some cancers are inherited. Retinoblastoma (REH tihn uh blah STOH muh) is a kind of eye cancer that is inherited by about 1 in 20 000 children. What does the word *inherited* mean as used in this description?

Chapter 22

Review

Summary

22:1 Mitosis

1. Living things grow and repair themselves in a process of cell reproduction called mitosis.
2. Before mitosis begins, chromosomes become doubled and form sister chromatids. Each chromatid is an exact copy of the original chromosome.
3. During mitosis, two identical cells form from one original cell. Mitosis takes place in four steps. It results in two new body cells, each with the same number of chromosomes as in the original cell.

22:2 Meiosis

4. Sex cells form through a process called meiosis.
5. Meiosis involves two cell divisions. It results in four sex cells, each with half the number of chromosomes as the original cell.
6. In human males, four sperm result from meiosis. In females, only one egg results from meiosis. Egg and sperm join together at fertilization. Each supplies half the new cell's chromosomes. The new organism develops by mitosis.

22:3 Changes in the Rate of Mitosis

7. Aging is due to a slowing down of mitosis in some cells of the body.
8. Cancer is the result of abnormal mitosis. Cells reproduce too rapidly.
9. Cancer can be caused by chemicals, radiation or viruses.

Key Science Words

body cell (p. 464)
cancer (p. 478)
meiosis (p. 471)
mitosis (p. 464)
ovary (p. 474)
polar body (p. 475)
puberty (p. 474)
sex cell (p. 471)
sister chromatids (p. 466)
testes (p. 474)

Testing Yourself

Using Words

Choose the word from the list of Key Science Words that best fits the definition.

1. cell reproduction that forms sex cells
2. abnormal cell mitosis
3. cell reproduction in which two identical cells are made
4. female sex organ
5. stage in life when sex cells start to undergo meiosis

Review

Testing Yourself *continued*

6. small cell in the female that dies after being formed by meiosis
7. organs that produce sperm

Finding Main Ideas

List the page number where each main idea below is found. Then, explain each main idea.

8. changes that take place in lung tissue with cancer
9. the role of the fibers during Steps Two and Three of mitosis
10. what is accomplished during meiosis
11. what the benefits of mitosis are
12. specific changes that take place between age 20 and age 70
13. how the chromosome grouping in Step Two of mitosis differs from that in Step Two of meiosis

Using Main Ideas

Answer the questions by referring to the page number after each question.

14. In what step of meiosis do sister chromatids separate? (p. 474)
15. Why can't the heart of an older person pump as much blood as that of a younger person? (p. 476)
16. Cats have 38 chromosomes in their body cells. How many chromosomes would their sex cells have? (p. 471)
17. How do each of the following compare for mitosis and meiosis? (pp. 469, 474)
    (a) the number of cells formed
    (b) numbers of chromosomes in the new cells compared with number of chromosomes in the original cell
    (c) type of cell formed
18. What are three main causes of cancer? (p. 478)

Skill Review 

*For more help, refer to the **Skill Handbook**, pages 704-719.*

1. **Classify:** Compare the following traits of eggs and sperm: size, number of chromosomes, process that makes them, polar bodies formed, tail present.
2. **Sequence:** What changes take place in the cell membrane and nuclear membrane before or during mitosis?
3. **Measure in SI:** A student makes 500 new bone cells and each is 0.015 mm in size. Assuming they all form in the same direction, how much did this student grow?
4. **Observe:** Examine a prepared slide of a lily anther and count the number of chromosomes in the young pollen cells.

Finding Out More

Critical Thinking

1. Through an error, part of a chromosome within a fertilized egg cell is lost. Will every body cell within the new organism show the same error? Explain.
2. How might tanning parlors and cigarettes be related to cancer?

Applications

1. Demonstrate how rapidly a living thing increases in size by calculating the number of cells that result from 10 generations of mitosis.
2. Build models of a cell's chromosomes showing the changes that take place during mitosis and meiosis.

CHAPTER PREVIEW

Chapter Content

Review this outline for Chapter 23 before you read the chapter.

Skills in this Chapter

The skills that you will use in this chapter are listed below.

- In **Lab 23-1,** you will form a hypothesis, experiment, and observe. In **Lab 23-2,** you will experiment, observe, measure in SI, and infer.
- In the **Skill Check,** you will understand science words.
- In the **Mini Lab,** you will experiment.

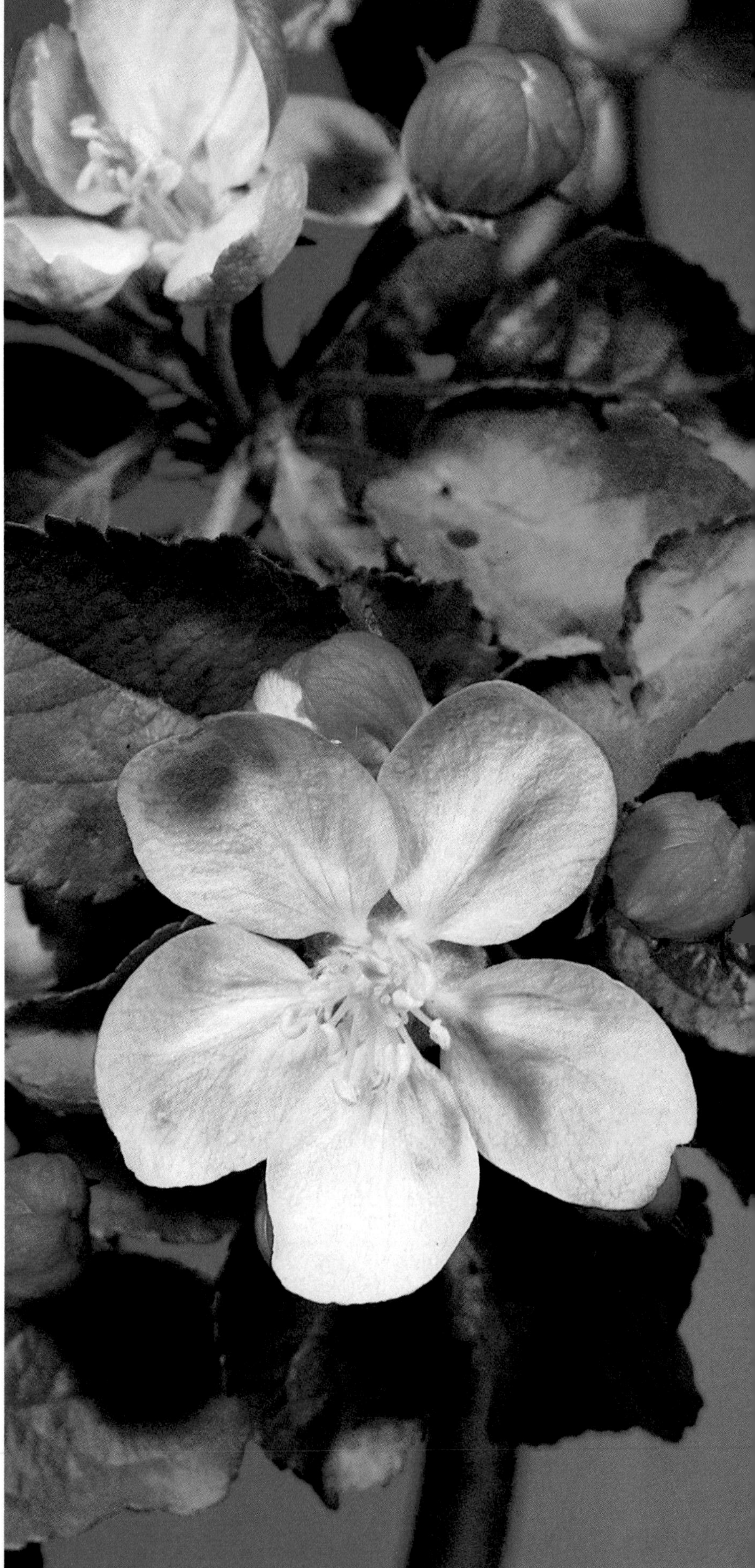

Chapter 23

Plant Reproduction and Development

You have read that two of the features of living things are that they grow and reproduce. How do flowering plants grow and reproduce? Look at the flower in the photo on the left. You may have noticed that insects are attracted to flowers. Insects have important relationships with many different flowering plants. Many insects help flowers reproduce.

Look at the photo below. This is a fruit that developed from the flower on the left. How did an insect help the flower turn into a fruit? The flower, the fruit, and the insect all play a role in the sexual reproduction of flowering plants. Flowering plants may also reproduce asexually from roots, stems, and leaves.

Try This!

Are fruits different from vegetables? Examine several different fruits and vegetables. Decide how these two plant parts differ. Form a definition of these two terms based on your observations.

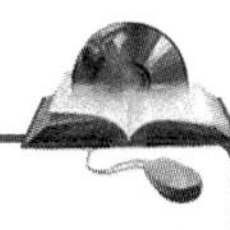

BIOLOGY Online

Visit the Glencoe Science Web site at science.glencoe.com to find links about **plant reproduction and development.**

A fruit develops from a flower.

23:1 Asexual Reproduction in Plants

Objectives

1. **Explain** how plants reproduce asexually from roots.
2. **Describe** how plants can reproduce asexually from leaves.
3. **Compare** asexual reproduction from runners, tubers, bulbs, cuttings, and grafting.

Key Science Words

tuber
runner
bulb
cutting
grafting

As you have read before, roots, stems, and leaves are plant parts with a variety of different jobs. One job of these plant parts is reproduction. Plant reproduction from roots, stems, and leaves is asexual reproduction. Sometimes they are also helpful in plant reproduction. Recall from Chapter 4 that reproduction from one parent is called asexual reproduction.

Asexual Reproduction by Roots

Many plants have roots that can grow into whole new plants. The sweet potatoes that you buy in the grocery store are roots. If you place a sweet potato in water, it will soon grow many small, new plants. Figure 23-1 shows how these changes take place. Notice that each new plant has its own roots, stems, and leaves. Each small plant can be pulled off the sweet potato and planted.

The sweet potato plants formed in this way are produced by asexual reproduction. The new plants have grown by mitosis. The original plant part didn't produce eggs and sperm, and there was no fertilization.

Many flowering plants reproduce asexually from roots. Dandelions and morning glory roots can be cut into many small pieces. Each piece of root can be planted and in time will form a new plant. Many trees, such as poplars and magnolias, are not killed when they are cut down. If the roots are left in the ground, they will send up new shoots.

Figure 23-1 A sweet potato is placed in water (a). Small plants begin to grow (b). Each new plant has its own roots, stems, and leaves (c).

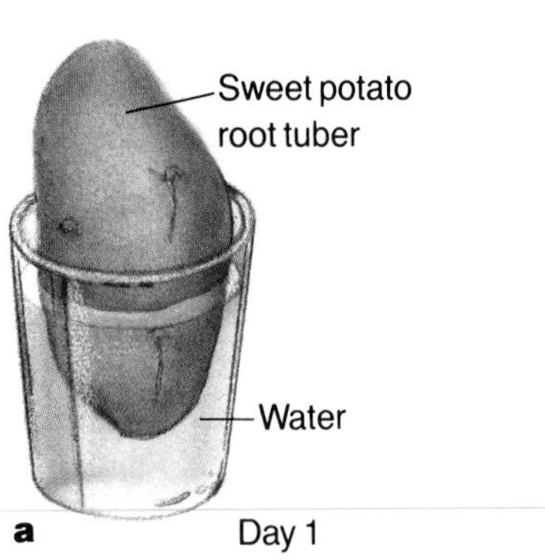

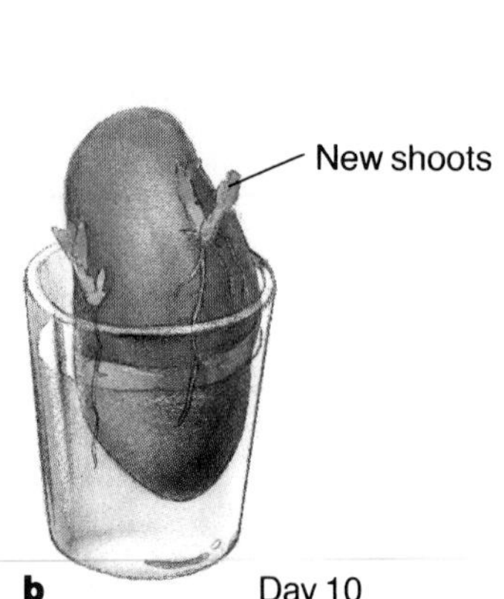

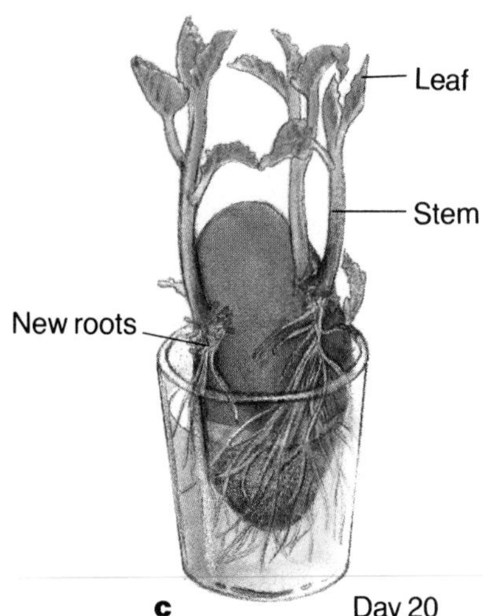

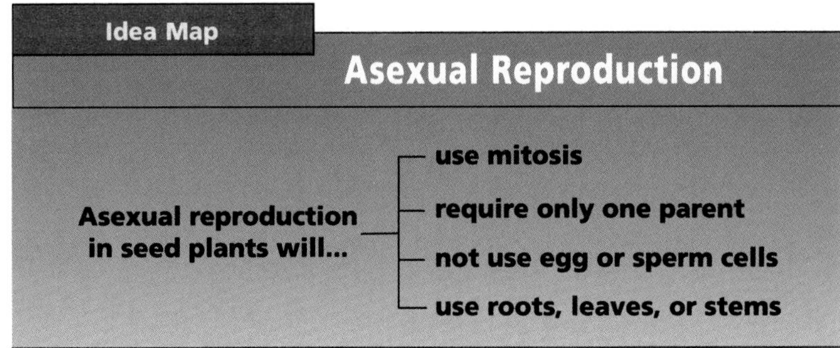

Idea Map
Asexual Reproduction
Asexual reproduction in seed plants will...
use mitosis
require only one parent
not use egg or sperm cells
use roots, leaves, or stems

Study Tip: Use this idea map as a study guide to asexual reproduction in plants. Which plant parts can reproduce asexually?

Other plants that reproduce asexually from roots include wild carrots, lilac bushes, rose bushes, and apple trees. They send up new shoots from their root systems underground. Many plants like the sweet potato have swollen roots called tubers. A **tuber** is an underground root or stem swollen with stored food. Begonias and dahlias, common garden flowers, also have root tubers. They grow new plants every year from the food stored in the tubers.

What kind of root is a sweet potato?

Asexual Reproduction by Leaves

How can an African violet reproduce asexually?

Some plants can reproduce asexually from leaves. Some leaves, when removed from the main plant, will grow into an entire new plant. For example, a leaf of an African violet plant can be placed with its stalk in the soil. An entire new plant will grow from this single leaf. Figure 23-2 shows the steps in the growth of a new plant from the leaf of a snake plant.

Remember that the process of making new plants from leaves, like roots, is asexual. There is only one parent. The new plants look just like the parent plants.

Figure 23-2 These are the steps for producing a new snake plant asexually from a leaf of a parent plant. Which plant part grows first?

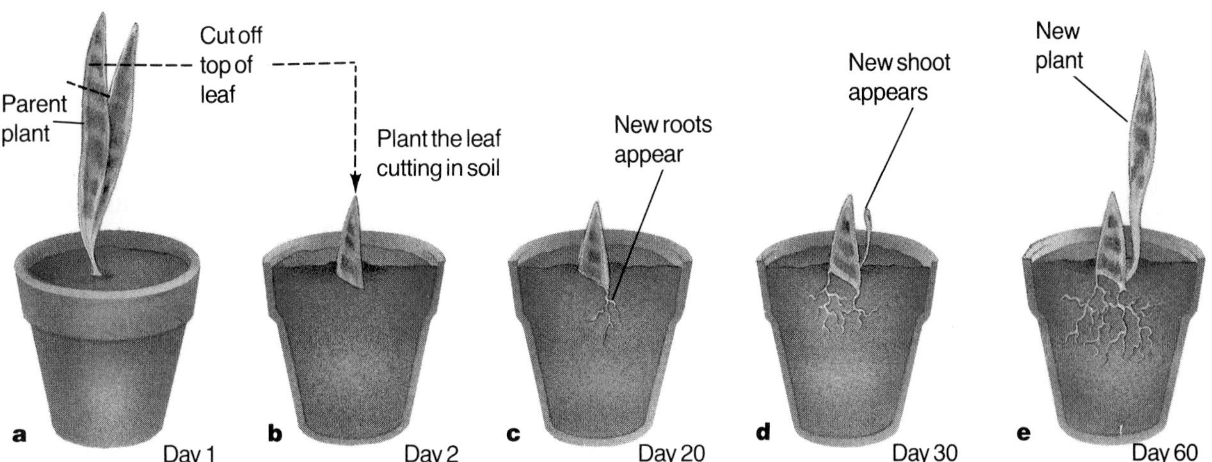

Cut off top of leaf
Plant the leaf cutting in soil

New shoot appears

Figure 23-3 New spider plants form at the end of runners (a). Potatoes are stem tubers (b). Both plants are reproducing asexually from stems.

a

b

Asexual Reproduction by Stems

What are two ways that plants can be reproduced from stems?

There are several ways that stems of flowering plants reproduce asexually. Two natural ways are by runners and underground stems.

Clover, strawberry, and spider plants grow small, new plants at the end of runners. A **runner** is a stem that grows along the ground or in the air and forms a new plant at its tip. Figure 23-3a shows an example of runners forming on a spider plant. Each new plant is a copy of the parent. In the forests of South Africa, where the spider plant grows naturally, these new plants would eventually break away from the parent plant. As houseplants, the small plants can be removed from the parent and placed in their own pots. They will then grow into new adult spider plants.

Many new plants grow asexually from underground stems. A potato is a stem tuber and an onion is a bulb. Both are underground stems. Have you ever found small plants growing from the sides of a potato? What you might have seen is shown in Figure 23-3b. Other plants with stem tubers are water lilies and caladium, a common houseplant with two-colored leaves shaped like arrowheads.

Figure 23-4 Notice the small stem at the base of an onion bulb.

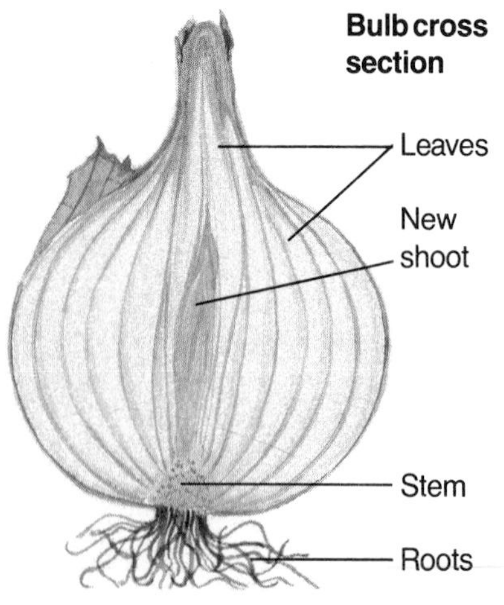

Many spring wildflowers have bulbs. A **bulb** is a short, underground stem surrounded by fleshy leaves that contain stored food. Look at the onion bulb that is cut in half, Figure 23-4. Notice the young shoot already formed at the center of the bulb. Garden flowers such as tulips, daffodils, and hyacinths (HI uh sihnths) are grown from bulbs.

Gardeners use their knowledge of this ability of stems to grow new plants to grow plants from cuttings or graftings. Have you ever tried to grow new geranium

① Cut off a small branch close to the stem.

② Place the stem in water.

③ New roots appear in a few days.

Parent plant

Cut here

a

Cutting

b

Figure 23-5 These are the steps for producing a new plant from a stem cutting.

plants using the method shown in Figure 23-5? If so, you have grown a plant from a cutting. A **cutting** is a small section of plant stem that has been removed from a parent plant and planted to form a new plant.

Grafting is a process of joining a stem of one plant to the stem of another plant. For example, the stem of a Delicious apple tree may be joined or grafted to the stem of a MacIntosh apple tree. The original Delicious apple stem will continue to form Delicious apples. The original MacIntosh stem will continue to form MacIntosh apples. The result is a tree that will form both kinds of apples.

Why is grafting important to gardeners, farmers, and plant scientists? A gardener can grow a variety of fruits from one tree. The most common reason for grafting is to protect a plant from disease. For example, some kinds of rose plants or apple trees are more resistant to disease than other kinds. The kinds that have better flowers or fruits are grafted onto the roots of similar kinds that are resistant to diseases, Figure 23-6.

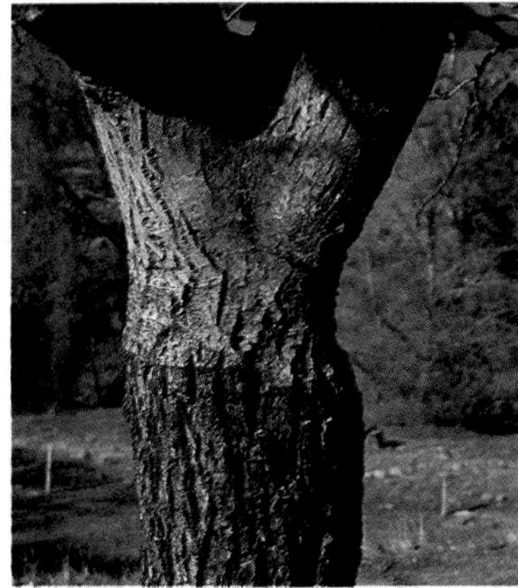

Figure 23-6 An old graft can be seen as a line where the two stems meet.

Check Your Understanding

1. How does a root tuber help a plant reproduce?
2. How can an African violet reproduce asexually?
3. Name five different ways that a plant stem can help in asexual reproduction and give an example of each.
4. **Critical Thinking:** How might asexual reproduction help a plant survive a forest fire?
5. **Biology and Reading:** What does the prefix *a-* mean when it is attached to the word *sexual*?

23:2 Sexual Reproduction in Plants

Objectives

4. **List** the parts of a flower and their functions.
5. **Describe** two methods of pollination.
6. **Sequence** the steps that lead to fertilization in flowering plants.

Key Science Words

sepal
petal
stamen
pistil
ovule
pollination
cross pollination
self pollination

Flowering plants make our world beautiful to live in. Flowers come in an enormous variety of colors, shapes, and sizes. But, plants don't produce flowers for our pleasure. Flowers are important to the survival of the plant. Without flowers, many plants wouldn't be able to reproduce sexually.

Flowers and Sexual Reproduction

Figure 23-7 is a diagram of a section through one kind of flower. Find the numbers on the diagram as you read the following statements that describe the different flower parts. In most flowers some flower parts are showy. These parts are neither male nor female. In most flowers some parts are female and other parts are male. Some flowers lack one or more of these parts. The flower in Figure 23-7 shows typical flower parts for a complete flower.

Parts 1 and 2 are sepals and petals. **Sepals** are often green, leaflike parts of a flower that protect the young flower while it is still a bud. **Petals** are often brightly colored and scented parts of a flower that attract insects to the flower.

What flower part forms pollen grains?

Part 3 is the male flower part. The **stamen** is the male reproductive organ of a flower. Usually there are one to many stamens in each flower. Each stamen has a saclike part at its top in which pollen grains are formed. Figure 23-8a shows pollen grains under an electron microscope. You read in Chapter 6 that pollen grains are male structures that contain sperm cells.

Study Tip: Use this idea map as a study guide to parts of a flower. Which structure in a flower contains ovules?

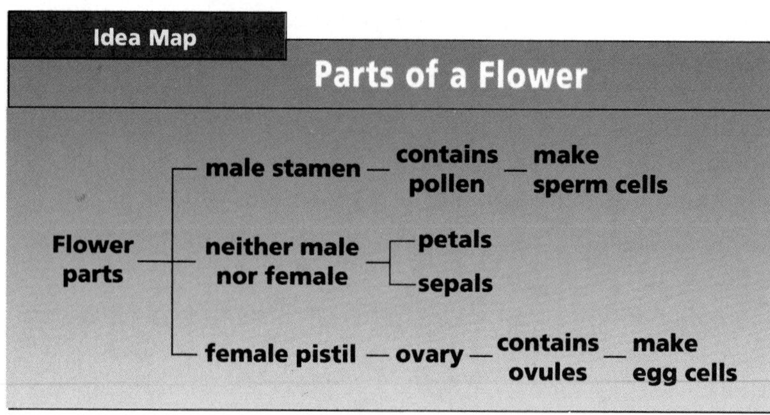

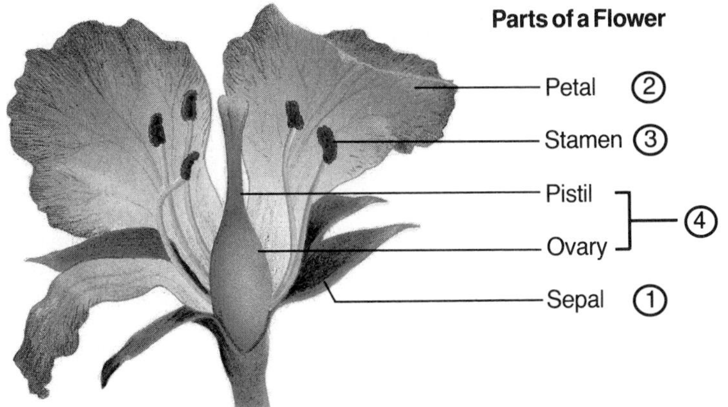

Figure 23-7 Follow the text to read the functions of each part of a flower.

Part 4 in Figure 23-7 is the female flower part. The **pistil** (PIH stul) is the female reproductive organ of a flower. Often there is only one pistil in the center of a flower. Each pistil has a large, round ovary at its base. In Figure 23-8b this is cut open so you can see the ovules inside. Above the ovary is a stalk with a sticky tip. The sticky tip helps to trap pollen grains. Inside the ovary are tiny, round parts called ovules. **Ovules** contain the egg cells of a seed plant.

If you ever looked closely at a flower, such as a dandelion, a rose, a carnation, or a daffodil, you may have noticed that none of these has the structure of our typical flower in Figure 23-7. Some flowers don't have sepals, some have petals that are fused to form a tube, some have many tiny flowers all clustered together in a head, and some don't have any stamens. Flowers, just like all other organisms, show an enormous range of variation.

Figure 23-8 Pollen grains under an electron microscope show wonderful surface patterns (a). A section through a pistil as seen under a light microscope shows the position of its ovules (b).

a

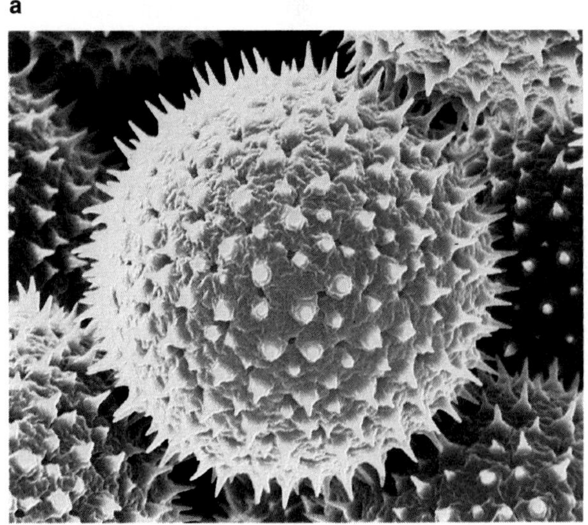

b

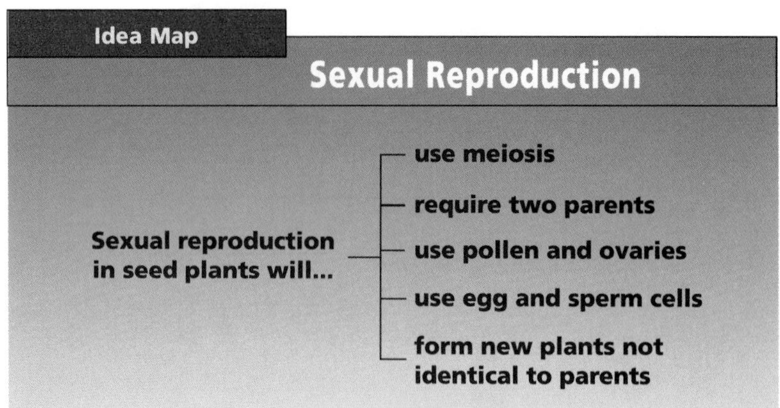

Idea Map
Sexual Reproduction
Sexual reproduction in seed plants will...
use meiosis
require two parents
use pollen and ovaries
use egg and sperm cells
form new plants not identical to parents

Study Tip: Use this idea map as a study guide to sexual reproduction in plants. What kind of cell division is necessary for sexual reproduction?

Pollination and Fertilization

You know that in sexual reproduction an egg and a sperm must join for fertilization. How does pollen get from the stamens of a plant to the pistil? Then, how do sperm cells inside the pollen reach the eggs within the ovules?

What is self pollination?

Pollen is transported from stamens to pistils by a process called pollination (pahl ih NAY shun). **Pollination** is the transfer of pollen from the male part of a seed plant to the female part.

In flowering plants, the design of the flower is very important in pollination. Many flowers with colorful petals have a strong scent that attracts insects. Insects climb over the flowers and pick up the pollen on their bodies. Many insects, such as bees, collect pollen for food. Some flowers also make a sugary chemical called nectar that bees and birds also use as food. Pollen catches on the hairs and feathers of their bodies while they feed on the nectar.

Some plants don't have large, colorful petals or attractive scents. These plants usually rely on wind to move the pollen from one plant to another. Grasses and many trees have this type of flower. Many trees, such as maples and willows, have long, hanging flowers that are yellow or green in color. These flowers are not attractive to insects. The pollen formed in these flowers is blown around by the wind.

Pollen is often carried by insects, birds, or wind between different plants. This is called cross pollination, Figure 23-9. **Cross pollination** is when pollen from the stamen of one flower is carried to the pistil of another flower on a different plant. Figure 23-9 also shows self pollination. **Self pollination** is when pollen moves from the stamen of one flower to the pistil of the same flower, or of another flower on the same plant.

Figure 23-9 Compare two types of pollination. How many plants are needed for self pollination?

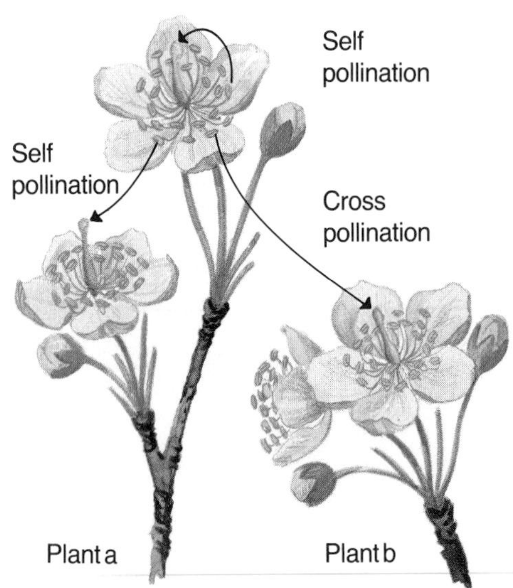

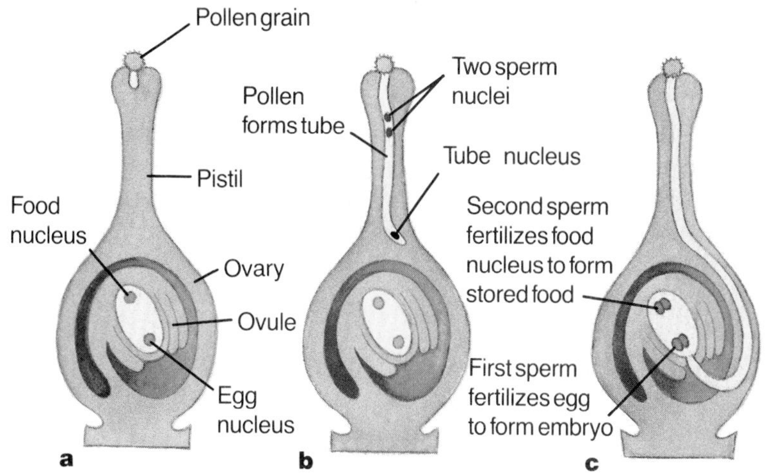

Figure 23-10 Follow the sequence of steps in plant fertilization as shown in these sections of a pistil. How many sperm are needed to fertilize one ovule?

How does sperm inside a pollen grain reach the egg inside an ovary? Follow the steps in Figure 23-10.

1. A pollen grain lands on the sticky tip of the pistil. It must somehow reach the egg within the ovule.
2. The pollen grain begins to grow a tube into the stalk of the pistil. This pollen tube grows all the way down to the ovary. Inside each pollen grain are three nuclei. One nucleus leads the way down the tube. The other two nuclei are sperm nuclei and these follow the first nucleus.
3. The tube reaches the ovule. A sperm nucleus inside the tube can now pass into the ovule and join with the egg. Fertilization takes place. The other sperm from the pollen tube joins with another nucleus in the ovule to form a food supply for the new seed.
4. A new plant or embryo now grows inside the ovule. As the ovule grows, it becomes a seed.

Bio Tip

Health: A person who is a vegetarian gets most of his or her needed protein from plant seeds.

How many nuclei pass down a pollen tube?

Check Your Understanding

6. List flower parts that are male, female, or neither.
7. What are two types of pollination?
8. How many sperm are needed to fertilize one ovule?
9. **Critical Thinking:** Hummingbirds and moths both feed on nectar of flowers. A hummingbird has a very long, thin beak and a moth has a very long mouthpart for feeding. What kinds of flowers would they most likely visit?
10. **Biology and Writing:** Are petals and sepals sexual parts in plants? Why or why not?

23:3 Plant Development

Objectives

7. **Describe** how fruits and seeds develop.
8. **Explain** how seeds are scattered and grow into new plants.
9. **Compare** asexual and sexual reproduction in plants.

Key Science Words

fruit
germination
clone

In addition to reproduction, two other features of living things are that they grow and develop. Once a seed is formed, it's ready to grow and develop into a new plant. What kinds of changes occur when the seed begins to grow?

Seeds and Fruits

After fertilization, changes take place in the flower. Use Figure 23-11 to follow these changes.

1. In this flower, many ovules have been fertilized by sperm cells.
2. Shortly after fertilization, the petals, stamens, and stalk of the pistil wither and die.
3. At the same time, the ovules and the ovary begin to grow.
4. Finally, the ovules mature into seeds. A seed is a plant part that contains a plant embryo. Remember from Chapter 6 that the embryo of a plant has a new young plant and stored food. The young plant and the stored food came from the joining of egg and sperm nuclei in the ovule.

As the ovary matures it becomes a fruit. A **fruit** is an enlarged ovary that contains seeds. You may have noticed that almost all the fruits you have ever eaten have contained seeds. If you use this definition of a fruit, can a tomato be a fruit? Is a squash, a green pepper, a cucumber, or a pumpkin a fruit? You must answer yes, because they all grew from ovaries and all have seeds inside.

Mini Lab

Will Seeds Grow Without Stored Food?

Experiment: Remove only the embryo from several soaked bean seeds. Place the embryos on moist paper in a closed petri dish and wait to see if they grow. *For more help, refer to the* ***Skill Handbook,*** *pages 704-705.*

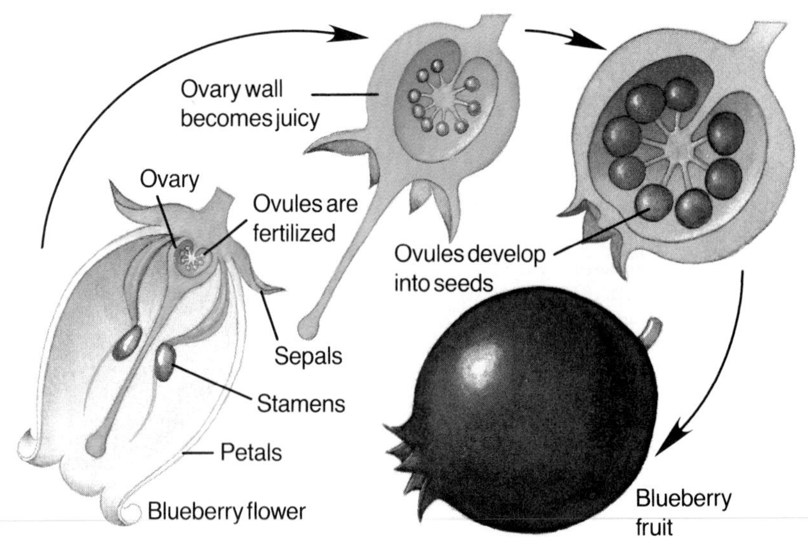

Figure 23-11 Follow the steps that occur after fertilization in the forming of seeds in a fruit.

TECHNOLOGY

Seed Banks

There are all kinds of banks. There are banks to hold your money, sperm banks, and even banks for body organs. Scientists also have formed a seed bank at a large laboratory in Fort Collins, Colorado. It is called the United States National Seed Storage Laboratory. The deposits made in this bank are exactly what you would expect from its name—seeds. In Fort Collins, seeds from all over the world are frozen in vats of liquid nitrogen. Someday, these living seeds will be removed from the vats and germinated.

Why do seeds need to be stored? Seeds are the link from one generation of plants to the next. One in 10 plant species is endangered. If a plant species becomes extinct, there is no way to get it back. Extinction is forever.

A seed bank is a means of saving plants that are endangered. Scientists of the future could cross-breed plants that have desirable traits with plants from the past.

Without plants for food, most living things can't survive. The "interest" the seed bank pays on seed deposits is that all living things benefit. The seed bank will help make the best seeds available for growing food plants for the future. Scientists will also be able to replace extinct plant species back into the wild when a new and suitable environment is found.

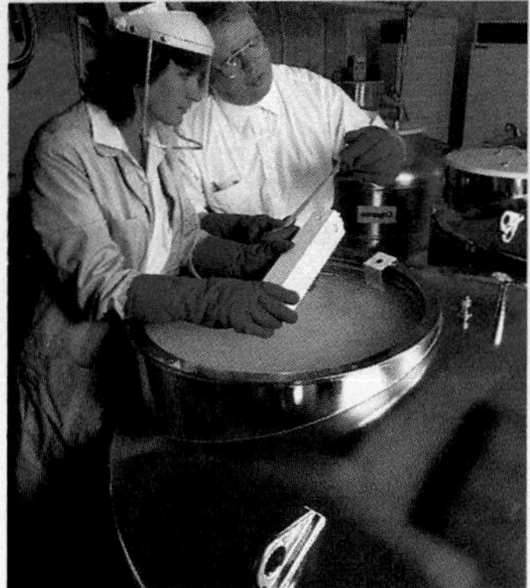

Seeds can be stored for a long time under controlled conditions.

If you could observe the development of a peanut or a walnut, you would conclude that they too were fruits. From these observations, it's clear that fruits are not just the juicy, fleshy fruits like oranges and grapes. Any plant part that develops from an ovary can be called a fruit. Fruits can also be nutty, and some aren't even edible by humans.

What is the ovule in a fruit?

One seed develops from each fertilized ovule in an ovary. How many ovules were inside a cherry flower before fertilization? The answer is one. How do we know? Because a cherry fruit has only one seed. How many ovules are inside an orange flower? At least ten because there are about ten seeds in an orange, Figure 23-12. Have you ever counted the seeds in a watermelon? If you have, then you will know how many ovules were fertilized.

Figure 23-12 The seeds of an orange develop from ovules. The juicy fruit and rind of the orange develop from the ovary.

Skill Check

Understand science words: germination. The word *germ* means a small part of living matter that can develop into a living organism. In your dictionary, find three words with the word part *germ* in them. *For more help, refer to the **Skill Handbook**, pages 706-711.*

Plant Development from Seeds

What happens to the fruit and seeds after they are fully ripe? If the fruit is brightly colored and juicy it may be eaten by an animal. The seeds pass through the animal's digestive system and when they are dropped, they are often far from the parent plant. In most cases, however, the fruit remains on the parent plant and dries up as the seeds mature. Sometimes, the fruits eventually burst open and the seeds are thrown away from the parent plant. Some seeds are carried away by wind. Larger seeds may be carried away by water. Sometimes, the fruits are carried away with the seeds held inside. Dandelion seeds are scattered by the wind while still held in their tiny dried fruits. These one-seeded fruits have hairs that act like a parachute. Have you ever seen a sticky seed that clings to the fur of an animal? The wall of the fruit is covered with tiny hooks that catch in fur or clothing. These seeds are carried away in an animal's coat. The new plants can then grow far from the parent plant where there is more space, light, and water that they need to survive.

How is the wind important to some seeds?

What happens next? If a seed lands on soil that has moisture and proper temperature, it may germinate. **Germination** (jur muh NAY shun) is the first growth of a young plant from a seed. In many flowering plants, the stored food is the main part of the seed. This food is available for the growth of the new plant.

Let's look at a bean seed as it germinates. First the root and then the stem grow from the seed. In beans, the two seed halves are made mostly of stored food for the new plant. The stem that comes out of the seed has a small pair of leaves on it. Once out of the seed, the stem and leaves soon turn green and begin to make food by photosynthesis. When the food supply of the two seed halves is used up, they drop off the plant.

Figure 23-13 Follow the sequence of steps in the development of a bean plant from a seed.

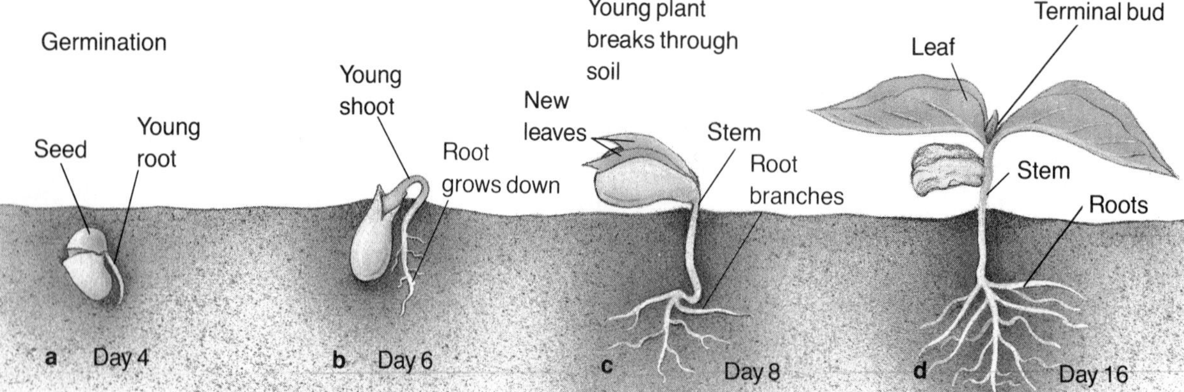

Lab 23–1 Germination

Problem: Will soaked seeds germinate faster than unsoaked seeds?

Skills

form a hypothesis, experiment, observe, interpret data

Materials

2 milk cartons	metric ruler
soil	rubber band
10 soaked corn seeds	small beaker
10 unsoaked corn seeds	pencil
water	tape
stapler	permanent marker
scissors	

Procedure

1. Copy the data table.
2. Prepare milk carton planters as shown in the figure.
3. Fill each carton with moist soil.
4. Wrap a rubber band around a pencil exactly 2 cm from the end of the eraser. Using your pencil as a measuring stick, poke a row of ten holes exactly 2 cm deep in both cartons as shown.
5. **Experiment:** To the first carton, add one soaked corn seed to each hole. Use the marker to label the carton Soaked Corn Seeds. Add your name and the date.

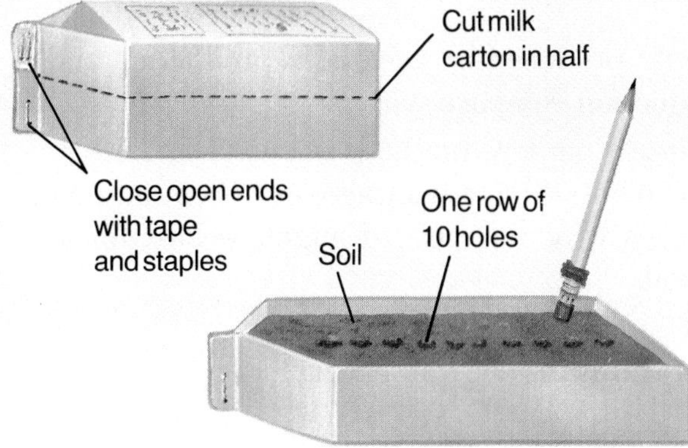

6. To the second carton, add one unsoaked corn seed to each hole. Label the carton Unsoaked Corn Seeds.
7. Gently cover the seeds with moist soil.
8. Read the problem question and write a **hypothesis** about it in your notebook.
9. **Observe:** Check each carton every day. Record the total number of plants that germinate each day.

Data and Observations

1. How long did it take for the first unsoaked corn seed to germinate?
2. How long did it take for the first soaked seed to germinate?

	NUMBER OF SEEDS GERMINATED					
Days from planting						
Soaked						
Not soaked						

Analyze and Apply

1. What is the purpose of this experiment?
2. (a) What is meant by a control?
   (b) Which group of seeds were the control?
3. (a) What is meant by a variable?
   (b) What was the variable?
4. **Check your hypothesis:** Is your hypothesis supported by your data? Why or why not?
5. **Apply:** How could you find out if your results would be the same using different kinds of seeds?

Extension

Design an experiment to test the effect of soaking the seeds for less than 24 hours.

Figure 23-14 Tulips of one variety all came from the same stock of bulbs and form a clone.

Comparing Kinds of Reproduction

You have read about two ways a plant can reproduce. Let's now compare the advantages of both asexual and sexual reproduction in plants.

Asexual reproduction

1. In asexual reproduction only one parent is needed. A new plant can form from a root, stem, or leaf by mitosis. Reproduction by this method produces adult plants in a short time.
2. Some plants, like bananas, do not form seeds. Thus, the only way for them to reproduce is asexually.
3. All the new plants from asexual reproduction are clones of the parent plant. A **clone** (KLOHN) is a group of living things that come from one parent and are identical to the parent. This means that any trait or feature that is useful for survival will be kept in all offspring. Useful traits include large flowers, bright colors, or juicy fruits.

What is a clone?

Sexual reproduction

1. Sexual reproduction requires two parents. Plants produced this way are made up of traits from both parents. No two plants will have the same combination of traits and so there will be a wide variety of traits in the offspring.
2. If proper growing conditions are not present, many seeds can survive for long periods of time. Some seeds have been known to germinate after hundreds of years of lying buried in the ground.
3. With a variety of traits, different plants will have a better chance to survive if conditions change.

Bio Tip

Leisure: You can make an indoor garden from kitchen wastes. Grow a pineapple plant from the leafy top of a pineapple, an orange tree from an orange seed, a carrot from a carrot top, and an avocado plant from an avocado seed.

Check Your Understanding

11. Describe what becomes of each of the following flower parts after fertilization: petals, ovules, ovary, stamens.
12. Which part of a plant grows first after a seed begins to germinate?
13. How will plants produced by asexual reproduction and sexual reproduction compare with their parents?
14. **Critical Thinking:** Compare methods of seed scattering from a parent plant. Give advantages for each method and give reasons why one method might be best for a plant's survival.
15. **Biology and Reading:** If you eat an orange or a tomato, which part of the plant are you eating?

Lab 23–2 Rooting

Problem: Can new plants be grown from roots and bulbs?

Skills

experiment, observe, measure in SI, infer

Materials

garlic bulb
carrot
labels
water
toothpicks
small beaker
metric ruler
razor blade
shallow dish

Procedure

1. Copy the data table.
2. **Experiment:** Stick three toothpicks into a garlic bulb as shown in the figure.
3. Fill a small beaker with water. Label the beaker with your name and the date.
4. Balance the garlic in the water as shown. Make sure that the pointed end of the garlic is sticking up out of the water.
5. **Observe:** Use the data table to draw what the garlic looks like today. Mark the date on the table.
6. **Measure in SI:** Use a metric ruler to measure about 2 cm from the top of a carrot. Use a razor blade to cut off a slice of a carrot where you measured it.

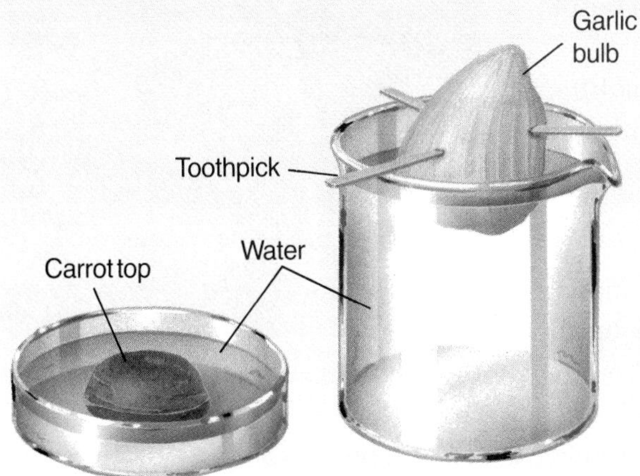

DATE				
garlic bulb				
carrot				

CAUTION: *Always cut away from yourself when using a razor blade.*

7. Place the carrot section into a shallow dish. The cut end should be facing down.
8. Label your dish. Add water to the dish.
9. Use the data table to draw what the root looks like today. Write in the data.
10. Check the bulb and root every day. Make a labelled diagram every four days on your data table.

Data and Observations

1. What plant part of the garlic was used? Of the carrot?
2. Describe what formed on the garlic and carrot slice after several days.

Analyze and Apply

1. Define asexual reproduction.
2. **Infer:** What did the parent plants of the garlic and carrot look like?
3. Using your data, how fast do garlic and carrot plants reproduce asexually?
4. **Apply:** How does asexual reproduction help a garlic plant survive?

Extension

Design an experiment to test if garlic bulbs will grow better in the light than in the dark.

Are These Seeds Alive?

Seeds serve as a food source for humans and other animals. They are also the most common method by which most plants reproduce. Farmers start the growing season by planting seeds for their spring and summer crops. They need to know what percentage of the seeds they plant are alive.

Identifying the Problem

You are a farmer who is about to plant soybeans for the coming season. You want to know what percentage of the beans planted will grow into new plants. Another way of looking at the problem is to ask whether all the seeds that you are about to plant are alive. Certainly, dead seeds will not grow into new plants. Therefore, you don't want to waste your time and energy planting a batch of seeds that contains many dead seeds. The problem is that you can't tell if a seed is alive just by looking at it. You want to experiment with your bean seeds to see how many in the package are alive. Thus, as a farmer, you can better predict how many future plants might grow once planted.

Technology Connection

As a farmer, you may not have the time to test samples of all the different kinds of seeds that you are about to plant. Scientists have a quick test that can tell if a seed is alive. A chemical called tetrazolium causes living seeds to turn pink or red. Dead seeds do not change color. Soak another 50 bean seeds as before. Open each into two equal halves, and throw away one half of each seed. Place the remaining half of each seed into a shallow dish with the flat, inner side facing down. Add a small amount of tetrazolium to the dish. Wait 30 minutes. Using forceps, turn over each seed half and examine the underside for a red or pink color. Record the number of living and nonliving seeds. How do the results of the tetrazolium test compare with your results in the original bean seeds?

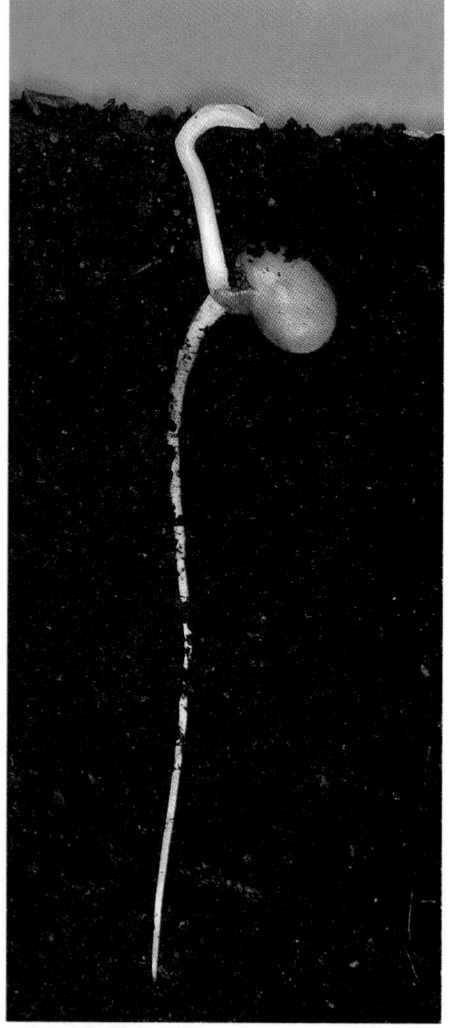

Collecting Information

Check the packages of beans that your teacher has supplied. Choose a variety that you want to test. Record the bean type you have selected. Examine the seeds and try to determine which seeds are alive. Record your observations. Try to estimate the percentage of seeds that are alive, and record this number. As a farmer or student, you should know that what you are doing is called sampling. You are checking a small sample of seeds rather than the entire batch to see if they are alive. Results from your small sample will tell you accurately what is true for the entire bag of seeds.

Carrying Out an Experiment   

1. Soak 100 bean seeds in a beaker of water. Soak the seeds for several hours.
2. Remove the seeds from the water, and wrap them completely in several layers of wet paper towels.
3. Slip the towels into a self-seal plastic bag. Label your bag with your name and the date.
4. After three days, open the bag, and examine the seeds. If a seed is alive, a root will be seen growing from it. If no root can be seen, the seed is not alive.
5. Count and record the number of living and nonliving bean seeds that you observe.

Career Connection

- **Farm Laborer** Carries out routine tasks such as caring for animals and crops; may be asked to clean pens, fertilize crops, or harvest crops when mature
- **Farm Office Worker** Gives clerical and technical support to farm owners; does accounting tasks such as payroll, handling billings, and mailings
- **Agronomist** Studies plants and soil in an effort to improve the process of growing farm crops; may work for large seed companies; aids in solving soil or crop problems

Assessing Your Results

Calculate and record the percentage of living bean seeds in your original 100-bean sample. Based on your experimental results, how would you answer the original question about whether all the seeds that you are about to plant are alive?

Chapter 23

Review

Summary

23:1 Asexual Reproduction in Plants

1. Asexual reproduction uses only one parent and produces identical offspring. Some plants reproduce from roots and root tubers.
2. Some plants reproduce asexually from leaves.
3. Some plants reproduce by underground stems, such as tubers and bulbs, or by runners. Some plants can be reproduced from cuttings or by grafting.

23:2 Sexual Reproduction in Plants

4. Sexual reproduction uses two parents and produces different offspring. The female part of the flower is the pistil. The male part is the stamen.
5. Pollination is the transfer of pollen from male to female flower parts. The two methods of pollination are self and cross pollination.
6. Pollen tubes grow down to ovules in the pistil. Sperm nuclei join with egg nuclei in the ovules.

23:3 Plant Development

7. After fertilization, a fruit forms from the ovary, and seeds form from ovules.
8. Seeds germinate and form new plants.
9. Asexual reproduction requires only one parent. Useful traits are kept in offspring. Sexual reproduction requires two parents and results in a wide variety of traits.

Key Science Words

bulb (p. 486)
clone (p. 496)
cross pollination (p. 490)
cutting (p. 487)
fruit (p. 492)
germination (p. 494)
grafting (p. 487)
ovule (p. 489)
petal (p. 488)
pistil (p. 489)
pollination (p. 490)
runner (p. 486)
self pollination (p. 490)
sepal (p. 488)
stamen (p. 488)
tuber (p. 485)

Testing Yourself

Using Words

Choose the word from the list of Key Science Words that best fits the definition.

1. exact copies of a living thing
2. underground stem of potato
3. the growth of a young plant from a seed
4. joining a stem of one plant to the stem of another
5. male reproductive organ of flower
6. enlarged ovary with seeds
7. female reproductive organ of flower
8. carrying of pollen from stamen of one flower to pistil of same flower

Review

Testing Yourself *continued*

Finding Main Ideas

List the page number where each main idea below is found. Then, explain each main idea.

9. how to grow a new plant from a leaf
10. why grafting is important
11. why squash, walnuts, and dandelion "seeds" are fruits
12. the differences between asexual and sexual reproduction in plants
13. how self pollination and cross pollination differ
14. the changes or development that take place in a seed after germination
15. how plants reproduce from root tubers
16. what the function of a pistil is
17. how sperm in pollen reach the egg
18. the roles of petals and sepals in a flower

Using Main Ideas

Answer the questions by referring to the page number after each question.

19. What are two advantages of sexual and two advantages of asexual reproduction to a plant? (p. 496)
20. What are the parts that form inside an ovary? (p. 492)
21. How do the color, structure, and scent of flowers aid pollination? (p. 490)
22. What happens to sperm and egg nuclei at fertilization? (p. 491)
23. How may some trees reproduce asexually after they have been cut down? (p. 484)
24. How does a sweet potato reproduce asexually? (p. 484)
25. What is the job of a stamen? (p. 488)
26. What are two ways leaves produce new plants? (p. 485)

Skill Review 

*For more help, refer to the **Skill Handbook,** pages 704-719.*

1. **Understand science words:** Which terms don't fit the definition of asexual reproduction? Explain why. clone, egg germination, mitosis, one parent
2. **Experiment:** What part of an experiment could be the control if you compared the rates of germination of wheat seeds and bean seeds?
3. **Infer:** How do plants get onto islands?
4. **Form a hypothesis:** If you noticed lots of young oak seedlings beneath an old oak tree, what would you hypothesize about how they got there?

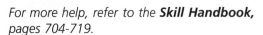

Finding Out More

Critical Thinking

1. *Aerodynamic* means designed for flight. What are some seeds that are carried away from the parent plant by wind with the help of aerodynamic fruits?
2. What is the advantage of a seed surviving for many years without germinating?

Applications

1. Examine pollen from different flowers under a microscope. Diagram what you see. How is pollen related to hay fever?
2. Prepare a list of 10 plants used by humans as food. Determine if these plants are grown by farmers using methods of asexual or sexual reproduction.

CHAPTER PREVIEW

Chapter Content

Review this outline for Chapter 24 before you read the chapter.

Skills in this Chapter

The skills that you will use in this chapter are listed below.

- In **Lab 24-1,** you will make and use tables, measure in SI, recognize and use spatial relationships, and make scale drawings. In **Lab 24-2,** you will make and use graphs, infer, and relate cause and effect.
- In the **Skill Checks,** you will sequence and understand science words.
- In the **Mini Labs,** you will use a microscope and observe.

Chapter **24**

Animal Reproduction

Look at the cloud of butterflies in the photo on the left. Where did they come from and where are they all going? The butterflies are migrating from Canada and the United States to Mexico. As the butterflies move southward, more and more butterflies join the flock. A flock of butterflies can number in the thousands. When the butterflies arrive in Mexico, they reproduce. In the spring, the butterflies will turn northward and head back to Canada and the United States.

How do butterflies reproduce? Like many other organisms, butterflies produce egg and sperm and have sexual reproduction. The joining together of the egg and sperm results in a new butterfly. In this chapter, you will study how different animals reproduce to form more of their own kind.

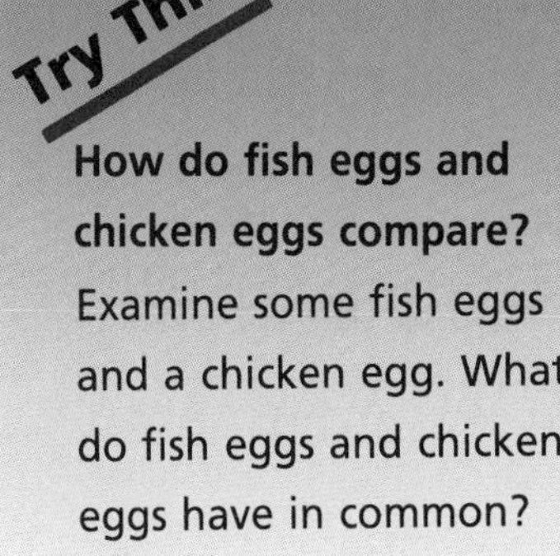

Try This!

How do fish eggs and chicken eggs compare? Examine some fish eggs and a chicken egg. What do fish eggs and chicken eggs have in common? How do they differ?

Visit the Glencoe Science Web site at science.glencoe.com to find links about **animal reproduction.**

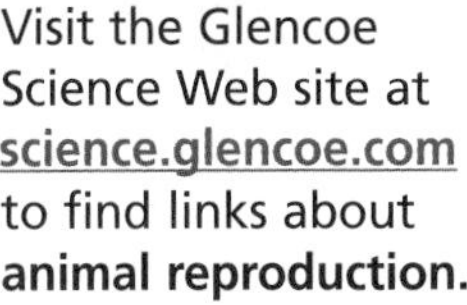

Adult monarch butterfly

24:1 Asexual Reproduction

Objectives

1. **Identify** the features of asexual reproduction.
2. **Describe** types of asexual reproduction.

Key Science Words

regeneration

All living things reproduce. Reproduction allows for the survival of different species of living things. For example, if butterflies didn't reproduce, there would be no more butterflies. The method or way in which living things reproduce differs from species to species. There are two main ways that animals reproduce. One of them is asexual reproduction.

Review of Asexual Reproduction

Sometimes only one parent is needed in order to have young. When reproduction requires only one parent, it is called asexual reproduction. You first studied asexual reproduction in Chapter 4 when you studied reproduction in bacteria. Let's review some features of asexual reproduction.

1. Egg and sperm cells are not used. Therefore, there is no meiosis.
2. Organs such as ovaries and testes are not used because no egg and sperm are produced in asexual reproduction.
3. Mitosis is the type of cell reproduction involved.
4. Since there are no egg or sperm certain body cells must undergo mitosis to form the offspring.
5. Offspring are identical to one another and the parent. They have the same kind of chromosome material and the same traits. The offspring are clones.

What type of cell reproduction occurs in asexual reproduction?

Study Tip: Use this idea map as you study asexual reproduction. Note the different forms of asexual reproduction—budding and regeneration.

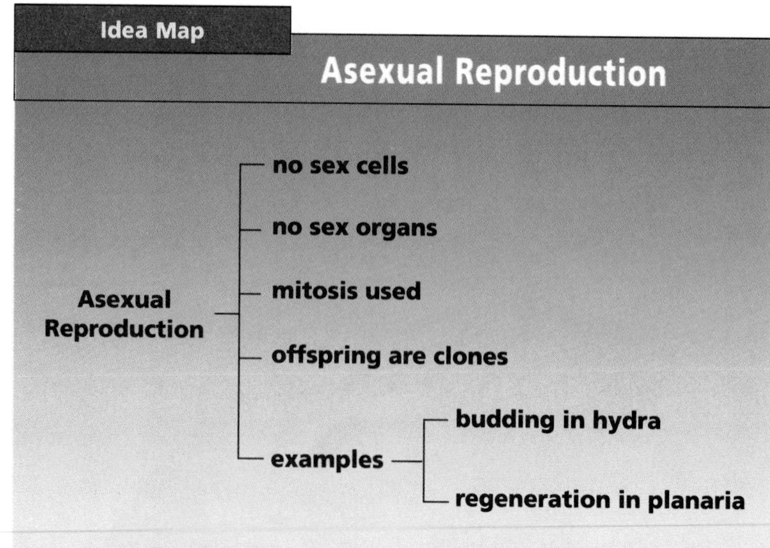

Asexual reproduction is sometimes said to be a simple form of reproduction. Since asexual reproduction is simple, do you suppose most animals use it? The answer is no. Simple animals, such as sponges, hydras, and flatworms, can reproduce this way. Complex animals, such as humans, birds, and fish, do not reproduce this way.

Methods of Asexual Reproduction

Animals are able to reproduce asexually by budding and regeneration (rih jen uh RAY shun). You learned about budding in Chapter 5. It is reproduction in which a small part of the body grows into a new organism. A hydra is a good example. Follow the steps in Figure 24-1 to see how budding works.

The new hydra, or bud, is at first smaller than the parent. Its other features, such as body shape and presence of tentacles, are the same as the parent's. This is an important feature of asexual reproduction. All offspring are exact copies, or clones, of the parent.

Figure 24-1 shows a photograph of a hydra forming a bud. The photo is enlarged about 50 times. Did the bud form by mitosis or meiosis? Why?

A few animals can reproduce asexually by regeneration. **Regeneration** is reproduction in which the parent separates into two or more pieces and each piece forms a new organism. A planarian, which is a flatworm, is an animal that undergoes regeneration.

Mini Lab

What Do Hydra Buds Look Like?

Use a microscope: Examine prepared slides of a hydra under low power. Diagram and label a hydra with buds. Does a hydra bud look like the parent? *For more help, refer to the* ***Skill Handbook,*** *pages 712-714.*

How is a bud of a hydra like the parent?

Figure 24-1 Hydra reproduce by budding.

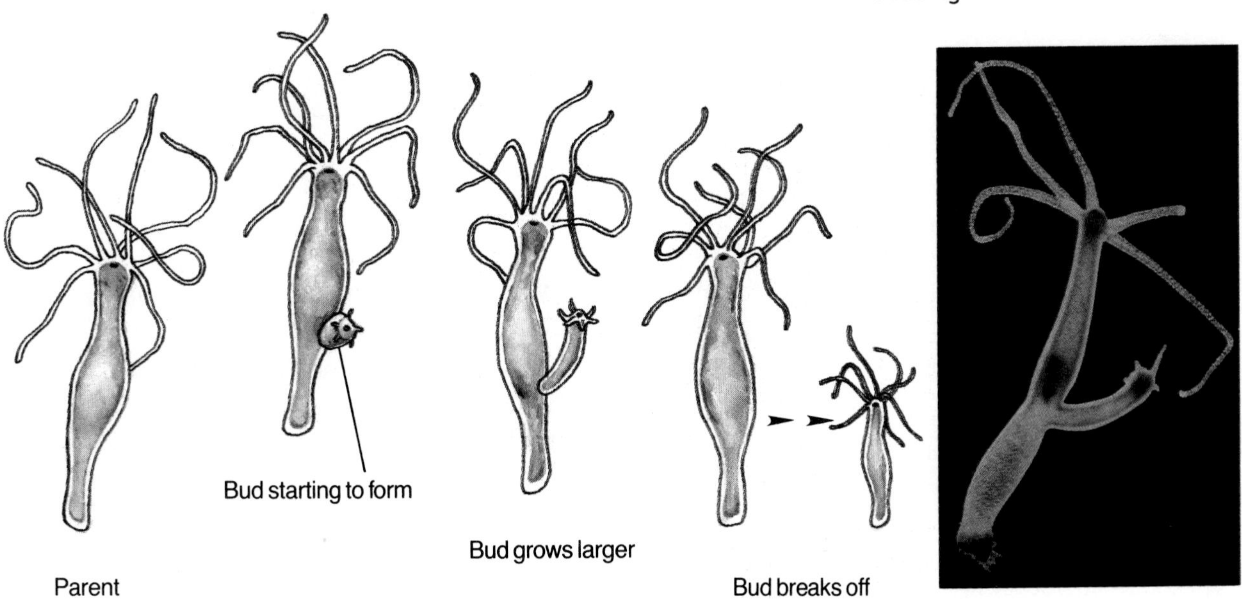

Figure 24-2 Planarians can reproduce by regeneration.

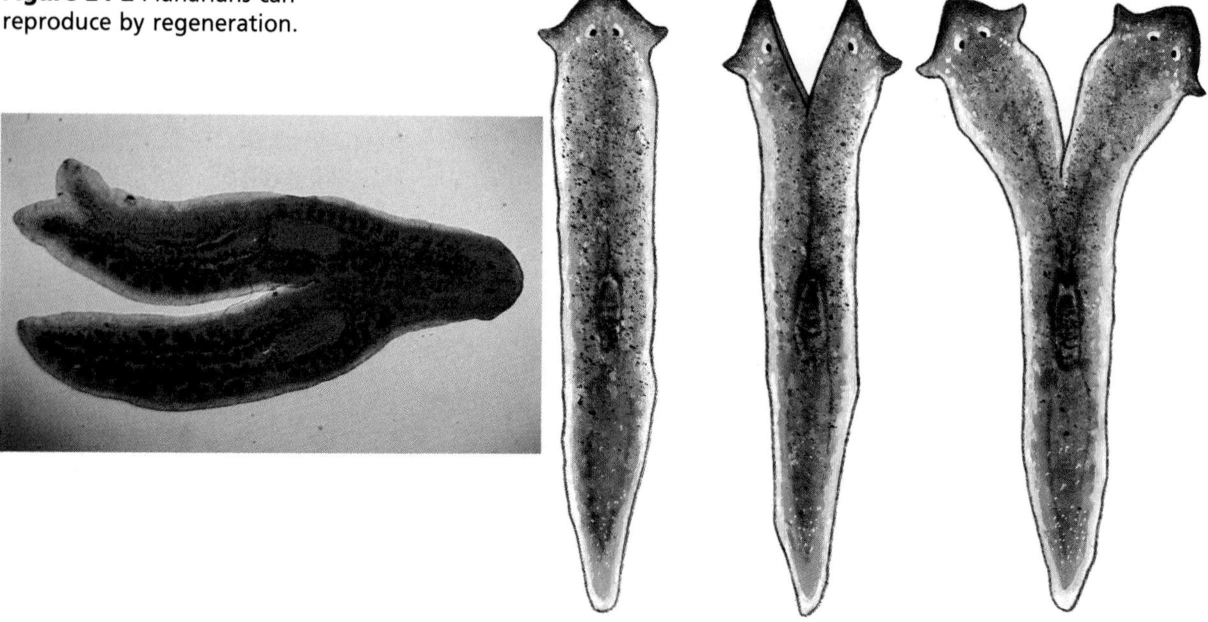

Figure 24-2 shows how a planarian regenerates. The planarian starts regenerating at its head or its tail end. Two heads or two tails form and then the animal begins to split along its body. Regeneration continues until what started out as a single animal ends up as two planarians.

The figure also shows a photograph of a planarian enlarged about 50 times. Note that the planarian appears to have two tail ends. In time, the two tails will separate to form two whole planarians.

Check Your Understanding

1. What is asexual reproduction?
2. What is regeneration?
3. What is budding?
4. **Critical Thinking:** What would be an advantage of asexual reproduction?
5. **Biology and Reading:** People who fish for a living consider starfish pests. When starfish are caught in the fishing nets, they are sometimes cut up and tossed back into the water. Each piece grows into a new starfish. What kind of reproduction is this?

24:2 Sexual Reproduction

Objectives

3. Describe the features of sexual reproduction.

4. Compare internal and external fertilization.

5. Discuss ways animals improve chances of fertilization.

Key Science Words

external fertilization
internal fertilization
estrous cycle

Can you name five animals that use sexual reproduction? You might name dogs, cats, horses, humans, and cattle. What about birds, insects, frogs, or fish? These animals also use sexual reproduction.

Most animals use sexual reproduction in forming offspring. Some simple animals like hydra and planarians can use sexual reproduction in addition to asexual reproduction.

Review of Sexual Reproduction

Let's review some of the features of sexual reproduction.

1. Sex cells, called egg and sperm, are used.
2. Organs are used to form the sex cells. Testes of the male form sperm cells. Ovaries of the female form egg cells.
3. Meiosis is the type of cell reproduction that forms the sex cells.
4. Mating may take place. Mating is the process of a male and female joining together to make sure that egg and sperm meet. Fertilization of egg by sperm is more likely to take place after mating.
5. Offspring formed by sexual reproduction will not look alike. They usually do not look exactly like either parent, although they may show some features of each parent.

Sexual reproduction uses what kind of cell reproduction?

Will offspring formed by sexual reproduction be clones? No. Sexual reproduction forms offspring with different features.

Idea Map

Sexual Reproduction

Sexual Reproduction —
- **sex cells**
- **sex organs**
- **meiosis used**
- **mating**
- **offspring different**

Study Tip: Use this idea map as you study sexual reproduction. Note that in sexual reproduction, eggs can be fertilized outside the body or inside the body.

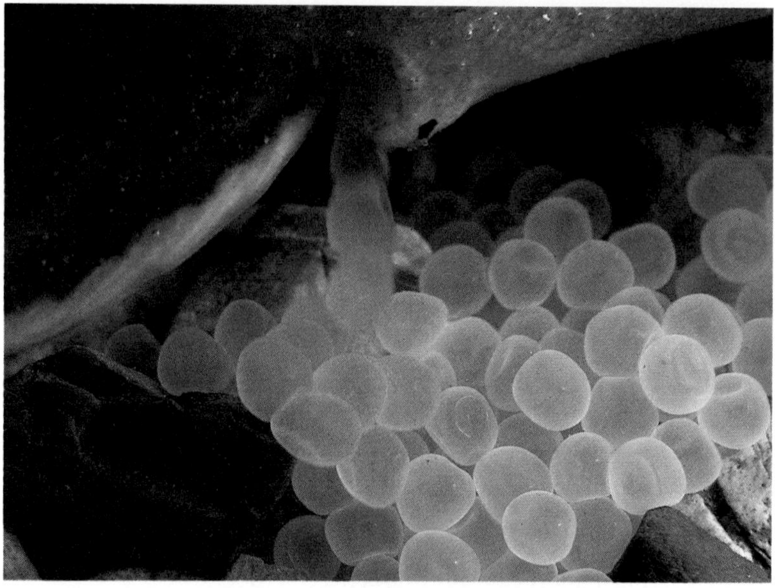

Figure 24-3 After the female fish deposits her eggs, the male will fertilize them.

External and Internal Fertilization

In some animals the sex cells join outside the body. How is this possible? Animals like the female fish in Figure 24-3 release their eggs in water. Meanwhile, male fish also release their sperm in water. Then sperm and egg meet, and fertilization takes place.

Where does external fertilization take place?

The events just described are called external fertilization. **External fertilization** is the joining of egg and sperm outside the body. It's easy to remember because *external* means outside. The fertilized egg then forms a new animal. The young usually grow without care from their parents. This lack of care by the parents seems to be true for many animals that have external fertilization.

Most animals with external fertilization live in water or deposit their sex cells in water. Sperm must be able to swim to the egg. They could not do so unless they were in water.

How can animals with external fertilization make sure that egg and sperm meet? They can't always. They can, however, improve the chances in two ways. First, these animals give off thousands of sex cells at one time. The large number of sex cells increases the odds that one of the eggs will be fertilized. Second, these animals gather in large groups. You may have heard large groups of frogs croaking on a warm, rainy spring night. They gather in large groups to reproduce. When the sex cells are given off, the chance that they will meet is greater.

Mini Lab

Are Sperm Attracted to Eggs?

Observe: Add a dropperful of sea urchin sperm to some sea urchin eggs. Observe what happens under the microscope. *For more help, refer to the* ***Skill Handbook,*** *pages 704-705.*

Fish, starfish, sponges, and frogs are animals with external fertilization. These animals are found in or near water. Frogs can hop around on land, but they usually reproduce only in water.

What are two ways that animals improve chances for external fertilization?

If *external* means outside, then what does *internal* mean? It means inside. Some animals have internal fertilization. **Internal fertilization** is the joining of egg and sperm inside the female's body.

When mating takes place during internal fertilization, sperm are released inside the female's body. Releasing the sperm this way increases the chance that an egg will become fertilized. It also allows reproduction to take place out of water. Animals with internal fertilization can reproduce on land. For example, insects, reptiles, birds, and mammals have internal fertilization. In addition to reproducing on land, animals with internal fertilization usually protect and care for their young.

Breeding Season

Most animals reproduce only during a certain time each year. For example, spring is the breeding season for frogs, Figure 24-4. A breeding season is a certain time of the year when particular kinds of animals reproduce.

The breeding season for frogs takes place when a lot of water is available and the temperature is warm enough. Why is water important for frog reproduction?

Figure 24-4 Frogs breed only at certain times of the year.

Figure 24-5 Large mammals, such as the deer, breed in the fall and have their young the following spring.

Most fish, birds, and many small mammals breed in the spring of the year. The young hatch from eggs, or are born during the late spring or summer of that same year. Many large mammals, such as the deer shown in Figure 24-5, breed in the fall of the year. The unborn young take longer to develop. The young are born during early spring of the next year. This breeding season helps make sure that the young are born in the warm part of the year when they have a better chance to survive.

Another way that animals improve the chance of fertilization is by having an estrous (ES trus) cycle. An **estrous cycle** is a cycle in which a female will mate only at certain times. It is only at these times that the female's eggs can be fertilized.

Mammals have estrous cycles. For example, rats are ready to mate every five days. Most dogs have two estrous cycles per year. Cows have a 20-day cycle.

Check Your Understanding

6. Which cells are used in sexual reproduction?
7. What is meant by external fertilization?
8. What is the difference between a breeding season and an estrous cycle?
9. **Critical Thinking:** Which animals do you think produce more offspring, those with external fertilization or those with internal fertilization? Why?
10. **Biology and Reading:** What kind of reproduction is it when starfish release sperm and eggs into the water?

Lab 24–1 Sex Cells

Problem: How alike are egg and sperm cells?

Skills

make and use tables, measure in SI, recognize and use spatial relationships, make scale drawings

Materials

microscope
prepared slide of starfish eggs
prepared slide of starfish sperm
metric ruler
petri dish for drawing circles

Procedure

1. Copy the data table.
2. Use your microscope to look at a prepared slide of starfish eggs under low power and high power.
3. Fill out the top row in your table.
4. Repeat step 2 with a prepared slide of starfish sperm.
5. Fill out the bottom row in your data table.
6. To measure the actual size of an egg cell, draw a circle on a sheet of paper. Use the petri dish as a guide.
7. **Recognize and use spatial relationships:** Using high power, draw one egg in the circle. Draw it to scale as seen through the microscope. Label it "starfish egg."
8. **Measure in SI:** Measure the diameter of your diagram in millimeters. Record this number in your table.
9. Multiply the number in step 8 by 0.004 to get the actual size of the cell. Record this number in your table.
10. Repeat steps 6 to 9 with the slide of sperm. Label your diagram "starfish sperm." Note that there are two parts to a sperm. Label the small, oval end "head." Label the long, threadlike part "tail."

Data and Observations

1. Compare the actual sizes of egg and sperm.
2. Compare the shapes of egg and sperm.

Analyze and Apply

1. Compare the following for starfish and humans:
   (a) type of reproduction—sexual or asexual
   (b) type of fertilization—internal or external
2. **Apply:** Explain how the shape of a sperm cell helps it swim.

Extension

Make scale drawings: Prepare a wet mount of a small piece of magazine print. Locate a period at the end of a sentence. Using the same procedures as in the lab, measure the diameter of the period under high power. Compare the sizes of the starfish egg and sperm to that of the period.

	SHAPE OF CELL	BODY OR SEX CELL	HAS A TAIL	SWIMS	NUMBER SEEN UNDER HIGH POWER	LENGTH OF DIAGRAM (mm)	MULTIPLY BY	SIZE OF CELL (mm)
Egg								
Sperm								

Menstrual Cycle

Lab 24–2

Problem: What causes uterus thickness to change during the menstrual cycle?

Skills

make and use graphs, infer, relate cause and effect

Materials

graph paper
colored pencil
ruler

Procedure

1. Obtain a sheet of graph paper.
2. Draw and label two graphs as shown in the figure. *For more help, refer to the* ***Skill Handbook,*** *pages 715-717.*
3. Use the data in the table to complete the top graph.
   (a) Plot the estrogen units on the graph.
   (b) Connect the points for estrogen with a line and label the line "estrogen units."
   (c) Plot the progesterone units.
   (d) Use a colored pencil to connect the points for progesterone and label the line progesterone units.
4. Use the data in the table to complete the bottom graph.
   (a) Plot uterus thickness on the graph.
   (b) Connect the points on the graph with a line.

DAY	AMOUNT OF HORMONE		THICKNESS OF UTERUS LINING (mm)
	ESTROGEN	PROGESTERONE	
1	50	5	.5
5	65	5	1.5
10	200	5	2.25
15	75	40	3.0
20	100	150	4.0
25	50	100	5.0
27	50	30	4.75
1	50	5	.5

Data and Observations

1. How do the amounts of estrogen and progesterone change throughout the menstrual cycle?
2. How does the thickness of the uterus change throughout the menstrual cycle?

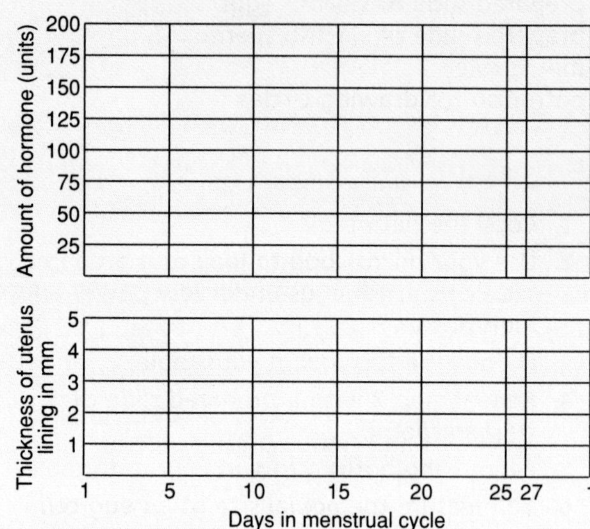

Analyze and Apply

1. **Infer:** How is the change in the thickness of the uterine lining related to the change in estrogen amount for days 1-10?
2. Compare the change in thickness of the uterine lining with the change in progesterone amount for days 10-27.
3. **Apply:** (a) What happens to the uterine lining between day 27 and day 1?
   (b) What is this process called?

Extension

Relate cause and effect: What happens to the egg on days 1-13, 14, and 15-27? How does this compare to the amounts of hormones and uterus thickness?

24:3 Reproduction in Humans

Humans have reproductive systems with many parts. The reproductive organs of males and females each do special jobs. Let's look at the human reproductive system to see how the different parts work.

Objectives

6. Identify the reproductive parts of humans and the stages of human reproduction.

7. Compare changes in the menstrual cycle with and without fertilization.

8. Discuss symptoms and problems of sexually transmitted diseases.

Key Science Words

reproductive system
scrotum
penis
vas deferens
oviduct
uterus
vagina
menstrual cycle
estrogen
menstruation
sexually transmitted disease

Human Reproductive System

The system used to produce offspring is called the **reproductive system.** What does this system look like in males and females?

Figure 24-6 shows the main parts of the male reproductive system. The job of each part is also listed. Study these jobs. Notice that this figure is a cut-away view from the side. Also notice that certain parts of the male reproductive system are also part of the excretory system studied in Chapter 13.

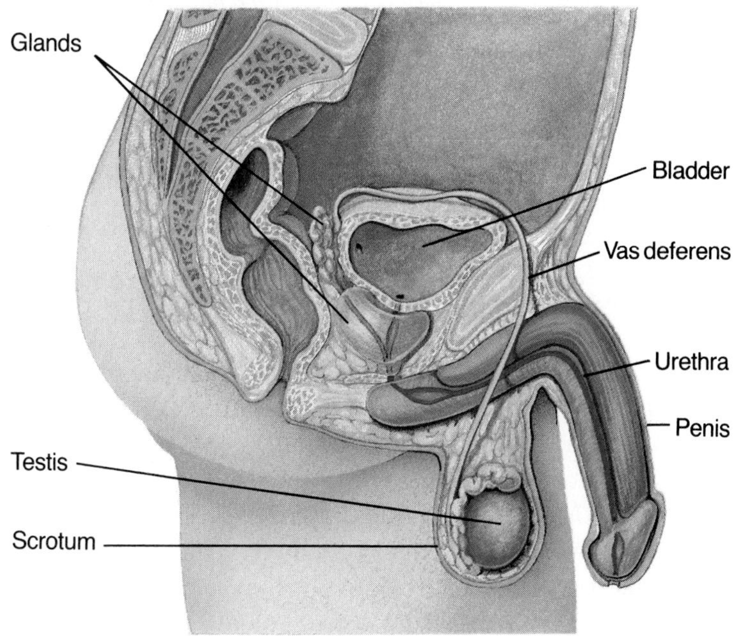

Figure 24-6 The male reproductive system is shown here in a cut-away view from the side.

PART	JOB
Testis	Produces sperm cells by meiosis
Scrotum (SKROH tum)	Sac that holds testes
Penis (PEE nus)	Places sperm in vagina during mating
Vas deferens (VAS·DEF uh runs)	Carries sperm from testes to urethra
Urethra (yoo REE thruh)	Carries sperm and urine out of body
Glands	Provides liquid in which sperm can swim

Figure 24-7 The female reproductive system is shown here in a cut-away view from the front.

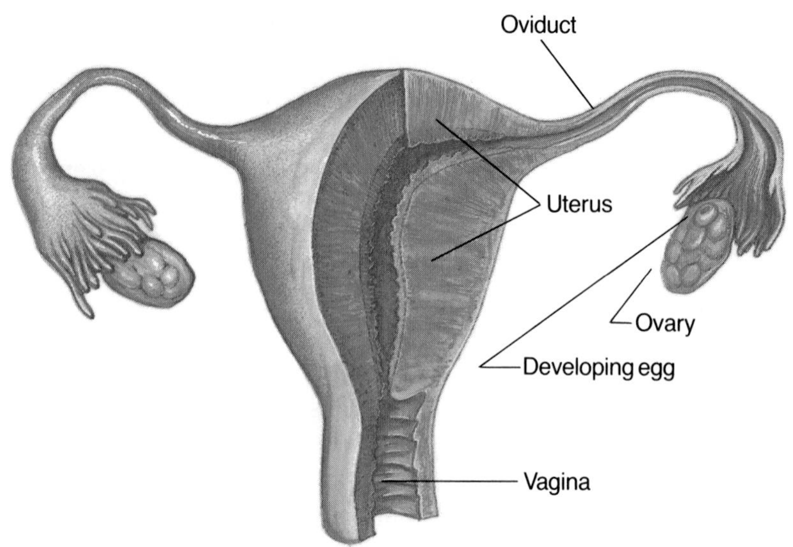

PART	JOB
Ovary	Produces egg cells
Oviduct	Carries egg from ovary to uterus
Uterus	Place where fertilized egg develops
Vagina	Receives penis of male during mating

The female reproductive system is shown in Figure 24-7. This figure shows a cut-away view from the front. Study the function of each part.

Stages of Reproduction

How does an egg become fertilized? What happens to the egg after it is fertilized? Look at Figure 24-8 and follow the numbers in the figure as you read.

1. Egg cells are formed in each ovary.
2. Each month, one ovary releases an egg. Usually, only one egg is released about every 28 days. The ovaries usually take turns releasing eggs. One ovary releases an egg one month, and the other ovary releases an egg the next month.

Figure 24-8 shows what happens next.

3. Once released from the ovary, the egg moves into a tube called an oviduct (OH vuh dukt). **Oviducts** are tubelike organs that connect the ovaries to the uterus. The **uterus** (YEWT uh rus) is a muscular organ in which the fertilized egg develops. The uterus is made of smooth muscle.

Figure 24-8 shows how fertilization takes place.

4. Sperm are released into the vagina during mating. The

Skill Check

Understand science words: oviduct. The word part *ovi* means egg. In your dictionary, find three words with the word part *ovi* in them. *For more help, refer to the **Skill Handbook,** pages 706-711.*

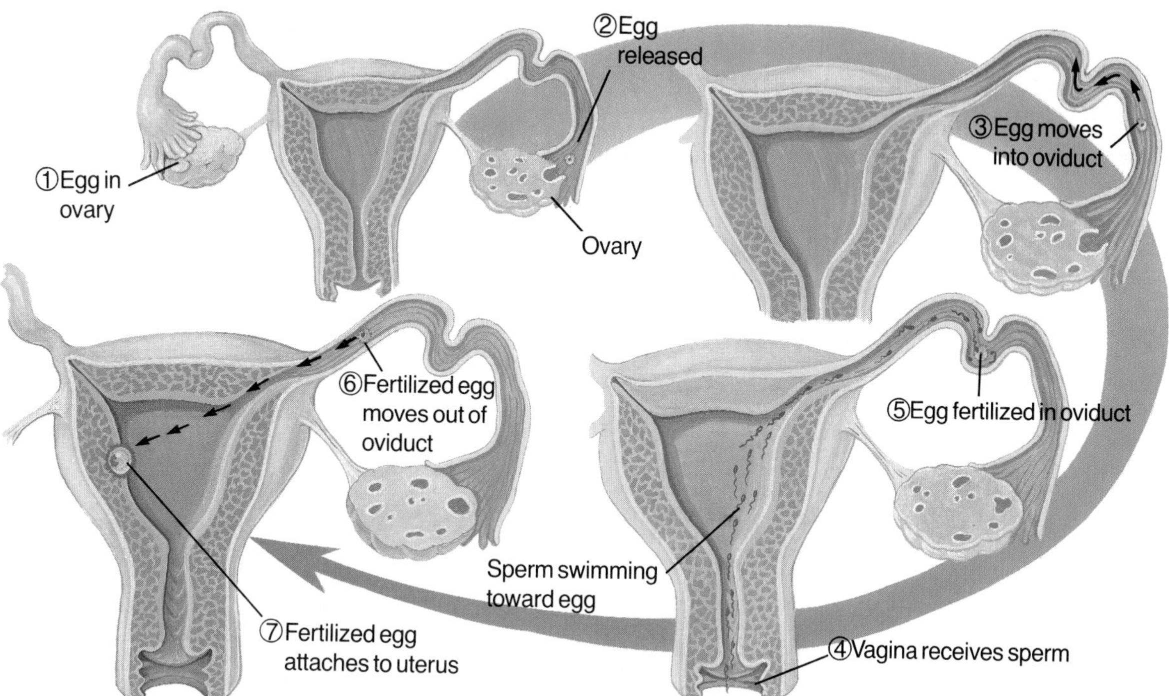

Figure 24-8 The stages of reproduction in the female are shown here. Where is the egg fertilized?

vagina (vuh JI nuh) is a muscular tube that leads from outside the female's body to the uterus. Sperm swim from the vagina into the uterus and then into the oviducts.

5. If an egg is present, fertilization takes place.

Figure 24-8 shows what happens after fertilization.

6. As the fertilized egg moves down the oviduct, it divides several times and becomes an embryo.
7. The embryo then attaches itself to the wall of the uterus. Once attached, it will remain there for nine months as it develops into a baby.

What happens to an egg after it is fertilized in the oviduct?

The Menstrual Cycle

You know that one egg is released by an ovary about every 28 days. What controls this timing? Many of the changes in the reproductive system are controlled by hormones. You learned in Chapter 15 that hormones are chemicals that affect certain body organs.

The monthly changes that take place in the female reproductive organs are called the **menstrual** (MEN struhl) **cycle.** The word *menstrual* comes from Latin and means monthly.

The menstrual cycle occurs about every month starting when a female is 10 to 13 years old. The monthly cycles continue for about 40 years.

Bio Tip

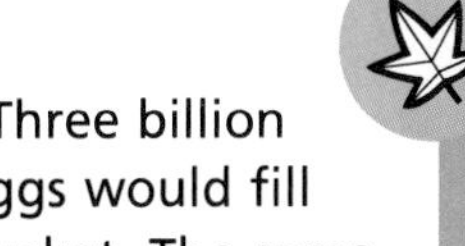

Science: Three billion human eggs would fill a small bucket. The same number of human sperm would fit into a thimble.

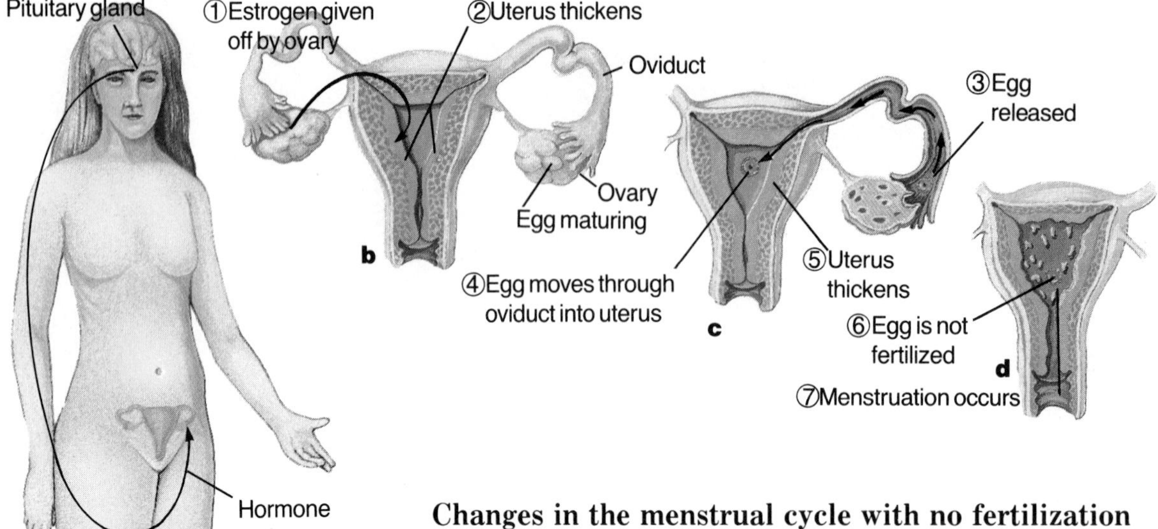

Figure 24-9 The changes shown here take place during the menstrual cycle if no egg is fertilized.

Changes in the menstrual cycle with no fertilization

Figure 24-9a shows events near the start of the menstrual cycle. The pituitary gland gives off a hormone. This hormone travels by way of the blood to the ovaries. It causes an egg in the ovaries to mature.

Figure 24-9b shows the changes that follow. The numbered steps in the diagram match the numbered statements.

1. The ovary itself forms estrogen (ES truh jun). **Estrogen** is a female hormone. It is responsible for the changes seen in a female as she enters puberty.
2. Estrogen also causes the lining of the uterus to increase in thickness. This increase in thickness makes it possible for a fertilized egg to attach to the uterine lining.

Figure 24-9c shows the next set of events. No sperm are present, so fertilization of the egg will not occur.

3. The ovary releases an egg. After the egg is released, the ovary begins to produce progesterone (pruh JES tuh rohn), a second female hormone.
4. The egg moves through the oviduct and enters the uterus.
5. Meanwhile, the uterine lining continues to thicken from the progesterone being made by the ovary.

Figure 24-9d shows what happens next.

6. The egg has not been fertilized. Therefore, it will not attach to the uterus.

What happens in the uterus when an egg is not fertilized?

7. The thick uterine lining is no longer needed. It begins to break apart. The cells of the thickened uterine lining break off and leave the body through the vagina along with the unfertilized egg and a small amount of blood. This loss of cells from the uterine lining, blood, and egg is called **menstruation** (men STRAY shun).

Science and Society

Sperm Banks

Technology has made it possible to store human sperm for years in sperm banks. Sperm are collected from donors and then frozen in liquid nitrogen. Chemicals are added to the sperm to keep them alive and prevent them from being damaged by the freezing temperatures. The sperm are kept frozen until they are needed.

How are the sperm banks used? For some couples, the husband is unable to produce sperm. These couples can receive sperm donated to the sperm bank by other males. Males who wish to donate sperm contact the sperm bank. The sperm is collected and stored. The stored sperm are used to fertilize one of the woman's eggs. The sperm are injected into the uterus with a syringe. This method of fertilizing eggs is called artificial insemination.

What Do You Think?

1. Couples who use sperm banks often don't know who donated the sperm. The name of the donor isn't revealed to them. Thus, the couple doesn't know who the biological father is. The donor also doesn't know who receives his sperm. How is keeping the identities of donors and couples who receive sperm a secret helpful? Would it be better to reveal the identities?

2. Couples who use sperm banks can look at catalogs that list the traits of different sperm donors. In this way, they can choose some of the genetic qualities of their future child. What if the couple wanted their child to be like Albert Einstein or Bruce Springsteen? Who would be responsible if the child were born with a genetic disorder instead of being like Einstein?

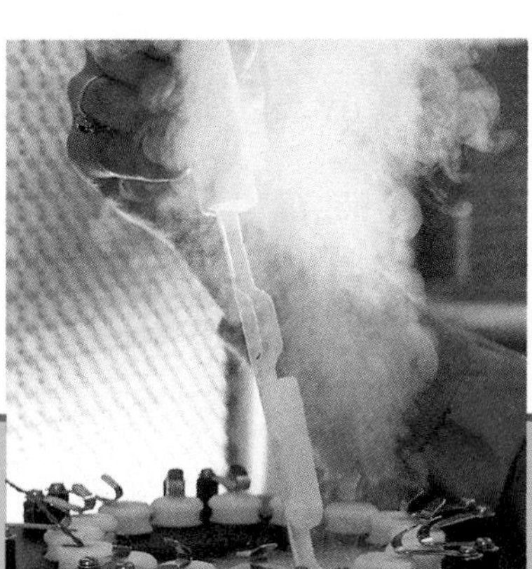

Sperm are stored in liquid nitrogen.

3. Imagine you could create a person with beautiful hair and eyes, a great personality, and lots of physical ability. This "super person" would have the traits you find desirable. Over time, a large group of "super people" could be produced. But, your friend likes blond hair and brown eyes. You prefer brown hair and blue eyes. Who decides which traits are the most desirable ones to have? What kinds of problems does society face with the ability to create "super people"?

Conclusion: Do the benefits of sperm banks outweigh the possible problems?

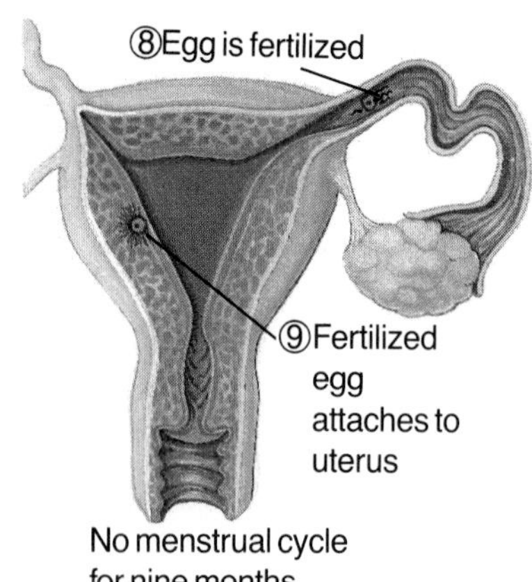
⑧Egg is fertilized
⑨Fertilized egg attaches to uterus
No menstrual cycle for nine months

Figure 24-10 If fertilization occurs, the fertilized egg attaches to the uterus.

TABLE 24–1. MENSTRUAL CYCLE		
Day in Cycle	**Event Occurring if Egg Is Not Fertilized**	**Diagram**
1-4	Menstruation—loss of egg, blood, uterus lining	24-9d
5-13	Thickening of uterus lining by estrogen	24-9a,b
14	Release of egg from ovary	24-9c
15-28	Continued thickening of uterus lining	24-9c
1-4	Menstruation if egg is not fertilized	24-9d

With menstruation, the cycle starts again. We could return to Figure 24-9a and follow the events for a second time. What event marks the beginning of a menstrual cycle? Table 24-1 shows you. It lists the days within the cycle when major events take place.

Changes in the menstrual cycle if fertilization occurs

What if fertilization does occur? Until step 5, the events are the same. After that, something different happens. Study Figure 24-10 as you read.

8. Sperm are present, and an egg is fertilized. Meanwhile, the uterine lining has thickened in preparation for receiving the embryo that develops from the fertilized egg.
9. The embryo attaches to the uterus. It remains attached for nine months while it develops into a baby. No new menstrual cycle will begin until the baby is born.

Skill Check

Sequence: Study Table 24-1 and Figure 24-9. Sequence the events that take place on days 5–28 of the menstrual cycle. *For more help, refer to the **Skill Handbook**, pages 706-711.*

Diseases of the Reproductive System

Like all other body systems, the human reproductive system can be invaded by microbes. This invasion can result in diseases being passed from person to person during sexual contact. Diseases transmitted through sexual contact are called **sexually transmitted diseases,** or STDs.

Some common STDs are listed in Table 24-2. These diseases can damage the reproductive system as well as other body systems. What kinds of microbes cause STDs? The diseases gonorrhea (gawn uh REE uh), chlamydia (kluh MIH dee uh), and syphilis (SIF uh lus) are caused by bacteria. AIDS and genital herpes are caused by viruses.

Gonorrhea, chlamydia, and syphilis can be treated and cured with antibiotics. Genital herpes and AIDS aren't curable.

TABLE 24–2. SEXUALLY TRANSMITTED DISEASES		
Disease	**Symptoms**	**Problems**
Gonorrhea	females–puslike discharge from vagina, painful urination, no symptoms in many women males–pus discharge from penis, painful urination	females–sterility, passed to newborn from infected mother during birth males–infection of the testes
Chlamydia	(see gonorrhea), no symptoms in 70% of females, 10% of males	females–sterility, problem pregnancies males–sterility
Syphilis	sore on vagina or penis followed by rash, sore throat, swollen glands	damage to heart and brain, passed from mother to fetus
Genital Herpes	blisterlike cut or sore on vagina or penis, may look like a rash	passed to newborn from infected mother during birth
AIDS	fever, night sweating, dry cough, weight loss, swollen glands, constant tiredness	destruction of the immune system, pneumonia, cancer, fatal after long illness

How can a person reduce the chance of getting an STD? The best way is to avoid sexual contact. A second way is to use a condom during sexual contact. A third way is to have sexual contact with only one partner.

Which STDs can be cured?

In addition to sexually transmitted diseases, the reproductive system can be affected by other types of diseases. In males, the gland that supplies the liquid in which sperm swim can become swollen or develop tumors. In females, the tissue that surrounds the opening to the uterus can develop cancer. Both of these problems can be cured if they are detected early. Regular visits to a doctor are the best prevention.

Check Your Understanding

11. Name the functions of the ovary and testes.
12. At about what age does the menstrual cycle first begin?
13. What kinds of organisms cause STDs?
14. **Critical Thinking:** How is it possible for a person to have a sexually transmitted disease for a long time without knowing it?
15. **Biology and Reading:** If nonidentical twins are produced, how many eggs must have been produced at the same time in the mother's ovaries? Is this normal?

Chapter 24

Review

Summary

24:1 Asexual Reproduction

1. Asexual reproduction does not require sex cells, mating, or sex organs. It does use mitosis.
2. Budding and regeneration are two types of asexual reproduction.

24:2 Sexual Reproduction

3. Sexual reproduction uses mating and egg and sperm that are formed during meiosis.
4. In external fertilization, egg and sperm join outside the bodies of the parents. In internal fertilization, egg and sperm join inside the body of the female.
5. Having breeding seasons or an estrous cycle increases the chance of fertilization.

24:3 Reproduction in Humans

6. Eggs are formed in the ovaries and released into the oviducts. Fertilization of an egg occurs in the oviduct. The fertilized egg attaches itself to the uterine lining after it has divided several times.
7. The menstrual cycle prepares the uterus for a fertilized egg. If fertilization does not take place, menstruation occurs.
8. Sexually transmitted diseases are caused by bacteria and viruses and are spread by sexual contact.

Key Science Words

estrogen (p. 516)
estrous cycle (p. 510)
external fertilization (p. 508)
internal fertilization (p. 509)
menstrual cycle (p. 515)
menstruation (p. 516)
oviduct (p. 514)
penis (p. 513)
regeneration (p. 505)
reproductive system (p. 513)
sexually transmitted disease (p. 518)
scrotum (p. 513)
uterus (p. 514)
vagina (p. 515)
vas deferens (p. 513)

Testing Yourself

Using Words

Choose the word from the list of Key Science Words that best fits the definition.

1. hormone found in females
2. certain time of the year when animals reproduce
3. loss of egg, uterine lining, and blood
4. organ in which a fertilized egg develops
5. cycle in which a female will mate only at certain times
6. egg and sperm joining outside body
7. tubelike organ connecting ovary to uterus
8. where sperm are released in the female
9. monthly changes in the female reproductive system

Testing Yourself *continued*

10. what it is called when the parent breaks into two or more pieces to reproduce

Finding Main Ideas

List the page number where each main idea below is found. Then, explain each main idea.

11. what the main features of asexual reproduction are
12. the pathway followed by sperm within the vagina
13. changes that take place during the different stages of the menstrual cycle
14. how animals with external fertilization improve the chances of fertilization
15. why animals with internal fertilization can reproduce on land
16. the roles of testes and ovaries

Using Main Ideas

Answer the questions by referring to the page number after each question.

17. What are three ways you can reduce the chances of getting an STD? (p. 519)
18. Why do animals with external fertilization almost always live or reproduce in water? (p. 508)
19. What happens on or about day 14 of the menstrual cycle? (p. 518)
20. How does mating help animals reproduce? (p. 509)
21. How does asexual reproduction compare to sexual reproduction in terms of (p. 504, p. 507):
    (a) the process of cell division used?
    (b) whether sex cells are needed?
22. How is budding different from regeneration? (p. 505)

Skill Review

For more help, refer to the ***Skill Handbook,*** *pages 704-719.*

1. **Make and use tables:** Make a table that compares the features of asexual and sexual reproduction.
2. **Sequence:** Sequence the steps involved in regeneration of a planarian.
3. **Understand science words:** What does the prefix *ex* mean in the word *external?*
4. **Make scale drawings:** An egg viewed under high power (430x) measures 30 mm when drawn to scale in a circle that has a diameter of 100 mm. What is the actual size of the egg?

Finding Out More

TECH PREP

Critical Thinking

1. As a hobby, you wish to grow identical planaria. Suggest a plan for carrying out your hobby.
2. Each ovary takes a turn releasing one egg every 28 days. Yet, people do have fraternal (unalike) twins. Explain how this is possible.

Applications

1. Explain how a condom, diaphragm, fertility awareness, birth control pill, vasectomy, and tubal ligation actually work as birth control methods.
2. Find out which mammals have menstrual or estrous cycles and how long these cycles are. Prepare a chart of your findings for room display.

CHAPTER PREVIEW

Chapter Content

Review this outline for Chapter 25 before you read the chapter.

Skills in this Chapter

The skills that you will use in this chapter are listed below.

- In **Lab 25-1,** you will measure in SI, use numbers, infer, and make and use graphs. In **Lab 25-2,** you will interpret data, measure in SI, and compare.
- In the **Skill Checks,** you will interpret diagrams and understand science words.
- In the **Mini Labs,** you will design an experiment.

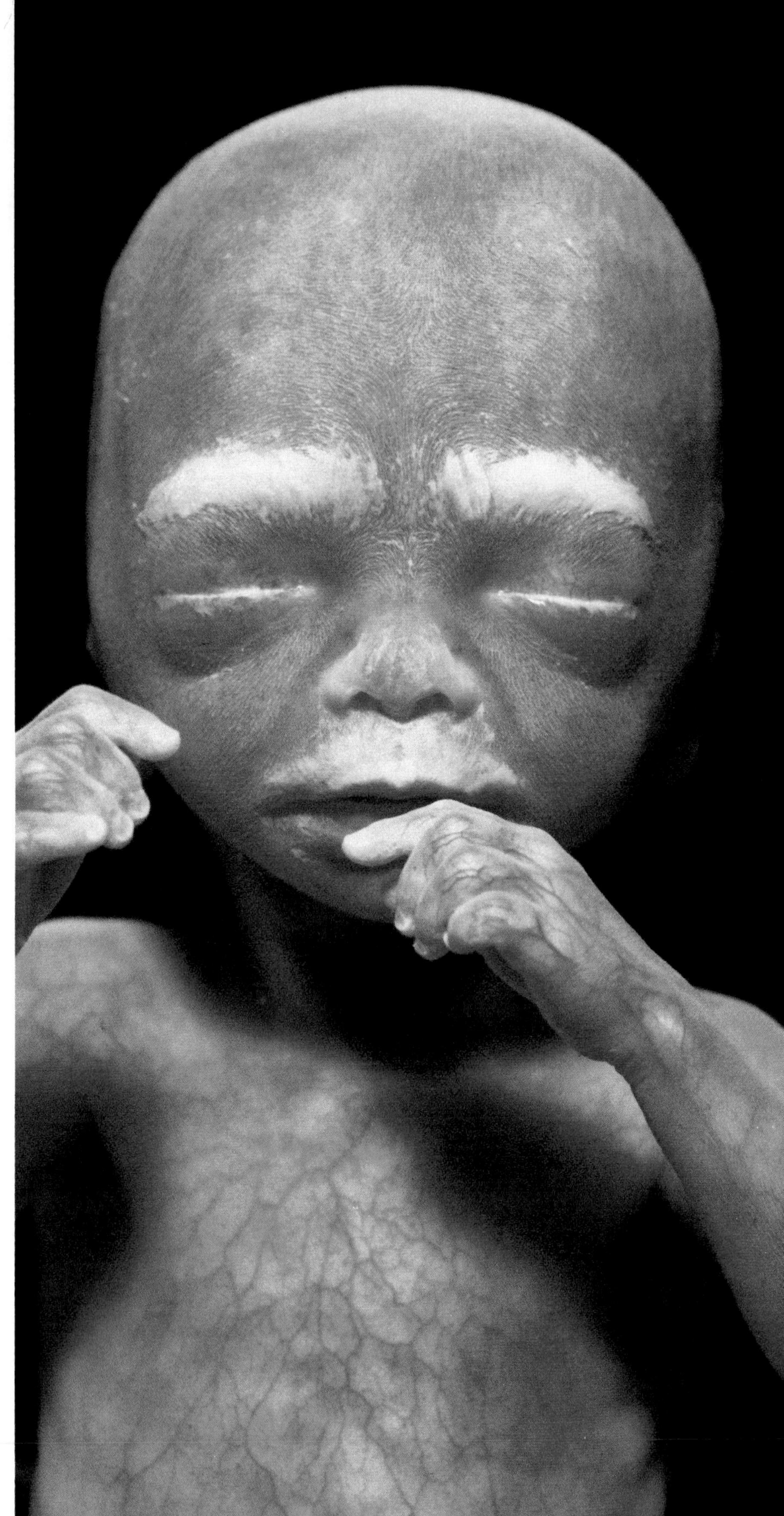

Six-month-old human fetus

Chapter 25

Animal Development

A human female carries an unborn child for about 266 days, approximately 9 months. Other animals carry their young different lengths of time. Rats carry their young for only 22 days. In cats and dogs, the length of time for carrying young is about 60 days. In horses, it is about 336 days. Some animals don't carry their young at all. The young develop outside the female's body. Birds, frogs, and insects are some animals that develop outside the female.

The photo on the left shows a six-month-old unborn human as it appears within the mother's body. Has it formed all the parts it will need at birth? How does it get food and oxygen? You will discover the answers to these questions as you read this chapter.

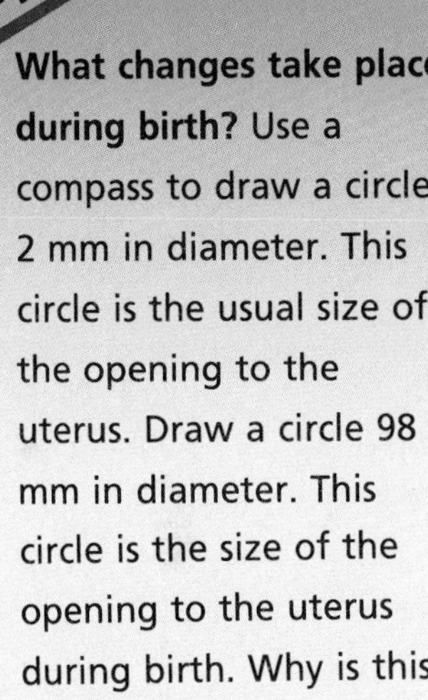

Try This!

What changes take place during birth? Use a compass to draw a circle 2 mm in diameter. This circle is the usual size of the opening to the uterus. Draw a circle 98 mm in diameter. This circle is the size of the opening to the uterus during birth. Why is this change in size needed?

BIOLOGY *Online*

Visit the Glencoe Science Web site at science.glencoe.com to find links about **animal development.**

25:1 Development Inside the Female

Objectives

1. **Sequence** the changes that take place in a fertilized egg until it attaches to the uterus.
2. **Explain** how the needs of an embryo are met as it develops.
3. **Describe** the stages of human development from the first month until birth.

Key Science Words

cleavage
amniotic sac
placenta
umbilical cord
fetus
labor
navel

You began as a fertilized egg and are now a young adult. You have gone through many changes. Many of these changes took place while you were still in your mother's body.

Early Stages of Development

All living things go through development. You learned in Chapter 2 that development is all the changes that occur in a living thing as it grows. In humans, development begins when a single fertilized egg changes into many different body parts or tissues. Development results in the forming of bone, muscle, hair, nerve, and skin tissues.

What are the changes that can be seen in humans as they develop? In order to answer this question we must start with a single fertilized egg. The fertilized egg undergoes changes that take it from a one-celled stage to a many-celled stage. These changes are largely the result of mitosis.

Most of the early changes take place while the egg is still within the female's oviduct. Follow the numbered steps of Figure 25-1 as you continue to read. Also pay attention to how long it takes for each change to take place.

1. An egg is fertilized.
2. The fertilized egg undergoes mitosis to make two cells. It is now an embryo. How long did this step take?

When does a fertilized egg become an embryo?

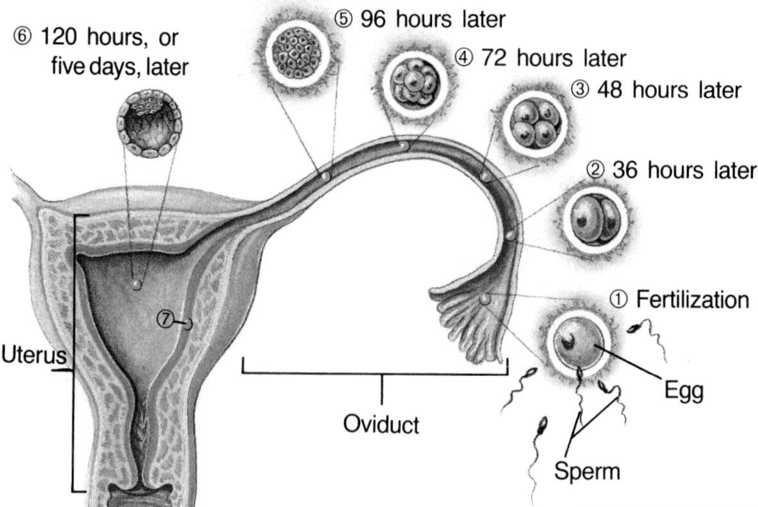

Figure 25-1 The fertilized egg goes through cleavage in the oviduct. How long does it take for the embryo to become a solid ball of cells?

3. Mitosis continues and two cells form four cells. Four cells then form eight. Eight cells will form sixteen cells.
4. Three days later, the embryo is sixteen cells in size.
5. Next, the embryo forms a solid ball of cells. Has it left the oviduct yet?
6. Finally, after five days (120 hours), the embryo moves out of the oviduct into the uterus. By this time, it is a hollow ball of cells.
7. Next, the embryo attaches itself to the lining of the uterus. This is where it develops for the next 37 weeks.

These early changes are called cleavage (KLEE vihj). **Cleavage** is the series of changes that take place to turn one fertilized egg into a hollow ball of many cells. The number of cells increases from one to about one hundred in just five days. The word *cleave* means to chop or divide. What divides during cleavage?

All the changes that take place during cleavage occur within the body of the female. Many animals, such as humans, cats, and whales, show this type of development.

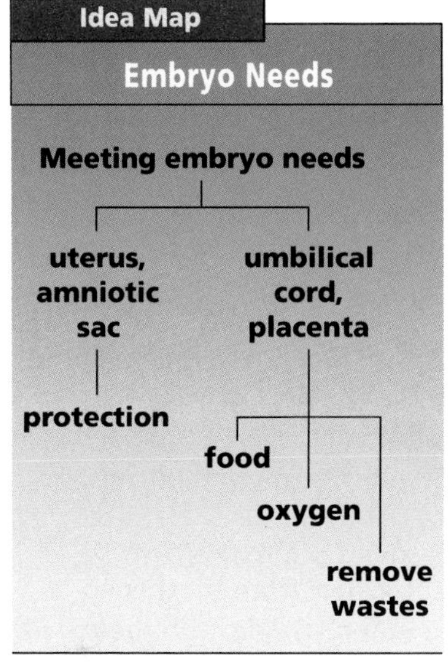

Study Tip: Use this idea map as a study guide to development inside the female. How are the needs of the embryo met?

Needs of the Embryo

Embryos, just like adults or newborns, have certain needs. These needs include protection, food, oxygen, and getting rid of wastes. How are these needs met?

Protection

The embryo must be protected against injury. An amniotic (am nee AHT ik) sac forms around it as shown in Figure 25-2. An **amniotic sac** is a tissue filled with liquid that protects the embryo. The liquid around the embryo cushions it in the same way as packing material cushions the contents of a package.

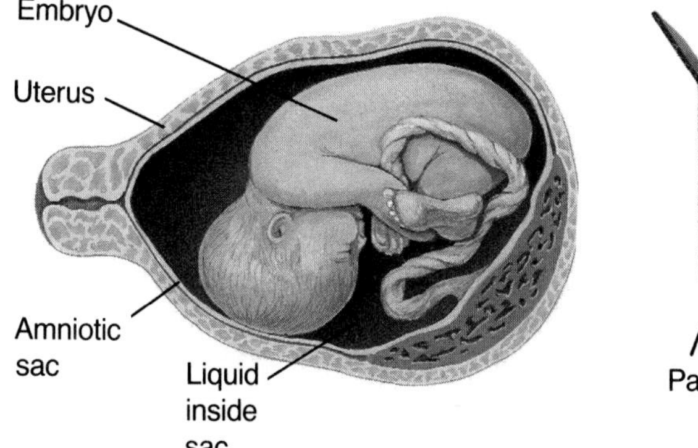

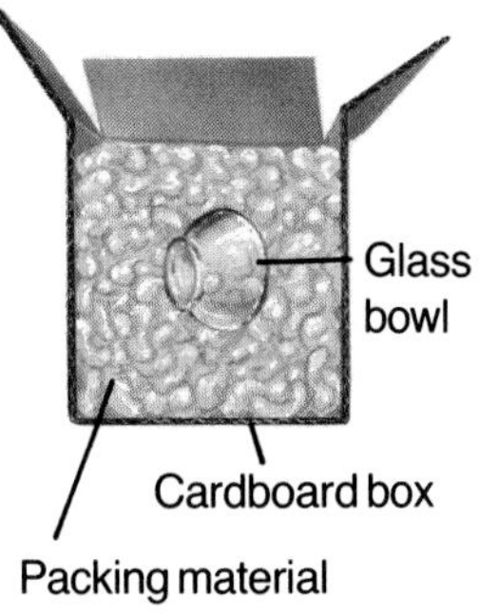

Figure 25-2 The amniotic sac protects the embryo, much like the packing material protects the bowl.

Figure 25-3 The embryo receives food and oxygen and gets rid of wastes through the placenta.

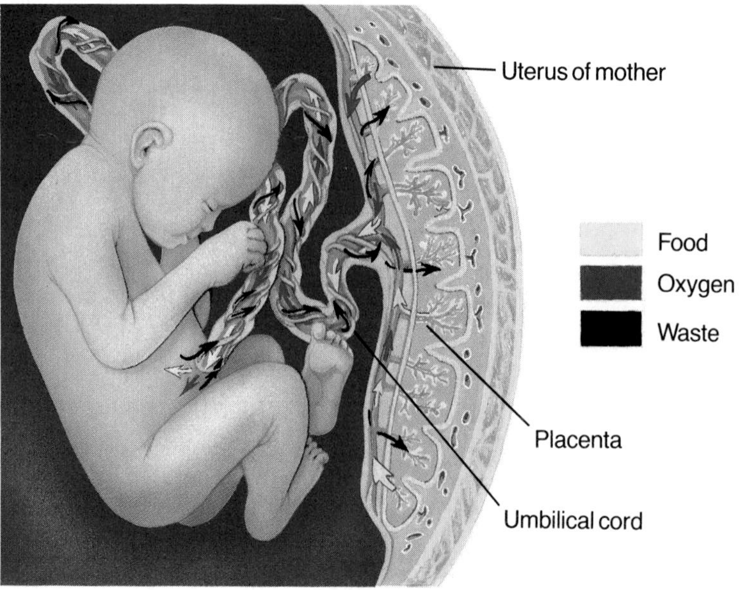

> **Skill Check**
>
> **Interpret diagrams:** Study Figure 25-3. What is the pathway of oxygen from mother to embryo? *For more help, refer to the* ***Skill Handbook,*** *pages 706-711.*

Food

How does an embryo get food? Certainly it doesn't eat while inside the amniotic sac. Yet, it must be supplied with food in order to grow.

Before a human is born, food is supplied by the mother. The embryo's food is in the mother's blood. This food comes from food that the mother eats.

How does the food get from the mother's blood to the embryo? Food passes from an organ called the placenta (pluh SENT uh) to the embryo. The **placenta** is an organ that connects the embryo to the mother's uterus. The placenta has many blood vessels in it. Food passes from the mother's blood through the placenta into a cord, called the umbilical (uhm BIL ih kuhl) cord. The **umbilical cord** is made of blood vessels that connect the embryo to the placenta.

Figure 25-3 shows the placenta and umbilical cord in a human embryo. Notice the yellow arrows. They show the direction of food from mother to embryo.

Oxygen

The embryo receives oxygen the same way as it receives food. Oxygen passes from the mother's blood in the uterus to the blood in the placenta. Once in the capillaries of the placenta, oxygen passes into the blood vessels of the umbilical cord. These blood vessels connect directly to the blood vessels of the embryo. Oxygen is then passed to the embryo. Figure 25-3 shows this pathway with blue arrows.

Career Close-Up

Ultrasound Technician

A woman is pregnant and her doctor suggests an ultrasound of the fetus. What actually is ultrasound? Who will take the ultrasound?

An ultrasound technician operates a machine that can take pictures of any organ inside the body. The machine uses sound waves that reflect off body organs. These sound waves are changed into pictures that allow a doctor to see inside the body. Unlike X rays, ultrasound pictures show movement. Ultrasound is also safer than X rays.

Ultrasound pictures will show if a woman is carrying twins. They help the doctor judge the age of the fetus. In this way, the doctor can give a better prediction as to when the child will be born. Abnormal development or birth defects may also be detected. Finally, the sex of the fetus can be determined.

To be an ultrasound technician a person needs a high school education. Two or more years of special training may be needed. After the proper education and training, most ultrasound technicians work in hospitals.

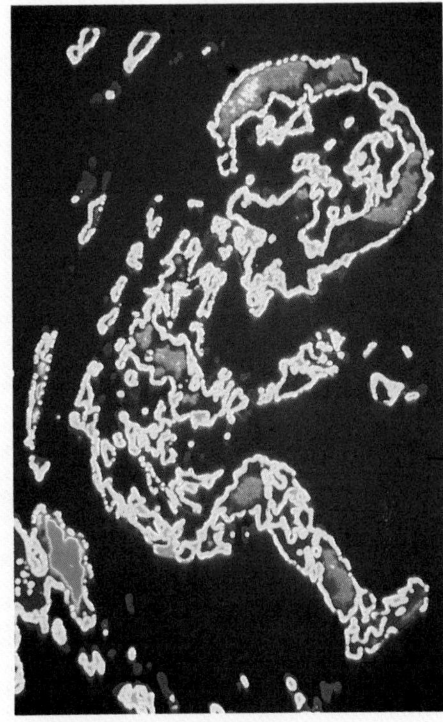

Ultrasound shows what a developing fetus looks like.

Wastes

The cells of the embryo produce wastes. Carbon dioxide and urea are two such waste chemicals. The cord and placenta help get rid of these wastes. The embryo's blood carries these wastes from the embryo to the placenta, where the wastes enter the mother's bloodstream. Figure 25-3 shows this pathway. In which direction are the black arrows that carry wastes going? Why?

How does an embryo get rid of wastes?

A pregnant female eats for herself and the developing embryo. She also breathes for herself and the embryo. She must get rid of her wastes and those of the embryo.

Now you see why proper nutrition is important during pregnancy. Good health is also very important at this time.

Human Development

What changes occur as the embryo continues to develop? When do these changes take place? Let's look at the human embryo to help answer these questions.

First Month

Many changes take place during the first month. Figure 25-4a shows what the embryo looks like at one month. The stomach, brain, and heart start to form. Liver and ears begin to appear. The thyroid gland forms. The heart is already beating. It can be seen as a bulge just in front of the arms.

> **Bio Tip**
>
> **Health:** A fetus born at 20 weeks of development has about a 5 percent chance of survival.

You can even see the start of eye development. The dark circle in the head will be the embryo's eye. Fingers can be seen. Can you see legs at this point in development?

How large is the embryo at this stage? To see how large the embryo is, measure the small figure in the box in millimeters. This figure is an embryo drawn to actual size. The embryo has a mass of 0.02 grams.

At what age does an embryo begin to move?

Second Month

During the second month, arms and legs form. Fingers and toes appear. The eyes have eyelids. The mouth has lips and the nose has nostrils. Figure 25-4b shows these parts. Muscles and bone form. The embryo can even move at this age.

Figure 25-4 A one-month-old embryo (a) and a two-month-old embryo (b) are shown enlarged and actual size.

a

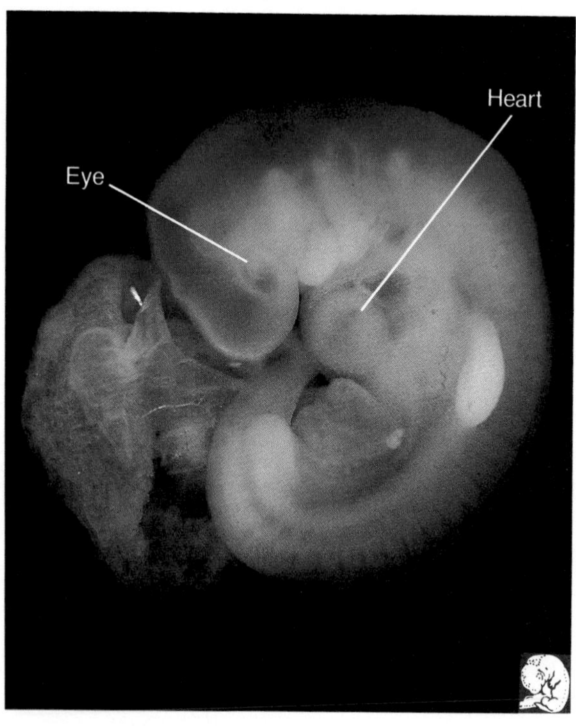

b

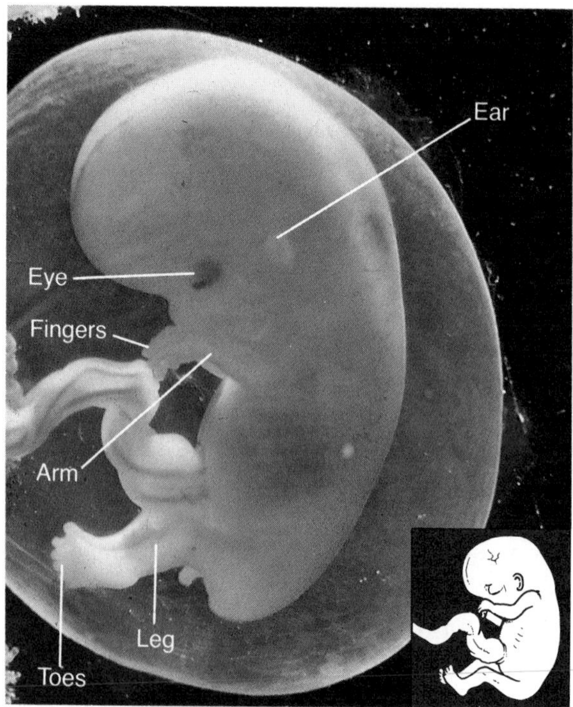

You cannot tell yet whether the embryo is a boy or a girl. Its mass is 1.0 gram. Measure the small figure in the box in millimeters. Has the embryo doubled in size from the first month?

Third Month

At the start of the third month, the embryo is called a fetus (FEET us). The **fetus** is an embryo that has all of its body systems. The organs are not complete, but they are all present. You can tell whether the fetus is a boy or a girl. Its mass is 14 grams. Figure 25-5 shows a life-size fetus at three months. Measure the length of the fetus in millimeters.

Fourth Month to Birth

By the fourth month, the fetus is rather active. It kicks, bends, and turns. The mother begins to feel it moving. By the end of the fourth month, the fetus is 160 millimeters long and has a mass of 100 grams.

Development goes on for the next five months. After a total of 38 weeks, or about 266 days, birth will take place. The fetus will be about 500 millimeters and 3000 grams. How much larger is it than it was at one month?

Not all animals develop in 38 weeks. Table 25-1 lists how long certain animals take to develop before they are born. Dogs take nine weeks to develop. Elephants take 84 weeks. Does length of time needed for development appear to be related to animal size? In what way?

TABLE 25–1. TIME FOR DEVELOPMENT

Animal	Weeks for Development
Rat	3
Rabbit	4
Cat	9
Dog	9
Sheep	21
Bear	28
Gorilla	36
Human	38
Horse	48
Elephant	84

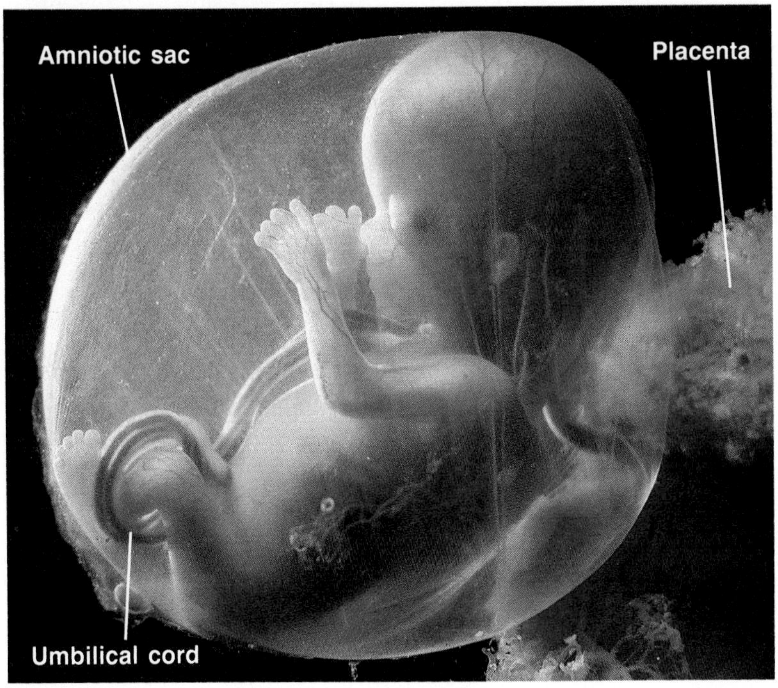

Figure 25-5 This three-month-old fetus is shown actual size.

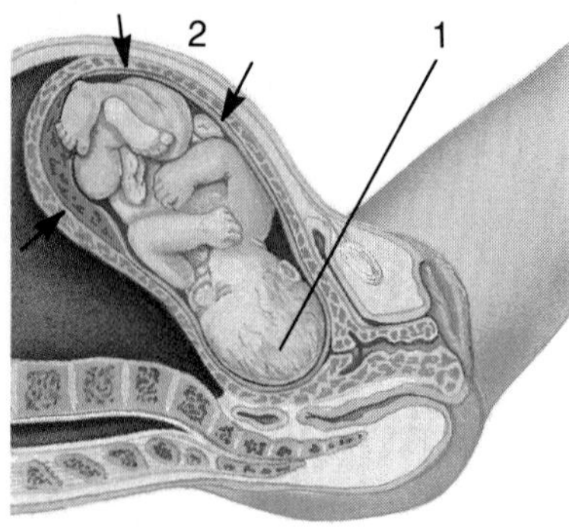

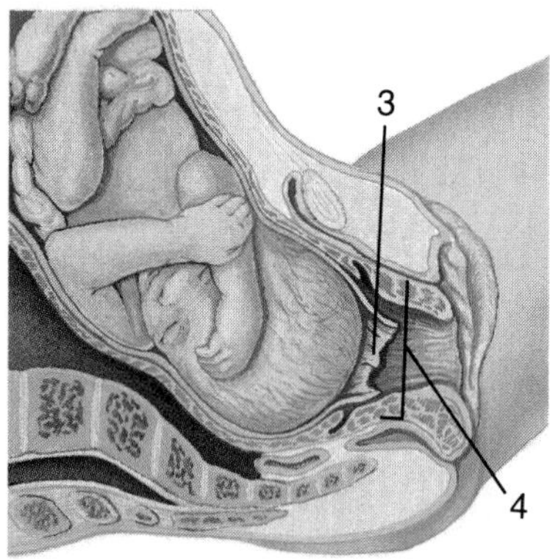

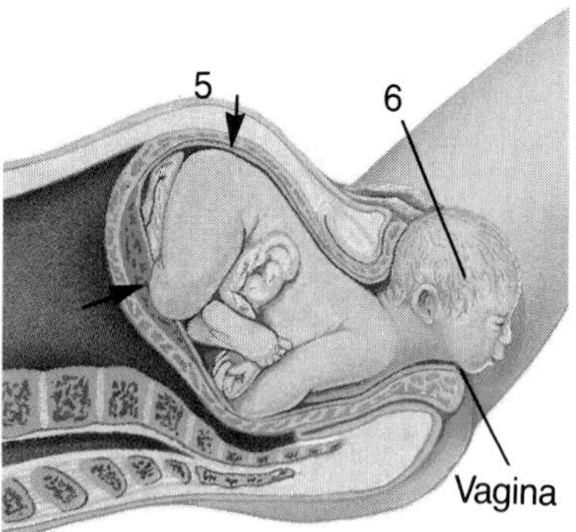

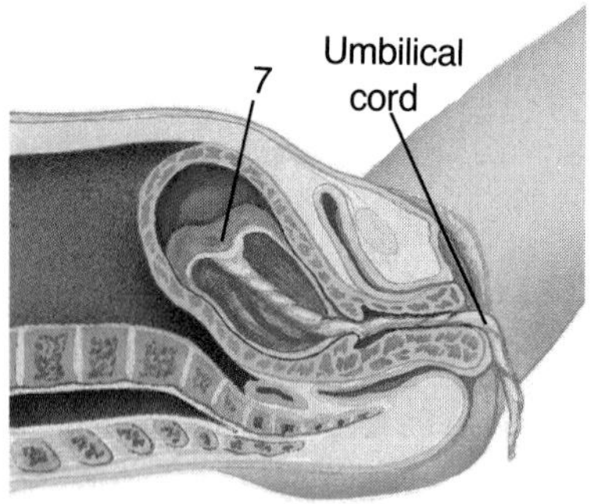

Figure 25-6 Study the steps of human birth shown here as you read the text.

Human Birth

Birth has one main purpose. It must push the fetus out of the uterus through the vagina. Follow the steps in Figure 25-6 as you read.

1. Before birth, the fetus usually turns head down. It will be born head first.
2. The uterus begins to contract. The uterus is smooth muscle. Is smooth muscle voluntary or involuntary? Contractions of the uterus are called **labor.**
3. The amniotic sac breaks. Liquid within the sac passes out through the mother's vagina.
4. The opening at the bottom of the uterus begins to widen. The fetus's head begins to move into this opening. Steps 2, 3, and 4 may take 8 to 12 hours.
5. More contractions take place. They help to push the fetus through the mother's vagina. Steps 4 and 5 might take several hours.
6. The fetus, or baby, is born.
7. The last step is getting rid of the placenta. The uterus contracts and pushes out the placenta and the amniotic sac. At this stage, the placenta is often called the afterbirth.

One evidence of your birth that still exists is your navel. Your **navel** is where the umbilical cord was attached to your body.

Check Your Understanding

1. (a) What is development?
   (b) Where does early development take place in humans?
2. What part protects the embryo?
3. At what age:
   (a) is an embryo called a fetus?
   (b) does the heart begin to beat?
   (c) can you tell if the embryo is male or female?
   (d) are all organs present?
4. **Critical Thinking:** If a pregnant woman smokes, carbon monoxide can pass through the placenta to the fetus. What effect would this have on the fetus? (HINT: In Chapter 13, you studied the effects of carbon monoxide on the body.)
5. **Biology and Reading:** What is the shape of the embryo at the time it moves out of the oviduct and into the uterus? What happens to the embryo after it leaves the oviduct?

Lab 25–1 Development

Problem: How do you judge the age of a human fetus?

Skills

measure in SI, use numbers, infer, make and use graphs

Materials

ruler

Procedure

1. Copy the data table below.
2. **Measure in SI:** The figure shows three fetuses 1/10 actual size. Measure the body length and foot length for each fetus in millimeters. Write the measurements in your data table.
3. Fill in all but the last column of your data table.
4. The data table on this page lists ages of a fetus. It also shows the average size of a fetus at each of these ages. Use the table and your measurements to determine the ages of fetuses A, B, and C.

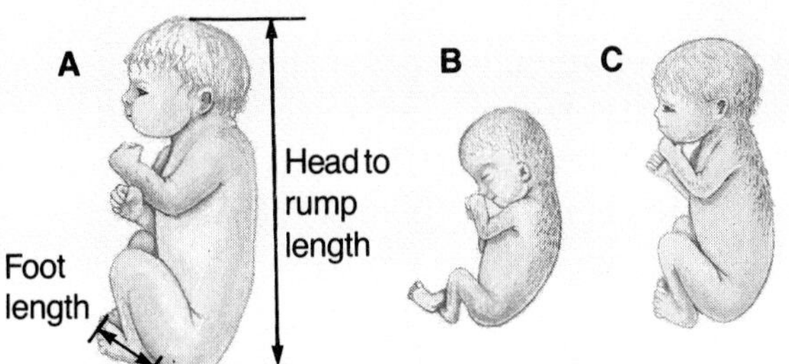

Data and Observations

1. At what age does the fetus open its eyes?
2. How much does the fetus grow between weeks 20 and 28?

AGE IN WEEKS	AVERAGE BODY LENGTH (mm)	AVERAGE FOOT LENGTH (mm)	EYES	HAIR
16	140	27	closed	none
20	210	40	closed	body, head
28	270	60	open	body, head
36	340	80	open	head

Analyze and Apply

1. How does a fetus obtain food and oxygen?
2. **Infer:** Could ultrasound tell you the sex of a fetus at six weeks of age? Why?
3. **Apply:** Assume you are an ultrasound technician and you report the age of fetus C to the doctor. The doctor tells you that the fetus should be 34 weeks old. The mother drinks a lot of alcohol. How would this affect the size of her fetus?

Extension

Make a line graph of the data for average body length.

FETUS	BODY LENGTH		FOOT LENGTH		EYES	HAIR	AGE OF FETUS (WEEKS)
	MEASURED	ACTUAL (×10)	MEASURED	ACTUAL (×10)			
A							
B							
C							

25:2 Development Outside the Female

Objectives

4. Compare development in frogs, birds, and reptiles.

5. Explain how the needs of a chick embryo are met.

All animals undergo development. The female, however, may not have a uterus. In this case, development must take place elsewhere.

Eggs That Are Laid

A female frog lays eggs outside her body. Birds and reptiles do the same thing. These eggs undergo development in a way similar to that in humans. One difference is that for frogs, birds, and reptiles, development takes place outside the female's body.

Where do frog eggs develop?

Fish, amphibians, and many invertebrates lay their eggs in water. These eggs undergo cleavage after they are fertilized, just as the fertilized human egg did. Figure 25-7 shows photos of cleavage taking place in a frog egg. Photo (a) is of one cell. Photo (b) is of two cells. How many cells can you see in photos (c) and (d)?

Bird and reptile eggs go through cleavage stages just as frog and human eggs do. Development in birds and reptiles takes place on land, not in water. Eggs of birds and reptiles can develop on land because they have shells. The shell keeps the embryo from drying out.

Needs of the Embryo

What are the needs of an embryo that undergoes development in an egg that is laid? Let's use a chicken egg as an example. A chick embryo must have food and oxygen. It must get rid of waste chemicals. It must also be protected against drying out and injury. How are these needs met? Look at Figure 25-8 as you read.

Figure 25-7 A frog egg undergoes cleavage.

a

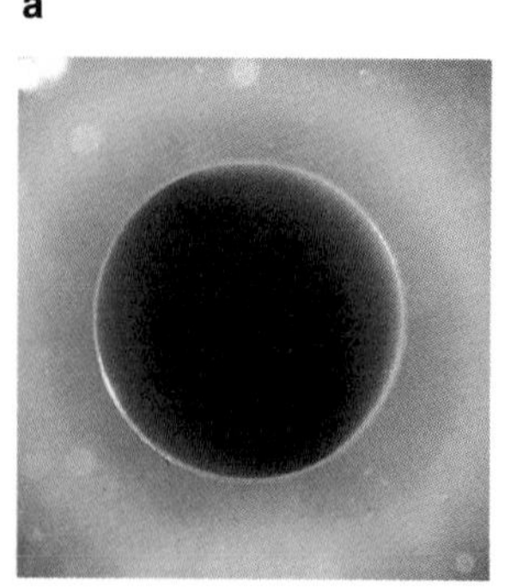

b

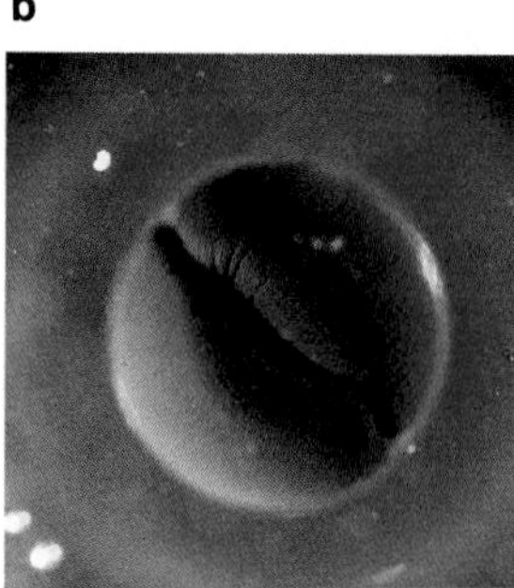

c

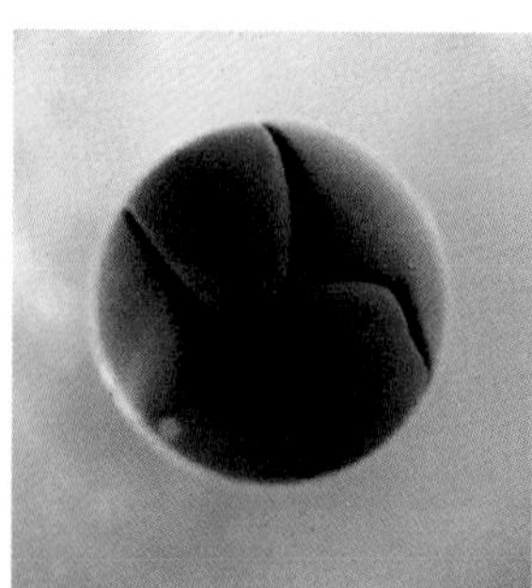

d

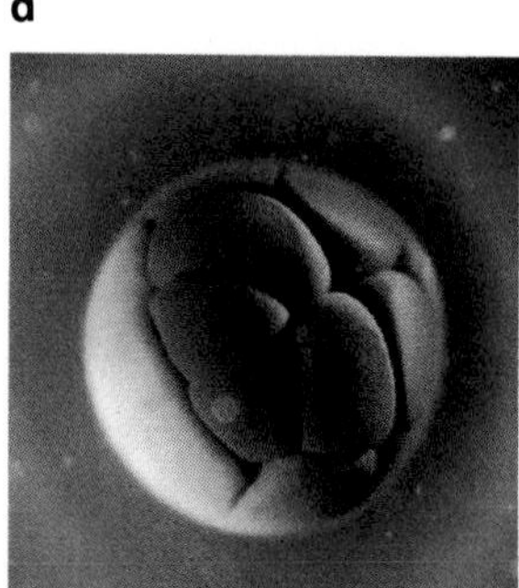

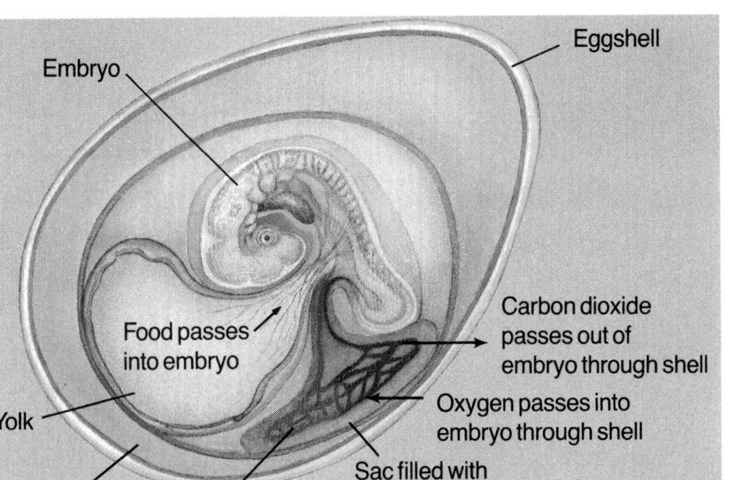

Figure 25-8 Chick embryo

Food Supply

You may not realize it, but you are familiar with the two parts that supply food to the embryo. They are the yolk and albumen (al BYEW mun). Yolk is the yellow part and is made up of protein and fat. Albumen is the white of the egg. Albumen also has protein in it. By hatching time, the chick has used up the yolk and albumen.

Oxygen Supply

Oxygen diffuses through the eggshell from the air outside. Blood vessels that lie just below the shell pick up the oxygen. The oxygen then passes to the embryo.

Waste Removal

Gas wastes leave the egg through the shell. Liquid wastes are stored in a sac within the egg until hatching.

Protection

The shell protects the chick embryo. It protects the embryo from water loss and injury.

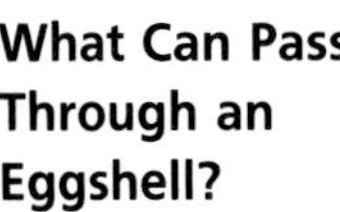

Mini Lab

What Can Pass Through an Eggshell?

Design an experiment: Design an experiment to test if glucose can pass through an eggshell. What is the control in your experiment? *For more help, refer to the **Skill Handbook,** pages 704-705.*

How is a chick embryo supplied with oxygen?

Check Your Understanding

6. Where do frog and bird development take place?
7. What does egg yolk supply to a developing chick?
8. What happens to wastes as a chick develops?
9. **Critical Thinking:** What is an advantage of developing outside the female's body? What is a disadvantage?
10. **Biology and Reading:** Which vertebrates lay eggs in water? Which vertebrates lay eggs on land?

Metamorphosis

Lab 25–2

Problem: How do complete and incomplete metamorphosis compare?

Skills

interpret data, measure in SI, compare

Materials

hand lens
grasshopper stages of metamorphosis
moth stages of metamorphosis
metric ruler

Procedure

Part A

1. Copy the data table.
2. Using a hand lens, examine the eggs of a grasshopper.
3. Complete the data table for this stage.
4. Repeat steps 2 and 3 for the nymph and adult stages.

Part B

1. Using a hand lens, examine the eggs of a moth.
2. Complete the data table for this stage.
3. Repeat step 2 for the larva, pupa, and adult stages. Measure only body length for the adult. Do not include the wings.

Data and Observations

1. For the grasshopper, list two ways in which
   (a) the egg differs from the nymph or adult
   (b) the nymph and adult are alike
2. For the moth, list two ways in which
   (a) the pupa differs from the adult
   (b) the larva and adult differ

Analyze and Apply

1. **Interpret data:** Which type of metamorphosis do grasshoppers show—complete or incomplete?
2. Which type of metamorphosis do moths show–incomplete or complete?
3. **Apply:** Metamorphosis has made it possible for insects to live in almost every type of environment and use many different materials for food. Why do you think this is so?

Extension

Compare the stages of metamorphosis for a crayfish to those of a grasshopper.

	STAGE	BODY LENGTH (mm)	LEGS PRESENT	NUMBER OF LEGS	WINGS PRESENT	CAN IT FLY?	MOUTH PARTS PRESENT?	CAN IT FEED?	CAN IT REPRODUCE?
Part A	Egg								no
	Nymph								no
	Adult								yes
Part B	Egg								no
	Larva								no
	Pupa								no
	Adult								yes

25:3 Metamorphosis

If you saw a puppy, you would be able to tell that it was a dog. If you saw a baby bird, you would be able to tell that it was a bird. Many young animals look like the adults, but some do not. These animals must go through another type of change before they look like an adult.

Objectives

6. Sequence the stages of frog metamorphosis.

7. Explain incomplete metamorphosis.

8. Describe complete metamorphosis.

Key Science Words

metamorphosis
larva
nymph
incomplete metamorphosis
complete metamorphosis
pupa

Frog Metamorphosis

Many animals hatch and go through a series of changes before they look like their parents. Changes in appearance and lifestyle that occur between the young and the adult stages are called **metamorphosis** (met uh MOR fuh sus).

You may have seen tadpoles swimming in a pond or stream. These tadpoles will undergo metamorphosis to become adult frogs. Follow the changes in Figure 25-9 to see how a young frog changes into an adult.

1. The adult female frog lays eggs in the water. Eggs are fertilized outside the body. Notice that frog eggs do not have shells. They have a jellylike covering that protects them.
2. Early development takes place for about 12 days.

Skill Check

Understand science words: metamorphosis. The word part *meta* means change. In your dictionary, find three words with the word part *meta* in them. *For more help, refer to the **Skill Handbook,** pages 706-711.*

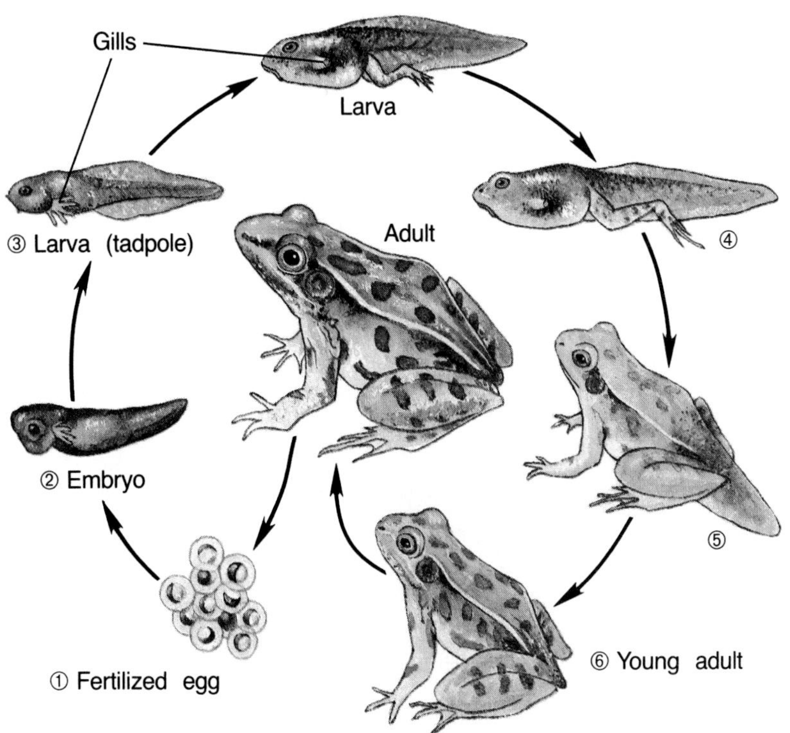

Figure 25-9 The stages of metamorphosis in a leopard frog are shown here.

3. A tadpole appears from the embryo. This tadpole is a larva (LAR vuh). A **larva** is a young animal that looks completely different from the adult. If you didn't know that a tadpole was a frog, you might mistake it for a small fish.

How does a tadpole differ from an adult frog?

4. Note how the larva differs from the adult. Look at its mouth, body, and tail shape. The larva has gills for obtaining oxygen. Does the adult have the exact same shape?

5. The larva continues to change. Front and hind legs form. Gills are replaced by lungs.

6. Finally the larva, or tadpole, no longer exists. It has changed into a young adult frog.

Insect Metamorphosis

Insects also undergo metamorphosis. Unlike frogs, insects have two types of metamorphosis. They are called incomplete and complete metamorphosis.

Incomplete metamorphosis

When some insects hatch from eggs, they almost look like miniature adults. There are many differences, however. These young are called nymphs (NIHMFS). A **nymph** is a young insect that looks similar to the adult. Note in Figure 25-10 that when a nymph first hatches, it does not have wings like the adult. It is also sexually immature.

The nymph grows and changes until it reaches adult size and appearance. This type of change is called incomplete metamorphosis. **Incomplete metamorphosis** is a series of changes in an insect from nymph to adult. Grasshoppers and crickets have incomplete metamorphosis.

Complete metamorphosis

Many insects, such as butterflies and moths, have complete metamorphosis. **Complete metamorphosis** is a series of changes in an insect in which the young do not look like the adult. Figure 25-11 shows these changes.

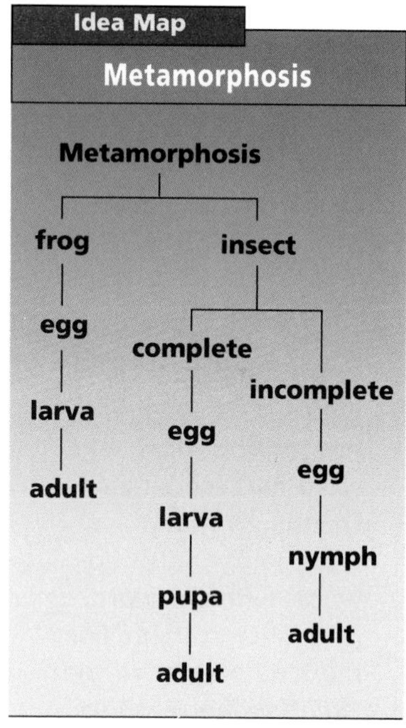

Study Tip: Use this idea map as a study guide to metamorphosis.

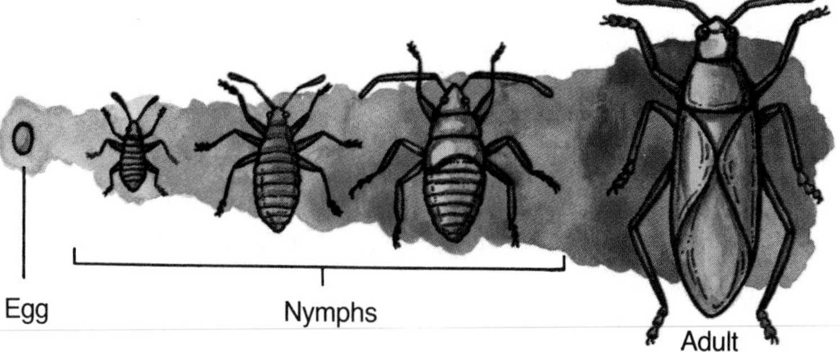

Figure 25-10 Incomplete metamorphosis in an insect includes egg, nymph, and adult stages.

Figure 25-11 Complete metamorphosis includes egg (upper left), larva (upper right), pupa (lower left), and adult (lower right) stages.

1. The adult female butterfly lays fertilized eggs that develop into embryos.
2. The embryo hatches into a small larva 10 to 15 days later. This larva is also called a caterpillar.
3. The larva moves about and feeds on plant matter.
4. The larva stops feeding and enters a pupa (PYEW puh) stage. A **pupa** is a quiet, non-feeding stage that occurs between larva and adult. The pupa is surrounded by a protective covering. Many changes take place in the pupa.
5. After about one week, the pupa opens and an adult butterfly emerges.

How do pupa and nymph stages differ?

Check Your Understanding

11. Name the stages of metamorphosis in a frog.
12. Arrange the following stages of development in their proper order: nymph, fertilized egg, adult.
13. How are frog eggs protected?
14. **Critical Thinking:** What form of a tapeworm is similar to the pupa stage of a butterfly? Why?
15. **Biology and Writing:** Is metamorphosis in a frog more like complete or incomplete metamorphosis? Write a paragraph that explains your answer.

Chapter 25

Review

Summary

25:1 Development Inside the Female

1. Cleavage is a series of changes from a fertilized egg to a hollow ball of cells.
2. A human embryo receives protection, food, and oxygen, and gets rid of wastes while in the uterus.
3. Most major changes to the human embryo take place within the first three months. Development continues through the ninth month.

25:2 Development Outside the Female

4. Development occurs outside the female's body in frogs, birds, and reptiles.
5. The chicken egg supplies the chick embryo with protection, food, and a way of obtaining oxygen and getting rid of gas wastes.

25:3 Metamorphosis

6. During metamorphosis, frogs change from an egg to a larva to an adult.
7. Incomplete insect metamorphosis has nymph stages that look like the adult.
8. Complete metamorphosis has larva and pupa stages that don't look like the adult.

Key Science Words

amniotic sac (p. 525)
cleavage (p. 525)
complete metamorphosis (p. 536)
fetus (p. 529)
incomplete metamorphosis (p. 536)
labor (p. 530)
larva (p. 536)
metamorphosis (p. 535)
navel (p. 530)
nymph (p. 536)
placenta (p. 526)
pupa (p. 537)
umbilical cord (p. 526)

Testing Yourself

Using Words

Choose the word from the list of Key Science Words that best fits the definition.

1. name given to the embryo at the start of the third month
2. blood vessels connecting the embryo to the placenta
3. young frog that looks completely different from the adult
4. young insect that looks similar to the adult
5. quiet, non-feeding stage during metamorphosis
6. series of changes that turn an egg into a hollow ball with many cells
7. organ connecting the embryo to the uterus
8. contractions of the uterus
9. liquid-filled tissue that protects the embryo
10. change in appearance that occurs between young and adult stages

Testing Yourself *continued*

Finding Main Ideas

List the page number where each main idea below is found. Then, explain each main idea.

11. how incomplete and complete metamorphosis are alike and how they differ
12. how the shell of an egg allows development out of water
13. the changes that take place during cleavage
14. how early development of a frog and bird are alike, how they differ
15. how contractions of the uterus help during birth
16. the four main needs of an embryo that develops inside the female and how each need is met
17. the main jobs of an eggshell
18. how a larva differs from an adult frog

Using Main Ideas

Answer the questions by referring to the page number after each question.

19. What changes happen to a human fetus from four months to birth? (p. 529)
20. What are the jobs of the uterus, placenta, and umbilical cord? (pp. 525, 526, 530)
21. How does human development compare with bird, amphibian, and reptile development? (pp. 524, 532)
22. How does a frog egg obtain protection? (p. 535)
23. What are the stages that take place during incomplete and complete metamorphosis? (pp. 536, 537)
24. What are the jobs of the shell, sac, and yolk in a bird egg? (p. 533)

Skill Review ✓

For more help, refer to the ***Skill Handbook,*** *pages 704-719.*

1. **Design an experiment:** Design an experiment to determine if flies will undergo metamorphosis faster if kept in the dark. What would be your control in this experiment?
2. **Infer:** An embryo measures 15 mm in length. Is it older or younger than 1 month in age?
3. **Make and use graphs:** Make a line graph that shows the number of cells in a human embryo after 36, 48, 72, 96, and 120 hours.
4. **Understand diagrams:** Study Figure 25-1. What changes does the fertilized egg undergo in the oviduct?

Finding Out More

TECH PREP

Critical Thinking

1. Babies are often born with diseases that the mother had during pregnancy. AIDS is an example. How might such a disease be passed to the fetus?
2. During pregnancy, the placenta sometimes pulls loose from the uterus. How might this affect the embryo?

Applications

1. How does the organ used for respiration differ between tadpoles and adult frogs? Relate this to where the tadpole and adult live.
2. Match the parts of a bird embryo with the following human parts and their jobs: kidney, skin, mouth, lungs. Explain your answer.

Biology in Your World

Growth, Development, Change

Development is not the same for all animals. Some animals undergo distinct stages, changing form entirely. Other animals just grow larger. The following examples compare human development with that of other organisms.

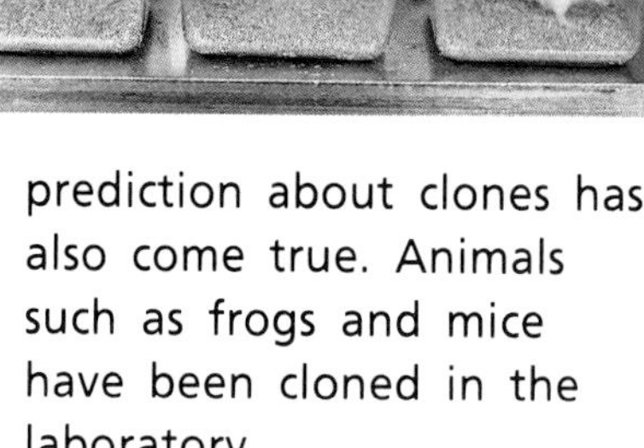

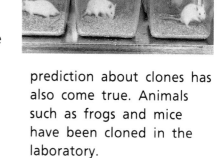

LITERATURE

A Crystal Ball?

In 1931, test-tube babies occurred only in science fiction. In *Brave New World,* author Aldous Huxley wrote about human eggs being fertilized outside the uterus. He also wrote about the eggs being made to divide into 96 identical embryos, or clones.

The process Huxley wrote about is a reality today. It is called *in vitro* fertilization. With *in vitro* fertilization, some childless couples are now able to have children. Huxley's prediction about clones has also come true. Animals such as frogs and mice have been cloned in the laboratory.

ART

Family Group, by Henry Moore

For all of history, our survival has depended on our ability to reproduce. However, just producing babies is not enough. Without care and attention, even well-fed babies will not gain weight.

Early artwork of females stressed fertility because of the importance of reproduction. Some artists have gone beyond just showing fertility. Henry Moore, a 20th century English sculptor, is one such artist. In 1949 he did a bronze sculpture called "Family Group," which shows a loving family environment. The parents are protecting and caring for the children. How does Moore's sculpture relate to what is known about growth and development?

LEISURE

See for Yourself!

In warm weather you can find frogs' eggs and tadpoles in ponds. The eggs look like small black beads surrounded by clear jelly. They can be collected using a long-handled dip net. The eggs will hatch in about one week at room temperature (don't leave them in the sun or near a heat source). Keep only eight tadpoles for each gallon of water. Use either pond water, bottled spring water, or tap water that has been sitting out for a few days. About a week after hatching, feed the tadpoles pond plants or boiled lettuce. In a few months, the tadpoles will develop hind legs. Front legs form later as the tails shorten. As the lungs develop, the young frogs will need a rock to rest on. If you plan to raise a frog, you must feed it live insects or mealworms. Good luck!

Return the frogs to the pond when you are finished observing them.

CONSUMER

The Luxury of Silk

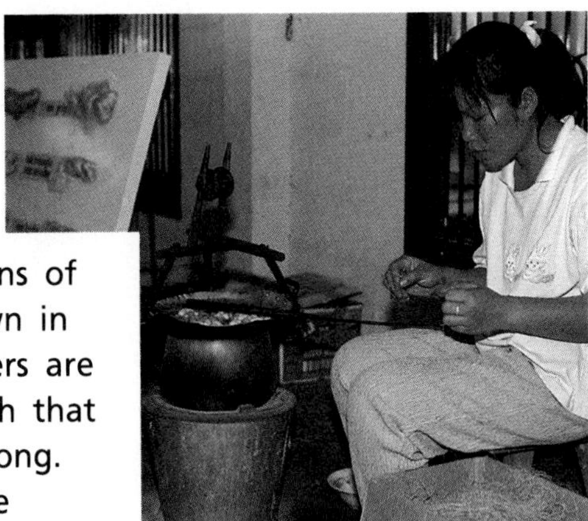

As you learned in this unit, a moth does not hatch from an egg completely formed. Instead, a caterpillar hatches from the egg. The caterpillar eventually forms a cocoon around itself. A moth then emerges from the cocoon.

Over 4000 years ago, the Chinese discovered that the cocoons of silk moths could be unwound into silk threads. Today, the caterpillars and cocoons of these moths are grown in factories. The silk fibers are woven into shiny cloth that is lightweight, yet strong.

Silk is used to make upholstery and fashionable clothing. Silk garments are often more expensive than those made from synthetic fibers. Why do you think silk is more expensive?

Unit 7

CONTENTS

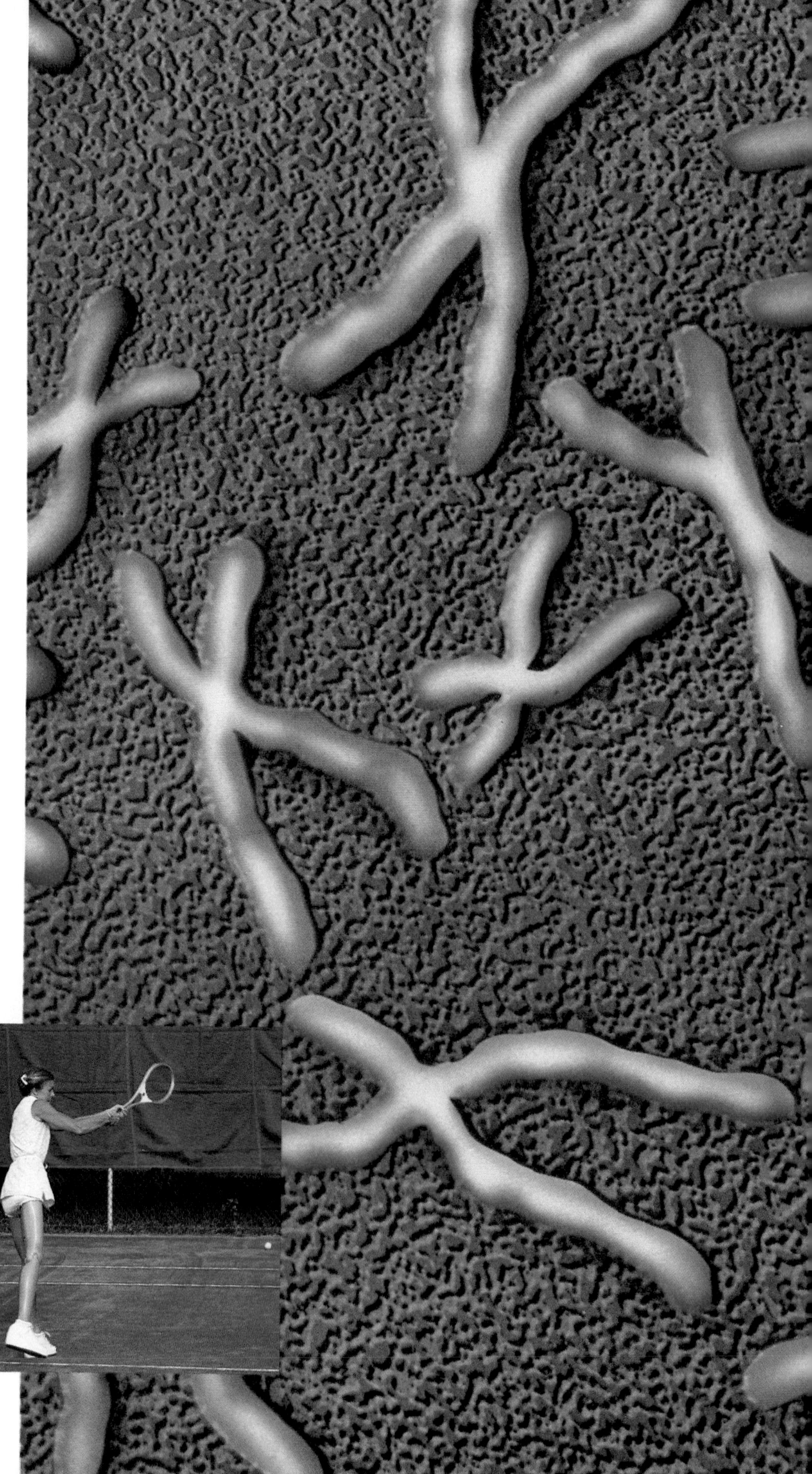

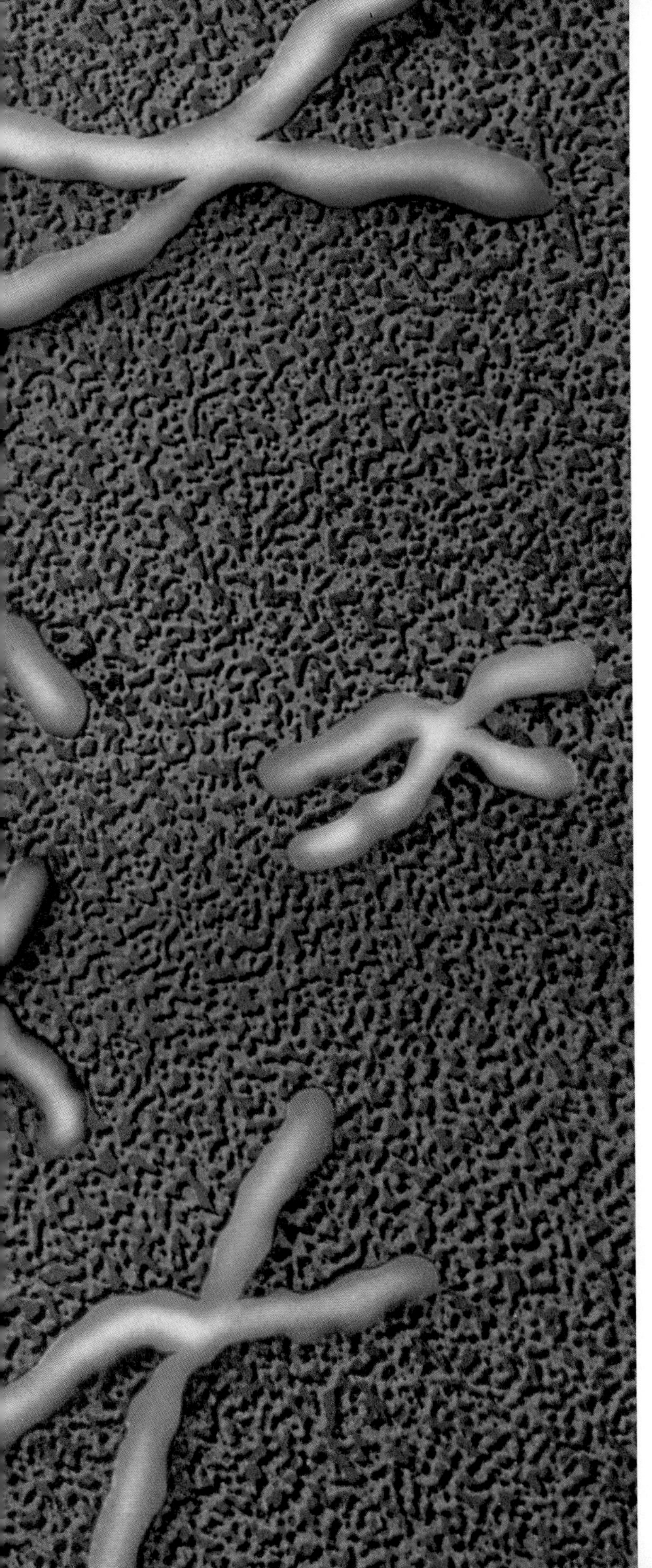

Traits of Living Things

What would happen if...

humans could regrow body parts that were damaged or cut off? People who lose a body part in an accident or during surgery would grow a replacement. An infected arm or leg could be removed to stop the spread of the infection. A new arm or leg would grow in its place.

The growing back of body parts is controlled by genes. In a young child, a finger tip could regrow if one were cut off. At your age, only the skin and a few other body tissues regrow. For some reason, most body tissues lose the ability to regrow as humans age.

Some organisms can regrow body parts. Scientists are studying them to learn how body parts regrow. Maybe they will learn how to make the genes that control the ability of body parts to regrow.

CHAPTER PREVIEW

Chapter Content

Review this outline for Chapter 26 before you read the chapter.

Skills in this Chapter

The skills that you will use in this chapter are listed below.

- In **Lab 26-1,** you will observe, predict, and interpret data. In **Lab 26-2,** you will observe, formulate a model, and form hypotheses.
- In the **Skill Checks,** you will calculate and understand science words.
- In the **Mini Labs,** you will classify and make and use tables.

Chapter 26

Inheritance of Traits

The small photo on this page shows two parent rabbits. The large photo on the opposite page shows some baby rabbits. Which of the baby rabbits do you think belong to the two parents? What traits did you use to help you answer the question? What traits did the parents pass to the offspring?

How about the all-white baby rabbit? Did someone who was caring for several litters of rabbits put the white rabbit in the wrong container? Probably not. Perhaps the parents passed the trait for white fur to the baby white rabbit. Can offspring show a trait that is not seen in the parents? You should be able to answer this question after you read this chapter.

Try This!

Are two peas in a pod alike? Open a pea pod and look carefully at the peas. Make a list of ways the peas are the same and ways they are different. Each pea in a pod is like each child in a family. Children in a family show different traits.

BIOLOGY Online

Visit the Glencoe Science Web site at science.glencoe.com to find links about **inheritance of traits.**

Parent rabbits

26:1 Genetics, How and Why

Objectives

1. **Compare** the number of chromosomes in sex cells and in body cells.
2. **Distinguish** between dominant and recessive genes.
3. **Describe** how different gene combinations result from fertilization and how traits are passed to offspring.

Key Science Words

genetics
gene
dominant gene
recessive gene
pure dominant
pure recessive
heterozygous

Genetics (juh NET ihks) is the study of how traits are passed from parents to offspring. Offspring usually show some traits of each parent. For a long time, scientists did not understand how this could happen. Later, they found that the traits of the parents were passed to offspring by sex cells.

Chromosomes

Before you learn how traits are passed from parents to offspring, let's review something about cells. Look at a cell in Figure 26-1. The large, round part in the center is the nucleus. It has two main jobs. One is to direct the actions of other cell parts. The other is to allow the cell to reproduce.

Inside the nucleus are long, threadlike parts called chromosomes. Chromosomes can be seen best when a cell is ready to reproduce. During cell reproduction, the chromosomes become short and thick.

Suppose we look closer at two kinds of cells. One kind is called a body cell. Remember that body cells are cells that make up most of the tissues and organs in your body. Look at Figure 26-1a. Notice that the body cell has two of each kind of chromosome. The chromosomes in body cells are paired.

Another type of cell is a sex cell. The sex cell can be a sperm cell or an egg cell. Notice in Figure 26-1b that there is only one of each kind of chromosome present in the sex cell. Sex cells, then, have half as many chromosomes as body cells.

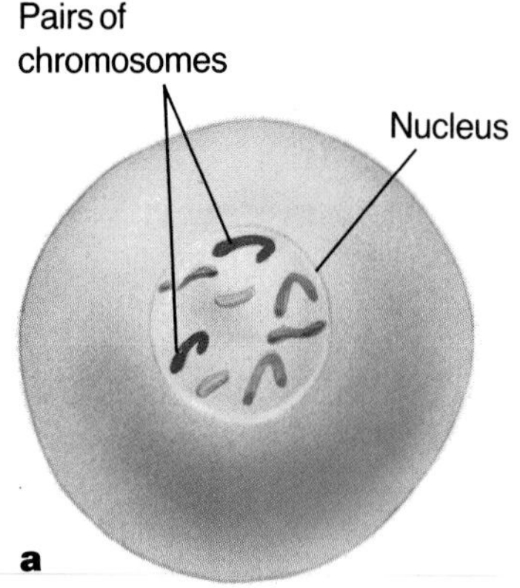

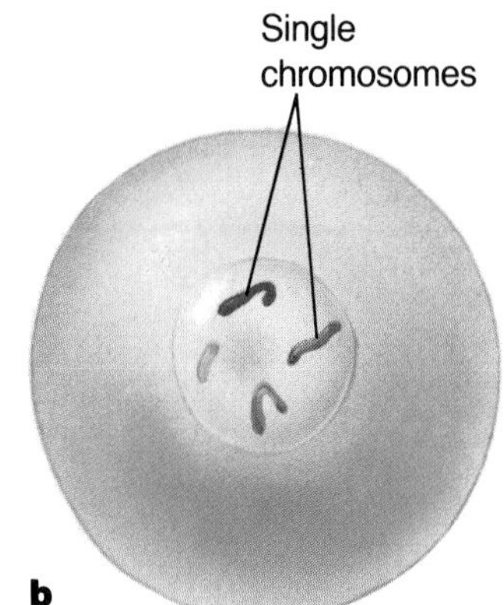

Figure 26-1 Chromosomes are found in the nucleus. Body cells have two copies of each chromosome (a). Sex cells have one copy of each chromosome (b).

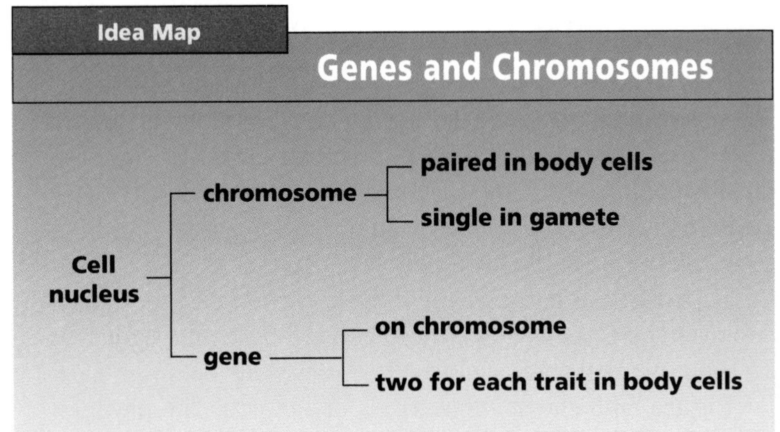

Study Tip: Use this idea map as a study guide to genes and chromosomes. Some features of each are shown.

Genes on Chromosomes

All chromosomes contain genes (JEENZ). A **gene** is a small section of chromosome that determines a specific trait of an organism. Examples of traits are eye color, hair color, and shape of body parts such as ears. Chemical processes inside the body, which cannot be seen, are also traits. Organisms have thousands of different traits. Genetics is really a study of the genes that control all of these traits. Note that the word *genetics* contains the word *gene.*

Genes are arranged on a chromosome, one next to another, much like beads on a necklace. Each chromosome has different kinds of genes that control different traits. Figure 26-2 shows drawings of human chromosomes. The locations of several genes are shown.

Remember that chromosomes in body cells are paired. The genes on chromosomes in body cells are paired, too. There is one gene of each gene pair on each chromosome of a chromosome pair. You can see this pairing in Figure 26-2. Each trait we study will have one gene pair, or two genes that represent it. The two genes representing the trait are located on chromosomes that make a pair.

Mini Lab

What Traits Do Some Seeds Share?

Classify: Look at the different dried beans found in the grocery store. Classify the beans into groups based on color, shape, and size. *For more help, refer to the **Skill Handbook,** pages 715-717.*

How many genes in a body cell represent each trait?

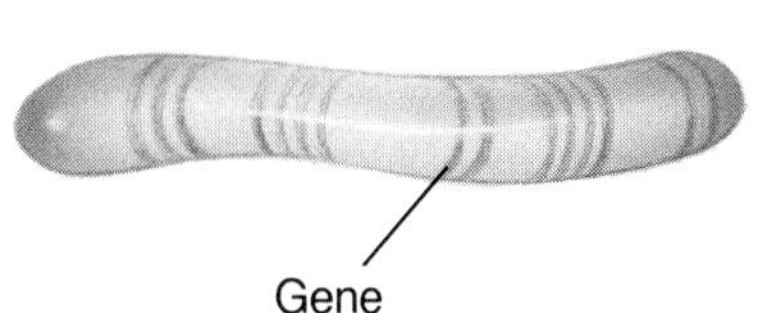

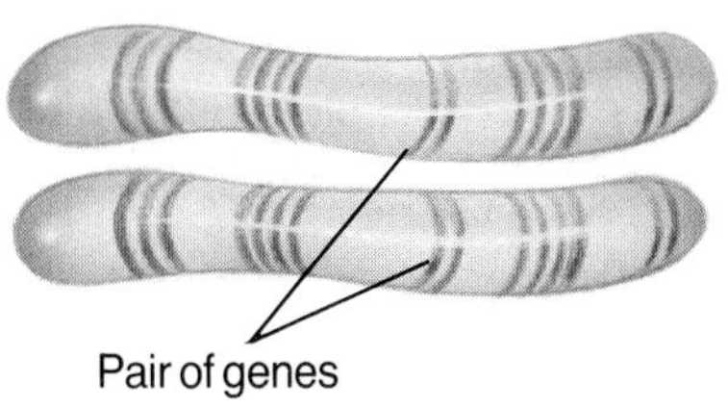

Figure 26-2 Genes are arranged next to each other on a chromosome. How many copies of a gene are in a sex cell?

Mini Lab

What Type of Earlobes Do Most Students Have?

Make and use tables: Make a table showing how many classmates have free or attached earlobes. Do more have the dominant or the recessive trait? *For more help, refer to the **Skill Handbook**, pages 715-717.*

Passing Traits to Offspring

How are traits passed from parents to their offspring? To answer this question, let's use the trait of earlobe shape in humans as an example. A person can have attached earlobes or free earlobes. Figure 26-3 shows what these traits look like. What genes would you expect to find in the body cells of each of these people?

Suppose the mother shows the attached earlobe trait. Suppose the father shows the free earlobe trait. Which trait would appear in their children?

Figure 26-3 shows what kinds of sex cells the parents can make. The mother has the gene pair for attached earlobes. She can make eggs that have this gene. The father has the gene pair for free earlobes. He can make sperm that have this gene. What genes for earlobe shape will their child have? The child will have one gene for each trait.

In fertilization, one sperm will join with one egg. Which sperm and egg will join? We don't know. We can see that it would make no difference in this example. Any child produced will have a gene for attached earlobes and a gene for free earlobes. From which parent did the child receive the gene for attached earlobes? From which parent did the child receive the gene for free earlobes?

Figure 26-3 Attached earlobe trait (a) and free earlobe trait (b). The trait for earlobe shape is passed from parents to their child.

a

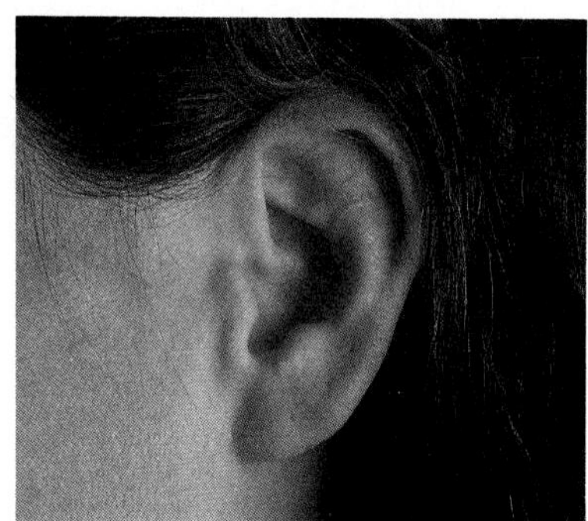

b

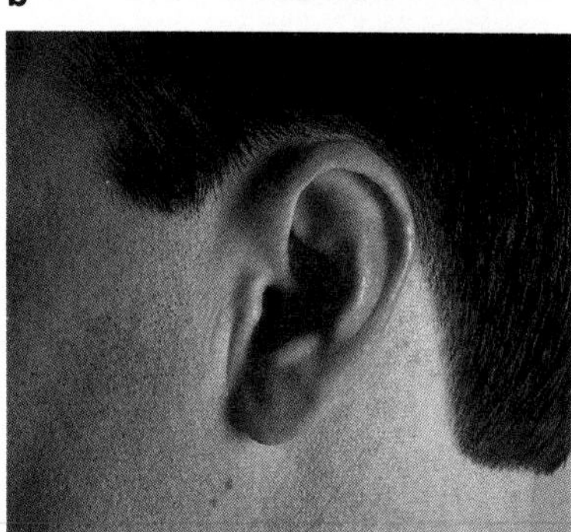

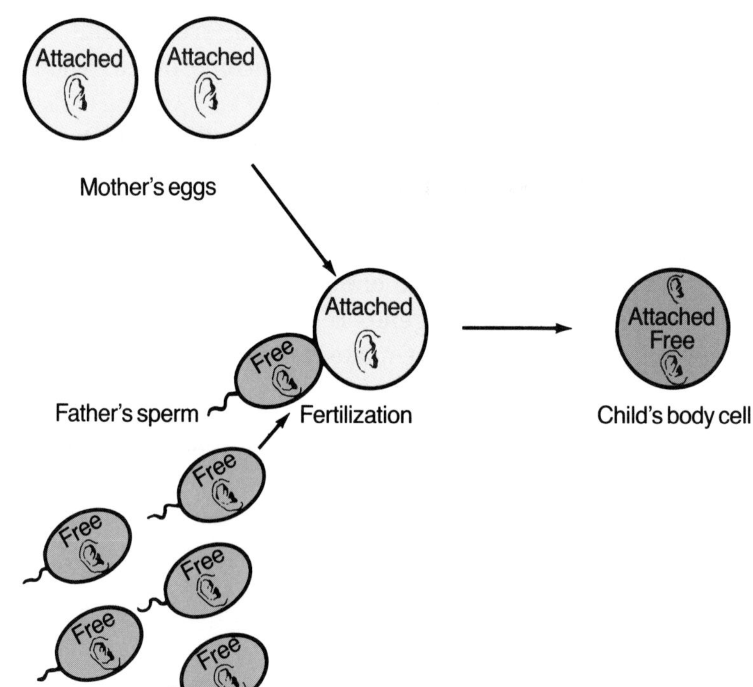

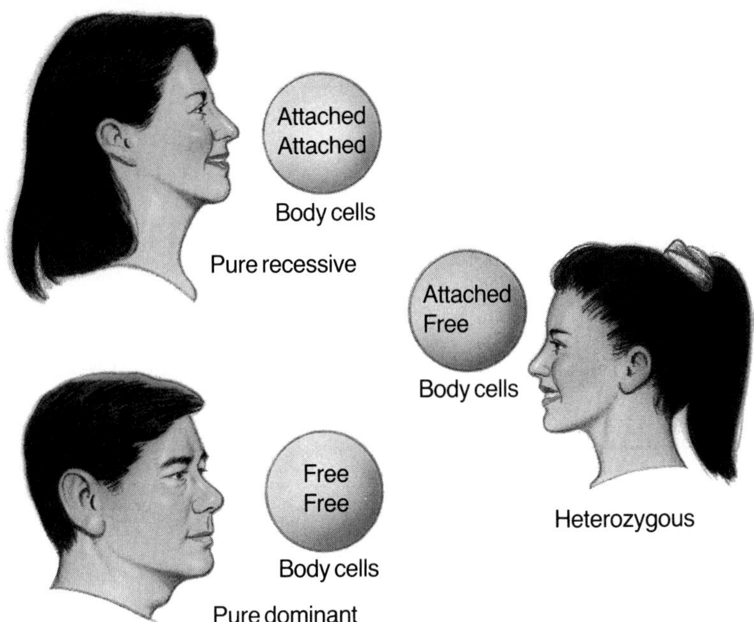

Figure 26-4 A child with one gene for attached earlobes and one gene for free earlobes will have free earlobes.

Skill Check

Understand science words: heterozygous. The word part *hetero* means different. In your dictionary, find three words with the word part *hetero* in them. *For more help, refer to the **Skill Handbook,** pages 706-711.*

Dominant and Recessive Genes

What does a child with one gene for attached earlobes and one for free earlobes look like? Figure 26-4 shows you. Children born with both genes will have free earlobes. Why? Some genes can keep others from showing their traits. Genes that keep other genes from showing their traits are called **dominant** (DAHM uh nunt) **genes.** The genes that do not show their traits when dominant genes are present are called **recessive** (rih SES ihv) **genes.** In this example, the gene for free earlobes is dominant. The gene for attached earlobes is recessive.

What is a recessive gene?

An organism with two dominant genes for a trait is said to be **pure dominant.** Using the word *pure* means that both genes are the same. In our example, the father is pure dominant for free earlobes.

An organism with two recessive genes for a trait is said to be **pure recessive.** In our example, the mother is pure recessive for attached earlobes.

The child with a gene for attached earlobes and a gene for free earlobes has two different genes. The child is heterozygous (HET uh roh ZI gus). A **heterozygous** individual is one with a dominant and a recessive gene for a trait. Even though the heterozygous individual has the recessive gene, the recessive trait does not show. The trait of the dominant gene shows.

Table 26-1 gives some traits in plants and animals. These traits are passed to offspring in the same way as the earlobe trait. They are dominant or recessive.

TABLE 26–1. TRAITS OF PLANTS AND ANIMALS

Trait Found in	Dominant Trait	Recessive Trait
flies	long wings	short wings
pea plants	red flowers	white flowers
humans	can roll tongue	can't roll tongue
corn plants	normal height	dwarf
dogs	short hair	long hair

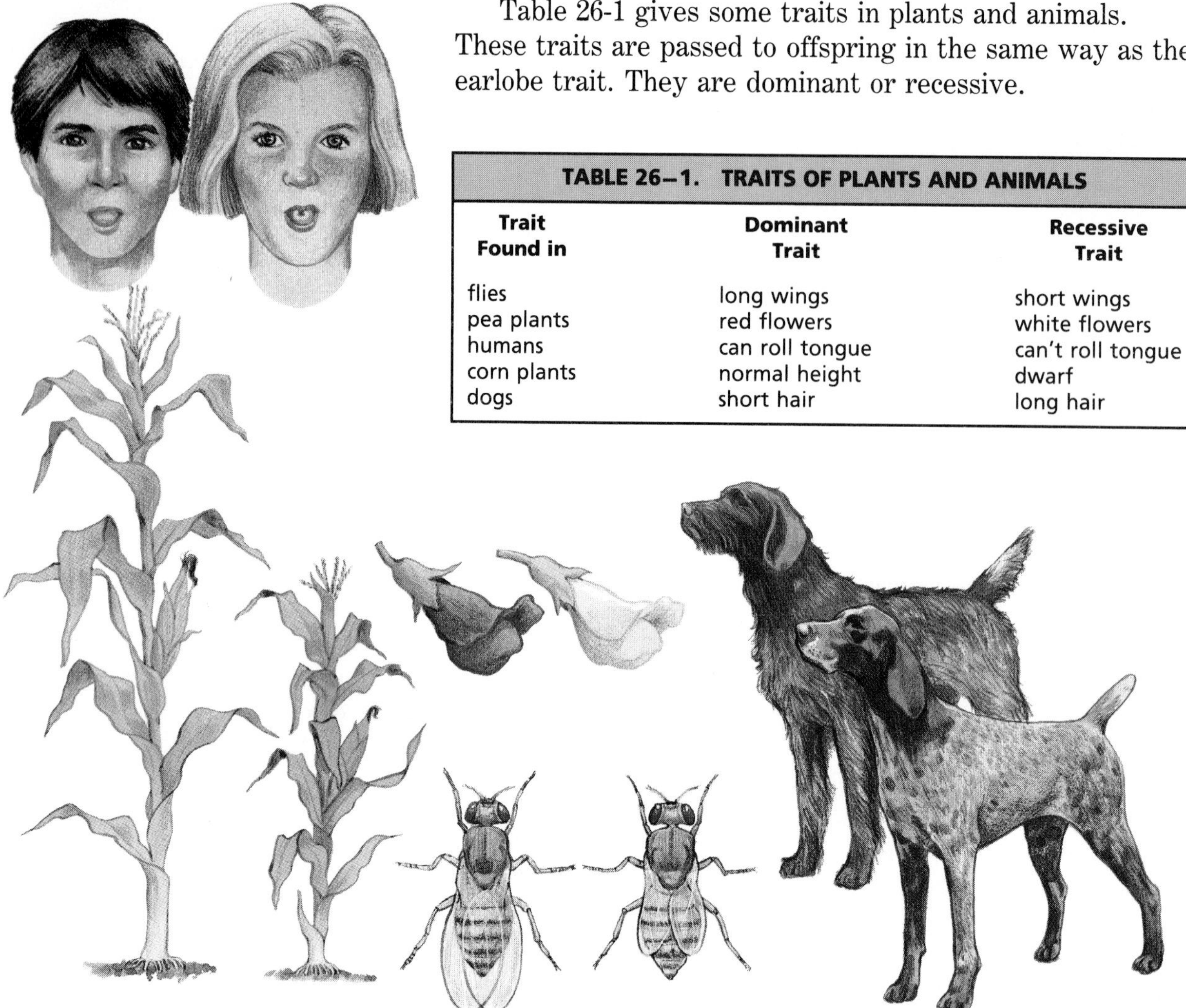

When Both Parents Are Heterozygous

When one parent had both genes for attached earlobes and the other parent had both genes for free earlobes, all of their children had a combination of genes. The children had free earlobes. Now, let's suppose both parents have a combination of genes. What could their children look like?

How many different kinds of sex cells can a heterozygous person make?

If the mother's body cells have both genes, she can make two kinds of eggs. Each egg will have a gene for attached earlobes or a gene for free earlobes, but not both. If the father's body cells have both genes, he can make two kinds of sperm. Each sperm will have the gene for attached earlobes or the gene for free earlobes, but not both. Figure 26-5 shows the possible combinations when egg and sperm join. There are four possible combinations.

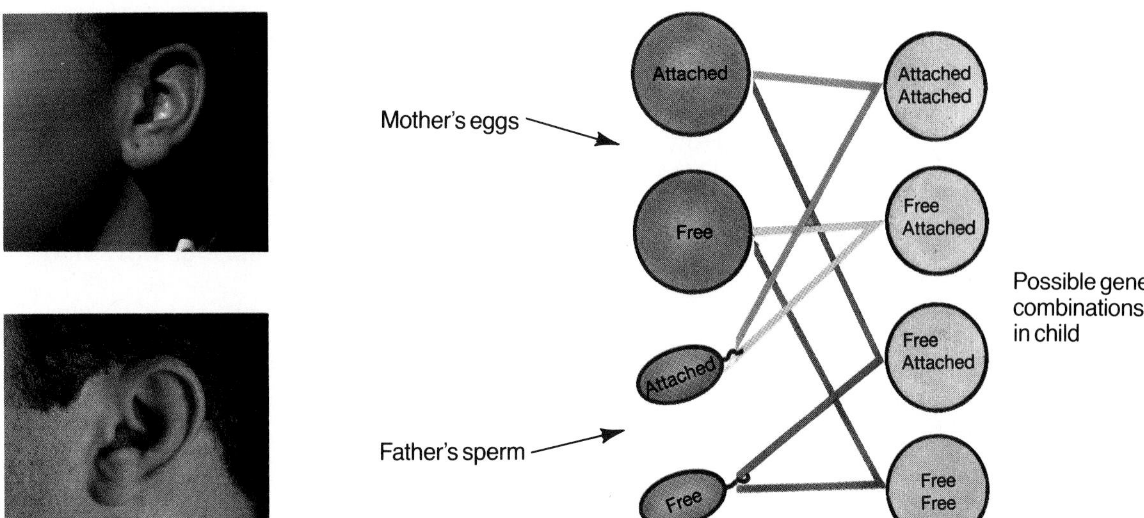

Figure 26-5 What genes can a child have if both parents are heterozygous?

Table 26-2 shows what a child with each combination would look like. There are three chances in four that a child would have free earlobes and one chance in four that a child would have attached earlobes. We call this a 3 to 1 ratio.

TABLE 26–2. GENE COMBINATIONS FOR EARLOBE TRAITS			
With these genes from the egg:	sperm:	The child is:	The child has:
free	free	pure dominant	free earlobes
free	attached	heterozygous	free earlobes
free	attached	heterozygous	free earlobes
attached	attached	pure recessive	attached earlobes

Check Your Understanding

1. How are chromosomes in sex cells different from those in body cells?
2. One person is pure dominant for a trait while another is heterozygous. How are their genes different?
3. How many different kinds of offspring could result if one parent is pure dominant and the other parent is pure recessive? How many different kinds of offspring could result if both parents are heterozygous?
4. **Critical Thinking:** Why does an individual usually carry only two genes for a certain trait?
5. **Biology and Math:** Why do living organisms have an even number of chromosomes in their body cells?

26:2 Expected and Observed Results

Objectives

4. **Discuss** the purpose of a Punnett square.
5. **Compare** expected results and observed results.
6. **Explain** the importance of Gregor Mendel's work.

Key Science Words

Punnett square

How can knowing the types of genes that each parent has be helpful? You can predict what traits their children could have. Sometimes, however, the combinations of genes that you expect do not appear in the offspring. Let's find out why this is true.

The Punnett Square

What does a Punnett square show?

We have seen how an egg and a sperm may combine to form an offspring. Each time, we figured out all the possible combinations of egg and sperm cells. There is an easier way to do this. This easy way is called the Punnett (PUH nuht) square. The **Punnett square** is a way to show which genes can combine when egg and sperm join. To make things easier, letters are used in place of genes. A large letter, such as ***F,*** is used for a dominant gene. The large ***F*** stands for free earlobe. A small letter, such as ***f,*** is used for a recessive gene. The small ***f*** stands for attached earlobe. A person with ***FF*** genes is pure dominant and has free earlobes. A person with ***Ff*** genes is heterozygous and has free earlobes. Notice that the large letter goes first in heterozygous organisms. A person with ***ff*** genes is pure recessive. What kind of earlobes does that person have?

Follow these six steps to determine the possible combinations of genes a child could have. In this example, both parents will be heterozygous. They will have a gene for free earlobes and a gene for attached earlobes. Use the letters ***Ff*** to stand for the gene pair in their body cells.

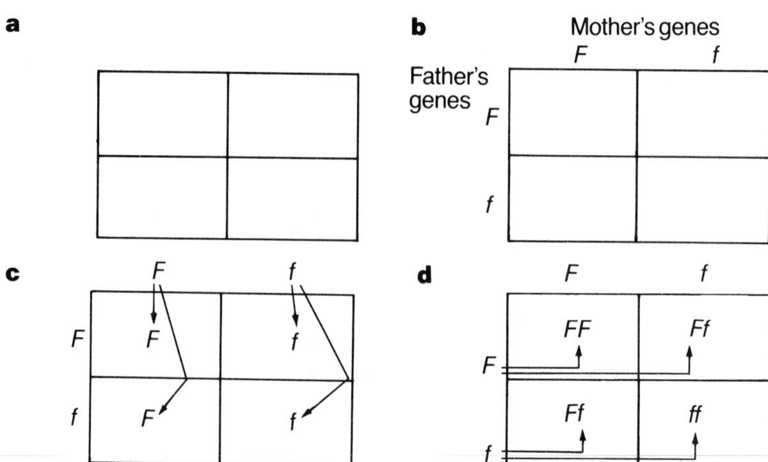

Figure 26-6 How to make and use a Punnett square

1. Draw a Punnett square as shown in Figure 26-6a. Each small box stands for one possible combination of genes that could show up in the offspring. Each combination of genes comes from a sperm cell fertilizing an egg cell.
2. Decide what kinds of genes will be in the sex cells of each parent. Write the letters that stand for the genes in the mother's egg cells across the top of the square. Figure 26-6b shows you how. In this case, put the two genes for eggs, ***F*** and ***f***, across the top.
3. Now, write the letters that stand for genes in the father's sperm along the side of the square. There are two possible genes for the sperm. They are ***F*** (free earlobes) and ***f*** (attached earlobes).
4. Copy the letters that appear at the top of the square into the boxes below each letter. Figure 26-6c shows how.
5. Copy the letters that appear at the side of the square into the boxes next to each letter. Figure 26-6d shows how.
6. Look at the small boxes in the Punnett square. They show the possible combinations of eggs and sperm. When egg and sperm combine, a new organism develops. The boxes also show what combinations of genes the organism could have. In our example, a child could have one of the following combinations: ***FF***, ***Ff,*** or ***ff.***

The Punnett square in Figure 26-7 shows you how the child could look. Remember, ***F*** is dominant over ***f.*** There are two different combinations of genes that would result in a child with free earlobes—***FF*** and ***Ff.*** Notice that three out of the four boxes contain either ***FF*** or ***Ff.*** Three out of four gene combinations of egg and sperm would result in a child with free earlobes.

There is only one combination of genes that would result in a child with attached earlobes—***ff.*** Only one out of four boxes in the Punnett square has ***ff.*** On average, one out of four gene combinations would result in a child with attached earlobes.

Bio Tip

Consumer: Sometimes an ear of white corn contains a yellow kernel of corn. A pollen grain carrying the dominant yellow gene was carried by the wind from a nearby garden. It fertilized the egg in one of the kernels of the white ear of corn.

Expected Results

The Punnett square that you drew shows what kinds of traits offspring can have. It shows what to expect when the sperm and egg of two parents join. Expected results are what can be predicted in offspring based on the genetic traits of parents. We predicted that one out of four children would have attached earlobes. We could do this because we knew what genes the parents had.

Figure 26-7 Three out of four gene combinations are for free earlobes.

	F	*f*
F	*FF* Free earlobes	*Ff* Free earlobes
f	*Ff* Free earlobes	*ff* Attached earlobes

Corn Genes

Lab 26–1

Problem: Do all corn seeds have genes for becoming green plants?

Skills

observe, predict, interpret data

Materials

20 corn seeds
paper towels
petri dish
water
wax marking pencil

Procedure

1. Copy the data table.
2. Look at the seeds carefully. Can you tell which seeds will grow into green plants and which will grow into white plants?
3. Moisten a paper towel with water.
4. Place the towel in the bottom half of the petri dish, Figure A. Fold it to fit.
5. Place the seeds on the towel. Cover the dish. Write your name on the cover.
6. Place the dish under a light and check it every day. Add more water if needed.
7. **Observe:** When the first seeds begin to sprout, record the leaf color. This is day 1.
8. Check your corn seeds for four more days. Record the numbers of new plants with green and with white leaves each day.

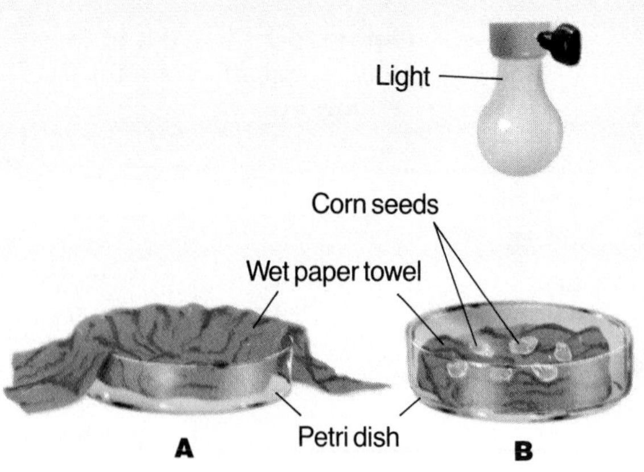

Data and Observations

1. What colors were the plants when they first sprouted?
2. Were there more green plants or white plants five days after sprouting?

	NUMBER OF CORN SEEDS SPROUTED	NUMBER WITH GREEN LEAVES	NUMBER WITH WHITE LEAVES
Day 1			
Day 2			
Day 3			
Day 4			
Day 5			
Total			

Analyze and Apply

Corn plants that remain white several days after sprouting have the albino trait. These plants soon die because they cannot make food. These plants are missing chlorophyll.

1. Could you tell before the seeds sprouted which would form green leaves? Explain.
2. **Predict:** What results would you see if the seeds came from heterozygous parents?
3. Why do albino plants not live as long as green plants?
4. **Apply:** Do you think most plants in fields and forests would be albino or green? Explain your answer.

Extension

Design an experiment to find out if the green corn plants in this lab carry the albino gene as a recessive trait.

Career Close-Up

Animal Breeder

An animal breeder works with a veterinarian or animal breeding specialist to keep records on animals and record what offspring of animals look like. Egg and milk production, size, and coat color are but a few of the genetic traits that animals show. Records of the traits of parent animals can help the animal breeder predict what traits can show in the offspring.

Many animal breeders work for state universities. Others may work on large farms that specialize in breeding. For example, farms that raise hundreds of cows may use an animal breeder to follow weight gain of cattle or record milk production. Animal breeders may work on horse ranches that specialize in race horses. They can also work at kennels where dogs are bred to be sold.

Wherever an animal breeder works, much of the training is learned on the job. A knowledge of biology and genetics is useful in this career.

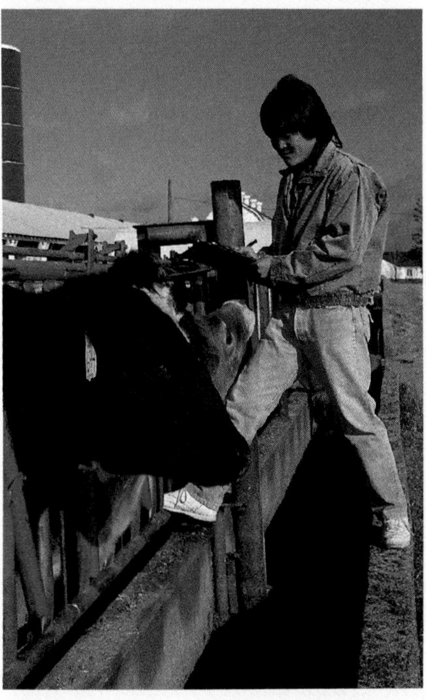

Some animal breeders record the milk production of cows.

The type of chin that a person has is also a genetic trait. A person with a cleft chin has a small indentation in the middle of the chin. A cleft chin is a dominant trait, Figure 26-8. A smooth chin is a recessive trait.

a

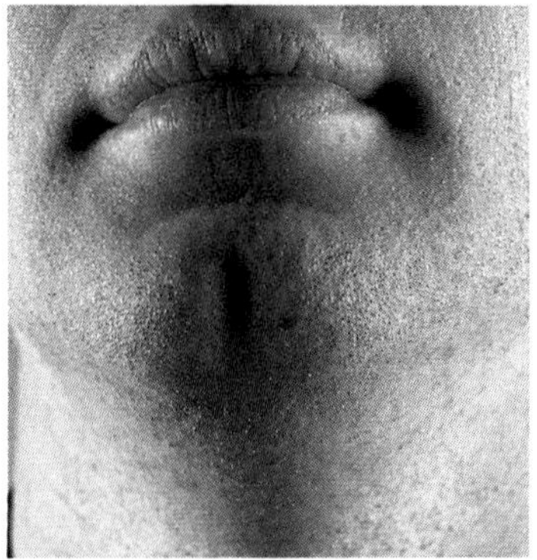

b

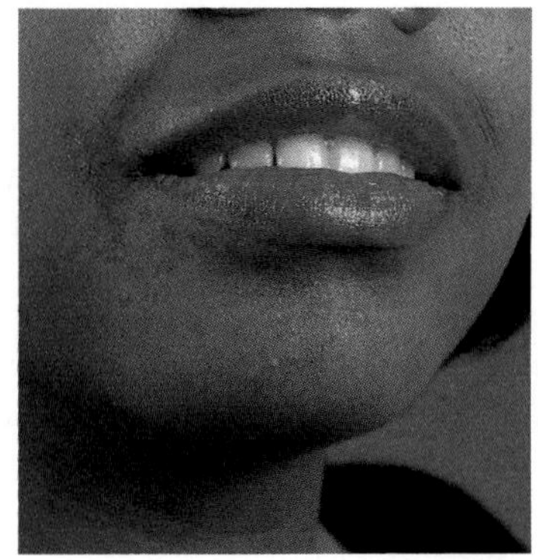

Figure 26-8 Person with cleft chin (a) and smooth chin (b)

Punnett Squares

Lab 26–2

Problem: What determines how offspring will look?

Skills

observe, formulate a model, form hypotheses

Materials

40 white beans
40 red beans
2 paper bags

Procedure

1. Copy the data table.
2. Place 20 white and 20 red beans into a paper bag. Label the bag female parent. These beans represent eggs.
3. Place 20 white and 20 red beans into a second paper bag. Label this bag male parent. These beans represent sperm.
4. Set up a Punnett square to show the parent gene types. Use ***R*** to stand for the dominant gene, red. Use ***r*** to stand for the recessive gene, white. Remember that each parent can produce sex cells carrying the red trait or the white trait. They are both heterozygous.
5. Complete your Punnett square to show the expected offspring of these parents.
6. Shake the bags.
7. Reach into each bag without looking and remove one bean. The two beans stand for the gene combination that results when sperm and egg join.
8. **Observe:** Look at the beans. Record the colors of the beans in your table next to trial 1. Use a check mark to record your results in the proper column.
9. Put the two beans back into the bags from which they came.
10. Write a **hypothesis** to show how many red/red ***(RR)***, red/white ***(Rr)***, and white/white ***(rr)*** pairs you will get in 40 trials. Use your Punnett square for help.
11. Repeat steps 6–9 for 39 more trials.
12. Using your table, find the totals of red/red (***RR***), red/white (***Rr***), and white/white (***rr***) combinations.

Data and Observations

1. What gene pairs result in red offspring?
2. What gene pairs result in white offspring?

	FEMALE PARENT		MALE PARENT		GENE PAIRS
TRIAL	RED BEANS	WHITE BEANS	RED BEANS	WHITE BEANS	
1					
2					
3					
4					
5					
40					

Analyze and Apply

1. Out of 40 offspring, how many did you expect to be pure dominant ***(RR)***, heterozygous ***(Rr)***, and pure recessive ***(rr)***?
2. Which of the three gene combinations did you expect in greatest number?
3. **Check your hypothesis:** Was your hypothesis supported by your data? Why or why not?
4. **Apply:** What determines how offspring will look?

Extension

Design an experiment to show that the more trials you do, the closer your observed results come to your expected results.

Let ***I*** stand for cleft chin and ***i*** stand for smooth chin. It might be easier to remember the symbols if you think of ***I*** standing for indentation. How might a child appear if one parent were ***Ii*** and the other were ***ii?*** Use the Punnett square in Figure 26-9 to get the answer. Two out of four gene combinations could be for children with cleft chins. Note that these children with cleft chins have ***Ii*** genes. The other two combinations could be for children with smooth chins, or children with ***ii*** genes. These results are what is expected. Would you expect these parents to have a pure dominant child? Why or why not?

	I	*i*
i	*Ii* Cleft chin	*ii* Smooth chin
i	*Ii* Cleft chin	*ii* Smooth chin

Figure 26-9 Children can have two possible chin types if one parent is heterozygous and the other is pure recessive.

Observed Results

We know that the results expected from the Punnett square do not always occur in every family. Look at Table 26-3. These data are from ten different families that show the cleft chin trait. In each family, one parent has the genes ***Ii*** and has a cleft chin. The other parent has the genes ***ii*** and has a smooth chin.

The table shows exactly what you would see if you looked at the children of these families. The traits actually seen in offspring when parents with certain genetic traits mate are the observed results. Using a Punnett square allows you to predict that half the children in these families could have cleft chins. Half the children could have smooth chins. The observed results, however, do not exactly match the expected results because you don't know which sperm and egg will join.

TABLE 26–3. OFFSPRING AND CHIN TYPES

Family	Number of Children in Family	Number with Cleft Chin	Number with Smooth Chin
A	2	0	2
B	1	1	0
C	5	3	2
D	4	2	2
E	2	1	1
F	3	1	2
G	1	1	0
H	6	4	2
I	2	1	1
J	3	1	2
Totals	29	15	14

Study Tip: Use this idea map as a study guide to expected and observed results. The idea map shows how they are different.

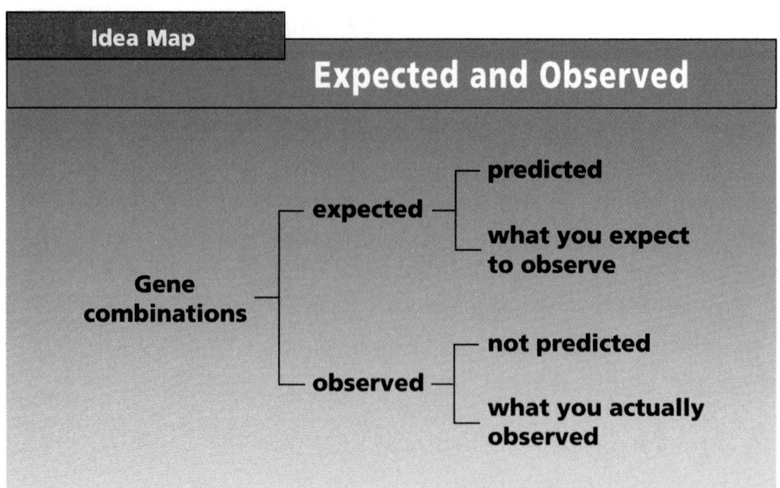

Skill Check

Calculate: Calculate how many heads and tails you would get if you were to flip a coin 10 times and then 50 times. Explain which you are able to calculate, expected or observed results. *For more help, refer to the* ***Skill Handbook,*** *pages 718-719.*

A little more than half the children have cleft chins. Nearly half the children have smooth chins. Each family by itself may not show the results you expect to see. If you add the results together, they are close to what you expect to see in a Punnett square. In fact, the more results you observe, the closer the observed results will be to the expected results.

Think about what happens when you flip a coin. You expect the coin to come up heads or tails. If you flip a coin two times, does it come up heads once and tails once? It may. It may come up heads twice or tails twice. Two heads or two tails are not what you expect. In fact, if you flip the coin four times you may even see four heads or four tails. Why?

When are observed results similar to expected results?

Each time you flip a coin, you are starting over. If you flip the coin once and see heads, it does not mean that the next flip must come up tails. The coin can land either way each time you flip it. With just a few flips, you are less likely to see the exact same number of heads as tails. If you flip the coin many, many times, the number of heads and tails should be about equal. The observed results will be closer to the expected results.

Mendel's Work

Let's go back in time to 1865. An Austrian monk named Gregor Mendel saw certain traits in the garden pea plants he grew in his garden. Mendel counted and recorded the

a

Figure 26-10 Gregor Mendel used a scientific method to study pea plants (a). Two of the traits he studied are shape and color of the seed (b).

traits he saw in the pea plants. He used a scientific method to do hundreds of experiments. Using the data he gathered and his knowledge of mathematics, he was able to explain some basic laws of genetics. Mendel explained what dominant and recessive traits were. He also showed how these traits passed from parent to offspring. How did he do it?

Why is Mendel's work important?

Mendel looked at several traits in the garden pea plant, Figure 26-10. One of those traits was height of the plant. He noticed that tall parent plants mated with short parent plants always produced tall plants, Figure 26-11. Remember that if ***T*** stood for the tall gene and ***t*** stood for the short gene, the parents would be ***TT*** and ***tt.*** What genes would the offspring have? Would the offspring be heterozygous or pure?

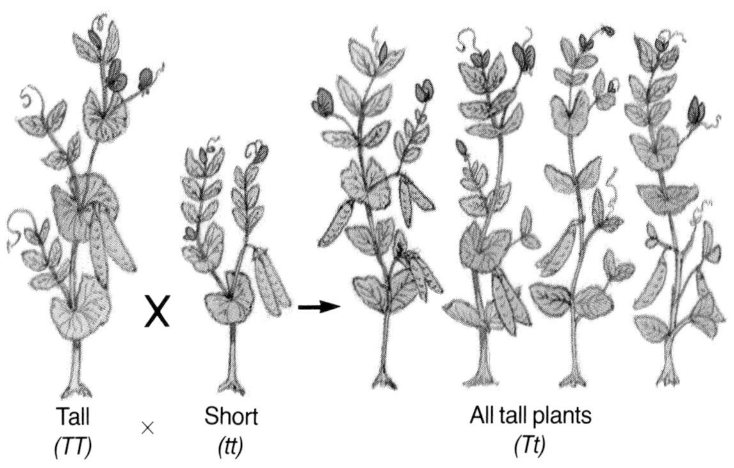

Figure 26-11 Mendel's results when he mated a tall parent plant with a short parent plant

Mendel then took the tall offspring and mated them with each other. He noticed that about three-fourths of the plants they produced were tall and about one-fourth were short. Look at Figure 26-12. Can you see that Mendel's results came very close to what we now expect to see when heterozygous plants are mated? Mendel concluded that his plants were heterozygous after he saw the results of his experiments.

Remember that a heterozygous individual has a dominant gene and a recessive gene. Mendel's heterozygous plants would be ***Tt.*** Study the Punnett square in Figure 26-12 to see what happens when two heterozygous plants mate.

	T	*t*
T	*TT* Tall	*Tt* Tall
t	*Tt* Tall	*tt* Short

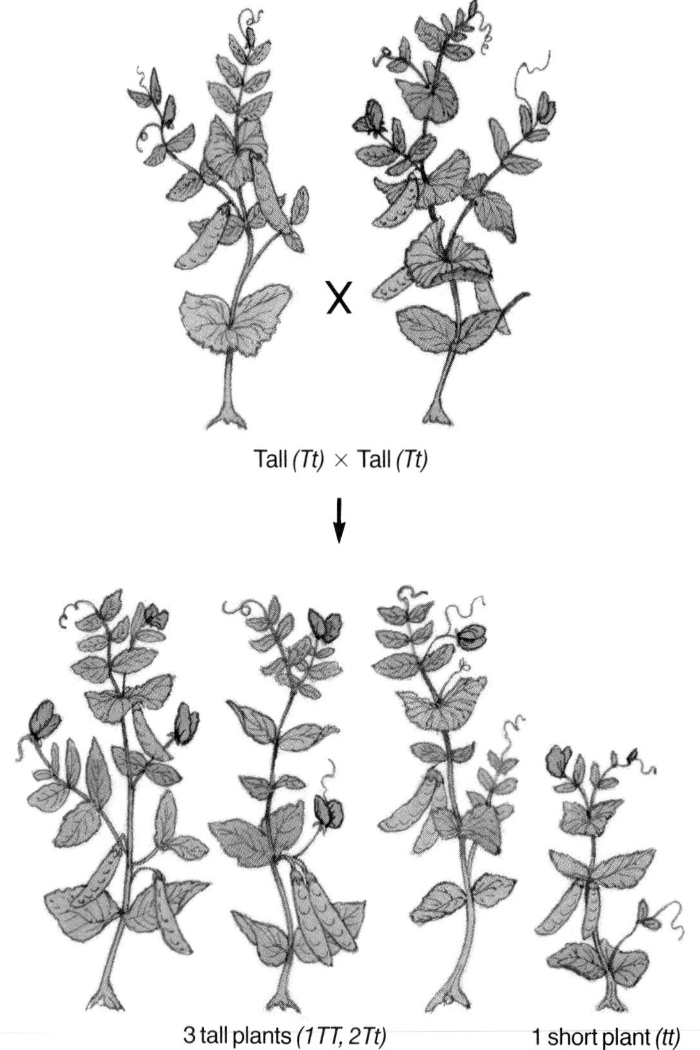

Figure 26-12 Mendel's results when he mated two heterozygous plants. How many gene combinations are possible?

TABLE 26–4. MENDEL'S RESULTS			
	Number of Offspring Observed		
Trait	**Dominant Trait**	**Recessive Trait**	**Total**
Color of pea pod	green–428	yellow–152	580
Shape of pea	round–5474	wrinkled–1850	7324
Color of pea	yellow–6022	green–2001	8023
Flower color	red–705	white–224	929

Mendel studied seven traits in pea plants. Some traits that he looked at were the color of the pea pod, shape of the peas, color of the peas, and color of the flower. Table 26-4 shows that he got the same results whenever he mated heterozygous plants. About three-fourths of the offspring had the dominant trait and one-fourth had the recessive trait.

Check Your Understanding

6. Why are observed results sometimes different from expected results?
7. Draw a Punnett square to show what genes a child would be expected to have if each parent were heterozygous for long eyelashes. Use ***L*** to stand for the dominant trait, long eyelashes. Use ***l*** to stand for the recessive trait, short eyelashes.
8. What laws of genetics did Mendel explain?
9. **Critical Thinking:** You mate a red-flowered plant with a white-flowered plant. You expect all the offspring to be red, but you find that half of them are white. Explain why.
10. **Biology and Writing:** Being an albino is a recessive trait in which no color is produced in the skin or eyes. Could an albino child be produced by two normally-pigmented parents? Explain.

Chapter 26

Review

Summary

26:1 Genetics, How and Why

1. Chromosomes are threadlike parts in the nuclei of body cells and sex cells. Body cells have twice as many chromosomes as sex cells.
2. Dominant genes keep recessive genes from showing their traits. An organism may have two dominant genes for a trait, two recessive genes, or one of each.
3. An individual that is pure for a trait can make one kind of sex cells for a certain trait. A heterozygous individual can make two kinds of sex cells for a certain trait. The kind of offspring produced depends on which sex cells combine.

26:2 Expected and Observed Results

4. The Punnett square helps to predict the combinations of genes offspring can receive.
5. Expected results are the predicted results. Observed results are what you actually see.
6. Mendel reported how traits were inherited in garden pea plants. He explained basic principles of genetics.

Key Science Words

dominant gene (p. 549)
gene (p. 547)
genetics (p. 546)
heterozygous (p. 549)
Punnett square (p. 552)
pure dominant (p. 549)
pure recessive (p. 549)
recessive gene (p. 549)

Testing Yourself

Using Words

Choose the word from the list of Key Science Words that best fits the definition.

1. has a dominant and a recessive gene for a trait
2. a gene that prevents a recessive gene from showing
3. has two dominant genes for a trait
4. having two recessive genes for the same trait
5. a small section of a chromosome that determines a trait
6. a way to show which genes combine when egg and sperm join
7. a gene that does not show when the dominant gene for the trait is present
8. study of how traits are passed from parents to offspring

Finding Main Ideas

List the page number where each main idea below is found. Then, explain each main idea.

9. how genes are arranged on a chromosome
10. how a Punnett square can be used to predict combinations of genes when sperm and eggs combine

Review

Testing Yourself *continued*

11. why recessive traits do not show in heterozygous individuals
12. what determines an organism's traits
13. how traits in the garden pea plant were studied
14. when chromosomes in a cell are seen best
15. what the four small boxes of a Punnett square represent
16. the traits that offspring actually have
17. that heterozygous parents can have children with a dominant or recessive trait
18. that each sex cell has one gene for a given trait

Using Main Ideas

Answer the questions by referring to the page number after each question.

19. What does a Punnett square predict? (p. 552)
20. Who discovered how garden pea plants inherit their traits? (p. 559)
21. What are genes that do not show even though they are present? (p. 549)
22. How are chromosomes and the nucleus related? (p. 546)
23. How is fertilization of an egg by a sperm like flipping a coin? (p. 558)
24. How are genes related to chromosomes? (p. 547)
25. What are two types of sex cells? (p. 546)
26. How do body cells and sex cells differ in chromosome number? (p. 546)
27. If Mendel crossed a heterozygous tall plant with a pure recessive short plant, what kinds of plants could he expect them to produce? (p. 553)

Skill Review ✓

*For more help, refer to the **Skill Handbook,** pages 704-719.*

1. **Calculate:** Use a calculator to find out how close Mendel's data (Table 26-4) are to showing a 3:1 ratio of dominant to recessive traits. Divide each trait by the total.
2. **Form hypotheses:** Form a hypothesis to explain why two offspring in a family have different traits.
3. **Interpret data:** Dogs with spots show a dominant trait and dogs without spots show a recessive trait. A dog breeder mates two spotted dogs. There are three dogs without spots in the first litter. Explain why.
4. **Classify:** Classify the following as cells or cell parts: sperm, gene, chromosome, egg, nucleus.

Finding Out More

TECH PREP

Critical Thinking

1. Suppose you planted flower seeds. The package stated that the plants grow two feet tall. You noticed that 30 percent of the plants grew less than one foot tall. What does this tell you about the tallness trait?
2. Explain why some children can resemble one parent more than the other.

Applications

1. Find out which members of your family can roll their tongues and which can't.
2. Find out how seed companies produce flower seeds that will grow into plants with certain traits.

CHAPTER PREVIEW

Chapter Content

Review this outline for Chapter 27 before you read the chapter.

Skills in this Chapter

The skills that you will use in this chapter are listed below.

- In **Lab 27-1,** you will make and use tables, formulate a model, calculate, and design an experiment. In **Lab 27-2,** you will observe, make and use tables, and interpret data.
- In the **Skill Checks,** you will formulate a model, and understand science words.
- In the **Mini Labs,** you will make and use tables.

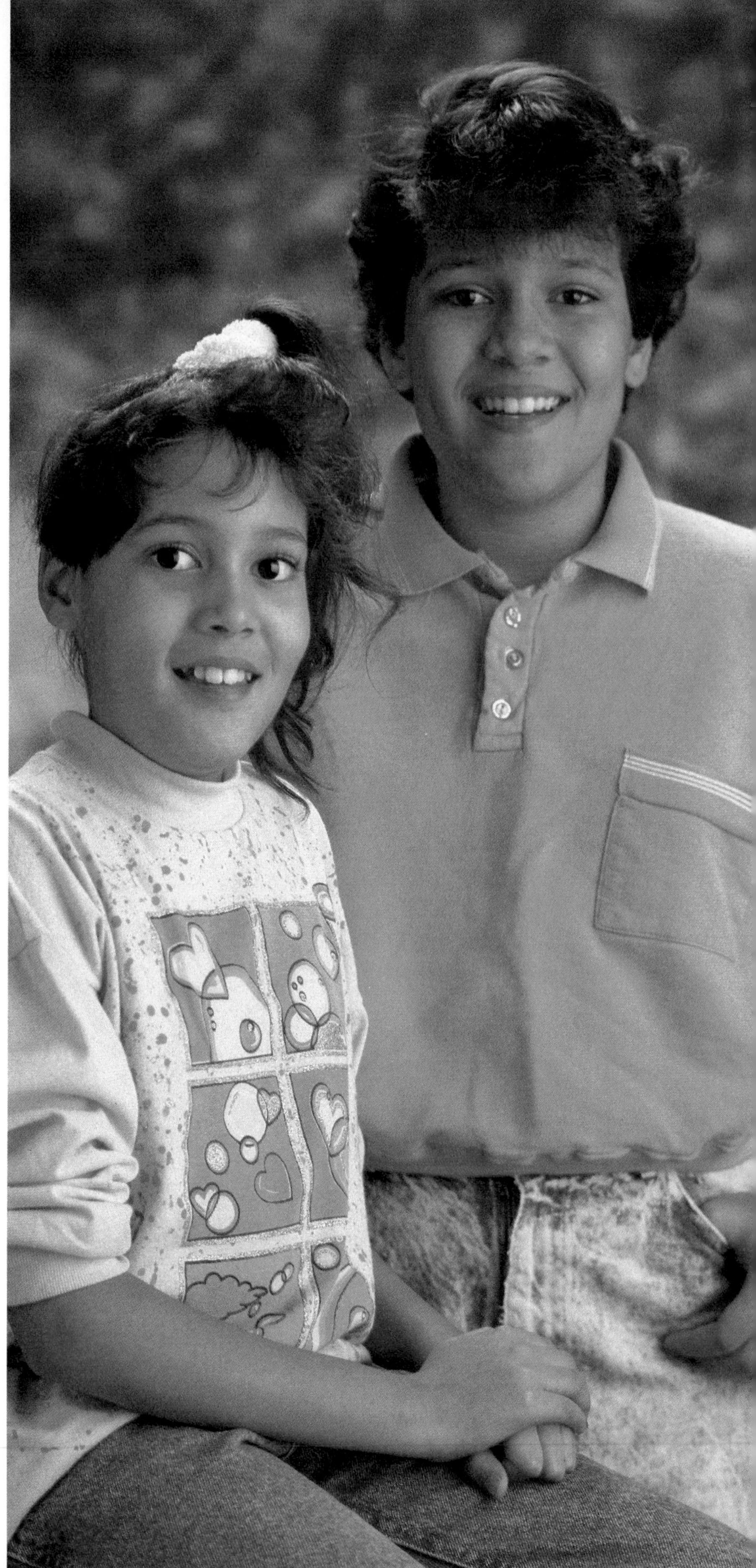

Chapter 27

Human Genetics

Are human traits different from traits of other animals? Look at pictures of several animals. Make lists to show which traits or body parts are shared by humans and other animals and which are different.

Did you ever notice how often the children in a family look alike? This is because they share many of the same traits they inherited from their parents. Some traits may not be the same in all the children, however. How can you explain these differences?

The girls in the large photo are sisters. What traits do they have in common? What traits can you see that are different in the two sisters? Can you tell anything about the traits of the parents without actually seeing them?

The girl in the smaller photo is a cousin of the two sisters. The father of the two sisters and the father of the cousin are brothers. What traits do the sisters share with the cousin? Why do sisters share more traits than two cousins?

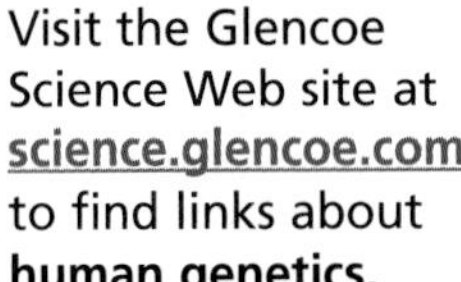
Visit the Glencoe Science Web site at science.glencoe.com to find links about **human genetics.**

27:1 The Role of Chromosomes

Objectives

1. **Compare** the chromosome numbers in body cells and sex cells.
2. **Describe** methods that doctors use to study chromosomes of a fetus.
3. **Compare** the chromosomes of males and females.

Key Science Words

amniocentesis
X chromosome
sex chromosome
Y chromosome
autosome

All living things can pass their traits to their offspring. You learned in Chapter 26 that these traits can be dominant or recessive. The genes that control the traits are found on sections of the chromosomes.

Chromosome Number

How are chromosomes arranged in body cells?

There are three things you should know about chromosome numbers in living things.
1. Each human sperm or egg has 23 chromosomes.
2. Each human body cell has 23 pairs of chromosomes. There are 46 chromosomes in each human body cell.
3. Different organisms have different numbers of chromosomes. The chromosomes in most living things are paired. A carrot plant has 18 chromosomes in each of its body cells. That's equal to nine pairs per cell. Table 27-1 gives the numbers of chromosomes in some other living things.

TABLE 27–1. CHROMOSOME NUMBERS

Animal or Plant	Number of Chromosomes in Body Cells	Number of Chromosome Pairs in Body Cells
Red clover	12	6
Pea	14	7
Onion	16	8
Corn	20	10
White pine	24	12
Cat	38	19
Rabbit	44	22
Chicken	78	39

Study Tip: Use this idea map as a study guide to the chromosomes in human cells. The numbers of chromosomes in body and sex cells are shown.

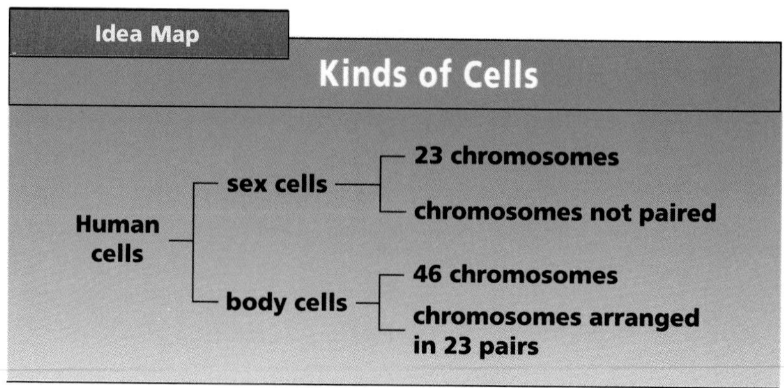

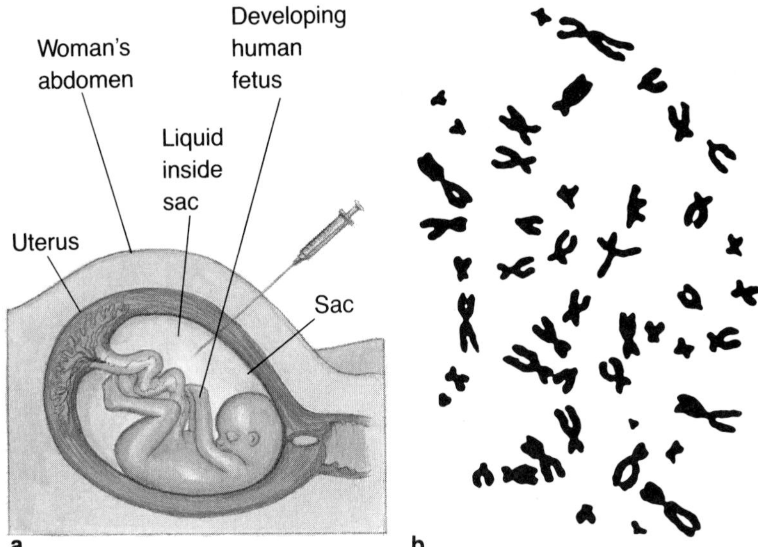

Figure 27-1 Amniocentesis (a). Chromosomes taken from the sac in which the fetus develops (b).

A Way to Tell Chromosome Number

Doctors today can tell if a child has the correct number of chromosomes even before it is born. They use a process called amniocentesis (am nee oh sen TEE sus). **Amniocentesis** is a way of looking at the chromosomes of a fetus.

Look at Figure 27-1a. A needle is inserted into the pregnant woman's abdomen. The needle enters the amniotic sac in which the fetus develops, and some liquid is removed through the needle. In this liquid are cells from the fetus's skin. The cells have rubbed off the fetus just as your skin cells do.

How does a doctor perform amniocentesis?

These cells are grown for about 10 days in a nutrient liquid that makes them divide. Remember that chromosomes can be seen best when the cell is dividing. The cells are then studied with a microscope and an enlarged photo like the one in Figure 27-1b is made. The chromosomes are cut out and the ones that match are arranged in pairs. A large chart is made showing the chromosome pairs. The lab technician looks at the chromosomes to make sure there are 46 of them and there are no missing or added parts.

Using a newer way to look at the chromosomes of a fetus, a doctor can take a small piece of the placenta surrounding the fetus. This area has cells with the same number and kind of chromosomes as the fetus. The chromosomes are counted and studied to see whether parts are missing or added.

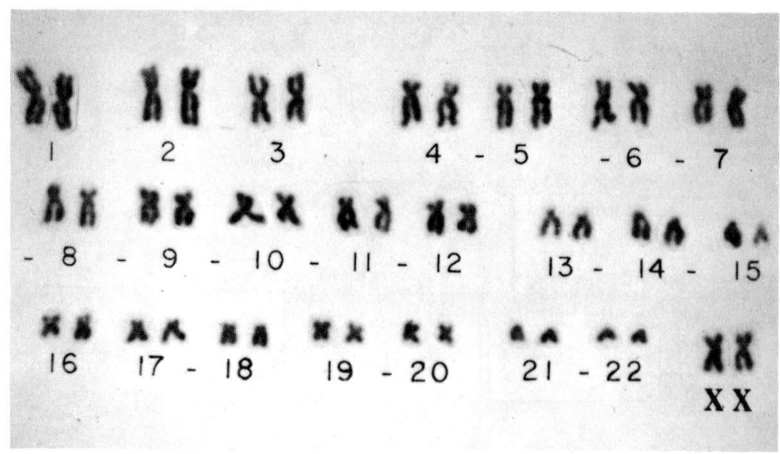

Figure 27-2 These photos of chromosomes have been cut out and arranged in pairs. How many chromosomes does a human body cell have?

Sex—A Genetic Trait

Whether you were born male or female depends on your chromosomes. In humans, a special pair of chromosomes determines sex.

What sex is a person with two X chromosomes?

Figure 27-2 shows the 23 pairs of chromosomes from a female. Notice that the chromosomes are numbered, except for those in the lower right corner. Those are marked XX. Each of the **X chromosomes** is a sex chromosome of the female. **Sex chromosomes** are chromosomes that determine sex. A human female has two X chromosomes in each body cell.

As you can see in Figure 27-3, the male's sex chromosomes are different from those of a female. The larger sex chromosome is an X chromosome. The smaller one is called a Y chromosome. The **Y chromosome** is a sex chromosome found only in males. A human male has one X and one Y chromosome in each body cell.

The numbered chromosomes in Figure 27-2 are called autosomes (AH toh sohmz). **Autosomes** are chromosomes that don't determine the sex of a person. They are also called nonsex or body chromosomes. The autosomes of a male are like those of a female.

Skill Check

Understand science words: autosome. The word part *auto* means self. In your dictionary, find three words with the word part *auto* in them. *For more help, refer to the* ***Skill Handbook,*** *pages 706-711.*

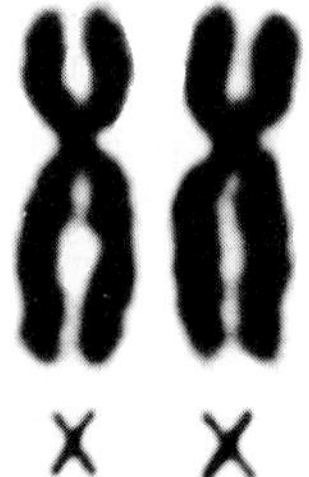

Figure 27-3 A female has two X chromosomes. A male has one X and one Y chromosome.

Science and Society

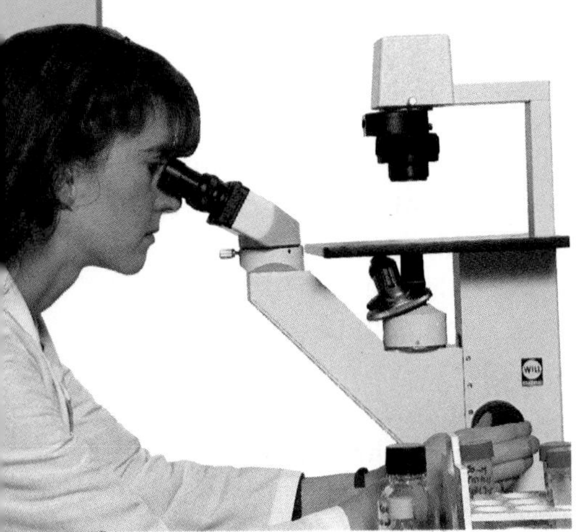

Can Couples Choose the Sexes of Their Children?

Suppose a couple had two girls and they wanted to have a boy. There is a 50 percent chance of having a baby boy. Is there any way to increase that chance? Researchers have found several ways to separate the sperm containing the Y chromosome from the sperm containing the X chromosome. Because the X chromosome is larger and heavier than the Y chromosome, these sperm can be separated by special filtering methods. When the sperm are separated, they are put into different tubes. Usually, one tube will have about three Y sperm for every X sperm. Thus, the chances of having a boy baby are increased from 50 percent to about 75 percent.

A doctor then inserts the sperm cells carrying the Y chromosome into the vagina of the female. If she becomes pregnant, she has a better chance of having a boy than a girl. However, she still has a 25 percent chance of having a girl.

What Do You Think?

1. About 200 genetic disorders are controlled by the X chromosome. Many of these are fatal or result in lifelong health problems. These disorders can be prevented if a couple has only girls. Is it acceptable for couples to choose the sex of their children to prevent health problems?

2. Certain cultures prefer children of one sex over children of the other sex. Suppose that over the next 10 years, all couples in Japan decided to have only boys. Predict some problems that could result in the country.

3. Choosing the sex of a child can be very expensive. Not all people can afford to choose the sex of their children. If choosing the sex of children becomes popular, should the government or health insurance companies pay the medical costs of couples who cannot afford to pay?

Conclusion: Under what conditions, if any, should people be allowed to choose the sex of their children? If you knew that these methods guaranteed the sex of a child instead of just increasing the chances, how would you feel about them?

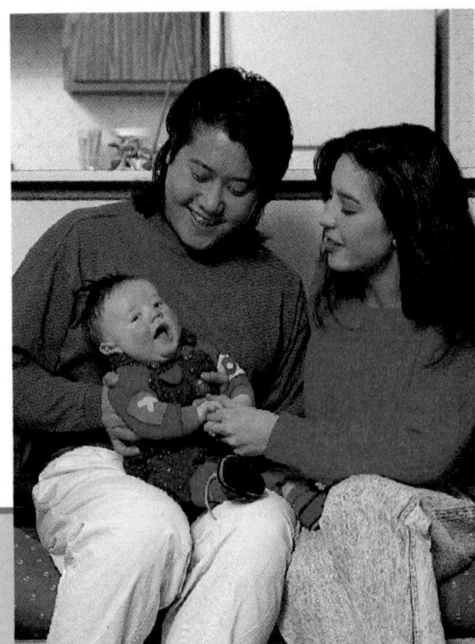

Some cultures prefer male children to female children.

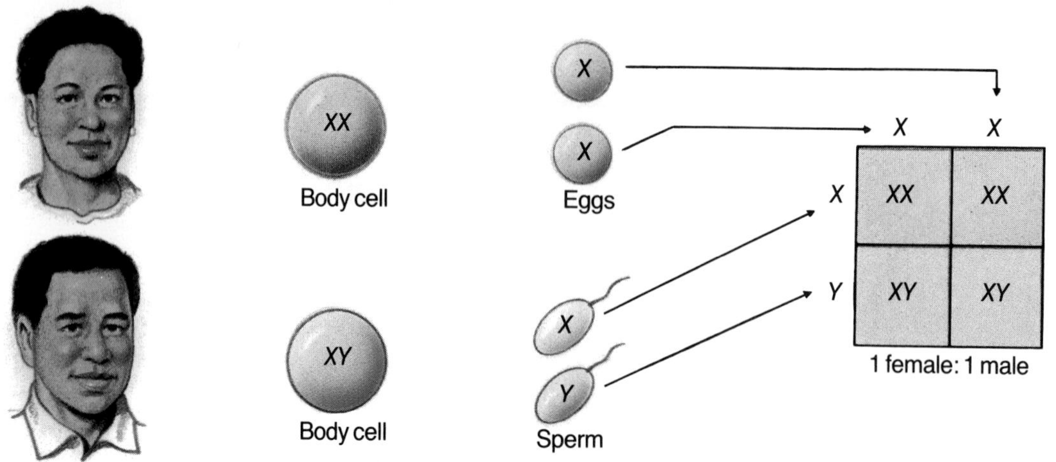

Figure 27-4 A child receives an X chromosome from the mother and an X or a Y chromosome from the father.

What kind of sex chromosomes, X or Y, are found in eggs and sperm? Because females have only one kind of sex chromosome, they can make only one kind of egg. Each egg of a female has one X chromosome. Males have two different sex chromosomes, so they can make two different kinds of sperm. Each sperm of a male has either an X chromosome or a Y chromosome. As you can see from Figure 27-4, a child can receive only an X chromosome from its mother. The child can receive an X or Y chromosome from its father.

Notice that the father determines the sex of the children. If the child receives the father's X chromosome, the child will be a girl. If the child receives the father's Y chromosome, what will be the sex of the child? This Punnett square shows that we can expect half the children to be females and the other half to be males.

Check Your Understanding

1. How do sex cells and body cells of humans differ in chromosome number?
2. What is amniocentesis?
3. How are the sex chromosomes of human males and females alike? How are they different?
4. **Critical Thinking:** What would happen if each sex cell had the same number of chromosomes as a body cell?
5. **Biology and Math:** Suppose a woman has three girls. What is the chance her next child will be a boy?

Lab 27–1

Sex Determination

Problem: In a family of four children, how many will be girls and how many will be boys?

Skills

make and use tables, formulate a model, calculate, design an experiment

Materials

2 coins
paper cup
masking tape

Procedure

1. Copy the data table 10 times.
2. Cover two coins with tape. Mark one coin with an X on each side. Mark the other coin with an X on one side and a Y on the other side.
3. Draw a Punnett square to show how many males and females you expect to see in one family with four children. Multiply the number of males by 10 and the number of females by 10. The totals are the numbers of males and females you would expect to see in ten families.
4. Put both coins in the paper cup. Shake the cup and drop the coins on your desk. Read the combination.
5. On table 1, make a check mark for your result on the first shake.
6. Shake, read, and mark your results three more times.
7. **Calculate:** Total the columns marked XX and XY.
8. Repeat steps 4 to 7 nine more times using a new table for each group of four shakes.
9. Add the number of females on all 10 tables. This is the total number of females you observed. Do the same for males.
10. Compare the observed totals in step 9 with the expected totals in step 3.

Data and Observations

1. How many male and female children do you expect in each family? Explain your answer using the words *chromosome, X, Y, sperm,* and *egg.*
2. How many males and how many females did you expect in 10 families?

	TABLE 1 RESULTS	
SHAKE	XX	XY
1		
2		
3		
4		
Total		

Analyze and Apply

1. Was the total number of males and females observed different from what you expected? If so, how can you explain the difference?
2. If you had made tables for 20 families, how many male children and how many female children would you expect?
3. If you had made tables for 100 families, would the observed results have been closer to the expected results? Explain your answer.
4. **Apply:** Suppose a doctor told a couple with three girls that their next child would be a boy. Is the doctor correct? Explain.

Extension

Design an experiment to show how often families with four children will have two boys and two girls.

27:2 Human Traits

Objectives

4. Compare recessive and dominant traits with incomplete dominance.

5. Describe different ways human traits can be inherited.

Key Science Words

incomplete dominance
sickle-cell anemia
color blindness

Have you ever gone "people watching"? The next time you are at a shopping center, movie theater, or football game, look at the people around you. They have many traits that you can see. These traits come from their parents. People also have many traits that you can't see. These traits direct things that go on inside the body.

Survey of Human Traits

The girl in Figure 27-5a has freckles. She also has dimples in her cheeks. Both traits are caused by dominant genes. In both of these traits, one or both of the parents shows the dominant trait. Remember, only one dominant gene is needed to make a dominant trait show up. Two recessive genes are needed to make a recessive trait show up.

What kinds of traits are recessive? Attached earlobes is a recessive trait. Neither of the parents needs to have attached earlobes for their child to show this trait. Set up a Punnett square to show why that is true.

Other recessive traits are straight hair and not being able to roll the sides of your tongue. Can you guess what the dominant traits are? The ability to roll your tongue is dominant. Curly hair is dominant to straight hair. Short eyelashes are recessive. What is the dominant trait?

a

b

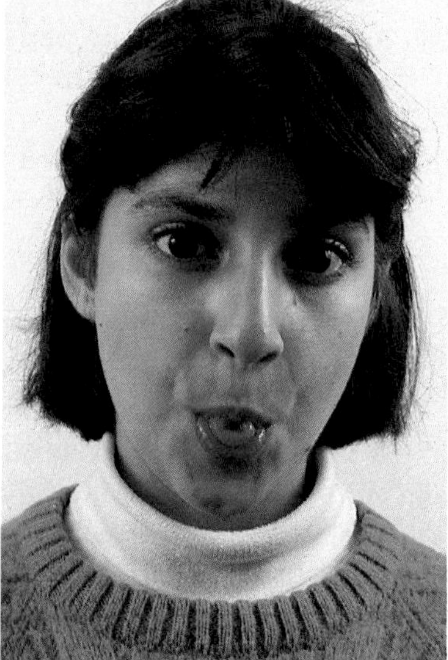

Figure 27-5 Freckles and dimples are dominant traits (a). The ability to roll the tongue is a dominant trait (b).

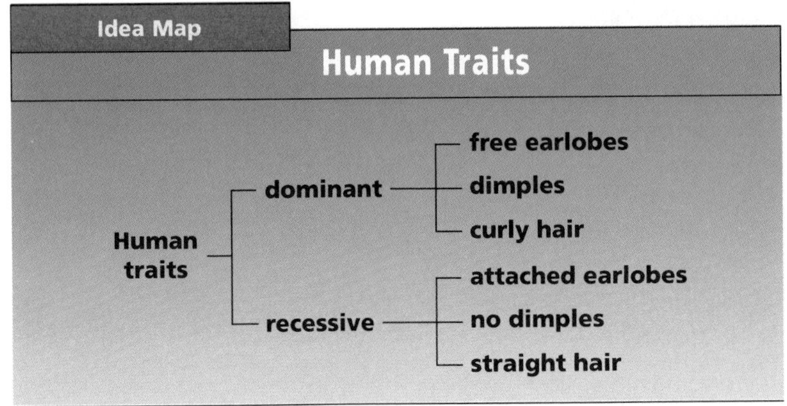

Study Tip: Use this idea map as a study guide to some human traits. How many other dominant and recessive traits can you think of?

Incomplete Dominance

Some traits are neither totally dominant nor totally recessive. Scientists know of several genes that don't show total dominance. A case in which neither gene is totally dominant to the other is called **incomplete dominance.** If you placed a piece of blue glass over a piece of yellow glass and held it up to the light, what would you see? Figure 27-6 shows that the glass would look green. If you took apart the pieces of glass, you would see that they were still blue and yellow.

What trait is seen in an individual that is heterozygous for an incompletely dominant trait?

The same kind of thing happens with incomplete dominance. Remember that when a dominant and a recessive gene are together in a heterozygous organism, only the dominant trait is usually seen. In incomplete dominance, we see a new trait that is a blend of the dominant and recessive traits. When pure dominant red snapdragons are mated with pure recessive white snapdragons, the heterozygous offspring are all pink. It is important to know that the genes themselves do not combine. When they are separated, we see the original traits again. If two pink snapdragons are mated, they produce red, white, and pink offspring.

Figure 27-6 In incomplete dominance, the combining of two genes gives a new trait, much as combining blue and yellow glass appears to make green glass (a). The combining of genes from red and white snapdragons gives offspring that are pink (b).

a

b

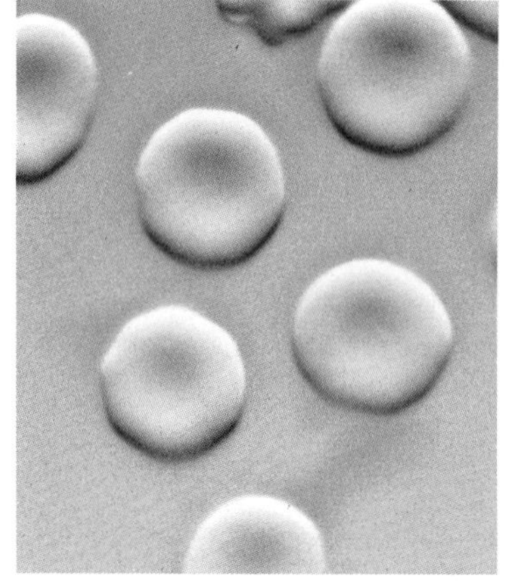

a

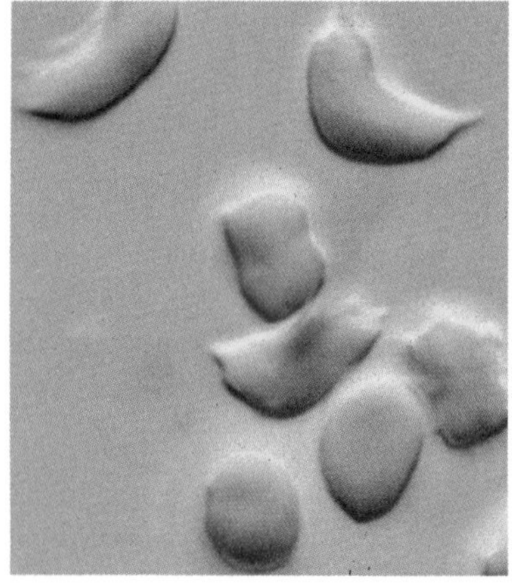

b

Figure 27-7 Normal round (a) and sickled (b) red blood cells. Which kind of cells does a heterozygous person have?

What human trait shows incomplete dominance?

Some traits in humans show incomplete dominance. One of these is the shape of red blood cells. In most people, red blood cells are round. Some people have red blood cells shaped like those in Figure 27-7b. These cells are called sickle cells. They are shaped like a sickle that is used to cut grass. Some people have both shapes of blood cells.

Let the letter ***R*** stand for the gene for round blood cells. Let ***R′*** stand for the gene for sickle cells. In incomplete dominance, two different letters or symbols are used for the genes controlling the trait. The ***R*** gene is not totally dominant over the ***R′*** gene. The ***R′*** gene is not totally dominant over the ***R*** gene. Capital letters can stand for both. A person with round cells has ***RR*** genes. A person with all sickle cells has ***R′R′*** genes. A person with both kinds of cells has ***RR′*** genes.

Someone with ***RR′*** genes usually doesn't have serious health problems. People who have some sickle cells, however, may not be as active as those who have all round cells. As you know, red blood cells carry oxygen to body cells. Sickle cells do not carry oxygen as well as round blood cells.

People with ***R′R′*** genes have sickle-cell anemia. **Sickle-cell anemia** is a genetic disorder in which all the red blood cells are shaped like sickles. This disorder is much more common among African-Americans than among people of other races. People with sickle-cell anemia have serious health problems. Their lives may be shortened because of the disorder. The sickle cells cannot carry enough oxygen. Body tissues can be damaged because they don't receive enough oxygen. Sickle cells do not move easily through the blood capillaries because their shape causes the capillaries to become clogged.

Mini Lab

What Traits Do Humans Share?

Make and use tables: Make a table of traits inherited from your parents. Compare your list with that of a classmate. How many traits do you have in common? *For more help, refer to the* ***Skill Handbook****, pages 715-717.*

If two parents have ***RR′*** genes, what would we expect the children to have? Look at the Punnett square in Figure 27-8. One out of four children is expected to have all round cells. Two out of four children are expected to have both round and sickle cells. The fourth child is expected to have all sickle cells. That child would have sickle-cell anemia.

	R	*R′*
R	*RR*	*RR′*
R′	*RR′*	*R′R′*

Figure 27-8 How sickle-cell anemia can be inherited when both parents are heterozygous. How many children can be expected to have sickle-cell anemia?

Blood Types in Humans

Remember from Chapter 12 that there are four blood types—A, B, AB, and O. Human blood type is controlled in part by dominance of genes. Although three genes control blood types, each person has only two of them. The two genes that you have control your blood type.

The three genes that control blood type are ***A, B,*** and ***O.*** Both ***A*** and ***B*** are dominant to ***O. A*** and ***B,*** however, are not dominant to each other. Look at Table 27-2 as you read the next three paragraphs.

If you have type O blood, you have ***OO*** genes. If you have type AB blood, you have one ***A*** gene and one ***B*** gene.

If you have type A blood, you have ***AA*** or ***AO*** genes. If you have ***AO*** genes, the ***A*** gene is dominant to the ***O*** gene.

If you have type B blood, you have ***BB*** or ***BO*** genes. If you have ***BO*** genes, the ***B*** gene is dominant to the ***O*** gene.

TABLE 27–2. BLOOD TYPES AND GENES

Blood Type	Genes
O	***OO***
AB	***AB***
A	***AA*** or ***AO***
B	***BB*** or ***BO***

Genes on the X Chromosome

Like other chromosomes, the sex chromosomes carry genes. Figure 27-9 shows some traits that are controlled by genes on the sex chromosomes. Notice that females have two genes for each of these traits. That is because females have two X chromosomes.

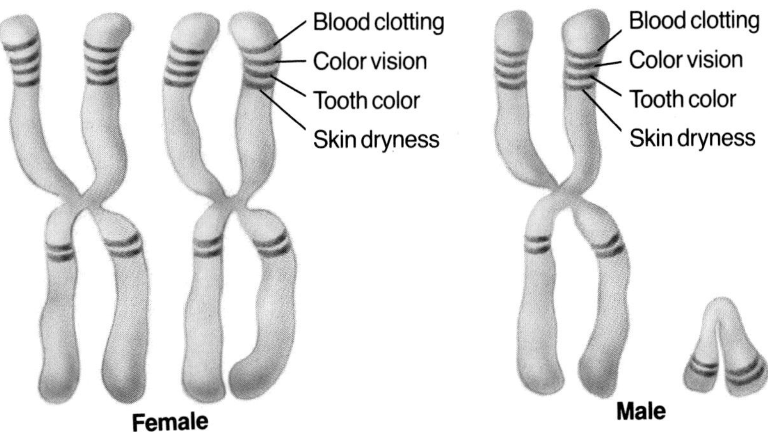

Figure 27-9 Certain genes on the X chromosome are missing from the Y chromosome.

Figure 27-10 To be tested for red-green color vision, a person may be shown a figure like this one.

Males have only one gene for each of these traits. The Y chromosome does not have the genes that are on the X chromosome. When the X and Y chromosomes are together, genes on the X chromosome control the traits.

Let's look at color blindness. **Color blindness** is a problem in which red and green look like shades of gray or other colors. Look at Figure 27-10. Someone who is color blind for red and green can't see the numbers in it.

Being able to see red and green as two separate colors is a dominant trait. Let's use ***C*** for this gene. Not being able to see red and green is a recessive trait. A ***c*** will be used for this gene.

A female who has ***CC*** genes will be able to see red and green. If she has ***Cc*** genes, she still will be able to see red and green. If she has ***cc*** genes, she will be color blind. Two of the three gene combinations for the female allow her to see red and green.

A male is more likely to be color blind. Why? Think about what genes he could have. Remember that the Y chromosome will not have a gene for the trait.

A male who has the ***C*** gene on his X chromosome will be able to see red and green. If he has the ***c*** gene on his X chromosome, he will not be able to see red and green. There are only two possible gene combinations for the male. One of them will allow him to see red and green. What would happen if a woman with ***Cc*** genes and a man with a ***c*** gene on his X chromosome had children? See if you can answer this question.

Check Your Understanding

6. What is incomplete dominance?
7. What two gene combinations can a person with type A blood have?
8. Why are there more color blind males than color blind females?
9. **Critical Thinking:** The gene for nearsightedness in humans is found on the X chromosome. A boy has a nearsighted father. Does this mean that the boy will become nearsighted? Why or why not?
10. **Biology and Reading:** The word part *hemi* is found in the word *hemisphere. Hemi* means half. The scientific term for genes on the X chromosome in a male is hemizygous. Why?

Lab 27–2 Human Traits

Problem: Which traits are most common?

Skills

observe, make and use tables, interpret data

Materials

pencil
paper

Procedure

1. Copy the data table.
2. **Make and use tables:** Have another student check you for each of the traits listed in the table. Record in your table with a check mark whether you show the dominant or recessive trait.
3. Reverse roles and check your partner for each trait.

TRAIT	DOMINANT	RECESSIVE
Skin color		
Extra fingers or toes		
Freckles		
Earlobe		
Tongue rolling		
Shape of hairline		
Hair on middle sections of fingers		
Chin shape		

Data and Observations

1. How many dominant traits do you have? How many recessive traits do you have?
2. How many other students in your class have the same traits as you?

Analyze and Apply

1. Which statement best explains this lab?
   (a) All students in the class show the same dominant and recessive traits.
   (b) A person may show some dominant and some recessive traits.
   (c) It is easier to recognize a dominant trait in someone than a recessive trait.
2. **Apply:** Were there more dominant or more recessive traits in your class? How can you explain this?

Extension

Design an experiment to determine if any of the traits are on the X chromosome.

TRAIT	DOMINANT	RECESSIVE
Skin color	Dark colors	Light colors
Extra fingers or toes	Six or seven fingers or toes	Five fingers and toes
Freckles	Freckles	No Freckles
Earlobe	Free	Attached
Tongue rolling	Can roll edges	Cannot roll edges
Shape of hairline	Pointed in middle	Not pointed in middle
Hair on middle sections of fingers	Hair	No hair
Chin shape	Indentation in middle	No indentation

27:3 Genetic Disorders

Objectives

6. **Describe** some genetic disorders in humans.
7. **Give examples** of how genetic counseling can help families.

Key Science Words

dyslexia
genetic counseling
pedigree

Each day, nearly 600 babies are born in the United States with some type of disorder. Some of the disorders are inherited. Let's look at a few of them.

Errors in Chromosome Number

Some people are born with more or fewer than 46 chromosomes. This happens when the sperm or egg cell does not have 23 chromosomes. Figure 27-12 shows how this can come about.

During meiosis, the sister chromatids are supposed to pull apart from each other. Notice in Figure 27-12 that the sister chromatids of the short chromosome pulled apart and were separated into different cells. The sister chromatids of the longer chromosome stuck together. Both of them went into the same cell. That cell now has an extra chromosome.

If these cells were eggs, the one on the bottom would have one fewer chromosome, or 22. If this egg joined a sperm, a child with only 45 chromosomes would result. The egg on the top has an extra chromosome. If this egg joined a sperm, a child with 47 chromosomes would result.

Not pulling apart can happen to almost any pair of chromosomes. When it happens to the sex chromosomes, certain traits show up. Figure 27-11 shows sex chromosome patterns that sometimes show up in people. The O stands for a missing sex chromosome. Notice the boy and the girl who have only one sex chromosome. The sex chromosome from one parent is missing. What are some of the problems that result from having one of these chromosome patterns?

Female	Male
XO	YO
	XXY

Figure 27-11 Sex chromosome patterns

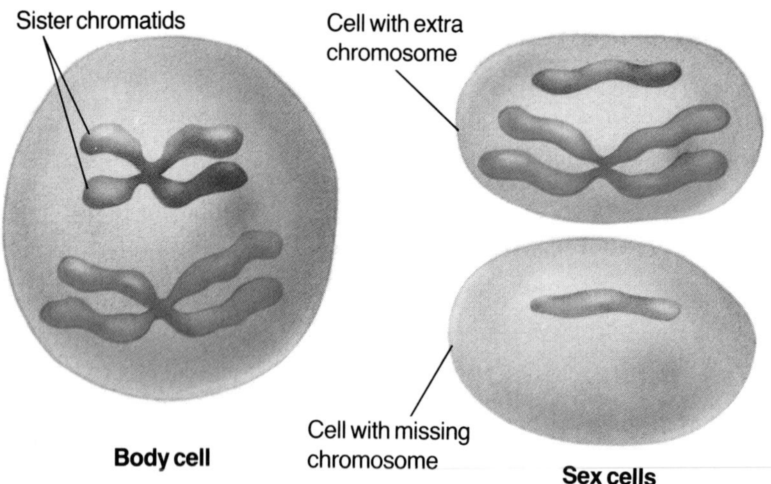

Figure 27-12 When sister chromatids do not pull apart during meiosis, there can be an error in chromosome number in the child.

A YO male will die before it is born. An XO female and an XXY male cannot make sex cells, but they can live.

The child in Figure 27-13 was born with an extra autosome. During meiosis, one of his autosomes did not pull apart from its copy. A doctor would say he has Down syndrome. The child learns more slowly than most children his age. Children with Down syndrome often have heart problems. Researchers are learning more about problems caused by having an extra chromosome.

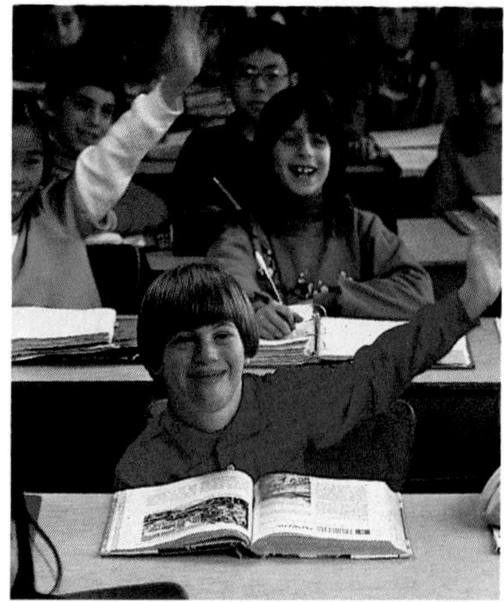

Figure 27-13 Child with Down syndrome

Genetic Disorders and Sex Chromosomes

You have studied traits controlled by genes on the sex chromosomes. The ability to see red and green was one of them. There are also serious genetic disorders controlled by genes on the X chromosome.

A certain recessive gene on the X chromosome can cause a rare disorder called hemophilia. Remember that hemophilia is a disorder in which a person's blood does not clot. Bleeding from a cut or bruise may take hours to stop. Hemophilia almost always shows up in male children. How would it be possible for a female child to inherit hemophilia?

Skill Check

Formulate a model: Make a model of the alphabet on a poster board showing how a person with dyslexia might see these letters. *For more help, refer to the **Skill Handbook**, pages 706-711.*

Genetic Disorders and Autosomes

Genetic disorders can also be caused by genes found on the autosomes.

Dyslexia (dis LEHK see uh) is a genetic disorder that is also called word blindness. It is caused by a dominant gene. People with dyslexia see and write some letters of the alphabet or parts of words backward. They have trouble learning to read.

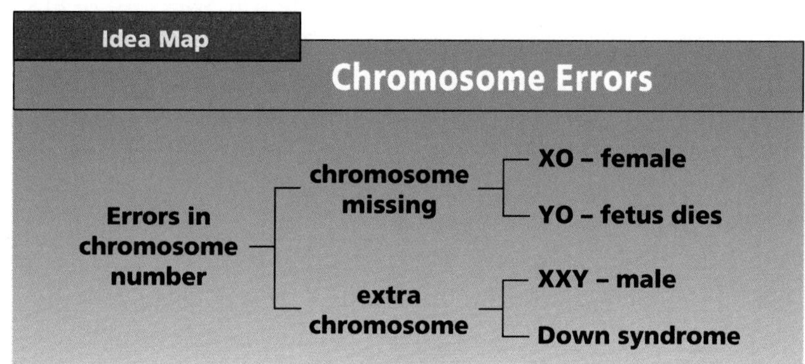

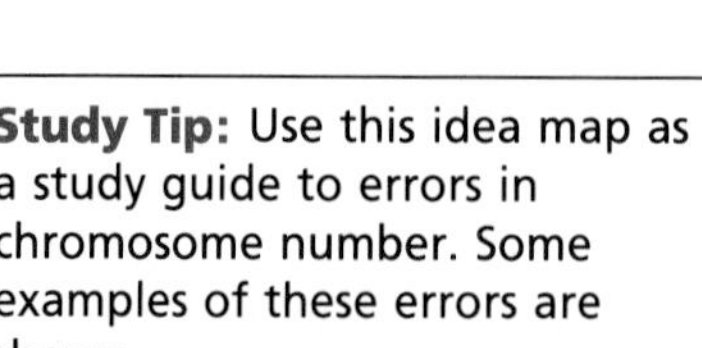
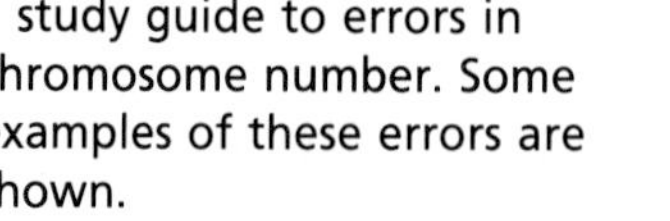

Study Tip: Use this idea map as a study guide to errors in chromosome number. Some examples of these errors are shown.

What is PKU?

Twenty years ago, a child born with PKU would spend its life in a hospital. PKU is a genetic disorder in which some chemicals in the body do not break down as they should. These chemicals can harm the brain cells. Today babies are tested for PKU soon after birth. A child born with PKU can be given a special diet, starting within the first few weeks of life. This diet has only small amounts of the chemicals that will not break down properly. The child can grow normally with this diet.

Bio Tip

Health: Cystic fibrosis patients lose large amounts of salt from their bodies, resulting in very salty sweat. For many years, having very salty sweat was used as a test for this disorder.

Genetic Counseling

"It runs in the family." Have you heard this comment before? It means that many family members have a certain trait. The trait could be found in parents, grandparents, brothers, sisters, or other relatives. Family members can have the same hair color, eye color, or shape of nose.

Genetic disorders run in families, too. How can a couple find out whether a disorder in their family will show up in their children? They can seek genetic counseling. **Genetic counseling** is the use of genetics to predict and explain traits in children. A genetic counselor can tell whether a problem is caused by genes.

A genetic counselor asks a lot of questions about many members of a family. Then the counselor makes some conclusions. A genetic counselor can help answer the following questions:

1. How did their baby get the disorder?
2. If the baby is healthy, does it have a problem gene?
3. Is the trait dominant or recessive?
4. What will happen to the baby's health as it gets older?
5. What are the chances that future children will have the trait?

Figure 27-14 A couple receiving genetic counseling

The baby of a young couple died when it was two years old. The doctor told the couple that their baby had cystic fibrosis (sis tik • fi BROH suhs). This is a genetic disorder in which the lungs and pancreas don't work the way they should. There are very serious breathing problems. Many children with this disorder die at an early age.

The baby's parents went to a genetic counselor, who told them that the disorder was caused by having two recessive genes, ***ff***. Both parents appeared normal, but they were each heterozygous for the dominant and recessive genes, ***Ff***.

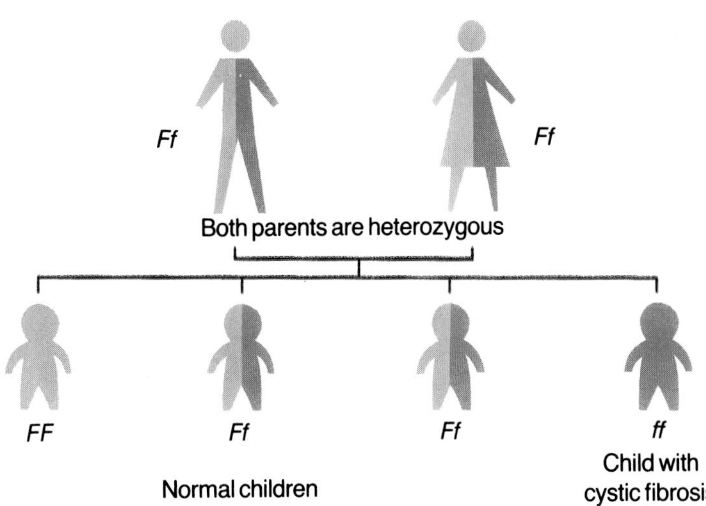

Figure 27-15 A family pedigree for cystic fibrosis

The counselor showed the parents a drawing of a family pedigree like the one in Figure 27-15. A **pedigree** is a diagram that can show how a certain trait is passed along in a family. A pedigree may be used to trace a family trait and to predict whether future generations will have the trait.

What is a pedigree?

The couple learned that there was a one in four chance of any of their children being born with the disorder and that tests such as amniocentesis can tell whether a fetus has the disorder. Talking to the counselor gave them an idea of what to expect if they decided to have more children.

Check Your Understanding

11. How can an error in meiosis cause a sex cell to have the wrong number of chromosomes?
12. Name three genetic disorders and tell how each one of them is controlled by genes.
13. How can a genetic counselor use a pedigree to help families with genetic disorders?
14. **Critical Thinking:** Why do you think an XO female can survive but a YO male will die before birth?
15. **Biology and Math:** A genetic counselor sees a couple who is worried that their unborn child may have hemophilia. Tests show that the fetus is female and that the mother is heterozygous for the recessive gene, which is on the X chromosome. Will the baby have hemophilia? Why or why not?

Chapter 27

Review

Summary

27:1 The Role of Chromosomes

1. Each human sex cell has 23 chromosomes. Each human body cell has 46 chromosomes arranged in pairs.
2. Amniocentesis is a way to look at the chromosomes of a fetus.
3. The sex of a person is determined by a pair of sex chromosomes. Human females have two X chromosomes. Human males have one X and one Y chromosome.

27:2 Human Traits

4. If a person has a dominant trait, one or both of the parents will also have the trait. If a person has a recessive trait, both of the parents will also carry the gene for the trait. Traits in which neither gene is totally dominant over the other show incomplete dominance.
5. Three genes control blood types in humans but each person has only two genes for this trait. Males show recessive traits located on the X chromosome more often than females do.

27:3 Genetic Disorders

6. If chromosome pairs do not pull apart during meiosis, errors in chromosome number result. Hemophilia, cystic fibrosis, and dyslexia are genetic disorders.
7. A genetic counselor uses knowledge of genetics to predict and explain disorders in children.

Key Science Words

amniocentesis (p. 567)
autosome (p. 568)
color blindness (p. 576)
dyslexia (p. 579)
genetic counseling (p. 580)
incomplete dominance (p. 573)
pedigree (p. 581)
sex chromosome (p. 568)
sickle-cell anemia (p. 574)
X chromosome (p. 568)
Y chromosome (p. 568)

Testing Yourself

Choose the word from the list of Key Science Words that best fits the definition.

1. chromosome not related to sex
2. way to observe cells from a fetus
3. word blindness
4. the use of genetics to predict problem traits in children
5. chromosome that determines sex of a child
6. female sex chromosome

Review

Testing Yourself *continued*

7. trait in which neither gene is totally dominant over the other
8. vision problem related to sex chromosome
9. male sex chromosome
10. diagram used to trace a family trait

Finding Main Ideas

List the page number where each main idea below is found. Then, explain each main idea.

11. how to see cells of a fetus
12. how a new trait appears when there is incomplete dominance
13. how errors in chromosome number occur
14. why people with sickle-cell anemia have serious health problems
15. what the X and Y chromosomes are
16. that sex cells contain half the number of chromosomes as body cells

Using Main Ideas

Answer the questions by referring to the page number after each question.

17. Which parent determines the sex of a child? (p. 570)
18. How can a child have cystic fibrosis when the parents do not? (p. 580)
19. How can two parents with type A blood have a child with type O blood? (p. 575)
20. A color-blind female marries a normal male. What could the color vision of their children be? (p. 576)
21. One parent is heterozygous for blood type A. The other is heterozygous for blood type B. What might the children's blood types be? (p. 575)

Skill Review ✔

*For more help, refer to the **Skill Handbook,** pages 704-719.*

1. **Make and use tables:** Make a table to show the numbers of baby boys and girls born in a local hospital in one month. How does the ratio of boys to girls compare with what you expect?
2. **Observe:** Find out how many students can roll their tongue and how many cannot. Explain why there are more students showing one trait.
3. **Interpret data:** Two parents have four children, each of which has a different blood type. One child is type A, one is type B, one is type O, and one is type AB. How can you explain this?
4. **Calculate:** If the body cells of an animal contain 78 chromosomes, how many chromosomes does a sperm cell from that animal contain?

Finding Out More

Critical Thinking

1. Explain why geneticists would want to spend more time finding out about genetic disorders in humans rather than how height, hair color, and eye color are inherited.
2. Why do most states require that newborn babies be tested for PKU soon after they are born?

Applications

1. Find out how people with Down syndrome lead useful lives.
2. Find out the treatment for cystic fibrosis.

CHAPTER PREVIEW

Chapter Content

Review this outline for Chapter 28 before you read the chapter.

Skills in this Chapter

The skills that you will use in this chapter are listed below.

- In **Lab 28-1,** you will formulate a model, interpret diagrams, and observe. In **Lab 28-2,** you will observe, infer, and interpret data.
- In the **Skill Checks,** you will understand science words.
- In the **Mini Labs,** you will formulate models.

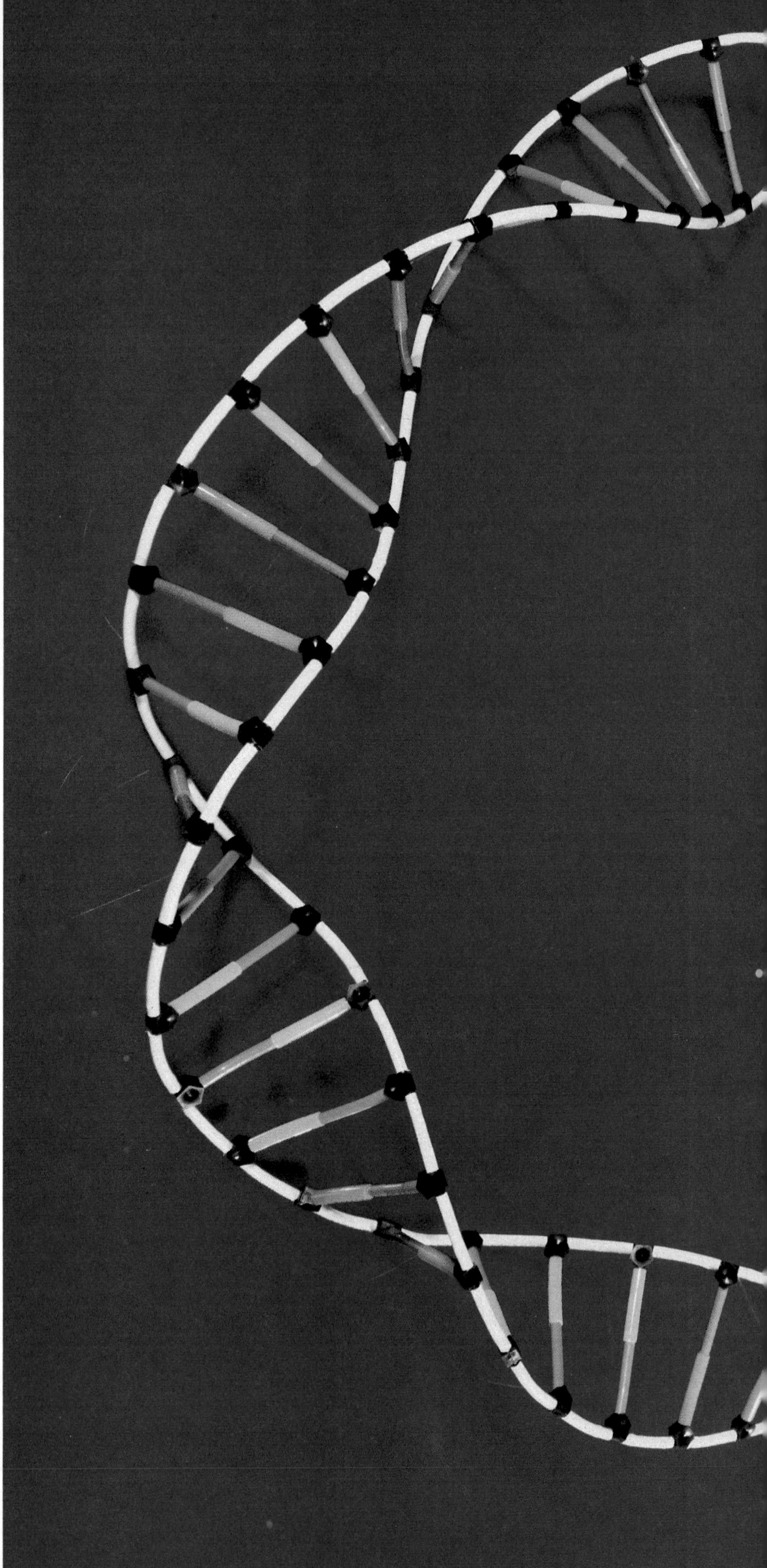

Chapter **28**

DNA—Life's Code

Many different kinds of codes are used for many different reasons. Braille is an example of a code. It is used by a visually impaired person to help him or her read. The photo on this page shows a book written in Braille. Another code is one that is found in each of your cells. It is a code that determines all of your traits. It directs your body to be you. What you see in the photo on the left is a model of a chemical molecule. How this molecule acts as a code to determine your body's traits is what this chapter is all about.

Try This!

How many words can you make with only four letters? Design an alphabet that has only four letters. Write as many words as possible using the four letters. You may use the same letter more than once.

BIOLOGY Online

Visit the Glencoe Science Web site at science.glencoe.com to find links about **DNA.**

Braille is a code.

28:1 The DNA Molecule

Objectives

1. **Describe** the structure of DNA.
2. **Explain** how DNA controls genetic traits.
3. **Describe** how DNA copies itself and works to make proteins.

Key Science Words

DNA
nitrogen base
RNA
genetic code

Within each of your cells is a chemical that controls life. It is what we have been calling *genes* in the past few chapters. Now you will find out how this chemical makes up the genes, how it determines all your traits, and how it passes an organism's instructions from generation to generation. You will see why this chemical has been called the code of life.

DNA Structure

The photograph on page 584 is a model of a DNA molecule. The letters DNA stand for *d*eoxyribo*n*ucleic (dee AHK sih ri boh noo klay ik) *a*cid. We shall use only the letters DNA each time we talk about this molecule. What is DNA? **DNA** is a molecule that makes up genes and determines the traits of all living things. Humans, birds, mushrooms, plants, protozoans, and bacteria have DNA. All living things contain DNA in their cells.

Scientists often use models to explain things that are very complex. We shall use a model of DNA to explain what DNA looks like and how it does its job.

Figure 28-1a shows a ladder. It is made up of two upright side pieces and many rungs. Figure 28-1b shows a small part of the DNA model. Note that it looks something like the ladder. An actual DNA molecule is much longer

Figure 28-1 DNA is often compared to a ladder (a). A model of DNA is shown (b).

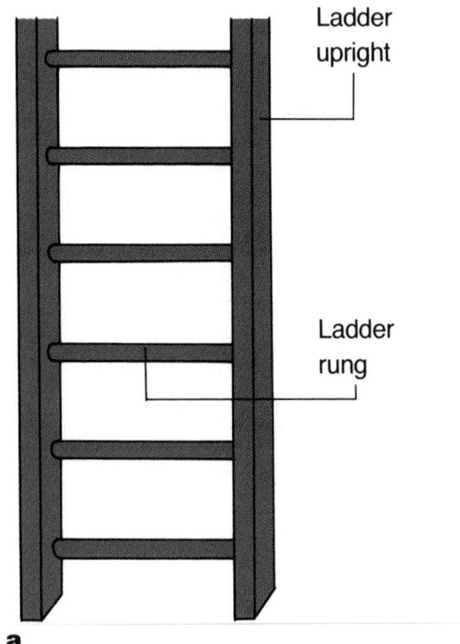

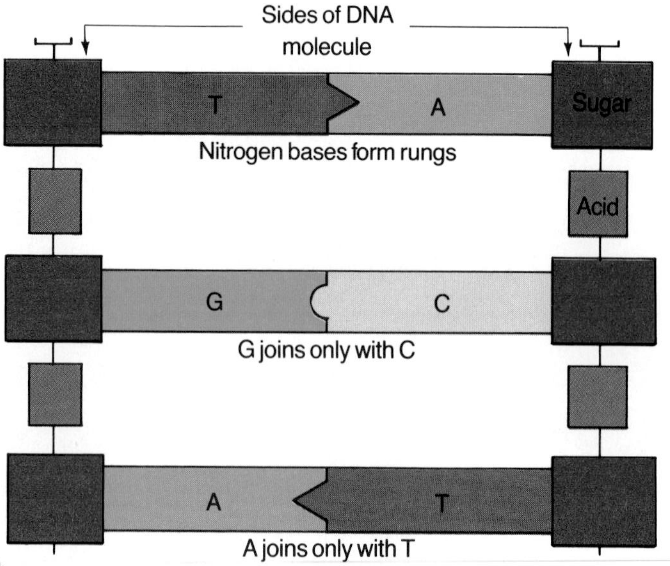

than the model. Scientists have estimated that a single DNA molecule in a human cell may contain about 100 million rungs.

What are the parts of a DNA molecule?

There are six features of the DNA model.

1. DNA has two main sides. These sides are like the upright parts of a ladder.
2. The sides are made of two different chemicals. One is a sugar. The other is an acid. These two chemicals alternate along each side.
3. There are parts that connect the two sides together. These parts look like the rungs of a ladder.
4. **Nitrogen bases** form the rungs of a DNA molecule.
5. There are four different nitrogen bases in DNA. The letters A, T, C, and G stand for the four bases.
6. The four bases join each other in certain ways to form the rungs. Like pieces of a puzzle, shape A will fit only shape T. Shape C will fit only shape G. As a result, base A joins only with base T. Base C joins only with base G.

You should know one more thing about the shape of DNA. Imagine holding each end of a ladder and twisting it. You would end up with a structure like the one in Figure 28-2. This figure shows that DNA is twisted into a spiral.

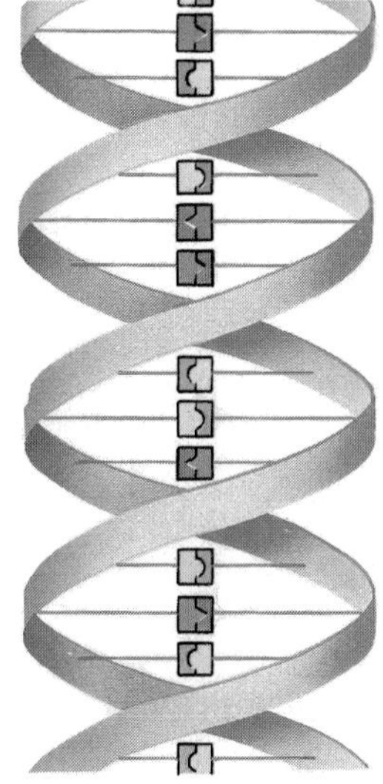

Figure 28-2 DNA is twisted into a spiral.

DNA and Chromosomes

DNA is in every cell in your body. It is also in every cell of all other living things. Where in these cells is it found? You probably know the answer already. DNA makes up parts of the chromosomes found in the nucleus.

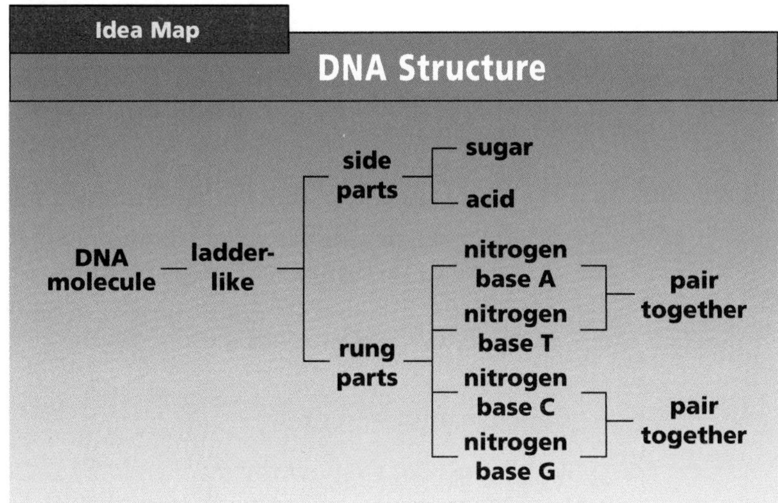

Study Tip: Use this idea map as a study guide to the structure of the DNA molecule. What is the shape of the DNA ladder?

DNA

Lab 28–1

Problem: What does a model of DNA look like?

Skills

formulate a model, interpret diagrams, observe

Materials

tracing paper
scissors
blank sheet of paper
heavy paper
red, blue, green, and yellow crayons
tape

Procedure

1. Trace the four parts shown in Figure A with tracing paper.
2. Cut out the four tracings and copy each part onto heavy paper four times. You should have 16 model parts.
CAUTION: *Always be careful when using scissors.*
3. Label each part with its correct letter (A, T, C, G).
4. Cut out the 16 model parts. Label the other side of each part to match the first side.
5. Color only the parts on your figures that are nitrogen bases. Use this coloring code: A = red, T = blue, C = green, G = yellow. Color both sides of each part.
6. Arrange eight of the figures in any order on a blank sheet of paper. Use Figure B to help you.
7. Tape the eight figures in place.
8. Arrange the remaining eight figures according to how they fit the first eight figures. You can turn the parts in any direction. Tape these figures in place.
9. Compare your model with those of your classmates.

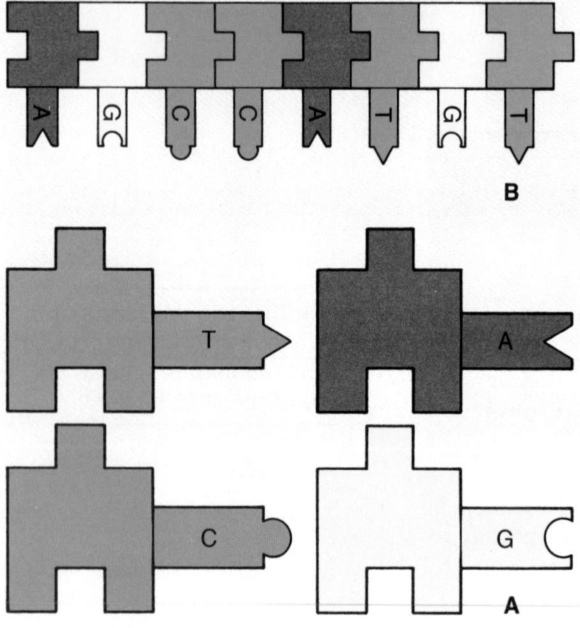

Data and Observations

1. What part, or letter, does the T fit? G? A? C? Explain these fits using your model.
2. In what ways is your model different from those of your classmates? In what ways is it alike?

Analyze and Apply

1. Where would you find this model if it were in a real cell?
2. Name the two chemicals that form the outer (uncolored) parts of your model.
3. **Interpret diagrams:** What chemical name describes the rungs that go across the middle?
4. **Apply:** Why do scientists use models?

Extension

Make a model to show the steps used by DNA as it copies itself. Use the parts from this lab.

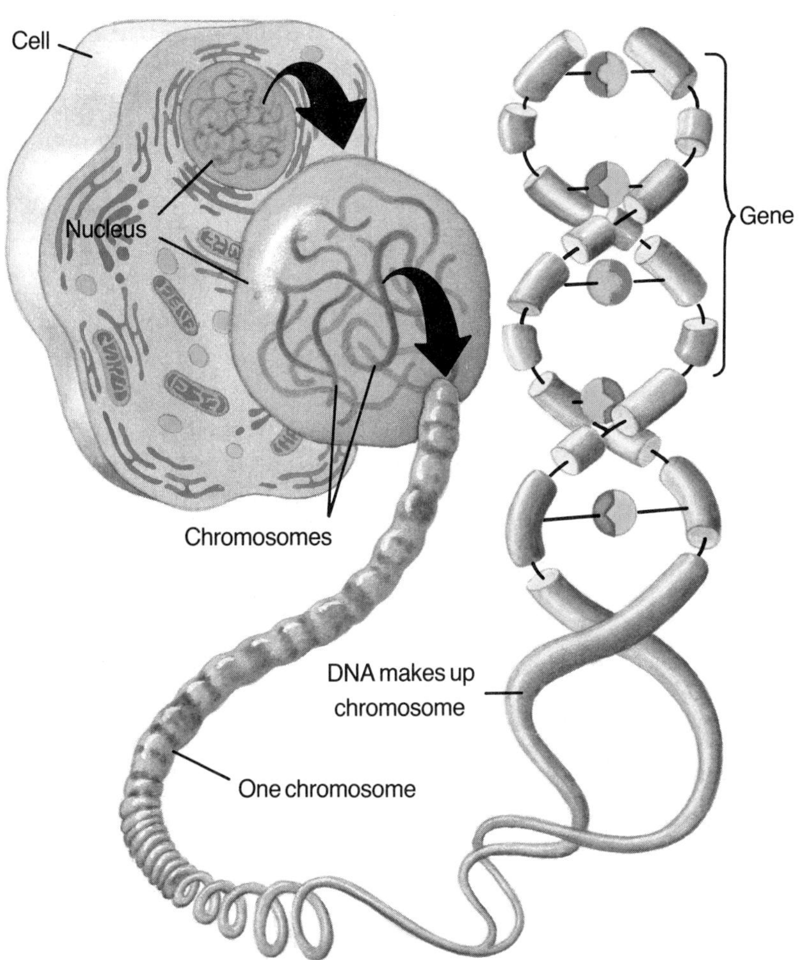

Figure 28-3 How the cell, nucleus, chromosomes, genes, and DNA are related

Look carefully at Figure 28-3. It shows how DNA is related to chromosomes. Keep in mind that this drawing is many times larger than the real thing.

Now think about this question. Where in the cell can we find genes? Genes are parts of chromosomes. We can describe a gene in two ways. First, it is a short piece of DNA. Second, it is a certain number of bases (rungs on the ladder) on the DNA molecule.

In Chapter 26, a gene was defined as a small section of chromosome that determines traits. All three definitions of a gene are correct.

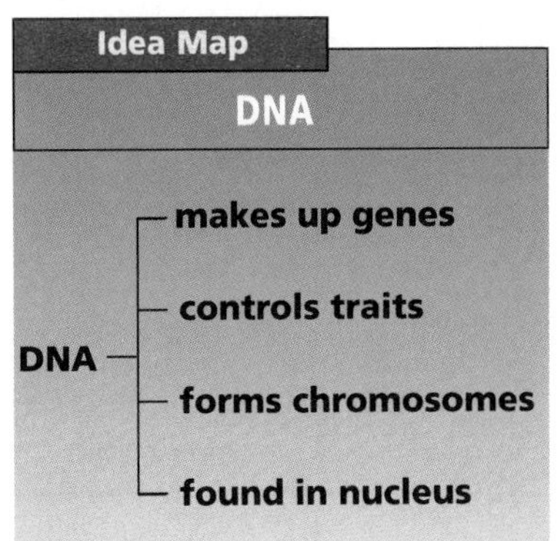

Study Tip: Use this idea map as a study guide to the DNA molecule. The features of DNA are shown.

Proof That DNA Controls Traits

How do scientists know that DNA controls our traits? To answer this question, let's look at some experiments done in 1928.

Figure 28-4 shows what happened in these experiments. Look at the first experiment, **a.** These results make sense. You would expect that living harmless bacteria would not kill a mouse. The results in experiments **b** and **c** also make sense. Living pneumonia bacteria would be expected to kill the mouse. Dead pneumonia bacteria would be expected to have no effect on the mouse. Do the results in experiment **d** make sense? They didn't at first. Remember from Chapter 1 that a scientist may have to form a new hypothesis if the experiment shows that the old hypothesis is not supported. The scientist who did this experiment had to form a new hypothesis to explain the results of experiment **d.** He hypothesized that the living harmless bacteria picked up a cell part from the dead harmful bacteria. He said that this cell part caused the harmless bacteria to change into harmful bacteria.

What cell part did the harmless bacteria pick up? This question was answered years later by another group of scientists. They injected a mixture of living harmless bacteria and the DNA from pneumonia bacteria into a mouse. The mouse died. Then they injected living harmless bacteria mixed with other chemicals from the pneumonia bacteria. When the harmless bacteria picked up DNA from the harmful bacteria, they were changed into harmful bacteria. When the harmless bacteria picked up any other

Figure 28-4 How scientists proved that DNA controls traits

	Experiment a	Experiment b	Experiment c	Experiment d
Scientist injects mouse with	Living, harmless bacteria	Living pneumonia bacteria	Dead pneumonia bacteria	Mixture of living, harmless bacteria and dead pneumonia bacteria
Results	 Mouse lives	 Mouse dies	 Mouse lives	 Mouse dies
Meaning	Harmless bacteria cannot kill a mouse	Harmful bacteria can kill a mouse	Dead, harmful bacteria cannot kill a mouse	? This combination should not kill a mouse

a. Person with normal blood cells

```
G A G T G A G G C T T C
| | | | | | | | | | | |
C T C A C T C C G A A G
```

b. Person with sickle cells

```
G A G T G A G G C T A C
| | | | | | | | | | | |
C T C A C T C C G A T G
```

Figure 28-5 A change in just one pair of DNA bases can change a living thing's traits.

chemical from the harmful bacteria, they did not change. The scientists concluded that only DNA caused the bacteria to change. DNA is the chemical that controls traits.

How DNA Works

How does DNA direct a plant to make chlorophyll? How does DNA direct the formation of sickle-shaped or round red blood cells? How does DNA direct all traits in all living things? Let's compare DNA and how it works to a computer. A computer reads electrical messages stored within its memory bank. These messages are stored in code form. DNA also has a code. The nitrogen bases in the DNA molecule spell out a message that is stored in code form.

What determines the coded message in DNA?

The order of the nitrogen bases, A, T, C, and G, is the coded message. Figure 28-5a shows the order of bases somewhere on a section of DNA. Let's suppose this order directs the formation of round red blood cells. Now, let's suppose we look at the same place on the DNA of another person. We see the message in Figure 28-5b. This order of bases directs the formation of sickle cells.

In this example, the orders of the bases in DNA are slightly different. This difference causes the traits to be very different. One person has normal round red blood cells. The other has sickle cells.

New orders of nitrogen bases can make new messages. Each message gives the cell instructions for a different trait. Think about our alphabet for a moment. The letters of the alphabet are always the same. We put the letters into different orders to form different words. The letters w, o, and l form the word *owl*. They also can form the word *low*. Each order codes for a different meaning. How many different letters are in the DNA alphabet?

Mini Lab

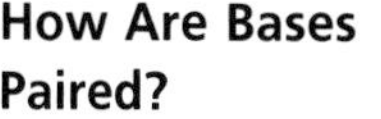

How Are Bases Paired?

Formulate a model: Mark each of 20 students with A, T, C, or G. Form a row of 10 students to show one side of DNA. Call up a student to pair with your letter. What pairs are possible? *For more help, refer to the* ***Skill Handbook,*** *pages 706-711.*

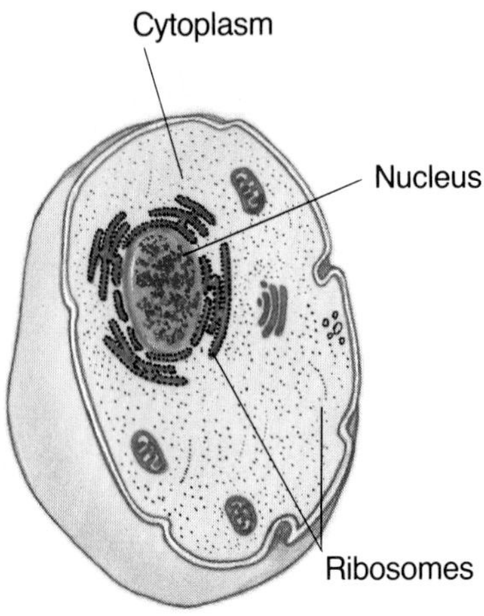

Figure 28-6 Ribosomes have a part in making proteins. Where are the ribosomes located?

Making Proteins

DNA directs the making of proteins in cells. How do cells make proteins? Remember from Chapter 2 that ribosomes are the cell parts where proteins are made. Look at Figure 28-6. Notice that the ribosomes are in the cytoplasm, not in the nucleus.

How can DNA in the nucleus control what goes on at the ribosomes? It has a helper molecule called RNA. These letters are the initials for *r*ibo*n*ucleic (ri boh noo KLAY ik) *a*cid. **RNA** is a chemical that acts as a messenger for DNA. RNA carries the coded DNA message from the nucleus to the ribosomes. The ribosome acts as a worktable for making proteins. When RNA arrives at the ribosomes, it carries a message that must be decoded before it can direct the formation of a protein. Think of the coded DNA message as being written in a certain language. This is the language of the nitrogen bases A, T, C, and G. A protein is written in another language. Somehow, the DNA language must be translated into the protein language. The **genetic code** is the code that translates the DNA language into the protein language. Once the message has been translated, the protein can be made. Figure 28-7 shows how proteins are made. Traits show up in a cell because of the kinds of proteins being made at the ribosomes. Thus, the genes on the chromosomes have sent their messages to the ribosomes to make certain traits appear.

What does RNA do in the making of proteins?

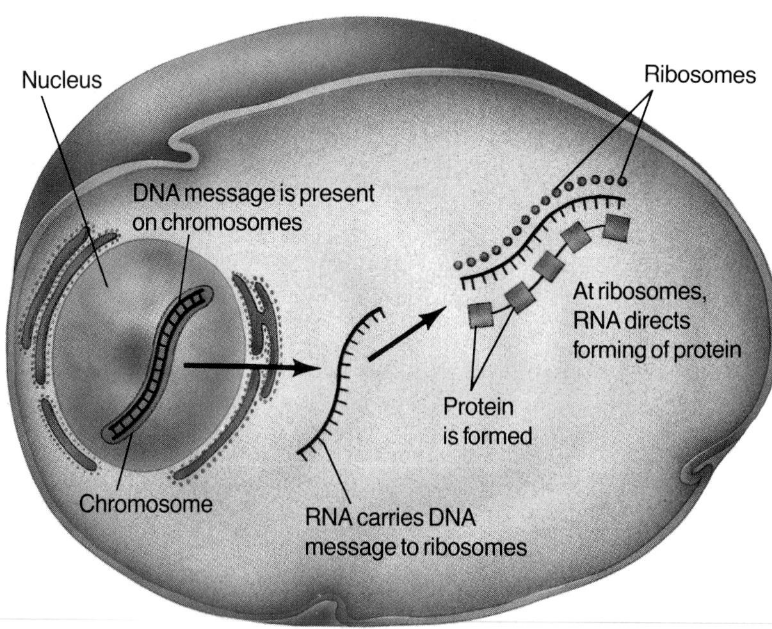

Figure 28-7 Steps needed in the cell to change the DNA message into a protein. Where are proteins made?

TECHNOLOGY

DNA Fingerprinting

In violent crimes such as rape or murder, physical evidence is often left at the scene. Blood, hair, or sperm cells may be the only evidence that the police have. A person can be arrested and found guilty long after a crime is committed. How do police know that this person is the rapist or murderer? They are able to identify the sperm or blood as his.

Every person has a special set of "fingerprints" from the moment he or she is born. These fingerprints are not on the fingertips. They are in each cell in the body. These fingerprints are the order of the bases in a person's DNA.

Today, it is possible to picture a person's DNA by using special laboratory tests. Any cell in the body can be used. In the case of a rapist, police are able to match the DNA of the sperm cells collected at the scene of the crime with the DNA of the arrested man's blood cells. Thus, police have proof that the sperm came from the man arrested. It is impossible for the sperm to have come from any other person because the order of the bases in DNA is different for each person. DNA fingerprinting will be used more and more in the future to solve crimes.

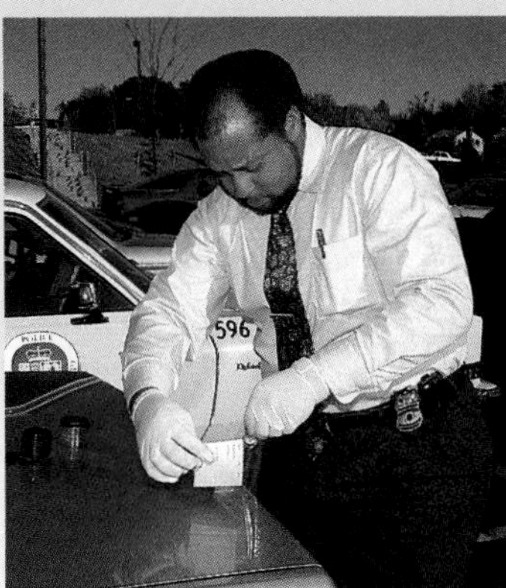

Mobile crime lab

How does the order of nitrogen bases in DNA fit into this story? The order of bases is the coded message that controls what kinds of proteins are made. The kinds of proteins you have determine what traits you have. Thus, all your traits and the traits of all living things result from the order of nitrogen bases in DNA.

How DNA Copies Itself

You have seen that DNA has many jobs. It carries the genetic messages for all living things. It controls the making of proteins. Remember that in cell reproduction, chromosomes make copies of themselves just before mitosis and meiosis. You know that chromosomes are made mostly of DNA. Now you know that DNA is what is copied. How does DNA copy itself? What is the story behind cell reproduction?

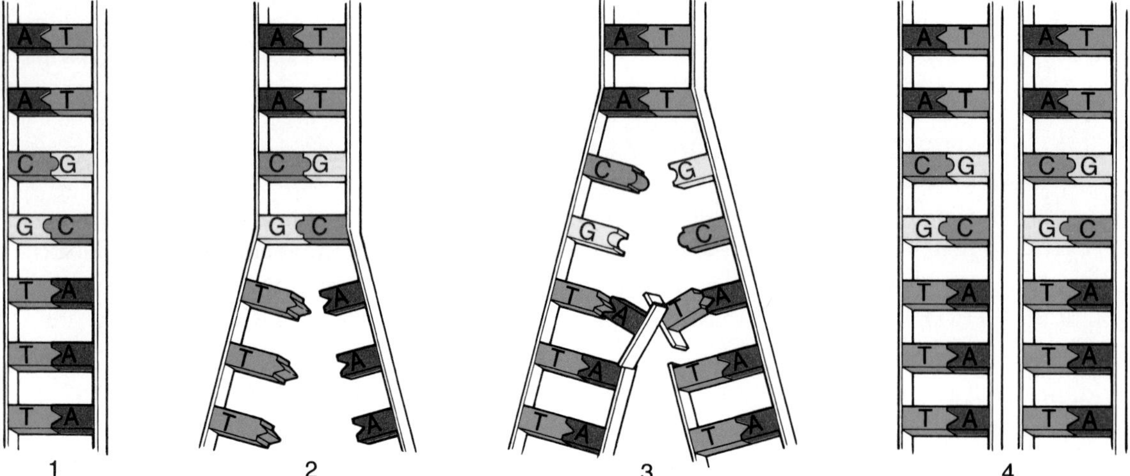

Figure 28-8 Steps used by DNA as it copies itself

Think about the ladder model of DNA again. Follow the steps in Figure 28-8 as you read how DNA copies itself.

1. DNA is ready to make a copy of itself.
2. The molecule begins to open up along its middle.
3. Loose nitrogen bases with the sugar and acid attached are present in the cell nucleus. These bases are not part of the DNA yet. They join the bases that are on the opened rungs. Look carefully and you will see that A joins only with T. C joins only with G. The loose nitrogen bases continue to join the bases on the DNA molecule.
4. Finally, two DNA molecules have formed.

Look closely at the two newly formed DNA molecules in step 4 of Figure 28-8. They are exactly alike. The order of nitrogen bases is also exactly the same as in the original molecule in step 1 of the figure. The genes on the two chromosomes are exactly the same. When a cell reproduces, the new cells that form have the same genetic message.

Check Your Understanding

1. Why is DNA important to all living things?
2. How does the order of nitrogen bases in DNA control traits?
3. How do RNA and ribosomes help DNA make proteins?
4. **Critical Thinking:** What features of the DNA structure are important if the genetic message is to be copied exactly?
5. **Biology and Reading:** Suppose one side of a piece of DNA has the bases A-A-G-C-T-C-C-T-G-C. What bases would the other side of the DNA molecule have?

28:2 How the Genetic Message Changes

In Section 28:1 you read that newly formed cells have the same genetic message as the cell from which they came. Sometimes they do not. What happens when the message changes? Can scientists change the message of living things?

Objectives

4. **Describe** how mutations occur.
5. **Describe** how cloning and breeding produce offspring with desired traits.
6. **Explain** how recombinant DNA and gene therapy can help humans.

Key Science Words

mutation
radiation
identical twins
fraternal twins
breeding
recombinant DNA
gene therapy

Mutations

Sometimes errors happen when chromosomes are copied. The bases A, T, C, and G may join incorrectly. Joining incorrectly results in a change called a mutation (myew TAY shun). A **mutation** is any change in copying the DNA message.

What happens if you hit the wrong key on a computer keyboard? Doesn't the computer get the wrong message? Cells are like the computer. A wrong base in the DNA gives the cell the wrong message. The result is that the wrong type of protein is made. This change may cause a different trait to appear.

Hemophilia is a serious blood disease that can start from a mutation. Look at the pedigree in Figure 28-9. You

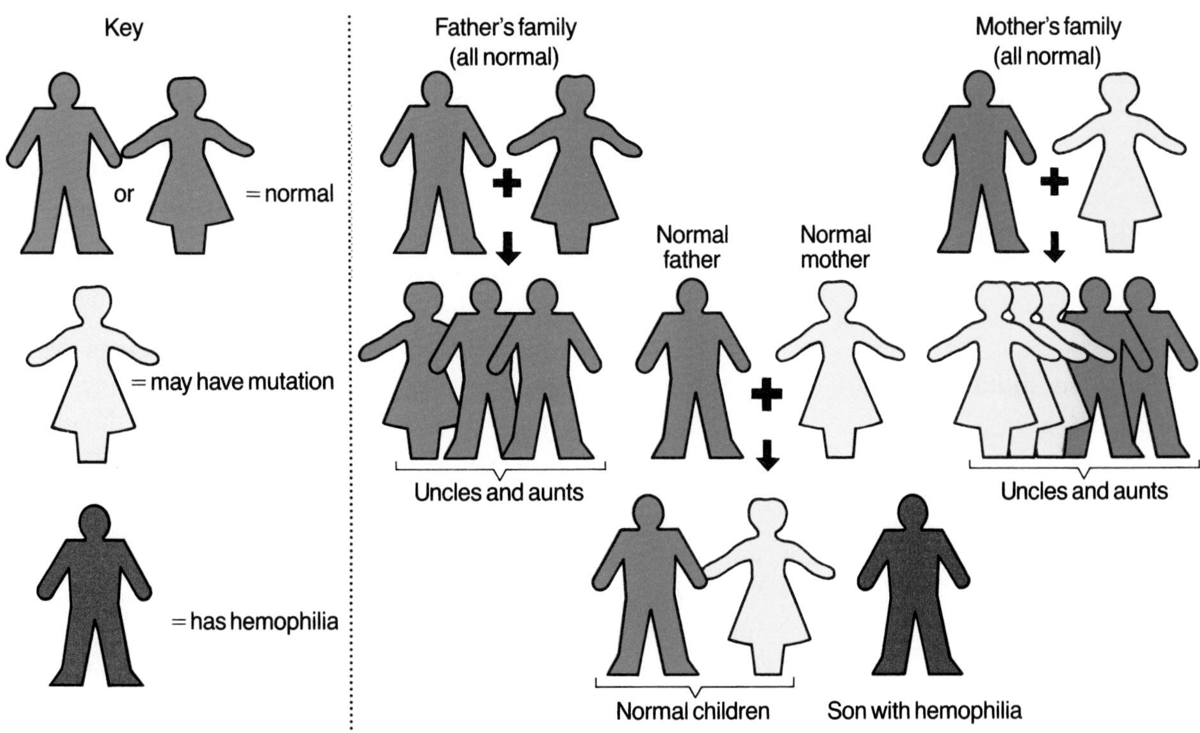

Figure 28-9 A family pedigree showing where a mutation for hemophilia may have occurred

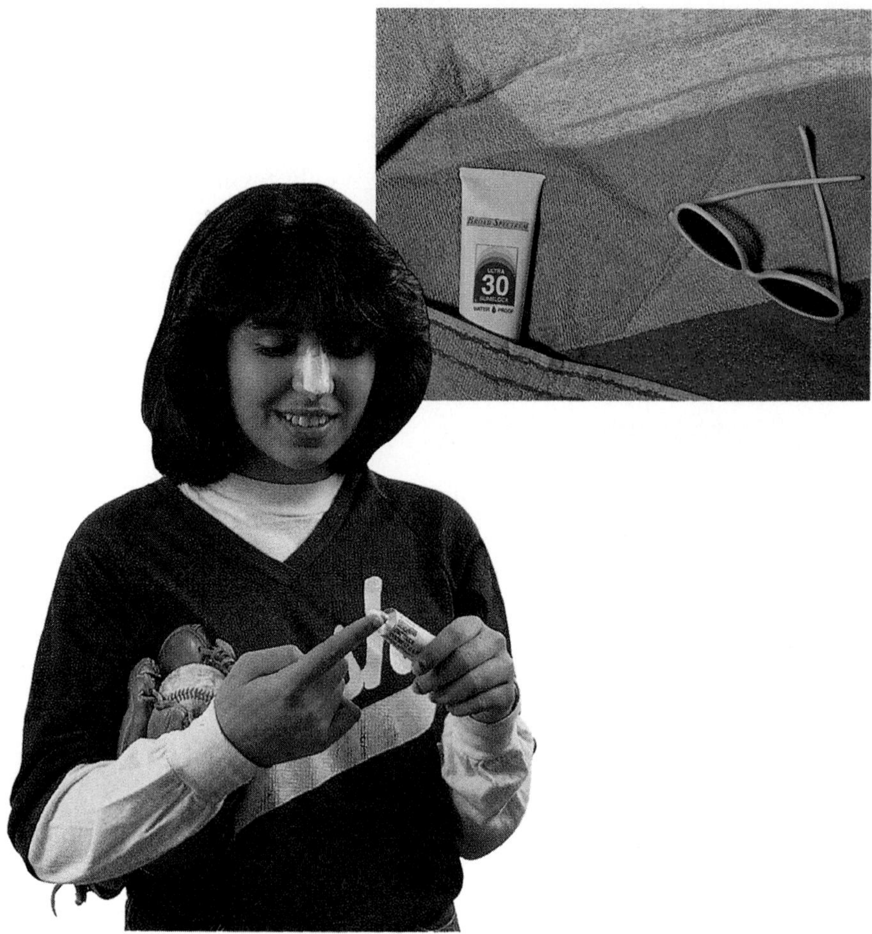

Figure 28-10 Sunscreens help protect the skin from the sun's radiation.

Mini Lab

How Does a Mutation Occur?

Formulate a model: Use parts like those in Lab 28-1 to build DNA as follows: AACGTA. Build a second model but show a mutation. Label the mutation. What can cause it? *For more help, refer to the* ***Skill Handbook,*** *pages 706-711.*

can see when the disease first appeared in this family. We expect traits to be passed from parents to children. In this family, the grandparents, uncles and aunts, and parents do not show the trait. It suddenly appears in the son. We can guess that a mutation took place when the sex cells of one of the parents were being made. It may have happened in the mother or grandmother, but it first appears in the child. A mutation causes a change in a child's trait only when it takes place in the parent's sex cells.

What causes mutations?

What causes mutations to occur? Many mutations are simply the results of copying mistakes. An error may take place in the pairing of bases when DNA is copied during cell reproduction. Other mutations may be caused by something from outside the cell. Certain chemicals and some forms of radiation (rayd ee AY shun) can cause mutations. **Radiation** is energy that is given off by atoms. The sun releases a large amount of radiation. X rays, ultraviolet light, and visible light are examples of kinds of radiation from the sun. Very powerful radiation, such as X rays and ultraviolet light, can cause mutations.

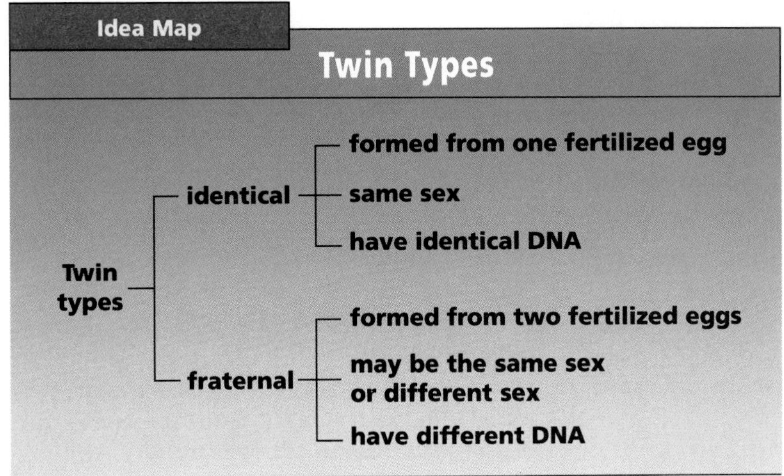

Study Tip: Use this idea map as a study guide to twin types. The features of identical and fraternal twins are shown.

Cloning

Do you have an identical twin? If you do, you have a clone. That's because **identical twins** are two children that form from the splitting of one fertilized egg. They have exactly the same genes. You know that two clones have the same genes. They are exact copies of each other. Since identical twins have the same genes, they must have the same DNA. See if you can tell which diagram in Figure 28-11 will result in identical twins. One of the diagrams shows how fraternal twins are formed. **Fraternal twins** are twins that form from two different fertilized eggs. They are not clones. Fraternal twins are no more alike than two other children in the same family would be. They don't even have to be the same sex.

Figure 28-11 Differences between identical and fraternal twins. Which diagram shows identical twins?

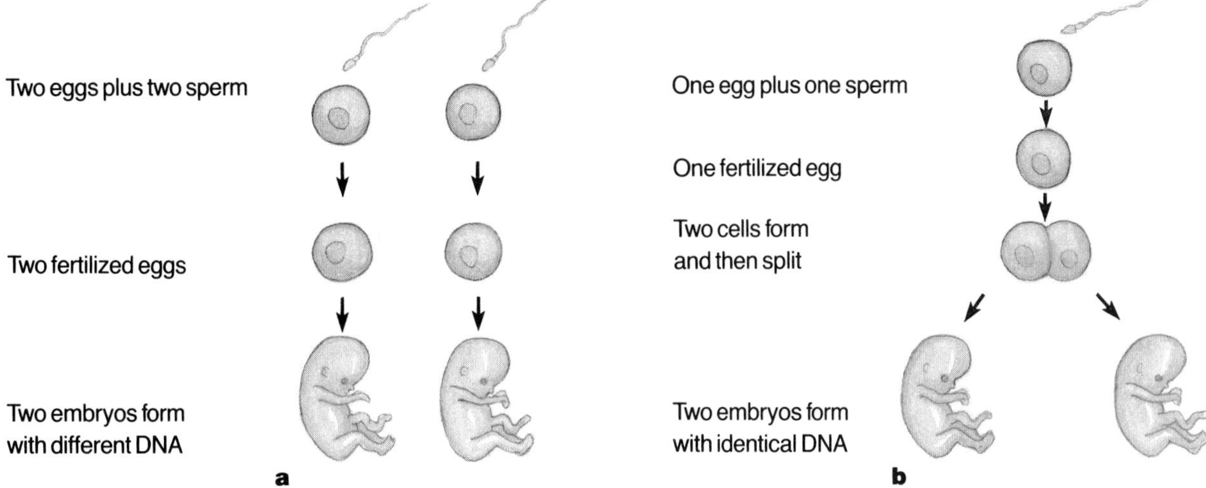

Cloning

Lab 28–2

Problem: Can a *Coleus* plant be cloned?

Skills

observe, infer, interpret data

Materials

clear plastic cup	soil
aluminum foil	water
scissors	label
Coleus plant	metric ruler

Procedure

1. Copy the data table.
2. Label a clear plastic cup with your name. Fill it with water.
3. Cover the top of the cup with foil.
4. Use scissors to punch a small hole in the center of the foil. **CAUTION:** *Always be careful when using scissors.*
5. Use scissors to cut off a small branch from a *Coleus* plant. The figure shows where to make the cut.
6. Insert the cut end of the branch through the hole in the foil. The bottom of the branch must dip into the water.
7. In your table, write today's date. Draw what the end of the cut branch looks like.
8. Record the number of students who cut a branch from the same plant.
9. **Observe:** Check the branch each day for the next few days. Look for roots that appear at the cut end.
10. When you see roots beginning to appear on your branch, record the date in your table and draw the cut end. Continue to check your branch and draw its appearance.
11. When roots are about two centimeters long, remove the plant from the cup. Pour out the water and fill the cup with soil.
12. Place your plant into the soil, being careful not to break its new roots. Moisten the soil with water.

Data and Observations

1. How many branches were cut from your original plant?
2. How many days did it take for roots to appear on your branch?

NUMBER OF STUDENTS USING SAME PLANT ____

Date ____	Date ____	Date ____
Date ____	Date ____	Date ____

Analyze and Apply

1. Explain the advantage of cloning plants.
2. **Infer:** How would the DNA in plants grown from the same original plant compare? Why?
3. **Apply:** Was the cloning method you used sexual or asexual reproduction? Explain.

Extension

Garden stores sell chemicals that are supposed to speed up root growth from the cut ends of plants. **Design an experiment** to test this.

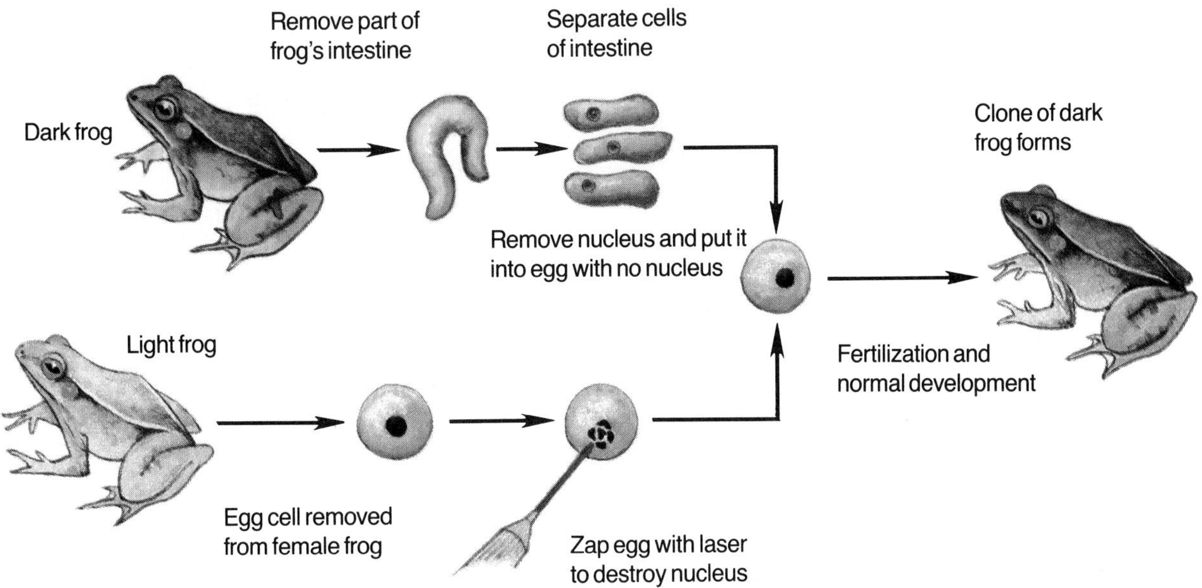
Remove part of frog's intestine
Separate cells of intestine
Dark frog
Clone of dark frog forms
Remove nucleus and put it into egg with no nucleus
Light frog
Fertilization and normal development
Egg cell removed from female frog
Zap egg with laser to destroy nucleus

Figure 28-12 Steps needed to clone a frog

Can humans clone animals? The answer is yes. Figure 28-12 shows how frogs are cloned. It is not science fiction. It has been done. Even a sheep has been cloned. Can humans be cloned in the laboratory? The technology makes it seem possible. More importantly, should it be done?

Plant and Animal Breeding

Cloning is a way of producing living things with identical desirable traits. How do we get a living thing with the desired trait in the first place?

In Chapters 26 and 27 you worked many genetic problems. They showed what to expect with certain matings. Knowing what to expect tells us which living things to breed for certain traits. **Breeding** is the bringing together of two living things to produce offspring.

Selective breeding can bring out the desired traits of living things. If breeders bring together parents with different desired traits, the result can be offspring with the best traits of both parents.

An example of this can be seen in cotton plants. A certain type of cotton plant grew in Africa, but a plant disease was destroying it. Another type of cotton plant did not get the disease, but this plant could not grow in Africa. It grew only in Central America. The two plants were bred with each other. The result was a new type of cotton plant that would grow in Africa and did not get the disease.

Bio Tip

Consumer: Humans have been breeding dogs for hundreds of years. The evidence is seen in the many different varieties that exist today.

How has breeding produced better cotton plants?

Figure 28-13 Scientists have selectively bred Brahman bulls for small size.

Brahman cattle are usually raised for meat, Figure 28-13. Recently, scientists have been able to breed smaller Brahmans. Each cow produces less meat than a larger size cow, but it also does not eat as much. This is a real help to those farmers who don't have much land for cattle to graze. Ten mini cows can graze on the same amount of grass as one large cow and can produce almost three times as much beef.

Many crops and livestock are the result of careful selective breeding. The fruits and vegetables that you buy at the grocery store are the products of selective breeding. Breeders are producing cattle that have less fat, and they are trying to breed chickens that lay low-cholesterol eggs.

Figure 28-14 Steps needed to splice a wire

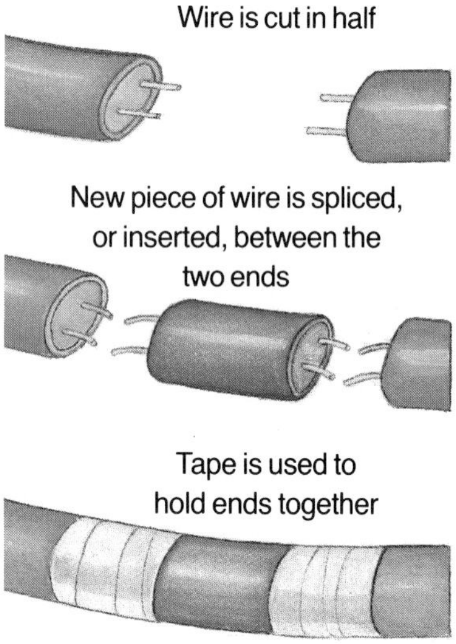

Wire is cut in half

New piece of wire is spliced, or inserted, between the two ends

Tape is used to hold ends together

Splicing Genes and Gene Therapy

Have you ever spliced a wire or broken film? If you have, then you know that the word *splice* means to insert, or join together. Figure 28-14 shows wires that have been joined together.

Today, scientists can splice genes. To do this, a bacterium receives a section of DNA from another organism, such as a human. The new DNA combination that is formed in the bacterium is recombinant (ree KAHM buh nunt) DNA. **Recombinant DNA** is the DNA that is formed when DNA from one organism is put into the DNA of another organism.

Gene splicing produces bacteria that can make certain chemicals. For example, scientists splice human genes that control the making of insulin into bacteria. The bacteria then make insulin in large amounts. Diabetics benefit

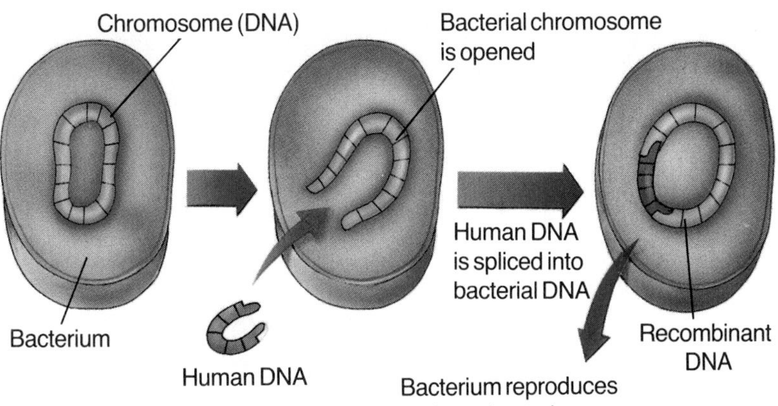
Chromosome (DNA)
Bacterial chromosome is opened
Human DNA is spliced into bacterial DNA
Bacterium
Human DNA
Recombinant DNA
Bacterium reproduces and new bacteria have the recombinant DNA

Figure 28-15 In gene splicing, a bacterium receives DNA from another organism.

because they can't make their own insulin.

Splicing of human genes into bacteria also has helped make human growth hormone. This hormone is given to children when their pituitary gland does not make enough of it. The real future of gene splicing may be in the ability to cure certain genetic diseases.

Gene therapy is the newest hopeful weapon against genetic disorders. **Gene therapy** is the adding of a healthy gene into the body of a person suffering from a disorder caused by a defective or mutated gene. This healthy gene then takes the place of the defective gene on a chromosome. How can healthy genes be delivered to chromosomes? Viruses can carry the healthy DNA or gene into cells. Once there, this healthy gene will begin to function as if it had been present in the defective cell from the start. Disorders that might be cured through gene therapy include cystic fibrosis, hemophilia, and certain muscle disorders.

Skill Check

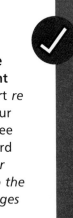

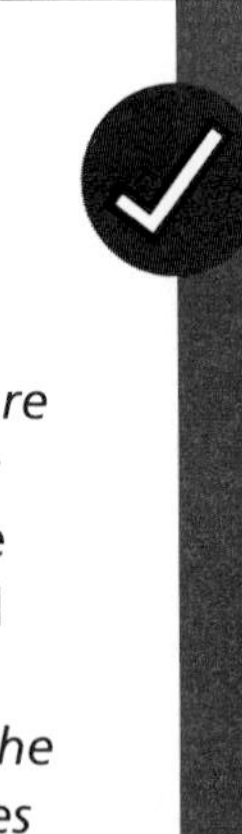

Understand science words: recombinant DNA. The word part *re* means again. In your dictionary, find three words with the word part *re* in them. *For more help, refer to the* ***Skill Handbook****, pages 706-711.*

Check Your Understanding

6. How does a mutation occur?
7. How could you produce an apple tree that was resistant to disease and produced large fruit?
8. What is being spliced to form recombinant DNA?
9. **Critical Thinking:** Could you breed an animal with exactly the combination of genes you wanted? Explain.
10. **Biology and Writing:** Do you think research on splicing genes should continue? Write a paragraph with at least three reasons that explain your answer.

Chapter 28

Review

Summary

28:1 The DNA Molecule

1. DNA is a molecule that makes up the chromosomes and genes of living things. The rungs of the ladder-shaped molecule are nitrogen bases.
2. Through experiments with bacteria, scientists showed that DNA controls traits.
3. The order of nitrogen bases A, T, C, and G forms a DNA message. RNA carries the DNA message to the ribosomes where specific proteins are formed. DNA is copied exactly to form new chromosomes just before mitosis and meiosis.

28:2 How the Genetic Message Changes

4. The DNA message may change if a mutation takes place. Mutations may be caused by some chemicals and radiation or by mistakes in copying.
5. Identical twins are clones. Clones have the same DNA and traits. Fraternal twins have different DNA. Selective breeding has produced new and better types of plants and animals.
6. Genes from one living thing can be spliced into the DNA of a different living thing. The result is recombinant DNA. Gene therapy adds healthy genes directly to an affected person to help in treating certain genetic disorders.

Key Science Words

breeding (p. 599)
DNA (p. 586)
fraternal twins (p. 597)
gene therapy (p. 601)
genetic code (p. 592)
identical twins (p. 597)
mutation (p. 595)
nitrogen base (p. 587)
radiation (p. 596)
recombinant DNA (p. 600)
RNA (p. 592)

Testing Yourself

Using Words

Choose the word from the list of Key Science Words that best fits the definition.

1. molecule that makes up genes
2. causing two living things to mate and produce desired offspring
3. form the rungs of the DNA molecule
4. a change in the DNA message
5. formed when DNA is spliced into the DNA of another organism
6. twins formed from the splitting of one fertilized egg
7. twins formed from two different fertilized eggs
8. molecule that acts as a cell messenger for DNA
9. translates the DNA language into the protein language
10. energy that is given off by atoms

Review

Testing Yourself *continued*

Finding Main Ideas

List the page number where each main idea below is found. Then, explain each main idea.

11. what is gained from selective breeding
12. what causes mutations to occur
13. how gene therapy works
14. how DNA in the nucleus can control what goes on at the ribosomes
15. the steps involved when DNA copies itself
16. what determines the DNA message
17. how identical and fraternal twins differ
18. what a model of DNA looks like
19. three ways to describe a gene
20. examples of how gene splicing has helped humans

Using Main Ideas

Answer the questions by referring to the page number after each question.

21. Which DNA bases can join with one another? Why? (p. 587)
22. How can one show that hemophilia in a family may result from a mutation? (p. 595)
23. What are the steps in gene splicing? (p. 600)
24. What are the steps in making protein, starting with DNA in the nucleus? (p. 592)
25. How has plant breeding offered a solution to the problem of diseased cotton in Africa? (p. 599)
26. What is copied when new chromosomes form just before mitosis and meiosis? (p. 593)
27. Where would you find DNA in a pine tree? Why does a pine tree have DNA? (pp. 587, 589)

Skill Review ✓

*For more help, refer to the **Skill Handbook,** pages 704-719.*

1. **Observe:** Look at Figure 28-4. Why do dead harmful bacteria injected into a rat not cause any problem to the animal?
2. **Interpret diagrams:** What happens to the original DNA molecule when it copies itself? Look at Figure 28-8 for help.
3. **Interpret data:** What part of a bacterium carries the genetic message? What is the evidence for this, as shown by the scientists who did the experiments described on page 590?
4. **Infer:** Why is it important that mutations be prevented from occurring in testes and ovaries?

Finding Out More

TECH PREP

Critical Thinking

1. Why would a mutation that appears in a person's body cell not appear in that person's offspring?
2. Should humans be cloned? Give reasons for your answer.

Applications

1. Write a report discussing the effects of the sun's radiation on the skin.
2. You wish to breed sheep in order to sell them as meat. What traits would you look for in the parents? Why?

CHAPTER PREVIEW

Chapter Content

Review this outline for Chapter 29 before you read the chapter.

Skills in this Chapter

The skills that you will use in this chapter are listed below.

- In **Lab 29-1,** you will form hypotheses, experiment, interpret data, infer, and calculate. In **Lab 29-2,** you will formulate a model, observe, and infer.
- In the **Skill Checks,** you will understand science words, define words in context, and interpret diagrams.
- In the **Mini Labs,** you will infer and make a bar graph.

Chapter **29**

Evolution

There are millions of different kinds of living things on Earth. Each and every kind of living thing is well suited to where it lives. For example, there are many different kinds of cacti. The smaller photograph below shows a close-up view of a cactus plant. Cactus plants have spines instead of leaves. The spines are an adaptation that helps to reduce water loss. Cactus plants are well suited for growing in areas where there is little water available. How do scientists explain the great variety of living things on the face of Earth? How do they explain the fact that living things are well suited to where they live? This chapter will answer these questions.

Try This!

What Does Opposable Mean? Tape your thumbs to the insides of your palms. Try to write, button a shirt, pick up a piece of paper, and turn the pages of this book. Most primates have an opposable thumb. What does opposable mean? How helpful is this trait?

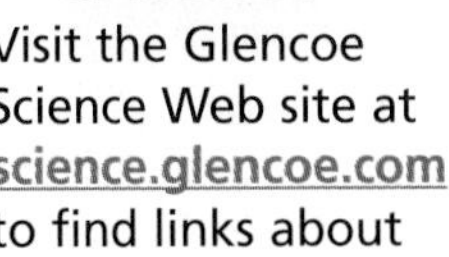

Visit the Glencoe Science Web site at science.glencoe.com to find links about **evolution.**

29:1 Changes in Living Things

Objectives

1. **Give examples** of how adaptations help organisms survive.
2. **Explain** how changes in life-forms occur.
3. **Describe** the classification and evolution of primates and humans.

Key Science Words

natural selection
species
fertile
primate
new-world monkey
old-world monkey

Living things are well suited to where they live. What exactly does this statement mean?

Adaptations

All the living things in the world have certain kinds of adaptations. An adaptation is a trait that makes a living thing able to survive in its surroundings. You first read about adaptations in Chapter 2.

Look at the foot of the bird shown in Figure 29-1a. The bird spends a great deal of its time perched in the long stalks of reeds. Note the bird's long toes. The long toes curl around small shoots and help the bird remain perched in the reeds. The bird shown in Figure 29-1b spends a lot of time in water. How does the webbing between the toes help this bird swim?

An Example of Survival

How do traits help organisms survive in their environments? What is the outcome when an organism with a certain trait is able to survive? The following example will answer these questions.

Figure 29-1 Many birds have feet adapted to perching on tree branches (a). Geese have feet adapted to swimming (b).

a

b

A group of mice lives in an area that has dark soil. Owls that eat mice also live in this area. Because dark mice blend well with the dark soil, owls cannot see them easily. Thus, the dark mice are better protected because they blend with the soil color, Figure 29-2. Their color is an adaptation, a trait that helps them survive.

Figure 29-2 Owls can easily see light-colored mice on dark soil.

Dark mice do not always have offspring that are also dark in color. Every now and then they have light-colored offspring. The light-colored offspring, in turn, have other light-colored mice. Light-colored mice are easy to spot against the dark soil. As a result of their color, the light-colored mice are usually the first to be eaten by the owls. Light-colored mice living on dark soil are poorly adapted to their surroundings. Few light mice survive. As a result, few light mice reproduce. The number of light-colored mice tends to remain low. The dark mice, however, survive and reproduce. The dark mice will continue to outnumber the light mice.

How do adaptations help living things survive?

Suppose chemical changes take place in the soil and cause it to change color. The soil becomes lighter. Now owls can spot the dark mice on the ground more easily than they can spot the light mice. As a result, owls eat more dark mice than light mice.

Dark mice are no longer adapted to these surroundings. However, light mice are adapted to the light-colored soil. Dark color has become an unfavorable trait. The balance between the two mouse types begins to change. More light mice survive and reproduce. Any dark mice that are born are more likely to be eaten. Few of them get a chance to survive and reproduce.

Adaptations

Lab 29–1

Problem: How is color an adaptation?

Skills

form hypotheses, experiment, interpret data, infer, calculate

Materials

white paper
black paper
30 white dots
30 black dots
damp paper towel

Procedure

1. Copy the data table. It should include space for 10 trials.
2. Spread 30 black and 30 white dots out on a page of white paper.
3. Write down a **hypothesis** as to whether you will be able to pick up more white dots or more black dots in 10 trials.
4. Moisten all your fingers with a damp paper towel. Quickly touch one dot with each finger. The dots should stick to your fingers. Do this as quickly as you can.
5. Count the number of black dots and white dots you picked up.
6. Record your totals as Trial 1.
7. Return all the dots to the white paper.
8. Repeat steps 4 through 7 nine more times.
9. Switch to the black background paper. Spread the dots on the black paper.
10. Write down a new **hypothesis.** Repeat steps 4 through 7 for ten trials.
11. Total each column in your table.

Data and Observations

1. Which color dots, black or white, were picked up more often on the white background?
2. Which color dots were picked up more often on the black background?

Analyze and Apply

1. **Check your hypotheses:** Are your hypotheses supported by your data? Why or why not?
2. **Infer:** What was the adaptation in this activity?
3. **Apply:** In nature, how does color help as an adaptation?

Extension

Calculate: Put the data for the class on the board. Calculate the averages for each category.

	WHITE BACKGROUND		BLACK BACKGROUND	
TRIAL	BLACK DOTS	WHITE DOTS	BLACK DOTS	WHITE DOTS
1				
2				
3				
9				
10				
Total				

Natural Selection

We can now ask an important question using the mouse and owl story. What determined which mouse was better adapted to its surroundings? The owls determined, by eating certain mice, which color was an adaptation for survival. Only mice that are not eaten survive to reproduce. On the dark soil, more dark mice survived because owls did not see them. On the light soil, more light mice escaped the owls. **Natural selection** is the process in which something in a living thing's surroundings determines if it will or will not survive to have offspring. In natural selection, something in nature does the selecting. In our example, the owls did the selecting. When the soil became lighter, the group of mice changed from mostly dark to mostly light. The change was the result of natural selection.

Living things that are suited to their surroundings survive. They will be the ones most likely to reproduce. Their traits will be passed on to their offspring. Living things that are not suited to their surroundings won't survive. They won't reproduce, they won't have offspring, and their traits won't be passed on.

Why wouldn't a bright red frog survive in a muddy pond? Can you explain why more mud-colored frogs are likely to survive in muddy ponds than red frogs? How can web-toed frogs survive in water better than frogs with toe pads?

Skill Check

Define words in context: Read this statement: "The dinosaurs became extinct due to natural selection." What does natural selection mean as used in this sentence? *For more help, refer to the* ***Skill Handbook,*** *pages 706-711.*

Mutations

Adaptations are traits that help living things survive in their environments. Recall from Chapter 26 that traits are controlled by genes. Thus, adaptations are controlled by genes.

What is a source for new traits that help living things survive? Many new traits come from mutations. Remember, a mutation is a change in the DNA code. Mutations may supply living things with sources of new traits. Thus, they may supply new adaptations.

Are all mutations helpful for survival? No, some are harmful. For example, a mutation causes a change in the gene that controls fur color. Inheriting this gene caused the deer in Figure 29-3 to have white fur. The deer no longer blends in with its surroundings. It can be seen more easily by its enemies and may be eaten.

Figure 29-3 A white deer may be more easily seen by its enemies than a brown deer.

Bio Tip

Consumer: Wearing a lead apron when you have dental X rays helps prevent mutations of your reproductive cells.

What if the deer had a mutation that gave it extra long legs? The long legs might help it run faster from its enemies and escape being eaten. This new trait would probably help the deer in its surroundings. The trait could be passed to the offspring, thus increasing the number of individuals with the trait.

Mutations are natural events. Mutations appear in every living thing. These changes in genes may be helpful, harmful, or have no effect at all.

Species Formation

Chapter 3 described a species as the smallest group of living things in classification. There is another way to describe a species. A **species** is a group of living things that can breed with others of the same species and form fertile offspring. **Fertile** means being able to reproduce by forming egg or sperm cells. For example, the plants shown in Figure 29-4c are very similar in appearance. Yet, they can't produce fertile offspring if they breed with one another. Another example is shown in Figures 29-4a and 4b. These macaws look very much alike. They belong to different species, however. If they breed with each other, they won't form fertile offspring.

Figure 29-4 The macaws (a,b) are different species, as are the plants (c).

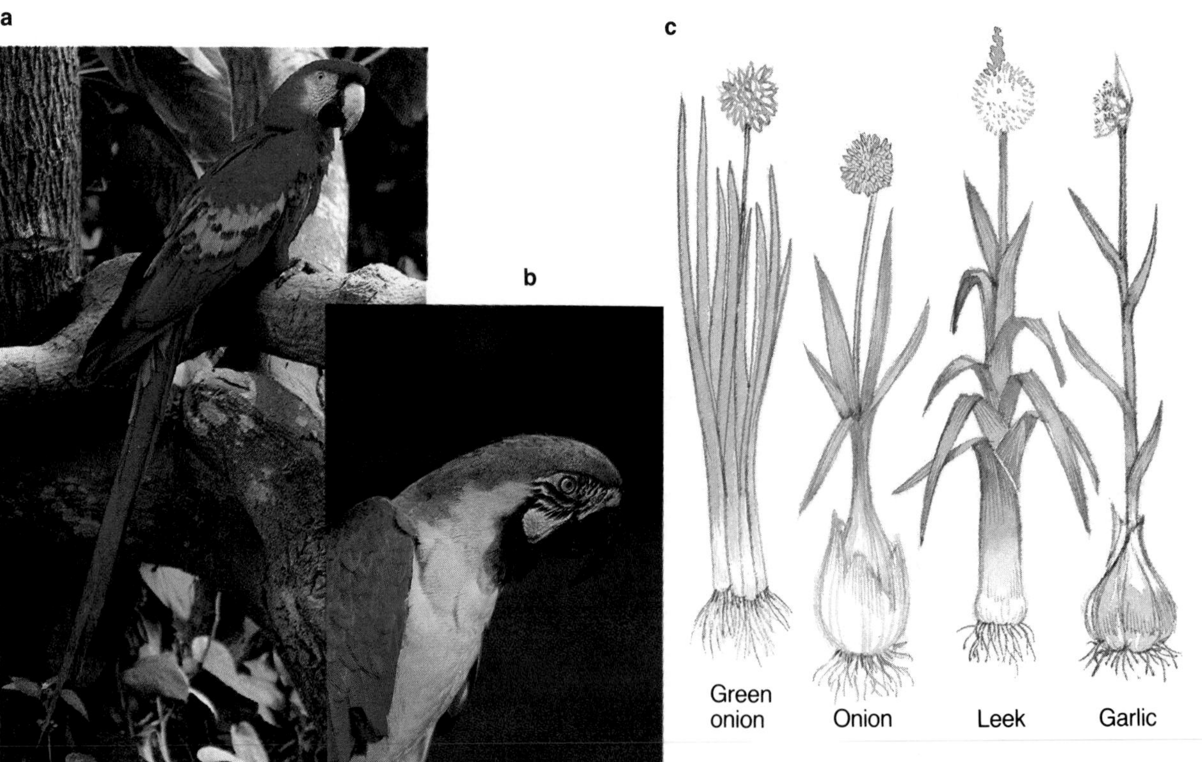

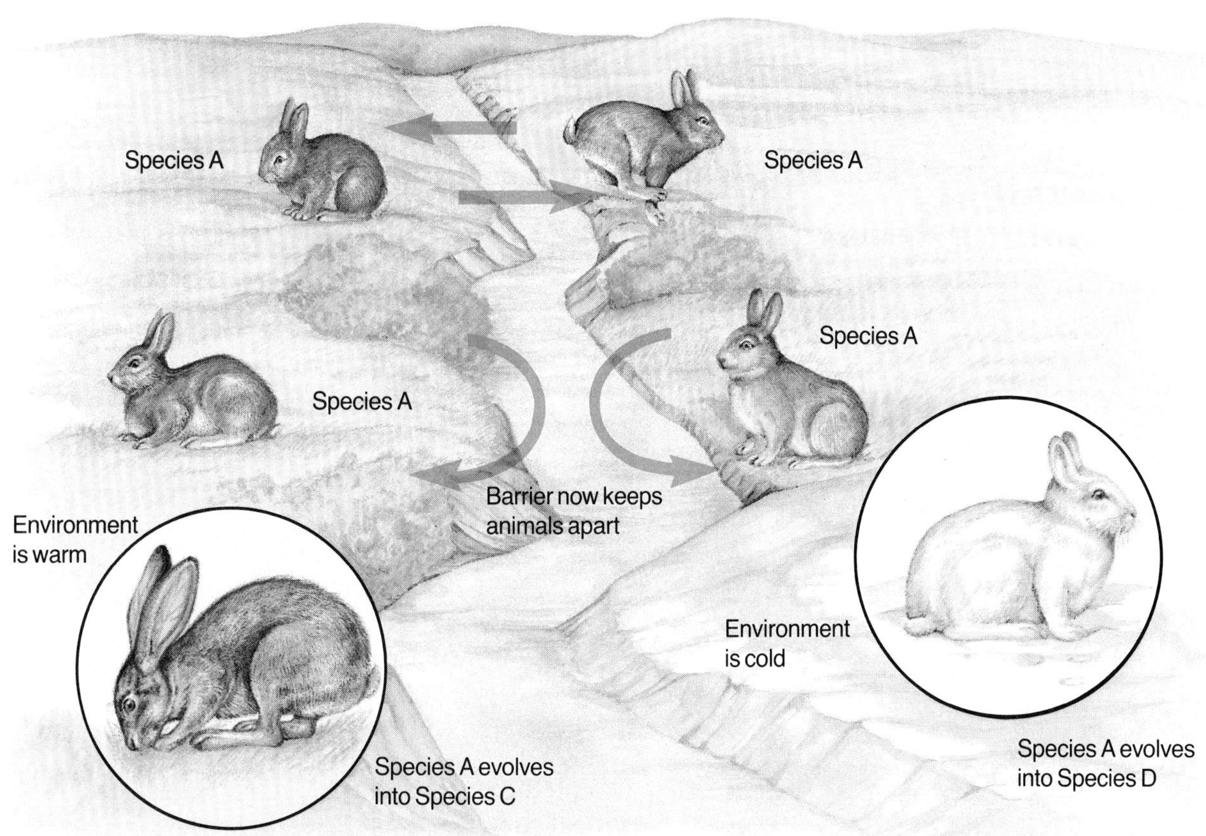
Species A
Species A
Species A
Species A
Barrier now keeps animals apart
Environment is warm
Environment is cold
Species A evolves into Species C
Species A evolves into Species D

Figure 29-5 A barrier can lead to the formation of new species.

Now we have a problem. Only members of the same species can breed and form offspring. Yet, new species are constantly appearing on Earth. How can new species appear? Follow the events shown in Figures 29-5 to see how this is possible. Let's start with a group of animals living on either side of a shallow stream. The animals at the top of Figure 29-5 are all of the same species. They can wade across the stream easily, but they are unable to swim. Due to a flood, the stream becomes a very wide river. It remains that way. The animals become separated into two groups. Being unable to swim, they are unable to cross the river. They continue to live apart for thousands of years. Note that the living conditions on each side of the river are different. During the time of separation, natural selection has taken place in each group. In each group, individuals lacking traits favorable for the new environment have died. Individuals with the favorable traits have survived and reproduced. The two groups gradually become different because their environments are different. In time, each group may become a different species.

How does the formation of a barrier lead to new species?

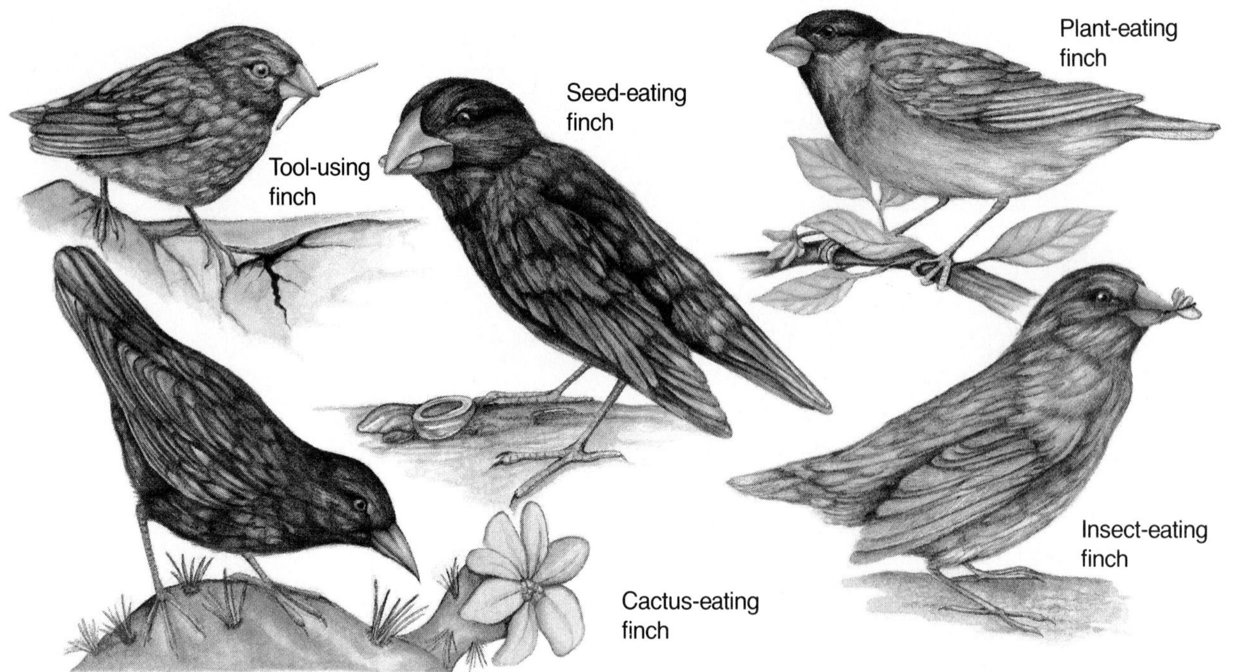

Figure 29-6 The finches on the Galapagos Islands are adapted to eating different kinds of food.

In the example just given, three events led to the development of two new species. First, a barrier formed that separated members of a species. The barrier could have been a river, ocean, new mountain, glacier, or a lava flow. Second, the animals found themselves living in different environments. Third, the groups began to show different traits as a result of natural selection. The two groups in time became two different species. As a result, they would not be able to breed and form fertile offspring if brought back together.

The finches that live on the Galapagos Islands are a well-known example of the forming of new species. On the Galapagos are several species of finch. Each species has a different beak shape. Some finches have thick beaks and are adapted to eating seeds. Some have small beaks and are adapted to eating insets. How did the finches come to live on the islands? The ancestor of the Galapagos finches probably flew to the islands from the mainland of South America. New species began to evolve when the finches spread out over the islands. The different groups of finches didn't come into contact with one another for a long time. Over time, the different groups became adapted to their new environments. They also became less like one another. A single finch ancestor had evolved into many different species.

Primate and Human Evolution

Fossils are very important when it comes to tracing the evolution of humans. They provide us with evidence of past life. Before looking at human evolution, a brief review of human classification will help. Some of the categories for humans are as follows: Phylum Chordate, Class Mammal, Order Primate.

The primate order is the one in which monkeys, apes, and humans are classified. **Primates** are mammals with eyes that face forward, a well-developed cerebrum, and thumbs that can be used for grasping. About 45 million years ago, primates evolved into two main groups, new-world monkeys and a second group. **New-world monkeys** have a tail that can grasp like a hand and nostrils that open upward. Howler and spider monkeys are examples of new-world monkeys. The second group that evolved is the ancestor of the old-world monkeys. It is also the ancestor of the group that evolved into apes and humanlike life-forms. **Old-world monkeys** can't grasp with their tails, if they have one, and their nostrils open downward. Baboons are old-world monkeys. Apes don't have tails and include gorillas and chimpanzees. Figure 29-7 shows the old-world monkeys, new-world monkeys, apes, and how they are related.

Skill Check

Understand science words: primate. The word *primate* comes from the word *prime* meaning first. In your dictionary, find three words with the word *prime* or word part *prima* in them. *For more help, refer to the* ***Skill Handbook,*** *pages 706-711.*

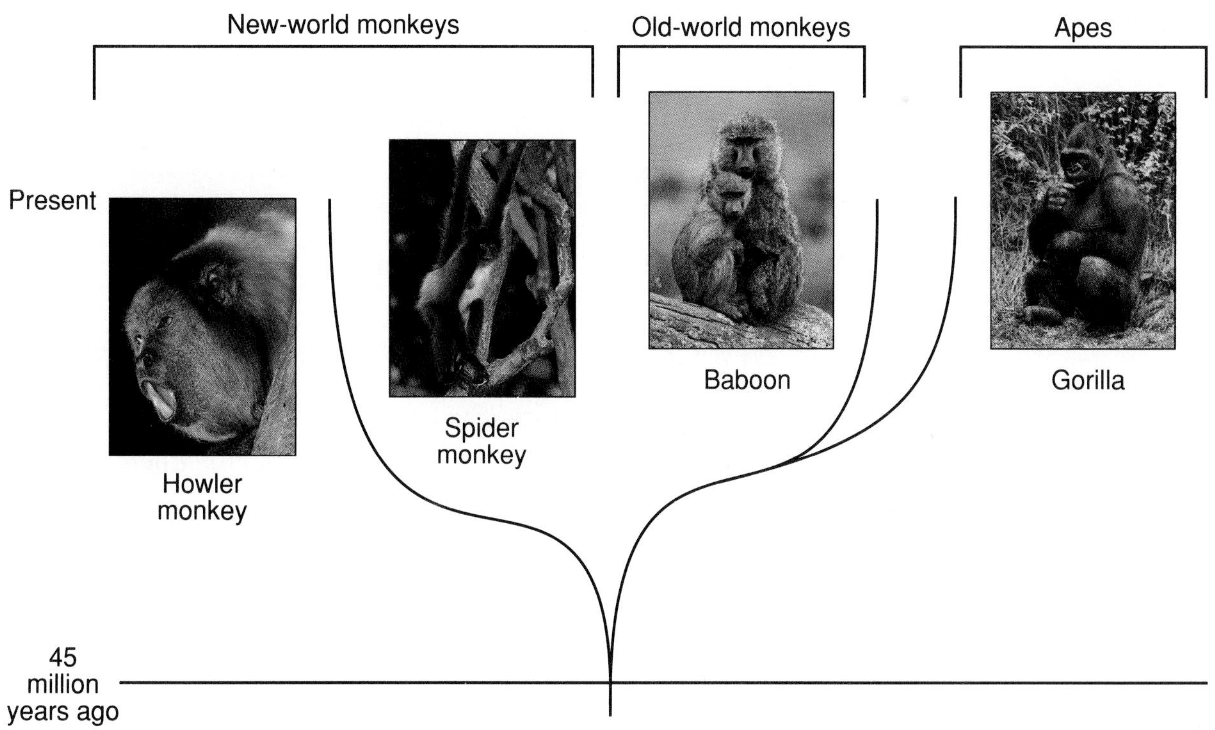

Figure 29-7 Primate groups include new-world monkeys, old-world monkeys, and apes.

Figure 29-8 Early humanlike forms include *Australopithecus* (ah stray loh PITH uh cus), *Homo habilis,* and Neanderthals. *Homo habilis* is called "handy man" because they used stone tools.

How are humans primates?

Humanlike ancestors first appeared around three million years ago. These ancestors walked upright, just as modern humans do. However, they were much shorter than modern humans. Several different humanlike groups evolved, but they all became extinct. Figure 29-8 shows these humanlike groups. Two groups of *Homo sapiens* have lived on Earth. *Homo sapiens* is the only human life-form alive today. One of the *Homo sapiens* groups was known as Neanderthal (nee AN dur thawl) man. Neanderthal man was shorter than modern humans and had thicker bones. Neanderthal man became extinct, and the second *Homo sapiens* group evolved into modern humans.

Check Your Understanding

1. How does color of a mouse affect its survival?
2. How can two species develop from one species?
3. How are new-world monkeys and old-world monkeys alike?
4. **Critical Thinking:** People who breed dogs sometimes end up with breeds that are different from any others. Are these dog breeds new species?
5. **Biology and Math:** In humans, the rate of mutation is about 1 in 100 000. You have about 100 000 genes. What is the chance that you have a gene that has a mutation and is, therefore, different from any of the genes your mother or father have?

29:2 Explanations for Evolution

How can we explain that life-forms have changed with time? What is the evidence that life-forms had a common beginning?

Objectives

4. **Communicate** Darwin's main ideas.
5. **Describe** evidence that supports evolution.

Key Science Words

variation
competition
evolution
fossil
extinct
sedimentary rock
vestigial structure

Darwin's Work

Much of what we have stated so far about adaptations and natural selection is not new. It was said over 100 years ago by Charles Darwin. When he was young, Darwin made a voyage around the world on a ship. During the trip, he observed many kinds of plants and animals and gathered examples of them. On some islands he collected living things not found anywhere else on Earth. Darwin saw that these living things were similar to life present in other parts of the world. For 20 years after his trip, Darwin studied the material he had collected. Finally, in 1859, he wrote a book explaining evolution (ev uh LEW shun) and his theory of natural selection.

Darwin made a number of important points in his book. We shall summarize several of his more important ideas.

1. Living things overproduce. More offspring are produced than survive. A single cottonwood tree forms thousands of seeds, Figure 29-9a. Frog eggs, shown in Figure 29-9b, are produced in the hundreds.

Figure 29-9 Cottonwood trees (a) and frogs (b) are examples of living things that produce more offspring than can survive.

a

b

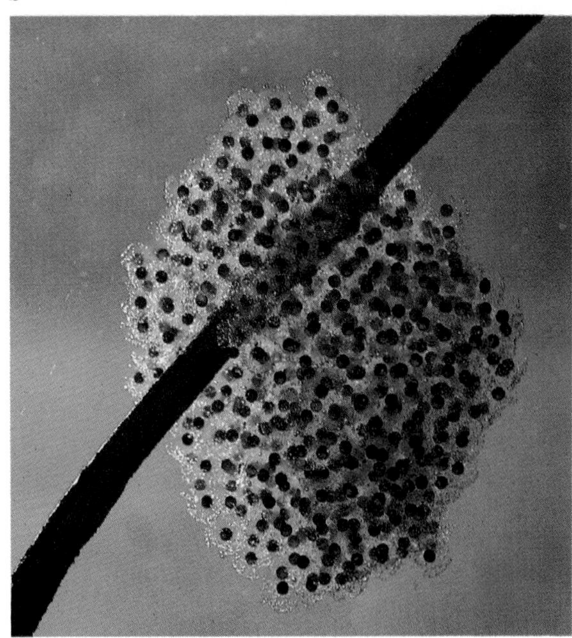

Figure 29-10 Tiger moths show variation in their wing patterns.

2. There is variation (ver ee AY shun) among the offspring. A **variation** is a trait that makes an individual different from others of its species. Each living thing does not appear exactly like all the others. Some of the differences between individuals are inherited. Figure 29-10 shows variations in color and pattern of tiger moth wings.

What does it mean that living things struggle to survive?

3. There is a struggle to survive. There are more living things than there are resources to go around. This results in competition. **Competition** is the struggle among living things to get their needs for life. Young pine trees compete for light, water, and soil nutrients. Rabbits compete with other rabbits for food, shelter, and mates.

4. Natural selection is always taking place. Individuals that have less desirable traits are less fit. They reproduce fewer offspring. Individuals that have desirable traits are more fit. They reproduce more offspring. The organisms alive today are the ones that are better suited to their surroundings. The traits, or variations, that make them more fit are the ones they inherited and will, in turn, pass to their offspring.

Darwin realized that species of organisms are always changing. He knew that the changes in species do not occur quickly. Darwin's studies led him to form the theory of natural selection to explain evolution. **Evolution** is a change in the hereditary features of a group of organisms over time. When a species changes through time, it is said to have evolved.

Mini Lab

How Do Peas Vary?

Make a bar graph: Find the average diameter of the peas in 10 pea pods. Make a bar graph of your data. How do your data relate to Darwin's statement about variation? *For more help, refer to the **Skill Handbook**, pages 715-717.*

Fossil Evidence

What evidence supports evolution? Some evidence of evolution comes from fossils. **Fossils** are the remains of once-living things from ages past.

A fossil may be a print of a leaf. It may be a footprint of an animal. It could even be a skeleton. A fossil could be an animal trapped and frozen in ice. Or a fossil could be an insect trapped in hardened plant sap.

When living things from the past are compared to living things today, we can see that change has occurred. For example, Figure 29-11 shows the fossil remains of an extinct (ihk STINGT) animal. An **extinct** life-form is one that no longer exists.

Fossils are found in Earth's crust. They are present in sedimentary rocks. **Sedimentary rocks** form from layers of mud, sand, and other fine particles. The mud, sand, and fine particles are called sediments. These sediments form at the bottom of seas. Many animals and plants die and settle to the bottoms of oceans, lakes, or ponds with the sediments. These sediments change into rock over millions of years. Fossils form within these sediment layers. Fossils give us a record of what types of living things were on Earth in the past. Scientists can tell how old fossils are by dating them. Being able to date fossils gives scientists an idea of the history of life on Earth.

Mini Lab

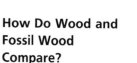

How Do Wood and Fossil Wood Compare?

Infer: Compare the density of a piece of wood with the density of a piece of fossil wood. Which has higher density? Why? *For more help, refer to the **Skill Handbook**, pages 706-711.*

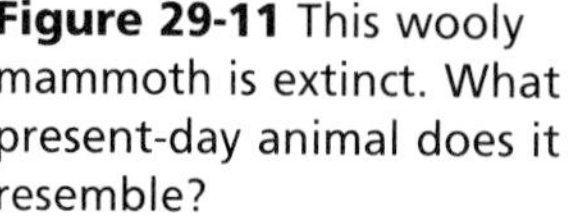

Figure 29-11 This wooly mammoth is extinct. What present-day animal does it resemble?

Fossil Prints

Lab 29–2

Problem: How is a fossil print made in rock?

Skills

formulate a model, observe, infer

Materials

plastic dish	warm water
100 mL graduated cylinder	plastic powder
leaf	metric ruler

Procedure

1. Copy the data table.
2. List as many traits as you can for the actual leaf. You may want to measure certain leaf parts with a ruler.
3. Measure 60 mL of plastic powder with a graduated cylinder. Pour the powder into a small dish as in the figure.
4. Slowly add 36 mL of warm water to the powder. Use your hands to mix the water and powder until it is claylike. The claylike substance represents soft rock.
5. Flatten the soft rock so that it covers the dish bottom. Smooth out the top surface.
6. Place a leaf onto the soft rock's surface.
7. Gently press the leaf into the soft rock.
8. Wait five minutes. Then remove the leaf.
9. Allow the soft rock to harden for a few more minutes.

10. Examine the leaf "fossil" print. Compare it to the original leaf.
11. **Observe:** List as many traits as you can for the leaf "fossil" in your data table.
12. Clean and return your materials to their proper place.

Data and Observations

1. How are the actual leaf and leaf "fossil" alike? How do they differ?
2. Match the following steps with the numbered steps in the procedure.
   (a) A soft soil such as clay is present.
   (b) A leaf falls onto the clay surface.
   (c) The leaf gets pressed down into the clay.
   (d) The clay hardens. The leaf breaks down and disappears.
   (e) A print of the leaf remains on the hard rock.

TRAITS	
Actual leaf	Leaf "fossil"

Analyze and Apply

1. How do fossils give information about past ages?
2. **Infer:** List several examples of what may become a fossil.
3. **Apply:** How long would steps 8 and 9 take if this were happening in nature?

Extension

Observe: Examine pieces of marble for fossils. How did the fossils get into the marble?

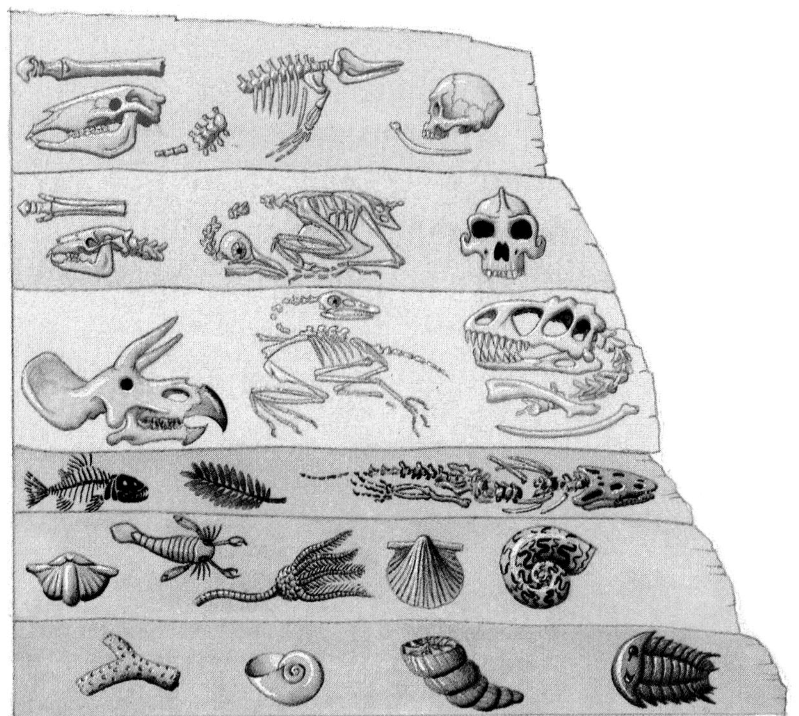

Figure 29-12 Fossils in the lower layers of sedimentary rock are older than those found near the top layers.

Figure 29-12 is a side view of sedimentary rocks in Earth's crust. Where is the oldest layer of rock located? Because it was the first layer to form, it is on the bottom. Younger layers settle on top of the oldest layer. So, as you move up the layers, the rocks get younger. This is similar to stacking newspapers. Suppose you always put the most recent newspaper on top of the stack. As long as the stack is left alone, the oldest paper will always be on the bottom. The newest paper will be on top. Fossils found in the lower layers of rock are older than those found in the upper layers.

Compare the fossils in the bottom layers to those near the top. How many fossil forms found toward the bottom and middle layers are still alive today? Not too many. How many fossil forms found in the top layers are still alive today? Certainly more than in the middle and bottom layers.

Skill Check

Interpret diagrams: Study Figure 29-12. In which rock layers are the more simple fossils found? *For more help, refer to the* ***Skill Handbook,*** *pages 706-711.*

Other Evidence

In Chapter 3, you studied that comparing the origins of body structures and comparing body chemistries are ways to determine relationships among different species. Each of these comparisons is evidence of evolution.

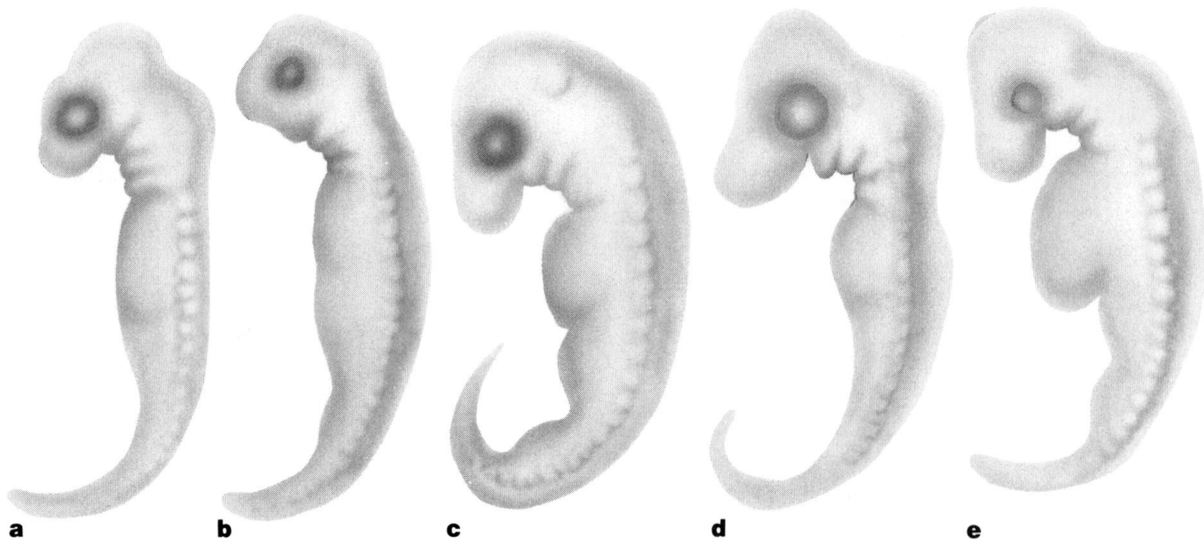

Figure 29-13 Embryos of a fish (a), frog (b), turtle (c), bird (d), and rabbit (e) look similar.

Another comparison also provides evidence of evolution. Figure 29-13 shows five different animal embryos. Fish, frog, turtle, bird, and rabbit embryos are shown. Which embryo is which? You really can't tell. They all look very much alike at this stage in their development. As they get older, you will be able to tell which embryo is which. How is the similar appearance of embryos evidence of evolution? All five animals are chordates. They have a common ancestry. The embryos share some of the same traits from their common ancestry. These traits result in all the embryos looking similar.

What other evidence is there of evolution? If later life-forms evolved from earlier ones, wouldn't the later forms have something in common with the earlier forms? Read the following examples for the answer to this question. Early life-forms are made of cells. So are later life-forms. Early life-forms have DNA as part of their chromosomes. So do later life-forms. The gene code in early life-forms is made of nitrogen bases, A, T, C, and G. Later life-forms have the same kind of gene code.

How are vestigial structures evidence of evolution?

Have you ever wondered what that little pink lump is in the corner of your eye? What does it do? What does your appendix do? Both of these body parts are called vestigial (vuh STIJ ee ul) structures. A **vestigial structure** is a body part that no longer has a function. How is a vestigial body part evidence of evolution? Most of these body parts do have jobs in other animals. For example, in many mammals the appendix helps digest food. Rabbits are examples of animals with an appendix that still works to digest food.

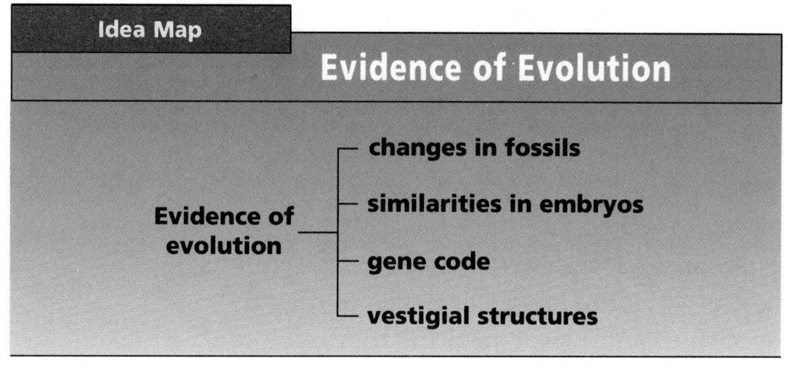

Idea Map
Evidence of Evolution
Evidence of evolution
changes in fossils
similarities in embryos
gene code
vestigial structures

Study Tip: Use the idea map as a study guide to the kinds of evidence that show that species evolve over time.

This part of a rabbit intestine helps to break down plant material, the rabbit's chief source of food. The pink lump in your eye is all that is left of a third eyelid. In other animals the third eyelid is usually very thin and covers the entire eye. Frogs and turtles are examples of animals with third eyelids. The third eyelid of these animals protects the eye while the animal is under water. Birds, fish, and reptiles also have third eyelids that protect the eye.

We may not have any use for our appendixes or what remains of our third eyelid. These structures, however, are still useful to related animals. Thus, the presence of vestigial structures is evidence of a common ancestor for us and related animals. We still have the genes for appendixes and third eyelids, even though we don't use these structures. Animals that are related to us also have the genes for these traits.

Check Your Understanding

6. List and describe the four major points of Darwin's theory of natural selection.
7. Give several examples of fossils and tell where they are formed.
8. Describe how the embryos shown in Figure 29-13 are alike.
9. **Critical Thinking:** An athlete breaks her leg in a high jump. Years later, she has a son who walks with a slight limp. Is this an example of evolution? Why or why not?
10. **Biology and Reading:** What kinds of evidence do scientists have for evolution?

APPLYING TECHNOLOGY

TECH PREP

A Fossil Hunt

Have you ever found a fossil? There is a certain thrill about finding evidence of life long gone. The presence of fossils in rock layers can be more useful than just showing us what life forms looked like millions of years ago. Fossils may also be used by scientists to judge the age of the rock layers in which they are found.

Identifying the Problem

You are a geologist who works for an oil company. After locating some fossils in sedimentary rock layers in Wyoming, you want to determine the age of these rock layers. Why would you want to know the age of a layer of rocks? As a geologist who is looking for rocks that may contain oil deposits, age is an important clue.

Collecting Information

One method for determining age of a rock layer is through the presence of index fossils. Index fossils are certain fossil types that are known to have lived during certain time periods on Earth. Thus, the age of these index fossils has been established. If a certain index fossil is found in a rock layer, the age of that layer can be determined by the presence of the index fossil buried in it.

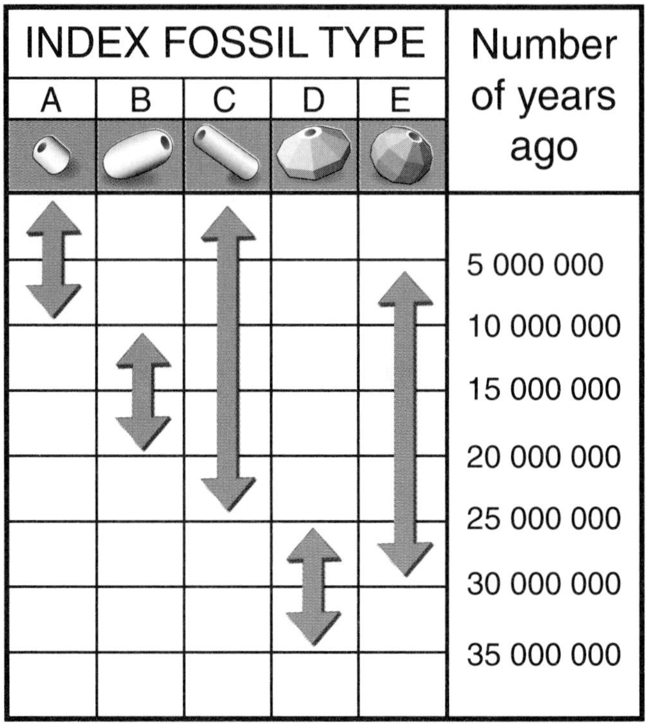

INDEX FOSSIL TYPE					Number of years ago
A	B	C	D	E	
					5 000 000
					10 000 000
					15 000 000
					20 000 000
					25 000 000
					30 000 000
					35 000 000

The table on this page shows the shapes of five beads. Each bead represents a different index fossil. The table gives the ages of the index fossils. If you find one, two, or even three index fossils in a rock layer, you should be able to determine the age of the rock from where the index fossils were found.

Technology Connection

Our dependence on oil as fuel to generate electricity may be decreasing. Solar cells have improved so much in the amount of energy they can trap that they can now compete with fossil fuels such as oil. Two large energy companies, Enron and Amoco, have joined together to build a solar plant in Nevada. This plant will generate enough energy to supply a city of 100 000 people.

Carrying Out an Experiment

You will be given three groups of "fossils" collected from three layers of rock. All the beads represent fossils, but only the special shapes shown in the table represent index fossils.

1. Make a data table in which you will record the number of your fossil group, the shape of the index fossils found, and the age of the rock layer from which the fossils were taken.
2. Examine the first group of fossils labeled #1. Look for the presence of index fossil shapes according to the table.
3. Record which index fossils are present in the group.
4. Using the rock ages on the right side of the table, record the age of the unknown rock sample from which the group of fossils was taken. For example, if you found index fossil shapes A, C, and E in the group, then the rock sample would be between 5 million and 10 million years old. Why? Because this is the only time period in which all three index fossils are found together.
5. Return all fossils to their container before starting to examine the next group of fossils.
6. Repeat steps 2-5 with two more groups of fossils marked #2 and #3.

Career Connection

- **Petroleum and Natural Gas Exploration Worker** Works for an oil company or an oil drilling company; may work on land or on an ocean platform
- **Field Assistant** Prepares, classifies, and sorts rock and fossil specimens; cleans, washes, and prepares samples for further study
- **Paleontologist** Studies fossil plants and animals in order to trace their evolutionary trends; categorizes fossils as to their age and location; looks for petroleum bearing rock formations

Assessing Your Results

Record the ages for rock samples #1, #2, and #3 in your data table. Explain why the age for rock sample #1 could *not* be between 5 million and 10 million years old. Explain how it might be possible for a rock sample to not contain an index fossil even though it is known that the fossil should be present.

Chapter 29

Review

Summary

29:1 Changes in Living Things

1. All living things have adaptations that help them survive.
2. Changes in life-forms are the result of natural selection, mutations, and changes in the environment.
3. Primates evolved about 45 million years ago into two main groups. One of these groups was the ancestor of apes and humans. Humans evolved about three million years ago.

29:2 Explanations for Evolution

4. Darwin described his ideas about evolution over 100 years ago. His four main points were: living things overproduce, variation occurs in living things, there is a struggle to survive, and those that survive to reproduce are the ones that have traits best suited for their surroundings.
5. Evidence of evolution comes from fossil remains, similarity of embryos, similarity between chemical makeup of early and present-day life-forms, and from the presence of vestigial structures.

Key Science Words

competition (p. 616)
evolution (p. 616)
extinct (p. 617)
fertile (p. 610)
fossil (p. 617)
natural selection (p. 609)
new-world monkey (p. 613)
old-world monkey (p. 613)
primate (p. 613)
sedimentary rock (p. 617)
species (p. 610)
variation (p. 616)
vestigial structure (p. 620)

Testing Yourself

Using Words

Choose the word from the list of Key Science Words that best fits the definition.

1. a change in hereditary features of a group over time
2. struggle among living things for their needs
3. species that is no longer living
4. when something in the environment determines whether or not a living thing survives
5. a trait that makes a living thing different from others in its species
6. a type of rock formed from layers of mud and sand
7. remains of any once-living thing
8. the appendix or the third eyelid of humans
9. mammal that has eyes that face forward and grasping thumbs
10. being able to reproduce by forming egg or sperm cells

Testing Yourself *continued*

Finding Main Ideas

List the page number where each main idea below is found. Then, explain each main idea.

11. why the oldest layer of sedimentary rock is found on the bottom
12. what Darwin meant when he said that living things overproduce
13. human classification and traits of the order to which they belong
14. the importance of chordate embryos looking alike
15. what may cause the balance to change in favor of the survival of light-colored or dark-colored mice
16. how natural selection works

Using Main Ideas

Answer the questions by referring to the page number after each question.

17. Why are dark-colored mice well adapted or not well adapted to living in areas with dark soil? (p. 607)
18. What are three events that lead to the formation of new species? (p. 612)
19. How are the following used as evidence of evolution?
    (a) comparison of fossil life-forms to present day life-forms (p. 617)
    (b) the DNA codes of all life-forms being made up of the same nitrogen bases (p. 620)
20. What did Darwin mean when he indicated that some individuals are more fit than others? (p. 616)
21. How are fossils formed? (p. 617)
22. How do monkeys and apes differ? (p. 613)
23. What are the three effects of mutations? (p. 609)

Skill Review 

*For more help, refer to the **Skill Handbook,** pages 704-719.*

1. **Infer:** Which flower would have the best chance of being pollinated by a moth, one that opens during the day or one that opens during the night? Why?
2. **Define words in context:** During the evolution of the horse, the number of toes decreased from four to three to one. What does the word evolution mean as used in this sentence?
3. **Interpret data:** During an experiment, a student reported the following distances that the frogs she was raising were able to jump: 1.2 m, 0.5 m, 0.9 m, 1.0 m, 0.6 m, 1.1 m. Are her data evidence of variation?
4. **Interpret diagrams:** Study Figure 29-13. How are the embryos alike?

Finding Out More

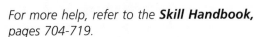

Critical Thinking

1. How might being white help a rabbit survive in Alaska?
2. Assume the trait in question 1 is a genetic trait. Will it be passed to future generations? Explain.

Applications

1. You wish to breed a chicken that produces many large eggs. Outline a breeding program that could accomplish this goal.
2. Use references to determine what the brain size, general appearance, and method of walking (two or four legs) were for early humanlike species.

Biology in Your World

Variety: The Spice of Life

In this unit, you learned that variations in plants and animals are due to changes in the genetic material. Studying fossils reveals the variety of living things from the past. Today, scientists can use the variations in living things to breed crops and animals with certain traits.

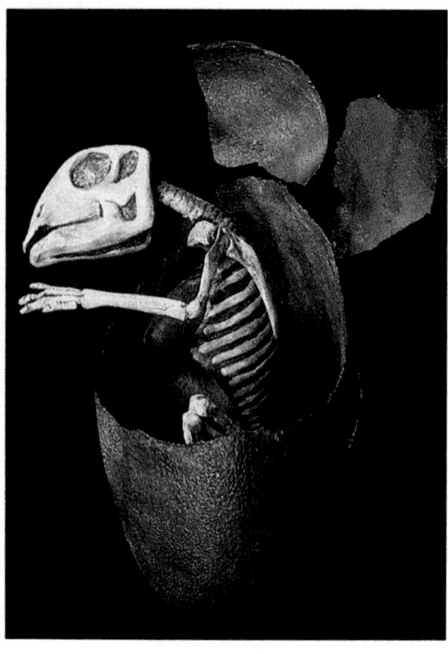

LITERATURE

Looking for Baby Dinosaurs

John Horner's interest in baby dinosaurs began with a can full of tiny dinosaur bones. He got the bones from the owners of a rock shop in Montana in 1978.

Horner spent the next six years looking for more bones at a site in Montana. He found the bones of baby dinosaurs in what appeared to be nests. The nests were close together and the babies were all the same age. Horner believes that these dinosaurs lived in groups and took care of their young.

Horner wrote *Digging Dinosaurs* about his findings. He also discusses why dinosaurs succeeded for 140 million years.

HISTORY

Putting the Pieces Together

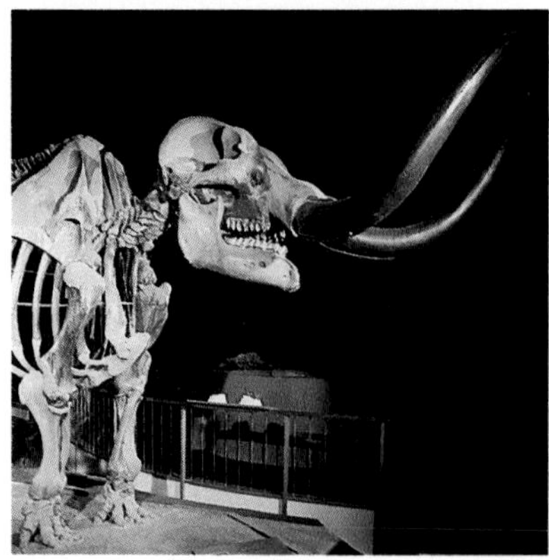

In 1801, farmers found some large bones in swamps near New York City. The bones were those of a mastodon. A mastodon is an extinct relative of the elephant. Mastodons lived in North America until about 10 000 years ago.

Charles Willson Peale was an artist and museum owner from Philadelphia who studied natural history. Hoping to find a complete mastodon skeleton, Peale organized an excavation at one of the sites. Most of the bones of an entire mastodon were found. In order to complete the skeleton, missing bones were carved from wood. The skeleton was displayed in Peale's museum.

LEISURE

Breeding Guppies

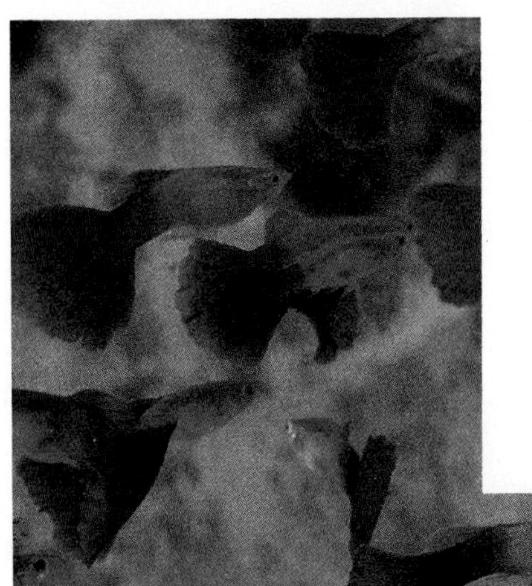

Female guppies are gray, ordinary-looking tropical fish. However, male guppies can be brightly colored. If you would like to see genetics in action, consider breeding guppies. You will need a tank and large jars to use as maternity wards. Check at a library or pet store for books on aquariums and breeding guppies. After each group is born, choose the largest and liveliest fish as the parents of the next generation. You can expect interesting results within a few generations. Keep records—you may succeed in producing a new strain of guppies!

CONSUMER

Delicious, Nutritious Bananas

Bananas are a very nutritious type of fruit. They are a good source of carbohydrates, phosphorus, and potassium. They are also a natural source of vitamins A and C. Bananas have very little fat.

Bananas grow in the tropics, where the plants produce fruit almost continuously. Thus, fresh bananas are almost always available.

The banana we eat is the Cavendish variety. It is being threatened by a fungus called black sigatoka (sihg uh TOH kuh). The fungus attacks the leaves of banana plants and can spread to the fruits. If not stopped, the fungus can kill the plants. At one time, chemical sprays were able to kill the fungus. But now, some of the fungi have become resistant to the chemicals.

Plant breeders are trying to develop a new hybrid banana that is resistant to black sigatoka. Some wild varieties of bananas are resistant to the fungus. The breeders are crossing these wild, nonedible varieties with the edible varieties.

Maybe in a few years you will see a new banana on your grocery store shelf.

Unit 8

CONTENTS

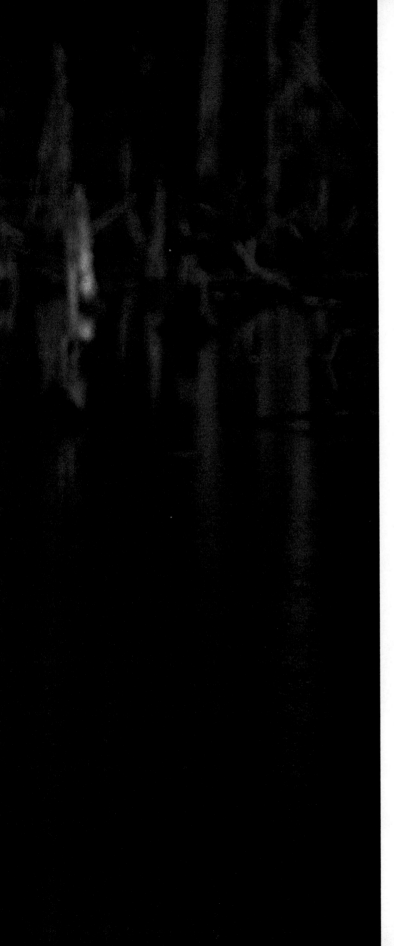

Relationships in the Environment

What would happen if...

there were no mosquitoes? You may have memories of hot, itchy summers when you thought the world would be better off without mosquitoes. But, would it? Fish depend on mosquito larvae for food. Many birds, in turn, depend on the fish. Without the mosquito, many fish would starve to death. Many birds would then starve. The balance of nature would be upset by the lack of mosquitoes.

What is the effect of spraying chemicals to kill mosquitoes? In many communities, trucks that spray to kill mosquitoes are a regular sight. The spray that kills mosquitoes kills honeybees, too. What would be the effect of killing honeybees? What is the cost to the environment of getting rid of mosquitoes? Is the cost too high?

CHAPTER PREVIEW

Chapter Content

Review this outline for Chapter 30 before you read the chapter.

Skills in this Chapter

The skills that you will use in this chapter are listed below.

- In **Lab 30-1,** you will form hypotheses, interpret data, infer, and design an experiment. In **Lab 30-2,** you will observe, infer, and design an experiment.
- In the **Skill Checks,** you will understand science words, make a line graph, and observe.
- In the **Mini Labs,** you will make and use tables.

Chapter 30

Populations and Communities

Some animals, like the caribou (KAR uh boo) on the left, live together in large groups called herds. There may be 50 to several hundred caribou in a herd. Each year the caribou migrate to a better feeding area. During migration, smaller herds combine to form a larger herd that may number in the thousands. What benefit might there be to living in a herd? What disadvantage is there to living in a herd?

Animals like the wolves shown below live together in small groups called packs. There may be 6 to 10 wolves in one pack. The pack may roam over an area of 25 to 50 square kilometers. Sometimes the wolves follow the caribou on their migration. How do wolves benefit from living in small packs? What might happen if the pack increased to 50 or 60 wolves?

Try This!

How important is each organism in a community? Write down the name of a plant. Now write the name of an organism that eats the plant. Continue to name an organism that eats the last organism you named. Your list shows relationships among living things.

BIOLOGY Online

Visit the Glencoe Science Web site at science.glencoe.com to find links about **populations and communities.**

30:1 Populations

Objectives

1. **Relate** the importance and methods of counting populations.
2. **Discuss** why populations change in size.
3. **Explain** how limiting factors affect a population.

Key Science Words

population
emigration
immigration
limiting factor

On your way to school each day, you see different kinds of living things. If you stopped to take a closer look, you might see many individuals of the same kind of organism. There could be dozens or hundreds of one kind of living thing in one place. These living things make up a population. A **population** is a group of living things of the same species that live in an area.

Population Size and Arrangement

Suppose you counted all of the daisies in the field in Figure 30-1. They would make up the daisy population of that field. You can't see the populations of spiders, earthworms, or field mice in the photo, but they are there as well.

A change in the size of one population often causes a change in the size of another population. For this reason, scientists usually want to know of any increases or decreases in populations they study. Finding the size of a population is not always easy. It may be hard to count the number of mice or earthworms in a population because these animals move. Some animals may be counted twice. Some may not be counted at all.

Counting the number of daisies in the small field would be easier because they do not move and the size of the field is small. Sometimes the area where the members of a population are found is very large. Finding out the size of the spruce tree population of Yellowstone National Park would be much harder due to the park's large size.

Skill Check

Understand science words: population. The word part *popula* means a group of people. In your dictionary, find three words with the word part *popula* in them. *For more help, refer to the* ***Skill Handbook,*** *pages 706-711.*

Figure 30-1 Some populations are easier to see and count than others.

a

b

c

Figure 30-2 Ear tags (a), leg bands (b), and radio transmitters (c) are ways of marking animals for population counts.

Populations are counted in different ways. Animals such as mountain goats can be marked with ear tags, Figure 30-2. Birds are marked by leg bands. Sometimes radio transmitters are put on animals. The animals are then followed to their nesting places and their offspring are counted. Trees may be marked with paint or ribbons. How is the human population in the United States counted?

How can an animal population be counted?

Populations are not spread out evenly. Figure 30-3 shows that the population of the United States is clumped in cities. Is there a high or low population where you live?

A clumped population can be a useful plan. Members of the population may help one another find food or shelter. Many species of animals live in groups for protection. What kinds of animals live in groups called herds, flocks, and packs? Plants may be helped by living in clumps. Trees may be protected from strong winds if they grow close together.

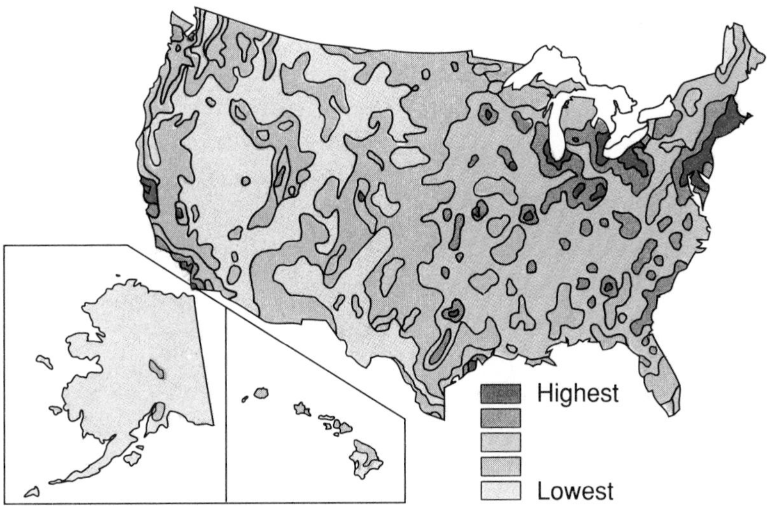

Figure 30-3 The human population in the United States is clumped. Which area has most people?

Skill Check

Make a line graph: Find out what the population size was for your state in 1950, 1960, 1970, 1980, and 1990. Plot the data on a line graph. Make a point showing where you think the population for 2000 will be. *For more help, refer to the* ***Skill Handbook,*** *pages 715-717.*

Population Changes

Let's examine how the human population has changed in the United States in the last forty years. You can see from the figures in Table 30-1 that the population is increasing.

TABLE 30–1. UNITED STATES POPULATION

Year	Number of People
1960	180 671 000
1970	205 052 000
1980	227 757 000
1990	249 600 000
1997 (est)	268 000 000

The graph in Figure 30-4 shows the growth in world population since the year 1500. Several things can cause a population to change in size. If the number of births goes up or the number of deaths goes down, a population will increase. The population will decrease if there are more deaths than births. Animal populations can increase or decrease in other ways. Animals move from place to place. The movement of animals out of an area is called **emigration** (em uh GRAY shun). Emigration causes a decrease in population numbers.

How does immigration affect a population?

Animals that move out of one population usually enter another population. The movement of animals into a population is called **immigration** (ihm uh GRAY shun). Immigration causes an increase in population numbers.

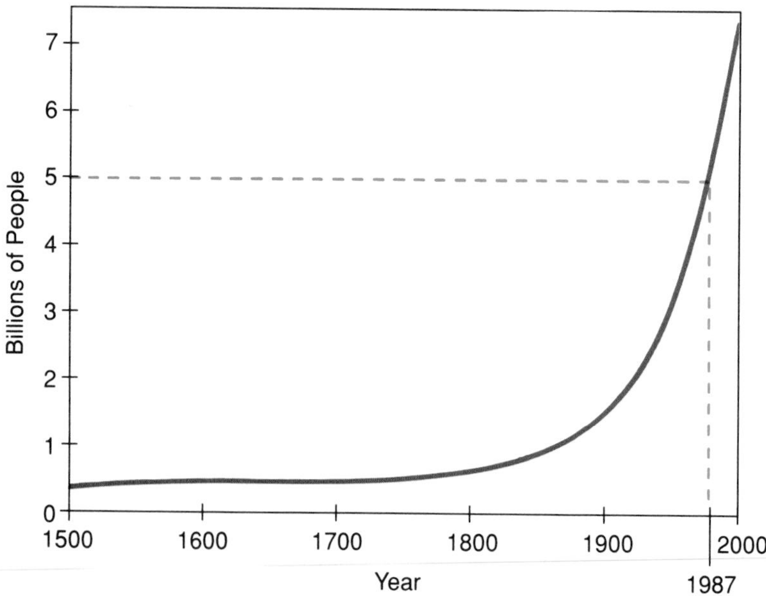

Figure 30-4 This graph shows how the human population in the world has grown since 1500. What is the population expected to be in the year 2000?

Science and Society

Problems with an Increasing Human Population

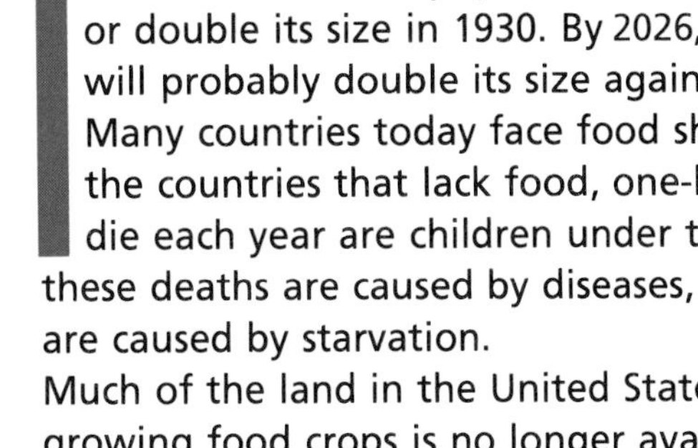

In 1975 the human population on Earth was four billion, or double its size in 1930. By 2026, the human population will probably double its size again, to almost 8.1 billion. Many countries today face food shortages each year. In the countries that lack food, one-half of the humans who die each year are children under the age of five. Some of these deaths are caused by diseases, but many of the deaths are caused by starvation.

Much of the land in the United States that was available for growing food crops is no longer available. Many farms have been sold, and the land is used for building homes and new industrial sites. Due to improved farming methods and better fertilizers, however, farmers can now grow more food on less land.

What Do You Think?

1. Thirty years ago, most countries produced their own food. Now, the United States and six other countries are the only ones in the world that produce enough food to feed all of their people. These countries have extra food that they store or sell to other countries.

Some people think that a country with extra food should give it to countries that do not have enough. If they give it away, who will pay the farmer?

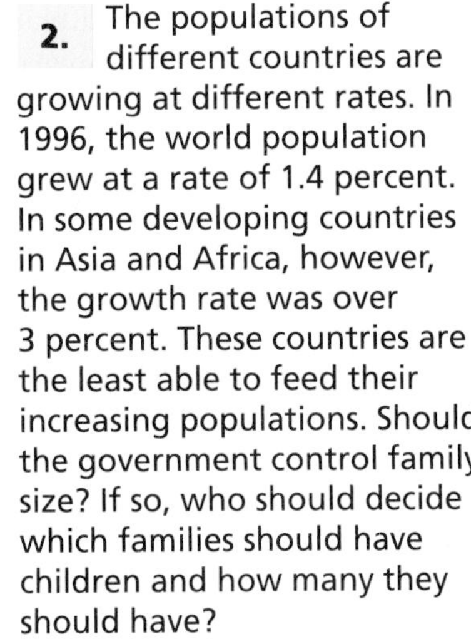

2. The populations of different countries are growing at different rates. In 1996, the world population grew at a rate of 1.4 percent. In some developing countries in Asia and Africa, however, the growth rate was over 3 percent. These countries are the least able to feed their increasing populations. Should the government control family size? If so, who should decide which families should have children and how many they should have?

3. Certain sprays that kill weeds and insects are banned in the United States because they are harmful. These sprays are sold to farmers in countries that have trouble raising enough food. If the sprays are not used, food production will decrease because some of the crops will be destroyed by insects or weeds. Should all countries pass laws banning the use of such sprays? Should the companies that make the chemicals be prevented from selling them?

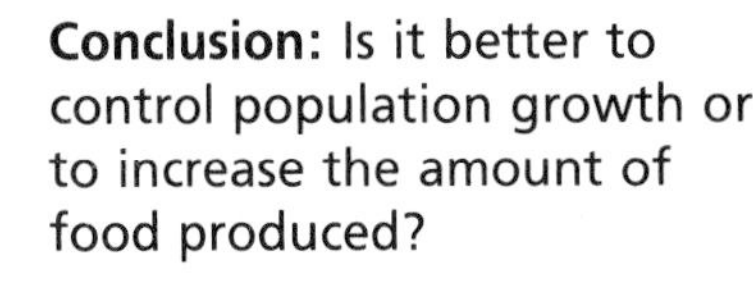

Conclusion: Is it better to control population growth or to increase the amount of food produced?

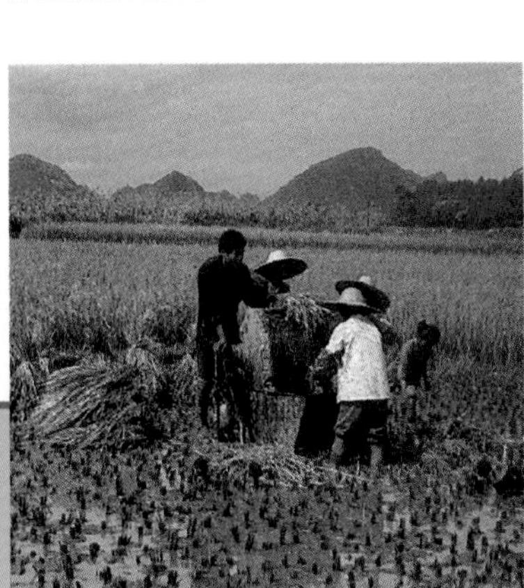

In some countries, people have trouble raising enough food to eat.

Figure 30-5 Competition affects the size of a population. What is the limiting factor in this photo?

Limits of Population Size

Why don't populations increase forever? What prevents the world from being overrun with all kinds of living things? Any condition that keeps the size of a population from increasing is called a **limiting factor.** Almost anything that affects the lives of organisms can be a limiting factor. Lack of light, space, water, or food are all limiting factors.

A population uses more sunlight, space, water, and food as it increases in size. Members of the same and similar species have the same needs. Remember from Chapter 29 that the struggle among organisms to get their needs for life is called competition. An increase in the population size causes more competition for the same materials. Some organisms may not get enough of these materials. The lack of needed materials causes population growth to slow down by decreasing the number of births and increasing the number of deaths.

What happens to plant growth when space is limited?

Plant roots growing close to each other compete for nutrients, water, and space. The stems and leaves of different plants compete for light. A decrease in the amount of space and light causes the growth of plants to slow down. Why do gardeners pull out some of their lettuce seedlings?

Look again at the human population graph in Figure 30-4. The line slants upward showing a rapid increase in population. What does this tell you about the limiting factors in a human population? Could you make a prediction about the population size in the year 2000? Do you think human populations are affected by limiting factors?

Check Your Understanding

1. Explain which is easier to count, plant populations or animal populations.
2. There are very few deaths in a deer population, yet the population is decreasing. Explain what might be happening to cause the decrease.
3. How can limiting factors cause a population change?
4. **Critical Thinking:** What might become limiting factors in a country whose population becomes too large?
5. **Biology and Math:** If there are 10 wolves in one pack roaming over an area of 20 square kilometers, what is the average density of wolves per square kilometer?

Lab 30–1 Competition

Problem: What happens when two animal species compete?

Skills

form hypotheses, interpret data, infer, design an experiment

Materials

graph paper
ruler

Procedure

In 1973, there were no houses in a large, old field. By 1988, there were 250 houses. A biologist found a population of dusky field mice living in the field. The biologist sampled this population from 1977 to 1992. Write a **hypothesis** in your notebook about what you think happened to the mouse population between 1977 and 1992.

1. Prepare a sheet of graph paper as follows. Mark the horizontal axis with the years 1977 to 1994. Label the horizontal axis "Years Sampled."
2. Mark the vertical axis from 0 to 90. Label the vertical axis "Size of Mouse Population per 1000 Square Meters."
3. Study the information in the table.
4. Plot the data for the field mouse population on the graph.
5. Connect the data points with a ruler to make a line graph. Continue the line to the right so you can estimate the size of the mouse population in 1994.

YEAR	NUMBER OF MICE PER 1000 SQUARE METERS	YEAR	NUMBER OF MICE PER 1000 SQUARE METERS
1977	89	1985	47
1978	85	1986	45
1979	82	1987	36
1980	79	1988	29
1981	78	1989	26
1982	51	1990	17
1983	48	1991	14
1984	42	1992	7

Data and Observations

1. Did the number of mice increase or decrease between 1977 and 1988?
2. Between which two years was the greatest change in population seen?

Analyze and Apply

1. **Infer:** What limiting factors are affecting the population size of the dusky field mouse?
2. With what other living thing do you think these mice are competing?
3. What might be happening to the birthrate of mice?
4. Suppose in 1984, 17 mice emigrated. What do you think would happen to the mouse population?
5. Predict what will happen to the number of mice in 1994.
6. List two factors that can increase the mouse population.
7. What effect would immigration have on the mouse population?
8. **Check your hypothesis:** Was your hypothesis supported by the data? Why or why not?
9. **Apply:** What happens when two animal species compete?

Extension

Design an experiment that would show how a limiting factor, such as size of habitat, affects a plant population.

30:2 Communities

Objectives

4. **Describe** different parts of a community.
5. **Explain** the importance of producers, consumers, and decomposers.

Key Science Words

community
habitat
niche
primary consumer
secondary consumer

The different populations in an area depend on each other. How can this be true? Look again at Figure 30-1. The daisies in the field can provide shelter for the spiders. The mice can eat the seeds of the daisy. What do you think the earthworm depends on? How do the mice and the earthworms help the daisies?

Parts of a Community

The populations of daisies, mice, and earthworms in Figure 30-1 make up a community. A **community** is all of the living things in an area that depend upon each other. Some communities, like the field of daisies, may be easy to see. Other communities may be harder to see. Is there a pond or lake near your home? What living things make up the pond community? Communities can be identified by the kinds of living things found there. What types of communities can you identify?

What is a habitat?

In a community, every living thing has a place to live that best suits it. You know that birds live in trees, fish live in ponds, and people live in houses. The place where a plant or animal lives is its **habitat** (HAB uh tat). A squirrel may use several different trees in the forest as its habitat, while a skunk may use one hollow log and the surrounding area as its habitat. Your skin is the habitat for the millions of bacteria that live there. Figure 30-6 shows a lichen attached to a rock. What do you think the lichen's habitat is?

Figure 30-6 Some lichens can grow on rocks.

Every living thing in a community also has a function, or job. The job of the organism in the community is its **niche** (NITSH). The job of most green plants is to produce food from sunlight, water, and carbon dioxide by photosynthesis. Animals have jobs, too. Earthworms help break up the soil so that plant roots can get water and nutrients. Bees pollinate flowers when they collect nectar to make honey.

You live in the community, too. Your habitat is your home, your school, and other places where you spend time. Your job is to study and learn. Your niche is that of a student. Think about other niches in your community. What do you think is the niche of an auto mechanic? How do other people depend on the mail carrier or the bus driver? How does your teacher depend upon you? How do you depend on your teacher?

a

b

c

Figure 30-7 In a pond community, water plants (a) may be the producers, fish (b) may be the primary consumers, and turtles (c) may be the secondary consumers.

Producers

All of the organisms in a community are related to each other through their jobs. These jobs may be divided into three types: producers, consumers, and decomposers.

Producers are the organisms that make food in a community. In almost all communities, green plants, algae, or blue-green bacteria are the producers. Producers use some of the food for themselves. The food they make will also help to feed all the animals in the community. Producers also make oxygen during photosynthesis. Other organisms need the oxygen to live.

What do producers do for a community?

Consumers

The animals in a community must have food. Animals can't make their own food, so they are consumers. Consumers are organisms that eat other organisms.

Many animals, such as mice and deer, get food by eating plants. Animals that eat only plants are **primary consumers.**

Some animals eat other animals. Owls eat mice, and turtles eat fish. These animals are secondary consumers. **Secondary consumers** are animals that eat other animals.

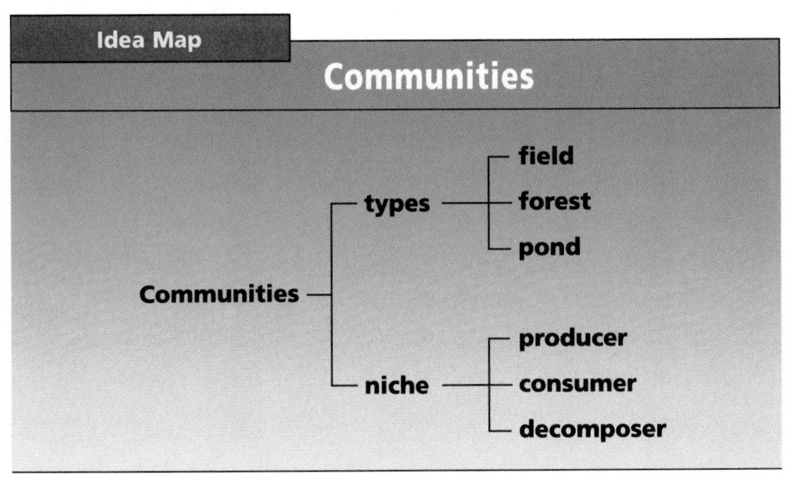

Study Tip: Use this idea map as a study guide to communities. Can you think of other types of communities?

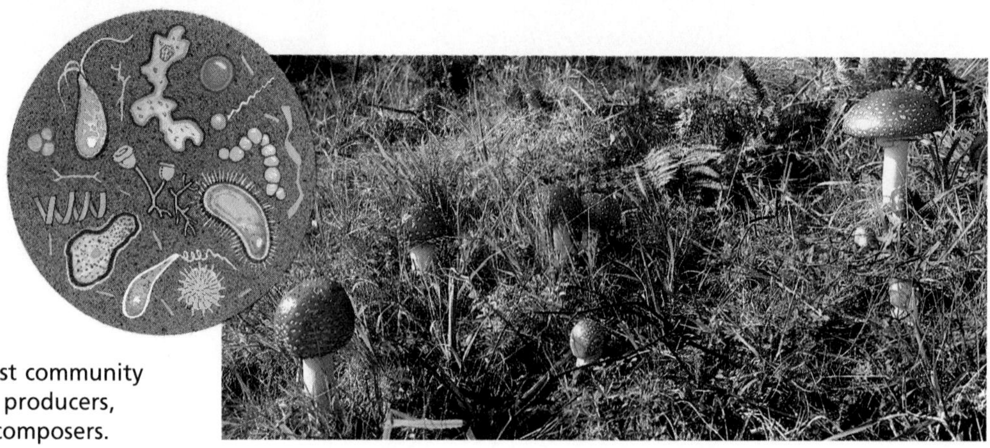

Figure 30-8 A forest community has many different producers, consumers, and decomposers.

Decomposers

In the pond and many other communities, decomposers form another group. Remember from Chapter 4 that decomposers are living things that get their food from breaking down dead matter. Bacteria and fungi, such as molds and mushrooms, are decomposers. Decomposers break down dead matter into simpler chemicals. Decomposers are very important to a community. They recycle nutrients so other organisms can use them. Without decomposers, Earth would soon be covered by dead material.

The forest community shown in Figure 30-8 has many different populations of producers, consumers, and decomposers. There are populations of ferns, oak trees, squirrels, and mushrooms. The upper left-hand corner of the figure shows a section of soil from the forest community. This soil section has been enlarged about 2000 times to show the small organisms found there. Bacteria, fungi, and protozoan populations fill the soil. Many of these organisms are decomposers.

Mini Lab

What's in a Community?

Make and use tables: Make a table listing 10 organisms you see in your area and whether each is a producer, consumer, or decomposer. Which appears in the largest number? *For more help, refer to the **Skill Handbook**, pages 715-717.*

Check Your Understanding

6. What is the habitat and niche of an earthworm?
7. How are producers different from consumers?
8. Why are decomposers important to a community?
9. **Critical Thinking:** Which of the following groups are necessary to life on Earth: producers, primary consumers, secondary consumers, decomposers?
10. **Biology and Reading:** Are humans primary consumers or secondary consumers? Explain your answer.

30:3 Energy in a Community

Recall that energy is the ability to do work. You learned in Chapter 2 that organisms get energy from food. This food energy can be traced back to the sun. How?

Producers, or green organisms, use light energy to make food. This light energy comes from the sun. To make food, green organisms change energy from the sun into chemical energy (sugar). This chemical energy is, in turn, used by consumers. A community's energy source is the sun.

Objectives

6. **Trace** the path of energy and materials through a community.
7. **Explain** how food chains are connected.

Key Science Words

food chain
food web
energy pyramid

Food Chains

When an animal eats a plant and is then eaten by another animal, you have a food chain. A **food chain** is a pathway of energy and materials through a community. Grass→grasshopper→bird is an example of a simple food chain. Each of the living things in it is a link in the food chain. Each depends on the other living things in the food chain. Figure 30-9 shows several important steps in a food chain.

What is an example of a simple food chain?

Figure 30-9 A food chain is a pathway of food through a community.

Figure 30-10 This food web shows the relationships among living things in a meadow community.

1. Food chains start with producers such as plants or other green organisms. They are the only organisms able to make their own food.
2. The next links in the food chain are the primary consumers (plant-eating animals).
3. Above the primary consumers are the secondary consumers (meat-eating animals). These animals eat primary consumers.
4. All living things die. They are food for the decomposers.

The arrows in the diagram show the direction in which the energy moves. In our example, plant leaves are eaten by grasshoppers, so an arrow points from the plant to the grasshopper. The next arrow shows that the grasshopper is eaten by a rat. Decomposers, in turn, feed on all organisms after they have died. Decomposers include mushrooms, bacteria, and protists.

> **Bio Tip**
>
> **Consumer:** On a farm, it takes 16 kilograms of plants to produce one kilogram of beef, but only five kilograms of plants to produce one kilogram of chicken. Think about this the next time you help with the shopping.

Several different primary consumers can eat the same producer. Several secondary consumers can eat the same primary consumer. The same producer or consumer may be part of several different food chains. Several food chains may be connected. Figure 30-10 shows how. Food chains connected in a community are called a **food web.** A food web shows how energy is moved through a community.

What is a food web?

Energy Flow in the Community

A producer gets energy from the sun and uses it to produce food and oxygen. Much of the food made through photosynthesis is used by the producer for growth and other cell processes. A large amount of the energy is lost as heat. Only a small amount of the food energy produced is available to the primary consumer that eats the producer. Therefore, primary consumers must eat large numbers of plants to get the energy they need for their cell processes and growth. When a secondary consumer eats a primary consumer, again only some of the food energy is available. The secondary consumer must eat several animals to get enough energy. You can see that each step of the food chain has fewer organisms than the step before it.

At each step of the food chain, from producers to primary consumers to secondary consumers, the amount of available energy becomes less and less. This loss of energy through a food chain can be shown in the shape of a pyramid. An **energy pyramid** is a diagram that shows energy loss in the food chain. Figure 30-11 shows an energy pyramid. The pyramid is wider at the bottom than at the top. There is more available energy at the bottom of the pyramid than at the top. Notice in Figure 30-11 that the number of primary consumers is smaller than the number of producers. Notice also that the number of secondary consumers is smaller than the number of primary consumers. There are fewer organisms at the top of the pyramid because less energy is available to them. Figure 30-11 also shows that all living things lose heat to their surroundings.

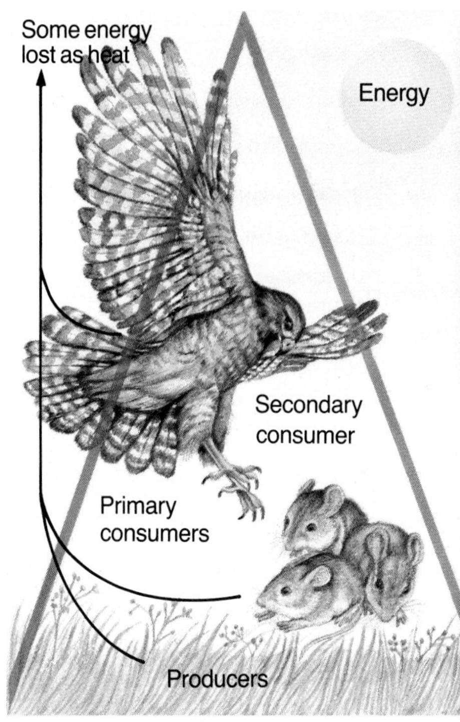

Figure 30-11 An energy pyramid shows energy loss in a food chain.

Check Your Understanding

11. Why must primary consumers eat large numbers of plants?
12. Which would be affected more by the loss of a species—a food chain or a food web? Why?
13. Why are plants necessary to the path of energy through a community?
14. **Critical Thinking:** What would happen to the food chains in an area if all plants were to die?
15. **Biology and Reading:** What happens to most of the energy at each level of a food chain?

30:4 Relationships in a Community

Objectives

8. **Give examples** of mutualism.
9. **Describe** commensalism.
10. **Compare** parasitism with predation.

Key Science Words

commensalism
parasitism
predation
predator
prey

There are many types of relationships within a community. Each relationship shows a different way that one living thing affects another. The relationship of one kind of organism to another may be helpful to both. In other relationships, one organism benefits while the second is harmed. There are still other cases where one organism is helped and the other is neither harmed nor helped. Communities could not exist without these different relationships.

Mutualism

Suppose there is a good rock concert in town and you would like to go. Right now, you do not have enough money for a ticket. Your best friend offers to buy you a ticket for this concert if you will give her a ride to the concert. It sounds like a good deal. Both you and your friend benefit from the arrangement.

What is mutualism?

This is an example of mutualism. Remember from Chapter 5 that mutualism is a relationship in which two organisms live in a community and depend on each other. Both organisms benefit from the relationship. Mutualism occurs among members of all five kingdoms. You first read about this kind of relationship in Chapter 5 when you read about lichens.

An example of a relationship between an animal and a protist can be seen in a dead log that contains termites, Figure 30-12. Termites are insects that eat wood. They are

Figure 30-12 The relationship among protists (a) living in the intestines of termites (b) is an example of mutualism.

Figure 30-13 The relationship between orchid plants and trees is an example of commensalism.

not able to digest wood. They could not use it for energy if it weren't for a protist that lives inside the termite's intestine. The protist can digest the wood for the termite so that the termite can use it for energy. The protist has a home inside the termite and uses some of the wood for its own energy.

Humans can show mutualism with other organisms. An example is the bacterium that lives in the human intestine. The bacterium makes vitamin B_{12} for the human, who provides a home and food for the bacterium.

> **Bio Tip**
>
> **Health:** The bacterium that makes vitamin B_{12} in the human intestine also makes vitamin K, which is needed for blood clotting. Newborn babies have to build up the population of these bacteria before they can get enough vitamin K in their bodies to clot their blood.

Commensalism

Let's return to the example of the rock concert tickets. Suppose your friend has a free ticket, but can't go to the concert. She gives the ticket to you. You benefit from the free ticket. Whether or not you use the ticket doesn't matter to your friend, does it? She got the ticket free. She won't lose anything whether you use the ticket or not.

What is commensalism?

Commensalism (kuh MEN suh lihz um) is seen in this example. **Commensalism** is a relationship in which two organisms live in a community, and one benefits while the other gets no benefit and is not harmed.

Commensalism occurs between many living things. For example, orchid plants and trees have this kind of relationship. Orchids receive more sunlight for photosynthesis if they grow high on tree branches rather than close to the ground, Figure 30-13. The orchid plants are helped by the tree. The tree gets nothing in return from the orchids.

Skill Check

Observe: A lamprey is a parasitic fish. Look at the photo of a lamprey on this page and see if you can tell how the lamprey gets its food. Does it have any body parts that are adapted to its way of life? *For more help, refer to the **Skill Handbook**, pages 704-705.*

Parasitism

Let's return to the rock concert tickets one more time. Suppose your friend offers to sell you some concert tickets. Later, you find out that she has charged you much more money than the tickets were worth. Your friend is making a profit at your expense.

This is an example of parasitism (PAR uh suh tihz um). **Parasitism** is a relationship between two organisms in which one is helped and the other is harmed. Remember that a parasite lives at the expense of another living thing. The living thing on which the parasite lives is called the host. The host is rarely killed in this kind of relationship. What would happen to the parasite if it killed its host?

Many organisms are parasites. Fleas are parasites with which you may be familiar. Fleas use the blood of dogs or other animals for food. The flea benefits and the host animal is harmed in the relationship.

There are other common examples of parasitism. When you get sick, chances are that bacteria or viruses are using your body as a host. At some time you may have had strep throat. This disease is caused by bacteria living in your throat and using your body as a host. The bacteria grow and reproduce. They give off chemicals that irritate your throat tissues and interfere with the function of other body systems.

Figure 30-14 The relationship between a flea (a) and a rabbit, or between a lamprey (b) and other fish is an example of parasitism.

a

b

Figure 30-15 A predator hunts and kills other animals for food.

Predation

There is another relationship in a community in which one organism benefits and the other is harmed. This is **predation,** or the predator-prey relationship. An animal that hunts, kills, and eats another animal is a **predator.** The **prey** is the animal that the predator kills and eats. Unlike parasitism, in predation the predator actually kills the prey.

What is a predator?

Predators are secondary consumers. They can limit the sizes of some populations. Rabbits living near a field of clover become prey for owls that live in the same area. Look at the graph of the rabbit population in Figure 30-16. The graph shows sudden increases and decreases in population size. What are some things that might cause the population to decrease?

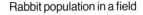

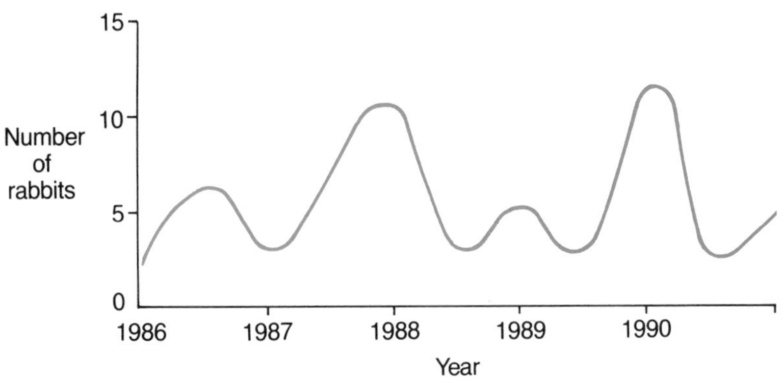

Figure 30-16 Predators can be a limiting factor for a prey population.

Study Tip: Use this idea map as study guide to community relationships. Can you give an example of each kind of relationship?

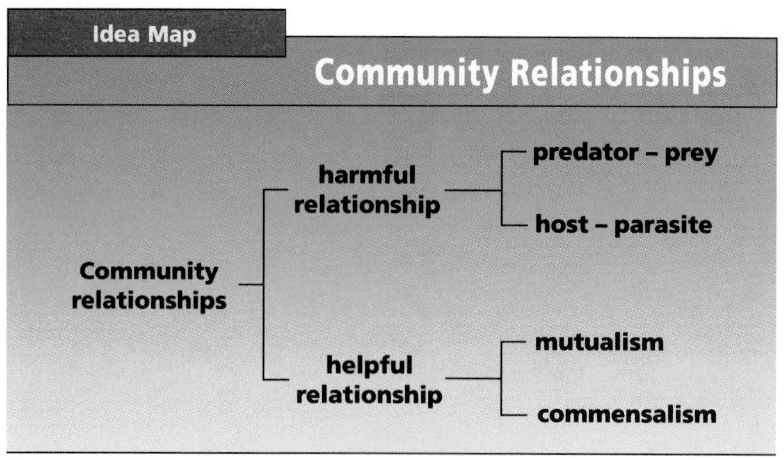

TABLE 30–2. RELATIONSHIPS		
Type of Relationship	**Organism 1**	**Organism 2**
Mutualism	helped	helped
Commensalism	helped	unaffected
Parasitism	helped	harmed
Predation	helped	harmed (killed)

Table 30-2 is a summary of the relationships that may occur among populations in a community. Now you can see why the statement "no living thing lives alone" is true.

Check Your Understanding

16. Give an example of mutualism involving humans.
17. How do orchid plants benefit by living high in trees?
18. What word describes the relationship between fleas and their hosts?
19. **Critical Thinking:** Explain how the relationship between orchid plants and a tree might change from commensalism to parasitism.
20. **Biology and Reading:** For each of the pairs of organisms below, write the word that best describes their community relationship.
    a. Ants protect a tree, which provides food for ants.
    b. A liver fluke lives inside a sheep, which develops liver disease.
    c. An insect eats the hairs that fall out of a deer's skin, but the deer is not affected.

Lab 30–2

Predation

Problem: What do owls eat?

Skills

observe, infer, design an experiment

Materials

water
bowl
forceps
glue
glass slide
coverslip
light microscope
owl pellet
cardboard

Procedure

Owl pellets are made of food, bones, and fur an owl has eaten but has not been able to digest. An owl forms pellets in its stomach and then coughs them up. Examining an owl pellet will give you some ideas about what an owl eats.

1. Copy the data table.
2. Place an owl pellet in the bowl. Add some water.
3. Use the forceps to remove the outside covering of the pellet.
4. Place a small piece of this covering on a glass slide and add a coverslip.
5. **Observe:** Examine the covering of the pellet on low power of your microscope. Make a drawing of what you see.
6. Continue to break the pellet open. Save all the parts present in the pellet.
7. Place the parts on a sheet of paper. Try to figure out how many living things the owl ate. Examine the bones and record your observations.
8. Try to assemble the bony parts into a skeleton. You can do this by gluing them to cardboard.

Data and Observations

1. What was the outside covering of the pellet? Where do you suppose it came from?
2. What types of things did you see in the pellet?

Analyze and Apply

1. An owl coughs up one pellet a day. How many animals did your owl eat each day?
2. Judging from the size of the remains in the owl pellet, name the kinds of living things that the owl ate.
3. Do owls eat plants or animals?
4. Is an owl a producer or a consumer? Explain your answer.
5. Is an owl a primary consumer or a secondary consumer?
6. **Apply:** Suppose an owl population increases from three to eight. What can happen to the population of organisms eaten by the owls? Why?

Extension

Design an experiment to show what happens to a predator population when the size of the prey population decreases.

NUMBERS OBSERVED				
Leg Bone	Rib	Mammal Skull	Bird Skull	Mammal Jawbone

Chapter 30 Review

Summary

30:1 Populations

1. There are many methods for counting the living things in a population.
2. A change in population size can be caused by a change in birth or death rates.
3. Limiting factors keep populations from increasing forever.

30:2 Communities

4. Each kind of organism in a community has its own habitat and niche.
5. Green organisms supply food in a community. Animals are consumers. Bacteria and fungi are decomposers.

30:3 Energy in a Community

6. The energy in a community comes originally from the sun. It is passed through the community as food.
7. Food chains are connected to form food webs.

30:4 Relationships in a Community

8. In mutualism, two living things depend on each other.
9. Commensalism is a relationship in which one organism benefits and the other is unaffected.
10. In parasitism, the host is harmed but not killed. A predator kills its prey.

Key Science Words

commensalism (p. 645)
community (p. 638)
emigration (p. 634)
energy pyramid (p. 643)
food chain (p. 641)
food web (p. 642)
habitat (p. 638)
immigration (p. 634)
limiting factor (p. 636)
niche (p. 638)
parasitism (p. 646)
population (p. 632)
predation (p. 647)
predator (p. 647)
prey (p. 647)
primary consumer (p. 639)
secondary consumer (p. 639)

Testing Yourself

Using Words

Choose the word from the list of Key Science Words that best fits the definition.

1. simple pathway of energy and materials through a community
2. relationship in which one member is helped and the other is harmed
3. keeps a population from increasing
4. place where an organism lives
5. a diagram that shows energy loss
6. animals moving into a population
7. eaten by a predator

Testing Yourself *continued*

8. animals moving out of a population
9. an organism's job in a community
10. a group of the same kind of living things in an area

Finding Main Ideas

List the page number where each main idea below is found. Then, explain each main idea.

11. examples of clumped populations
12. the job of decomposers in a community
13. what a food chain shows
14. a relationship in which two organisms depend on each other
15. the importance of plants to a community
16. what the second link of a food chain is
17. what methods are used to count populations
18. what the limiting factors for animal populations are
19. the difference between primary consumers and secondary consumers
20. what an energy pyramid shows

Using Main Ideas

Answer the questions by referring to the page number after each question.

21. If a predator population in an area decreases, what could happen to the size of the prey population? (p. 647)
22. What role does the sun play in a community? (p. 643)
23. What are two examples of decomposers? (p. 640)
24. Why are radio transmitters sometimes used to count a population? (p. 633)
25. What factors can increase a population? (p. 634)
26. What relationship is shown by orchids growing on a tree? (p. 645)

Skill Review 

*For more help, refer to the **Skill Handbook,** pages 704-719.*

1. **Make a line graph:** Find out from the school office how many students have been in your school each year during the past ten years. Make a line graph to show these data. Is the school population increasing or decreasing?
2. **Infer:** Infer what happens to two different organisms in a community if they have the same needs for food.
3. **Form hypotheses:** Form a hypothesis to explain how you could tell if owls eat small plant-eating animals.
4. **Make and use tables:** Make a table listing 10 organisms as producers, primary consumers, and secondary consumers. List how each organism gets food.

Finding Out More

Critical Thinking

1. What causes a prey population to increase again after it has been decreased by predators?
2. What would happen in a food web if one kind of organism in the web were all killed by a disease?

Applications

1. Make a list of 6 to 10 different organisms that live near your school or home. Construct a food web using the organisms on your list.
2. Find out why fungi and bacteria are useful organisms in a farm field.

CHAPTER PREVIEW

Chapter Content

Review this outline for Chapter 31 before you read the chapter.

Skills in this Chapter

The skills that you will use in this chapter are listed below.

- In **Lab 31-1,** you will observe, form hypotheses, interpret data, infer, and design an experiment. In **Lab 31-2,** you will observe, form hypotheses, interpret data, and design an experiment.
- In the **Skill Checks,** you will understand science words and sequence.
- In the **Mini Labs,** you will interpret diagrams and experiment.

Chapter 31

Ecosystems and Biomes

Look at the photo on the left. What part of the environment is causing the trees to bend toward one side? Why are the limbs growing more on one side than on the other? The photo on this page shows a field. Do you see any animals in the field? Where could they be? What happened to the green plants in the field?

In the last chapter, you learned that populations of living things interact with each other in a community. Within a community, living things also interact with their environment. In this and the next chapter, you will see how living things interact with their environments.

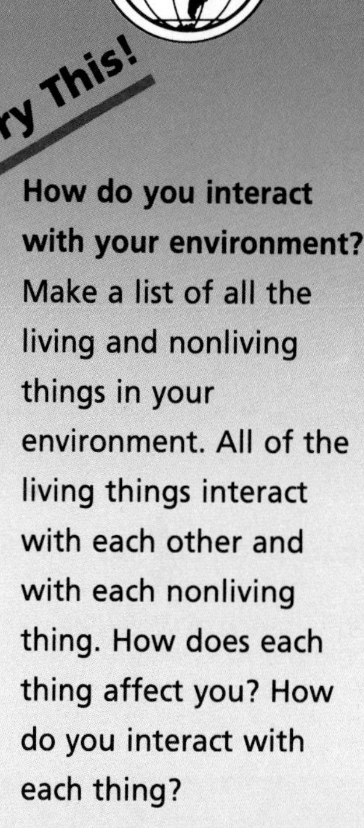

Try This!

How do you interact with your environment? Make a list of all the living and nonliving things in your environment. All of the living things interact with each other and with each nonliving thing. How does each thing affect you? How do you interact with each thing?

Visit the Glencoe Science Web site at science.glencoe.com to find links about **ecosystems and biomes.**

31:1 Ecosystems

Objectives

1. **Describe** the parts of an ecosystem.
2. **Describe** how the water cycle affects an ecosystem.
3. **Explain** the nitrogen cycle and the oxygen-carbon dioxide cycle.

Key Science Words

ecosystem
ecology
nitrogen cycle
water cycle

A community interacting with the environment is an **ecosystem.** An ecosystem can be as small as a roadside ditch or as large as one of the Great Lakes.

Parts of an Ecosystem

What two parts make up an ecosystem?

How do communities differ from ecosystems? A community includes the living things in an area. When you study a community, you study only how the living things affect each other. An ecosystem includes the nonliving parts as well as the living parts in an area. When you study an ecosystem, you study how the nonliving and living parts affect each other.

The study of how living things interact with each other and with their environment is called **ecology** (ih KAHL uh jee). Ecologists study ecosystems to find out how the different parts interact. They observe living things in nature and in the laboratory. They collect data by making measurements and by carrying out carefully controlled experiments. Sometimes it is difficult or impossible to set up the same conditions in a lab that exist in nature. Then, ecologists may set up a model of an ecosystem on a computer, Figure 31-1a. However, they must realize when they make conclusions based on lab experiments or models that the data did not come from nature.

Look at the seashore ecosystem in Figure 31-1b. The living parts of this ecosystem are the plants, animals, and algae you see, as well as the bacteria, protozoans, and fungi that are too small to be seen. All these organisms are the producers, consumers, and decomposers you read about in Chapter 30. Suppose you could remove the living parts

Figure 31-1 Ecologists sometimes try to reproduce natural conditions by setting up a computer model (a). A seashore (b) is an ecosystem.

a

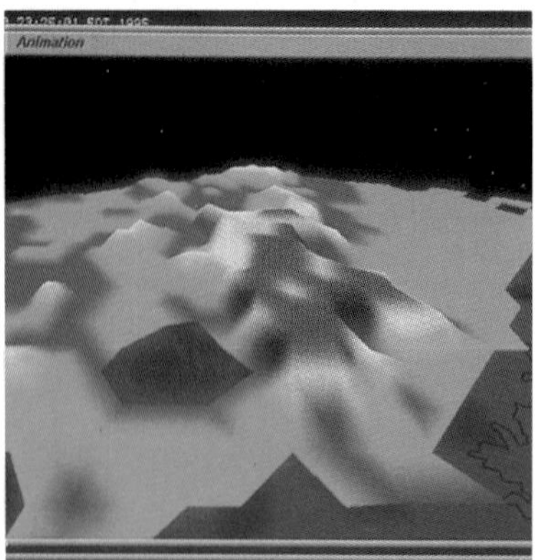

b

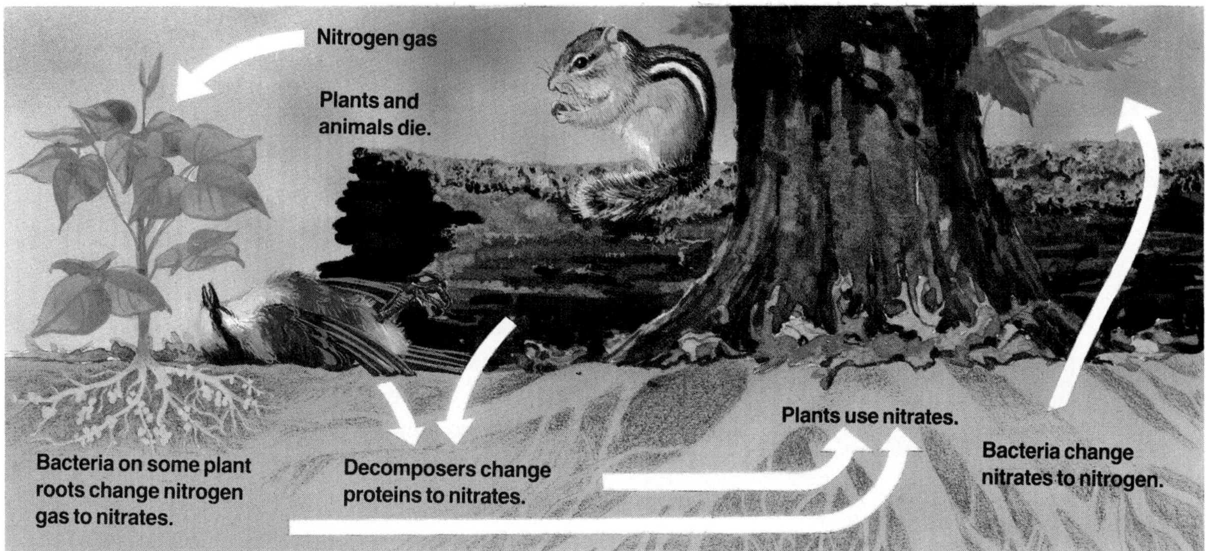
Nitrogen gas
Plants and animals die.
Bacteria on some plant roots change nitrogen gas to nitrates.
Decomposers change proteins to nitrates.
Plants use nitrates.
Bacteria change nitrates to nitrogen.

Figure 31-2 The nitrogen cycle. How is nitrogen gas returned to the air?

from Figure 31-1b. What would you see? You would see the nonliving parts. The nonliving parts of an ecosystem include the soil, air, water, light, and temperature. Nonliving parts help to determine where organisms can live.

Soil and the Nitrogen Cycle

The soil is an important nonliving part of an ecosystem. Soil comes from rocks that have broken down. It has many jobs. Soil serves as a place for plants to anchor their roots. Soil serves as a home for many living things. Ants and earthworms moving through the soil make holes for air and water. Moles, shrews, and other animals that burrow help mix and loosen the soil as they tunnel in all directions.

Soil holds other nonliving parts such as water, nutrients, and gases. Some of the most important soil nutrients needed by plants are called nitrates (NI trayts). Nitrates are chemicals that contain nitrogen. The way that nitrates get into soil is an example of the way living and nonliving parts of the ecosystem work together. Use Figure 31-2 to trace how nitrogen moves through an ecosystem.

1. Plants and animals need nitrogen to make protein. Air is mostly nitrogen, but most living things can't make protein from the nitrogen in the air. Bacteria living on the roots of some plants change the nitrogen in air to a form that plants can use to make protein.
2. When living things die, they are decomposed by bacteria and fungi. Some of the nitrogen in their bodies is changed into nitrates.

Skill Check

Understand science words: ecology: The word part *eco* means environment. In your dictionary, find three words with the word part *eco* in them. *For more help, refer to the* ***Skill Handbook,*** *pages 706-711.*

Career Close-Up

Wildlife Technician

Wildlife technician observing animal populations

A wildlife technician is a person who usually assists a wildlife biologist. Wildlife technicians count the numbers of animals in populations such as deer, turkeys, squirrels, and rabbits.

A wildlife technician may be expected to speak to people about wildlife management. During the hunting season, he or she may examine the animals killed by hunters to check for diseases and determine the ages of the animals. These data are used to predict how many surviving animals of each type are in the area. This information will tell if the number of hunting permits issued must be changed or if an area must be restocked.

A wildlife technician usually works for a state wildlife agency. The individual must have an interest in all kinds of living things and some knowledge of populations, communities, and ecosystems. Some wildlife technicians get an Associate of Arts degree to prepare for their jobs.

Bio Tip

Consumer: Manure rebuilds soil better than chemical fertilizers do. Manure broken down by decomposers provides nitrogen for plants, becomes part of the soil, and helps retain water in the soil. Dry manure can be bought in garden shops.

3. The nitrates are left in the soil. They can be used by plants as a source of nitrogen for making protein. Some plants are then eaten by animals. These animals can use nitrogen compounds found in plants to make proteins. When the plants or animals die, the nitrogen in their bodies is again returned to the soil.
4. Some nitrates in the soil are changed back into nitrogen gas by bacteria. The nitrogen gas is released into the air. The cycle repeats itself over and over again. The reusing of nitrogen in an ecosystem is called the **nitrogen cycle.**

The nonliving parts of the nitrogen cycle are nitrogen and soil. The living parts are the plants, animals, bacteria, and fungi. In every ecosystem, the nonliving and living parts interact with each other. Many other chemical materials are cycled through the ecosystem just as nitrogen is cycled.

Water

Water is a nonliving part of an ecosystem that is needed by all living things. The cells of all living things are mostly water. Without water, the cells would die. Plants can't make food without water. What would happen to an ecosystem if plants couldn't make food?

Water, like nitrogen, is cycled through the ecosystem. The path that water takes through an ecosystem is called the **water cycle.** Use Figure 31-3 to help you understand the steps of the water cycle.

1. Water in the air falls to Earth as rain or snow.
2. Some water runs off the land into rivers, ponds, lakes, or oceans.
3. Water from rain may soak into the soil.
4. Some of this water is taken up by plants through their roots. Animals may drink some of the water on the ground. The rest flows into underground lakes and rivers.
5. Excess water passes out of plants through their leaves. Animals lose water through body openings or as wastes are removed. The water evaporates (ee VAP uh raytz) into the air. Evaporate means to change from a liquid to a gas. Water is present in the air as a gas.
6. Water from lakes, rivers, and oceans also evaporates into the air.

Mini Lab

What Are the Parts of the Water Cycle?

Interpret diagrams: Look at Figure 31-3. What parts of an ecosystem are important in the water cycle? What role does each part play? Are all living things part of the water cycle? *For more help, refer to the* ***Skill Handbook,*** *pages 706-711.*

In what form is water found in the air?

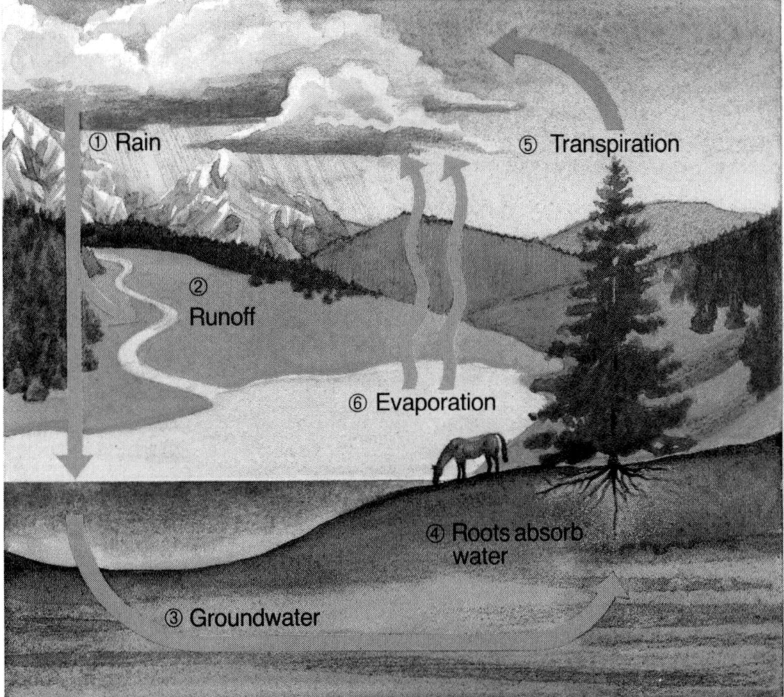

Figure 31-3 The water cycle. How is water returned to the air?

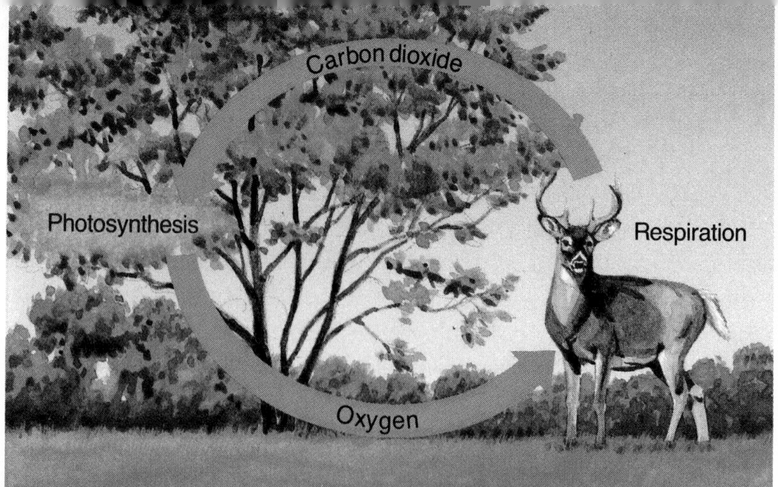

Figure 31-4 The oxygen-carbon dioxide cycle. What is given off during photosynthesis?

Oxygen-Carbon Dioxide Cycle

Why do living things need oxygen?

Air is needed in an ecosystem. Air is made up of several gases that are important to living things. You have learned about nitrogen. Two other gases in the air are oxygen and carbon dioxide. Why are these gases important? Without oxygen and carbon dioxide, respiration and photosynthesis can't take place.

Trace the path of oxygen and carbon dioxide through the ecosystem in Figure 31-4. Oxygen given off by producers during photosynthesis is needed by all living things for cellular respiration. Carbon dioxide given off during respiration is used by producers for making food. Oxygen and carbon dioxide can be cycled.

Producers and consumers in a water ecosystem, such as a lake, also have an oxygen-carbon dioxide cycle. Water has air mixed into it. Oxygen in the water is used by fish, plants, and other water life for respiration. Carbon dioxide in the water is used by water plants, blue-green bacteria, and algae for photosynthesis.

Check Your Understanding

1. What two parts make up an ecosystem?
2. Draw a diagram of the water cycle.
3. Why do plants and animals need nitrogen?
4. **Critical Thinking:** If a large forest was destroyed, how would cycling of materials be affected?
5. **Biology and Reading:** What is meant by the word *cycle* when discussing the nitrogen, water, and oxygen–carbon dioxide cycles?

Lab 31–1 Carbon Dioxide

Problem: Do plants use carbon dioxide or give off carbon dioxide?

Skills

observe, form hypotheses, interpret data, infer, design an experiment

Materials

4 test tubes
marking pencil
4 stoppers
lamp
blue test liquid
green test liquid
2 *Elodea* sprigs
2 test-tube racks

Procedure

1. Read the problem. In your notebook, write a **hypothesis** that answers the question.
2. Copy the data table.
3. Number the tubes 1 to 4 with the marking pencil.
4. Fill tubes 1 and 2 nearly full with green test liquid. Put an *Elodea* sprig in tube 2. Stopper both tubes. Place them in a rack.
5. Record the color of the liquid in each tube.
6. Place both tubes in front of the lamp and turn the lamp on.
7. Keep both tubes in the light overnight.
8. Fill tubes 3 and 4 nearly full with blue test liquid. Put an *Elodea* sprig in tube 4. Stopper both tubes. Place them in a rack.
9. Record the color of the liquid in each tube.
10. Place both tubes in the dark overnight.
11. **Observe:** Record the color of the liquids in all tubes the next day.
12. Dispose of the materials as directed by your teacher.

Data and Observations

1. In which tubes, 1 or 2, did the green test liquid change color?
2. In which tubes, 3 or 4, did the blue test liquid change color?

TUBE	PLANT ADDED?	WHERE PLACED (DARK OR LIGHT)	COLOR OF TUBE FIRST DAY	COLOR OF TUBE SECOND DAY
1				
2				
3				
4				

Analyze and Apply

1. What was the purpose of tube 1?
2. What was the purpose of tube 3?
3. If the green liquid turns blue, carbon dioxide was used in the tube. Which tube, 1 or 2, showed evidence that carbon dioxide was used overnight?
4. If the blue liquid turns green, carbon dioxide was given off in the tube. Which tube, 3 or 4, showed evidence that carbon dioxide was given off overnight? Explain.
5. Do plants use or give off carbon dioxide?
6. **Check your hypothesis:** Was your hypothesis supported by your data? Why or why not?
7. **Apply:** What process occurs in green plants that gives off carbon dioxide? What process in plants uses carbon dioxide?

Extension

Design an experiment to test whether other water plants, such as hornwort, use and give off carbon dioxide.

31:2 Succession

Objectives

4. Describe succession in a land community.

5. Describe succession in a water community.

Key Science Words

succession
climax community

What is succession?

The living and nonliving parts of an ecosystem may change over a period of time. For example, as trees grow taller in a forest ecosystem, they change the amounts of light, temperature, and water in the forest. Forests that have large trees are cooler, darker, and wetter than forests made up of small trees. Decaying leaves make the soil more acid. Some forest organisms cannot live in these new conditions. They are replaced by organisms that can live in the new conditions. When we refer to the changes in an ecosystem, we often refer just to the changes in the living part, the community. The changes that take place in a community as it gets older are called **succession.** All communities go through succession. You might not notice the changes unless you observe the community for many years, because they happen slowly.

Succession in a Land Community

To understand the stages of succession in one type of land community, we will start with a plowed field. In Figure 31-5a, there is nothing but bare soil.

After a few weeks or months, weed seeds are carried to the field by wind and animals. Weeds begin to grow over the soil. Worms and grasshoppers may be among the first animals to arrive. Beetles and ants arrive soon after. Figure 31-5b shows the changes that have occurred in the community.

As plants and animals die and decompose, their remains add nutrients to the soil. The soil becomes better suited to a greater variety of larger plants. Bushes and small trees begin to grow. Animal populations also change. Rabbits,

a

b

Figure 31-5 The first stage of succession on land is bare soil (a). The next stage is the growth of producers (b).

a

b

Figure 31-6 A wide variety of plants provide food for many different animals (a). A climax community is usually a forest (b).

mice, foxes, skunks, and hawks may be found in the community shown in Figure 31-6a.

As more plants and animals die in the community, more nutrients are added to the soil. The soil itself becomes looser and better suited to the growth of larger plants. Now large trees are able to grow. If the community is not disturbed, a forest begins to develop. There are more kinds of animals in the forest than there are in a vacant field.

The last stage in the succession of this community is a forest like the one in Figure 31-6b. We call the forest the climax community. A **climax community** is the last or final stage of succession in a community. It is the final stage because it is stable and can replace itself with little change from then on. It may take 150 years or more for an area to become a climax community.

Mini Lab

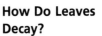

How Do Leaves Decay?

Experiment: Fill a jar half full with dead leaves and soil. Moisten with water and cover with plastic wrap. Punch holes in the wrap. Record the odor weekly. What causes the change? *For more help, refer to the* ***Skill Handbook,*** *pages 704-705.*

Succession in a Water Community

Could the place you are sitting right now have been a pond or lake a few hundred years ago? It may seem impossible when you think of the living things that make up a water community. Once, however, there might have been a water community where you are sitting.

Let's look at the stages of succession for a pond that changes into a land community. The bottom and sides of a new pond are bare. The only green organisms in the pond are algae floating on the water surface. Snails and a few small fish can be found. After several years, the pond becomes more shallow because dead plants and algae have piled up on the bottom. Larger fish, tadpoles, crayfish, and frogs are now present.

Why does a pond become more shallow over time?

Figure 31-7 Succession in a pond leads to dry land.

Over the years, more dead plants pile up and soil washes into the pond. The pond becomes more shallow and plants begin to grow around the edge of the pond, Figure 31-7c. Eventually, there will not be enough water for the fish to survive. Animals that can live out of water part of the time, such as turtles and frogs, will increase in number.

A hundred years later, the ground looks nothing like a pond. It may look like what you see in Figure 31-7d. Grass, shrubs, and small trees grow where the old pond was. Much of the animal life is completely different from when there was a young pond. Some day the area the pond occupied may contain a hardwood forest. A forest is not the only type of climax community that can develop. Depending upon soil, water, air, and sunlight, the climax community for an area might be a prairie or grassland, a desert, or a tropical rain forest. What kind of climax community can you identify where you live?

Check Your Understanding

6. What is succession?
7. If weeds are the first things to appear in a land community, what will the first animals probably be?
8. What animals are found in an older pond that is very shallow? Why?
9. **Critical Thinking:** How would succession be affected if animals did not return to an area after a fire?
10. **Biology and Reading:** Is the climax community for a lake a watery environment? Why or why not?

31:3 How Living Things Are Distributed

The nonliving parts of an ecosystem usually control what organisms can live in a certain area. Temperature, sunlight, and water are very important factors in determining which organisms live in a place.

Objectives

6. Explain how climate helps determine what living things live in an area.

7. Describe the major land biomes and water ecosystems.

Key Science Words

precipitation
climate
biome

Climate

Light and temperature are two nonliving parts of an ecosystem that are not cycled. Light from the sun is used by plants to make food. Producers transfer energy to other living things through food chains.

Temperature and light are often related. The soil in a field is warmed by the sun. The soil in a forest is cool because the leaves of trees prevent most of the sun's light from warming the ground.

The temperature of an ecosystem helps determine what organisms live there. Polar bears can live in very cold ecosystems, while lions, elephants, and palm trees live only in warmer ecosystems. Temperatures that are too low for some living things are just right for others.

The water cycle is also related to temperature and light. The amount of sunlight can affect the rate of evaporation. Water evaporates faster at warm temperatures. The temperature of the air affects the type of precipitation (prih sihp uh TAY shun) falling to Earth. **Precipitation** is water in the air that falls to Earth as rain or snow. Figure 31-8 shows temperatures and precipitation for the land biomes. You will learn about biomes on the next page.

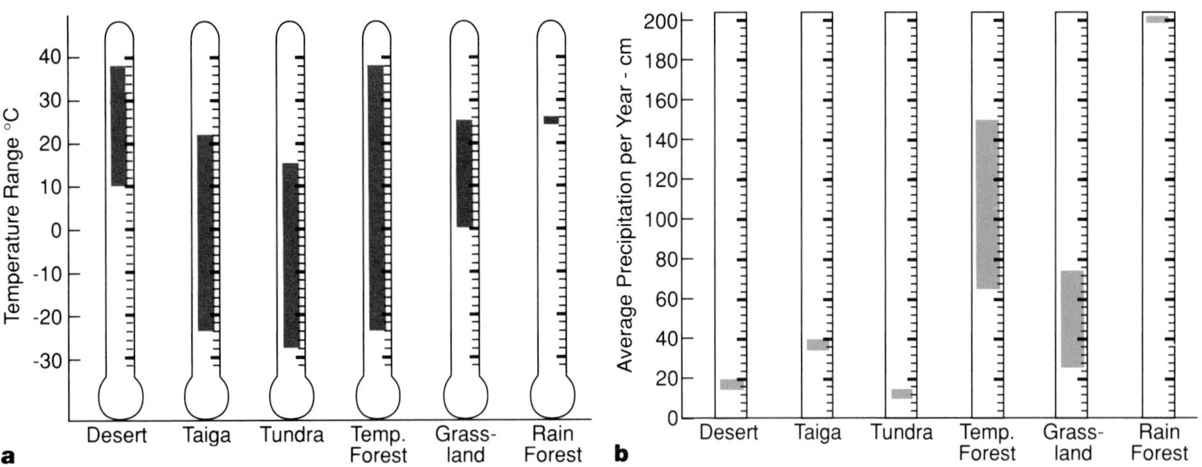

Figure 31-8 Each biome has a distinct temperature range (a) and amount of precipitation (b).

What is climate?

All of these factors—light, temperature, and precipitation—taken over many years is called the **climate** of an area. Some climates are wet while others are dry. Some are cold while others are warm or hot. The climate of an area helps determine what kinds of plants and animals can live there. You can see how living and nonliving factors act together to make an ecosystem.

Land Biomes

There are large areas on Earth that have similar climates and climax communities. A land area with a distinct climate and with specific types of plants and animals is called a **biome** (BI ohm). Each biome has its own distinct producers, consumers, and decomposers. A biome is made up of all the ecosystems on Earth that have similar climates and organisms. For example, there are many deserts on Earth. They all have very dry climates and similar organisms. All the deserts together make up the desert biome. Figure 31-10 shows the major biomes and some of the living things found in each one. Biomes include tropical rain forests, grasslands, deserts, temperate forests, taiga (TI guh), and tundra (TUN druh).

A biome usually covers parts of several continents. Each continent has several biomes. Look at the map in Figure 31-9. How many different biomes are there in the continent of North America?

Skill Check

Sequence: List the six major land biomes in order from (a) least to most precipitation and (b) from the lowest to highest temperature. Which biome is the wettest and warmest? Which biome is coolest and driest? *For more help, refer to the **Skill Handbook**, pages 706-711.*

Figure 31-9 The major land biomes

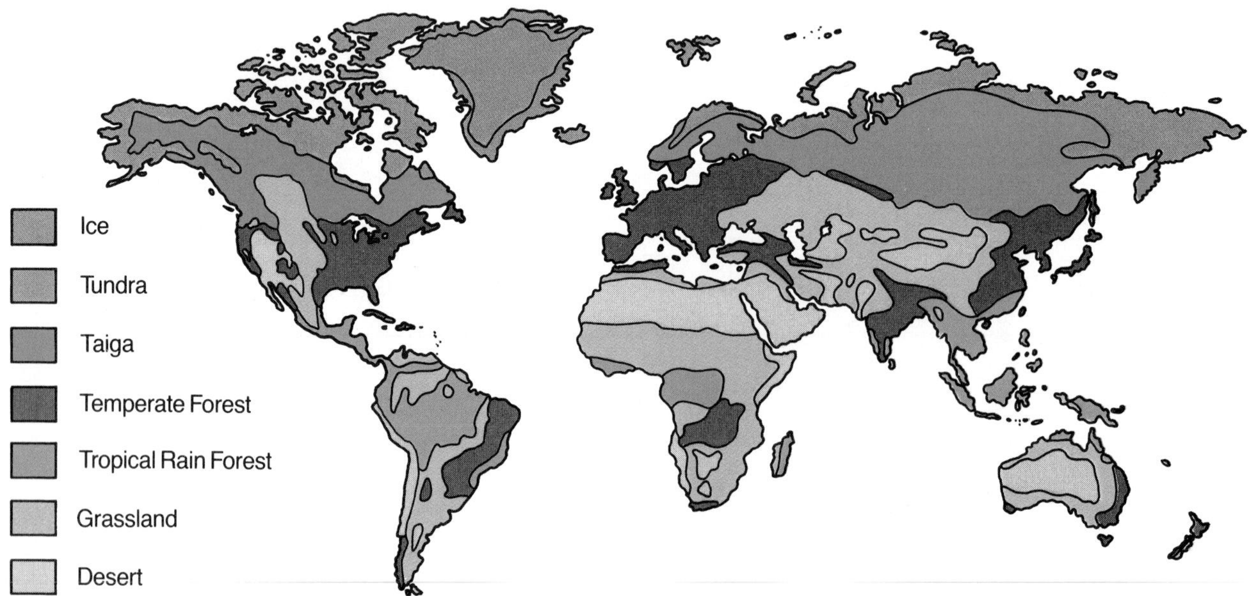

Figure 31-10 Biomes found on land

COMMON PLANTS	BIOME	COMMON ANIMALS
Vines Palm trees Orchids Ferns	 Tropical rain forest	Tree frogs Birds Insects Monkeys
Grasses	 Grassland	Antelope Gophers Rabbits Prairie dogs
Cacti Small bushes	 Desert	Lizards Snakes Scorpions Mice
Trees: Hickory Maple Beech Oak	 Temperate forest	Deer Black bears Squirrels Insects
Evergreen trees: Spruces Conifers Firs	 Taiga	Moose Weasels Mink
Lichens Mosses Grasses Small bushes	 Tundra	Caribou Musk oxen Polar bears

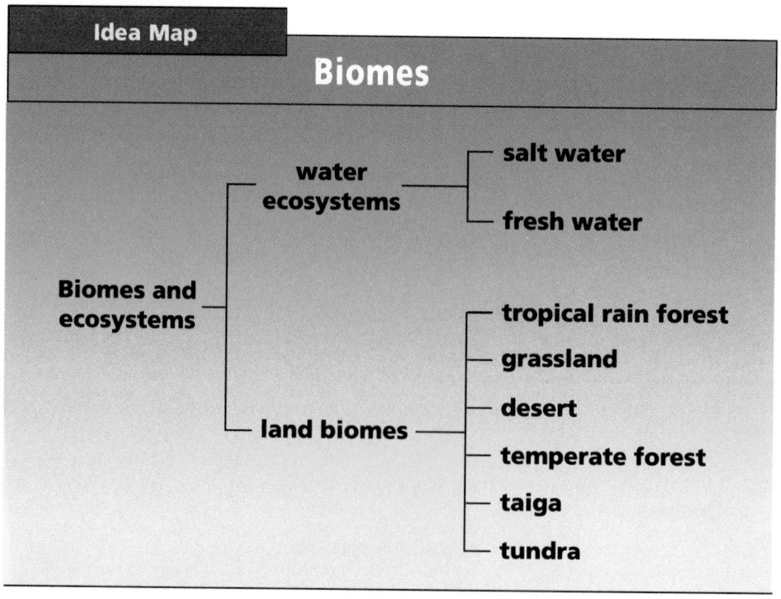

Idea Map

Biomes

- Biomes and ecosystems
  - water ecosystems
    - salt water
    - fresh water
  - land biomes
    - tropical rain forest
    - grassland
    - desert
    - temperate forest
    - taiga
    - tundra

Study Tip: Use this idea map as a study guide to biomes and ecosystems. Give an example of an animal that lives in each one.

Water Ecosystems

Water ecosystems are divided into fresh water and salt water. Streams, rivers, and lakes make up the freshwater ecosystem. Streams and rivers may start as water draining from a pond or lake and eventually flow into an ocean. The oceans make up the saltwater ecosystem because they contain large amounts of salts. Most freshwater organisms would die if placed in an ocean.

What is Earth's largest ecosystem?

The ocean ecosystem is Earth's largest ecosystem. It can be divided into smaller ecosystems. These include tidal pools, salt marshes, seashores, coral reefs, and the open ocean.

Check Your Understanding

11. Why can't palm trees live in the tundra?
12. What biome has the lowest average precipitation? Name three organisms that live there.
13. What major difference between freshwater and saltwater ecosystems determines what organisms will live there?
14. **Critical Thinking:** What might be an effect on today's biomes if all climates became warmer?
15. **Biology and Writing:** Write a paragraph of at least five sentences that describes the biome you live in.

Lab 31–2 Brine Shrimp

Problem: How does salt affect the growth of brine shrimp?

Skills

observe, form hypotheses, interpret data, design an experiment

Materials

flat toothpick
brine shrimp eggs
weak salt solution
strong salt solution
distilled water
3 small plastic cups
marking pen
hand lens

Procedure

1. Copy the data table.
2. Label three cups 1, 2, and 3 with a marking pen.
3. Fill the cups nearly full with the following:
   cup 1—distilled water
   cup 2—weak salt solution
   cup 3—strong salt solution
4. Dip a toothpick into the brine shrimp eggs. Brine shrimp are tiny animals that live in salt water. Place the eggs that cling to the toothpick into cup 1.
5. Repeat step 4 with cups 2 and 3. The eggs will float on the water.
6. Place the cups on a shelf in the classroom.
7. Write a **hypothesis** in your notebook to explain which cup will have the most brine shrimp at the end of 3 days.
8. **Observe:** Observe the cups for the next three days. Use the hand lens to look for small orange-colored animals swimming in a jerking motion. These objects are the brine shrimp that hatched from the eggs. Note the cup in which you see the first shrimp and the number of shrimp that hatch in each cup on each day. Record your data in the data table.

CUP	AMOUNT OF SALT	DAY WHEN MOST SHRIMP HATCHED	CUP WITH MOST BRINE SHRIMP

Data and Observations

1. How many days did it take for you to see the first brine shrimp?
2. In which of the three cups did they hatch first?
3. In which of these three cups did the most brine shrimp hatch? In which did they grow best?
4. In which of the three cups did the brine shrimp not hatch at all?

Analyze and Apply

1. What would you call the ecosystems in which you were trying to grow brine shrimp?
2. In which type of ecosystem do you think brine shrimp hatch and grow best?
3. **Check your hypothesis:** Was your hypothesis supported by your data? Why or why not?
4. **Apply:** How does salt affect the growth of brine shrimp?

Extension

Design an experiment to find out if there is a maximum strength of salt water at which brine shrimp will grow.

TECH PREP APPLYING TECHNOLOGY

Taking a Sidewalk's Temperature

Communities can be found right "under your nose." A parking lot contains a community of living things. So does a concrete sidewalk and the grassy area alongside it. Your backyard, a vacant lot, or a school football field also houses a community. What, then, makes a community different from an ecosystem? Once you begin to study both the living and the nonliving things that are present in a community, you are looking at the bigger picture—an ecosystem.

Identifying the Problem

What information can you gain by examining the sidewalk outside your school and any grassy areas next to the sidewalk? You could determine temperature changes that occur during the day. You could also conduct an inventory of the animal and plant types that are present. You might want to see if the humidity of the sidewalk ecosystem changes during the day. In this activity, you will explore the temperature changes that occur during the day and survey the plant and animal life in a nearby grassy area. Form a hypothesis about how temperature changes may differ between the sidewalk and the grassy areas during the day.

Technology Connection

An important ecosystem worldwide is the land farm. However, farmers are now able to move beyond raising crops only on land. Through advances in technology, one of the newest farming practices is aquaculture. Aquaculture is the farming of certain plants and animals in large tanks of water (see photo). Salmon, shrimp, catfish, and tilapia (a freshwater fish) are animals being raised through this new technology.

Collecting Information

Using references, look up and define the meaning of the terms *abiotic factors* and *biotic factors.* Give examples of each. Describe some of the biotic and abiotic factors that you might study in a sidewalk or grassy area ecosystem. Describe the biotic and abiotic factors that you actually will be studying in this activity.

Carrying Out an Experiment

1. Prepare a table in which you will record the temperature of the sidewalk and the grassy area nearby. You should record the data your group collects and the average data from all the groups in your class. Other classes will collect data at different times throughout the day. You will complete your data table the following day by adding the data collected by these classes.
2. Place a thermometer onto the chalk-marked area of the school sidewalk. Wait five minutes, and record the temperature and time.
3. Place the thermometer in the grassy area near the sidewalk. Wait five minutes and record the temperature.
4. Record and describe any animal life present on the sidewalk and in the grassy area.
5. Record and describe any plant life present on the sidewalk (or in any cracks) and in the grassy area.
6. Calculate class averages for the sidewalk temperature and grassy area temperature. Record these numbers for the time that your class meets.
7. At the next class session, complete your data table by adding the information from all other classes.
8. Prepare a graph that plots temperature changes for the sidewalk and grassy area against time in hours.

Career Connection

- **Fisher** Fishes the oceans or lakes; may also dig for clams and oysters near shore; helps set and empty lobster traps
- **Ecologist** Regulates, controls, and prevents air, land, and water pollution; may work for the federal government or local agencies; solves environmental problems
- **Park Ranger** Makes surveys of plant and animal life; works for the federal government in national parks, forests, and wildlife refuges; studies relationships among living things

Assessing Your Results

Explain why the sidewalk and the grassy area are ecosystems. How did the temperature of the sidewalk change during the day? How did the temperature of the grassy area change during the day? Which area showed a greater temperature change? Offer an explanation for why this may have been observed. How might this change affect the types of life forms that live there? Was your hypothesis supported by your data?

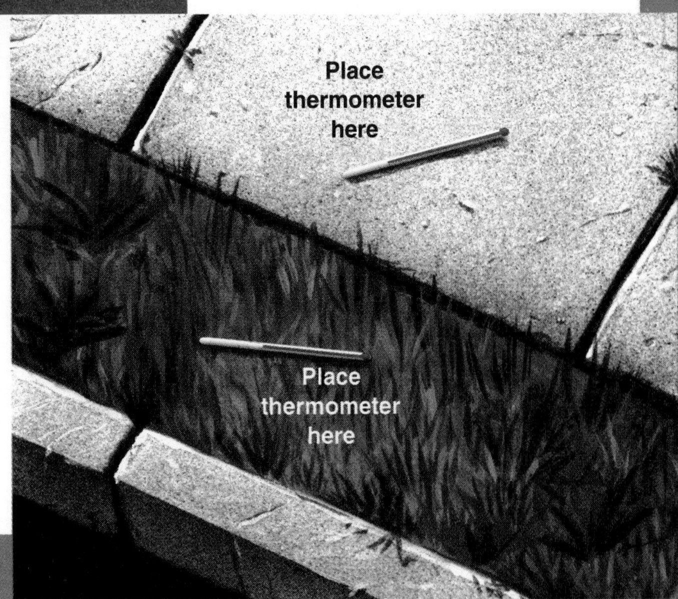

Chapter 31

Review

Summary

31:1 Ecosystems

1. Ecosystems are made up of living and nonliving parts.
2. Water is a nonliving part that is cycled through the ecosystem and is necessary for all living things.
3. The nitrogen cycle returns nitrates to the soil so they can be used for plant growth. The oxygen-carbon dioxide cycle provides carbon dioxide for plant photosynthesis and oxygen for plant and animal respiration.

31:2 Succession

4. The changes that take place in the living parts of a community as it gets older are called succession. Each stage of succession is represented by different plants and animals.
5. Succession in a water community results in formation of dry land.

31:3 How Living Things Are Distributed

6. Light, temperature, and precipitation make up an area's climate. Climate determines what living things live in an area.
7. Biomes are large land areas with distinct climate and living things. Water ecosystems include fresh water and salt water.

Key Science Words

biome (p. 664)
climate (p. 664)
climax community (p. 661)
ecology (p. 654)
ecosystem (p. 654)
nitrogen cycle (p. 656)
precipitation (p. 663)
succession (p. 660)
water cycle (p. 657)

Testing Yourself

Using Words

Choose the word from the list of Key Science Words that best fits the definition.

1. path of water through an ecosystem
2. has a distinct climate and certain plants and animals
3. community interacting with the environment
4. the last stage of succession
5. one factor that makes up the climate of an area
6. the average of temperature and precipitation over many years
7. the reusing of nitrogen in an ecosystem
8. the study of how living things interact with each other and their environment
9. changes in a community over time

Review

Testing Yourself *continued*

Finding Main Ideas

List the page number where each main idea below is found. Then, explain each main idea.

10. how water is recycled in an ecosystem
11. where ecosystems can be found
12. why the soil in a vacant field has a different temperature than the soil in a forest
13. what uses oxygen in an ecosystem
14. whether soil is a living or a nonliving part of an ecosystem
15. what the different biomes are in North America
16. what the first stage of succession on land is
17. why oxygen and carbon dioxide are important gases in an ecosystem
18. what happens to the depth of a pond as the community changes

Using Main Ideas

Answer the questions by referring to the page number after each question.

19. Which of the following are recycled through an ecosystem: oxygen, water, light, nitrogen, carbon dioxide? (pp. 655, 657, 658, 663)
20. How is the oxygen-carbon dioxide cycle important to an ecosystem? (p. 658)
21. How does climate affect plants and animals? (p. 663)
22. How can the soil be affected by the death of plants and animals? (p. 655)
23. Which has more salts, an ocean or river? (p. 666)
24. Why does a pond become more shallow as it ages? (p. 662)
25. What kinds of things separate one biome from another? (p. 664)

Skill Review ✓

For more help, refer to the ***Skill Handbook,*** *pages 704-719.*

1. **Sequence:** List 5 ecosystems in the area where you live. Sequence these ecosystems from smallest to largest.
2. **Infer:** How can boiling water become part of the water cycle?
3. **Infer:** Explain why leaves that fell on the forest floor in autumn disappear during the next summer.
4. **Design an experiment:** Design an experiment to show how you could tell if you are giving off carbon dioxide when you breathe out.

Finding Out More

TECH PREP

Critical Thinking

1. Suppose you set up a terrarium with plants, soil, decomposers, and water. You sealed it so it was air-tight. Could the plants live in this sealed environment? What is the one factor you would have to supply?
2. Use the map of biomes on page 664 to find out which biome you live in. What would happen if your biome got half the amount of precipitation it now gets?

Applications

1. Visit a park or area near your school. Record evidence for how nonliving things affect the living things.
2. Find out about Biosphere II. Describe some of the problems scientists had to solve while planning and building this project.

CHAPTER PREVIEW

Chapter Content

Review this outline for Chapter 32 before you read the chapter.

32:1 Resources and Human Activities
- Wildlife and Plants
- Soil
- Water
- Fossil Fuels

32:2 Problems from Pollution
- Air Pollution
- Water Pollution
- Acid Rain
- Land Pollution

32:3 Working Toward Solutions
- Conserving Our Resources
- Keeping Our Environment Clean

Skills in this Chapter

The skills that you will use in this chapter are listed below.

- In **Lab 32-1,** you will observe, experiment, interpret data, and classify. In **Lab 32-2,** you will observe, form hypotheses, interpret data, and design an experiment.
- In the **Skill Checks,** you will calculate, experiment, and understand science words.
- In the **Mini Labs,** you will outline and classify.

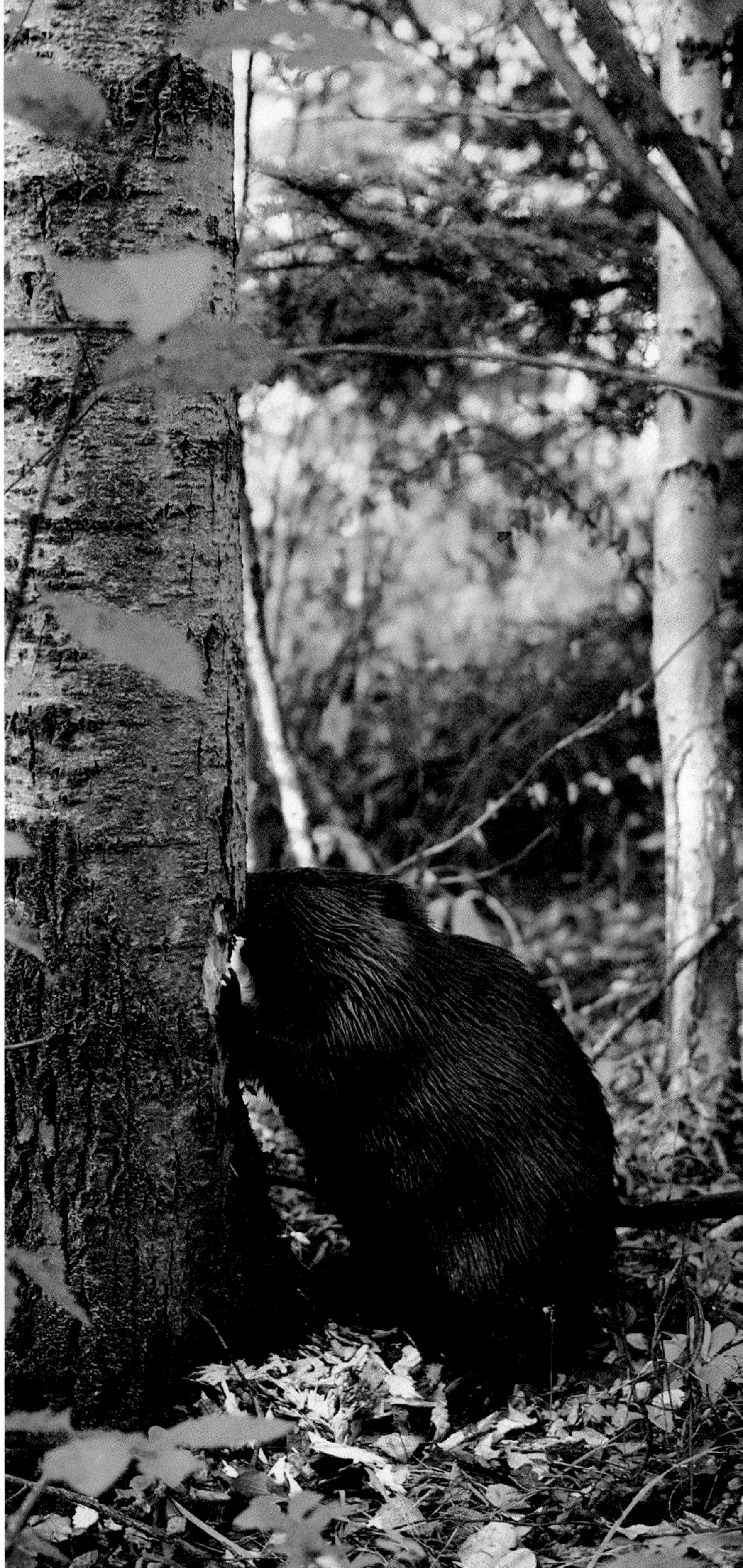

Chapter **32**

Solving Ecological Problems

The increasing human population has brought about vast changes in the environment. Humans have changed their environment by building factories, parks, shopping malls, and by farming. Other living things change the environment, too, but not as fast as humans do. The beavers on the opposite page cut down trees and dam up streams. This activity creates ponds where the beavers can build their dens. The river in the photo on this page has been littered with trash thrown away by humans. Both of these activities have changed the environment.

In this chapter, you will read about the changes humans have made to the environment. Think about some of the ways you may have changed your environment.

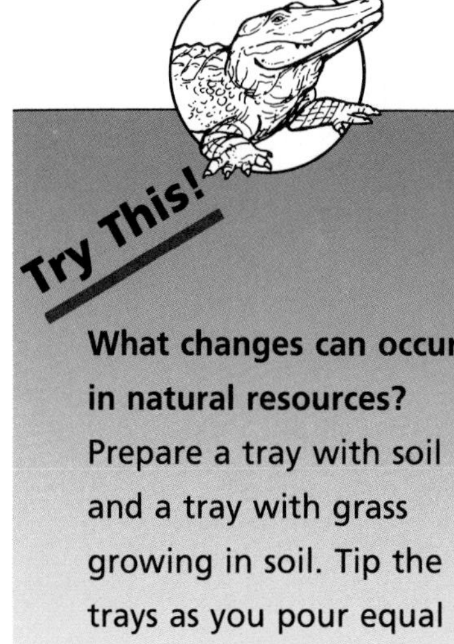

Try This!

What changes can occur in natural resources? Prepare a tray with soil and a tray with grass growing in soil. Tip the trays as you pour equal amounts of water over each. Collect the water as it runs off the soil. Compare the amount of soil that has washed out of each tray. What effect did the grass have?

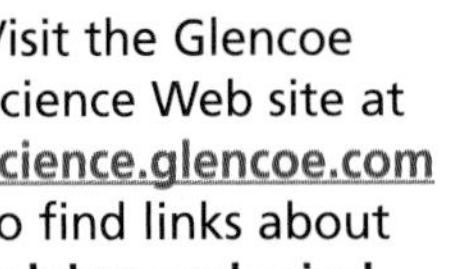

BIOLOGY Online

Visit the Glencoe Science Web site at science.glencoe.com to find links about **solving ecological problems.**

32:1 Resources and Human Activities

Objectives

1. **Explain** how wildlife and plants are affected by humans.
2. **Describe** how water and soil can be lost to the environment.
3. **Explain** why fossil fuels are being used up.

Key Science Words

natural resource
endangered
threatened
erosion
sediment
pollution
fossil fuel

Air, water, soil, plants, and animals are some of our natural resources. A **natural resource** is any part of the environment used by humans. If these resources are used unwisely, the lives of many things can be harmed. For instance, cutting down a forest will destroy many of the animal populations that live there. See Figure 32-1.

Wildlife and Plants

Wildlife and plants are important resources for humans. They provide food, fibers, and building materials. Nearly half the medicines being used today come from living things. Plants make oxygen, which all living things use during respiration. Soil that is covered with plants absorbs and holds water from precipitation. Thus, the soil is not worn away. Animals may be considered a renewable resource. A renewable resource is a resource that can be replaced within a person's lifetime. Animals can reproduce, keeping the population stable. Plants are a renewable resource, too. A field can regrow in about five years. A forest may take 25 to 100 years or more to regrow.

What is a renewable resource?

As the human population gets bigger, it needs more space to grow food and to build homes and factories. Where do we get the space? We use the land around us. As we take over this land, we use more natural resources, make more trash, and take away habitats of animals and plants.

Figure 32-1 Cutting down a forest destroys many of the plant and animal populations that live there.

a

b

c

Figure 32-2 The green pitcher plant (a) and whooping crane (b) are endangered species. Elephants (c) are killed illegally so their tusks can be carved into jewelry.

What happens to animal populations that are forced from their habitats? They must find new places to live. If they can't find enough space or food, they die. Unlike animals, plants can't move to a new area when their habitats are used for land development. These plants usually die. When a species of living thing no longer exists anywhere on Earth, it is extinct. Over the past three and one-half centuries, nearly 200 animal species have become extinct in the United States alone!

In the world today, over 1200 animal and plant species are endangered. **Endangered** means that a species is in danger of becoming extinct. Two endangered species are shown in Figure 32-2. Some living things are threatened. **Threatened** means that a species of a living thing is close to being endangered. There are almost 250 threatened species of plants and animals in the world.

How do animals and plants become threatened or endangered? Illegal hunting is one way. For example, African elephants are killed for their ivory tusks. Reptiles such as alligators and crocodiles are hunted illegally so their skins can be made into shoes, belts, and purses.

The rain forests of South America, Africa, and Asia are being cut down at an extremely high rate to make room for farmland. Every year, 200 000 square kilometers of rain forest are destroyed. This is an area equal to 76 football fields being destroyed every minute! Decreasing the forests means there are fewer plants releasing oxygen into the air and more carbon dioxide building up in the air. The ecological balance is upset. Why is this a problem for humans and other living things?

Scientists estimate that over 100 species of plants in these forests are becoming extinct each day. Extinction has always happened as a part of evolution, but by cutting down the rain forests, we speed up the process.

Mini Lab

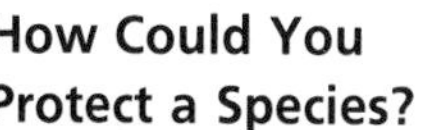

How Could You Protect a Species?

Outline: Pick a plant or animal and assume that it has become endangered. Make an outline showing what you would do to protect this species. What needs does this species have? *For more help, refer to the **Skill Handbook**, pages 706-711.*

Mini Lab

What Does Sediment Look Like?

Classify: Add soil to a jar of water. Cover and shake. Let the soil settle for three days. How many layers of sediment do you see? Describe the layers. Is the water clear? *For more help, refer to the* ***Skill Handbook****, pages 715-717.*

Soil

Soil is an important natural resource. The roots of plants hold soil in place. When plants are removed and the soil is not protected, erosion can take place. **Erosion** is the wearing away of soil by wind and water. When soil erodes, it is carried away faster than it can be replaced.

Soil that washes into streams and rivers can cause problems. As the current in a stream slows, the floating soil particles settle to the bottom. Material that settles on the bottom of the stream is called **sediment.** Sediment covers the habitats of water plants and animals. Fish eggs on the river bottom die because they don't get enough oxygen.

How does dry soil erode?

When bare soil dries out, it can erode quickly. Why? The wind easily separates the small particles of soil and blows them away. Erosion carries away good topsoil. The subsoil layer under the topsoil is exposed. Plants don't grow well in subsoil because it is packed down, contains little oxygen, and lacks the nutrients found in topsoil.

Water

Water is the most abundant resource on Earth. It is a renewable resource because it is cycled in the environment and can be reused. Water covers over 70 percent of Earth's surface. Over 70 percent of your body is water. You could not go without water for more than a few days at a time.

Water in the air is an important factor in climate and weather. The more water there is in the air, the wetter the climate will be in an area. Water is used to irrigate crops, for industry, and for cooling in electric power plants.

Figure 32-3 Soil washes into streams and becomes sediment (a). Wind can erode soil by blowing it away (b). Water is used to irrigate crops (c).

a

b

c

A major problem with lakes and ponds is rapid aging. Remember that lakes and ponds normally go through succession and become dry land. Producers living in the ponds and lakes need nutrients to live. Sometimes more nutrients are added to the water than are needed by these producers. The extra nutrients are in fertilizers washed from nearby farms, and in untreated sewage. These nutrients cause the pond or lake to go through succession faster than normal. Review the stages of aging in a lake as shown in Figure 31-7 on page 662. Aging is normally a slow, natural process, but it is speeded up because of pollution. **Pollution** is anything that makes the surroundings unhealthy or unclean.

Skill Check

Calculate: The United States has 1/22 of the world's population, but uses 1/3 of the world's energy resources. If the population of the United States is 268 million, what is the world population? *For more help, refer to the* ***Skill Handbook,*** *pages 718-719.*

Fossil Fuels

Coal, oil, and natural gas are all fossil fuels that humans use to run their cars, heat their homes, and produce electricity. A **fossil fuel** is the remains of organisms that lived millions of years ago. These remains were compressed over long periods of time at tremendous pressure. Because they take so long to form, fossil fuels are not renewable.

Why are fossil fuels not renewable?

Americans use fossil fuels for about 85 percent of our energy needs. As the human population continues to grow, more energy is needed to run larger cities and more cars, homes, and factories. The supply of fossil fuels is being used up at an alarming rate. When coal, oil, and natural gas supplies are gone, there will be no more. New energy sources must be found. Governments must help to save our fossil fuel supplies by passing laws limiting their use.

Check Your Understanding

1. What is the difference among extinct, endangered, and threatened species?
2. How is aging of ponds and lakes affected by pollution?
3. Why must new sources of energy be found?
4. **Critical Thinking:** Why should people who live in cities be concerned about rain forests being destroyed?
5. **Biology and Math:** It is estimated that 100 species of plants become extinct in rain forests each day. How many plant species become extinct in rain forests in a year?

32:2 Problems from Pollution

Objectives

4. **Discuss** the causes of air pollution and acid rain.
5. **Discuss** problems that come from nonbiodegradable chemicals that pollute water.

Key Science Words

toxic
smog
greenhouse effect
ozone
radon
pesticide
biodegradable
acid
base
acid rain

What is the effect of pollution on our bodies and on our environment? A number of health problems are now known to be caused by pollutants in our environment. Plants and animals are dying from pollutants. Humans are affected by pollution, too.

Air Pollution

Most air pollution is caused by burning coal, oil, gasoline, or natural gas. In North America, cars, lawn mowers, and other gasoline engines burn a total of 14 million gallons of fuel each hour. Gasoline produces several harmful gases when burned in an engine. These pollutants are given off in toxic amounts. **Toxic** means poisonous.

Many large cities, such as New York and Los Angeles, suffer from smog emergencies. **Smog** is a combination of smoke and fog. It is made thicker by chemical fumes. Most of the smog that blankets our cities is produced when the harmful gases from industry and auto exhausts react with the energy in sunlight. Trace how smog develops, using Figure 32-4b. This reaction produces new chemicals that are irritating to the eyes, nose, lungs and throat. A smog emergency occurs when the air is very still for a long time, allowing dangerous amounts of these gases to build up.

Factories and power plants that burn coal also make a lot of air pollution. Coal contains a small amount of sulfur.

Figure 32-4 Air pollution is caused by the burning of fossil fuels in engines (a). Smog is formed when harmful gases from burning fossil fuels react with sunlight (b).

a

b

a

b

Figure 32-5 Factories and power plants that burn coal cause air pollution (a). Chemicals from spray cans and plastic foam containers (b) destroy Earth's ozone layer.

When coal burns, the sulfur combines with oxygen in the air to form sulfur dioxide. Sulfur dioxide can kill humans. Power plants that burn coal also give off tiny particles of soot or dust. These particles block sunlight and get into our lungs when we breathe.

When fossil fuels are burned by factories and power plants, large amounts of carbon dioxide are made and released into the air. We don't usually think of carbon dioxide as being a pollutant, but in large amounts it can be. This carbon dioxide forms a layer around Earth and traps heat so it can't escape into the atmosphere. The trapped heat may cause temperatures on Earth to rise slowly, a process know as the **greenhouse effect.** If temperatures on Earth rise even a few degrees, polar ice may melt and the habitats of plants and animals may change. Deserts can form in areas that were once forests.

You may have heard or read that Earth's ozone layer is getting thinner. **Ozone** is a molecule made of three oxygen atoms. It forms a layer high above Earth's surface that keeps harmful radiation from reaching Earth.

Scientists have learned that ozone is being destroyed by certain chemicals found in spray cans, refrigerators, and air conditioners. The same chemicals are used to make plastic foam food containers. The chemicals destroy ozone, and harmful radiation can reach Earth's surface.

Some air pollution problems can be found inside your home. You may have heard of **radon**, a gas that gives off radiation. Radon is found naturally in the ground. In large amounts, radon may cause cancer. Radon can build up to dangerous levels in some buildings. Fans can be added to these buildings to blow the polluted air outside.

Skill Check

Experiment: Wash and rinse a plastic cup in distilled water. Use the cup to collect rain water, then evaporate the water. Look for a film of tiny particles on the side of the cup. How do particles get into the cup? *For more help, refer to the **Skill Handbook,** pages 704-705.*

How is ozone destroyed?

TECHNOLOGY

The Greenhouse Effect — A Computer Model

Many scientists are certain that the hotter than usual weather of 1988 was due to the greenhouse effect. They say that the gases from factories, home fuels, and car exhaust are causing the temperature on Earth to rise. These gases trap the heat energy from the sun. Any increase in the amounts of these gases allows more heat to be trapped in the atmosphere.

People who study weather and climate patterns have developed a computer model to explain the hot, dry summer of 1988. Worldwide climate data from several years were placed into the computer, which showed what climate patterns had looked like in the past.

The scientists added more atmospheric gases to the computer model. The gases were added in the amount in which they had been increasing. The computer showed the expected changes for the next 30 years.

What changes were predicted based on the increase in gases? These changes included warmer winters around the world, increased melting of glaciers and polar ice, a six-foot rise in sea level, some desert biomes becoming wetter, and rain forests becoming drier.

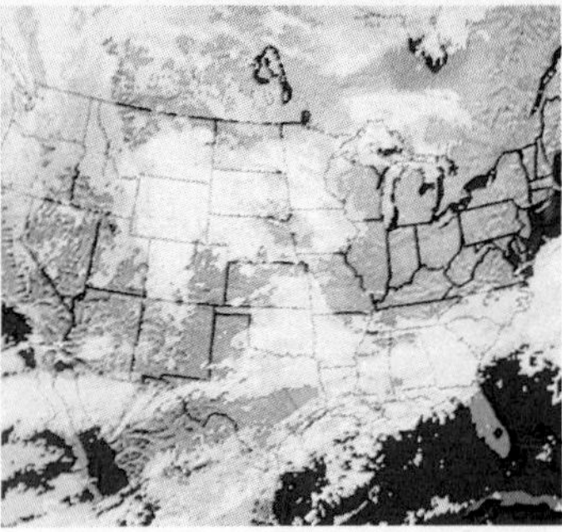

Not all scientists agree on the results of the computer model. Some say that the model has not included the effects of plants, clouds, soil moisture, and mountains on global weather. They also claim that it is too difficult to predict weather for more than a few weeks at a time. They agree that Earth is warming, but not as fast or as much as the model shows. Which group of scientists is correct? More work using the models to predict weather patterns may give us the answers.

Scientists can use computers to model weather patterns.

Water Pollution

Why are pesticides a problem to the environment?

Water from streams and rivers supplies this country with about half of its drinking water. However, much of it contains toxic wastes. Our water also contains untreated sewage, and fertilizers and pesticides washed from farmlands. **Pesticides** are chemicals used to kill unwanted pests such as rodents or insects. When pesticides are washed off farmlands into rivers and streams, they pollute the water. Pesticides can cause damage to soil and wildlife. Because pesticides are so dangerous, many of them are no longer allowed to be used in this country.

The best known of all pesticides is DDT. DDT is a chemical spray used to kill insects. It has been shown to cause cancer. DDT also affects the shells of bird eggs. The shells of birds that have eaten DDT in their food are thin and break easily. These eggs don't hatch, so no offspring are produced. Because of DDT, some bird species have become endangered. DDT is so harmful that the Environmental Protection Agency (EPA) ruled in 1973 that DDT could no longer be used in the United States. However, it is still widely used in other countries.

A serious problem today is that other pesticides similar to DDT are still being used. Many of these chemicals are not biodegradable (bi oh duh GRAYD uh bul.) **Biodegradable** means that something can be broken down by microbes into harmless chemicals and used by other living things. Because many pesticides are not biodegradable, they remain in an ecosystem. DDT, other pesticides, and weed killers also move through food chains. This means that humans and other meat-eating animals can have very high amounts of these chemicals in their bodies.

We use chemicals to rid our crops of pests, to wash our clothes, and to manufacture the many products we use. Many dangerous wastes are left over when chemicals are made. The waste products from chemical manufacturing are often toxic. These wastes may cause pollution.

A big problem with toxic wastes is how to get rid of them. One solution has been to bury them at dump sites. We now know that buried chemical wastes can escape into the air, the surrounding soil, and nearby water. When these wastes escape into the water, the water is polluted. Therefore, burying wastes is not a good solution.

Skill Check

Understand science words: pesticide. The word part *pest* means something that annoys. In your dictionary, find three words with the word part *pest* in them. *For more help, refer to the **Skill Handbook**, pages 706-711.*

a

b

clothing. If swallowed, DO NOT make person vomit. Call a doctor immediately. **Note to Physicians:** Emergency Information—call (415) 233-3737. This product contains a light petroleum solvent and may present an aspiration problem. Diazinon is an organophosphorus cholinesterase inhibitor. If signs and symptoms of cholinesterase inhibition develop, atropine is antidotal; 2-PAM may also be given in conjunction with atropine but should never be used alone.

ENVIRONMENTAL HAZARDS: This product is toxic to fish and wildlife. Birds feeding on treated areas may be killed. Keep out of any body of water. Do not contaminate ornamental fishponds. Do not contaminate water by cleaning of equipment or disposal of wastes. This pesticide is highly toxic to bees exposed to direct treatment or to residues remaining on treated areas. Do not apply when bees are actively visiting the crop. Application should be timed to provide the maximum possible interval between treatment and the next period of bee activity. **PHYSICAL OR CHEMICAL HAZARDS:** Do not use or store near heat or open flame.

DIRECTIONS FOR USE: It is a violation of Federal law to use this product in a manner inconsistent with its labeling.

READ ENTIRE LABEL. USE STRICTLY IN ACCORDANCE WITH LABEL PRECAUTIONARY STATEMENTS AND DIRECTIONS.

HOW TO USE: Use 2 teaspoonfuls (⅓ fl. oz.) per gal. of water. Begin spraying after blooming or when pests first appear and repeat at 7-10 day intervals as necessary. **Numbers in ( ) indicate days required between last application and harvest.**

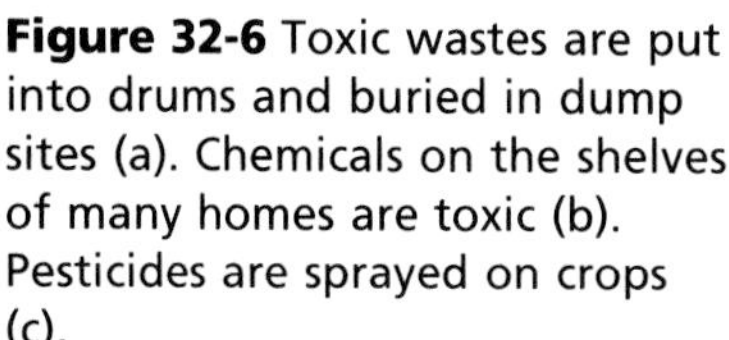

Figure 32-6 Toxic wastes are put into drums and buried in dump sites (a). Chemicals on the shelves of many homes are toxic (b). Pesticides are sprayed on crops (c).

c

pH

Lab 32–1

Problem: How does the pH of rainwater compare with that of some household products?

Skills

observe, experiment, interpret data, classify

Materials

pH paper
pH chart
forceps
marking pencil
7 glass slides
7 droppers
ammonia
baking soda
cola
vinegar
2 samples of rainwater
distilled water

Procedure

1. Copy the data table.
2. Label 5 glass slides 1 to 5.
3. Put a drop of liquid on each slide, as listed below. **CAUTION:** *If solutions are spilled on skin, rinse with water at once and notify your teacher.*
   Slide 1—baking soda
   Slide 2—distilled water
   Slide 3—vinegar
   Slide 4—cola
   Slide 5—ammonia
4. Pick up a piece of pH paper with the forceps.
5. Touch the pH paper to the liquid on slide 1 and remove the paper.
6. **Observe:** Compare the color of the wet end of the paper with the pH color chart.
7. Record the pH of the liquid on slide 1.
8. Discard the pH paper and rinse the forceps in tap water.
9. Test the liquids on slides 2 through 5 by repeating steps 4 through 8.
10. Put a drop of the first rainwater sample on a clean slide. Put a drop of the second rainwater sample on a second slide.
11. Test the pH of each sample of rainwater. Follow steps 4 through 8.
12. Record the pH of each sample.

SLIDE	CHECK CORRECT ANSWER			
	pH	ACID	NEUTRAL	BASE
1. Baking soda				
2. Distilled water				
3. Vinegar				
4. Cola				
5. Ammonia				
6. Rainwater				
7. Rainwater				

13. **Classify:** Finish filling in the data table by checking the correct column.

Data and Observations

1. Which of the materials tested was the strongest acid? How do you know?
2. Which of the materials tested was the strongest base? How do you know?

Analyze and Apply

1. Were your rainwater samples acidic or basic?
2. Why did you handle the pH paper with forceps?
3. What household product had a pH closest to your rainwater samples?
4. **Apply:** Why is acid rain a problem?

Extension

Experiment: Sample several bodies of water in your community and determine if the pH of the water is above or below 5.5.

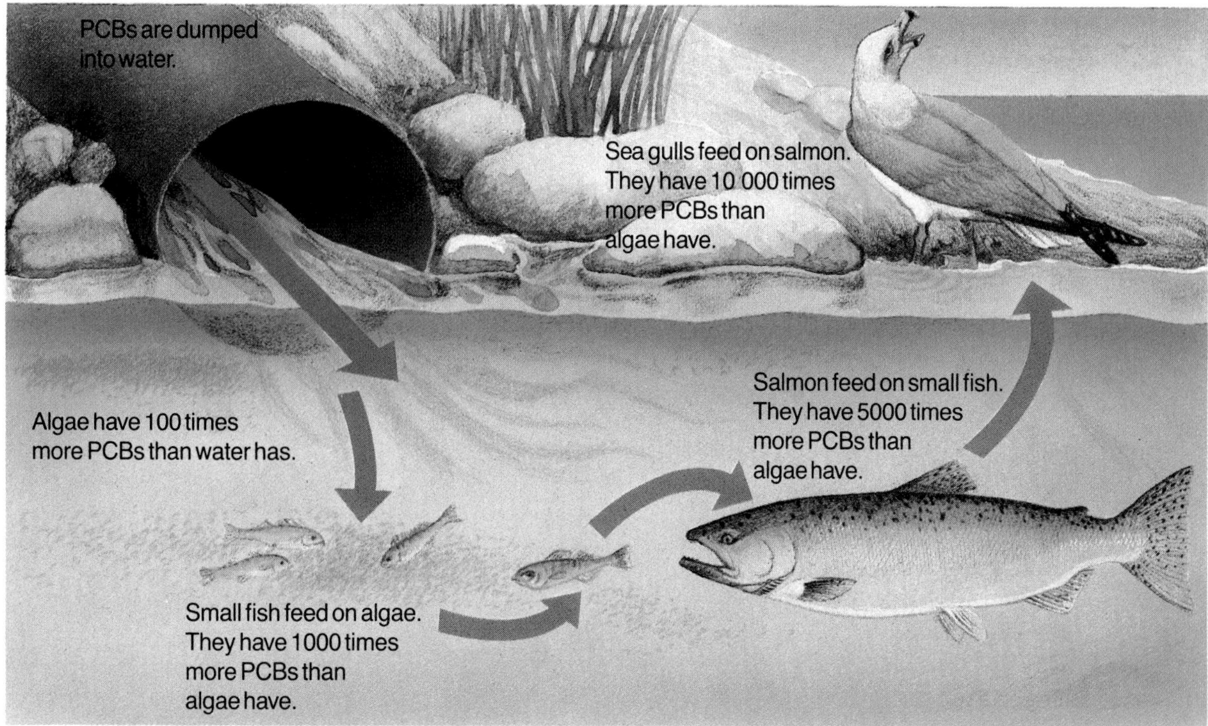
PCBs are dumped into water.
Sea gulls feed on salmon. They have 10 000 times more PCBs than algae have.
Algae have 100 times more PCBs than water has.
Salmon feed on small fish. They have 5000 times more PCBs than algae have.
Small fish feed on algae. They have 1000 times more PCBs than algae have.

Figure 32-7 PCBs increase in the bodies of animals at every step of a food chain. Why does this happen?

One class of dangerous chemicals present in water is PCBs. PCBs are toxic wastes produced when paints and inks are made. PCBs cause many serious health problems in people. Many lakes polluted with PCBs have signs posted around them warning people not to eat fish taken from the lake. Fish have large amounts of PCBs in their bodies because of their place in the food chain. Figure 32-7 shows what happens to PCBs in a food chain. Animals can't excrete PCBs from their bodies. At every step up in a food chain, the amount of PCBs in each animal increases.

Other pollutants found in water are heavy metals such as lead and mercury. These metals come from factories that dump their wastes into rivers or lakes. These chemicals, too, can build up in fish and the animals that eat fish. Lead is also a problem in paint found in old houses. Sometimes small children eat this paint and get sick.

Acid Rain

A very serious problem related to pollution is acid rain. A substance can be an acid, a base, or neutral. Scientists have devised a scale called the pH scale to measure how strong an acid or a base is. The pH scale is numbered from 0 to 14. See Figure 32-8.

Figure 32-8 pH scale

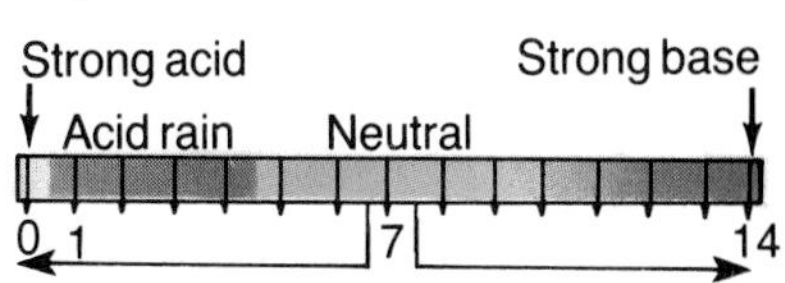
Strong acid
Strong base
Acid rain
Neutral
0
1
7
14

Figure 32-9 Acid rain damages buildings (a) as well as forests (b). What gas causes acid rain?

a

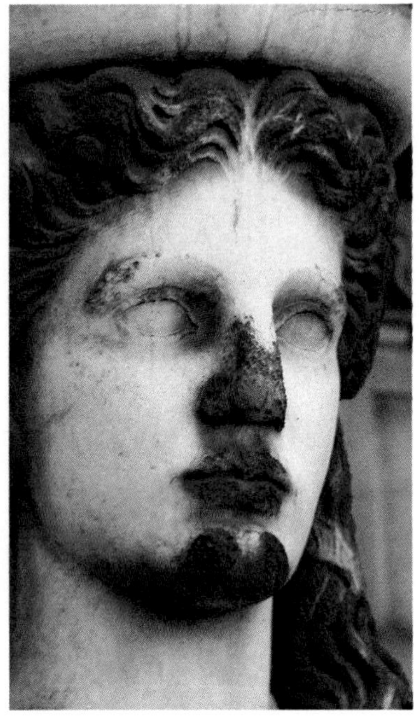

Acids, such as vinegar and lemon juice, are liquids that have pH values lower than 7. The stronger an acid is, the lower its pH. Pure water is neutral, neither acid nor base, and has a pH of 7. **Bases,** such as ammonia and lye, are liquids that have pH values greater than 7. The stronger a base is, the higher its pH. The pH of rain is normally above 5.5, or almost neutral. **Acid rain** is rain that has a pH between 1 and 5.5.

b

How is it possible for rain (or fog or snow) to be acid? Acids form in the air when gases such as sulfur dioxide react with water. Remember that these gases are formed by burning fossil fuels. As acid rain falls onto Earth, it damages forests, crops, soil, and buildings. See Figure 32-9. Acid rain has turned lakes and ponds into bodies of water with a pH value below 4.5. It has been estimated that 15 percent of all Minnesota lakes are now too acid for most living things. Over 200 lakes in New York State are dead due to acid rain. Nothing can live in them.

Acid rain seems to be worse for the east coast of the United States and Canada. Much of the countries' industry is in the east. Winds also carry acid rain from the western and central parts of these countries toward the east.

Bio Tip

Ecology: Each person in the United States produces more than four pounds of garbage each day. Eighty percent of this garbage can be recycled by using modern technology.

Land Pollution

We live in a plastic and chemical age. Think about all the items we use each day that are made of plastic. You probably have heard that plastic is not biodegradable . Plastics can't be broken down.

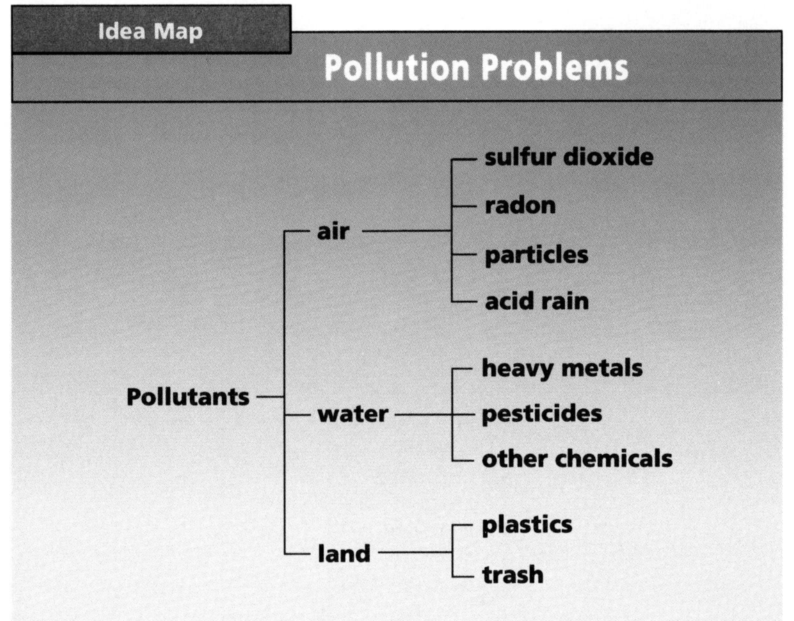

Study Tip: Use this idea map as a study guide to pollutants. Can you think of any other pollutants?

Suppose you set a bag filled with bits of food, paper scraps, and plastic jugs and bottles on the curb. The trash collectors forget to come to your house. In a few days, the bits of food begin to rot. The plastic containers, however, remain unchanged. After several weeks, the food and paper have rotted away. The plastic is still there. In fact, if you were to move away and then return in a few years, the plastic would still be unchanged. Think of all the plastic items people throw away each day. None of them are biodegradable. Dumps and landfills are becoming filled with these items. We are running out of places to put them.

Check Your Understanding

6. What is a major source of air pollution?
7. In what parts of the food chain do pesticides build up?
8. What problems arise from the use of plastics?
9. **Critical Thinking:** Bald eagles almost became extinct because DDT prevented their eggs from hatching. The eagles got the DDT from the fish they ate. How did the fish get the DDT?
10. **Biology and Writing:** In this section, you learned about damage to humans from air, water, and land pollution. Tell about another kind of pollution that damages humans.

Pollution

Lab 32–2

Problem: What is the effect of pollutants on yeast?

Skills

observe, form hypotheses, interpret data, design an experiment

Materials

4 test tubes
marking pencil
test-tube rack
5 droppers
blue test liquid
dead yeast mixture
live yeast mixture
detergent
hydrogen peroxide
water

Procedure

Yeast cells are living organisms. In this lab, they will be treated with different pollutants to see if they are killed.

1. Copy the data table.
2. Number four test tubes 1 to 4.
3. Add the following to each tube: Tube 1—10 drops of dead yeast mixture and 10 drops of water; Tube 2—10 drops of live yeast mixture and 10 drops of water; Tube 3—10 drops of detergent and 10 drops of live yeast mixture; Tube 4—10 drops of hydrogen peroxide and 10 drops of live yeast mixture.
4. Fill all tubes almost full with blue test liquid.
5. **Observe:** Record the color in each tube. This is day 1.
6. Write a **hypothesis** to state whether the yeast will be killed by detergent or hydrogen peroxide.
7. Leave the tubes in the rack overnight.
8. Record the color in each tube on day 2.
9. Dispose of your materials as directed by your teacher.

TUBE	CONTENTS	COLOR IN TUBE DAY 1	COLOR IN TUBE DAY 2	YEAST ALIVE OR DEAD
1	Dead yeast and water			
2	Live yeast and water			
3	Detergent and live yeast			
4	Hydrogen peroxide and live yeast			

Data and Observations

1. What color was each tube on day 1?
2. What color was each tube on day 2?

Analyze and Apply

If the tube contents change from blue to green overnight, the yeast cells are alive. If the tube contents stay blue, the yeast cells are dead. Complete the last column of the data table, then answer the questions.

1. Which tubes contained live organisms the first day? Which tubes contained live organisms the second day?
2. **Interpret data:** The blue liquid changes to green if carbon dioxide gas is present. How is this gas related to living things?
3. What was the purpose of tubes 1 and 2?
4. **Check your hypothesis:** Did your data support your hypothesis? Why or why not?
5. **Apply:** From this lab, what would you say might be the effect of hydrogen peroxide and detergent on some other living things?

Extension

Design an experiment to test the effect of an acid and a base on yeast.

32:3 Working Toward Solutions

Objectives

6. Explain methods of conserving resources.

7. Discuss ways to keep the environment clean.

Key Science Words

recycling

Our environment is in trouble. It is being damaged by pollutants in the air, soil, and water. We are using up our resources and making more and more wastes as our population increases. Many resources can't be replaced.

Look at the area where you live. You probably can find examples of water, land, and air pollution. What can be done to solve pollution problems before they begin to limit population growth? How can we use resources wisely?

Conserving Our Resources

There are several solutions to the problem of endangered plants and animals. One depends on the United States government. Another is up to you as an active, caring individual.

Today, the government has set aside over 500 areas called National Wildlife Refuges. A refuge is an area that protects or shelters living things from being harmed by humans.

What is the Endangered Species Act?

In 1973, the United States Congress passed the Endangered Species Act. This law states that anyone found guilty of killing, capturing, or removing any endangered species from its environment can be fined up to $20 000 and jailed for one year. The law also protects the habitat of any endangered species. In order for government refuges and laws to work, everyone must cooperate. This means that everyone must obey the law and make every effort not to reduce our wildlife populations.

Erosion of soil can be slowed by planting crops across the slope of the land, rather than up and down, as shown in Figure 32-10. Strips of grass can be planted between fields to hold the soil. Rows of trees can be planted as windbreaks so the wind doesn't blow away dry soil.

Figure 32-10 Erosion of soil can be slowed by planting crops across a slope.

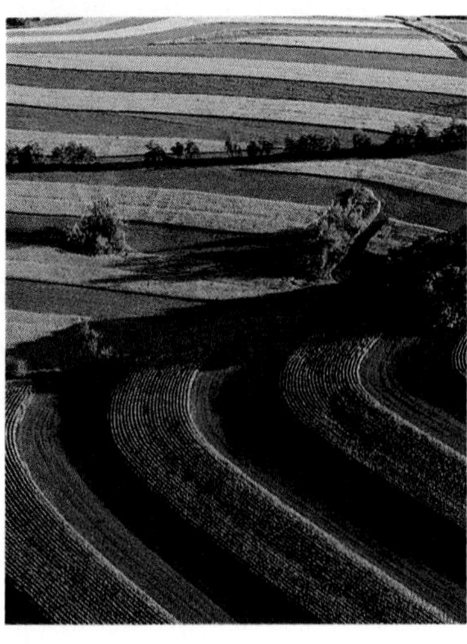

Water can be reused if wastes and harmful chemicals are first removed. Conserving water would also guard the supply of this precious resource. Each time an average-size lawn is watered, up to 2000 gallons of water is used. That's enough water to fill your bathtub 40 times! Now, imagine thousands of people watering their lawns daily. You can see that a lot of water is used on lawns.

Do you allow the water to run as you brush your teeth? If you do, you are sending about five gallons of water down the drain each time you brush. Turning off the tap is a simple solution to this problem.

Keeping Our Environment Clean

Why do some power plants use scrubbers?

How can we clean up our environment and keep it clean? Many steps have been taken to improve our air and water. New power plants that use coal must have scrubbers in their smokestacks. The scrubbers help to remove sulfur dioxide and particles from the smoke before it is released into the air. More improvements in air quality could be made if cleaner sources of energy were used. Solar and nuclear energy can be used to make electricity, instead of using coal or oil.

How has water quality improved? Several large bodies of water in the United States have been cleaned up in the last 20 years. Parts of Lake Erie were once so polluted with chemicals that people could not eat Lake Erie fish. Today people can swim and fish in Lake Erie again.

Wise use of technology also can help solve the pollution problem. Scientists have found that insect pests can be destroyed by their natural enemies. Aphids are tiny insects that suck the juices from plants. The ladybird beetle is a natural predator of aphids. Ladybird beetles can be released in an area where there are aphids. The aphids will be destroyed in a natural way without the use of pesticides.

Bacteria and viruses also can destroy some kinds of insects without hurting other living things. For example, bacteria can be used to control gypsy moths. The gypsy moth caterpillar has destroyed many oak forests. A bacterium found in the soil is able to kill this caterpillar. Biologists grow this type of bacterium in large tanks. The bacteria are then sprayed on tree leaves. The gypsy moth caterpillars die several hours after eating leaves with bacteria on them. Biologists are discovering ways to solve problems in our environment without creating new ones.

Figure 32-11 Soil bacteria can interrupt the life cycle of the gypsy moth (a). A ladybird beetle eats aphids (b).

a

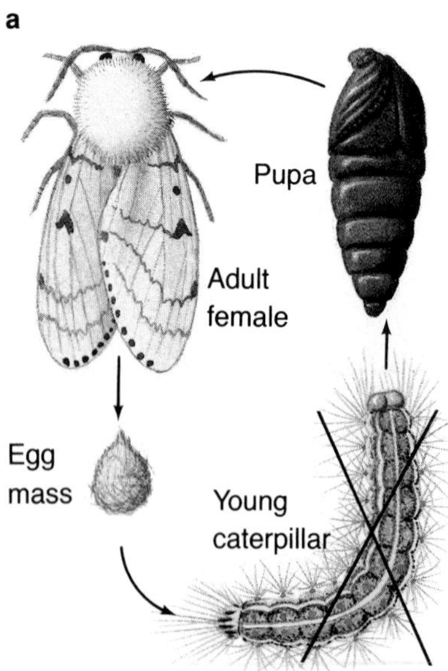

b

a

b

Figure 32-12 Recycling (a) can reduce the amount of trash that goes into landfills.
Over-packaged products (b) cause landfills to fill up. Why?

There are many things that people can do to help prevent pollution. **Recycling** is the reusing of resources. There are recycling centers for glass, plastic, and aluminum. Used paper and cardboard can be made into new paper. Over 500 000 trees could be saved every week if everyone recycled just the Sunday newspapers.

Being a wise consumer can also help. You can buy products wrapped in paper or packed in cardboard or glass. You can avoid buying individually wrapped items such as cheese slices. Every item that is packed in something other than plastic will help to reduce landfills.

How can you help reduce pollution? You can walk or bike short distances instead of driving, shop wisely and carefully for items that won't pollute the environment, recycle, clean up litter, turn off lights and water, and use natural pesticides. The answers to pollution problems are neither easy nor inexpensive. To have a good quality of life, everyone must work to keep the environment clean.

Bio Tip

Ecology: Using one metric ton of recycled paper for printing is equal to cutting down 17 trees to make into paper pulp.

Check Your Understanding

11. List one way to conserve each of the following: animals, water, trees.
12. How have laws helped in cleaning up our environment?
13. How can gypsy moths be controlled naturally?
14. **Critical Thinking:** How can producing more fuel-efficient cars help reduce air pollution?
15. **Biology and Reading:** What are at least two steps that have been taken to keep the air clean?

Chapter 32 Review

Summary

32:1 Resources and Human Activities

1. Wildlife and plants are natural, renewable resources. They become endangered when they die faster than they can renew themselves.
2. Soil erosion occurs when soil is carried away by wind and water faster than it can be replaced. Water is a valuable resource. Pollution can cause ponds and lakes to go through a rapid aging process.
3. Fossil fuels are a nonrenewable source of energy. They are rapidly being used up.

32:2 Problems from Pollution

4. Air pollution is caused by burning fossil fuels in large amounts.
5. Heavy metals and pesticides can pollute water and build up in fish and other animals. Plastics and other materials that are not biodegradable fill up landfills and pollute Earth's surface.

32:3 Working Toward Solutions

6. Resources can be conserved by effective laws and by each person using resources wisely and carefully.
7. Keeping the environment clean depends on industry, government enforcement of laws, the development of clean energy sources, and the actions of all people.

Key Science Words

acid (p. 684)
acid rain (p. 684)
base (p. 684)
biodegradable (p. 681)
endangered (p. 675)
erosion (p. 676)
fossil fuel (p. 677)
greenhouse effect (p. 679)
natural resource (p. 674)
ozone (p. 679)
pesticide (p. 680)
pollution (p. 677)
radon (p. 679)
recycling (p. 689)
sediment (p. 676)
smog (p. 678)
threatened (p. 675)
toxic (p. 678)

Testing Yourself

Using Words

Choose the word from the list of Key Science Words that best fits the definition.

1. has a pH value less than 7
2. poisonous or harmful
3. particles deposited in a stream
4. wearing away of soil
5. close to being endangered
6. a chemical that kills unwanted organisms
7. a gas that gives off radiation
8. a combination of smoke and fog
9. ability of a chemical to be broken down by microbes
10. process that causes heat to be trapped near Earth's surface

Testing Yourself *continued*

Finding Main Ideas

List the page number where each main idea below is found. Then, explain each main idea.

11. what happens to a pond when extra nutrients are washed into it
12. what can happen to toxic waste buried in dump sites
13. what the purpose of the Endangered Species Act is
14. how fossil fuels are formed
15. why the ozone layer is being destroyed
16. how acid rain is formed
17. why bare soil is easily moved about by the wind
18. why plants and animals are renewable resources
19. how water can be conserved

Using Main Ideas

Answer the questions by referring to the page number after each question.

20. What are two ways that people make living things endangered or threatened? (p. 675)
21. What are two energy sources that are cleaner than fossil fuels and are renewable? (p. 688)
22. What is radon? (p. 679)
23. What effects does DDT have on living things? (p. 681)
24. How can bacteria be used to control plant pests naturally? (p. 688)
25. Where does sulfur dioxide in the air come from? (p. 679)
26. How do pollutants get into our water supply? (p. 680)
27. Why do you suppose water at a water treatment plant is checked for mercury and lead? (p. 683)

Skill Review

*For more help, refer to the **Skill Handbook,** pages 704-719.*

1. **Outline:**Make an outline of the different kinds of pollution discussed in this chapter.
2. **Calculate:** Assume there are 4800 species of animals in your state. Pollution is causing 30 species to die each week. Calculate how many years it will be before all animals are gone.
3. **Classify:** Classify the following items as to whether each is biodegradable: banana peel, plastic bottle, newspaper, DDT, plastic foam container, sandwich.
4. **Design an experiment:** Design an experiment to show whether newspapers are biodegradable.

Finding Out More

Critical Thinking

1. Marigolds produce chemicals that drive away insect pests. How could you use this knowledge to protect your garden?
2. Some people think that zoos and refuges are a waste of time and money. Give reasons for and against this statement.

Applications

1. Tape cardboard coated with a thin layer of vaseline to various locations in your community. After a few days, remove the cards and examine the particles trapped in the vaseline. Try to find out where they came from.
2. Write a report describing why Love Canal in New York State caused problems in the environment.

Biology in Your World

Disrupting a Delicate Balance

As you have seen in this unit, relationships are important in a community. If the survival of a single species is threatened, it may result in a chain reaction. We live in a rather fragile environment. Even well-meaning interference can be disastrous.

LITERATURE

Environmental Awakening

In 1962, biologist Rachel Carson wrote about the dangers of spraying with pesticides in her book, *Silent Spring.* She pointed out that robins died after eating earthworms that had picked up DDT residues. DDT had been sprayed on the leaves of elm trees. It was used to kill the beetles carrying the fungus that caused Dutch Elm disease. Carson also warned that pesticides could contaminate human food supplies.

Because of Carson's book, the use of certain pesticides was restricted in the United States.

HISTORY

A Ticking Time Bomb

In the 1920s, Love Canal was a partly-dug canal in the city of Niagara Falls, New York. It became a disposal site for household trash. Nearby chemical plants also dumped wastes in it. When the site was covered in 1958, about 20 000 metric tons of waste had been buried. A school was built on one edge of the canal. Homes were built on the other side. After very heavy rains in the mid 1970s, chemicals seeped out of the canal. These chemicals have been known to cause cancer, birth defects, miscarriages, and liver damage. Dangerous levels of these chemicals were found in the basements of many homes. In 1978-79, families were told to move from their homes. In 1990, the government allowed people to move into the homes.

LEISURE

Making Your Own Paper

Recycle old newspapers yourself. First, tear the paper into small pieces. Put the pieces into a blender with five cups of water. Blend until you get a watery pulp. Put a piece of window screen into a dishpan with about three centimeters of water. Pour one cup of the paper pulp over the screen and spread it evenly with your fingers. Lift the screen and drain. Put the screen and pulp between several sheets of newspaper, and press with a board to squeeze out the extra water. Remove the pulp from the screen and leave it on dry newspaper overnight. When the pulp is dry, carefully peel off your handmade paper.

CONSUMER

Stop the Throw-away Lifestyle

Landfills are overflowing. Some resources are being used up. What can you do to help? Instead of creating more waste, you can make some smart choices as a consumer.

Try not to buy products that are packaged in single-serving sizes or in unnecessary plastic. Whenever you can, buy things in large containers or in bulk. If possible, take your own containers and bags with you when you shop.

Set up a system that encourages recycling. Separate aluminum cans, glass containers, and newspapers and recycle them if you have a recycling center near you.

When you buy a new sweater to replace one that does not fit anymore, do not put the old one in the trash. Give it to someone who can use it or add it to the rag bag.

Substitute paper for plastic whenever you can. The plastic rings connecting a six-pack of canned soft drinks take a long time to decay in a landfill. Also, birds can get caught in the plastic rings and be strangled. Some types of plastic give off toxic fumes when they are burned.

Do Earth a favor. Think of the long-term effects before you buy something.

APPENDIX A

Measuring in SI

Careful measurement is an important part of science. But even if you measure very carefully, your figures will have no meaning if your measuring unit means something different each time you use it. To be practical, a measurement unit must always mean the same thing to everybody. A *standard* unit is a definite amount used by everyone when measuring.

Most people in the world and all scientists use the International System (SI) of units. This is a modern form of the metric system.

In SI, all of the units are related to each other by the same set of prefixes. It is easy to change to smaller or larger units simply by multiplying or dividing by ten.

The main units used for measuring are the meter–for distance, gram–for mass, and liter–for volume.

Below is a table of the SI prefixes, their meanings, and rules for changing from one unit to another.

I. Rules for expressing units of measurement

A. Changing meters
   1. Meters can be changed into smaller units such as decimeters (dm), centimeters (cm), or millimeters (mm).
   2. Meters can be changed into larger units such as dekameters (dam), hectometers (hm), or kilometers (km).

B. Changing grams
   1. Grams can be changed into smaller units such as decigrams (dg), centigrams (cg), or milligrams (mg).
   2. Grams can be changed into larger units such as dekagrams (dag), hectograms (hg), or kilograms (kg).

C. Changing liters
   1. Liters can be changed into smaller units such as deciliters (dL), centiliters (cL), or milliliters (mL).
   2. Liters can be changed into larger units such as dekaliters (daL), hectoliters (hL), or kiloliters (kL).

II. Rules for changing from one unit to another

A. When changing from a *smaller unit to a larger unit* you must *divide.* (This type of change is shown by the arrow that points upward in the table.)

SI UNITS

Prefixes	Symbol	Meaning		Larger unit to Smaller unit ↓	Smaller unit to Larger unit ↑
kilo	k	1000	thousand		
hecto	h	100	hundred		
deka	da	10	ten		
gram, meter, liter	g,m,L		main unit		
deci	d	0.1	tenth		
centi	c	0.01	hundredth		
milli	m	0.001	thousandth		

1. If you move up the table from any prefix to
   one above it, then divide by 10.
   two above it, then divide by 100.
   three above it, then divide by 1000.
   four above it, then divide by 10 000.
   five above it, then divide by 100 000.
   six above it, then divide by 1 000 000.
2. Examples
   Change 13.2 grams (g) to hectograms (hg). (*Hecto* is two places above gram in the table. Divide by 100.)
   13.2 g ÷ 100 = 0.132 hg

   Change 2.6 decimeters (dm) to kilometers (km).
   (*Kilo* is four places above *deci* in the table. Divide by 10 000.)
   2.6 dm ÷ 10 000 = 0.00026 km

   Change 14 milliliters (mL) to liters (L). (Liter is three places above *milli* in the table. Divide by 1000.)
   14 mL ÷ 1000 = 0.014 L

B. When changing from a *larger unit to a smaller unit* you must *multiply.* (This type of change is shown by the arrow that points downward in the table.)
1. If you move down the table from any prefix to
   one below it, then multiply by 10.
   two below it, then multiply by 100.
   three below it, then multiply by 1000.
   four below it, then multiply by 10 000.
   five below it, then multiply by 100 000.
   six below it, then multiply by 1 000 000.

2. Examples
   Change 25 hectograms (hg) to dekagrams (dag). (*Deka* is one place below *hecto* in the table. Multiply by 10.)
   25 hg × 10 = 250 dag

   Change 126 meters (m) to millimeters (mm). (*Milli* is three places below meter in the table. Multiply by 1000.)
   126 m × 1000 = 126 000 mm

   Change 0.08 kiloliters (kL) to dekaliters (daL).
   (*Deka* is two places below *kilo* in the table. Multiply by 100.)
   0.08 kL × 100 = 8.0 daL

APPENDIX B

Classification of Living Things

Scientists recognize five kingdoms of living things: the Moneran Kingdom, Protist Kingdom, Fungus Kingdom, Plant Kingdom, and Animal Kingdom. These five kingdoms are divided into smaller groups. Scientists do not always agree about the grouping of living things within a kingdom, or that there should be five kingdoms. This appendix lists the classification groups that are generally accepted by most scientists. Use this appendix as you study Chapters 3-8.

MONERAN KINGDOM

One-celled; no nucleus; cell wall and cell membrane present.

TRUE BACTERIA

Includes several phyla; may be round, rod-shaped, or spiral-shaped; cannot make food. Example: bacteria that cause sore throats.

Blue-green Bacteria Phylum

Contain colored pigments, usually blue-green; can make food. Example: *Anabaena.*

PROTIST KINGDOM

Nucleus present; one-celled or multicellular.

ANIMAL-LIKE PROTISTS

One-celled; most move about; classification based on type of movement; cannot make food.

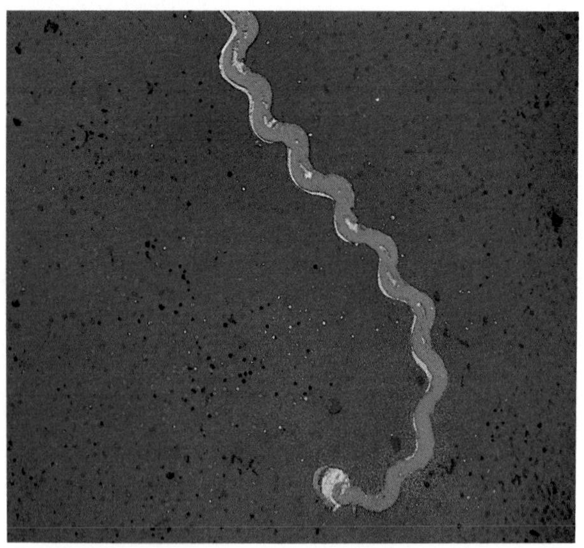

Amoeba Phylum

Move with false feet. Example: *Amoeba proteus.*

Ciliate Phylum

Move with cilia. Example: *Paramecium.*

Flagellate Phylum

Move with flagella. Example: *Trypanosoma.*

Sporozoan Phylum

Do not move. Example: *Plasmodium.*

PLANTLIKE PROTISTS

Some can move about; one-celled or multicellular; can make food.

Euglena Phylum

Move with flagella; one-celled; no cell wall. Example: *Euglena gracilus.*

Diatom Phylum

One-celled; golden-brown color; has two-part glass covering. Example: *Navicula.*

Dinoflagellate Phylum

One-celled; red or brown color. Example: *Gongaulax.*

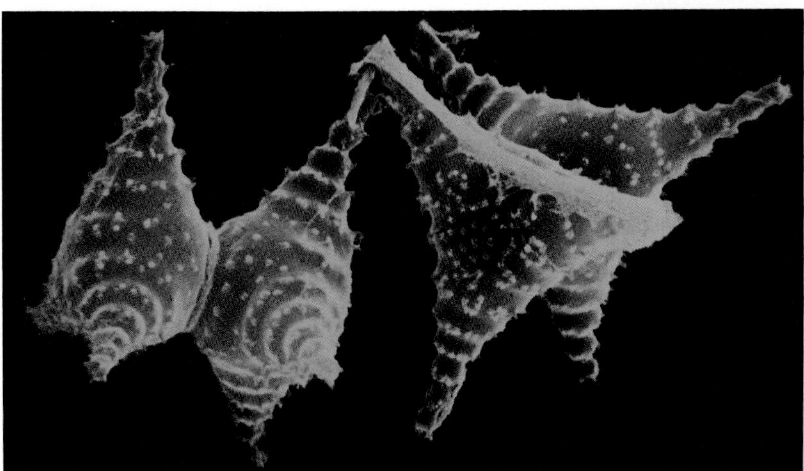

Green Algae Phylum
One-celled or multicellular. Example: *Spirogyra* (spi ruh GI ruh).

Red and Brown Algae Phyla
Multicellular. Example: kelp.

FUNGUSLIKE PROTISTS
Have a life cycle that includes a moving, slimy mass, an amoebalike stage, and a spore-forming stage; cannot make food. Examples: two phyla that include the slime molds.

FUNGUS KINGDOM
Body consisting of hyphae; reproduce by spores.

Sporangium Fungus Phylum
Cannot make food; spores formed in sporangia. Example: bread mold.

Club Fungus Phylum
Cannot make food; spores formed in club-shaped structures. Examples: mushrooms, smuts.

Sac Fungus Phylum
Cannot make food; spores formed in saclike structures. Examples: yeast, *Penicillium.*

Lichens
Can make food; combination of fungus and organism with chlorophyll. Examples: British soldier lichen, reindeer moss.

PLANT KINGDOM
Green; chlorophyll in cells; can make food; cell walls present.

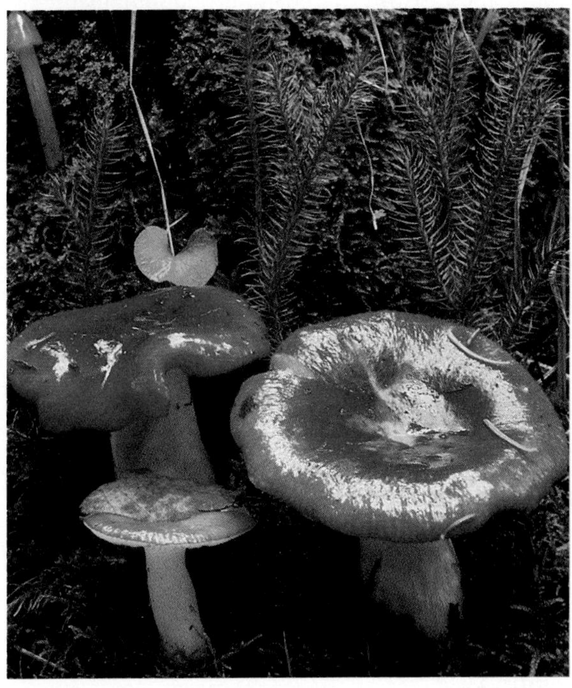

NONVASCULAR PLANTS
No tubes for carrying materials throughout plant; reproduce by means of spores; no roots, stems, leaves; includes several phyla (botanists use the term *division* instead of *phylum*). Examples: mosses and liverworts.

VASCULAR PLANTS
Tubes for carrying materials throughout plant; roots, stems, and leaves present; includes several phyla.

Fern Phylum
Reproduce with spores. Example: Boston fern.

Conifer Phylum
Reproduce with seeds; seeds in cones. Examples: pines, firs.

Flowering Plant Phylum
Reproduce with seeds; seeds in flowers. Examples: maple tree, daisy, wheat.

ANIMAL KINGDOM
Cannot make food; can move about; multicellular.

INVERTEBRATES
Animals without backbones; includes several phyla.

Sponge Phylum
Two cell layers; no symmetry; body with pores and canals; no tissues or organs; reproduce sexually and asexually. Examples: glass sponges, bath sponges.

Stinging-cell Animal Phylum
Two cell layers; radial symmetry; one opening (mouth) leading into a hollow body; mouth surrounded by tentacles; stinging cells; first phylum to show tissues. Examples: coral, sea anemone, *Hydra.*

Flatworm Phylum
Three cell layers and body organs; flattened body; bilateral symmetry; parasitic or free living; one opening (mouth) if present; first phylum to show a nervous system. Examples: *Planaria,* tapeworms.

Roundworm Phylum
Round body; bilateral symmetry; parasitic or free living; mouth and anus; sexual reproduction only. Examples: hookworm, *Ascaris.*

Segmented Worm Phylum
Segmented body; bilateral symmetry; free living; mouth and anus; heart and circulatory system. Examples: earthworm, clamworm, leech.

Soft-bodied Animal Phylum
Soft body covered by a fleshy mantle; movement by a muscular foot; bilateral symmetry; free living; mouth and anus; reproduce sexually.

Snail Class
Large muscular foot; one shell or no shell on outside of body. Examples: snails, slugs.

Clam Class
Large muscular foot; two shells. Examples: clams, oysters, scallops.

Squid Class
Muscular foot divided into tentacles and arms; shell (if present) inside body. Examples: squid, octopus.

Jointed-leg Animal Phylum
Body in sections; bilateral symmetry; free living; mouth and anus; first phylum to show a skeletal system (exoskeleton).

Crayfish Class
Body in two sections; five pairs of legs; two pairs of antennae. Examples: crabs, crayfish, shrimp.

Spider Class
Body in two sections; four pairs of legs; no antennae. Examples: spiders, scorpions.

Centipede Class
Body in two sections; one pair of poisonous claws; body segmented with one pair of legs per body segment. Example: centipedes.

Millipede Class
Body in two sections; body segmented with two pairs of legs per body segment. Example: millipedes.

Insect Class
Body in three sections; three pairs of legs; one pair of antennae; wings usually present. Examples: bees, ants, moths, beetles.

Spiny-skin Animal Phylum
Body with five part design; radial symmetry; covered with spines; tube feet. Examples: sea urchin, starfish.

VERTEBRATES
Animals with backbones; includes one phylum.

Chordate Phylum
Tough, flexible rod along the back sometime during life; skeleton inside body (endoskeleton); sexual reproduction; bilateral symmetry.

Jawless Fish Class
No scales or jaw; skeleton of cartilage; fins not paired; no gill covering; cold blooded; parasitic or free living. Example: lamprey.

Cartilage Fish Class
Toothlike scales; jaw; skeleton of cartilage; paired fins; no gill covering; cold blooded. Examples: sharks, rays.

Bony Fish Class
First to show bony skeleton; bony scales; paired fins; covered gills; most have a swim bladder; cold blooded. Examples: trout, salmon, goldfish.

Amphibian Class
Moist, scaleless skin; young live in water, adults on land; four legs; cold blooded. Examples: frog, toad, salamander.

Reptile Class
Dry, scaly skin; first to show ability to lay eggs outside of water; egg with shell and protective membranes; four legs (except snakes); cold blooded. Examples: turtle, snake, alligator.

Bird Class
Feathers; wings; scales and claws on legs; beak present; no teeth; warm blooded. Examples: swan, robin, sparrow, owl.

Mammal Class
Hair; mammary glands for nursing young; young develop within body of mother; warm blooded. Examples: human, dog, lion, whale.

APPENDIX C

Rootwords, Prefixes, and Suffixes

Many words can be broken down into root words, prefixes, and suffixes. A prefix is a word part that, when added to the beginning of a word, changes the meaning of the word. A suffix is a word part that, when added to the end of a word, changes the meaning of the word. Knowing the meanings of certain root words, prefixes, and suffixes can help you understand biology words, and therefore, biology better. Use the table on these two pages to help you understand words used in biology.

P = prefix
R = root
S = suffix

Word Part	Word	Meaning	Word Example	Definition
R	amphi	double	amphibian	double life
P	anti	against	antibiotic	against life
P	arthro	join/joint	arthritis	disease of joints
R	atrium	entrance	left atrium	entrance to heart
R	bio	life	biodegradable	change by life forms
			biology	study of life
R	carb	carbon	carbohydrate	having carbon and water
			carbon dioxide	having carbon and oxygen
P	cardio	heart	cardiac muscle	muscle found in heart
R	chloro	green	chlorophyll	green pigment in leaf
			chloroplast	green body in plant cell
R	chromo	colored	chromosome	colored body in cell nucleus
P	cyto	cell	cytoplasm	liquid within cell
R	dermis	covering	dermis	skin
			epidermis	outer skin covering
P	dia	through	diaphragm	through the middle of body
P	endo	inside	endoskeleton	skeleton inside body
			endospore	spore formed inside bacteria
P	epi	upon	epidermis	outer skin covering
			epiglottis	upon the glottis, or throat
P	ex	out of	exhaling	push air out of lungs
			extinct	life forms no longer on Earth
P	exo	outer	exoskeleton	skeleton on outside of body
R	fertile	to bear	fertilization	process that occurs before bearing young
			fertilizer	helps soil bear better plants
R	flagellum	whip	flagellum	cell part that looks like a whip
P	gen	create	genetics	how life traits are created
R	genos	race	genus	classification group

Word Part	Word	Meaning	Word Example	Definition
P	hemo	blood	hemophilia	disease in which the blood does not clot properly
			hemoglobin	protein in red blood cells
R	herb	grass	herb	small grass-like plant
			herbivore	plant-eating animal
R	hydr	water	hydroponics	growing plants in water
P	hypo	below	hypothesis	to put under or suppose
S	logy	study of	biology	study of life
			ecology	study of where life forms live
P	micro	small	microscope	instrument used to look at small objects
P	mito	thread	mitochondrion	thread-like parts in cytoplasm
			mitosis	changes in cell chromosomes (threads) during reproduction
R	morph	shape	metamorphosis	changes in body shape
P	multi	many	multicellular	having many cells
R	nephros	kidney	nephron	filter unit of kidney
R	neuro	nerve	neuron	nerve cell
P	non	not	noncommunicable	disease that is not catching
R	ov	egg	ovary	organ that makes egg cells
			oviduct	organ that carries egg from ovary
R	para	beside or near	parasite	living beside or inside
R	pherein	to carry	pheromone	carries insect messages
P	photo	light	photosynthesis	use of light by plants to make food
			phototropism	plants turning toward light
R	phyl	tribe	phylum	group name used to classify
R	phyll	leaf	chlorophyll	cell part that gives leaves their green color
R	pneuma	breath	pneumonia	disease of lungs that makes breathing difficult
P	post	after	posterior	hind part of animal
P	pre	before	prenatal	before birth
			prescription	writing before
P	re	back	regenerate	to grow back again
P	semi	half	semilunar	half moon-shaped valve
R	scop	to look	microscope	instrument you look through
S	soma	body	chromosome	colored body in cells
R	spor	seed	spore	cell that acts like a seed in reproduction
			sporozoan	protozoan that forms spores during reproduction
R	stoma	mouth	stomate	pore in leaves like a mouth
			stomach	sac connected to mouth
S	trop	turning	phototropism	turning toward light
			thigmotropism	turning toward something that touches
P	zoo	animal	zoology	study of animals

Safety in the Laboratory

The biology laboratory is a safe place to work if you are aware of important safety rules and if you are careful. You must be responsible for your own safety and for the safety of others. The safety rules given here will protect you and others from harm in the lab. While carrying out procedures in any of the **Labs,** notice the safety symbols and caution statements. The safety symbols are explained in the chart on the next page.

1. Always obtain your teacher's permission to begin a lab.
2. Study the procedure. If you have questions, ask your teacher. Be sure you understand all safety symbols shown.
3. Use the safety equipment provided for you. Goggles and a safety apron should be worn when any lab calls for using chemicals.
4. When you are heating a test tube, always slant it so the mouth points away from you and others.
5. Never eat or drink in the lab. Never inhale chemicals. Do not taste any substance or draw any material into your mouth.
6. If you spill any chemical, wash it off immediately with water. Report the spill immediately to your teacher.
7. Know the location and proper use of the fire extinguisher, safety shower, fire blanket, first aid kit, and fire alarm.
8. Keep all materials away from open flames. Tie back long hair.
9. If a fire should break out in the classroom, or if your clothing should catch fire, smother it with the fire blanket or a coat, or get under a safety shower. **NEVER RUN.**
10. Report any accident or injury, no matter how small, to your teacher.

Follow these procedures as you clean up your work area.

1. Turn off the water and gas. Disconnect electrical devices.
2. Return materials to their places.
3. Dispose of chemicals and other materials as directed by your teacher. Place broken glass and solid substances in the proper containers. Never discard materials in the sink.
4. Clean your work area.
5. Wash your hands thoroughly after working in the laboratory.

FIRST AID IN THE LABORATORY

Injury	Safe response
Burns	Apply cold water. Call your teacher immediately.
Cuts and bruises	Stop any bleeding by applying direct pressure. Cover cuts with a clean dressing. Apply cold compresses to bruises. Call your teacher immediately.
Fainting	Leave the person lying down. Loosen any tight clothing and keep crowds away. Call your teacher immediately.
Foreign matter in eye	Flush with plenty of water. Use eyewash bottle or fountain.
Poisoning	Note the suspected poisoning agent and call your teacher immediately.
Any spills on skin	Flush with large amounts of water or use safety shower. Call your teacher immediately.

SAFETY SYMBOLS

DISPOSAL ALERT
This symbol appears when care must be taken to dispose of materials properly.

BIOLOGICAL HAZARD
This symbol appears when there is danger involving bacteria, fungi, or protists.

OPEN FLAME ALERT
This symbol appears when use of an open flame could cause a fire or an explosion.

THERMAL SAFETY
This symbol appears as a reminder to use caution when handling hot objects.

SHARP OBJECT SAFETY
This symbol appears when a danger of cuts or punctures caused by the use of sharp objects exists.

FUME SAFETY
This symbol appears when chemicals or chemical reactions could cause dangerous fumes.

ELECTRICAL SAFETY
This symbol appears when care should be taken when using electrical equipment.

SKIN PROTECTION SAFETY
This symbol appears when use of caustic chemicals might irritate the skin or when contact with microorganisms might transmit infection.

ANIMAL SAFETY
This symbol appears whenever live animals are studied and the safety of the animals and the students must be ensured.

RADIOACTIVE SAFETY
This symbol appears when radioactive materials are used.

CLOTHING PROTECTION SAFETY
This symbol appears when substances used could stain or burn clothing.

FIRE SAFETY
This symbol appears when care should be taken around open flames.

EXPLOSION SAFETY
This symbol appears when the misuse of chemicals could cause an explosion.

EYE SAFETY
This symbol appears when a danger to the eyes exists. Safety goggles should be worn when this symbol appears.

POISON SAFETY
This symbol appears when poisonous substances are used.

CHEMICAL SAFETY
This symbol appears when chemicals used can cause burns or are poisonous if absorbed through the skin.

SKILL HANDBOOK

1 Practicing Scientific Method

Scientists use orderly methods to learn new information and solve problems. The methods used in this process of understanding the world include observing, forming a hypothesis, separating and controlling variables, interpreting data, and designing an experiment. Practice scientific methods by following the examples explained below.

Skill: Observing

When you use your senses, you are observing. What you observe can give you a lot of information about events or things. This information helps you make sense of the world around you.

Learning the Skill
There are two kinds of observations, qualitative and quantitative.

(a) Qualitative observations describe something without using numbers. You might use words such as sweet or sour, good or poor.

(b) In quantitative observations, numbers are used to describe something. Measurements are often made. Your height, age, and mass are quantitative observations.

Example

1. Qualitative observations for an animal may be any of the following: large or small, tall or short, smooth or furry, brown or black, long ears or short ears, naked or hairy tail.
2. Quantitative observations for the same animal may be: mass–459 g; height–27 cm; ear length–14 mm; age–283 days.

Skill: Forming a Hypothesis

A hypothesis is a possible explanation.

Learning the Skill

1. First, you need to recognize the problem or event that needs explaining. You can then begin to form a hypothesis to explain it.
2. You make a hypothesis in the form of a statement that contains the words *if* and *then.*

Example
You are raising African violets and want to know if salt water will kill them. A possible hypothesis is: *If* I give salt water to my African violets, *then* they will continue to grow as usual. Another hypothesis could be: *If* I give salt water to my African violets, *then* they will die.

Skill: Separating and Controlling Variables

To be sure you know which variable caused a particular outcome in your experiment, you must test only one variable at a time.

1. Determine what variables can affect the outcome of an experiment.
2. Make sure that only one variable is allowed to change during the experiment.
3. Set up an experimental group and a control group. The experimental group is the one in which you expect to see a change. The control group is the one in which you don't expect to see a change.

Example

1. You wish to know whether or not caffeine increases heart rate. First, you must find out what other things (variables) might increase heart rate. Is activity important? Is amount of caffeine consumed or time of day important?

2. Keep all variables but one constant. For example, you could:
   (a) have all your subjects sit during the entire experiment.
   (b) give all your subjects the same amount of caffeine.
   (c) collect your data the same time each day.

   The variable that remains constant is the giving of caffeine to all the subjects. Changing any other variable can cause misleading results. For example, you allow your subjects to walk around as they wish. You won't be able to tell if any changes in heart rate are due to the caffeine or the different activities.
3. Compare your results with those from a control group. In the control group, all variables are identical to those in the experimental group with one exception. The members of the control group drink plain water.

Skill: Interpreting Data

The word *interpret* means to explain the meaning of something. When you interpret data, you explain what the data mean.

Learning the Skill
Check the data:
   (a) Compare the control group and the experimental group. Compare them for qualitative and quantitative differences.
   (b) Decide if the variable being tested had any effect. If there is no difference between the control group and the experimental group, then the variable being tested probably had no effect.

Example
1. You wish to find out if fertilizer affects plant growth. Your hypothesis is: If I add fertilizer to the soil, then my plants will grow larger.
2. The table shows the data you collect.
3. You compare group A (control group) to groups B and C (experimental groups). In groups B and C, the plants are taller.
4. You decide that adding fertilizer to the soil did have an effect on plant growth. It caused the plants in groups B and C to grow taller. Also, the amount of fertilizer had an effect because the plants in group C grew taller than the plants in group B.

EFFECT OF FERTILIZER ON PLANTS

GROUP	TREATMENT	HEIGHT	AFTER 3 WEEKS
A	no fertilizer added to soil	16.5 cm	17 cm
B	3 g fertilizer added to soil	16.5 cm	31 cm
C	6 g fertilizer added to soil	16.5 cm	48 cm

Skill: Designing an Experiment

A successful experiment follows the steps of the scientific method.

Learning the Skill
1. Make observations. You observe that your neighbor fertilizes her chrysanthemums and that they are taller than yours.
2. Recognize the problem. Does fertilizer cause plants to grow larger?
3. Research the problem to see what kinds of things could cause plants to grow larger.
4. Form a hypothesis: If I fertilize my plants, then they will grow larger.
5. Run the experiment.
   (a) Gather your materials. For this experiment, you would need fertilizer, water, plants, soil, a ruler, and a light source.
   (b) Set up the procedure. The only difference between the control group and the experimental group should be whether or not fertilizer is received.
   (c) Collect the data in a table.
6. Interpret the data. Decide if the fertilizer had an effect on plant growth.
7. Reach a conclusion. Is your original hypothesis supported by your data?
8. If your original hypothesis is not supported, form another hypothesis and test it.

2 Reading Science

As you use this textbook, you will find it useful to practice the following reading skills. Processes in nature generally occur in particular sequences. Learning the skill of sequencing will help you remember these processes. The skills of outlining and summarizing main ideas, understanding science words, and interpreting diagrams will help you study the major concepts in this text. As you learn about biology, you will find the skills of inferring and making models will help you apply your knowledge.

Skill: Sequencing

Sequencing is the arranging of facts or ideas into a series. The order in which the facts or ideas are arranged is important because one fact must lead to the next in the series.

Learning the Skill

1. List all items to be sequenced.
2. Decide which item is the first in the series.
3. Continue to list all items in their proper order.

Examples

Place the following steps of sharpening a pencil in their proper sequence. Remember that the steps must be in the proper order for the activity to be carried out successfully. Each item must lead to the next.

(a) Turn the handle of the sharpener several times.
(b) Locate the pencil sharpener.
(c) Insert the pencil into the opening of the sharpener.
(d) Check the pencil point to see if it is properly sharpened.
(e) Remove the pencil from the sharpener.

In this example, the correct sequence is (b), (c), (a), (e), (d).

Skill: Outlining and Summarizing Main Ideas

Being able to outline the material in your textbook will help you learn and understand it and to see how ideas are connected.

Learning the Skill

1. Look over the material to be outlined and summarized. Locate the major headings or ideas. These are the numbered sections in each chapter. Use a Roman numeral for each numbered section and write the title of the section next to the Roman numeral.
2. Write a capital letter for each subsection. Subsections are indicated by the headings that follow each numbered section in a chapter.
3. Write the main idea of the subsection next to the capital letter. Leave several lines between all capital letters for supporting ideas.
4. List several supporting ideas below each main idea. These ideas should be in the form of numbered statements.
5. To summarize the main ideas, write the outline in the form of complete sentences.

Example

1. Outline numbered Section 22:1.

I. Mitosis (title of numbered section)
   A. Body Growth and Repair (title of subsection)
      1. cell reproduction is called mitosis (supporting idea)
      2. body growth occurs because of mitosis (supporting idea)
      3. body repair occurs because of mitosis (supporting idea)
   B. An Introduction to Mitosis (title of subsection)
      1. one cell forms two during mitosis (supporting idea)
      2. mitosis can be compared to making a photocopy (supporting idea)

2. Summarize the information in the outline. Mitosis is the process of cell reproduction that occurs as our bodies grow and repair themselves. During this process, one cell reproduces to form two. Mitosis can be compared to the making of a photocopy.

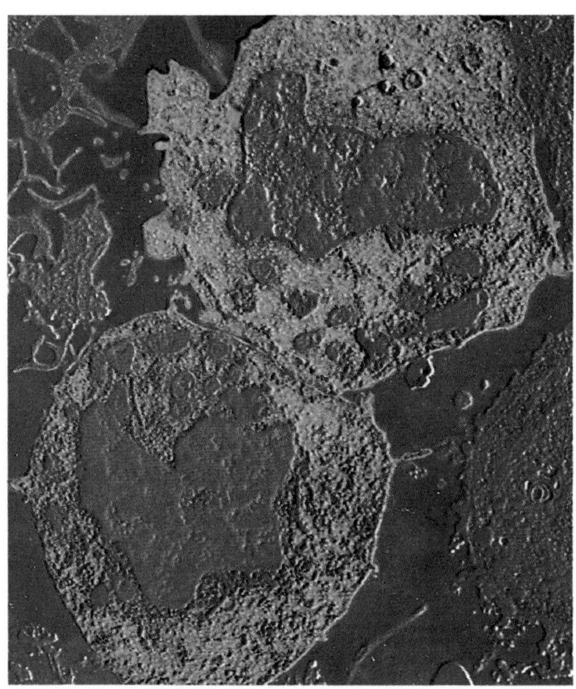

Skill: Understanding Science Words

There are two ways you can understand science words better. One way is to define the word in context. The way the word is used gives you a clue as to its meaning. A second way to understand science words is to look at the parts that make up the word. Each word part can give you a clue as to the meaning of the whole word.

Learning the Skill: Defining Words in Context

1. First, read to see if the word is defined directly in the sentence.
2. If the word is not defined directly, read several sentences beyond the one in which the word first appears. These sentences may provide information about the definition of the word.
3. If possible, define the word based on your own past knowledge. You may have learned the word in an earlier chapter, or you may be familiar with it because you hear it day to day.
4. Figure out the meaning of the word by how it is used in the sentence and by the sentences around it.

Examples

Find the definitions of the underlined words.

1. *Biology* is the study of life. The word *is* gives you a clue that the word is defined directly in the sentence.
2. A cat is a *mammal.* A mammal is an animal that has body hair, provides its young with milk from mammary glands, and is warm-blooded. The second sentence contains the definition of mammal.
3. All living things can *reproduce. Reproduce* is a familiar word. It is defined in Chapter 2. It is also one you probably hear from day to day.
4. Green plants carry out *photosynthesis.* Thus, they are able to make their own food. These sentences tell you that photosynthesis is the ability of a plant to make food.

Learning the Skill: Understanding Word Parts

1. Look at the word to see how many word parts you think it has. The word may have one or more word parts.
2. You may recognize parts of the word from previous lessons. Or, you may recognize parts of the word from other familiar words. Try to define each word part if you can. Then define the whole word.
3. Look for root words and prefixes or suffixes. A root word is the main part of the word. A prefix is a word part added to the front of a root word to change its meaning. A suffix is a word part added to the end of a root word to change its meaning.

Examples

1. What does the word *microorganism* mean? The word *microorganism* has two word parts, *micro* and *organism.* You remember the word *microscope* and that *micro* means small. You also remember that the word *organism* means a living thing. Therefore the word *microorganism* means a small living thing.
2. Examples of root words are:
   emia—blood
   vertebrate—animal with a backbone
   bio—life
   zoo—animal
3. The prefixes *an* and *in* mean without. What do the words *anemia* and *invertebrate* mean?
   anemia—having too little blood
   invertebrate—animal without a backbone
4. The suffix *logy* means the study of. What is biology? Zoology?
   biology—the study of life
   zoology—the study of animals

Skill: Interpreting Diagrams

When you look at a diagram, note first the orientation and symmetry of the organism. Knowing the orientation tells you where the front end is, where the tail end is, where the front side is, and where the back side is. Knowing the symmetry tells you how many sides the organism has. If it is an inside view of an organism, note whether it shows the whole organism or only part of one. Usually, inside views, also called sections, show only parts of organisms. If the diagram is of a process, note the order in which steps of the process occur.

Learning the Skill: Orientation

1. First, note whether the diagram, as in Figure 1, shows an internal or external view.
2. Locate the anterior, or head end, posterior, or tail end, dorsal, or back side, ventral, or belly side.

FIGURE 1

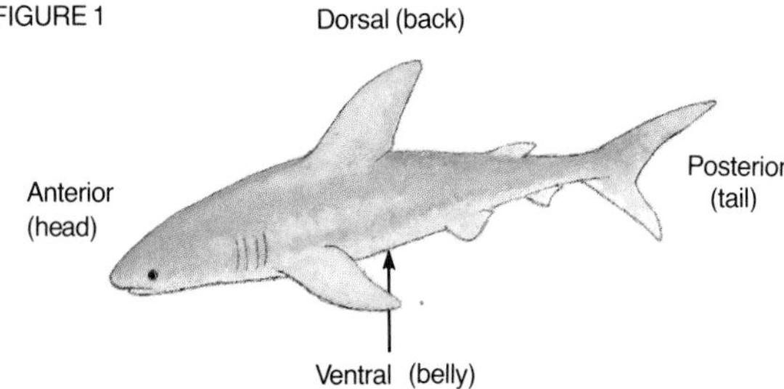

Learning the Skill: Symmetry

There are two types of symmetry, radial and bilateral. Bilateral means having two sides alike. Radial means spreading out from the center.

1. Draw an imaginary line through the center of the animal from its anterior end to its posterior end.
2. If the animal forms two mirror images, it probably has bilateral symmetry. You must do step 3 to find out for sure.
3. Draw a second line at right angles to the first. If the animal forms four equal parts, it has radial symmetry. If it doesn't, it has bilateral symmetry.

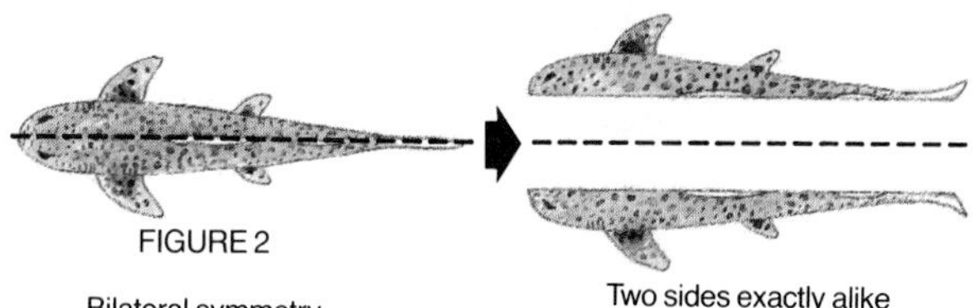

FIGURE 2

Bilateral symmetry

Example

1. Study the figure of the shark. A shark has bilateral symmetry. Note that the two sides are mirror images of one another.
2. Study the figure of the bicycle wheel and the sea anemone. A sea anemone has radial symmetry. The parts of the animal spread out from the center, much like the spokes of a wheel.

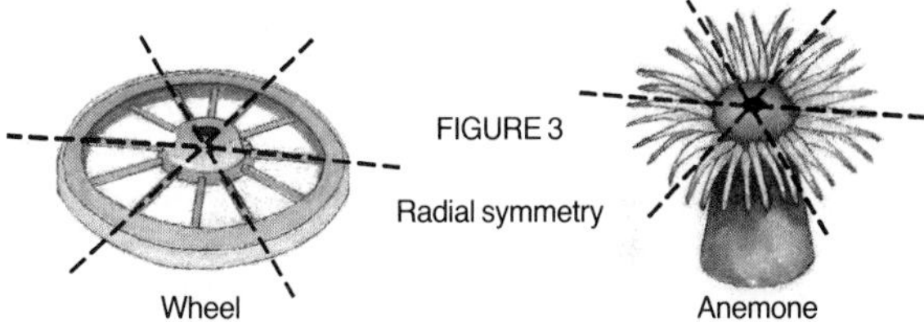

FIGURE 3

Radial symmetry

Learning the Skill: Sections

There are two kinds of sections, long sections and cross sections.

1. Cross sections are slices made at right angles to the axis. A slice through your waist would give a cross section of you.
2. Long sections are slices that run along the axis. A slice from the top of your head to the floor would give a long section of you.

Example

Study the figure of the squash. This figure shows a cross section of the squash. A cross section across the top part shows only the fleshy part of the squash. A cross section across the lower part of the squash shows an interior chamber filled with seeds.

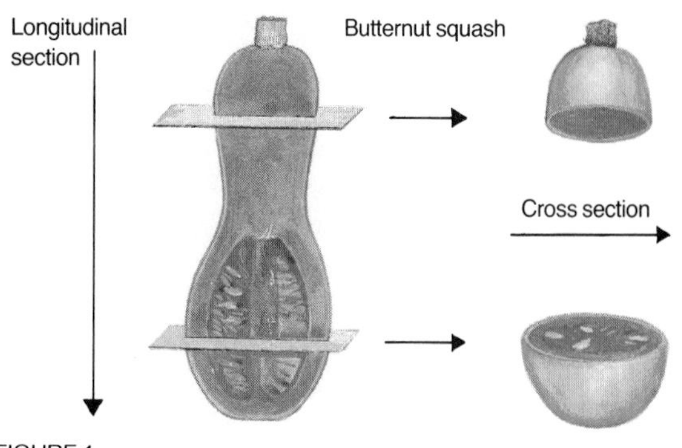

FIGURE 4

Skill: Inferring

To infer means to make an assumption based on facts, observations, or experimental data. You make an inference each time you say "I'll bet the school cafeteria is serving pizza for lunch because I can smell it."

Learning the Skill

1. Define the question. Sometimes you infer something without thinking about the question behind the inference. In the pizza example, what is the question? The question might be "what are we having for lunch today?" Sometimes, the question is asked directly, as in the example that follows.
2. Study the available information. In the pizza example, the available information is the odor of pizza in the air.
3. Look for connecting ideas. Make comparisons. Again in the pizza example, what else could you be smelling? Could it be sub sandwiches? The odor is similar, but your cafeteria has never served subs before.
4. Answer the question. Based on your responses to steps 2 and 3, you infer that pizza is on the menu.
5. Support your answer using the available information.

Example

1. The following information is about human population growth. One question that might come up regarding this information is "which country might have a food shortage by 2020?"
2. Study the table. What information does it give? It gives information about expected population growth in Japan and Pakistan.
3. Compare the expected increase in population for Japan with the expected increase in population for Pakistan.
4. Based on the data, you could infer that Pakistan will have a food storage by 2020.
5. Supporting evidence is that the population of Pakistan is increasing at a much faster rate than the population of Japan. The larger a population becomes, the more food it requires. What additional information could change your inference? If Japan had to import most of its food, and Pakistan was the largest producer of grain in the world, you might infer that Japan would have a food shortage, or that neither country would have a food shortage.

POPULATION GROWTH

COUNTRY	1996	2020 (ESTIMATED)
JAPAN	125 million	124 million
PAKISTAN	129 million	199 million

Skill: Formulating Models

In a model, ideas in biology are represented by familiar objects. Models help to simplify processes or structures that are often difficult to understand.

Learning the Skill

1. Recognize the process or structure that is to be modeled.
2. Research the process or structure before you start your model. Find out how a structure such as an artery is put together. Find out what takes place during a process, such as breathing in and out.
3. Think of a simple way of showing this same process or structure using materials that are readily available.
4. Construct your model. Note if it operates correctly and shows the idea you wish it to show.

Example

1. Formulate a model that shows how the sizes of arteries, veins, and capillaries differ.
2. Read what you can about arteries, veins, and capillaries. Arteries are usually round and are the thickest of the three blood vessel types. Veins are flat and thinner than arteries. Capillaries are very small in size. All three vessels are hollow tubes.
3. Materials that you could use include straws to represent capillaries, mostaccioli noodles to represent veins, and manicotti noodles to represent arteries.
4. To construct the model, you might want to cook the noodles first. Cooking will make them softer, and make their texture more like that of arteries and veins. Cut the noodles and straws into short sections and glue them to thick paper so that the openings are facing outward. Label each structure appropriately.

3 Using a Microscope

There are many opportunities in this biology text for you to make your own observations and experimentations. In many labs, you will be able to practice your skills of using a compound microscope, making a wet mount, calculating magnification, and making scale drawings.

Skill: Using a Compound Microscope

The compound microscope allows you to magnify objects 100, 400, and 1000 X (times) their natural size. The word *compound* refers to the fact that a compound microscope has several lenses for magnifying objects.

Steps Needed to Acquire Skill

1. Review the parts of the compound microscope shown in Figure 1.
2. Place the glass slide and object to be viewed onto the stage. Hold the slide in place with the stage clips.
3. Move the slide so that the object to be viewed is directly over the stage opening.
4. Turn the low-power objective into place until you hear a click. Look to the side and turn the coarse adjustment until the low-power objective is almost touching the slide.
5. Turn on the microscope (if electric) or adjust the mirror toward a light source. Never use direct sunlight.
6. Look through the eyepiece and adjust the diaphragm until you see a bright light. The mirror may also have to be readjusted.
7. Use the coarse adjustment to raise the body tube until you can see the object.
8. Use the fine wheel adjustment to bring the object into sharp focus.
9. To see the object under high power, rotate the objectives until you hear another click. Use only the fine adjustment to sharpen the focus.

Troubleshooting

Problem: Object can't be seen clearly or found under low or high power.

Solution: Make sure that the eyepiece, objectives, slide, and coverslip are clean. Check to be sure that the water is not covering the coverslip and if it is, make a new wet mount. If the objective or eyepiece lens is dirty, clean it with lens paper. Check the coarse adjustment. Check the diaphragm opening. If there is too much light, transparent objects, such as amoebas, will be washed out.

Problem: Object can't be found under high power.

Solution: Check to be sure the object is centered over the stage opening. You may need to return to low power first.

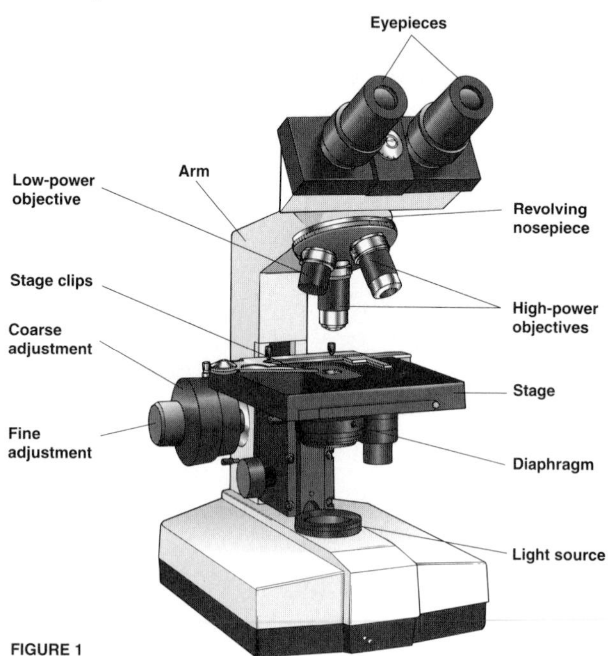

FIGURE 1

Skill: Making a Wet Mount

All the slides you prepare for observing under the microscope are called wet mounts. They are called wet mounts because the object to be viewed is prepared or mounted in water.

Learning the Skill

1. Add a drop or two of water to the center of a clean microscope slide.
2. Place the object to be viewed in the drop of water as shown in Figure 1.
3. Pick up a coverslip by its edges. Do not touch the surface of the coverslip. Stand the coverslip on its edge next to the drop of water.
4. Slowly lower the coverslip over the drop of water and the object to be viewed as shown in Figure 2.
5. Make sure that the object is totally covered with water. If it is not, remove the coverslip, add more water, and replace the coverslip.

Troubleshooting

1. Figures 3 and 4 show correctly prepared wet mounts.
2. The directions that follow tell you what to do if problems occur when you are making wet mounts.
   (a) Figure 5 shows what happens when there is not enough water on the wet mount or the coverslip is lowered too quickly. The water doesn't cover the entire object, or air bubbles may form.
   (b) Figure 6 shows what happens when too much water is used or the coverslip is dirty. If too much water is on the slide, place the tip of a paper towel at the edge of the coverslip to absorb some of the excess water. If there is dirt on the coverslip, replace it. If these steps don't work, clean the slide, dry the slide and coverslip, and make a new wet mount.
   (c) Figure 7 shows what happens when the object being viewed is too thick. Prepare a thinner object for viewing and make a new wet mount.

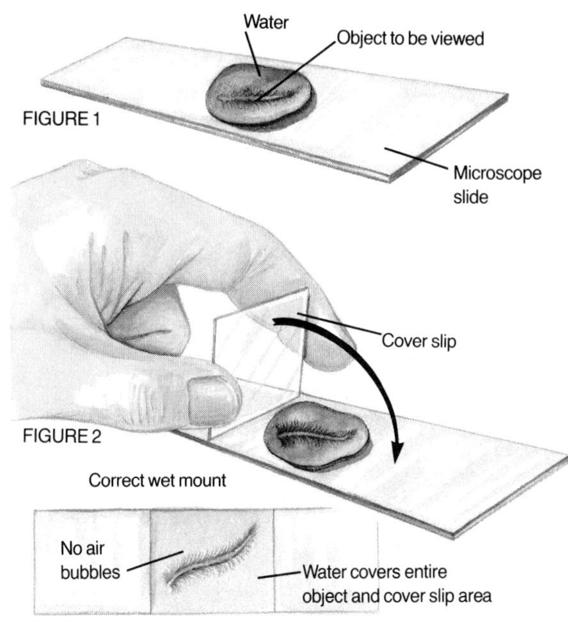

FIGURE 1

FIGURE 2

FIGURE 3

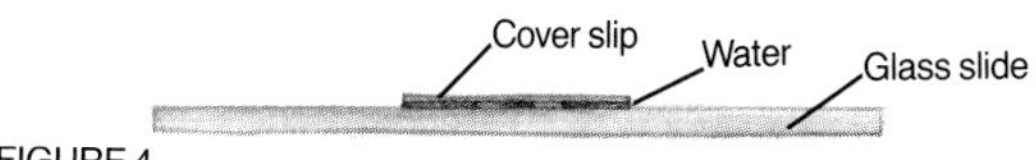

FIGURE 4

Incorrect wet mounts

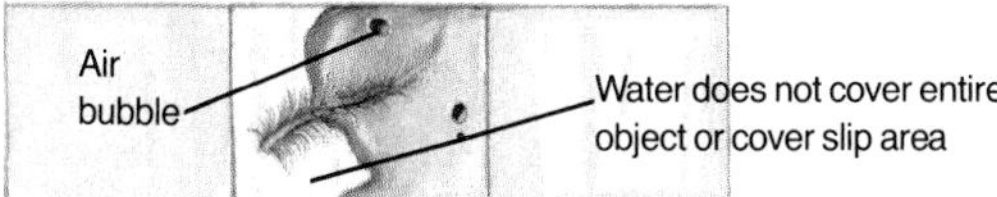

FIGURE 5

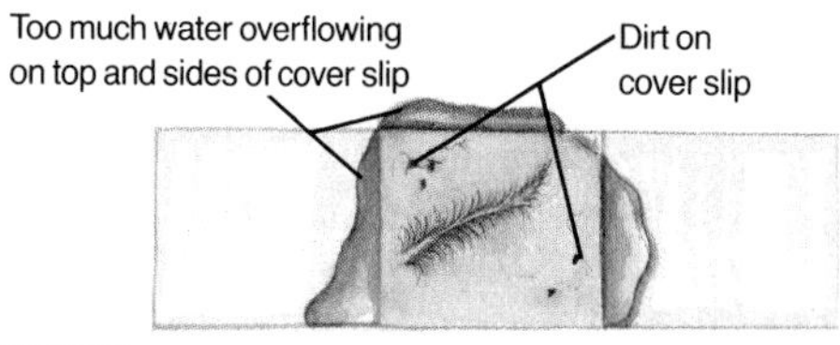

FIGURE 6

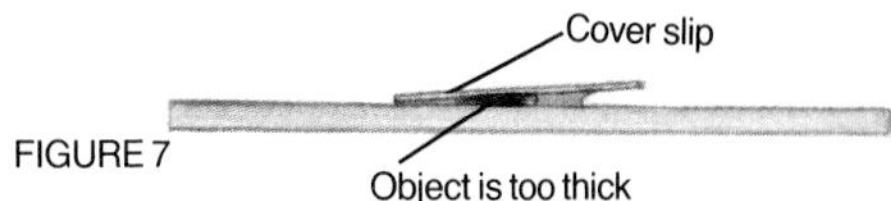

FIGURE 7

Skill: Calculating Magnification

Objects viewed under the microscope appear larger than normal because they are magnified. Total magnification describes how much larger an object appears when viewed through the microscope.

Learning the Skill

1. Look for a number marked with an X on the following:
   (a) eyepiece
   (b) low-power objective
   (c) high-power objective
   The X stands for how many times the lens of the microscope part magnifies an object.
2. To calculate *total* magnification, multiply the number on the eyepiece by the number on the objective.

Example

If the eyepiece magnification is 7×, the low-power objective magnification is 10×, and the high-power objective magnification is 40×:
(a) then total magnification under low power is 7× for the eyepiece × 10× for the low-power objective = 70× (7 × 10 = 70).
(b) then total magnification under high power is 7× for the eyepiece × 40× for the high-power objective = 280× (7 × 40 = 280).

Skill: Making Scale Drawings

When you draw objects seen through the microscope, the size that you make your drawing is important. Your drawing should be in proportion to the size the object appears to be when viewed through the microscope. This is called drawing to scale. Drawing to scale allows you to compare the sizes of different objects. It also allows you to form an idea of the actual size of the object being viewed, Figure 1.

Learning the Skill

1. Draw a circle on your paper. The circle may be any size.
2. Imagine the circle divided into four equal sections, as in Figure 2.

FIGURE 1

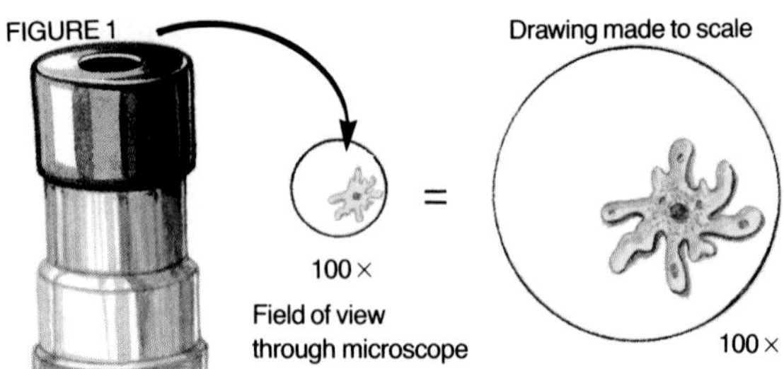

3. Locate an object under low or high power of the microscope. Imagine the field of view also divided into four equal sections.
4. Note how much of the field of view is taken up by the object. Also note what part of the field of view the object is in.
5. Draw the object in the circle. Position the object in about the same part of the circle as it appears in the field of view. Also, draw the object so that it takes up about the same amount of space within the circle as it actually takes up in the field of view, Figure 2.

FIGURE 2

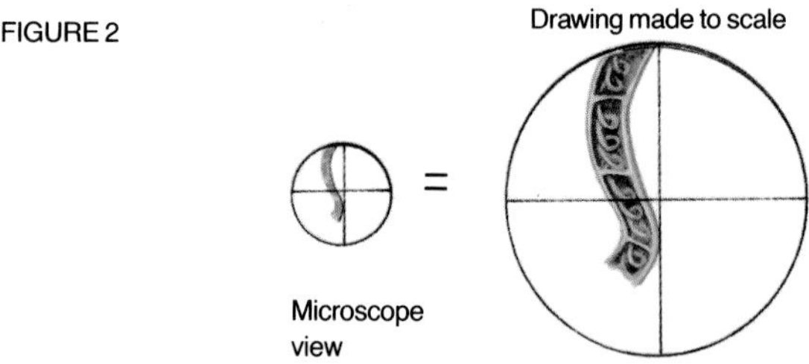

Troubleshooting

Figure 3 shows objects that haven't been drawn to scale correctly. One object has been drawn too small. One object has been drawn too large. The object in Figure 2 is drawn to the correct size and in the correct position.

FIGURE 3

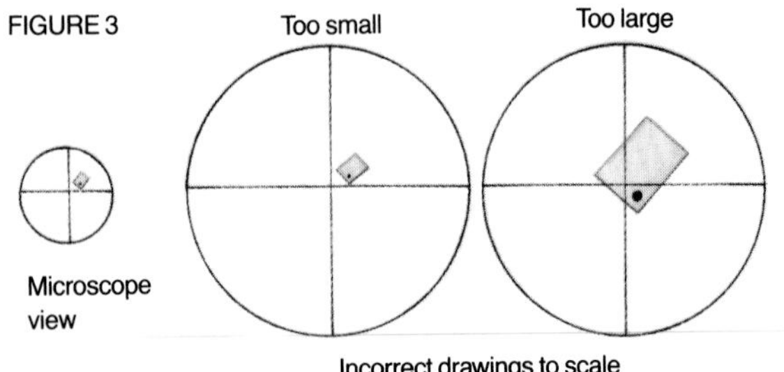

4 Organizing Information

As a scientist, you will be expected to collect data and draw conclusions. The skills of classifying, making and using tables, making line graphs, and making bar graphs will help you organize and interpret the information that you gather.

Skill: Classifying

Everyone, including scientists, groups or classifies to show similarities and differences among things and to put things in order.

Learning the Skill

1. Study the things to be classified.
2. Look at their traits and note traits that are similar and those that are different.
3. Pick out one major trait that can be used to separate the objects into at least two (or even three) separate groups.
4. Separate the things to be classified into these two groups. Then, taking each group at a time, look for other traits that can be used to separate each of the two groups into subgroups.
5. Continue to separate each subgroup into smaller and smaller groups until each group contains only one object.

FIGURE 1

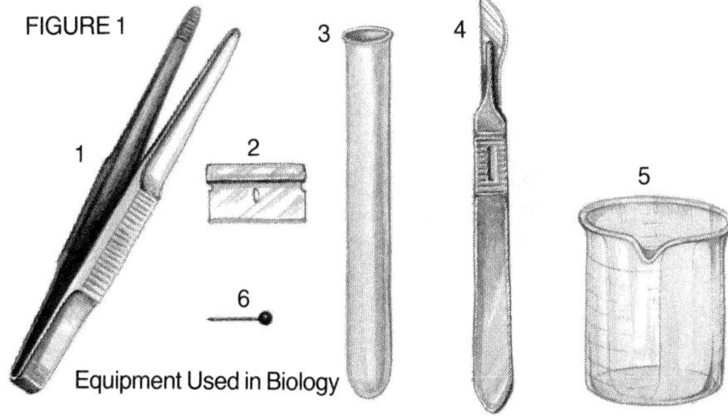

Equipment Used in Biology

Example

1. Classify equipment used in biology labs.
2. First, separate the equipment into two groups, possibly glass and metal.
3. Separate the glass items into subgroups. Separate metal items into subgroups. The diagram shows some possible subgroups for classifying the equipment.

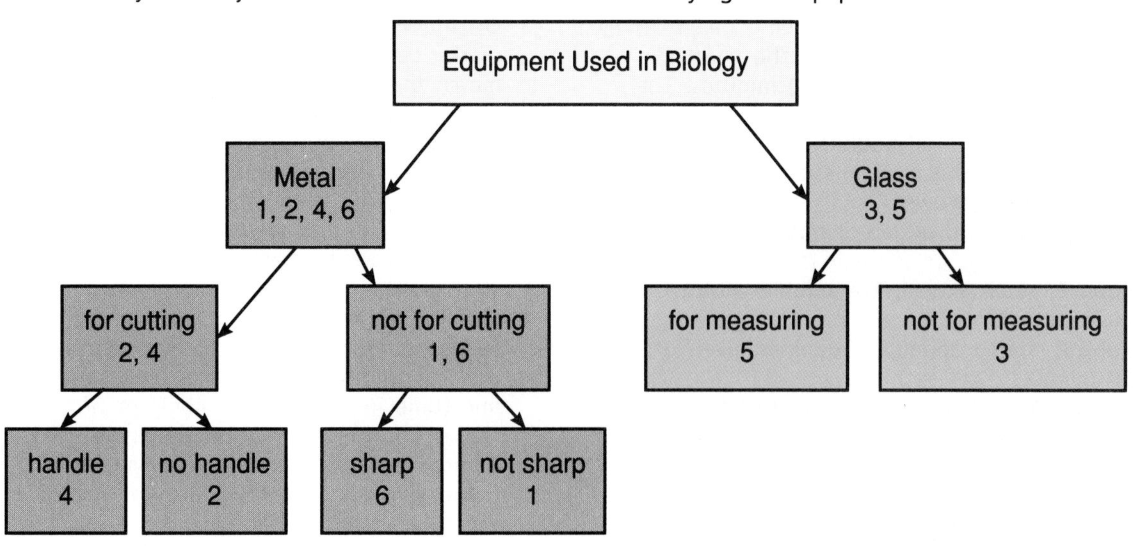

Skill: Making and Using Tables

Tables organize information in a way that makes it easy to read and interpret. Tables may be used to show qualitative information or quantitative information. (See page 704 of the Skill Handbook.) Tables have two uses. One, they are used to record data. Two, they provide information.

Learning the Skill

Tables have the following characteristics:

1. a title that tells what kind of information appears in the table
2. columns that run up and down
3. rows that run left to right
4. a heading for each column that tells what type of information is in that column

Example

An experiment was run to see if yeast cells give off different amounts of carbon dioxide gas when provided with different kinds of food. Starch, a food source, was put into a test tube containing yeast. Sugar, a second food source, was put into another test tube containing yeast. A third test tube that contained yeast, the control, had no food added. The test tubes were labeled A (no food), B (sugar added), and C (starch added). For the next 20 minutes, the experimenter counted the number of bubbles given off in each test tube. The number of bubbles was used as a measurement of the amount of carbon dioxide gas being given off by the yeast. The results were that the yeast in tube A gave off 3 bubbles in 20 minutes. The yeast in tube B gave off 104 bubbles. The yeast in tube C gave off 58 bubbles.

The results of this experiment could be summarized as follows:

Tube A, no food added, 3 bubbles given off in 20 minutes.

Tube B, sugar added, 104 bubbles given off in 20 minutes.

Tube C, starch added, 58 bubbles given off in 20 minutes.

Note that in this experiment, quantitative observations were made.

Another way to summarize the results is in a table. Study the table. Note how the information gathered in the yeast experiment is organized.

BUBBLES GIVEN OFF BY YEAST USING DIFFERENT FOOD TYPES

TUBE	FOOD ADDED	BUBBLES GIVEN OFF IN 20 MINUTES
A	none	3
B	sugar	104
C	starch	58

Skill: Making Line Graphs

A line graph is used to show how quantities of things change. It shows these changes in picture form. Very often, the data that are shown in a line graph are collected in a table first. Then, the data in the table are graphed.

Learning the Skill

1. Look at Figure 1 as you follow these directions. On a piece of graph paper draw two lines, one horizontal (A) and one vertical (B). The lines must meet in the lower left corner.
2. On the horizontal line, make regularly spaced marks (C). Label the marks. For example, if data were gathered every hour, then the first mark on the horizontal line would be labeled 1 hour, the second mark 2 hours, and so on.
3. On the vertical line, make regularly spaced marks. Label the marks. These labels must also match the data to be graphed. If the numbers to be graphed are the numbers of yeast cells, then the first mark could be labeled 1 cell, the second mark 2 cells, and so on. The way you label your graph depends on the data. If the quantities you want to graph are large, you will want to start with larger numbers on your graph.
4. Plot the data. Look at the data in the following table. One set of numbers represents the information that goes on the horizontal line (time in hours). The other set of numbers represents the information that goes on the vertical line (number of cells). First, plot the number of hours (zero). Place your pencil on the horizontal line at zero hours. Next plot

REPRODUCTION IN YEAST	
NUMBER OF YEAST CELLS	TIME IN HOURS
5	0
10	1
18	2
45	3
30	4
8	5

the number of cells. Move your pencil up to the mark labeled 5. Make a dot (D). Continue to plot the data until all the rows of numbers have been plotted.

5. If you have trouble determining where to place your dot, draw dotted lines from the horizontal and vertical lines. Where the dotted lines intersect is where you place your dot (E).
6. Connect each dot with a smooth line (F).
7. If you are graphing two sets of data, be sure to make the two lines different colors so you can tell them apart. Or, you can use a solid line for one and a dashed line for the other. Include a key that tells which color or line represents which set of data (G).

Example

Let's see how the data in the table would look in a line graph.

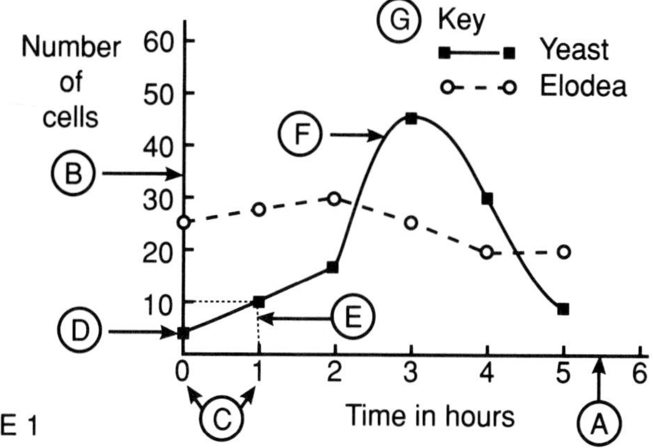

FIGURE 1

Skill: Making Bar Graphs

A bar graph is used to compare quantities of different kinds of things. It shows these comparisons in picture form.

Learning the Skill

1. Repeat steps 1-3 under Making Line Graphs. Label the horizontal and vertical lines with the appropriate labels.
2. Plot the data from the table below.

BREATH HOLDING ABILITY	
AMOUNT OF TIME ABLE TO HOLD BREATH (MINUTES)	ANIMAL
20	Gray Seal
10	Walrus
50	Blue Whale
15	Beaver

Place your pencil on the horizontal line where it is marked gray seal. Plot the number of minutes a gray seal can hold its breath by moving your pencil up to the appropriate mark on the vertical line. Place a small mark there. Then, draw a vertical bar up to the mark. Continue to plot the data until the values for all animals have been plotted.

Example

1. Let's see how the data in the table would look in a graph.

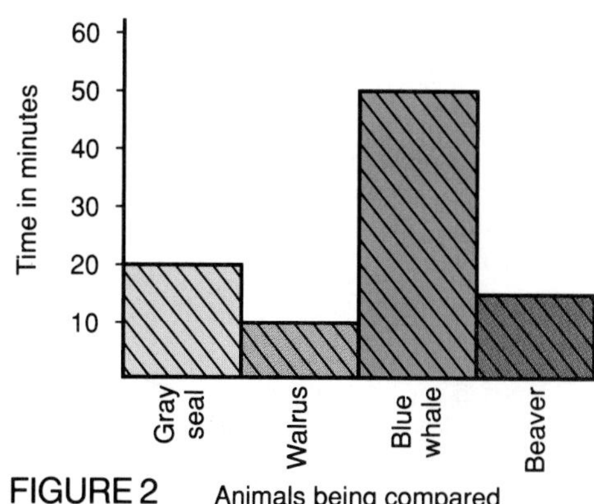

FIGURE 2

5 Using Numbers

Two important methods of science are the comparing of information and the repeating of experiments. All scientists use the skills of measuring volume, length, and mass in SI, and calculating in order to confirm hypotheses, results, and conclusions.

Skill: Measuring in SI—Volume

Volume measurements are made in milliliters (mL) or liters (L). 1000 milliliters = 1 liter. The equipment used for measuring volume is a graduated cylinder.

Learning the Skill

1. Locate the units on the graduated cylinder in Figure 1.
2. To read the volume of liquid in a graduated cylinder, do the following:
   (a) Place the cylinder on a flat surface.
   (b) Make sure you read the volume of liquid at eye level.
   (c) The surface of the liquid will form a curve. Find the lowest point of the curve and read the volume.

Troubleshooting

1. Figure 1 shows a correct volume reading.
2. The reading in Figure 2 is incorrect because:
   (a) The cylinder is not on a flat surface.
   (b) The top of the liquid is not at eye level.
   (c) The lowest point of the curve is not being used to read the volume.

Skill: Measuring in SI—Length

Measurements of length, distance, and height are made in millimeters (mm), centimeters (cm), and meters (m). The equipment used for measuring length is a metric ruler.

Learning the Skill

1. First decide which units you want to measure in. The units you choose will depend on the size of the object or distance being measured.
2. Note the divisions on the ruler. Figure 3 shows these divisions. One meter is divided into 100 centimeters. Each centimeter is divided into 10 millimeters.
3. Place one end of the object you are measuring next to the zero mark on the ruler. Note where the other end stops and read the number of units shown there.

Example

1. Measure the length of a leaf.
2. Place one end of the leaf at the zero mark as shown in Figure 4.
3. The leaf in the figure is 35 mm.

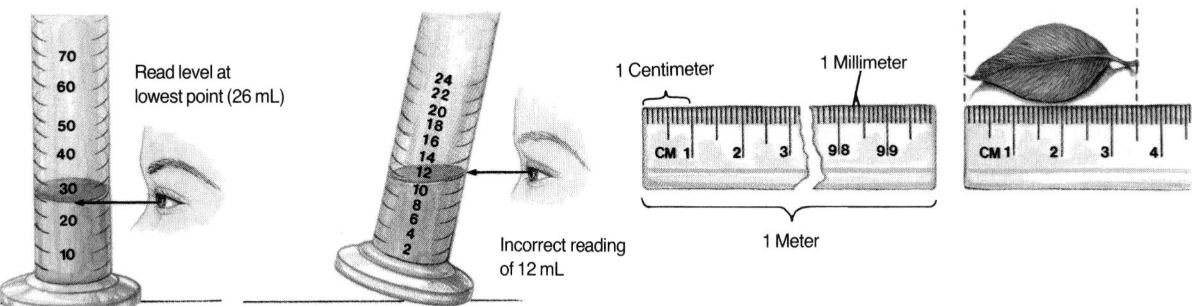

FIGURE 1 FIGURE 2 FIGURE 3 FIGURE 4

Skill: Measuring in SI—Mass

Mass measurements are made in grams (g), milligrams (mg), or kilograms (kg). The equipment used is a balance.

Learning the Skill

1. Place the object to be massed onto the pan. Note: liquids must be in a container; powders or solids must be placed on paper. The mass of the container or paper must later be subtracted from the total mass.
2. Slide all the riders to the left and release the locking mechanism.
3. Figure 1 shows that the back beam measures mass in 10 gram units. The middle beam measures mass in 1 gram units. The front beam measures mass in 0.1 and 0.01 gram units.
4. Starting with the rider on the back beam, move the rider to the right until the pointer drops below the zero point. Then move it back one notch. Repeat this process with the rider on the middle beam. Move the front rider until the pointer lines up with the zero point (center line).
5. The mass of the object will be the total of all the numbers indicated by the riders of the three beams.

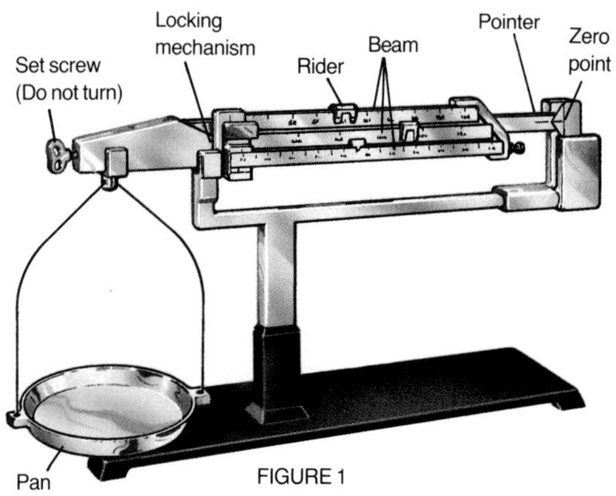

FIGURE 1

Skill: Calculating

Calculations may involve averaging numbers, comparing sets of numbers, or calculating area or volume.

Learning the Skill: Averaging

1. Add together the numbers to be averaged.
2. Divide the total by the number of figures used. The resulting number is the average.

Learning the Skill: Comparing Numbers

1. To find out how many times larger something is, *divide the smaller number into the larger.*
2. To find out how many times smaller something is than something else, *divide the larger number into the smaller.*

Learning the Skill: Calculating Area

1. Multiply the length times the width.
2. The result is the area in *square* units, for example, square meters or square centimeters.

Example

1. Calculate the area of the rectangle shown below.

3 cm

5 cm

2. Length = 3 cm
   Width = 5 cm
   Area = length × width = 15 square centimeters
   Square units may be written cm^2, m^2, or mm^2.

Learning the Skill: Calculating Volume

1. Multiply length times width times height.
2. The result is the volume in *cubic* units, such as cubic meters or cubic centimeters.

Example

1. Calculate the volume of the cube shown below.

15 mm

35 mm

15 mm

2. Length = 35 mm
   Width = 15 mm
   Height = 15 mm
   Volume = length × width × height = 7875 cubic millimeters
   Cubic units may be written cm^3, m^3, or mm^3.

GLOSSARY

PRONUNCIATION KEY

a . . . b**a**ck (bak)
ay . . . d**a**y (day)
ah . . . f**a**ther (fahth ur)
ow . . . fl**ow**er (flow ur)
ar . . . c**ar** (car)
e . . . l**e**ss (les)
ee . . . l**ea**f (leef)
ih . . . tr**i**p (trihp)
i(i+con+e) . . . **i**dea, l**i**fe (i dee uh, life)
oh . . . g**o** (goh)
aw . . . s**o**ft (sawft)
or . . . **or**bit (or but)
oy . . . c**oi**n (coyn)
oo . . . f**oo**t (foot)
ew . . . f**oo**d (fewd)
yoo . . . p**ure** (pyoor)
yew . . . f**ew** (fyew)
uh . . . comm**a** (cahm uh)
u(+con) . . . flow**er** (flow ur)
sh . . . **sh**elf (shelf)
ch . . . na**t**ure (nay chur)
g . . . **g**ift (gihft)
j . . . **g**em (jem)
ing . . . s**ing** (sing)
zh . . . vi**si**on (vihzh un)
k . . . ca**k**e (kayk)
s . . . **s**eed, **c**ent (seed, sent)
z . . . **z**one, rai**s**e (zohn, rayz)

A

Acids: liquids that have pH values lower than 7 (p. 684)

Acid rain: rain that has a pH between 1 and 5.5 (p. 684)

Acquired Immune Deficiency Snydrome, or **AIDS:** a disease of the immune system (p. 259)

Adaptation: a trait that makes a living thing better able to survive (p. 29)

Algae (AL jee): plantlike protists (p. 97)

Alveoli (al VEE uh li): the tiny air sacs of the lungs (p. 268)

Amniocentesis (am nee oh sen TEE sus): a way of looking at the chromosomes of a fetus (p. 567)

Amniotic (am nee AHT ik) **sac:** a tissue filled with liquid that protects the embryo (p. 525)

Amphibian (am FIHB ee un): an animal that lives part of its life in water and another part of its life on land (p. 169)

Anemia: a condition in which there are too few red blood cells in the blood (p. 248)

Animals: organisms that have many cells, can't make their own food, and can move (p. 62)

Annual (AN yul): a plant that completes its life cycle within one year (p. 447)

Annual ring: each ring of xylem in a woody stem (p. 423)

Antacid (ant AS ud): a drug that changes acid into water and a salt (p. 382)

Antennae (an TEN ee): appendages of the head that are used for sensing smell and touch (p. 159)

Antibiotics (an ti bi AHT iks): chemical substances that kill or slow the growth of bacteria (p. 84)

Antibodies (ANT ih bohd eez): chemicals that help destroy bacteria or viruses (p. 257)

Antigens (ANT ih junz): foreign substances, usually proteins, that invade the body and cause diseases (p. 257)

Antihistamine (ANT i HIHS tuh meen): a drug that reduces swelling of tissues by stopping the leaking of blood plasma from capillaries (p. 380)

Anus (AY nus): an opening through which undigested food leaves the body (p. 145)

Aorta (ay ORT uh): the largest artery in the body (p. 230)

Appendage (uh PEN dihj): a structure that grows out of an animal's body (p. 158)

Appendix (uh PEN dihks): a small, fingerlike part of the digestive system found where the small and large intestines meet (p. 212)

Artery (ART uh ree): a blood vessel that carries blood away from the heart (p. 226)

Arthritis (ar THRIT us): a disease of bone joints (p. 299)

Asexual (ay SEK shul) **reproduction:** the reproducing of a living thing from only one parent (p. 80)

Atria (AY tree uh): the small top chambers of the heart (p. 225)

Auditory (AHD uh tor ee) **nerve:** a nerve that carries messages of sound to the brain (p. 341)

Autosomes (AH toh sohmz): chromosomes that do not determine the sex of a person (p. 568)

Axon (AK sahn): the part of a neuron that sends messages to surrounding neurons or body organs (p. 313)

B

Bacteria: very small, one-celled monerans (p. 79)

Balanced diet: a diet with the right amount of each nutrient (p. 192)

Ball-and-socket joint: a joint that allows you to twist and turn the bones in a circle where they meet (p. 291)

Bases: liquids that have pH values greater than 7 (p. 684)

Behavior: the way an animal acts (p. 352)

Bicuspid (bi KUS pud) **valve:** the valve between the left atrium and the left ventricle (p. 227)

Biennial (bi EN ee ul): a plant that produces seeds at the end of the second year of growth and then dies (p. 447)

Bile: a green liquid that breaks large fat droplets into small fat droplets (p. 210)

Biodegradable (bi oh duh GRAYD uh bul): something that can be broken down by microbes into harmless chemicals and used by other living things (p. 681)

Biology: the study of living and once-living things (p. 5)

Biome (BI ohm): a land area with a distinct climate and with specific types of plants and animals (p. 664)

Biotechnology (bi oh tek NAHL uh jee): the use of living things to solve practical problems (p. 85)

Blade: the thin, flat part of a leaf (p. 398)

Blood pressure: the force created when blood pushes against the walls of vessels (p. 232)

Blue-green bacteria: small, one-celled monerans that contain chlorophyll and can make their own food (p. 87)

Body cells: cells that make up most of the body, such as the skin, blood, bones, and stomach (p. 464)

Bone marrow: the soft center part of the bone (p. 248)

Bony fish: fish that have skeletons made mostly of bone (p. 168)

Brain: the organ that sends and receives messages to and from all body parts (p. 314)

Breeding: the bringing together of two living things to produce offspring (p. 599)

Bronchi (BRAUN ki): two short tubes that carry air from the trachea to the left and right lung (p. 267)

Budding: reproduction in which a small part of the parent grows into a new organism (p. 106)

Bulb: a short, underground stem surrounded by fleshy leaves that contain stored food (p. 486)

C

Caffeine: a stimulant found in coffee and tea, cocoa, chocolate, and some soft drinks (p. 385)

Calorie: a measure of the energy in food (p. 193)

Cambium: a thin layer of cells that divide to form new phloem on the outside and new xylem on the inside (p. 421)

Cancer: a disease in which body cells reproduce at an abnormally fast rate (p. 478)

Capillary: the smallest kind of blood vessel (p. 234)

Capsule: a sticky outer layer produced by bacteria (p. 80)

Carbohydrates (kar boh HI drayts): nutrients that supply you with energy (p. 185)

Carbon monoxide (CO): an odorless, colorless gas sometimes found in the air (p. 272)

Cardiac (KAR dee ac) **muscle:** the muscle that makes up the heart (p. 294)

Cartilage (KART ul ihj): a tough, flexible tissue that supports and shapes the bodies of some fish and some animal parts (p. 167)

Cartilage fish: jawed fish in which the entire skeleton is made of cartilage (p. 167)

Cell: the basic unit of all living things (p. 27)

Cell membrane: the cell part that gives the cell shape and holds the cytoplasm (p. 32)

Cell wall: the thick, outer covering outside the cell membrane (p. 35)

Cellular respiration: the process by which food is broken down and energy is released (p. 27)

Celsius (SEL see us): a scale with which scientists measure temperature (p. 14)

Centrioles (SEN tree ohlz): cell parts that help with cell reproduction (p. 35)

Cerebellum (ser uh BEL uhm): the brain part that helps make your movements smooth and graceful, rather than robotlike (p. 317)

Cerebrum (suh REE brum): the brain part that controls thought, reason, and the senses (p. 316)

Chemical change: turns food into a form that cells can use (p. 206)

Chlorophyll (KLOHR uh fihl): a chemical that gives plants their green color and traps light energy (p. 114)

Chloroplasts (KLOR uh plasts): cell parts that contain the green pigment, chlorophyll (p. 35)

Cholesterol (kuh LES tuh ral): a fatlike chemical found in certain foods (p. 237)

Chordate (KOR dayt): an animal that, at some time in its life, has a tough, flexible rod along its back (p. 165)

Chromosomes (KROH muh sohmz): cell parts with information that determines what traits a living thing will have (p. 32)

Cilia: short, hairlike parts on the surface of a cell (p. 96)

Circulatory system: the body system made up of your blood, blood vessels, and heart (p. 222)

Class: the largest group within a phylum (p. 54)

Classify: to group things together based on similarities (p. 48)

Cleavage (KLEE vihj): the series of changes that take place to turn one fertilized egg into a hollow ball of many cells (p. 525)

Climate: the average light, temperature, and precipitation in an area taken over many years (p. 664)

Climax community: the last or final stage of succession in a community (p. 661)

Clone (KLOHN): a living thing that comes from one parent and is identical to the parent (p. 496)

Club fungi: fungi with club-shaped parts that produce spores (p. 105)

Cocaine: a controlled drug used for its stimulant effects (p. 384)

Cochlea (KAHK lee uh): a liquid-filled, coiled chamber in the ear that contains nerve cells (p. 341)

Cold-blooded: having a body temperature that changes with the temperature of the surroundings (p. 166)

Colony: a group of similar cells growing next to each other that do not depend on each other (p. 79)

Color blindness: a problem in which red and green are not seen as they should be (p. 576)

Commensalism (kuh MEN suh lihz um): a relationship in which two organisms live together, and one benefits while the other gets no benefit and is not harmed (p. 645)

Communicable (kuh MYEW nih kuh bul) **diseases:** those that can be passed from one organism to another (p. 83)

Community: all of the living things in an area that depend upon each other (p. 638)

Competition: the struggle among living things to get their needs for life (p. 616)

Complete metamorphosis: a series of changes in an insect in which the young do not look like the adult (p. 536)

Compound eyes: eyes with many lenses (p. 159)

Cones: nerve cells that can detect color (p. 336)

Conifer (KAHN uh fur): a plant that produces seeds in cones (p. 124)

Consumers: living things that eat, or consume, other living things (p. 27)

Control: a standard for comparing results (p. 18)

Cork: the outer layer of a woody stem (p. 421)

Cornea (KOR nee uh): the clear outer covering at the front of the eye (p. 335)

Coronary (KOR uh ner ee) **vessels:** the blood vessels that carry blood to and from the heart itself (p. 237)

Cortex: a food storage tissue in plants (p. 420)

Courting behaviors: behaviors used by males and females to attract one another for mating (p. 359)

Cross pollination: when pollen from the stamen of one flower is carried to the pistil of a flower on a different plant (p. 490)

Cutting: a small section of plant stem that has been removed from a parent plant and planted to form a new plant (p. 487)

Cyst (SIHST): a young worm with a protective covering (p. 143)

Cytoplasm (SITE uh plaz um): the clear, jellylike material between the cell membrane and the nucleus that makes up most of the cell (p. 32)

D

Data: the recorded facts or measurements from an experiment (p. 18)

Day-neutral plants: plants in which flowering doesn't depend on day length (p. 441)

Decomposers: living things that get their food from breaking down dead matter into simpler chemicals (p. 81)

Dendrites: parts of the neuron that receive messages from nearby neurons (p. 313)

Dependence: needing a certain drug in order to carry out normal daily activities (p. 383)

Depressant (dih PRES unt): a drug that slows down messages in the nervous system (p. 377)

Dermis: a thick layer of cells that form in the inner part of the skin (p. 343)

Development: all the changes that occur as a living thing grows (p. 26)

Diabetes mellitus (di uh BEET us · MEL uht us): a disease that results when the pancreas doesn't make enough insulin (p. 324)

Diaphragm: a sheetlike muscle that separates the inside of your chest from the intestines and other organs of your abdomen (p. 270)

Diffusion (dif YEW zhun): the movement of a substance from where there is a large amount of it to where there is a small amount of it (p. 38)

Digestion: the changing of food into a usable form (p. 205)

Digestive system: a group of organs that take in food and change it into a form the body can use (p. 204)

DNA: a molecule that makes up genes and determines the traits of all living things (p. 586)

Dominant (DAHM uh nunt) **genes:** those that keep other genes from showing their traits (p. 549)

Dosage: how much and how often to take a drug (p. 372)

Drug: a chemical that changes the way a living thing functions when it is taken into the body (p. 370)

Drug abuse: the incorrect or improper use of a drug (p. 383)

Dyslexia (dis LEHK see uh): a genetic disorder in which the person sees and writes some letters or words backward (p. 579)

E

Eardrum: the membrane that vibrates at the end of the ear canal (p. 340)

Ecology (ih KAHL uh jee): the study of how living things interact with each other and with their environment (p. 654)

Ecosystem: a community interacting with the environment (p. 654)

Egg: a female reproductive cell (p. 118)

Embryo (EM bree oh): an organism in its earliest stages of growth (p. 124)

Emigration (em uh GRAY shun): the movement of animals out of an area (p. 634)

Emphysema (em fuh SEE muh): a lung disease that results in the breakdown of alveoli (p. 273)

Endangered: a species of living thing that is in danger of becoming extinct (p. 675)

Endocrine (EN duh krin) **system:** the body system made up of small glands that make chemicals that carry messages through the body (p. 320)

Endodermis: a ring of waxy cells that surrounds the xylem in roots (p. 429)

Endoskeleton (EN doh skel uht uhn): a skeleton on the inside of the body (p. 165)

Endospore (EN duh spor): a thick-walled structure that forms inside a bacterial cell (p. 81)

Energy pyramid: a diagram that shows energy loss in the food chain (p. 643)

Enzymes (EN zimes): chemicals that speed up the rate of chemical change (p. 206)

Epidermis (ep uh DUR mus): the outer layer of cells of a plant (p. 401)

Epiglottis (ep uh GLAHT uhs): a small flap that closes over the windpipe when you swallow (p. 267)

Erosion: the wearing away of soil by wind and water (p. 676)

Esophagus (ih SAHF uh gus): a tube that connects the mouth to the stomach (p. 209)

Estrogen (ES truh jun): a female hormone (p. 516)

Estrous (ES trus) **cycle:** a cycle in which a female will mate only at certain times (p. 510)

Ethyl (ETH ul) **alcohol:** a drug formed from sugars by yeast that is found in all alcoholic drinks (p. 386)

Evolution (ev uh LEW shun): a change in the hereditary features of a group of organisms over time (p. 616)

Excretory (EK skruh tor ee) **system:** the body system made up of those organs that rid the body of liquid wastes (p. 274)

Exoskeleton (EK soh skel uht uhn): a skeleton on the outside of the body (p. 159)

Experiment: testing a hypothesis using a series of steps with controlled conditions (p. 16)

External fertilization: the joining of egg and sperm outside the body (p. 508)

Extinct (ihk STINGT): a life-form that no longer exists (p. 617)

F

Family: the largest group within an order (p. 54)

Farsighted: being able to see clearly far away but not close up (p. 345)

Fats: nutrients that are stored by your body and used later as a source of energy (p. 185)

Fern: a vascular plant that reproduces with spores (p. 121)

Fertile: being able to reproduce by forming egg or sperm cells (p. 610)

Fertilization: the joining of the egg and sperm (p. 118)

Fertilizer: a substance made of minerals that improves soil for plant growth (p. 449)

Fetus (FEET us): an embryo that has all of its body systems (p. 529)

Fibrous roots: many-branched roots that grow in clusters (p. 428)

Fission (FIHSH un): the process of one organism dividing into two organisms (p. 80)

Fixed joints: joints that don't move (p. 291)

Flagellum (fluh JEL um): a whiplike thread used for movement by bacteria (p. 80)

Flatworms: the simplest worms, they have flattened bodies (p. 142)

Flower: the reproductive part of a flowering plant (p. 128)

Flowering plant: a vascular plant that produces seeds inside a flower (p. 128)

Food chain: a pathway of energy and materials through a community (p. 641)

Food web: food chains connected in a community (p. 642)

Fossil fuel: the remains of organisms that lived millions of years ago (p. 677)

Fossils: the remains of once-living things from ages past (p. 617)

Fraternal twins: twins that form from two different fertilized eggs (p. 597)

Fruit: an enlarged ovary that contains seeds (p. 492)

Fungi: organisms that have cell walls and absorb food from their surroundings (p. 61)

G

Gallbladder: a small, baglike part located under the liver that stores bile (p. 210)

Gene (JEEN): a small section of chromosome that determines a specific trait of an organism (p. 547)

Genetic code: the code that translates the DNA language into the protein language (p. 592)

Genetic counseling: the use of genetics to predict and explain traits in children (p. 580)

Genetics (juh NET ihks): the study of how traits are passed from parents to offspring (p. 546)

Genus (JEE nus): the largest group within a family (p. 55)

Germination (jur muh NAY shun): the first growth of a young plant from a seed (p. 494)

Gill: a structure used by fish and some other animals to breathe in water (p. 166)

Grafting: the process of joining a stem of one plant to the stem of another plant (p. 487)

Gravitropism (grav uh TROH pihz um): the response of a plant to gravity (p. 444)

Greenhouse effect: the trapping of heat near Earth's surface by a layer of carbon dioxide (p. 679)

Guard cells: green cells that change the size of the stomata in a leaf (p. 402)

H

Habitat (HAB uh tat): the place where a plant or animal lives (p. 638)

Heart attack: the death of a section of heart muscle (p. 237)

Hemoglobin (HEE muh gloh bun): a protein in red blood cells that joins with oxygen and gives the red cells their color (p. 248)

Hemophilia (hee muh FIHL ee uh): a disease in which a person's blood won't clot (p. 252)

Herbaceous (hur BAY shus) **stems:** soft, green stems (p. 420)

Heterozygous (HET uh roh ZI gus): an individual with a dominant and a recessive gene for a trait (p. 549)

Hibernation: the state of being inactive during cold weather (p. 170)

Hinge joints: joints that allow bones to move only back and forth (p. 291)

Hookworm: a roundworm that is a parasite of humans (p. 145)

Hormones: chemicals made in one part of an organism that affect other parts of the organism (p. 320)

Host: an organism that provides food for a parasite (p. 73)

Hydrochloric (hi druh KLOR ik) **acid:** a chemical often called stomach acid (p. 210)

Hypertension (HI pur ten chun): occurs when blood pressure is extremely high (p. 236)

Hyphae (HI fee): threadlike structures that make up the bodies of most fungi (p. 102)

Hypothesis (hi PAHTH uh sus): a statement that can be tested (p. 16)

I

Identical twins: two children that form from the splitting of one fertilized egg (p. 597)

Immigration (ihm uh GRAY shun): the movement of animals into a population (p. 634)

Immune system: the body system made up of proteins, cells, and tissues that identify and defend the body against foreign chemicals and organisms (p. 256)

Immunity: the ability of a person who once had a disease to be protected from getting the same disease again (p. 258)

Incomplete dominance: a case of inheritance in which neither gene is totally dominant over the other (p. 573)

Incomplete metamorphosis: a series of changes in an insect from nymph to adult (p. 536)

Inhalant: a drug breathed in through the lungs in order to cause a behavior change (p. 379)

Innate behavior: a way of responding that does not require learning (p. 353)

Instinct: a complex pattern of behavior that an animal is born with (p. 353)

Insulin (IHN suh lun): a hormone that lets your body cells take in glucose, a sugar, from your blood (p. 323)

Interferon (ihnt ur FIHR ahn): a chemical substance that interferes with the way viruses reproduce (p. 77)

Internal fertilization: the joining of egg and sperm inside the female's body (p. 509)

International System of Units: a measuring system based on units of 10 (p. 11)

Invertebrates (in VERT uh brayts): animals without backbones (p. 134)

Involuntary muscles: muscles you can't control (p. 294)

Iris: a muscle that controls the amount of light entering the eye (p. 334)

J

Jawless fish: fish that have no jaws and are not covered with scales (p. 167)

Jointed-leg animal: an invertebrate with an outside skeleton, bilateral symmetry, and jointed appendages (p. 158)

K

Kilogram (kg): an SI unit of mass (p. 14)

Kingdom: the largest group of living things (p. 53)

Koch's postulates (KAHKS · PAHS chuh lutz): steps for proving that a disease is caused by a certain microscopic organism (p. 82)

L

Labor: contractions of the uterus (p. 530)

Large intestine: a tubelike organ at the end of the digestive tract (p. 212)

Larva (LAR vuh): a young animal that looks completely different from the adult (p. 536)

Lateral buds: buds along the sides of a stem that give rise to new branches, leaves, or flowers (p. 423)

Learned behaviors: behaviors that must be taught (p. 356)

Lens: a clear part of the eye that changes shape as you view things at different distances (p. 335)

Lens muscle: a muscle that pulls on the lens and changes its shape (p. 335)

Leukemia (lew KEE mee uh): a blood cancer in which the number of white blood cells increases at an abnormally fast rate (p. 250)

Lichen (LI kun): a fungus and an organism with chlorophyll that live together (p. 107)

Ligaments (LIGH uh munts): tough fibers that hold one bone to another (p. 289)

Light microscope: light passes through the object being looked at and then through two or more lenses (p. 7)

Limiting factor: any condition that keeps the size of a population from increasing (p. 636)

Liver: the largest organ in the body, it makes a chemical called bile (p. 210)

Long-day plants: plants that flower when the day length rises above 12 to 14 hours (p. 441)

M

Mammal (MAM ul): an animal that has hair and feeds milk to its young (p. 173)

Mammary (MAM uh ree) **glands:** body parts that produce milk (p. 173)

Mantle: a thin, fleshy tissue that covers a soft-bodied animal (p. 149)

Medulla (muh DUL uh): the brain part that controls heartbeat, breathing, and blood pressure (p. 317)

Meiosis (mi OH sus): a kind of cell reproduction that forms eggs and sperm (p. 471)

Menstrual (MEN struhl) **cycle:** the monthly changes that take place in the female reproductive organs (p. 515)

Menstruation (men STRAY shun): a loss of cells from the uterine lining, blood, and egg (p. 516)

Metamorphosis (met uh MOR fuh sus): the changes in appearance and lifestyle that occur between the young and the adult stages (p. 535)

Meter: an SI unit of length (p. 11)

Midrib: the main vein of a leaf (p. 399)

Migration: a kind of behavior in which animals move from place to place in response to the season of the year (p. 363)

Minerals: nutrients needed to help form different cell parts (p. 190)

Mitochondria (mite uh KAHN dree uh): cell parts that produce energy from food that has been digested (p. 34)

Mitosis (mi TOH sus): cell reproduction in which two identical cells are made from one cell (p. 464)

Molting: shedding an exoskeleton (p. 159)

Monerans: one-celled organisms that don't have a nucleus (p. 61)

Moss: a small, nonvascular plant that has both stems and leaves but no roots (p. 117)

Mucus (MYEW kus): a thick, sticky material that protects the stomach and intestinal linings from enzymes and stomach acid (p. 216)

Multicellular: an organism having many different cells that do certain jobs for the organism (p. 97)

Muscular dystrophy (MUS kyuh lur · DIHS truh fee): a disease that causes the slow wasting away of skeletal muscle tissue (p. 300)

Muscular system: all the muscles in your body (p. 292)

Mutation (myew TAY shun): any change in copying the DNA message (p. 595)

Mutualism: a living arrangement in which both organisms benefit (p. 107)

N

Natural resource: any part of the environment used by humans (p. 674)

Natural selection: the process in which something in a living thing's surroundings determines if it will or will not survive to have offspring (p. 609)

Navel: where the umbilical cord was attached to the body (p. 530)

Nearsighted: being able to see clearly close up but not far away (p. 344)

Nephron (NEF rahn): a tiny filter unit of the kidney (p. 276)

Nerve: many neurons bunched together (p. 312)

Nervous system: the body system made up of cells and organs that let an animal detect changes and respond to them (p. 310)

Neurons (NOO rahnz): nerve cells (p. 312)

New-world monkeys: those that have a tail that can grasp like a hand and nostrils that open upward (p. 613)

Niche (NITSH): the job of the organism in the community (p. 638)

Nicotine: a stimulant found in tobacco (p. 385)

Nitrogen bases: the chemicals that form the rungs of a DNA molecule (p. 587)

Nitrogen cycle: the reusing of nitrogen in an ecosystem (p. 656)

Nonvascular (nahn VAS kyuh lur) **plants:** plants that don't have tubelike cells in their stems and leaves (p. 116)

Nuclear membrane: a structure that surrounds the nucleus and separates it from the rest of the cell (p. 32)

Nucleolus (new KLEE uh lus): the cell part that helps make ribosomes (p. 32)

Nucleus (NEW klee us): the cell part that controls most of the cell's activities (p. 32)

Nutrients (NEW tree unts): the chemicals in food that cells need (p. 184)

Nutrition (new TRISH un): the study of nutrients and how your body uses them (p. 184)

Nymph (NIHMF): a young insect that looks similar to the adult (p. 536)

O

Old-world monkeys: those that can't grasp with their tail, if they have one, and whose nostrils open downward (p. 613)

Olfactory (ohl FAK tree) **nerve:** a nerve that carries messages from the nose to the brain (p. 339)

Optic (AHP tihk) **nerve:** a nerve that carries messages from the retina to the brain (p. 336)

Order: the largest group within a class (p. 54)

Organ: a group of tissues that work together to do a job (p. 41)

Organ system: a group of organs that work together to do a certain job (p. 41)

Organism: a living thing (p. 41)

Osmosis (ahs MOH sus): the movement of water across the cell membrane (p. 39)

Ovaries (OHV uh reez): the female sex organs that produce eggs (p. 474)

Over-the-counter drug: one that you can buy legally without a prescription (p. 371)

Overdose: the result of too much of a drug in the body (p. 374)

Oviducts (OH vuh dukts): tubelike organs that connect the ovaries to the uterus (p. 514)

Ovules: tiny, round parts of a seed plant that contain the egg cells (p. 489)

Ozone: a molecule made of three oxygen atoms that forms a layer high above Earth's surface (p. 679)

P

Palisade layer: the layer of long, green cells below the upper epidermis of a leaf (p. 402)

Pancreas (PAN kree us): an organ located below the stomach that makes three different enzymes (p. 210)

Parasite (PAR uh site): an organism that lives in or on another living thing and gets food from it (p. 73)

Parasitism (PAR uh suh tihz um): a relationship between two organisms in which one is helped and the other is harmed (p. 646)

Parental care: a behavior in which adults give food, protection, and warmth to eggs or young (p. 365)

Pasteurization (pas chuh ruh ZAY shun): the process of heating milk to kill harmful bacteria (p. 85)

Pedigree: a diagram that can show how a certain trait is passed along in a family (p. 581)

Penis (PEE nus): places sperm in vagina during mating (p. 513)

Perennial (puh REN ee ul): a plant that doesn't die at the end of one or two years of growth (p. 448)

Pesticides: chemicals used to kill unwanted pests (p. 680)

Petals: often brightly colored and scented parts of a flower that attract insects to the flower (p. 488)

Pheromones (FER uh mohnz): chemicals that affect the behavior of members of the same species (p. 360)

Phloem (FLOH em): cells that carry food that is made in the leaves to all parts of the plant (p. 121)

Photosynthesis (foht oh SIHN thuh sus): the process in which plants use water, carbon dioxide, and energy from the sun to make food (p. 114)

Phototropism (foh toh TROH pihz um): the growth of a plant in response to light (p. 443)

Phylum (FI lum): the largest group within a kingdom (p. 54)

Physical change: occurs when large food pieces are broken down into smaller pieces (p. 205)

Pistil (PIH stul): the female reproductive organ of a flower (p. 489)

Pituitary (puh TEW uh ter ee) **gland:** an endocrine gland that forms many different hormones (p. 321)

Placenta (pluh SENT uh): an organ that connects the embryo to the mother's uterus (p. 526)

Planarian (pluh NAIR ee un): a common freshwater flatworm that is not a parasite (p. 144)

Plants: organisms that are made up of many cells, have chlorophyll, and can make their own food (p. 62)

Plasma (PLAZ muh): the nonliving, yellow liquid part of blood (p. 247)

Platelets (PLAYT lutz): cell parts that aid in forming blood clots (p. 251)

Pneumonia (noo MOH nyuh): a lung disease caused by bacteria, a virus, or both (p. 273)

Polar body: a small cell formed during meiosis in a female (p. 475)

Pollen: the tiny yellow grains of seed plants in which sperm develop (p. 125)

Pollination (pahl ih NAY shun): the transfer of pollen from the male part of a seed plant to the female part (p. 490)

Pollution: anything that makes the surroundings unhealthy or unclean (p. 677)

Population: a group of living things of the same species that live in an area (p. 632)

Pore: a small opening in a sponge through which water enters (p. 137)

Precipitation (prih sihp uh TAY shun): water in the air that falls to Earth as rain or snow (p. 663)

Predation: the predator-prey relationship (p. 647)

Predator: an animal that hunts, kills, and eats another animal (p. 647)

Prescription drug: one that a doctor must tell you to take (p. 371)

Prey: the animal that the predator kills and eats (p. 647)

Primary consumers: animals that eat only plants (p. 639)

Primates: mammals with eyes that face forward, a well-developed cerebrum, and thumbs that can be used for grasping (p. 613)

Producers: living things that make, or produce, their own food (p. 27)

Proteins (PROH teenz): nutrients that are used to build and repair body parts (p. 185)

Protists: mostly single-celled organisms that have a nucleus and other cell parts (p. 61)

Protozoans (proht uh ZOH uhnz): one-celled animal-like organisms with a nucleus (p. 94)

Psychedelic (si kuh DEL ihk) **drug:** one that alters the way the mind works and changes the signals we receive from our sense organs (p. 378)

Puberty (PEW bur tee): the stage in life when a person begins to develop sex cells (p. 474)

Pulmonary (PUL muh ner ee) **artery:** an artery that carries blood away from the heart to the lungs (p. 229)

Pulmonary veins: veins that carry blood from the lungs to the left side of the heart (p. 229)

Punnett (PUH nuht) **square:** a way to show which genes can combine when egg and sperm join (p. 553)

Pupa (PYEW puh): a quiet, non-feeding stage that occurs between larva and adult (p. 537)

Pupil: an opening in the center of the iris through which light enters the eye (p. 334)

Pure dominant: an organism with two dominant genes for a trait (p. 549)

Pure recessive: an organism with two recessive genes for a trait (p. 549)

R

Radiation: energy that is given off by atoms (p. 596)

Radon: a gas found in the ground that gives off radiation (p. 679)

Recessive (rih SES ihv) **genes:** those that do not show their traits when dominant genes are present (p. 549)

Recombinant DNA: the DNA that is formed when DNA from one organism is put into the DNA of another organism (p. 600)

Recommended daily allowance: the amount of each vitamin and mineral a person needs each day to stay in good health (p. 189)

Recycling: the reusing of resources (p. 689)

Red blood cells: cells in the blood that carry oxygen to the body tissues (p. 247)

Reflexes: quick, protective reactions that occur within the nervous system (p. 318)

Regeneration (rih jen uh RAY shun): reproduction in which the parent separates into two or more pieces and each piece forms a new organism (p. 505)

Reproduce: to form offspring similar to the parents (p. 26)

Reproductive system: the body system used to produce offspring (p. 513)

Reptile: an animal that has dry, scaly skin and can live on land (p. 170)

Respiratory (RES pruh tor ee) **system:** the body system made up of body parts that help with the exchange of gases (p. 264)

Retina (RET nuh): a structure at the back of the eye made of light-detecting nerve cells (p. 335)

Ribosomes (RI buh sohmz): cell parts where proteins are made (p. 34)

RNA: a chemical that acts as a messenger for DNA (p. 592)

Rods: nerve cells that detect motion and help us to tell if an object is light or dark (p. 336)

Root hairs: threadlike cells of the epidermis that absorb water and minerals for a plant (p. 429)

Roundworms: worms that have long bodies with pointed ends (p. 145)

Runner: a stem that grows along the ground or in the air and forms a new plant at its tip (p. 486)

S

Sac fungi: fungi that produce spores in saclike structures (p. 106)

Saliva: a liquid that is formed in the mouth and contains an enzyme (p. 208)

Salivary (SAL uh ver ee) **glands:** three pairs of small glands located under the tongue and behind the jaw (p. 208)

Saprophytes (SAP ruh fites): organisms that use dead materials for food (p. 81)

Scientific method: a series of steps used to solve problems (p. 15)

Scientific name: the genus and species names together (p. 59)

Sclera (SKLER uh): the tough, white outer covering of the eye (p. 334)

Scrotum (SKROH tum): the sac that holds the testes (p. 513)

Secondary consumers: animals that eat other animals (p. 639)

Sediment: material that settles on the bottom of a body of water (p. 676)

Sedimentary rocks: those that form from layers of mud, sand, and other fine particles (p. 617)

Seed: a part of a plant that contains a new, young plant and stored food (p. 124)

Segmented worms: worms with bodies divided into sections called segments (p. 146)

Self pollination: when pollen moves from the stamen of one flower to the pistil of the same flower or of another flower on the same plant (p. 490)

Semicircular canals: inner ear parts that help us keep our balance (p. 341)

Semilunar (sem ih LEW nur) **valves:** valves located between the ventricles and their arteries (p. 227)

Sense organs: parts of the nervous system that tell an animal what is going on around it (p. 332)

Sepals: often green, leaflike parts of a flower that protect the young flower while it is still a bud (p. 488)

Sex cells: reproductive cells produced in sex organs (p. 471)

Sex chromosomes: chromosomes that determine sex (p. 568)

Sexual reproduction: the forming of a new organism by the union of two reproductive cells (p. 118)

Sexually transmitted diseases: those transmitted through sexual contact (p. 518)

Short-day plants: plants that flower when the day length falls below 12 to 14 hours (p. 441)

Sickle-cell anemia: a genetic disorder in which all the red blood cells are shaped like sickles (p. 574)

Side effect: a change other than the expected change caused by a drug (p. 372)

Sister chromatids (KROH muh tidz): two strands of a doubled chromosome (p. 466)

Skeletal muscles: muscles that move the bones of the skeleton (p. 293)

Skeletal system: the framework of bones in your body (p. 286)

Slime molds: funguslike protists that are consumers (p. 99)

Small intestine: a long, hollow, tubelike organ where most of the chemical digestion of food takes place (p. 210)

Smog: a combination of smoke and fog (p. 678)

Smooth muscle: involuntary muscle that makes up the intestines, arteries, and many other body organs (p. 295)

Social insects: insects that live in groups, with each individual doing a certain job (p. 362)

Soft-bodied animals: animals with a soft body that is usually protected by a hard shell (p. 149)

Solid bone: the very compact or hard part of a bone (p. 289)

Species: the smallest group of living things (p. 55); a group of living things that can breed with others of the same species and form fertile offspring (p. 610)

Sperm: a male reproductive cell (p. 118)

Spinal cord: the body part that carries messages from the brain to body nerves or from body nerves to the brain (p. 315)

Spiny-skin animal: an invertebrate with a five-part body design, radial symmetry, and spines (p. 163)

Sponges: simple invertebrates that have pores (p. 137)

Spongy bone: the part of a bone that has many empty spaces, much like those in a sponge (p. 289)

Spongy layer: the layer of round, green cells directly below the palisade layer of a leaf (p. 402)

Sporangia: structures, found on the tips of hyphae, that make spores (p. 104)

Sporangium (spuh RAN jee uhm) **fungi:** fungi that produce spores in sporangia (p. 104)

Spores: special cells that develop into new organisms (p. 96)

Sporozoans (spor uh ZOH uhnz): protozoans that reproduce by forming spores (p. 96)

Sprains: injuries that occur to your ligaments at a joint (p. 300)

Stamen: the male reproductive organ of a flower (p. 488)

Stereomicroscope (STER ee oh MI kruh skohp): used for viewing large objects and things through which light cannot pass (p. 8)

Stimulant (STIHM yuh lunt): a drug that speeds up body activities that are controlled by the nervous system (p. 376)

Stimulus: something that causes a reaction in an organism (p. 353)

Stinging-cell animals: animals with stinging cells and hollow, sock-shaped bodies that lack organs (p. 139)

Stoma (STOH muh): a small pore or opening in the lower epidermis of a leaf (p. 402)

Stomach: a baglike, muscular organ that mixes and chemically changes protein (p. 209)

Succession: the changes that take place in a community as it gets older (p. 660)

Symmetry (SIH muh tree): the balanced arrangement of body parts around a center point or along a center line (p. 134)

Synapse (SIN aps): a small space between the axon of one neuron and the dendrite of a nearby neuron (p. 313)

T

Taproot: a large, single root with smaller side roots (p. 428)

Tapeworm: a kind of flatworm that has a flattened, ribbonlike body divided into sections (p. 142)

Taste buds: nerve cells in the tongue that detect chemical molecules (p. 339)

Technology: the use of scientific discoveries to solve everyday problems (p. 20)

Tendon (TEN duhn): a tough, fibrous tissue that connects muscle to bone (p. 295)

Tentacles (TENT ih kulz): armlike parts of stinging-cell animals (p. 139)

Terminal bud: the bud at the tip of a stem (p. 423)

Testes (TES teez): the male sex organs that produce sperm (p. 474)

Theory: a hypothesis that has been tested again and again by many scientists, with similar results each time (p. 19)

Thigmotropism: a plant growth response to contact (p. 444)

Threatened: a species of living thing that is close to being endangered (p. 675)

Thyroid (THI royd) **gland:** an important endocrine gland that is found near the lower part of your neck and produces thyroxine (p. 322)

Thyroxine (thi RAHK sun): the hormone that controls how fast your cells release energy from food (p. 322)

Tissue: a group of similar cells that work together to carry out a special job (p. 40)

Toxic: poisonous (p. 678)

Trachea (TRAY kee uh): a tube about 15 centimeters long that carries air to two shorter tubes called the bronchi (p. 267)

Trait: a feature that a thing has (p. 48)

Transpiration (trans puh RAY shun): the process of water passing out through the stomata of leaves (p. 403)

Tricuspid (tri KUS pud) **valve:** the valve between the right atrium and the right ventricle (p. 227)

Tropism (TROH pihz um): a movement of a plant caused by a change in growth as a response to a stimulus (p. 443)

Tube feet: parts of a starfish that are like suction cups and help the starfish move, attach to rocks, and get food (p. 163)

Tuber: an underground root or stem swollen with stored food (p. 485)

U

Umbilical (uhm BIL ih kuhl) **cord:** made of blood vessels that connect the embryo to the placenta (p. 526)

Urea (yoo REE uh): a waste that results from the breakdown of body protein (p. 274)

Ureter (YOOR ut ur): a tube that carries wastes from a kidney to the uninary bladder (p. 276)

Urethra (yoo REE thruh): a tube that carries liquid wastes from the urinary bladder to outside the body (p. 276)

Urinary bladder: a sac that stores liquid wastes removed from the kidneys (p. 276)

Urine: the waste liquid that reaches the ureter (p. 277)

Uterus (YEWT uh rus): a muscular organ in which the fertilized egg develops (p. 514)

V

Vaccines: substances made from weakened or dead viruses that protect you against certain diseases (p. 78)

Vacuole (VAK yuh wol): a liquid-filled space that stores food, water, and minerals (p. 35)

Vagina (vuh JI nuh): a muscular tube that leads from outside the female's body to the uterus (p. 515)

Valves: flaps in the heart that keep blood flowing in one direction (p. 227)

Variable: something that causes the changes observed in an experiment (p. 18)

Variation (ver ee AY shun): a trait that makes an individual different from others of its species (p. 616)

Vas deferens (VAS · DEF uh runs): carries sperm from testes to urethra (p. 513)

Vascular (VAS kyuh lur) **plants:** plants that have tubelike cells in their roots, stems, and leaves to carry food and water (p. 116)

Vein: a blood vessel that carries blood back to the heart (p. 226)

Vena cava (VEE nuh · KAY vuh): the largest vein in the body (p. 229)

Ventricles (VEN trih kulz): the large bottom chambers of the heart (p. 225)

Vertebrates (VERT uh brayts): animals with backbones (p. 134)

Vestigial (vuh STIJ ee ul) **structure:** a body part that no longer has a function (p. 620)

Villi (VIHL i): the fingerlike parts on the lining of the small intestine (p. 214)

Virus: a chromosome-like part surrounded by a protein coat (p. 72)

Vitamins: chemical compounds needed in very small amounts for growth and tissue repair of the body (p. 188)

Vitreous (VI tree us) **humor:** a jelly-like material inside the eye (p. 335)

Volume: the amount of space a substance occupies (p. 13)

Voluntary muscles: muscles you can control (p. 294)

W

Warm-blooded: having a body temperature that is controlled so that it stays about the same no matter what the temperature of the surroundings (p. 171)

Water cycle: the path that water takes through an ecosystem (p. 657)

White blood cells: the cells in the blood that destroy harmful microbes, remove dead cells, and make proteins that help prevent disease (p. 249)

Wilting: when a plant loses water faster than it can be replaced (p. 403)

Woody stem: a nongreen stem that grows to be thick and hard (p. 421)

X

X chromosomes: sex chromosomes of the female (p. 568)

Xylem (ZI lum): cells that carry water and dissolved minerals from the roots to the leaves (p. 121)

Y

Y chromosome: a sex chromosome found only in males (p. 568)

INDEX

C

D

E

F

G

H

I

J

K

L

M

N

O

Q

R

S

T

U

V

W

X

Y

PHOTO CREDITS

Cover photo: Claude Steelman/Tom Stack & Associates
iv, Walter Cunningham; **v**(t) Larry Lefever/Grant Heilman, (inset) Rod Plank/Tom Stack & Associates, (c) Alvin Staffan, (b) Steve Lissau; **vi**(tl) Johnny Johnson/Animals, Animals, (tc) Walt Anderson/Tom Stack & Associates, (b) Aaron Haupt/Glencoe; **vii**(t) David Dennis, (c) a, Larry Lipsky/Tom Stack & Associates, b, Hans Pfletschin/Peter Arnold, Inc., c, National Audubon Society/Photo Researchers, d, Fred Bavendam/Peter Arnold, Inc., e, Spencer Swanger/Tom Stack & Associates, f, Aaron Haupt/Glencoe, g, Tim Courlas, h, H. Gritscher/ Peter Arnold, Inc., (b) Lynn Stone; **ix**(tr) a, Donald Dietz/Stock Boston, b, Tim Davis/Photo Researchers, (tr), Hank Morgan/Science Source/Photo Researchers, (bl) a, John Radcliffe Infirmary/SPL/Photo Researchers, b, Lennart Nilsson, (br) First Image; **x**(t) Cobalt Productions, (b) Alan Carey; **xi**(t) R. Feldman/Dan McCoy Rainbow, (b) Harris Biological Supply Company; **xii**(b) W. Gregory Brown/Animals, Animals, (t) Lynn Stone; **xiii**(b) Gerard Photography, (t) Dr Tony Brian/SPL/Photo Researchers; **xiv**(t) David Brownell, (b) Pictures Unlimited; **xv**(b) Grant Heilman/Grant Heilman, (t) Elaine Shay; **xvi,** James Westwater; **xvii**(t) Pictures Unlimited, (b) Alan Carey; **xviii**(t) William Weber, (b) John Giannicchi/SPL/Photo Researchers; **xix**(t) Elaine Shay, (b) David Frazier; **xx**(t) Frank Balthis, (b) Ted Rice; **xxi,** David Dennis; **xxii**(t) Comstock, (b) Bob McKeever/Tom Stack & Associates; **xxiii**(t) Larry Lefever/Grant Heilman, (b) G.C. Lockwood; **xxiv,** Pictures Unlimited; **1,** M. Woodbridge Williams/National Park Service; **2-3,** JPL/NASA, (inset)NASA; **4,** Bob Daemmrich; **5,** Tom Courlas; **6**(l) USDA, (r) Doug Martin; **8**(l) E. Herzog/Photo Researchers, (c) Michael Abbey/Photo Researchers, (r) Martin Rodgers/Taurus Photos; **11**(t) David Dennis, (b) Tom Courlas; **13,** NASA; **20,** Hank Morgan/Photo Researchers; **21**(t) F. Stuart Westmorland/Tom Stack & Associates, (b) Lionel Deleingne/Stock Boston; **24,** Johnny Johnson/Animals, Animals; **25,** Walt Anderson/Tom Stack & Associates; **27**(l) David M. Dennis/Tom Stack & Associates, (r) Dwight Kuhn; **29**(t) John Kaprieliam/Photo Researchers, (b) Larry Hamill; **31,** Cabisco/Visuals Unlimited; **35**(l) K.R. Porter/Photo Researchers, (r) Biophoto Associates/Science Source/Photo Researchers; **37**(tl) John Shaw/Tom Stack & Associates, (bl) Oxford Scientific Films/Animals, Animals, (tr) Fritz Prenzel/Animals, Animals, (br) Juan M. Renyito/Animals, Animals; **42** Studiohio; **43** Grant Heilman/Grant Heilman Photography; **46,** Brian Parker/Tom Stack & Associates; **47,** Peter Parks/Oxford Scientific Films/Animals, Animals; **48,** Bud Fowle, **50,** First Image; **54**(l) Jeffrey W. Lang/Photo Researchers, (tr) John Serrao/Photo Researchers, (br) Soames Summerhay/Photo Researchers; **56,** First Image; **57**(t) Comstock, (c) Lynn Stone, (b) Dave R. Fleethan/Tom Stack & Associates; **59**(l) Anup Shah/Animals, (c) Ralph A. Beinhold/Animals, Animals, (r) Rod Planckt/Tom Stack & Associates, (tr) Jerry L. Ferrana/Photo Researchers; **60**(l) Alan Carey, (c) Michael Collier, (r) David Zosel; **61,** Phil Farnes/Photo Researchers; **66,** Steve Schapiro/Sygma; **67,** Doug Martin; **68-69,** Roger & Donna Aitkenhead/Animals, Animals, (inset) Runk-Schoenberger/Grant Heilman; **70,** Dr. Gopal Murti/Science Photo Library/Custom Medical Stock Photo; **71,** CNR/Science Photo Library/Photo Researchers; **72**(l) Omikron/Science Source/Photo Researchers, (c) CNRI/SPL/Photo Researchers, (r) R. Feldman-Dan McCoy/Rainbow; **73,** Omnikon/Photo Researchers; **74**(t) Prof. Luc Montagnier/CNRI/Science Photo Library/Photo Researchers, (b) James E. Lloyd/Animals, Animals; **78**(l) Pictures Unlimited, (r) Martin Rodgers/Stock Boston; **79**(tl) Dr. Tony Brain & David Parker/Science Photo Library/Photo Researchers, (bl) CNRI/Science Photo Library/Photo Researchers, (r) CNRI/Science Photo Library/Photo Researchers; **81,** John Kohout/Root Resources; **83**(l) NIH/Science Photo Library/Photo Researchers, (r) CNRI/Science Photo Library/Photo Researchers; **84**(l) Doug Martin, (c),(r) First Image; **85,** First Image; **86**(t) Ken W. Davis/Tom Stack & Associates, (b) First Image; **88,** Brian Parker/Tom Stack & Associates; **89**(l) Biophoto Associates-Science Source/Photo Researchers, (r) Arthur M. Siegelman; **92,** Rod Planck/Tom Stack & Associates; **93,** Ward's Natural Science Establishment; **94**(l) Eric Grave/Science Source/Photo Researchers, (c) Oxford Scientific Films/Animals, Animals, (r) John Shaw/Tom Stack & Associates; **96,** Eric Grave/Photo Researchers; **98**(t) Jan Hinsch/SPL/Photo Researchers, (b) Ward's Natural Science Establishment; **99,** Joyce Photographics/Photo Researchers; **102**(r) David M. Dennis/Tom Stack & Associates, (l) Jack Swenson/Tom Stack & Associates; **105,** Cary Wolinsky/Stock Boston; **106**(l) A. McClenaghan/Photo Researchers, (r) Biophoto Associates/Photo Researchers; **107,** Breck Kent; **112,** Ray Bishop/Stock Boston; **113,** Al Staffan; **114,** Courtesy: Ray F. Everett; **117**(l) Rod Planck/Tom Stack & Associates, (r) Jack Wilburn/Earth Series; **118,** Aaron Haupt/Glencoe; **122,** Breck P. Kent/Earth Scenes; **124**(l) Rich Brommer, (c) John Gerlach/Tom Stack & Associates, (r) Doug Sokell/Tom Stack & Associates; **126**(t) Eric Carle/Stock Boston, (b) Bob Daemmrich; **127**(t) Donald Dietz/Stock Boston, (b) Tim Davis/Photo Researchers; **128**(t) Kodansha/Science Software System, (bl) Jack Wilburn/Earth Scenes, (br) Zig Leszczynski/Earth Scenes; **129**(t) Patti Murray/Earth Scenes, (b) First Image; **132,** Nancy Sefton/Photo Researchers; **133,** Brian Parker/Tom Stack & Associates; **137,** Cindy Garoutte/Tom Stack & Associates; **139**(l) William Curtsinge/Photo Researchers, (c) Dave Fleetham/Tom Stack & Associates, (r) Lynn Funkhouser/Peter Arnold, Inc.; **142,** Michael Klein/Peter Arnold, Inc.; **146**(l) Alvin Staffan, (r) Breck Kent; **149**(l) George Holton/Photo Researchers, (r) Garry G. Gibson/Photo Researchers; **150**(t) F.E. Unverhav/Animals, Animals, (b) Ed Robinson/Tom Stack & Associates; **152**(l) Larry Lipsky/Tom Stack & Associates, (lc) Hans Pfletschinger/Peter Arnold, Inc., (rc) National Audubon Society/Photo Researchers, (r) Fred Bavendam/Peter Arnold, Inc.; **153**(rc) Tim Courlas, (r) H. Gritscher/Peter Arnold, Inc., (l) Spencer Swanger/Tom Stack & Associates, (lc) Aaron Haupt/Glencoe; **156,** Mark Newman/Tom Stack & Associates; **157,** Colin Milkens/Oxford Scientific Films/Animals, Animals; **158,** Jack Swenson/Tom Stack & Associates; **159**(t) George Harrison/Grant Heilman, (b) Donald Specker/Animals, Animals; **160,** Tom McHugh/Photo Researchers; **163**(t) Ed Robinson/Tom Stack & Associates, (c) Larry Tackett/Tom Stack & Associates, (r) Runk-Schoenberger/Grant Heilman, (b) Ed Robinson/Tom Stack & Associates; **164,** Brian Parker/Tom Stack & Associates; **165,** Animals, Animals; **168**(t) Fred McConnaughey/Photo Researchers, (b) Al Staffan; **169,** Breck Kent/Animals, Animals; **170**(t) Alvin Staffan, (b) David Dennis; **173,** Wendy Neefus/Animals, Animals; **174**(l) Dave Watts/Tom Stack & Associates, (r) Dave Watts/Tom Stack & Associates; **175**(t) Tom McHugh/Photo Researchers, (b) P.F. Gero/Sygma; **178**(t) Peter Veit/DRK Photo, (b) Jim Zuckerman/Stock Concepts; **179**(t) Runk-Schoenberger/Grant Heilman, (b) Bud Fowle; **180-181,** Doug Martin, (inset) Bob Daemmrich; **182, 183,** ML Uttermohlen; **185,** Tim Courlas; **191,** Studiohio; **196,** Light Source; **197,** Bob Daemmrich; **198,** Mark Burnett; **199,** Glencoe file photo; **202,** Calvin Larsen/Photo Researchers; **203,** Davis Scharf/Peter Arnold, Inc.; **211,** Doug Martin; **217,** ML Uttermohlen; **220,** Steve Elmore/Tom Stack & Associates; **221,** Lennart Nilsson; **225,** David Gifford/SPL/Photo Researchers; **237,** Randy Schieber; **238,** Dan McCoy/Rainbow; **239**(l) Sygma, (r) Dan McCoy/Rainbow; **242,** CNRI/SPL/Photo Researchers; **243,** Aaron Haupt/Glencoe; **244,** Doug Martin; **245,** Geri Murphy; **246,** Carolina Biological Supply Co.; **247**(l) Doug Martin, (r) E.R. Degginger/Bruce Coleman; **249**(l) Runk-Schoenberger/Grant Heilman, (r) Lennart Nilsson; **250**(t) John Radcliffe Infirmary/SPL/Photo Researchers, (r) Lennart Nilsson; **251**(rt) Hickson-Bender, (rc) S.I.U./Bruce Coleman, (l) Lennart Nilsson, (rb) L. Moskowitz/Medichrome; **252,** Doug Martin; **259,** Lennart Nilsson; **262,** Dave B. Fleetham/Tom Stack & Associates; **263,** Doug Martin; **272,** Young/Hoffhines; **273,** Susan McCartney/Photo Researchers;

280, Owen Franklin/Stock Boston; **281**(t) Doug Martin, (b) Hank Morgan/Science Source/Photo Researchers; **284,** Roy Bishop/Stock Boston; **285,288,** First Image; **292,293** Bob Daemmrich; **294,** M. Huberland/Science Source/Photo Researchers; **299,** Comstock, (inset) Alexander Tsiaras/Stock Boston; **300,** Julie Simon/Discover Magazine; **301,** First Image; **308,** Aaron Haupt/Glencoe; **309,** SPL/Science Source/Photo Researchers; **304**(t) Doug Martin, (b) Scala/Art Resource; **305,** Doug Martin; **306-307,** David Parker/SPL/Photo Researchers, (insert) Sheila Terry/SPL/Photo Researchers; **326,** MRI-Cross section of body; **327,** Blair Seitz/Photo Researchers; **330,** Holt Studios, Ltd/Animals, Animals; **331,** Thomas Eisner; **333,** W. Perry Conway/Tom Stack & Associates; **334,** Tim Courias; **336,** Ralph Eagle, Jr. MD/Science Source/Photo Researchers; **339,** Omnikron/Science Source/Photo Researchers; **342,** Doug Martin; **350,** ML Uttermohlen; **351,** Anthony Mercieca/Photo Researchers; **356,** David Frazier; **359**(l) Nik Kleinberg/Stock Boston, (r) J.H. Robinson/Photo Researchers; **361**(l) Cary Wolinsky/Stock Boston, (r) Stephen Krasemann/Peter Arnold, Inc.; **362,** Gary Milburn/Tom Stack & Associates; **363**(l) Stephen Krasemann/Peter Arnold, Inc., (r) K. Brink/Vireo; **364,** John David Brandt/Animals, Animals; **365,** Mac Albin; **368,** Pictures Unlimited; **369,** Aaron Haupt/Glencoe; **370,** Doug Martin; **371**(l) Tim Courlas, (r) Aaron Haupt/Glencoe; **378,** Tom HcHugh/Photo Researchers; **380,** Doug Martin; **381,** ML Uttermohlen; **382,** Pictures Unlimited; **383,** Scott Camazine/Photo Researchers; **384**(l) Christopher Brown/Stock Boston, (r) Hugh Patrick Brown/Sygma; **385**(l) Norm Thomas/Photo Researchers, (r) Pictures Unlimited; **386,** ML Uttermohlen; **388,** Larry Kolvoord/The Image Works; **389**(t) Bob Daemmrich/Stock Boston, (b) Bob Daemmrich/The Image Works; **392**(t) Historical Picture Service, (b) Tate Gallery/Art Resource; **393,** Doug Martin; **394-395,** Michael P. Gadomsli/Photo Researchers, (inset) Sam Bryan/Photo Researchers; **396,** Randy Schieber; **397,** Aaron Haupt/Glencoe; **399**(l) Jack Fields/Photo Researchers, (r) G.I. Bernard/Earth Scenes; **403**(t) Jerry Bauer, (b) Studiohio; **404,** Steve Lissau; **408,** Pictures Unlimited; **411**(l),(c) G.R. Roberts, (r) John Gerlack/Tom Stack & Associates; **412,** Aaron Haupt/Glencoe; **413**(t) G.I. Bernard/Earth Scenes,` (b) Aaron Haupt/Glencoe; **414**(l) Pictures Unlimited, (r) John Shaw/Tom Stack & Associates; **418,** H. Reinhard/Okopia/Photo Researchers; **419,** Doug Martin; **421,** Jeff Lepore/ Photo Researchers; **422,** Doug Martin; **426,** Kjell B. Sandved/Photo Researchers; **427,** First Image; **428**(l) John Kaprielian/Photo Researchers, (r) Michael P. Gadomski/Photo Researchers; **431**(l) Bob Daemmrich, (r) Alan & Sandy Carey; **433**(l) G.R. Roberts, (r) Dwight Kuhn; **435**(l) Breck P. Kent/Earth Scenes, (r) Elaine Braithwaite/Peter Arnold, Inc.; **438,** Lefever-Grushow/Grant Heilman; **439,** Norm Thomas/Photo Researchers; **440,** Pat Lynch/Photo Researchers; **443,** G.R. Roberts; **444,** John Colwell/Grant Heilman; **446**(l) Charles C. Johnson, (r) Nuridsany et Perennou/Photo Researchers; **447**(l) Richard Kolar/Earth Scenes, (r) Richard Schiell/Earth Scenes; **449**(l) David M. Dennis, (c) Greg Vaughn/Tom Stack & Associates, (r) First Image; **450**(l) First Image, (c) Rod Planck/Tom Stack & Associates, (r) Kenneth Murray/Photo Researchers; **451**(l) Alan & Sandy Carey, (r) Bob Daemmrich; **452**(l) Michael S. Thompson/Comstock, (c) Kathy Merrifield/Photo Researchers, (r) G.R. Roberts; **453**(l) Patti Murray/Animals, Animals, (r) John Gerlach/Tom Stack & Associates; **454**(t) Mark Thayer; **454**(b) Morton/White; **455** Doug Martin; **458**(t) Zig Leszczynski/Earth Scenes, (b) Illustrated London News; **459**(t) USDA, (b) David Frazier; **460-461,** Grant Heilman/Grant Heilman, (inset) Jeff Heger/Houston Convention & Visitor Bureau; **462,** IFA/Peter Arnold, Inc., **463,** D.R. Specker/Earth Scenes; **465,** ML Uttermohlen; **467,** Grant Heilman/Grant Heilman; **470,** Photo Researchers; **474,** David Scharf/Peter Arnold, Inc., **478,** Oscar Auerbach, M.D.; **479**(l) Aaron Haupt/Glencoe, (r) Brian Yablonsky/duomo; **482,** Breck Kent/Animals, Animals; **483,** Frank Cezus; **486,** Pictures Unlimited; **487,** Biological Photo Service; **489**(l) Dr. Jeremy Burgess/SPL/Photo Researchers, (r) Runk-Schoenberger/Grant Heilman; **493**(t) Agricultural Research Service/USDA, (b) First Image; **496,** LLT Rhodes/Earth Scenes; **498,** Stephen J. Krasemann/Photo Researchers; **499,** Glencoe file photo; **502,** Peter Menzel/Stock Boston; **503,** John Gerlach/Animals, Animals; **505,** Walker England/Photo Researchers; **506,** Carolina Biological Supply Company; **508,** Mark Stouffer/Animals, Animals; **509,** David M. Dennis; **510,** John Shaw/Tom Stack & Associates; **517**(t) Francis Leroy/Biocosmos/Science Photo Library/Photo Researchers, (b) Hank Morgan/Science Source/Photo Researchers; **522,** Lennart Nilsson; **523,** Studiohio; **527,** Howard Sochurek/Medical Images, Inc.; **528**(l) Lennart Nilsson, (r) Hans Pfletschinger/Peter Arnold, Inc.; **529,** Lennart Nilsson; **532,** Hans Pfletschinger/Peter Arnold, Inc.; **537**(tl), (bl) John Shaw/Tom Stack & Associates, (tr) Michael Fogden/Animals, Animals, (br) Don and Pat Valenti/Tom Stack & Associates; **540**(t) Latent Image, (b) David Frazier; **541**(t) Zig Lesczynski/Animals, Animals; **542-543,** Omnikon/Science Source/Photo Researchers, (inset) Catherine Ursillo/Photo Researchers; **544,545;** Studiohio; **548,** Aaron Haupt/Glencoe; **551**(t) Mavournea Hay/Daemmrich Associates, (b) Aaron Haupt/Glencoe; **555**(t) Doug Martin, (b) First Image; **559,** Bettmann Archives; **564,565,** Tim Courlas; **567,** American Cancer Society; **568**(t) File Photo, (b) Biophoto Associates/Photo Researchers; **569,** Doug Martin; **572**(l) Studio Productions, (r) First Image; **573,** Tim Courlas; **574**(l) Biological Photo Service, (r) Murayama/Biological Photo Service; **576,** First Image; **579,** Richard Hutching/Science Source/Photo Researchers; **580,** Doug Martin; **584,** Runk-Schoenberger/Grant Heilman; **585,593,** First Image; **596**(t) Aaron Haupt/Glencoe, (b) First Image; **600,** John Colwell/Grant Heilman; **604,** D. Cavagnaro/Peter Arnold, Inc.; **605,** First Image; **606**(l) Richard R. Hansen/Photo Researchers, (r) Cary Wolinsky/Stock Boston; **609,** Richard P. Smith/Tom Stack & Associates; **610**(l) Tom McHugh/Photo Researchers, (r) Kenneth W. Fink/Photo Researchers; **613**(l) Tom McHugh/Photo Researchers, (lc) Tom McHugh/Photo Researchers, (rc) Stephen J. Krasemann/Photo Researchers, (r) Tom McHugh/Photo Researchers; **615**(l) Aaron Haupt/Glencoe, (r) John Shaw/Tom Stack & Associates; **616**(l) Ray Coleman/Photo Researchers, (r) Jeff Lepore/Photo Researchers; **619,** National Museum of Natural History; **619,** Bohdan Hrynewych/Stock Boston; **623,** Francois Gohier/Photo Researchers; **626**(t) E.R. Degginger/Earth Scenes, (b) National Museum of Natural History; **627**(t) Studiohio, (b) Bud Fowle; **628-629,** Larry Lefever/Grant Heilman, (inset) Rod Plank/Tom Stack & Associates; **630,** Stephen J. Krasemann/Photo Researchers; **631,** Thomas Kitchin/Tom Stack & Associates; **632,** Tim Cullinan; **633**(l) Joe McDonald/Animals, Animals, (c) Larry Brock/Tom Stack & Associates, (r) Richard P. Smith/Tom Stack & Associates; **635**(t) Bruno J. Zehnder/Peter Arnold, Inc., (r) Eugene Gilliam; **636,** John Shaw/Tom Stack & Associates; **638,** Gerald Corsi/Tom Stack & Associates; **639**(l) C.C. Lockwood/Earth Scenes, (c) E.R. Degginger/Animals, Animals, (r) Zig Leszcynski/Animals, Animals; **640,** David Thompson/Oxford Scientific Films/Earth Scenes; **644**(l) Eric V. Grave/Photo Researchers, (r) Dwight R. Kuhn, **645,** Jacques Jangoux/Peter Arnold, Inc.; **646**(l) Animals, Animals/G.I. Bernard, (r) Zig Leszczynski/Animals, Animals; **647,** Charles G. Summers, Jr/Tom Stack & Associates; **652,** Kevin Schafer/Tom Stack & Associates; **653,** Sydney Thomson/Earth Scenes; **654**(l) Matt Meadows; (r) David C. Fritts/Animals, Animals; **656,** Bob Daemmrich; **660**(l) Doug Martin, (r) Pictures Unlimited; **661**(l) Michael P. Gadomski/Photo Researchers, (r) Michael P. Gadomski/Photo Researchers; **665**(a) Brian Parker/Tom Stack & Associates, (b) Richard R. Thom/Tom Stack & Associates, (c) Michael & Barbara Reed/Earth Scenes, (d) Michael P. Gadomski/Photo Researchers, (e) Charlie Ott/Photo Researchers, (f) Tom Stack/Tom Stack & Associates; **669,** Carl Purcell/Photo Researchers; **672,** Pat & Tom Leeson/Photo Researchers; **673,** David Dennis;

674, Richard Kolar/Earth Scenes; **675**(l) Jeff Lepore/Photo Researchers, (c) Cary Wolinsky/Stock Boston, (r) Sven-Olofinblad/Photo Researchers; **676**(l) Gregory G. Dimijian, MD/Photo Reserachers, (c) Owen Franken/Stock Boston, (r) Tom Kitchin/Tom Stack & Associates; **678,** Doug Martin; **679**(l) Frederick D. Bodin/Stock Boston, (r) ML Uttermohlen; **670,** Doug Martin; **681**(l) Tom Kitchin/Tom Stack & Associates, (c) Pictures Unlimited, (r) David Frazier; **684**(l) Don & Pat Valenti/Tom Stack & Associates, (r) John Shaw/Tom Stack & Associates; **687,** Cary Wolinsky/Stock Boston; **688,** Dr. Jeremy Burgess/SPL/Photo Researchers; **689,** ML Uttermohlen; **692**(t) Arthur C. Twomey/Photo Researchers, (b) Michael J. Philippot/Sygma; **693**(t), Studiohio, (b) Thomas Kitchin/Tom Stack & Associates; **696**(b) Omnikon/Science Source/Photo Researchers, (t)Manfred Kage/Peter Arnold, Inc.; **697**(t) Richard Kessel, (b)Stephen J. Krasemann/DRK Photo; **698,** Pictures Unlimited; **699**(t) Geri Murphy, (b)Lynn Stone; **706,** Doug Martin; **707,** CNR/SPL/Photo Researchers; **710,711,** Aaron Haupt/Glencoe.